Modern Multidisciplinary Applied Microbiology

Edited by
Antonio Mendez-Vilas

Related Titles

Coico, R.
Current Protocols in Microbiology
Set
2006
ISBN 0-471-72924-8

Schumann, W.
Dynamics of the Bacterial Chromosome
Structure and Function
2006
ISBN 0-527-30496-7

Hogg, S
Essential Microbiology
2005
ISBN 0-471-49753-3

Singleton, P.
Bacteria in Biology, Biotechnology and Medicine
2004
ISBN 0-470-09026-X

Dale, J. W., Park, S.
Molecular Genetics of Bacteria
2004
ISBN 0-470-85084-1

Modern Multidisciplinary Applied Microbiology

Exploiting Microbes and Their Interactions

Edited by
Antonio Mendez-Vilas

WILEY-VCH Verlag GmbH & Co. KGaA

The Editor

Dr. Antonio Mendez-Vilas
Formatex
Zurbaran 1
Planta 2a, Local 1
06001 Badajoz
Spain

Library of Congress Card No.: applied for

British Library Cataloguing-in-Publication Data
A catalogue record for this book is available from the British Library.

**Bibliographic information published by
Die Deutsche Bibliothek**
Die Deutsche Bibliothek lists this publication in the Deutsche Nationalbibliografie; detailed bibliographic data is available in the Internet at <http://dnb.ddb.de>.

© 2006 WILEY-VCH Verlag GmbH & Co. KGaA, Weinheim

Printing and Bookbinding   Markono Print Media Pte Ltd, Singapore
Cover   Adam Design, Weinheim

Printed in Singapore
Printed on acid-free paper

ISBN-13: 978-3-527-31611-3
ISBN-10: 3-527-31611-6

Preface

This book contains selected papers related to contributions that were presented in short during the 1st International Conference on Environmental, Industrial and Applied Microbiology (BioMicroWorld-2005), held on March 15–18th, 2005 in Badajoz (Spain). http://www.formatex.org/biomicroworld2005, and it is intended to give an overview on the current state-of-the-art of the field.

Focus of the Conference. While Microbiology is about the study of microorganisms (bacteria, viruses, algae, fungi, and protozoa) and of related topics such as microbes interactions, the immune response and molecular genetics, Applied Microbiology is quite interdisciplinary, overlapping aspects of several other academic branches, some of them traditionally near, such as cell biology, molecular and cell biophysics, physiology, parasitology, biochemistry, genetics, medicine, pharmacology, and medical technology, and other not so near as physics, physical (bio)chemistry, materials science, nanotechnology, computer science, information technology, instrumentation, but collaboration with which is resulting in extraordinary advances in this post-genomic world. Thus, cross-disciplinary cooperation in Microbiology has made possible that microbiologists can not only study traditional microorganisms, but also aspects of molecular genetics, biosensors, cancer, aging, immunodeficiency diseases, animal and plant cell cultures, and microscopy, among others. Modern microbiology includes a broad variety of scholarly approaches which lead to a better understanding of all living things at the micrometer-scale/cellular and nanometer-scale/molecular level, and which produce beneficial applications in medicine, agriculture, industry, and ecology.

In this context, the Conference called for papers reporting interdisciplinary resear-chers, relating Microbiology with other Sciences as Physico/Chemistry, Environ-mental Science, Genetics, Pharmacology, Nanoscience, Microscopy/Imaging Science, etc. In other words, we are specially (but not exclusively) interested in reports applying the techniques, the training, and the culture of Microbiology to research areas usually associated with other scientific and engineering disciplines. Over 750 participants from over 60 countries attended the Conference, 15% of which participated with a grant from the conference organization. Over 1100 works were presented during the different oral and posters sessions. Good examples of modern interdisciplinary applied microbiology were the works represented by the three Plenary Speakers:

David C. White, Director of the Center for Biomarker Analysis, University of Tennessee, USA
Lecture: *Biomarkers to define Interactions in the Environment and Health*

Alexander Steinbüchel, Institut für Molekulare Mikrobiologie und Biotechnologie, Münster, GERMANY
Lecture: *Unspecific Microbial Enzymes for Template-independent Biosynthesis of Poly-oxoesters, Polythioesters and Polyamides*

Timo Lövgren, Department of Biochemistry and Food Chemistry/Biotechnology, University of Turku, FINDLAND
Lecture: *Novel time-resolved Fluorescence based Immunoassays and Real-time PCR assays in Microbiological Applications*

I am are very grateful to all members of the Organizing Committee for the hard work done in the Conference preparation (which began over a year before the conference) and for the good job that made the Conference so successful that there are already several candidates to host next edition of BioMicroWorld. We would also like to thank the members of the International Scientific Advisory Committee, as well as the reviewers, for their advice, which has certainly helped to improve the quality, accuracy and relevance of this conference Program and publications.

Finally, we would like to thank *BIOMEDAL S.L.-Advances for the Postgenomic Era* (*http://www.biomedal.es*) for sponsoring the Conference.

Hoping that this edition will stimulate further meetings focusing on the interdisciplinary nature of current relevant applied research, we hope that readers will find the content fruitful and interesting.

Antonio Mendez-Vilas
Editor

FORMATEX, Badajoz, Spain
June, 2006

List of Contents

Industrial Microbiology – Future Bioindustries

Food Microbiology

Agriculture, Soil, Forest Microbiology

Bioremediation

Microbial Biotechnology

Microfactories – Microbial Production of Chemicals and Pharmaceuticals. Biopolymers

Microbial Physiology, Metabolism and Gene Expression

Medical Microbiology

Analytical Techniques, Imaging Techniques, Microscopy

Methods in Basic and Applied Microbiology. Microbiology Education

Environmental Microbiology, Marine Microbiology, Water/Aquatic Microbiology, Geomicrobiology

A new potential indicator of virological contamination of surfaces

M. Verani[1], **B. Casini**[1], **E. Rovini**[1], **P. Paone**[1], **A. Mansi**[2], **R. Lombardi**[2], and **A. Carducci**[1]

[1] Department of Experimental Pathology, Medical Biotechnologies, Infectivology and Epidemiology, University of Pisa, Via S. Zeno 35 56127 Pisa, ITALY
[2] Institute of Occupational Safety and Prevention, Department of Occupational Hygiene, Rome, ITALY

Keywords: TTVirus, Viral indicator, environmental monitoring

1 Introduction

TT virus is a widespread infectious agent of humans identified in 1998 (1). The studies carried out till now have evidenced a large diffusion of the virus in the world population with a persistent viremia in infected people characterized by a low pathogenic potential. TTV is present in blood, serum, faeces, pharyngeal and nasal swabs and other biological fluids, it has a high environmental resistance and several ways of elimination (2,3). All this characteristics, linked to the technical problems and great limits that present the direct research of pathogenic viruses in environment, suggest the utilization of TT virus as a new possible indicator of presence of haematic and entero-oral transmission viruses (4). TTV could be researched in particular settings like hospitals, clinical laboratories, etc. in witch the presence, at high concentration, of these pathogenic viruses represent a real risk for health-care workers (5).

2 Materials and methods

The study was carried out in two stages:

2.1 Estimating the sensitivity limits of the technique for TTV detection on surfaces artificially contaminated.

2.2 Environmental monitoring utilising the moist sensible technique.

2.1.1 Preparation of artificial samples

The serum from a TTV virus-infected subject with viral title of 46×10^6 copies/ml has been diluted from 10^{-1} to 10^{-4} in bovine fetal serum negative for TTV-DNA presence. 10 microliters of the whole serum and the dilutions, in addition to a negative control were spread on a sterilized stainless steel plate. The plate was then dried for about 10 minutes.

2.1.2 Estimation of sensitivity limits of sample purification and DNA extraction methods

Two analytical protocols have been compared on artificial samples:

a) Two eluents were tested contemporary: beef extract (BE) 3% at pH 9 and bovine serum albumin (BSA) 1% with NaCl 0.85%. For elution trials, the surface was repeatedly wiped with a cotton swab impregnated with 1 ml of eluent. The swab was then dipped in a test tube containing 1 ml of the same eluent, kept in a refrigerator for 2 hours at 4°C during which time it was shaken every 15 minutes, and finally mixed with vortex. The eluates were then recovered and extracted using a commercial kit "QIAamp DNA Mini Kit" (Qiagen). The recovery test was repeated three times.

b) Application of a commercial kit (DNA IQ^{TM} SYSTEM), generally used in forensic field for the detection and purification of DNA present in several biological samples, on the body and on objects for personal use; it was modified in order to achieve the present work using the Lysis Buffer of the kit as eluent to wash the artificially contaminated surface (250 µl) and cutting the swab of 1 cm long to allow its introduction in a spin basket put in a centrifuge test-tube containing 50 µl of the same eluent. Subsequently the swab was treated as indicated by the protocol of the commercial kit. Also in this case the recovery tests were repeated four more times.

2.1.3 Qualitative detection and quantitative determination of the TTV genome by PCR

The qualitative detection of the TTV genome was carried out by "nested PCR", which uses specific primers drawn by the UTR region of the TTV genome (6). For every reaction negative and positive controls were inserted. In order to quantify the TTV genome the TaqMan-Applied Biosystems Prism 7700 (Poster City, California) system has been used, drawing the primers and probe from the UTR region (7).

2.2 Application of the selected method for the environmental monitoring

The most sensible technique is actually used for an environmental monitoring in different sites of an hospital associated with research of an other biological indicator: haemoglobin (Table 1). A total of 74 selecting points were sampled, chosen mainly for the high probability of becoming contaminated, such as work benches, centrifuges, biosafety cabinet, and other instrumentation.
For TTV and haemoglobin detection, cotton swabs soaked in the relative eluent were wiped repeatedly on area of 36 cm squares. For the detection of haemoglobin a kit used for the blood detection hidden in faeces (Kit OC-Hemocard-Alfabiotech®-Wasserman) was modified and applied to the purpose (8).

Table 1 Environmental monitoring

ENVIRONMENT	ANALYZED SAMPLES
Clinical Lab.	12
Surgery	31
Cardiac Unit Intensive Terapy	16
Rianimation	9
Surgery passage	5
TOT	74

3   Results and discussion

In all recovery tests, the elution technique using both eluents was able to reveal the TTV genome up to the dilution of 10^{-2}, while in those carried out by DNA IQ^{TM} System protocol (Promega Corporation USA) the maximum dilution testing positive was 10^{-3} (Table 2).

The medium viral titres measured by Real-time PCR (TaqMan) at the highest positive dilution for the tests using beef extract and serum bovine albumin for the elution, resulted respectively 8100 copies/ml and 7600 copies/ml (Table 3). Given the viral content of the original solution (46.000 copies/ml, for 10^{-2} dilution) it was estimated that these eluents gave respectively 17,6% and 16,5% recovery efficiency. With the DNA IQ^{TM} System protocol the medium viral titre measured at the dilution of 10^{-3}, was 3.900 copies/ml. This value represents 84,8% of recovery frequency in comparison with the expected 4.600 copies/ml.

Therefore the technique of recovery with DNA IQ^{TM} System has resulted the most efficient and it has been chosen for the following environmental monitoring.

Table 2 Results of recovery tests

Dilution of TTV serum	Presence of TTV in eluates		
	BE	SA	LB
Whole	3/3	3/3	3/3
10-1	3/3	3/3	3/3
10-2	3/3	3/3	3/3
10-3	0/3	0/3	3/3
10-4	0/3	0/3	0/3

BE: Beef Extract 3% pH 9
SA: bovine serum albumin 1% with NaCl 0,85%
LB: Lysis Buffer "DNA IQ SYSTEM" kit (Promega)

Table 3 Recovery percentage for each TTV dilution

DILUTION OF TTV SERUM	EXPECTED COPIES NUMBER	COPIES NUMBER/ml RECOVERED FOR EACH TEST		
		BE	SA	L B
Whole	46 x 105	380.000	356.000	1.900.000
10-1	46 x 104	132.000	99.000	105.000
10-2	46 x 103	8.100	7.600	68.000
10-3	46 x 102	NP	NP	3.900
% of recovery		17,6	16,5	84,8

The samples analysed until now are 74: 5 of these resulted positive for TTV-DNA (7%), while 14 for haemoglobin (19%) (Figure 1). Only in 2 samples is possible to evidence the contemporary presence of these two potential indicators.

6

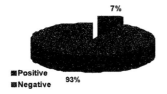
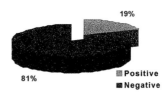

Fig. 1 Results of environmental monitoring

4 Conclusion

Results obtained from tests on artificially-contaminated samples show the technique applied, using for this purpose the DNA IQ^{TM} System kit (Promega Corporation USA), was of sufficient sensitivity for the TTV detection on surfaces in health settings. The sensitivity level is comparable to those obtained in the previous HCV studies, using the elution technique with beef extract (8). Despite this high sensitivity the study on field permitted to detect the TTV in only 5 samples. Nevertheless from the artificial samples to the study on field, is essential consider that the molecular techniques used were standardised for clinical materials such as serum or plasma (7), while the environmental samples besides to have inferior viral concentrations can contain many factors with interfere with the reactions of amplification. Then to exclude false negativity it would be necessary to set a system of internal positive controls (9).

At present the short number of samples doesn't permit to establish a correlation among haemoglobin and TTV, but environmental monitoring permitted to reveal critical points for the sanitization.

References

[1]   T. Nishizawa, H Okamoto., K Konishi., H Yoshizawa.,. Y Miyakawa, and M. Mayum. A novel DNA virus (TTV) associated with elevated transaminase levels in posttranfusion hepatitis of unknown etiology. *Biochem. Biophys. Res. Commun.* **241**:92-97 (1997).

[2]   K. Abe, T. Inami, K. Asano, C. Miyoshi, N. Masaki, S Hayashi, K. I. Ishikawa, Y. Takebe, K. M. Win, A. R. EI-Zayadi, K. H. Han, D. Y. Zhang. TT virus infection is widespread in the general populations from different geographic regions. *J. Clin. Microbiol.* **37**, 2703-2705 (1999).

[3]   T. Ishikawa, Y. Hamano, H. Okamoto Frequent detection of TT virus in throat swabs of pediatrie patients. *Infection* **27**, 298 (1999).

[4]   A. Carducci, C. Gemelli, L. Cantiani, B. Casini, E. Rovini. Assessment of microbial parameters as indicators of viral contamination of aerosol from urban sewage treatment plants. *Lett. In Appl. Microbiol.* **28 (3)**, 207-210 (1999).

[5]   A. Carducci, M. Verani, B. Casini, R. Lombardi, F. Maggi Un nuovo possibile indicatore di rischio virale occupazionale in ambienti sanitari . *Riv. It. Ig.* Printing (2004).

[6]   T. P. Leary, J.C. Erker, M.L. Chalmers, S.M. Desai, I.K. Mushahwar. Optimized PCR assay for the detection of TT virus. *J. Virol. Meth.* **82**, 109-112 (1999).

[7]   F. Maggi, C. Fornai, L. Zaccaro, A. Morsica, ML Vatteroni, P. Isola, S. Marchi, A. Ricchiuti, M. Pistello, M. Bendinelli. TT Virus (TTV) loads associated with different peripheral blood cell types and evidence for TTV replication in activated mononuclear cells. *J. Med. Virol.* **64,** 190-194 (2001).

[8]   A. Carducci, M. Verani, B. Casini, A. Giuntini, F. Mazzoni, E. Rovini, A. Passaglia, L. Giusti, A. Valenza, R. Lombardi. Detection and potential indicators of the presence of hepatitis C virus on surfaces in hospital settings. *Lett. In Appl. Microbiol.* **34**: 189-193 (2002).

[9]   J. Hoorfar, N. Cook, B. Malorny, M. Wagner, D. De Medici, A. Abdulmawlood and P. Fach. Making Internal Amplification Control Mandatory for Diagnostic PCR. *J. of Clin. Microbiol.* **41,** 5835 (2003).

A novel mechanism for bacterial acid resistance: a carbon dioxide-dependent system

L. Sun, O. Sutoh, Y. Kurokawa, T. Fukamachi, H. Saito, and H. Kobayashi*

Graduate School of Pharmaceutical Sciences, Chiba University, 1-8-1, Inohana, Chuo-ku, Chiba 260-8675, Japan

*Corresponding author: e-mail: hiroshi@p.chiba-u.ac.jp, Phone: +81-43-226-2890, Fax: +81-43-226-2892.

It has been reported that decarboxylation of amino acids such as glutamate, lysine and arginine induces bacterial acid resistance (AR). In the present study, expression of lysine and glutamate decarboxylase genes, *cadBA*, *gadA* and *gadB*, was found to be repressed by the CO_2 supply. If amines produced by decarboxylation of amino acids induced AR, the AR should decrease with the CO_2 supply, although the supply induced AR in *Escherichia coli* (Sun *et al.*, Lett. Appl. Microbiol. in press). These results led us to propose a novel mechanism by which amino acid decarboxylases supply CO_2 which is required for bacterial survival under extreme acidic conditions.

Keywords: carbon dioxide, *cadBA*, *gadA*, *gadB*, acid resistance.

1 Introduction

Environmental pH is an important factor for bacterial growth and survival, but bacterial habitats encompass a wide range of pH from 1 to 11 [1]. To infect animals and cause disease, it is important that bacteria overcome acidic stress to survive in the gastric stomach. It is known that environmental factors such as nutrient level, temperature and pH affect the induction of acid resistance (AR) in *Escherichia coli*. In addition, glutamate, lysine and arginine decarboxylases can help it to survive during extreme acidic exposure [2–6].

Amino acid decarboxylases produce amine and carbon dioxide. It has been proposed that amines increase the acidic pH of both the cytoplasm and the outside medium, and that this neutralization is important in inducing AR. However, adaptation with amino acids induced AR even if the change in medium pH during the adaptation stage was small [7]. Recently, it was shown that the internal pH is not a primary factor for acid tolerance [8].

E. coli cells adapted at pH 5.5 before the acid challenge were more resistant at pHs below 3 [9]. It has been demonstrated recently that CO_2 strongly induced the AR of *E. coli* and the CO_2 supply was more effective in the adaptation stage at pH 5.5 [7]. In this study, CO_2 was found to repress expression of lysine and glutamate decarboxylase genes in *E. coli* at acidic pH, making it less likely that amines produced by these enzymes have an essential role in the induction of acid resistance. Based on these results, a novel mechanism for acid resistance has been proposed; amino acid decarboxylases provide CO_2 for CO_2-dependent metabolism that is required for survival under acidic environments.

2   Materials and methods

2.1 Bacterial strains and growth conditions

The strains used in this study are listed in Table 1. *E. coli* strains were grown at 37°C with vigorous shaking or in a tightly capped tube using media M655, M675 and M680 [10]. When CO_2 was supplied, the culture tube was connected with a rubber balloon containing CO_2 (Fig. 1). Kanamycin (25 µg/ml), tetracycline (10 µg/ml), hygromycin (100 µg/ml), and chloramphenicol (20 µg/ml) were used. The *hemA* mutant was grown in the presence or absence of 80 µg/ml 5-aminolevulinic acid (ALA). Bacterial growth was monitored by the absorbance of the culture media at 600 nm. The gene deletions of all strains used in this study were checked by PCR.

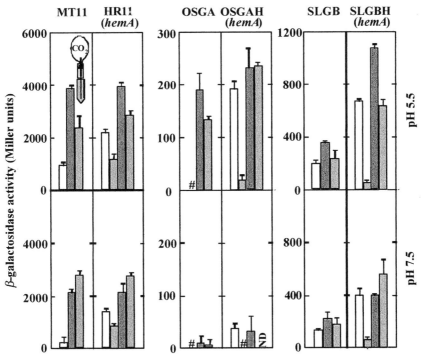

Fig. 1 Effect of medium pH and CO_2 on expression of *cadBA*, *gadA* and *gadB*. *E. coli* strains were grown in medium M655 (pH 5.5) and M675 (pH 7.5) containing glucose (1%). When MT11 and HR11 were grown, lysine (20 mM) was added. β-galactosidase activity was assayed in duplicate in at least four independent experiments. Each point represents the mean ± S.D. Symbols: white bars, cultured with shaking; striped bars, cultured with shaking in the presence of ALA (80 µg/ml); gray bars, cultured without shaking; dotted bars, cultured without shaking with the CO_2 supply. #, activities less than 20 Miller units. ND, not determined.

2.2 Other methods and chemicals

The promoter regions of *gadA* (from -1434 to +20) and *gadB* (from -1212 to +296) amplified with PCR were fused with *lacZ*, and strains containing *gadA-lacZ* or *gadB-lacZ* fusion genes were constructed, as described previously [11]. Various genes were transferred with P1-mediated transduction [12]. β-galactosidase activity of strains harvested in the mid-logarithmic phase was measured as described previously [13]. Enzyme activity was expressed in Miller units. Reagents used were of analytical grade.

Table 1. Bacterial strains used in this study

Strains	Relevant genotype, phenotype, or description	Origin or reference
MC4100	*araD139Δ (lac) U169 str A thi*	[25]
MT11	ST11 Φ (*cadB-lacZ*)(Km^r)	[10]
OSGA	MC4100 Φ (*gadA-lacZ*)	This study
OSGAH	OSGA *hemA*::Km^r	This study, OSGAxP1(H500[26])
SLGB	MC4100 Φ (*gadB-lacZ*)	This study
SLGBH	SLGB *hemA*::Km^r	This study, SLGBxP1(H500[26])
HR11	MT11 *hemA*::Tn*10*(Tet^r)	This study, MT11xP1(ME8366)
SL1102	MT11 *lrp*::Tn*10*(Tet^r)	This study, MT11xP1(BE1[27])
SL1103	MT11 *rpoS*::Tn*10*(Tet^r)	This study, MT11xP1(EF362[28])
SL1104	MT11 *crp*::Cm^r	This study, MT11xP1(EF528[28])
SL1111	MT11 *hns*::*hph*(Hy^r)	This study, MT11xP1(YK4124[29])

Km^r, Kanamycin resistant; Tet^r, Tetracycline resistant; Cm^r, Chloramphenicol resistant; Hy^r, hygromycin resistant. ME8366 (*hemA*::Tet^r) was obtained from Genetic Stock Center of National Institute of Genetics, Mishima, Japan.

3 Results

3.1 Effect of CO_2 on the expression of *cadBA*, *gadA* and *gadB*

Expression of the *cad* operon (*cadBA*), *gadA* and *gadB* was induced strongly under O_2-limiting conditions at pH 5.5. Expression of these genes was repressed by the CO_2 supply under such conditions (Fig. 1). The CO_2 supply again decreased expression of *gadA* and *gadB* at pH 7.5, although the expression level was low (Fig. 1). These results suggest that expression of these genes is regulated by CO_2.

3.2 Deletion of the respiratory chain increased the expression of *cadBA*, *gadA* and *gadB*

The main metabolic pathway to produce CO_2 is the TCA cycle under aerobic growth conditions. The activity of the TCA cycle is dependent on respiratory activity; the production of CO_2 decreases with reduced respiratory activity. The deletion of the *hemA* gene encoding the enzyme to synthesize ALA increased the expression of the *cad* operon under aerobic growth conditions at pH 5.5 and 7.5 (Fig. 1). The increase in *gadA* expression was more dramatic at pH 5.5 (Fig. 1). When ALA was added, the expression levels of these genes decreased (Fig. 1). These results suggest that the genes for lysine and glutamate decarboxylases were expressed when respiration was less active due to a limited supply of CO_2.

3.3 Effect of the global regulators Lrp, RpoS, CRP and H-NS on the expression of the *cad* operon

Lrp (leucine responsive regulatory protein), RpoS (sigma38), CRP (cAMP receptor protein) and H-NS (histone like protein) are global regulatory proteins in *E. coli*, which affect the expression of numerous genes at the transcriptional level [14, 15]. It was shown that no significant expression was detected in *lrp*, *rpoS* or *crp* mutants under aerobic conditions at alkaline conditions of pH 7.5 (Fig. 2). In contrast, mutations lacking *lrp*, *rpoS* and *crp* had no significant effect on the *cad* operon expression at pH 5.5 (Fig. 2).

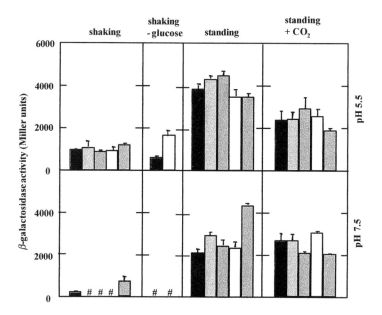

Fig. 2 Lrp, RpoS and CRP had no essential effect on expression of the *cad* operon at acidic pH. *E. coli* MT11 (wild type, black bars), SL1102 (*lrp*⁻, gray bars), SL1103 (*rpoS*⁻, hatched bars) and SL1104 (*crp*⁻, open bars) were grown in medium M655 (pH 5.5) and M675 (pH 7.5) containing lysine (20 mM) and glucose (1%). SL1111 (*hns*⁻, dotted bars) was grown in the same media of pH 5.5 and 8.0. SL1104 was also grown without the addition of glucose. β-galactosidase activity was assayed and is represented as described in the legend of Fig. 1.

The deletion of *hns* increased the *cad* operon expression only at pH 8.0 (Fig. 2), in agreement with a previous report [16]. No such increase was observed when CO_2 was supplied (Fig. 2). Deletion of *crp* stimulated the expression of the *cad* operon in the presence of a limited amount of glucose at pH 5.5 (Fig. 2). It should be noted that tryptone contains approximately 7 % carbohydrate [17].

4   Discussion

E. coli has amino acid decarboxylase systems such as glutamate, lysine, arginine and ornithine decarboxylases to overcome extremely acidic conditions [18]. It has been accepted for a long time that the conversion of these extracellular amino acids to amines neutralizes acidic media and that medium neutralization is the major factor inducing AR. However, the present results showed this possibility less likely, because CO_2 repressed amine production via lysine and glutamate decarboxylation (Fig. 1), but induced AR under the same conditions [7]. Based on these results, we propose that the production of CO_2 via amino acid decarboxylases participates in the induction of AR and that their genes are expressed with a decrease in the CO_2 level.

CO_2 is supplied mainly from the TCA cycle coupled with respiration under aerobic growth conditions. The present study demonstrated that the expressions of *cadBA*, *gadA* and *gadB* were stimulated by a deficiency in the respiratory chain, even if enough air is supplied. CO_2 is again supplied from air under aeration. The respiratory activity decreased at acidic pH [19]. Simultaneously, the medium level of carbonate decreased at low pH, because a high percentage of carbonate converts to CO_2 at pH below pKa (6.3). The level of carbonate plus CO_2 decreases at low pH. Therefore, the CO_2 supply could be limited under conditions of external low pH and anaerobiosis. The genes encoding lysine and glutamate decarboxylases were expressed under these conditions [20, 21].

The AR induced under aerobic oxidation was repressed by glucose [4]. It is generally accepted that glucose metabolism represses respiratory activity and that CRP has some participation in the regulation [22]. Our data demonstrated that the deletion of *crp* increased the *cad* operon expression in the presence of a limited amount of glucose at acidic pH (Fig. 2).

It remains unclear how CO_2 protects bacteria against cell death at low pH. Adaptation treatment was suggested to cause transient synthesis of some key acid shock proteins to protect cells against extreme acidic pH [6, 23]. CO_2 is a substance for biosynthesis of carbamoylphosphate, the first metabolite of the pyrimidine nucleotide synthetic pathway. Thus, the maintenance of some types of metabolism at a high level may be required for survival under extremely acidic conditions. Carbonate is secreted in the human esophagus and from the salivary glands [24], resulting in an increase in the CO_2 level in the stomach. CO_2 is also supplied from food. Therefore, it would be physiologically significant if metabolism using CO_2 has some role in supporting bacterial life under acidic conditions.

Acknowledgements

We are grateful to John W. Foster, Rowena G. Matthews, Akiko Nishimura and Yasunobu Kano for the generous gift of strains.

References

[1] T. A. Langworthy, In Microbial Life in Extreme Environments, Kushner, D. J. (ed.). London: Academic Press, pp. 279 (1978).

[2] E. F. Gale, Adv. Enzymology 6, 1 (1946).

[3] D. E. Guilfoyle, and I. N. Hirshfield, Lett. Appl. Microbiol 22, 393 (1996).

[4] J. Lin, Lee, S. I. J. Frey, J. L. Slonczewski, and J. W. Foster, J. Bacteriol 177, 4097 (1995).

[5] J. Lin, M. P. Smith, K. C. Chapin, H. S. Baik, G. N. Bennett, and J. W. Foster, Appl. Environ. Microbiol 62, 3094 (1996).

[6] J. W. Foster, J. Bacteriol 173, 6896 (1991).

[7] L. Sun, T. Fukamachi, H. Saito, and H. Kobayashi, Lett. Appl. Microbiol 40, 397 (2005).

[8] H. Richard, and J.W. Foster, J. Bacteriol 186, 6032 (2004).

[9] J. W. Foster, and H. K. Hall, J. Bacteriol 172, 771 (1990).

[10] M. Takayama, T. Ohyama, K. Igarashi, and H. Kobayashi, Mol. Microbiol 11, 913 (1994).

[11] T. Shijuku, T., Yamashino, H. Ohashi, H. Saito, T. Kakegawa, M. Ohta and H. Kobayashi, Biochim. Biophys. Acta 1556, 142 (2002).

[12] E. S. Lennox, Virology 1, 190 (1955).

[13] J. H. Miller, Cold Spring Harbor Laboratory, Cold Spring Harbor, N.Y. (1972).

[14] F. Colland, M. Barth, R. Hengge-Aronis, and A. Kolb, EMBO. J 19, 3028 (2000).

[15] Z. Ma, H. Richard, and J. W. Foster, J. Bacteriol 185, 6852 (2003).

[16] X. Shi, B. C. Waasdorp, and G. N. Bennett, J. Bacteriol 175, 1182 (1993).

[17] E. Yohannes, D. M. Barnhart, and J. L. Slonczewski, J. Bacteriol 186, 192 (2004).

[18] C. W. Tabor, and H. Tabor, Microbiol. Rev 49, 81(1985).

[19] E. Padan, D. Zilberstein, and H. Rottenberg, Eur. J. Biochem 63, 533 (1976).

[20] S. Y. Meng, and G. N. Bennett, J. Bacteriol 174, 2659 (1992a).

[21] S. Y. Meng, and G. N. Bennett, J. Bacteriol 174, 2670 (1992b).

[22] T. Inada, K. Kimata, and H. Aiba, Genes. Cells 1, 293 (1996).

[23] J. W. Foster, J. Bacteriol 175, 1981 (1993).

[24] C. M. Brown, C. F. Snowdon, B. Slee, L. N. Sandle, and W. D. Rees, Dig. Dis. Sci 40, 1642 (1995).

[25] M. J. Casadaban, J. Mol. Biol 104, 541 (1976).

[26] T. Nakayashiki, K. Nishimura, R. Tanaka, and H. Inokuchi, Mol. Gen. Genet 249, 139 (1995).

[27] B. R. Ernsting, M. R. Atkinson, A. J. Ninfa, and R. G. Matthews, J. Bacteriol 174, 1109 (1992)

[28] M. P. Castanie-Cornet, T. A. Penfound, D. Smith, J. F. Elliott, and J. W. Foster, J. Bacteriol 181, 3525 (1999).

[29] K. Yasuzawa, N. Hayashi, N. Goshima, K. Kohno, F. Imamoto, and Y. Kano, Gene 122, 9 (1992).

An individual based model to study the main groups of microbes active in composting process

A. Gras[*1], **C. Prats**[2], and **M. Ginovart**[3]

[1] College of Agriculture of Barcelona
[2] Department of Physics and Nuclear Engineering
[3] Department of Applied Mathematics III
Technical University of Catalonia, Campus Baix Llobregat, Avda. Canal Olímpic s/n, 08860 Castelldefels (Barcelona), SPAIN

[*]Corresponding author: e-mail: anna.gras@upc.edu, Phone: +34 935521224, Fax: +34 935521001

The computer code INDISIM, which share the philosophy of the individual based models, has been adapted to several microbial systems with success. INDISIM -COMP is an extension of INDISIM that enables us to study microbial activity in the composting process. This first individual based, discrete aproach takes into account the evolution and behaviour of different types of organic and mineral compounds, and three different prototypes of microorganisms.

Keywords Composting process; Individual-based Model

1 Introduction

Composting has been defined as "the biological decomposition and stabilization of organic substrates under such conditions that allow development of thermophilic temperatures as a result of biologically produced heat, with a final product sufficiently stable for storage and application to land without adverse environmental effects". This definition differentiates composting from the mineralization of dead Organic Matter (OM) taking place in nature above the soil. It describes the compost pile as a man-made microbial ecosystem. Nevertheless, the microbiology of composting is somehow related to soil microbiology, turnover of OM in nature and formation of humic substances. The microbiology of self-heating of moist OM has been studied in the case of agricultural products. This phenomenon led to the concept of heat generation as part of microbial metabolism. If the composting process is carried out as a batch culture it proceeds in various phases which are recognised superficially by the stages of temperature rise and decline. These temperature phases are the reflection of the activity of successive microbial populations performing the degradation of labile OM and increasingly more stable OM [1]. Despite the many facts describing this degradation qualitatively, there are no standard mathematical or computational models of composting processes. There are, however, many ways in which relevant knowledge and hypotheses can be expressed in models [2,3]. Composting and soil processes share a variety of simultaneously assimilated substrates and microorganisms.
A discrete methodology which shares the philosophy of the Individual based Models (IbM) has been developed to study bacterial cultures and microbial activity in the mineralization and immobilization of soil carbon and nitrogen and nitrification process [4,5]. The aim of this work is to develop and to present an IbM to deal with the composting process, a simu -lator code called INDISIM-COMP. This takes into account the evolution and behaviour of different types of organic and mineral compounds, and three different prototypes of microorganisms (mesophilic bacteria, termophilic actinomycetes and mesophilic fungi).

The cell model controls the following actions for each individual of this microbial population. The medium model considers mass and heat transfer processes, hydrolysis reactions of the complex to simple compounds, output/input flows of gaseous compounds, and diffusion of soluble organic and mineral compounds. Simulation results are presented in order to discuss specific aspects related with the individual behaviour of the cells and the microbial successions through the composting process. The individual approach is a bottom-up approach which starts at the bottom level of the systems investigated, i.e. at the individual level, and then tries to understand how the system's properties emerge from the interaction among these parts. To our knowledge this is the first study of the composting process using individual-based simulations. This paper presents a bottom-up approach for the composting microbial system involved in the OM degradation. The formulation of these kinds of models is both a feasible alternative and a valuable tool to investigate complex systems, the dynamics of which are mainly driven by microorganisms. Such dynamics emerge mainly from the actions of microbial cells and their interactions with the environment and with each other.

2   Simulation model

2.1 General outline of the simulation model INDISIM-COMP

A computer code called INDISIM (INDividual DIScrete SIMulations) was developed by Ginovart et al. [4] specifically to study bacterial cultures, and it was devised to deal with systems in which bacterial activity is one of the fundamental parts of the system. These simulation models consider each element individually, each following its own set of rules. The use of random variables is required to define them, which enables the introduction of variability in the individual behaviour. Since the simulated systems are built from both biotic and abiotic elements, the differences between microorganisms and spatial heterogeneity can be readily accounted for. INDISIM-SOM, an extension of INDISIM to deal with the Soil Organic Matter (SOM), is an individual simulation model that enables us to study microbial activity in soil, dealing with the mineralization and immobilisation of C and N, and the integration of the nitrification process in this context [5,6]. INDISIM-SOM is focused on the microbial activity and assumes two different prototypes of cells: decomposer microorganisms and nitrifier bacteria. It also takes into account the role of C and N during their microbial lives, thus linking the C and N cycles. INDISIM-SOM has been calibrated using experimental data from incubations of different soils [5,6]. INDISIM-COMP, another extension of INDISIM, is focused on the microbial activity in the composting process and shares some of the ideas developed and implemented in INDISIM-SOM. INDISIM-COMP is discrete in space and time. The physical domain is subdivided into squared spatial cells. The time evolution of the system is divided into equal intervals that we identify with time steps. The behaviour of microbial population is specifically considered, taking into account their motion, uptake, metabolism, reproduction, viability and lyses of each individual that makes up the system. At each time step and for each microorganism the simulator controls its own time dependent properties: biomass, position in the spatial domain, reproduction biomass, state of its cellular cycle, and its internal C/N ratio. These microorganisms and different substrate particles are evolving in a two-dimensional spatial grid. Diverse considerations make it possible to set specific rules for the simulated microbial population and for the spatial domain where this population evolves. Figure 1 shows the sketch of the INDISIM-COMP model. After modelling each

16

part of the system and implementing the overall model in a computer code, we follow the activity of the microbial populations and the different pools of matter acting together. At each time step we have a complete temporal and spatial description of the simulated system.

2.2 Medium and space model

NDISIM-COMP controls different types of substrates, six of them are organic substrates namely: polymerised carbon (C_P), polymerised nitrogen (CN_P), labile carbon (C_L), labile nitrogen (CN_L), resistant carbon (C_R) and stable organic matter (CN_{MOE}), and the others are mineral compounds. It controls compounds in the gaseous phase of the compost like CO_2 or O_2, and water vapour (($H_2O)_{vapour}$), and mineral compounds in the liquid phase like water $(H_2O)_{liquid}$ and ammonium (NH_4^+). The organic substrates in the simulation model have been chosen according to their bioavailability, their C/N ratio, and whether they were monomeric or polymeric compounds. It is assumed that: (i) Soluble and/or labile compounds represent glucose and fatty acids, substances without N in their molecules (C_L) and aminoacids like substance with N in them (CN_L), (ii) Polymerised compounds are polymers of compounds described before (C_P and CN_P), (iii) C_R represents the structural matter of the mass compost added to improve the structure and facilitate the aeration, (iv) CN_{MOE} is the end organic product of the composting process. The   simulation   model considers: (i) the output flow of CO_2 and $(H_2O)_{vapour}$ from the system, (ii) the input flow of O_2, (iii) the diffusion of the soluble organic labile forms and mineral compounds in the medium, (iii) the mass transfer reactions from complex to labile compounds in each spatial cell and at each time step, the hydrolysis processes rates are related to the actinomycets and fungi biomass, (iv) the heat transfer processes which are driven mainly by the evaporation of water and the aeration of the compost. These processes play an important role to keep the mass between the optimal ranges of temperature that allow a good composting progress.

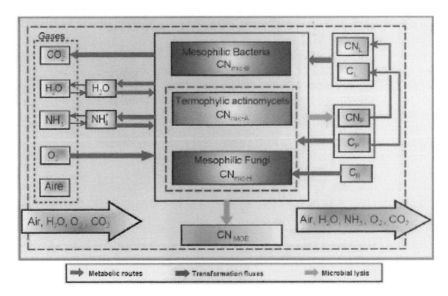

Fig. 1 Sketch of the composting process model INDISIM-COMP.

2.3   Microbial cell model

We consider in the simulated system three microorganims groups: the mesophilic bacteria, the thermophilic actinomycets and the mesophilic fungi. They differ on their microbial biomass composition defined by their C/N ratio, their optimum temperatures to develop and their metabolisms. All microorganisms are modelled with heterotrophic metabolisms. The difference between them is the capability they have to metabolise polimerized and/or resistant organic compounds. For each microbial cell, with a set of time dependent variables, we control the individual properties, its biomass; its position in the spatial domain; its reproduction biomass; its state of cellular cycle and its biomass C/N ratio. At each time step, a microorganism may perform the following actions controlled by our computer code: (a) Motion. The cell has a position in the space and it can move at random from one spatial cell to another adjacent spatial cell. (b) Uptake. The cell uptakes some of the substrate particles surrounding it, from the spatial cell that it is occupying. The individual maximum uptake rate depends on temperature following Ratkowsky model [7]. The availability of substrates in the medium increases with temperature until a maximum is reached. (c) Metabolism. The substrate particles the cell has uptaken are metabolised following specific metabolic pathways and as a consequence of these reactions the heat dissipation takes place. (d) Reproduction. The bipartition model used in INDISIM has been assumed [4]. (e) Death and lysis. When the cell has no possibility of obtaining its maintenance requirements, neither from the external source nor from its own biomass, cellular lyses occurs, its biomass is released to the medium and it is transferred to CN_{MOE}, C_P, CN_P and N_{NH4} in order to balance carbon and nitrogen.

3   Results

The simulation results obtained with the computer code INDISIM-COMP and presented in this work must be regarded as preliminary and interesting. The first simulations investigated the interactions between different types of microorganisms and substrates, microbial biomass growth, substrate concentration and temperature. Microbial activity can be represented, in broad outline, by CO_2 production. The simulated system develops in a changing environment mainly due to the diverse effects that the microbial activity has in the medium. The succession of the different microbial prototypes is clearly observed in figure 2 that shows the temporal evolution of the mesophilic bacteria, the thermophilic actinomycets and the mesophilic fungi, the three microbial populations considered in the model. Figure 3 shows the cumulative CO_2 produced by the microbial activity and directly related to the individual cell metabolism.

18

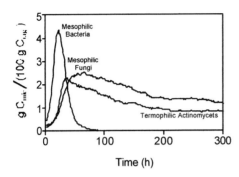

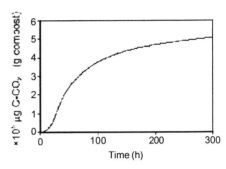

Fig. 2 Simulated temporal evolution of the three microbial populations.

Fig. 3 Simulated cumulative CO_2, produced by the microbial populations.

It is possible to further develop INDISIM-COMP so as to explore a wide range of phenomena in microbial composting process activity, and that is precisely what we are committed to do.

Acknowledgements

The financial supports of the Ministry of Education and Science of Spain Grant REN2000-0049-P4-04, the DURSI Generalitat de Catalunya 2003ACES00064, and the Ministry of Science and Education Plan Nacional I+D+i CGL 2004-01144.

References

[1]   J. Klein and J. Winter, Biotechnology, Volume 11c, Environmental Processes III (2003).
[2]   J. Kaiser, Ecological Modelling **91**, 25 (1996).
[3]   F. Solé, J. Illa and X. Flotats, Proceedings of RAMIRAN'04, 11th International Conference of the FAO ESCORENS Network on Recycling of Agricultural, Municipal and Industrial Residues in Agriculture, Murcia, Spain, (2004).
[4]   M. Ginovart, D. López and J. Valls, Journal of Theoretical Biology **214**, 305 (2002).
[5]   A. Gras and M. Ginovart, in: Proceedings of the IASTED International Conference on Applied Simulation and Modelling, Rhodes, Greece, 2004, (ACTA Press), pp. 125-129.
[6]   M. Ginovart, D. López and A. Gras. Nonlinear Analysis: Real World Applications (in press).
[7]   D.A Ratwosky et al. Journal of Bacteriology **154**, 1222 (1983).

Biotechnological approach for treatment of textile wastewater – A case study

R.A. Pandey*, M.v. Lakshane, K.V. Padoley, T.V. Subbarao and S.N. Kaul

National Environmental Engineering Research Institute, Nagpur, India

*Corresponding author: e-mail: rapandey@hotmail.com; Ph. +91-0712- 2240097, Fax No : +91-712-2249900

An existing full-scale biological treatment plant for treating textile wastewater comprising of an anaerobic up-flow packed bed reactor followed by aerobic completely mixed activated sludge (CMAS) system was evaluated with reference to physico-chemical and biological parameters. The full-scale anaerobic biological treatment plant, operated with an average COD load of 8.46 Kg $COD/m^3/d$, could remove only 48 % of COD from textile wastewater at a HRT of 5.3 hours. The CMAS unit of full-scale treatment system was operated at optimal conditions with a load of 0.189 Kg COD/Kg MLSS/d with residual COD and resulted in the removal of 54.23 % of COD at a HRT of 23.6 hours. The overall COD removal of the full-scale treatment system was 76.20 %. The biomass of the full-scale CMAS treatment system contained the dominant microorganisms of the genera of *Nocardia, Bacillus, Bacteriodes* and *Cardiobactrium.*

A bench-scale unit having similar set up as that of full-scale, using the textile wastewater from the same industry, was investigated. The bench-scale up-flow anaerobic packed bed system operated with an average loading rate of 12 Kg $COD/m^3/d$, could remove 53.4 % of the COD at a HRT of 3.8 hours. Further, the bench-scale CMAS unit operated at optimal conditions with an average load of 0.39 Kg COD/Kg MLSS/d and at a HRT of 23.6 hours could remove 78.5 % of COD from the partially treated textile wastewater. The overall COD removal of the bench-scale treatment system was 89.97% .The biokinetics of CMAS system have also been evaluated.

Topic + Keywords: Applied Environmental Microbiology, Textile wastewater, Biological treatment, Activated Sludge Process, Anaerobic treatment, Decolorization, Microorganisms

1   Introduction

The textile industries in India are generally of three types, based on the scale of operation; viz; small, medium and large-scale textile industries. Most of the textile industries have problem in the treatment of textile wastewater using existing effluent treatment plant. Several processes have been cited in the literature for treatment of textile wastewater and these include physical, chemical and biological and sometimes combination of these processes (**Table 1**). In order to understand the deficiency in the treatment of textile wastewater, an investigation was carried out by selecting a full-scale operating biological treatment plant for textile wastewater vis-à-vis a bench-scale biological treatment unit having similar stages of the treatment.

2   Materials and methods

The schematics of a selected full-scale biological treatment system and a bench scale system consisted of a packed bed upflow anaerobic (PBUA) reactor and a completely mixed activated sludge (CMAS) were used in this investigation are presented in **Figs. 1** and **2** respectively and operational parameters are given in **Table 2**. The textile wastewater was supplemented with dipotassium phosphates and potassium hydrogen phosphate (K_2HPO_4

20

and KH_2PO_4) along with ammonium sulphate as nutrients and pH was adjusted in the range of 7.3 –7.5, using mineral acid / sodium hydroxide and this wastewater was fed to the biological treatment system.

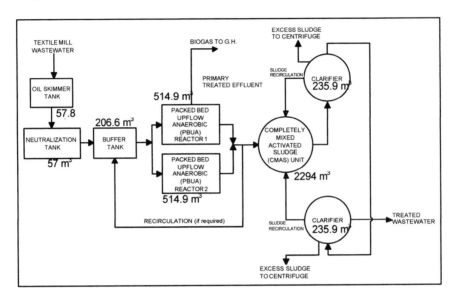

Fig. 1 Schematics of full-scale biological treatment system for textile wastewater

Table 1 Different physico-chemical and biological treatment options cited in literature for the treatment of textile wastewater

Treatment options	Reference
Adsorption on activated carbon	Faria et al. (2004);
Adsorption on modified TiO_2 surface	Andrezejewska (2004)
Ozonation	Muthukumar and Selvakumar (2004)
Fenton's treatment	Meric et al. (2004)
Coagulation / carbon adsorption combined treatment	Papi et al. (2004)
Advanced oxidation process (photocatalysis)	Lim et al. (2004)
Activated sludge process	Tokal (2002)
Anaerobic treatment	Islk and Sponza (2004)
Anaerobic – aerobic system	Libra et al. (2004); Panswad and Luangdilok (2000)
Anaerobic / oxic system	Qian et al. (1996)
Anoxic + anaerobic / aerobic SBR activated sludge process	Panswad and Techovanich (2002);
Anaerobic / aerobic sequential process	Rott and Minke (1989)

Table 2 Operational parameters of full-scale and bench scale biological treatment system for textile wastewater at optimal conditions

Parameters	Full scale system		Bench scale system	
	PBUA	CMAS	PBUA	CMAS
Flow rate	2340.4 m³/day		21.6 L/day	---
Dissolved oxygen (mg/L)	---	0.6 – 0.8	---	1.2 – 1.5
HRT (hours)	5.3	23.6	3.8	23.6
Loading kg COD/m³/day	8.46	0.987	12	0.9
kg COD/kg MLSS/day	---	0.189	---	0.39
Load removal rate for PBUA reactor (kg/m³/day)	0.48	---	0.53	---
Load removal rate for CMAS unit (kg COD / kg MLSS / day)	---	0.54	---	0.78
MLSS (mg/L)	---	5000	---	2280
MLVSS (mg/L)	---	3000	---	1800
Total Viable count (Nutrient agar) CFU/ml	---	6×10^4	---	30×10^7
Specific count (Wastewater – as a carbon source-agar) CFU/ml	---	5×10^3	---	18×10^6

Biokinetic constants	Maximum reaction rate $(V_{max}) = 6.82$ kg COD / m³ / day; Half velocity concentration $(k_s) = 0.464$ g/L	Yield coefficient (Y) = 0.22 mg biomass / mg/ COD; Decay coefficient (kd) = 0.02 hr⁻¹; Half saturation rate constant (ks) = 284.74 mg/L as COD; Maximum specific growth rate $(\mu_{max}) = 0.34$ hr⁻¹

Contd…

MLSS: Mixed Liquor Suspended Solids; MLVSS: Mixed Liquor Volatile Suspended Solids; CFU/ml: Colony forming units/ml; All values are average of five sets of observations

As mentioned in the preceeding section a full-scale existing biological wastewater treatment system of a textile industry was selected and investigated for evaluation of the performance parameters. Grab samples were collected on hourly basis and composited for 24 hrs and analysed by adopting Standard Procedures (APHA.AWWA.WEF, 1995). Thirteen pure cultures were isolated, from the biomass of the existing full-scale aerobic treatment plant, by adopting Standard Procedures and were designated as C1 - C13.

The sludge from the full-scale PBUA treating textile wastewater was used as a starter seed for the bench scale PBUA unit. The textile wastewater was recycled in a closed loop till the performance of anaerobic system in terms of COD removal was consistent. The isolated potential cultures were used for development of biomass (2500 MLSS) in the CMAS unit through fill and draw method using textile wastewater. The combined PBUA – CMAS reactor system was operated in series at different hydraulic retention times (HRTs) and COD loadings and optimized for the process parameter.

3   Results and discussion

Evaluation of the existing full-scale biological treatment system

The textile industry selected for investigation produced 80,000 meter of cloth/day, leading to the generation of wastewater to the magnitude of $2400 \pm 200 m^3$/day. The performance data under optimal conditions for both the systems are presented in **Tables 3**. The average optimal volumetric loading rate to the PBUA unit was 8.46 kg COD/m^3/day with a HRT of 5.3 hrs, resulting in 48% of COD removal. The CMAS unit of the existing system was operated at 0.189 kg COD/kg MLSS/day with a HRT of 23.6 hrs at optimal condition with a COD removal efficiency of 54.23%. The overall removal of COD and BOD in the existing full-scale treatment system for textile wastewater at the industry was found to be 76.2 % and 80 % respectively. The dissolved oxygen (DO) level in the contents of CMAS of the full-scale system was found to be 0.6-0.8 mg/L, which was due to improper installation and inadequate number of fixed aerators in the system. This resulted in poor removal of COD from the textile wastewater. As per the conventional treatment system, the minimum DO level in the CMAS must be 1-1.5 mg/L   (Metcalf and Eddy, 1996). The total bacterial count was observed to be $6x10^4$ CFU/ml, while the average specific bacterial count using, wastewater as a carbon source, indicated a magnitude of $5x10^3$ CFU/ml.

Table 3 Performance of full-scale and bench-scale biological treatment systems operating at optimal conditions

Parameter	Full-scale system				Bench-scale system				*Standards for discharge into water bodies
	PBUA Influent	PBUA Effluent	CMAS Influent	CMAS Effluent	PBUA Influent	PBUA Effluent	CMAS Influent	CMAS Effluent	
pH	6.9 – 7.2	7.2 – 7.5	7.2 – 7.5	7.9 – 8.0	7.0±0.25	7.3±0.2	7.3±0.2	7.7±0.2	5.5 – 9.0
COD (mg/L)	1800±50	900±20	500±30	380 – 430	1800±50	800±30	800±30	180±20	250
BOD (mg/L)	850±50	400±20	200±30	76 – 86	770±50	350±50	350±50	16±10	30

* Guidelines for discharge of effluent, Central Pollution Control Board (CPCB), Ministry of Environment and Forests, Govt. of India, New Delhi (1996)

Isolation, characterization and identification of microorganisms present in the biomass of the CMAS of full-scale treatment system

The biomass of the full scale CMAS unit showed the presence of microorganisms belonging to genus *Brevicatena, Cardiobacterium, Bacteriode, Bacillus subtilis, Nocardia brasiliensis* and *Nocardia brevicatena* respectively (Bergey's Manual, 1984, 1986). Out of the thirteen isolates, three cultures showed the potential for effective biodegradation of the textile wastewater (C11 - *Bacillus subtilis*, C12 - *Nocardia brasiliensis* and C13 - *Nocardia brevicatena*).

Bench scale treatment of textile wastewater in a combined PBUA – CMAS System at optimal condition

As discussed earlier the PBUA bench-scale system was started in batch mode of operation. After achieving the consistent COD removal efficiency, both the treatment units were operated in series. After achieving the psuedo-steady state level with reference to the COD removal in a continuous feed mode, the system was evaluated for the ascertainment of optimal HRT and loading. The COD removal in the PBUA - CMAS system at different HRTs is presented in **Figs. 3 and 4** respectively. The results indicate that a HRT of 3.8 hours and a COD load of 12 kg COD / m^3 / day with a COD removal efficiency of 53.4% for the PBUA reactor **(Fig. 3 and 4),** while a corresponding HRT of 23.60 hours with a load of 0.39 kg COD / kg MLSS / day is optimal for CMAS unit with a COD removal efficiency of 78.5% **(Fig. 5 and 6)**. The overall COD reduction in the bench scale unit was found to be 89.97%. Under optimal conditions of HRT and loading, a total and specific count of 30×10^7 CFU/ml and 18×10^6 CFU/ml respectively was obtained.

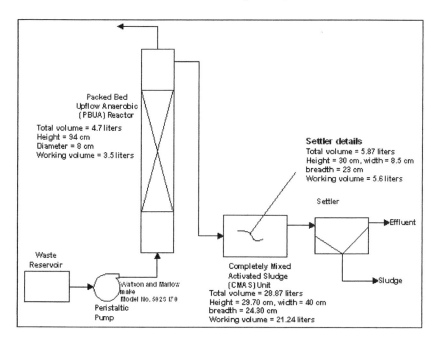

Fig. 2 Schematics of bench-scale biological treatment system for textile wastewater

On comparing the performance of bench-scale and full-scale PBUA – CMAS treatment systems for textile wastewater in the present investigation, it can be seen that the DO level in the CMAS of bench-scale unit is higher than that of full-scale CMAS unit. The biomass content of the full scale CMAS unit is in the range of 4500 – 5500 mg/L, while the biomass in the CMAS of the bench-scale unit is in the range of 2200 – 2500 mg/L. However, the total and specific viable counts of microorganisms present in the biomass of the bench-scale CMAS unit are higher than those of the full-scale CMAS unit. Thus, the low level of biomass with maximum viable potential microorganisms for effective biodegradation of constituents of textile wastewater has resulted in improved efficiency in the treatment of the textile wastewater. However, the characterization of the biomass of the CMAS indicated the presence of *Bacillus* and *Nocardia* sps. as dominant cultures (**Plate 1**). Rott and Minke (1999) and Sponza and Isik (2002) have reported the treatment of textile wastewater using combined anaerobic-aerobic system. The COD removal of 40% -60% in the anaerobic reactor was obtained at a COD load in the range of 5 kg COD/m^3.d to 25 kg COD/m^3.d. The residual COD was removed with an efficiency of 85-90% in the aerobic reactor. Qian et al. (1996) have reported 68% removal efficiency for textile wastewater using combined anaerobic-aerobic treatment system at an organic loading rate of 5.3 Kg COD/m^3/d. The HRT of anaerobic reactor was 6-10 hours, while that of the aerobic unit was 6.5 hours.

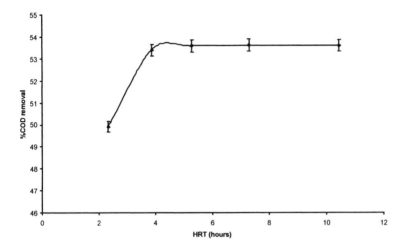

Fig. 3 Assessment of HRT for PBUA on bench scale system for treatment of textile wastewater

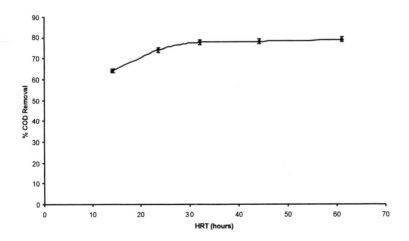

Fig. 4 Evaluation of HRT for CMAS on bench scale system for treatment of textile wastewater

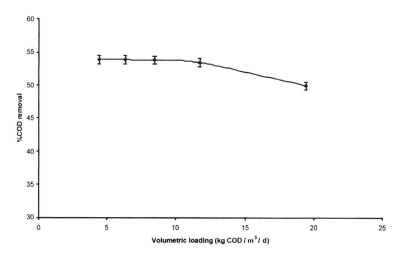

Fig. 5 Selection of COD load for PBUA on bench scale system at optimal HRT for treatment of textile wastewater

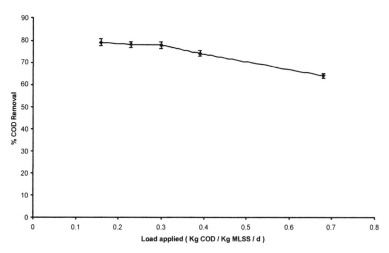

Fig. 6 S election of load for bench scale CMAS system at optimal HRT for treatment of textile wastewater

Evaluation of biokinetic constants for PBUA - CMAS bench-scale system

The experimental results of the PBUA using mixed biomass and COD as growth substrate, were used to determine the kinetic constants, viz. maximum reaction rate (V_{max}) and half velocity constant (K_s) by using the Monod hyperbolic relationship between the reaction rate and residual COD in the treated effluent. The computed biokinetic constants for PBUA are presented in **Table 2** Results on biokinetic constants indicate maximum reaction rate (V_{max}) of 6.82 Kg COD / m^3 / day and the half velocity constant (K_s) of 0.4652 g/L. The experimental results with textile wastewater in a CMAS have been used for establishing a relationship between the growth of biomass and the substrate utilization rate as per the Monod equation for computation of biokinetic constants for the system. The biokinetics of CMAS system indicate the yield coefficient (Y=0.22 mg biomass/mg COD), decay coefficient (K_d=0.027/hour), half saturation rate constant (K_s=284.74 mg/L as COD) and specific growth rate constant (μ_{max}=0.34/hour) (**Table 2**). The biokinetic constants observed for the present CMAS system are comparable with Tokal (2002).

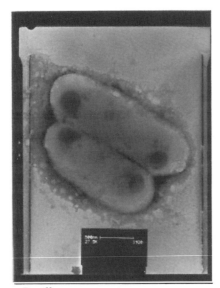
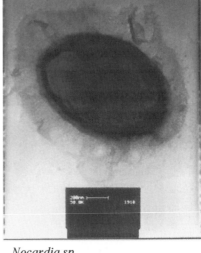

Bacillus sp.   *Nocardia sp.*

Plate 1. Transmission electron micrograph of the culture isolated from the bench scale CMAS unit treating textile wastewater

4   Conclusions

The full-scale and bench-scale PBUA and CMAS combined systems were evaluated for treatment of textile wastewater. The performance of bench scale system for the treatment of textile wastewater is found to better (99.97% COD removal) than the full scale treatment system (96.20% COD removal) owing to the high level of DO in the CMAS and better and efficient operation of the CMAS. The biomass of the full-scale CMAS treatment system contained the dominant microorganisms of the genera of *Nocardia, Bacillus, Bacteriodes, Cardiobactrium,* and *Brevicatena*, while the biomass of the bench-scale unit indicated the presence of dominant microorganisms of the genera *Nocardia* and *Bacillus* only. The biokinetic for PBUA reactor and CMAS unit have shown the comparable results with the literature reports

References

[1]   American Public Health Association (APHA), 1995. Standard Methods for the Examination of Water and Wastewater. (19th ed). APHA, AWWA and WEF, Washigton, D.C.

[2]   Andrzejewska A., Krysztafkiewicz A. and Jesionowski T., 2004, Adsorption of organic dyes on the aminosilane modified TiO2 surface, Vol. 62, pp 121-130

[3]   Bergeys Manual of Systematic Bacteriology, 1984, Volume 1, Krieg, N.R. and Holt, J.G. (eds.), Williams and Wilkins, USA.

[4]   Bergeys Manual of Systematic Bacteriology, 1986, Volume 2, Sneath, P.H.A., Mair, N.S., Sharpe, M.E. and Holt, J.G. (eds.), Williams and Wilkins, USA.

[5]   Faria P.C.C., Orfao J.J.M. and Pereira M.F.R., 2004, Adsorption of anionic and cationic dyes on activated carbons with different surface chemistries, Water Research, Vol. 38, Iss. 8, pp 2043-2052

[6]   Feitkenhauer *H. and Meyer U., 2003, Anaerobic digestion treatment of textile wastewater with high alcohol sulfate (surfactant) contents, Institute for chemical -and Bioengineering

[7]   Islk M. and Sponza D.T., 2004, A batch kinetic study on decolorization and inhibition of Reactive Black 5 and Direct Brown 2 in an anaerobic mixed culture, Chemosphere, Vol. 55, Iss.1, pp 119-128

[8]   Lim B.R., Hu H.Y., Ahm K.H., Fujie K., 2004, Oxidative treatment characteristics of biotreated textile dyeing wastewater and chemical agents used in a textile dyeing process by advanced oxidation process, Wat. Sci. and Tech., Vol. 49, No. 5-6, pp 137-143

[9]   Metcalf and Eddy, 1996. Wastewater Engineering, Treatment, Disposal, Reuse. (3rd ed.). Tata McGraw-Hill, Inc. India, pp. 370.

[10]   Muthukumar M. and Selvakumar N., 2004, Studies on the effect of inorganic salts on decolouration of acid dye effluents by ozonation , Dyes and Pigments ,Vol. 62 ,Iss. 3, pp 221-228

[11]   Meric S., Kaptan D., and Olmez T., 2004, Color and COD removal from wastewater containing Reactive Black-5 using Fenton's oxidation process, Chemosphere, Vol. 54, Iss.3, pp 435-441

[12]   Papi S., 2004, Removal of some reactive dyes from synthetic wastewater by combined Al(III) coagulation / carbon adsorption process, Vol. 62, Iss. 3, pp 291-298

[13]   Panswad T., Techovanich A., 2002, Comparison of dye wastewater treatment by normal anoxic + anaerobic / aerobic SBR activated sludge processes. Presented at Ist IWA congress, Paris

[14]   Rott U., and Minke R., 1999, Overview of wastewater treatment and recycling in the textile processing industry, Water Science & Technology, Vol. 40, No. 1, pp 137-144

[15]   Sponza D.T., and Isik M., 2002, Ultimate azo dye degradation in anaerobic / aerobic sequential processes, J. Water Science & Technology, Vol. 45, No.12, pp 217 – 278

[16]   Tokal E., 2002, Colour removal from cotton textile industrial wastewater in an activated sludge system with various additives, J. Wat. Res. Vol. 36, Iss. 11, pp 2920-2925

[17]   Qian A.H., Gu X., Tang W.Z., 1996, Biotreatment of dye wastewater using an anaerobic / oxic system. Chemosphere, Vol. (33), No. 12, pp 2533-2542

[18]   Wang A., Qu J., Liu H., and Ge J., 2004, Degradation of azo dye Acid Red 14 in aqueous solution by electrokinetic and electrooxidation process, Chemosphere, Vol. 55, Iss. 9, pp 1189-1196

Catalytic performance of lignin peroxidase hosted in AOT reverse micellar medium

Xirong Huang[*,1,2], **Wenjuan Zhang**[1], **Yuezhong Li**[2], **Yinbo Qu**[2], and **Peiji Gao**[2]

[1] Key Lab for Colloid and Interface Chemistry of the Education Ministry of China, Shandong University, Jinan 250100, P.R. CHINA
[2] State Key Lab of Microbial Technology of China, Shandong University, Jinan 250100, P.R. CHINA

[*]Corresponding author: e-mail: xrhuang@sdu.edu.cn, Phone: +86 531 8365433, Fax: +86 531 8564750

Studies on the catalytic and kinetic properties of lignin peroxidase (LiP) hosted in AOT reverse micelle were reported in this paper. Results showed that LiP could express its activity in the AOT/isooctane/toluene/water reverse micelles, but its activity depended, to a great extent, on the composition of the reverse micelles. Optimum activity occurred at a molar ratio of water to AOT ($?_0$) of 11, a pH value of 3.6. The dependence of LiP activity on the volume fraction of water in the medium indicated that veratryl alcohol (VA) was mainly solubilized in the pseudophase of the reverse micelle. Its partition coefficient between the pseudophase and the organic solvent phase was determined to be 35.8 based on the pseudo-biphasic model and the corresponding kinetic method. The steady state kinetics of the model reaction in a component fixed AOT reverse micellar medium indicated that the reaction mechanism was the same as that in bulk aqueous medium, but the kinetic parameters were quitely different in the two media. This difference was caused by the microheterogeneity of the reverse micelle.

Keywords lignin peroxidase; reverse micelle; veratryl alcohol; catalytic performance

1   Introduction

The lignin-degrading white rot fungus *Phanerochaete chrysosporium* is capable of degrading many aromatic pollutants such as chlorinated phenols, polycyclic aromatic hydrocarbons (PAHs) etc. and attracts a lot of attention in the research field of environment cleanup [1, 2]. It is believed that the non-specific extracellular hemoprotein peroxidases play an important role in the biodegradation [3]. Lignin peroxidase (LiP) is one of the enzymes secreted by the fungus under ligninolytic conditions. Due to its high redox potential, many attempts have been made to degrade aromatic pollutants with LiP [4]. Because many of environmentally persistent aromatic pollutants such as PAHs are poorly soluble in water, an organic solvent seems to be required to increase the degradation efficiency [5]. Therefore, an active LiP system in hydrophobic solvents is of great significance. A reverse micelle is a surfactant-stabilized aqueous microdroplet suspended in an organic solvent. A water soluble enzyme resides in the aqueous microdroplet, while the substrate can be solubilised in the aqueous microdropt and /or in the organic solvent, depending on its solubility in water and the organic solvent. Because the direct contact of enzyme with organic solvent is avoided, an enzyme entrapped in a reverse micelle is usually active. In addition, high dispersion of the catalyst and fast exchange between the solutes in reverse micelles result in great improvement of the catalytic performance of an enzyme. It follows that it is of great interest to study the catalytic and kinetic properties of LiP in a reverse micellar medium.

Veratryl alcohol (VA), a secondary metabolite of *Phanerochaete chrysosporium*, is a optimum substrate of LiP. Studies in aqueous solution on the enzymatic properties of LiP were usually based on the LiP-catalyzed oxidation of VA by H_2O_2. In addition, it has been reported that VA has a mediating effect on the oxidation of xenobiotics [6]. So in our research VA was selected as the substrate of LiP. The reverse micellar medium was composed of AOT, isooctane, toluene and water.

2   Materials and methods

Chemicals   Sodium bis(2-ethylhexyl)sulfosuccinate (AOT) was purchased from Sigma and used without further purification; veratryl alcohol (VA) was obtained from Fluka; all other chemicals were of analytical grade. The triply distilled water was used throughout.

Lignin peroxidase production and purification   Culture conditions used to produce lignin peroxidase (LiP) by *Phanerochaete chrysosporium* and its purification were described in our previous paper [7].

Saturated solution of hydrogen peroxide in toluene   Preparation of the saturated solution of hydrogen peroxide in toluene and calibration of its concentration were made based on the procedure described in the literature [8].

Measurement of an initial reaction rate   The initial reaction rate of the LiP-catalyzed oxidation of VA by H_2O_2 in AOT reverse micellar medium was measured based on the absorbance change with time at 310nm, which was recorded on a Shimadzu UV-240 spectrophotometer at 30°C. The detailed procedure was as follows: into a 1 cm long cell were added aliquots of a stock solution of AOT in isooctane (500 mM), isooctane, a stock solution of VA in toluene (30 mM), toluene, and a stock solution of the citrate buffer, respectively. The mixture was shaken until it became transparent. After that, 20µl LiP solution was added and mixed gently, giving a LiP containing reverse micellar solution. Two minutes later, 100µl saturated solution of H_2O_2 in toluene was added to the resulting solution to initiate the oxidative reaction. A plot of absorbance (*A*) at 310nm versus the reaction time (*t*) was recorded immediately after the initiation. The initial rate ($?_0$ in µmol•l^{-1}•min^{-1}) was then calculated based on the slope of the linear portion of the *A~t* curve via an absorbance/concentration conversion.

3   Results and discussion

3.1   Effect of the composition of AOT reverse micelle on the activity of LiP

3.1.1 Size of reverse micelles

To make LiP express high activity, several $?_0$ values at a fixed value of the volume fraction of water in the medium ($?$) were tested, and a bell-shaped curve was observed for the activity of LiP as a function of $?_0$. With the increase of $?_0$, the size of reverse micelles increased, and the rigidity of the conformation of LiP decreased accordingly, resulting in an increase in the activity. At higher $?_0$, more flexible conformation was vulnerable to attack by AOT, a denaturant. Figure 1 indicated that the optimum $?_0$ value was ca. 11.

3.1.2 Buffer pH

The effect of aqueous pH on the activity of LiP in the reverse micelles was investigated by varying the pH of the citrate buffer (0.1M). A bell-shaped curve was observed (see Fig. 2).

The optimum pH was 3.6, which was higher than that (3.2) in bulk aqueous solutions. This phenomenon may be caused by the favorable electrostatic interactions of H^+ with the anionic AOT headgroups, which made the protein sense a more acidic environment.

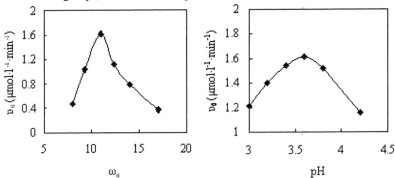

Fig. 1(left) Effect of ω_0 on the activity of LiP. $T=30°C$; $V_{isooctane}:V_{toluene}=7:1;\theta=0.0286;pH=3.6$; [VA]=1.6 mM; [LiP]= 0.20μM;[H₂O₂]= 80μM.

Fig.2(right) Effect of aqueous pH on the activity of LiP. $T=30°C$; $\omega_0=11;V_{isooctane}:V_{toluene}=7:1$; $\theta=0.0286$;[VA]=1.6mM;[LiP]= 0.20μM;[H₂O₂]=80μM.

3.2 Partitioning of VA

Figure 3 shows the variation of the initial rate of the LiP catalyzed oxidation of VA with respect to θ at several VA concentrations. As we can see, the initial rate decreases with the increase of the θ values. This result indicates that VA is dissolved mainly in the pseudophase. Based on the pseudo-biphasic model and the corresponding kinetic method [9], the partition coefficient of VA, P, was determined to be 35.8.

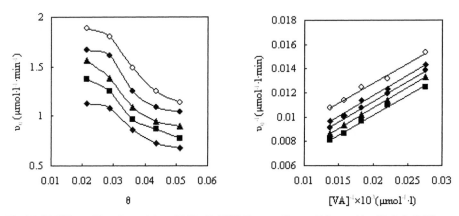

Fig.3(left) Effect of θ on the activity of LiP. $T=30°C$;$V_{isooctane}:V_{toluene}=7:1$; $\omega_0=11$; pH=3.6; [LiP]= 0.20μM;[H₂O₂]= 80μM; [VA]×10⁻³=1.0(◆), 1.2(■), 1.4(▲), 1.6(●), 1.8(O) μM.

Fig.4(right) Double reciprocal plot of the initial rate versus the concentration of VA at different levels of H₂O₂ in AOT reverse micelles. $T=30°C$ $V_{isooctane}:V_{toluene}=7:1$; $\theta=0.0286$; $\omega_0=11$; pH=3.6; [LiP]=7.0 μM; [H₂O₂]⁻¹×10³=0.580(■), 0.696(▲), 0.870(●), 1.16(◆), 1.74(○) μmol⁻¹•l.

3.3 Reaction mechanism and kinetic parameters

Figure 4 showed the double reciprocal plots of the initial rate versus the concentration of VA at several different fixed concentrations of H_2O_2 (the concentrations, including the initial rate, were expressed as true not apparent values). For each given concentration of H_2O_2, the data of $?_0^{-1}$ and $[VA]_0^{-1}$ were best-fit to a linear line, and, moreover, these linear lines drawn at different levels of H_2O_2 were almost parallel, suggesting that the LiP catalyzed oxidation of VA by H_2O_2 in the AOT reverse micelles followed a ping-pong mechanism Based on the data in Fig.4, the true kinetic parameters, k_{cat}, $K_{m,VA}$, and $K_{m,H2O2}$ were calculated to be 59.6 min^{-1}, 139 mM , and 948 μM, respectively.

For comparison, the steady state kinetics of the model reaction in bulk aqueous solution (pH=3.6) was also studied. The k_{cat}, $K_{m,VA}$, and $K_{m,H2O2}$ were determined to be 142 min^{-1}, 113 μM, and 103 μM, respectively. As far as k_{cat} is concerned, the k_{cat} value in AOT reverse micelles is approximately half of that in the aqueous solution, i.e., the reverse micellar medium has an adverse effect on the expression of the activity of LiP. Our previous studies showed that in an aqueous solution, AOT was a strong inhibitor for LiP; it deactivated LiP completely at 1 mM AOT level. In the AOT reverse micellar medium, the apparent AOT concentration was as high as 140 mM, and, moreover, the LiP molecule was surrounded by AOT monolayer, but LiP still retained 50% of the activity in the aqueous solution without AOT. This result indicated that the effect of AOT on the activity of LiP related to its aggregate structure or the microheterogeneity of the midia.

H_2O_2 should be solubilized in the aqueous microdroplets. The fact that the $K_{m,H2O2}$ in the reverse micelles increased as compared with that in the aqueous solution may be ascribed to the conformational change of LiP after encapsulation. This change altered the peripheral microenvironment of the substrate binding site and hence decreased the affinity between native LiP and H_2O_2.

In the case of $K_{m,VA}$, its value in the reverse micelle was much higher than that in the aqueous solution; i.e., the affinity between VA and LiP(I) and LiP(II) decreased greatly when LiP was entrapped in the reverse micelles. This result had much to do with the loci of VA and LiP in the pseudophase in addition to the changes, caused by encapsulation, of the conformation of LiP and the reactivity of VA. LiP was located in the inner core of the pseudophase, while VA was mainly solubilised in the interfacial membrane. This localization impeded the formation of the substrate-enzyme complex.

To sum up, the above mentioned difference in the kinetic parameters was caused by the microheterogeneity of the reverse micelle.

Acknowledgements

The authors gratefully acknowledge the financial supports from K. C. Wong Education Foundation, Hong Kong; the Middle-aged and Youthful Excellent Scientists Encouragement Foundation of Shandong Province; the Provincial Natural Science Foundation of Shandong; and the National Natural Science Foundation of China.

References

[1]   D.P. Barr, and S.D. Aust, Environ Sci Technol **28**(2), 78A(1994).
[2]   A. Paszczynski, and R.L. Crawford, Biotechnol Prog **11**, 368(1995).
[3]   M.D. Cameron, S. Timofeevski, and S.D. Aust, Appl Microbiol Biotechnol **54** (6), 751(2000).

32

[4]   D.K. Joshi, and M.H. Gold, Eur J Biochem **237**, 45(1996).

[5]   J. Michizoe, S. Okazaki, M. Goto, and S. Furusaki, Biochem Eng J **8,** 129(2001).

[6]   X. Huang, D. Wang, C. Liu, M. Hu, Y. Qu, and P. Gao, Biochem Biophys Res Commun **311**, 491(2003).

[7]   Y.Z. Li, P.J. Gao, and Z.N. Wang, Acta Microbiol Sin **34**, 29(1994).

[8]   P.A. Mabrouk, J Am Chem Soc **117**(8), 2141(1995).

[9]   Y.L. Khmelnitsky, I.N. Neverova, V.I. Polyakov, V.Y. Grinberg, A.V. Levashov, and K. Martinek, Eur J Biochem **190**, 155(1990).

Characterization of ß-lactam-resistant genes from a metagenomic library of cold-seep sediments in deep-sea

Jae Seok Song[1], Jeong Ho Jeon[2], Jung Hun Lee[1, 2], Jung-Hyun Lee[2], and Sang Hee Lee[*1]

[1] Department of Biological Sciences, Myongji University, San 38-2 Namdong, Yongin, Kyunggido, 449-728, SOUTH KOREA
[2] Marine Biotechnology Center, Korea Ocean Research & Development Institute, Ansan P.O. Box 29, Seoul 425-600, SOUTH KOREA

[*]Corresponding author: e-mail: sangheelee@mju.ac.kr, Phone: +82 31 3306195, Fax: +82 31 3358249

To determine prevalence and genotypes of ß-lactamases among clones of a metagenomic library from the cold-seep sediments of Edison seamount (10,000 years old), we performed pulse-field gel electrophoresis, antibiotic susceptibility testing, pI determination, and DNA sequencing analysis. Among the 8,823 clones of the library, thirty clones produced ß-lactamases and had the high genetic diversity. According to minimum inhibitory concentration patterns, we found that five (16.7%) of thirty clones produced an extended-spectrum ß-lactamase. Banding patterns of PCR amplification with the designed primers showed that 837- and 259-bp fragments specific to bla_{TEM} genes were amplified. TEM-1 was the most prevalent ß-lactamase and conferred the resistance to ampicillin, piperacillin, and cephalothin. TEM-116 had a spectrum that was extended to ceftazidime, cefotaxime, and aztreonam. The resistance levels conferred by the pre-antibiotic era alleles of TEM-type ß-lactamases were essentially the same as the resistance levels conferred by the TEM-type alleles that had been isolated from clinically resistant strains of bacteria after antibiotic era. Our first report on the TEM-type ß-lactamases before the antibiotic era indicates that TEM-type ß-lactamases paint a picture in which most of the diversity of the enzymes may not the result of recent evolution but that of ancient evolution.

Keywords ß-lactamases; deep-sea; evolution; metagenomic library; TEM-1; TEM-116

1 Introduction

The ß-lactamases (EC 3.5.2.6) produced by bacteria are known to protect against the lethal effect of ß-lactams (penicillins, cephalosporins, carbapenems, or monobactams) on cell-wall synthesis. The production of ß-lactamase is the single most prevalent mechanism responsible for resistance to ß-lactams among clinical isolates of the family *Enterobacteriaceae* [1]. A variety of ß-lactamases has been classified into class A, B, C, and D according to their amino acid homology [2]. The most common ß-lactamases among the *Enterobacteriaceae* are the plasmid-born class A TEM (named for a patient called Temoniera) and SHV (named for sulfhydryl variable) ß-lactamases [3]. TEM-1 (classical TEM-type ß-lactamase), first reported in 1965 [4], confers a high level of resistance to penicillins (ampicillin and piperacillin) and early cephalosporins (cephalothin) but little to oxyiminocephalosporins (cefotaxime and ceftazidime), carbapenems (imipenem and meropenem), and monobactams (aztreonam) [3]. Beginning in the early 1980s, extended-spectrum ß-lactamases (ESBLs) driven from TEM-1 began to appear following the introduction of ß-lactam antibiotics into medical use. ESBL are ß-lactamase inhibitor-susceptible enzymes conferring broad resistance to penicillins, aztreonam, cefotaxime, and ceftazidime [5]. ESBLs are often plasmid mediated, and most are mutants of the TEM-1 enzymes, with one or more amino acid substitutions around the active site [6]. These changes allow hydrolysis of extended-spectrum cephalosporins (ceftazidime and cefotaxime) and monobactams (aztreonam), which are stable to TEM-1 enzymes [7]. The rapid appearance of new TEM variants active against those ß-lactams has strongly affected public health policy and has conditioned thinking about the evolution of antibiotic resistance. Hall and Barlow [8] estimated that TEM ß-lactamases diverged around 400 million years ago (before the antibiotic era). Lacking a fossil record, the values for the divergence times are fairly speculative. Therefore, it is necessary to investigate ß-lactamase gene from microorganisms before the antibiotic era. Deep-sea surface sediments harboring ancient microorganisms were estimated to range from 5800 to about 180,000 years [9]. A metagenome is defined as the collective genomes of all microorganisms in a given habitat, so it is the starting material for culture-independent microbial genomic analysis [10]. Thus, metagenomics can be used to address the challenge of studying microorganisms in the environment such as cold-seep sediments that are, as yet, unculturable and which represent more than 99% of the organisms there [11].

In this study we detected TEM-type ß-lactamases from a metagenomic library of the cold-seep sediments of deep-sea Edison seamount (about 10,000 years old) [12] and characterized ß-lactamases. We suggested that ß-lactam resistance in microorganisms was likely to have been present prior to the modern antibiotic era, paying particular attention to understanding the rapid appearance of strains that express ß-lactamase genes following the introduction of ß-lactams into medical and agricultural use.

2   Materials and methods

2.1 Bacterial strains

A total of 8,823 clones of a metagenomic library were constructed. Thirty clones were selected as ß-lactamase-producing clones. *E. coli* EPI 300 strain (Epicentre, Madison, WI, USA) was used as a host of the metagenomic library construction. *E. coli* ATCC 25922 was used as the minimum inhibitory concentration (MIC) reference strain.

2.2 Study site and metagenomic library construction

Cold-seep sediment samples (depth, 1,450 m) were obtained at the summit of deep-sea Edison seamount (GTVA 31 site; 03°19′ S, 152°34′ E) (Figure 1), south of Lihir Island in the New Ireland Fore-arc, near Popua New Guinea, from the R/V Sonne during the 2002research cruise SO-166 (September 2002). Samples of the sediment consisted of foraminiferal carbonate ooze, minor smectite, and disseminated Femonosulfide. The intact sediments devoid of contamination were taken from inside of sediments with a camera-guided grab sampler. The sediment samples were aliquoted into sterile conical tubes and preserved at 4 °C during transportation for 2 weeks before storage at -80 °C. Community DNA was extracted from the sediment samples by the method using a cation-exchange resin [13].

Fig. 1 Map of the New Ireland fore-arc region with Lihir Island. Sampling was performed at south of Lihir Island (closed rectangle).

The extracted DNA was end-repaired by end-repair enzyme mix (Epicentre) which made the DNA blunt-ended and 5′-phosphorylated. The end-repaired DNA was separated by low-melting-temperature agarose gel (SeaPlaque, Cambrex, Baltimore, MD, USA) electrophoresis at 35 V for 13 h. The regions containing approximately 40 kbp DNA fragments were cut from the gel. A metagenomic library was constructed using the approximately 40 kbp end-repaired DNA fragments and CopyControl™ Fosmid Library Production Kit (Epicentre). CopyControl cloning system was based on technology developed by Wild et al. [14]. The recombinant fosmid containing the end-repaired DNA fragments was purified by alkaline lysis method [15].

2.3 Screening of β-lactamase-producing clones from the metagenomic library

In order to screen β-lactamase-producing clones from the metagenomic library, a total of 8,823 fosmid clones were transferred to Luria-Bertani (LB, Difco Laboratories, Detroit, MI USA) plate containing 100 μg/ml ampicillin. The ampicillin-resistant clones were selected for further analyses. Unless otherwise stated, molecular biological reagents and restriction enzymes were obtained from Sigma-Aldrich (St Louis, MO, USA). To test genetic diversity of fosmid clones, the purified fosmid was digested with *Bam*HI and was analyzed by pulse-field gel electrophoresis (PFGE). *Bam*HI-digested fosmid DNA was prepared according to the instruction of Bio-Rad (Hercules, CA, USA) and fragments were separated for 12 h at 6 V/cm at 11 °C using a CHEF-DRII system (Bio-Rad), with initial and final switch times of 0.05 and 0.46 s, respectively. DNA fingerprints were interpreted as recommended by Tenover et al. (1995).

2.4 Susceptibility to β-lactams

Minimum inhibitory concentrations (MICs) of antimicrobial agents were determined by the agar dilution method [17]. Antibiotics were from the following suppliers: ampicillin and cephalothin (Sigma-Aldrich); cefotetan (Merck Sharp and Dohme-Chibret, West Point, PA, USA); cefotaxime (Handok Pharmaceuticals, Seoul, Korea); ceftazidime and ampicillin-sulbactam (Hanmi Pharmaceuticals, Hwasung, Korea); aztreonam and cefepime (BMS Pharmaceutical Korea, Seoul, Korea); cefoxitin, imipenem, and meropenem (Choongwae Pharmaceuticals, Seoul, Korea); piperacillin (Yuhan, Seoul, Korea); and piperacillin-tazobactam (Wyeth Korea, Seoul, Korea). The results of antimicrobial susceptibility tests were interpreted according to the National Committee for Clinical Laboratory Standards' criteria [18].

2.5 Isoelectric focusing (IEF)

Crude cell extracts containing β-lactamases from fosmid clones were prepared by the osmotic shock method detailed in pET system manual (Novagen, Madison, WI, USA). IEF was performed in Ready Gel Precast IEF Polyacrylamide Gel containing Ampholine with a pH range of 3.5 to 9.5, placed into a Mini-Protein 3 Cell, as described by the manufacturer (Bio-Rad, Hercules, CA, USA). Gels were developed with 0.5 mM nitrocefin (Oxoid, Basingstoke, United Kingdom) as a chromogenic substrate (cephalosporin) of β-lactamases.

2.6 Molecular studies

The purified fosmid DNA was used as template DNA in polymerase chain reaction (PCR). The primers for PCR amplification were designed by selecting consensus sequences in multiple-nucleotide alignment of 60 TEM-type β-lactamase genes (bla_{TEM}), 27 SHV-type β-lactamase genes (bla_{SHV}), and 5 CMY-type β-lactamase (class C) genes (bla_{CMY}) by using the Primer Calculator program (Williamstone Enterprises, Waltham, MA, USA). The primers were described previously (Lee et al. 2000): T1, T2, T3, and T4 were used for bla_{TEM}; S1, S2, S3, and S4 were used for bla_{SHV}; C1, C2, C3, and C4 were used for bla_{CMY}. PCR amplifications were carried out as described previously [19]. DNA sequencing was performed by the direct sequencing method with an automatic sequencer (ABI PRISM3100; Applied Biosystems, Weiterstadt, Germany), as previously described [20]. DNA sequence analysis was performed with DNASIS for Windows (Hitachi Software Engineering America Ltd., San Bruno, CA, USA). Database similarity searches for both the nucleotide sequences and deduced protein sequences were performed with National Centre for Biotechnology Information BLAST server (http://www.ncbi.nlm.nih.gov/BLAST/) [21].

3 Results

3.1 Construction of metagenomic library

The environmental DNA from sediments of deep-sea Edison seamount was used to construct a metagenomic library using the fosmid vector, pCC1FOS (Epicentre). ES01 was the first metagenomic library which consisted of 8,823 clones and were stored at -80ºC in 96-well microplate. The average insert DNA size was 32.3 kbp (Figure 2b). It was estimated that the library ES01 contained approximately 284 Mbp of DNA which corresponded to about 95 bacterial genomes based on the average bacterial genome of 3 Mbp. 3.2 Screening of β-lactamase-producing clones

Thirty of 8,823 clones grew on LB plate containing 100 μg/ml ampicillin (Figure 2a). The cell extracts of thirty clones changed the pale yellow colour of nitrocefin into a pink colour of cephalosporanic acid (the end product hydrolyzed from nitrocefin by β-lactamases), which indicated all thirty clones produced β-lactamases. The isoelectric point (pI) value of β-lactamases produced by thirty clones was 5.4 (Table 1).
PFGE analysis generates a characteristic genomic fingerprinting which can be used to reveal intra- and inter-species genotypic variations. All thirty clones generated distinct bands ranging in size 0.5 to 28.1 kbp and varied from 3 to 11 bands by the *Bam*HI macrorestriction analysis (Figure 2b). Distinct PFGE patterns indicated that all thirty clones showed different genotypes.

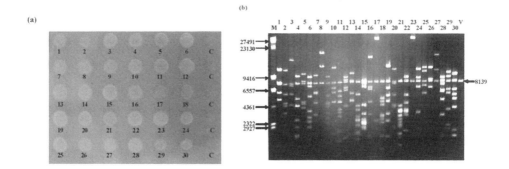

Fig. 2 Restriction analysis (b) by *Bam*HI of thirty clones (a) producing ß-lactamase. Thirty clones (a) grow on LB plate containing 100 µg/ml ampicillin but negative controls (*E. coli* EPI 300 host strain, C) was not grown. Each number of (a) indicates each clone identical to that of (b). Lanes of (b): M, ? *Hind*III molecular mass marker [sizes in base pairs (bp) are indicated on the left edge of the gel]; V, Fosmid vector pCC1FOS™ (8139 bp); 1, ES01003A02; 2, ES01003G06; 3, ES01008A11; 4, ES01008C04; 5, ES01008E12; 6, ES01008F07; 7, ES01013B05; 8, ES01013E09; 9 ES01013F06; 10, ES01014A10; 11, ES01014D12; 12, ES01014G12; 13, ES01016A08; 14, ES01016A11; 15, ES01016B02; 16, ES01016C11; 17, ES01019B07; 18, ES01020A11; 19, ES01023A03; 20, ES01023A09; 21, ES01023B06; 22, ES01023H02; 23, ES01023H07; 24, ES01024D03; 25, ES01024E08; 26, ES01027F09; 27, ES01031E08; 28, ES01031F09; 29, ES01033B01; 30, ES01046B06.

Table 1 Profiles of ß-lactamase-producing clones, MIC reference strain (*E. coli* ATCC 259220), and host strain (*E. coli* EPI 300) for transfection.

Clones or strains	MIC(µg/ml) of the following ß-lactams[a]:														pI	ß-lactamase
	AMP	A/S	PIP	P/T	CEP	FOX	CTT	CAZ	CTX	FEP	IPM	MER	AZT		
ES01003A02	256	8	>256	16	16	0.25	0.25	0.12	0.12	0.12	0.25	0.12	0.5	5.4	TEM-1
ES01003G06	256	16	>256	16	32	0.5	0.25	0.5	0.25	0.25	0.12	0.12	0.5	5.4	TEM-1
ES01008A11	128	16	>256	32	16	0.25	0.25	0.5	0.25	0.25	0.25	0.25	0.5	5.4	TEM-1
ES01008C04	256	16	>256	32	32	0.5	0.5	0.5	0.25	0.25	0.12	0.12	0.25	5.4	TEM-1
ES01008E12	256	8	>256	16	32	0.25	0.5	0.12	0.25	0.12	0.25	0.12	0.5	5.4	TEM-1
ES01008F07	>256	16	>256	32	128	0.5	0.5	64	64	0.5	0.25	0.12	32	5.4	TEM-116
ES01013B05	256	16	>256	32	32	0.5	0.5	0.25	0.25	0.12	0.12	0.25	5.4	TEM-1
ES01013E09	>256	8	>256	32	128	0.5	0.25	64	64	0.25	0.25	0.12	32	5.4	TEM-116
ES01013F06	256	8	>256	16	32	0.25	0.5	0.25	0.25	0.25	0.25	0.12	0.25	5.4	TEM-1
ES01014A10	256	8	>256	32	32	0.5	0.5	0.25	0.25	0.25	0.25	0.12	0.25	5.4	TEM-1
ES01014D12	256	16	>256	32	32	0.5	0.25	0.25	0.25	0.12	0.12	0.12	0.5	5.4	TEM-1
ES01014G12	128	8	>256	16	16	0.5	0.12	0.25	0.25	0.12	0.25	0.25	0.25	5.4	TEM-1
ES01016A08	256	8	>256	32	16	0.5	0.25	0.25	0.25	0.12	0.25	0.12	0.25	5.4	TEM-1
ES01016A11	>256	16	>256	32	64	0.5	0.5	32	64	0.5	0.25	0.12	32	5.4	TEM-116
ES01016B02	256	16	>256	16	16	0.5	0.25	0.5	0.5	0.5	0.12	0.12	0.25	5.4	TEM-1

	AMP	A/S	PIP	P/T	CEP	FOX	CTT	CAZ	CTX	FEP	IPM	MER	AZT	pI	
ES01016C11	256	8	>256	16	32	0.5	0.5	0.25	0.25	0.25	0.25	0.25	0.25	5.4	TEM-1
ES01019B07	256	8	>256	32	16	0.5	0.25	0.5	0.5	0.25	0.25	0.25	0.25	5.4	TEM-1
ES01020A11	256	8	>256	32	32	0.5	0.25	0.25	0.5	0.25	0.25	0.12	0.25	5.4	TEM-1
ES01023A03	256	8	>256	16	32	0.25	0.5	0.25	0.25	0.25	0.25	0.12	0.25	5.4	TEM-1
ES01023A09	>256	16	>256	32	64	0.25	0.25	32	64	0.25	0.5	0.25	32	5.4	TEM-116
ES01023B06	128	8	>256	16	16	0.5	0.25	0.25	0.5	0.12	0.25	0.12	0.5	5.4	TEM-1
ES01023H02	256	16	>256	32	16	0.5	0.5	0.5	0.5	0.25	0.25	0.06	0.25	5.4	TEM-1
ES01023H07	256	16	>256	32	16	0.25	0.5	0.5	0.5	0.5	0.06	0.06	0.25	5.4	TEM-1
ES01024D03	256	8	>256	16	32	0.25	0.25	0.5	0.25	0.12	0.12	0.12	0.25	5.4	TEM-1
ES01024E08	256	16	>256	16	32	0.25	0.25	0.5	0.25	0.25	0.25	0.12	0.25	5.4	TEM-1
ES01027F09	>256	16	>256	32	128	0.5	0.25	64	64	0.5	0.12	0.25	32	5.4	TEM-116
ES01031E08	128	4	>256	16	16	0.5	0.5	0.25	0.25	0.12	0.25	0.12	0.5	5.4	TEM-1
ES01031F09	256	8	>256	16	32	0.5	0.5	0.25	0.25	0.25	0.12	0.12	0.5	5.4	TEM-1
ES01033B01	256	16	>256	16	32	0.25	0.25	0.5	0.25	0.25	0.25	0.25	0.25	5.4	TEM-1
ES01046B06	256	8	>256	16	16	0.25	0.12	0.25	0.25	0.12	0.25	0.12	0.25	5.4	TEM-1
ATCC 25922	0.5	0.5	0.5	0.5	0.5	0.12	0.12	0.12	0.25	0.12	0.06	0.06	0.12		
E. coli EPI 300	0.5	0.5	0.5	0.5	0.5	0.12	0.12	0.12	0.12	0.12	0.06	0.06	0.12		

[a]Abbreviation: AMP, ampicillin; A/S, ampicillin/sulbactam; PIP, piperacillin; P/T, piperacillin/tazobactam; CEP, cephalothin; FOX, cefoxitin; CTT, cefotetan; CAZ, ceftazidime; CTX, cefotaxime; FEP, cefepime; IPM, imipenem; MER, meropenem; AZT, aztreonam.

3.3 Susceptibility to ß-lactams

ß-lactam MICs and pI for thirty clones were listed in Table 1. According to the National Committee for Clinical Laboratory Standards' criteria, all clones were resistant to ampicillin, piperacillin, and cephalothin but were sensitive to cefoxitin, cefotetan, cefepime, imipenem, and meropenem. All clones showed sixteen-fold to more than thirty two-fold decrease in the MIC of ampicillin and piperacillin by ß-lactamase inhibitors (sulbactam and tazobactam, respectively). Five clones (ES01008F07, ES01013E09, ES01016A11, ES01023A09, and ES01027F09) were resistant to ceftazidime, cefotaxime, and aztreonam. The resistance phenotypes of these five clones indicated that they produced ESBL(s). All clones produced ß-lactamases with a pI of 5.4.

3.4 Genotyping of ß-lactamases

Taking into account the resistance phenotypes of these thirty clones, the resistance genotypes of these clones were analysed. Using the designed primers, we detected ß-lactamase genes by PCR. The electrophoretic analysis of PCR product showed that TEM-type ß-lactamases genes (bla_{TEM}) were exactly amplified and the fragment sizes of PCR product were the same as indicated in Material and methods (Figure 3). SHV-type ß-lactamase genes and CMY-type ß-lactamase genes were not detected. The resistance genotypes of thirty clones were analyzed by direct sequencing of the PCR-amplified bla_{TEM} genes. Only one large open reading frame was found, which corresponds to a putative protein of 286 amino acids for TEM-type ß-lactamase. Two different TEM-type (bla_{TEM-1} and $bla_{TEM-116}$) ß-lactamase sequences were found (Table 1). Table 1 showed that five clones (ES01008F07, ES01013E09, ES01016A11, ES01023A09, and ES01027F09) among thirty clones produced TEM-116. The remaining twenty five clones produced TEM-1.

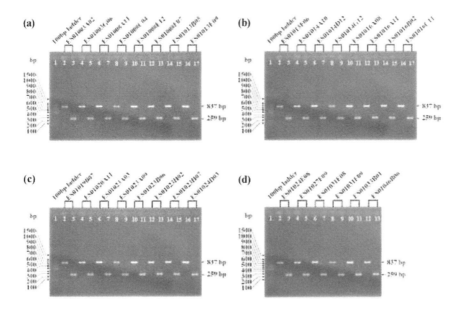

Fig. 3 Banding patterns of PCR products of *bla*TEM β-lactamase genes. Lane 1 of each figure: DNA molecular mass marker [100 bp ladder, sizes in base pairs (bp) are indicated on the left edge of the gel]. Even lanes in (a), (b), (c), and (d): PCR products (837 bp) of *bla*TEM with primer pair of T1 and T2. Odd lanes in (a), (b), (c), and (d): PCR products (259 bp) of *bla*TEM with primer pair of T3 and T4.

4 Discussion

Traditional methods of culturing microorganisms that grow under laboratory conditions limit analysis of real microbial diversity and their function [11, 22]. In order to investigate ß-lactamases from an ancient habitat, we constructed a metagenomic library (ES01) from the cold-seep sediments of deep-sea Edison seamount (about 10,000 years old, Schmidt et al. 2002). PFGE analysis of thirty clones producing ß-lactamases from the ES01 metagenomic library indicated that they were not closely related but had the high genetic diversity. Our partial results of the end-sequencing of the ES01 library and previous data of microbial diversity in the cold-seep sediments of Edison seamount [23] showed that microorganisms identified in the sediments were primarily affiliated with one of four groups: the ?-, d-, and e-subdivisions of *Proteobacteria*, and *Cytophaga-Flavobacterium-Bacteroides*.

40

Table 2 Amino acid substitutions of TEM-1 and TEM-116 ß-lactamases.

ß-lactamase	pI	Residue (coding triplet) at amino acid:	
		84	184
TEM-1	5.4	Val (GTT)	Ala (GCA)
TEM-116	5.4	Ile (ATT)	Val (GTA)

On the basis of the pI of ß-lactamases, their resistance phenotypes against ß-lactams, their profiles of inhibition by sulbactam and tazobactam, and DNA sequencing of PCR-amplified bla_{TEM} genes, the ß-lactamase produced by five clones (ES01008F07, ES01013E09, ES01016A11, ES01023A09, and ES01027F09) was a TEM-type ESBL (TEM-116). The ß-lactamase produced by the remaining twenty five clones was a classical TEM-type ß-lactamase (TEM-1). These ß-lactamases had the same MIC patterns against ß-lactams, amino acid sequences, and pI as those previously reported in clinical isolates [24, 25]. The prevalence of TEM-1 and TEM-116 in sediments of Edison seamount was 0.3% (25 of 8,823) and 0.06% (5 of 8,823). The TEM-1 ß-lactamase exists at high frequencies in antibiotic-resistant pathogens across the globe [3, 26]. While TEM-1 has a spectrum that is limited to penicillins and early cephalosporins, it has given rise to more than ninety descendent alleles such as TEM-116 that confer resistance to most modern ß-lactam antibiotics (the website of The Hall Laboratory of Experimental Evolution, http://www.rochester.edu/Colledge/BIO/labs/HallLab/AmpC-Phylo.html). The deduced amino acid sequence of TEM-116 had two amino acid substitutions from TEM-1 (Table 2). Two schemes for in vivo evolution were reported: (i) from TEM-1 to TEM-19, from TEM-19 to TEM-20, and from TEM-20 to TEM-52 and (ii) from TEM-1 to TEM-116 [25]. The TEM descendants might evolve from TEM-1 since the beginning of the antibiotic era, about 1950. However, TEM-116 as well as TEM-1 ß-lactamase genes were detected in sediments of deep-sea Edison seamount which was about 10,000 years old [12]. Therefore, it might be not the case that antibiotic resistance genes have evolved exclusively as a response to the modern clinical use of antibiotics such as ß-lactams. There is growing evidence that resistance genes were capable of conferring resistance long before the discovery of modern antibiotics. For one evidence, two strains of *Citrobacter freundii* that were collected in the pre-antibiotic era (the 1920s) carried *ampC* ß-lactamase genes whose products were fully as active toward as wide a range of ß-lactams as were the AmpC ß-lactamases (class C) that were found on plasmid after the antibiotic era [27]. Our first report on the TEM-type ß-lactamases (class A) before the antibiotic era indicates that TEM-type ß-lactamases paint a picture in which most of the diversity of the enzymes may be the result of ancient evolution. According to these results, we should not be surprised by the rapid appearance of strains that express those genes following the introduction of ß-lactam antibiotics into medical and agricultural use because classical ß-lactamases and ESBLs have been present before the antibiotic era.

Acknowledgements

This work was founded through BioGreen 21 Program (to S.H. Lee), Rural Development Administration, Republic of Korea, and in part by in-house program (PE87200 to J.-H. Lee) of Korea Ocean Research & Development Institute (KORDI).

References

[1]   C. C. Sanders and W. E. Jr. Sanders, Clin. Infect. Dis. **15**, 824 (1992).

[2]   R. P. Ambler, Philos. Trans. R. Soc. Lond. B Biol. Sci. **289**, 321 (1980).

[3]   A. A.Medeiros, Clin. Infect. Dis. **24**, S19 (1997).

[4]   N. Datta and P. Kontomichalou, Nature **208**, 239 (1965).

[5]   D. M. Livermore, Clin. Microbiol. Rev. **8**, 557 (1995).

[6]   D. L. Paterson, W. -C. Ko, A. V. Gottberg, J. M. Casellas, L. Mulazimoglu, K. P. Klugman, R. A. Bonomo, L. B. Rice, J. G. McCormack, and V. L. Yu, J. Clin. Microbiol. **39**, 2206 (2001).

[7]   S. K. DuBois, M. S. Marriott, and S. G. B. Amyes, J. Antimicrob. Chemother. 35, 7 (1995).

[8]   B. G. Hall and M. Barlow,. Drug Resist. Update. **7**, 111 (2004).

[9]   C. Raghukmar, S. Raghukumar, G. Sheelu, S. M. Gupta, B. N. Nath, and B. R. Rao,. Deep-Sea Res. Part I-Oceanogr. Res. Pap. **51**, 1759 (2004).

[10]   J. Handelsman, M. R. Rondon, S. F. Brady, J. Clardy, and R. M. Goodman, Chem. Biol. **5**, 245 (1998).

[11]   R. I. Amann, W. Ludwig, and K. H. Schleifer, Microbiol. Rev. **95**, 143 (1995).

[12]   M. Schimidt, R. Botz, K. Winn, P. Stoffers, O. Thiessen, and P. Herzig, Chem. Geol. **186**, 249 (2002).

[13]   R. A. Hurt, X. Qiu, L. Wu, Y. Ro, A. V. Palumbo, J. M. Tiedje, and J. Zhou, Appl. Environ. Microbiol. **67**, 4495 (2001).

[14]   J. Wild, Z. Hradecna, and W. Szybalski, Genome Res. **12**, 1434 (2002).

[15]   J. Sambrook and D. W. Russell (eds.), Molecular Cloning: a laboratory manual (Cold Spring Harbor Laboratory, New York, 2001), pp 1.40-1.41.

[16]   F. C. Tenover, R. D. Arbeit, R. V. Goering, P. A. Micklsen, B. E. Murray, D. H. Persing, and B. Swanminathan, J. Clin. Microbiol. **33**, 2233 (1995).

[17]   S. H. Lee, J. Y. Kim, S. H. Shin, Y. J. An, Y. W. Choi, Y. C. Jung, H. I. Jung, E. S. Sohn, S. H. Jeong, and K. J. Lee, J. Clin. Microbiol. **41**, 2477 (2003).

[18]   National Committee for Clinical laboratory Standards (NCCLS), Methods for dilution antimicrobial susceptibility tests for bacteria that grow aerobically, fifth approved standards, (NCCLS, Villanova, Pennsylvania, 2000), pp M7-A5.

[19]   S. H. Lee, J. Y. Kim, S. K. Lee, W. Jin, and K. J. Lee, Lett. Appl. Microbiol. **31**, 307 (2000).

[20]   S. H. Lee, J. Y. Kim, S. H. Shin, S. K. Lee, M. M. Choi, I. Y. Lee, Y. B. Kim, J. Y. Cho, W. Jin, and K. J. Lee, FEMS Microbiol. Lett. **200**, 157 (2001).

[21]   S. F. Altschul, T. L. Madden, A. A. Schäffer, J. Zhang, Z. Zhang, W. Miller, and D. J. Lipman, Nucleic Acids Res. **25**, 3389 (1997).

[22]   P. Hugenholtz and N. R. Pace, Trends Biotechnol. **14**, 190 (1996).

[23]   S. H. Lee, H. -R. Oh, J. -H. Lee, S. -J. Kim, and J. -C. Cho, J. Microbiol. Biotechnol. **14**, 906 (2004).

[24]   P. Giakkoupi, E. Tzelepi, P. T. Tassios, N. T. Legakis, and L. S. Tzouvelekis, J. Antimicrob. Chemother. **45**, 101 (2000).

[25]   S. H. Jeong, I. K. Bae, J. H. Lee, S. G. Sohn, G. H. Kang, G. J. Jeon, Y. H. Kim, B. C. Jeong, and S. H. Lee, J. Clin. Microbiol. **42**, 2209 (2004).

[26]   J. J. Yan, S. M. Wu, S. H. Tasi, J. J. Wu, and I. J. Su, Antimicob. Agents Chemother. **44**, 1438 (2000).

[27]   M. Barlow and B. G. Hall, Antimicob. Agents Chemother. **46**, 1190 (2002).

Cobalt-induced stimulation and inhibition of cytochromes synthesis and extracellular nitrite release in *Paracoccus denitrificans*

Ezzatollah Keyhani[1,2], **Shokoofeh Golkhoo**[3], and **Jacqueline Keyhani**[*,2]

[1] Inst. Biochem. Biophys., University of Tehran, P.O. Box 13145-1384, 13145 Tehran, IRAN
[2] Laboratory for Life Sciences, Saadat Abade, Sarve Sharghi 34, 19979 Tehran, IRAN
[3] Faculty of Sciences, Azzahra University, Tehran, IRAN

[*]Corresponding author: e-mail: keyhanie@ibb.ut.ac.ir, Phone: +98-21-207-4804, Fax: +98-21-640-4680

When Co^{2+} was added to the culture medium, *Paracoccus denitrificans* yield after 24 h increased progressively with increasing Co^{2+} concentrations, reaching a maximum at 1 mg/l Co^{2+} where it almost doubled. Thereafter, the cell yield progressively decreased until 50 mg/l Co^{2+} where no growth was observed. Assay of cytochromes by difference spectrophotometry of cell suspensions showed that, after 24 h culture in the presence of 1 mg/l Co^{2+}, the amount of cytochrome *c* was twice the control value while that of cytochromes *b* and *c* oxidase were, respectively, 1.5 times and 1.1 times the control value. The cytochromes content then decreased with increasing Co^{2+} concentrations to reach approximately 42% of the control value for cytochrome *c* and, respectively, approximately 47% and 26% of the control value for cytochromes *b* and *c* oxidase, at 50 mg/l Co^{2+}. No significant amount of nitrite was released in the culture medium with less than 2 mg/l Co^{2+} in the medium. As Co^{2+} concentration increased, the amount of nitrite detected in the culture medium increased rapidly to reach a maximum at 5 mg/l Co^{2+}.

Keywords cobalt; cytochromes; hypoxia-anoxia; nitrite; *Paracoccus denitrificans*

1   Introduction

The transition metals iron, copper, cobalt, zinc and nickel are essential nutrients for all living organisms [1, 2]. These metals are required as trace elements at nanomolar concentrations. However, higher concentrations of the same metals, in the micro- or millimolar range, are extremely toxic, producing a wide range of diseases ultimately leading to death [3, 4]. In a number of bacteria and some other unicellular organisms, resistance, which is mostly plasmid-borne, has evolved in response to metal toxicity [5–7]. Cobalt is a transition metal that plays a critical role in many biological functions. Besides a key role played by cobalt in vitamin B_{12}-dependent enzymes, eight noncorrin-cobalt-containing enzymes have been reported [8]. Recently it was demonstrated that cobalt induced heme oxygenase-1 expression by a hypoxia-inducible factor independent mechanism [9], but also inhibited the interaction between hypoxia-inducible factor a and von-Hippel-Lindau protein by direct binding to hypoxia-inducible factor a [10].
Bacteria that often live in stringent microenvironments such as presence of heavy metals, high temperatures, etc., have developed a flexible metabolic pathway to adapt to such situations. *Paracoccus denitrificans* is one of these bacteria that developed a highly sophisticated branched respiratory network to ensure optimum amount of energy for its survival in various environments.
In this work, the effect of increasing cobalt concentrations on *Paracoccus denitrificans* growth, cytochromes biosynthesis and on the release of metabolites (nitrite) in the culture medium was investigated. The results showed that, in concentrations up to 1 mg/l, cobalt

induced an enhancement of cytochromes biosynthesis, but that in higher concentrations it produced an inhibition of cytochromes biosynthesis. When the cytochromes biosynthesis was inhibited, nitrite was released into the extracellular medium, suggesting that an alternate electron transport chain became operative.

2   Materials and methods

Paracoccus denitrificans (strain ATCC 17741) was cultured in enriched medium containing 16 different cobalt chloride (Co^{2+}) concentrations (0.05 mg/l to 50 mg/l). Cells were harvested after 24 h, washed twice in phosphate buffer 0.1 M, pH 7 and used for cytochromes quantification. The amount of cytochromes *c* oxidase, *b* and *c* present in the cells was deduced from dithionite-reduced minus air-oxidized difference spectra [11] obtained with a DW-2 Aminco spectrophotometer. The extinction coefficient for cytochrome *c* oxidase was 16, for cytochrome *b*, 18, and for cytochrome *c*, 19 [11]. The amount of nitrite released in the culture medium was measured according to [12].

3   Results and discussion

Figure 1 shows that upon addition of increasing concentrations of cobalt to the culture medium , the cell yield obtained after 24 h increased progressively, reaching a maximum at 1 mg/l Co^{2+} where it almost doubled. Thereafter the cell yield progressively decreased reaching zero growth at 50 mg/l Co^{2+}. Reduced-minus-oxidized difference absorption spectra of *Paracoccus denitrificans* intact cells grown in the presence of various Co^{2+} concentrations are shown in Fig. 2. The spectrum of the control exhibited absorption peaks at 608 nm, 560 nm and 552 nm due to cytochrome *c* oxidase, cytochrome *b* and cytochrome *c*, respectively. Two ? absorption bands were detected, one at 444 nm due to cytochrome *c* oxidase and one at 428 nm due to the other cytochromes. After growth in the presence of 0.4 mg/ml cobalt, cytochrome *b* and cytochrome *c* oxidase exhibited a decrease in their absorption band, while cytochrome *c* exhibited an increase in its absorption band. Higher concentrations of cobalt, 20 and 50 mg/l, produced a decrease in the absorption band of all cytochromes. Figure 3 shows the pattern of cytochromes biosynthesis when *Paracoccus denitrificans* was cultured in the presence of 16 different concentrations of cobalt chloride (from 0.05 to 50 mg/l). It showed a biphasic pattern of enhancement of cytochromes biosynthesis. In up to 0.2 mg/l cobalt, there was a moderate increase in the level of cytochrome *c* oxidase, cytochrome *b* and cytochrome *c*, followed by a trough, and again an enhancement of all cytochromes biosynthesis. Thus for 1 mg/l cobalt, there was a maximum increase of cytochromes (2-fold for cytochrome *c*, 1.5-fold for cytochrome *b* and 1.1-fold for cytochrome *c* oxidase). This showed that the response of *Paracoccus denitrificans* to cobalt induction of protein synthesis, was different for all three cytochromes, with cytochrome *c* exhibiting the highest amount compared to cytochromes *b* and *c* oxidase. Incidently, 1 mg/l cobalt induced a maximum cytochromes biosynthesis, corresponding to a maximum cell yield (Fig. 1). In high concentrations of cobalt (10 to 50 mg/l), the amount of cytochromes decreased maximally, but remained constant over the cobalt concentrations range of 20-50 mg/l (Fig. 3).

44

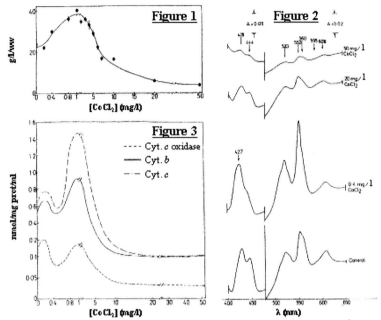

Fig. 1 Growth of *Paracoccus denitrificans* in the medium with increasing Co^{2+} concentrations. Approximately 2-g inoculum was cultured in a 2-liter flask containing 500 ml culture medium and various cobalt concentrations.

Fig. 2 Dithionite-reduced minus H_2O_2-oxidized difference spectra of *Paracoccus denitrificans*.

Fig. 3 The amount of cytochrome *c* oxidase, cytochrome *b* and cytochrome *c* for cells grown with increasing Co^{2+} concentrations .

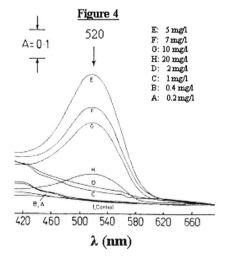

Fig. 4 Nitrite release in the medium as a function of cobalt concentration. A–E represent cobalt concentrations expressed in mg/l.

The determination of nitrite release in the medium was shown in Fig. 4. A small amount of nitrite was released by cells grown in up to 2 mg/l cobalt. The maximum amount of nitrite was released by cells grown in 5 mg/l cobalt. Thus high amounts of nitrite were released into the medium by cells grown in 5–10 mg/l cobalt in the order 5 > 7 > 10 mg/l cobalt. This corresponded to the period 1) when the cell yield reached almost no growth (Fig. 1) and when the biosynthesis of cytochromes was also severely repressed (Fig. 3). These observations suggested that when cytochromes biosynthesis was inhibited by cobalt, there was a repression of the cell yield. Consequently, in order to survive, the cells developped (or activated) an alternate electron transport chain, using the nitrate present in the medium as electron acceptor, thus releasing nitrite into the medium. Repression by Co^{2+} of cytochromes biosynthesis and ensuing hypoxia led the cells to adjust their metabolic activity to survive. The adaptative response to hypoxia and to low oxygen tension is a signalling event for cell adaptation [13, 14] and cytochrome c oxidase is likely the sensor hemoprotein [13]. Moreover, cobalt is considered a chemical inducer of hypoxia-inducible factor 1 [15]. Nitrate respiration is well studied in both *Escherichia coli* and *Paracoccus denitrificans* [16]. The nitrate respiration under cobalt stress induced the relase of nitrite into the medium as reported in this paper.

References

[1]   H. L. Ehrlich, Appl. Microbiol. Biotechnol. **48**, 687 (1997).
[2]   D. C. Rees, Annu. Rev. Biochem. **71**, 221 (2002).
[3]   M. Friedman (ed.), Protein-Metal Interactions, Advances in Experimental Medicine and Biology Vol. 48 (Plenum Press, New York 1974).
[4]   D. H. Nies, Appl. Microbiol. Biotechnol. **51**, 730 (1999).
[5]   L. Diels, and M. Mergeay, Appl. Environ. Microbiol. **56**, 1485 (1990).
[6]   N. El Solh, and S. D. Ehrlich, Plasmid 7, 77 (1982).
[7]   H. Horitsu, K. Yamamoto, S. Wachi, K. Kawai, and A. Fukuchi, J. Bacteriol. **165**, 334 (1986).
[8]   M. Kobayashi, S. Shimizu, Eur. J. Biochem. **261**, 1 (1999).
[9]   P. Gong, B. Hu, D. Stewart, M. Ellerbe, Y.G. Figueroa, V. Blank, B. S. Beckman, and J. Alam, J. Biol. Chem. **276**, 27018 (2001).
[10]   Y. Yuan, G. Hillard, T. Ferguson, and D. Millhorn, J. Biol. Chem. **278**, 15911 (2003).
[11]   E. Keyhani, and J. Keyhani, Biochim. Biophys. Acta **717**, 355 (1982).
[12]   S. Finegold, and W. J. Martin, in: Diagnostic Microbiology, Bailey and Scott, eds, (C.V. Mosby, 1990), p. 113.
[13]   R. S. Zitomer, and C. V. Lowry, Microbiol. Rev. **56**, 1 (1992).
[14]   K. E. Kwast, P. V. Burke, and R. O. Poyton, J. Exptl. Biol. **201**, 1177 (1998).
[15]   J.-P. Piret, D. Mottet, M. Rae, and C. Michiels, Ann. N. Y. Acad. Sci. **973**, 443 (2002).
[16]   W. G. Zumft, Microbiol. and Mol. Biol. Rev. **61**, 533 (1977).

Communication of microbial risk to workers in a wastewater treatment plant

A. Calamusa[1], M. Verani[1], and A. Carducci[1]

[1] University of Pisa - Health Observatory Communication – Via S. Zeno 35 56127, Pisa ITALY

1   Introduction

The prevention of microbiological hazard in working setting necessarily implies a training for the operators, who should become able to adopt right behaviours and procedures within their competence and activity. The Italian Decree Law n. 626/94 expects training as an integral part of the hazards risks and control process from the employer (1).

A proper communication about this topic has a great importance because of the present lack of specific knowledge in the most part of working settings not related with healthcare that could lead both to underestimate dangers and to an excessive alarmism, due to media as well.

It is evidently necessary an integrated training programme, with a specific methodology taking in account quality standards for the construction of messages (correction-completeness, reliability, usefulness, incitement to attention, comprehensibility, balance, independence, accessibility, coherence, cultural suitability, scientific facts, continuity-repetitiveness, timeliness and for the formulation of communicative strategy) (2,3).

2   Methodology

Before embarking on any programme of prevention-aimed information, it is necessary to know the knowledge level, the behaviours and the hazard perception of people involved. The actual intervention has therefore been preceded by a questionnaire aimed at estimating some basic knowledge about microbiological hazard among the workers and its perception (4). The questionnaire has been divided into four sections (Fig. 1).

Giving the questionnaire created the occasion to stimulate the workers attention and to make them thinking over these issues, thus creating the premises of an involvement in the training process further step. The information obtained have been used for the production of explanatory material, with a particular care for the textual and graphical formulation. Such material (leaflets, plasticized pocket cards, posters) has been given to the workers during some meetings that aimed at illustrating these issues through the workers active involvement in the discussion, in order to facilitate the attention and a quick understanding.

| Brief introduction of questionnaire to workers | General information (age, level of education, etc…) | Information received about microbiolo- gical hazard and their sources | Definitions (definition of infection, microbiological hazard, etc.) | Evaluations giving a value of importance from 1 to 10 |

Fig. 1 Questionnaire

3   Results

The questionnaire has been given to 20 workers assigned to particularly hazardous jobs. The 71% had already received information about microbiological hazard, mainly from the media (31%), then from the qualified doctor (28%) and from the workers representatives for safety (17%). The definitions of infection and of microbiological hazard are known respectively to the 67% and the 46% of workers, whereas, in a scale from 1 to 10, the majority of the interviewed (25%) gave to such hazard an 8 importance level; for the 20%, the level to be assigned to it should be 10. The interviewed proved to know the main features distinguishing chemical from microbiological agents (Figure 2), even if they showed some biases.

As for the transmission routes, the respiratory one turns to be the most important, followed by the ones associated to injuries, objects, to the ingestion and, lastly, to the contact with people. Therefore enteric transmission is considered not so important, despite the hazard of the sewage (Figure 3).

As to confirm this perception, the precautionary measures regarded as the most important are the use of gloves, then face guards and cleaning one's face, whereas the prohibition to eat and drink on the working facility comes only at the fourth place (Figure 4).

As a consequence of these results, informative material has been produced in order to create a greater awareness about microbiological hazard and precautionary measures to be adopted. The meetings during which such material was given led to a greater awareness of these issues, checked through the discussion.

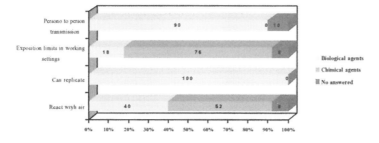

Fig. 2 QUESTION N. 5 "Related the following definition with microbiological or chemical agents"

48

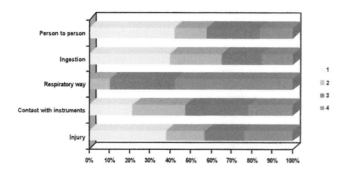

Fig. 3 QUESTION N. 9 "Give a value of importance (from 1 to 4) to the following way of microbiological agents transmission

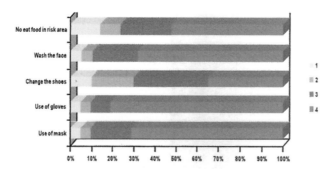

Fig. 4 QUESTION N. 10 "Give a value of importance (from 1 to 4) to the following methods of prevention

4   Conclusions

For the accomplishment of this intervention, a methodology has been used which is based upon quality standards. The use of a simple language and the active involvement of workers: it produced an evident awareness of the dangers connected to the working activity. The workers themselves began to suggest working procedures, protection and prevention interventions overcoming organizational difficulties. Even though a complete evaluation of the intervention would be possible only after a lapse of time, it is however sensible to affirm that the methodology used produced good results.

References

[1]   E.B Arkin.. Translation of risk information for the public: message development in effective risk communication. Ed by V.T. Covello, D.B. 1989.

[2]   A. Carducci., A. Calamusa, A. Giuntini, F. Mazzoni, B. Casini, M. Verani, E. Rovini – La comunicazione sanitaria attraverso i mass media: il modello AIDS. – Difesa Sociale, LXXXI (1): 23-40, 2002.

[3]   A. Carducci, A. Calamusa, e gruppo "Leggere e ascoltare la salute". – La comunicazione di massa sulla salute. Proposta di un sistema di valutazione della qualità del processo di produzione dell'informazione sulla salute. – Educazione sanitaria e promozione della salute, 26 (1): 67-79, 2003.

[4]   A. Carducci, A. Calamusa, P. Manfredi, J. Williams, F. Romano, A. Giuntini, S. Marcantonio, R. Bianchi Bendinelli, R. Piz, F. Tarini, M. Verani, G. Privitera – Indagine sugli stili di vita e la salute degli studenti universitari dell'Ateneo Pisano. – Annali di Igiene, 16 (5): 673-684, 2004.

De novo protein production under long term starvation in *Flexibacter chinensis*

Jamshid Raheb[1]*, **Shamim Naghdi**[1] and **Ken P. Flint**[2]

[1]National Institute for Genetic Engineering and Biotechnology (NIGEB), P.O.BOX: 14155-6343, Tehran, IRAN
[2]Department of Biological Sciences University of Warwick, Coventry CV4 7AL, UK
*Corresponding author: e-mail: jam@nrcgeb.ac.ir, Fax: 0098 21 44580399

The molecular and physiological effects of starvation on *Flexibacter chinensis* have been investigated. The survival of the organism under starvation conditions was temperature dependent with the longest survival occurring at 4°C and shortest at temperatures above 30°C. The cell size of *F. chinensis* was reduced during starvation and this reduction was also temperature-dependent. The changes in protein profile at the starvation process were investigated using two-dimensional gel electrophoresis.

Keywords: *Flexibacter chinensis*, stress, protein profile

1   Introduction

One of the remarkable features of bacterial species is their capacity for rapid growth when nutrients are available and the appropriate conditions for growth are provided and, probably, a more remarkable characteristic is their ability to remain viable under starvation conditions [1]. De novo protein synthesis especially, in the first few hours, after the onset of starvation is essential for the longevity of starving *E. coli* cultures. These starvation proteins have different temporal profiles of synthesis; some being synthesized very transiently during the starvation period, whereas others have a broader peak of synthesis and persist for longer period of time.

The synthesis of these new proteins gives bacteria the ability to scavenge more efficiently during times of nutrient limitation. They also confer no bacteria the ability to overcome stress conditions and confer a more stress-resistant phenotype on the bacteria and also confers the ability on bacteria to survive prolonged periods of stress or starvation [2].

The synthesis and degradation of proteins have a role in the production of free amino acids. Therefore degradation of these peptides is required to establish a pool of amino acids which could be utilized for the synthesis of starvation specific proteins [3, 4]. The complete loss of any protein which is required for protein synthesis should be lethal, for example inactivation of elongation factor G in a temperature-sensitive mutant of *Salmonella typhimurium* leads to a rapid loss of viability. A similar loss in viability would be expected if ribosomes, any of the soluble protein factors involved in translation or amino acyl-tRNA *synthesis*, or RNA polymerase were completely lost from the bacterial cell. Loss of some ribosomes is one example of the third mechanism described above. Loss of ribosomes is a well-known major response of bacteria to starvation which provided nucleotides and amino acids, and possibly energy in the form of nucleotide diphosphates if RNA is degraded by polynucleotide phosphorylase [5].

2 Materials and methods

2.1 Bacterial Strain

The bacterial strain used in this study was *Flexibacter chinensis* obtained from Dr. Flint (Warwick university, UK).

2.2 Bacterial Growth Media and Conditions

All bacterial strains were routinely grown in Luria Broth (10g/l Bacto tryptone, 5g/l yeast extract, 5g/l NaCl, pH 7.2) or on Luria Agar (10g/l bacto tryptone, 5g/l yeast extract, 5g/l NaCl, 15g/l Agar). Plates were incubated at 30°C.

2.3 Protein purification method

F. chinensis was grown in a suitable liquid medium for 24 hr at 30°C. The cells were harvested by centrifugation and sonicated in 1 mM Tris buffer (pH 8.0). The sonicated cells were centrifuged at 1,500 g for 20 min at 4°C. The pellet was discarded and the supernatant centrifuged at 48,400 g for 60 min at 4°C to harvest the membrane proteins and leave the soluble proteins in the supernatant. The membrane proteins were resuspended in 3 ml Tris buffer (pH 8.0) and stored at -20°C.
The supernatant was concentrated using an Amicon ultrafiltration apparatus to about 5 ml. An equal amount of saturated ammonium sulphate was added to the concentrated supernatant fraction to give a final ammonium sulphate concentration of 50%, and left overnight to allow the proteins to precipitate. The precipitant was pelleted by centrifugation at 35,000 g for 30 min, resuspended in 5 ml of sterile distilled water and dialyzed overnight against at least 4 liter of distilled water to remove any ammonium sulphate from the samples.

2.4 Two-dimensional gel electrophoresis

This technique was done according to the method described be Ofarrell [6].

3 Results

3.1. A comparison of the changes in protein profiles between cells of *F. chinensis* in the stationary phase and subjected starvation.

The changes in the protein profiles in cells in various stages of growth and in cells subjected to long term starvation in sterile distilled water at 15°C were compared.
Figure shows the protein profiles obtained for cells harvested in the exponential phase, early stationary phase, and late stationary phase compared with the protein profiles for cells starved for 7 days, 14 days and 28 days. Each protein which showed a change was given an identifying number.
Investigation about the appearance and disappearance of different major proteins identified on the gels showed differences at different stages of starvation. The proteins numbered 3, 6,

8, 9, and 15 were found in all the Figures. The proteins numbered 4 and 5 appeared from early stationary phase and then persisted until the end of the starvation period. The protein numbered 16 appeared in early stationary phase and was stable until 7 days of starvation but disappeared afterwards. The proteins numbered 1, 7, and 2 were found from the exponential phase until 14 days starvation but disappeared before 28 days starvation.

The proteins numbered 12 and 17 appeared after 14 days starvation and were then stable until the end of the starvation period. The proteins numbered 13 and 14 were only present after 7 days starvation. The proteins numbered 18 and 19 were only present in late stationary phase and were also found under carbon and nitrogen limitation. The protein numbered 20 was only present in cells starved for 7 days. All these results show that changes in the protein profiles are occurring constantly even though these cells are subjected to extreme starvation conditions.

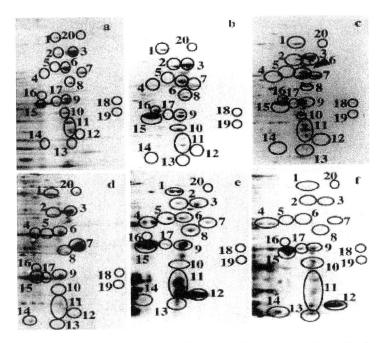

Fig. 1 Two dimensional electrophoresis analysis. Figure a shows the protein profile for cells harvested in the exponential phase, Figure b cells harvested in early stationary phase, Figure c cells harvested in late stationary phase, Figure d cells starved for 7 days. Figure e cells starved for 14 days, and Figure f cells starved for 28 days. The proteins which show the major changes are numbered from 1–20.

4   Discussion

One of the most remarkable features of bacterial endospores is their high resistance to environmental stress especially starvation condition. Sporulation is triggered by environmental stimuli and, in different bacteria, leads to the formation of a series of proteins which are only produced in response to starvation conditions. Starvation induces the development of a more resistant state even in bacteria which cannot produce endospores.

Kramer and Singleton [7] reported that in *Vibrio furnissii* in response to nutrient depletion, Cell mass and rRNA decreased rapidly while the total cell counts remained stable. This reduction in cell mass is seen as miniaturization of cells, similar to the phenomenon observed in other starved *Vibrio species*. The overall rate of protein synthesis in *Vibrio S14* after 24 hr starvation was about 10% of that of cells growing exponentially in a complex medium and the global rate of protein synthesis was reduced by 95% in *E. coli* [8]. In both cases the proteins which were being synthesized were the ones specifically associated this starvation stress and the development of resistance to stress in these bacteria. The rate of total protein synthesis decreased immediately at the onset of starvation and after 48 hr was less than 1% of the rate of synthesis during growth [9].

During the transition on the stationary phase at temperatures around the optimum for growth, the concentration of proteins which are involved in transcription and translation are greatly reduced. However, some proteins which have roles in energy metabolism are greatly increased in *E. coli* [10]. To eliminate some of the changes in protein profiles in starvation medium which could be induced by high temperature, all the experiments in starvation medium were conducted at 15°C which is within the normal temperature range for growth of *F. chinensis* but is less than the optimum temperature for growth of this organism, to obtain only the starvation proteins in absences of the effect of high temperature.

Lim [11] reported that, in *Aeromonas hydrophila* the sequential induction, repression and increased production of individual proteins was likely to be a process of adaptation of the cells for survival. He showed that starvation-specific proteins were sequentially induced during the whole period of starvation at temperatures for 4°C to 37°C. Some proteins showed only a transient induction phase after a specific period of starvation and were subsequently repressed. Reeve *et al.* [12] reported that, in *E. coli*, protein synthesis during the first 9 hr of starvation was the most crucial for survival and the hypothesis that these proteins play a role in long term starvation is supported by the observation that the addition of chloramphenicol or amino acid at the beginning of the starvation period had a negative effect on culture viability.

This study has shown that there are many proteins synthesized in *F. chinensis*. Some of these proteins must play a major role in the stability of the cell under starvation stress. Most of the new proteins which are synthesized in response to starvation appeared during the early stationary phase than in any other phase. This shows the importance of the new proteins produced in this phase of growth to the maintenance of the cell during the onset of starvation. The continued presence of these proteins in cells subjected to long term starvation shows that the proteins have a continued role in the survival of the organisms under extreme nutrient limitation.

The final analysis of a single protein found to be synthesized under starvation stress conditions revealed it to have some homology to a known chaperonin from *E. coil*.

References

[1]   Deborah, A. S., and R. Kolter.(1992). Minireview. Life after log. *Journal of Bacteriology*, 174, 345-384.

[2]   Ozkanca, R. (1993). Survival and physiological status of Escherichia coli in lake water under different nutrient conditions. *PhD. Thesis. University of Warwick. U.K.*

[3]   Reeve, C. A., A. T. Bockman, and A. Matin. (1984). Role of protein degradation in the survival of carbon-starved Escherichia coli and Salmonella typhimurium. *Journal of Bacteriology*, 157, 758-763.

54

[4]   Matin, A., E. A. Auger, P. H. Bluem, and J. E. Schultz. (1989). Genetic basis of starvation survival in non-differentiating bacteria. *Annual Review of Microbiology*, 43, 293-315.

[5]   Albertson, N. H., T. Nystrom, S, and Kjelleberg. (1990). Macromolecular synthesis during recovery of the marine Vibrio sp. S14 form starvation. *Journal of General Microbiology*, 136, 2201-2207.

[6]   OFarrell, P. H. (1975). High resolution two-dimensional electrophoresis of proteins. *Journal of Biological Chemistry*, 250, 4007-4021.

[7]   Kramer, J. K., and F. L. Singleton. (1992). Variations in rRNA content of marine Vibrio spp-during starvation-survival and recovery. *Applied and Environmental Microbiology*, 58, 201-207.

[8]   Davis, D. B., S. M. Luger, and P. C. Tai. (1986). Role of ribosome in the death of starved Escherichia coil cells. *Journal of Bacteriology*, 166, 439-445.

[9]   Flardh, K., P. S, Cohen, and S. Kjelleberg. (1992). Ribosomes exist in large excess over the apparent demand for protein synthesis during carbon starvation in marine *Vibrio* sp-strain CCUG 15956. *Journal of Bacteriology*, 174, 6780-6788.

[10]   Herendeen, S. L, R. A. Vanbogelen, and F. C. Neidhardt. (1979). Level of major proteins of Escherichia coli during growth at different temperatures. *Journal of Bacteriology*, 139,185-194.

[11]   Lim, C. H. (1995).The effect of environmental factors on the physiology of *Aeromonas hydrophila* in lake water. *PhD thesis. University of Warwick. U.K.*

[12]   Reeve, C. A., P. S. Amy, and A. Matin. (1984). Role of protein synthesis in the survival of carbon-starved Escherichia coil K-12. *Journal of Bacteriology*, 160, 1041-1046.

Development of a biotechnology tool using New Zealand white-rot fungi to degrade pentachlorophenol in soil – A summary

Monika Walter and Kirsty S. H. Boyd-Wilson

HortResearch, Environment and Risk Management Group, P. O. Box 51, Lincoln, NEW ZEALAND

We describe the research undertaken in order to select isolate(s) for bioremediation of pentachlorophenol (PCP) in the field using a proof-of-concept field biopile. This includes screening approaches, growth and survival studies, growth substrate development and evaluation of bioremediation potential in the laboratory.

Keywords *Trametes versicolor*, PCP, laboratory studies, field remediation.

1   Introduction

In New Zealand the fungicide pentachlorophenol (PCP) was used extensively by the forestry industry from the late 1940s to prevent sapstaining of pine. Until the 1970s, New Zealand was a heavy user of industrial grade PCP and there are approximately 800 contaminated sites as a legacy [1].

International research has shown that soils contaminated by xenobiotics may be ameliorated using white-rot fungi. Because of restrictions on the importation and use of overseas microorganisms our research has focused on the use of New Zealand white-rot fungi to degrade PCP contaminated soils (Fig. 1). Native white-rot isolates were collected, selected for their ability to degrade PCP and studied for their mechanisms and pathways of degradation. Organic waste materials were also evaluated for their suitability to serve as a growth substrate for fungal augmentation to the polluted soil. Selected organisms were evaluated for PCP loss and breakdown in soil. Soil limiting factors affecting colonisation of augmented isolates were identified. These included soil type, moisture, temperature and pollutant concentration. These results then were transferred into the field and PCP degradation studied using proto-type biopiles.

2   Experimentation

2.1. Screening experiments [2]

From a pool of 481 white-rot fungi native to New Zealand (over 77 genera), isolates were screened for their PCP bioremediation potential. Fungi were tested for their ligninolytic activity using dye discoloration and wood decay assays, temperature growth curves, tolerance to PCP, PCP degradation *in vitro*, and enzyme expression. Of the isolates tested, 26% showed a discolouration in the polymeric dye assay, but all caused wood decay on willow cuttings. Some 18% and 40% did not survive incubation at 35 and 40°C, respectively. In the PCP resistance tests, 9% were able to grow on 200 mg/L PCP amended agar, of which 20 isolates were further studied for laccase expression and PCP degradation in liquid. All 20 isolates reduced ($P<0.05$) PCP in the liquid fraction in the absence or

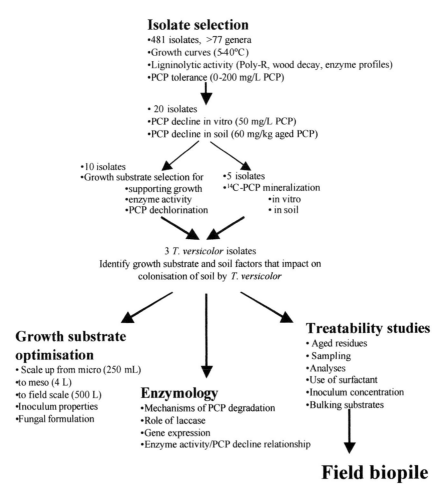

Isolate selection
- 481 isolates, >77 genera
- Growth curves (5-40°C)
- Ligninolytic activity (Poly-R, wood decay, enzyme profiles)
- PCP tolerance (0-200 mg/L PCP)

- 20 isolates
- PCP decline in vitro (50 mg/L PCP)
- PCP decline in soil (60 mg/kg aged PCP)

- 10 isolates
- Growth substrate selection for
  - supporting growth
  - enzyme activity
  - PCP dechlorination

- 5 isolates
- ^{14}C-PCP mineralization
  - in vitro
  - in soil

3 *T. versicolor* isolates
Identify growth substrate and soil factors that impact on colonisation of soil by *T. versicolor*

Growth substrate optimisation
- Scale up from micro (250 mL)
- to meso (4 L)
- to field scale (500 L)
- Inoculum properties
- Fungal formulation

Enzymology
- Mechanisms of PCP degradation
- Role of laccase
- Gene expression
- Enzyme activity/PCP decline relationship

Treatability studies
- Aged residues
- Sampling
- Analyses
- Use of surfactant
- Inoculum concentration
- Bulking substrates

Field biopile

Fig. 1   The pathway for selecting white-rot fungi for bioremediation of PCP in the field.

presence of laccase. Five of the isolates produced no detectable levels of PCP. None of the screening tests were predictive for PCP degradation in liquid.

2.2 Growth substrate selection and mineralization experiments [3]

Nine New Zealand native white-rot fungi were studied for their ability to grow and survive on different substrates formulated from bark, wheat straw, sawdust, apple pomace and maize products. We identified their PCP biodegradation potential and selected a fungal carrier for bioaugmentation of polluted soils. Isolates were also evaluated for their ability to mineralize ^{14}C-PCP in liquid culture and in soil. An American isolate of *Phanerochaete chrysosporium* outgrew the native fungi on the substrates tested, but the high colonisation did not result in PCP dechlorination, as measured by chloride release and PCP mineralisation. Virtually no pentachloroanisole (PCA), a toxic metabolite of PCP, was captured in the volatile fraction of *T. versicolor* isolates, whereas 75% of the volatile fraction of *P. chrysosporium* consisted of PCA. This indicates that different pathways and mechanisms of degradation are involved.

2.3 PCR detection [4]

DNA was isolated from three *Trametes versicolor* isolates in soil using bead beat extraction, combined with extract clean-up. This was amplified by PCR (polymerase chain reaction) using ITS-1 and ITS-2 primers. The detection limit of mycelium in soil was dependent on isolate, soil type and mycelium type (fresh versus freeze-dried). Nested PCR greatly enhanced *T. versicolor* detection compared to single PCR amplifications. DNA from fresh mycelium was more readily amplified (approximately 10 fold) compared to freeze-dried mycelium (based on mycelial dry weights).

2.4 Soil factors and fungal inoculum properties [4, 5]

Soil colonisation was significantly dependent on isolate and soil type, ranging from sparse to complete colonisation for the three different *T. versicolor* isolates studied. There was a significant interaction between soil type and fungal isolate. Linear correlation studies between soil properties (11 soils) and colonisation by *T. versicolor* showed that the only significant correlation that occurred for all three isolates was between mycelial cover and percentage base saturation. Soil colonisation was also affected by fungal inoculum properties, such as growth substrate composition, age, inoculum concentration, particle size and inoculation method. Other parameters that significantly affected fungal growth in soil were concentration of non-colonised organic matter or bulking substrate, temperature, pH, soil moisture level and pollutant concentration.

2.5 Treatability studies [6]

Laboratory studies were conducted to standardise treatability assessments using aged, field contaminated PCP residues. The main finding in these experiments was that similar levels of aged PCP residues in soils are more toxic to white-rot fungi than an equivalent spiked PCP contamination. Also, pollutant degradation followed different patterns depending on whether the contaminant is aged, or added fresh. The upper limit for PCP bioremediation by *T. versiolor* was in the order of 1300-1800 mg/kg aged PCP.
Working with aged residues, high variations between samples occurred. These were overcome by the stratified sampling of combining sub-samples, homogenising and re-sampling. This halved the analytical costs. This also improved the demonstration of treatment effects.

2.6 Field remediation [7]

Engineered soil cells were designed to develop proof-of-concept biopiles for white-rot bioremediation of aged PCP contaminated soil from a former timber treatment site. Soil cells were constructed to allow for forced aeration, irrigation, leachate collection and monitoring of temperature and soil moisture content. We studied the effect of a New Zealand *Trametes versicolor* isolate on PCP degradation, the effect of fungal inoculum concentration on PCP degradation and reproducibility of the experiments. PCP degradation and fungal survival were monitored at regular intervals for 2.5 years. The experiments were set up in January 2000. There was no effect of inoculum concentration. Treatment effects were reproducible. PCP residue levels declined from 800-1000 mg/kg to 0-9.4 mg/kg in a first order like decay curve. Irrigation was not required during the 2.5 years of the study, nor did any leaching occur. The soil cell temperatures did not exceed daily maximum temperatures of 35°C.

3   Conclusions

All isolates showed ligninolytic activity by decaying willow cuttings. The polymeric dye indicated hydrogen peroxidase activity. The laccase assay identified isolates that produced certain lignin-modifying enzymes. However, neither ligninolytic activity, nor growth rate, nor pollutant tolerance were indicative of biodegradation potential. This demonstrates that isolates need to be characterised and tested vigorously prior to augmentation into polluted soil sites.

The soil microcosm studies, using contaminated soil from a timber treatment site, clearly showed that the New Zealand *T. versicolor* isolates mineralized PCP. Degradation of PCP in non-sterile soil was higher in the presence of white-rot fungi than in soil without white-rot fungus. Viable white-rot fungus is necessary for significant PCP degradation. *T. versicolor* isolates showed PCP remediation potential. Wheat straw and SCS could be suitable carriers for New Zealand native *T. versicolor* isolates for bioremediation of PCP polluted soil sites, however, in New Zealand, sawdust is more readily available and cheaper.

For successful bioremediation in the field, good colonization of the polluted soil is necessary. Rate-limiting factors for colonization and bioremediation were identified. The variable nature of contaminated soil requires reliable assessments (treatability studies) to optimise the bioremediation process in the field. Our treatability studies were designed to select fungal isolates, to determine optimum fungal inoculum concentration and maximum contaminant concentration, and to predict degradation over time. Isolate specific effects highlight the importance of having a detailed database of growth and survival characteristics under specific conditions in order to select isolates for transfer to the field.

Our proof-of-concept biopiles demonstrated that a New Zealand native white-rot isolate is capable of degrading PCP from 800 mg/kg to less than 50 mg/kg in 74 weeks. Additional 'natural attenuation' for 2 more years further reduced the PCP levels to less than 4 mg/kg. In our screening tests, all isolates survived temperatures of between 0 and 30°C. The temperatures within our soil cells stayed within this range, so biopile temperatures would not have limited growth of these isolates. However, it is important to control temperature and prevent biogenic heat production to avoid killing the fungal mycelium by over-heating. The results clearly show that the white-rot fungus *T.versicolor* can biodegrade PCP from aged soils in the field.

Acknowledgements

We would like to thank the Foundation for Research, Science and Technology for funding. Thanks to all colleagues and staff involved in the research to date. Thanks to Forest Research, Landcare Research, AgResearch, WRONZ, URS (formerly Woodward-Clyde), Utah State University (USA), Massey University and University of Canterbury for isolates and/or research collaboration. Special thanks to Environment Canterbury for providing the field site and to all regional/local councils and industry partners for their support and interest in the project.

References

[1]   T.W. Finnbogason and O.N.C. St. Quintin, Review and assessment of available pentachlorophenol (PCP) and dioxin treatment technologies. (Envirochem Special Projects Inc. North Vancouver, B.C. 1994) p.123.

[2]   M. Walter, J. M. Guthrie, S. Sivakumaran, E. Parker, A. Slade, D. McNaughton and K.S.H. Boyd-Wilson, Bioremediation J. **7** (2), 119 (2003).

[3]   M.Walter, L. Boul, R. Chong, and C. Ford, J. Environ. Manage. **71**, 361 (2004).

[4]   M. Walter, Towards optimisation of white-rot fungi bioremediation. Ph.D. Thesis (University of Canterbury, Christchurch, NZ, 2004) p. 222.

[5]   K.R. Schmidt, S. Chand, P.A. Gostomski, K.S.H. Boyd-Wilson, C. Ford and M. Walter, Biotech.Prog. **21**, 377 (2005).

[6]   M. Walter, K.S.H. Boyd-Wilson and D. McNaughton, Int. Biodeter. Biodeg. **55**, 121 (2005).

[7]   M.Walter, K.S.H. Boyd-Wilson, L. Boul, C. Ford, D. McFadden, R. Chong and J. Pinfold, Int. Biodeter. Biodeg. **56**, 51 (2005).

Ecological interactions among protozoan parasites and their avian hosts: an approach

A. Marzal[*1], **M.I. Reviriego**[1], **C. Navarro**[1], **S. Díaz**[1], **F. De Lope**[1] and **A.P. Møller**[2]

[1] Department of Animal Biology, University of Extremadura, Avda. Elvas s/n, E-06071 Badajoz, Spain
[2] Lab. Ecologie Evolutive Parasitaire, CNRS, Univ. Pierre et Marie Curie, 7 quai St. Bernard, F-75252 Paris, France

[*] Corresponding author: e-mail: amarzal@unex.es, Phone/ Fax: +34 924 289412

Ecologist and evolutionary biologist virtually ignored parasites, but nowadays scientists recognize the benefits of using data on the prevalence and distribution of avian haemosporidian in ecological, evolutionary and behavioural studies. We present three different researches on ecological interactions among protozoan parasites and their avian hosts focus on answering whether parasites regulate host populations; and whether parasites act as selective agents that results in adaptative responses in their hosts. Results demonstrate that protozoan malarial parasites can have dramatic effects on their hosts because they are involved in different host's life history traits such as senescence, reproductive success, survival and predation-prey interaction.

Keywords: Ecological interaction, Protozoa, malarial parasites, senescence, predation, reproductive success.

Parasitic animal are one of the major cause of the infection diseases which affect man and his domestic stock, and their effects are enormous in terms of mortality, chronic diseases and economic loss. To prevent this many advances in microbiology and parasitology are required with an obvious applied significance because they throw light upon a wide variety of researches, such as immunology, conservation biology, behavioural ecology, biogerontology and evolutionary biology. So, modern researches are directed towards the understanding of parasite's effects on their hosts. Here we present three different researches that allow us to advance in knowledge about parasites and immune responses in wild birds. Most aspects of the life history of hosts have been hypothesised to be affected by parasites, like age of maturity, clutch size and offspring size, but our knowledge about causal relationship is still rudimentary due to a scarcity of experimental manipulation. Malaria is supposed to have strong negative effects of host fitness because this group of intra-cellular parasites cause dramatic effects over the efficiency of metabolism. However, the fitness consequences of malarial infections are generally poorly known. So far only a single study has experimentally treated malarial infections in birds showing a direct effect of avian malaria on reproductive output [1]. But there are more questions that must be answered. Do malarial parasites have effects on clutch size? If so, do these early effects of malarial infection during the reproductive cycle have disproportionately large effects on seasonal reproductive success? So early treatment of individuals host would then be expected to reduce or maintain the levels of infection at a time in the reproductive cycle when infections are normally rapidly increasing.

We experimentally reduced levels and intensity of blood parasites infection, by randomly treating birds (House martin *Delichon urbica*) with an anti-malarial drug (Primaquine. Sigma, St.Louis, Mo) at the beginning of the breeding season [2]. Results showed that clutch size was on average 18% larger in treated birds, while the difference increased to 39% at hatching and 42% at fledging. Our study is the first to demonstrate that there indeed is an improvement in clutch size as a response to a removal of a parasite. The possible mechanisms generating these effect could be a direct impact on foraging ability and therefore rate of level of resource acquisition necessary for production of eggs, or it also would be possible that reduction of clutch size was due to a energy draining by the parasites. In addition, carotenoids used by the immune system for fighting serious infection may also play a role in egg formation [3-5]. If there is a trade-off between use of carotenoids and other antioxidants for egg production and immunity, we hypothesise that control females laid eggs with reduced levels of yolk antioxidants. That could directly lead to reduced hatching success [6]. Alternatively, adult house martins may have been affected by malarial infections, causing a reduction in the efficiency of incubation and provisioning of offspring. To sum up, these findings demonstrate that malarial parasites can have dramatic effects on clutch size and other demographic variables.

Predation is one of the most important causes of natural selection [7]. Although predation often is studied in isolation, several pieces of evidence suggest that different kinds of interespecific interaction may affect each other, so parasites may play an important role in predation-prey interaction. In another study, House sparrow (*Passer domesticus*) were either exposed to a predator (owl) or a control (pigeon) while development of malaria infections was recorded during the following six weeks [8]. We tried to test how perceived predation risk affects the ability of potential prey to produce a response to an immune response (T-cell mediated immune response to a challenge with phytohemaglutinin, PHA), and if there were a long term effects of reduced immune response on malarial infection. The main findings of this study were that predation has significant depressing effect on T-cell response because it was 20 % smaller in the owl (predation) than in pigeon (control) treatment. Figure 1 shows that an increased predation risk was associated with an increase prevalence and intensity of *Haemoproteus* malarial blood parasite.

To the best of our knowledge, this provides the first experimental demonstration of a direct relationship between predation risk and immune function, and the first successful experimental manipulation of infection with a blood parasite, and this could result in an increased risk of mortality [9-12]. In conclusion, predators may interact indirectly with parasites through their effects on immune function in hosts.

Senescence is the progressive lost of functionality accompanied by fertility decreasing and increasing in risk of mortality as time goes by [13]. Immune system deteriorates as ageing, so it causes an ability reduction to fight against parasites and to prevent diseases [14].

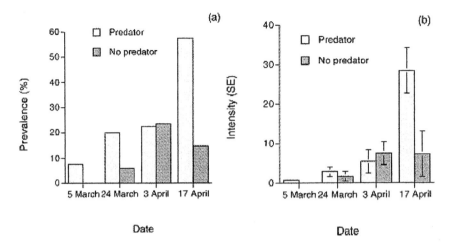

Figure 1. Prevalence (a) and intensity (b) of infection with *Haemoproteus* blood parasites in relation to date and treatment. Error bars are SE. Sample sizes are 40 for predator exposure and 34 for control treatment.

This is the reason why parasitism have been involved in the expression on the senescence for a long time [15, 16]. So, must we expect higher blood parasites infection in wild birds as ageing? Many studies tried to show an increasing in parasites infection in wild populations of migratory birds. The main problem of all these researches was the impossibility to study the same individual consecutive years. In the third study we made a comparative test for prevalence of malarial infection between two avian species (House martin and Barn swallow), capturing the same individual every year. As we can see at figure 2 we found significance differences in the number of infected adult birds between years in both species.

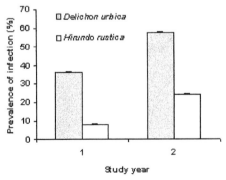

Figure 2. Prevalence of infection with *Haemoproteus* and *Trypanosoma* blood parasites in relation to study year. Error bars are SE. Sample sizes are 127 for martins (*Delichon urbica*) and 50 for swallows (*Hirundo rustica*).

Moreover, many individuals significantly change their status from non infected to infected year after year. This could be explained because the host's ability loss to fight against parasites as growing old [17], a spring relapse in blood parasites [18], a higher vector exposition [19] or a immune function deterioration associated to senescence [20, 21]. In resume, protozoan parasites are related to the expression of their hosts senescence in wild populations of avian species because blood parasites infection levels increase as their hosts grow old.

All these findings experimentally demonstrate that protozoan malarial parasites affects their hosts in many different ways, such as senescence, reproductive success, immune function and mortality. Thus, the study of parasites is a field that must to pay attention because it gives many answers to unsolved questions in behavioural ecology and evolutionary biology.

Acknowledgements

This study was supported by grants from the Spanish Ministry of Science and Technology BOS 2003-01713 for FdL and AM. AM and MIR were supported by predoctoral grants from Junta de Extremadura (FIC01A043) and from Spanish Ministry of Science and Technology (BES 2004-4886), respectively.

References

[1]   S. Merino, J. Moreno, J.J. Sanz, & E. Arriero. Proc. R. Soc. Lond. B **267** p.2507 (2001)
[2]   A. Marzal, F. de Lope, C. Navarro and A. P. Møller Oecología **142** p.541(2005)
[3]   J.D. Blount, D.C. Houston, P.F. Surai & A.P. Møller Proc R Soc Lond Ser B **271** [Suppl] p. 79 (2004)
[4]   N. Saino, R. Ferrari, M. Romano, R. Martinelli & A.P.Møller. Proc R Soc Lond Ser B **270** p.2485 (2003)
[5]   D. Wakelin, D. Immunity to parasites: How parasitic infections are controlled. Cambridge University Press, Cambridge, U. K.(1996)
[6]   P.F. Surai. Natural antioxidants in avian nutrition and reproduction. Nottingham University Press, Nottingham, U. K. (2003)
[7]   J.A. Endler. Natural selection in the wild. Princeton, Princeton University Press (1986)
[8]   C. Navarro, F. de Lope, A. Marzal, and A. P. Møller Behavioural Ecology **15** p. 629 (2004)
[9]   P. Christe, A.P. Møller, N. Saino, F. de Lope. Heredity **85** p. 75 (2000)
[10]   G. González, G. Sorci & F. de Lope. Behavioral Ecology and Sociobiology **46** p.117 (1999)
[11]   S. Merino, A.P. Møller, F. de Lope. Oikos **90** p.327 (2000)
[12]   M. Soler, M. Martín-Vivaldi, J.M. Marín & A.P. Møller. Behavioral Ecology **10** p.281 (1999)
[13]   T.B. Kirkwood & S.N.Austad. Nature **408** p. 233 (2000)
[14]   R.A. Miller. Science **273** p. 70 (1996)
[15]   P.B. Medawar. An Unsolved Problem in Biology. H. K. Lewis, London (1952)
[16]   G.C. Williams. Evolution **11** p.398 (1957)
[17]   A.P. Møller & F. de Lope. Journal of Animal Ecology **68** p.163 (1999):
[18]   C. Atkinson. & C. van Riper III. In Bird-Parasite interacctions (Loyle & Zuk eds.) Oxford University Press. Oxford pp. 19-48 (1991)
[19]   K. Allander & G.F. Bennet. Journal of Avian Biology **25** p.69 (1994).
[20]   M. Cichon, J. Sendecka, L. Gustafsson. Journal of Evolutionary Biology **16** (6) p. 1205 (2003)
[21]   N. Saino, R. P. Ferrari, M. Romano, D. Rubolini & A. P. Møller Journal of Evolutionary Biology **16** (6) p. 1127 (2003)

Effects of hydrodynamics on biofilm formation

Fouad M. Qureshi [1,*] **, Nuzhat Ahmed** [1]**, John A. Finlay**[2]**, and Lynne E. Macaskie** [2]

[1] Centre for Molecular Genetics, University of Karachi, Karachi 75270, Pakistan
[2] School of Bioscience, University of Birmingham, Birmingham B15 2TT, United Kingdom

*Corresponding author: e-mail: qureshi@icgeb.org, Tel.: +92-21-4966045

A large number of factors can play an important role in the establishment of successful biofilms. An ideal biofilm for the purpose of environmental bioremediation should be thick, should hold fast to the support, and should have a large surface area and a density optimal for the bioremediation process. With the aim to achieving these properties in biofilms the present study focused on using quartz sand and polyvinyl chloride (PVC) support materials in pure and consortium cultures. The effect of varying hydrodynamics has been studied under the SEM.

1 Introduction

Bacterial adhesion to surfaces is a major factor in infections of the hospital, biomaterial implants, and catheters etc. [1–4]. On the other hand, bacterial adhesion to support materials is a requirement in certain environmental bioremediation applications [5–15], especially those making use of bioaccumulation or biosorption phenomena. The thicker the biofilm, more the cells per unit volume, the better is the efficiency of the bioreactor. Bioreactors typically make use of biomass immobilized in the reactor vessel. Biofilms have been used extensively for immobilisation of cells in bioreactors for a wide range of applications. In nature very complex and very extensive biofilms can be found existing on rocks submerged in streams, on river banks, on rocks under waterfalls, on trees, in caves, tooth plaque etc. [2]. However, obtaining optimal biofilms in the laboratory is not always possible, primarily due to poor understanding of the structure and the functions of different components of this micro-community of cells in the closed artificial microenvironment.

Biofilms both natural and cultured have recently been understood to behave as multicellular organisms [16]. With different communal sections within a biofilm; differing with respect to their function. Cell layers closest to the support are usually sessile and hold fast to the surface. Higher level layers more and more planktonic in nature. Similarly, higher layers are aerobic while deeper layers are relatively anaerobic [17] and the methane producing consortia biofilms drive energy from methanogenesis [2, 18]. This is typically so among the naturally occurring multi-species biofilms and also among some of the laboratory-cultured biofilms composed of carefully selected member-microbes of a microbial consortium. However, many pure-culture biofilms are regulated by a quorum sensing mechanism [19], which becomes the biofilm proliferation limiting factor in such microbes. Despite this degree of understanding, cultured biofilms are usually thin films of cells on a support. Depending on application this may be sufficient but some applications, such as bioremediation, benefit from highly proliferant biofilms. Biofilm proliferation can depend on a number of different factors such as the quality and quantity of nutrition, ambient conditions, etc. The present study was aimed to obtain as dense a biofilms as possible and focused on the effects of hydrodynamic factors during biofilm culture.

2 Materials and methods

2.1 Culture Medium

All of the biofilms were cultured in acetate-wastewater medium (as described in Qureshi et al. [11]) that was prepared in two parts. Each part was autoclaved separately at 121°C for 15 min, allowed to cool down to room temperature and mixed to constitute the working solution. Part 1 consisted of 1.0 g of NH_4Cl, 0.2 g of $MgSO_4 \cdot 7H_2O$, 0.01 g of $FeSO_4 \cdot 7H_2O$, 0.01 g of $CaCl_2 \cdot 2H_2O$, 5.0 g of $CH_3COONa \cdot 3H_2O$, and 0.5 g of yeast extract dissolved in 800 ml of real-time or synthetic wastewater as indicated. With the pH adjusted to 7.0 with 2 N NaOH the volume was raised to 990 ml. Part 2 consisted of 0.5 g of K_2HPO_4 in 10 ml final volume in distilled water and with the pH adjusted to 7.0.

The real-time wastewater used for preparing the medium consisted of parts-per-million As (0.045), Br (<0.01), Cd (0.01), Cl (1650), Co (1.2), Cr (<0.01), Cu (0.026), F (1), Fe (<0.01), N_{TOTAL} (6.1), Na (2050), Ni (0.88), Sb (0.023), SO_4^{2-} (1880), Zn (0.1), biological oxygen demand (<5), chemical oxygen demand (<10), and solid particles (5.2). Wherever the synthetic wastewater was used, it was modelled after the real-time wastewater (to the best availability of ingredients) and consisted of parts per million: As (0.045), Br (0.01), Cd (0.01), Cl (1,650), Co (1.2), Cr (0.01), Fe (0.01), Na (2,050), Ni (0.88), SO_4^{-2} (1,880), and Zn (0.1).

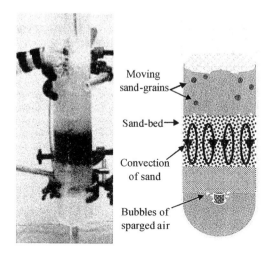

Moving sand-grains
Sand-bed
Convection of sand
Bubbles of sparged air

Fig. 1. Fluidized bed reactor in an aerated tower configuration used for biofilm production on sand grains. The schematic on right indicates fluid dynamics. Red arrows indicate the direction of convection currents observed in the sand-bed.

2.2 Biofilm on quartz sand

Two biofilms were cultured on quartz sand grains; (1) pure culture biofilm of *Serratia* sp. N14 and (2) a biofilm of a consortium of nine bacterial strains.

2.2.1 Monoculture biofilm: Serratia sp. N14

The *Serratia* sp. N14 biofilm on quartz sand grains was cultured in a fluidized bed reactor in an aerated tower configuration (Fig. 1) in acetate-wastewater medium prepared in real-time wastewater. The setup was initiated in batch culture mode for the first 24 hrs. It was then switched over to chemostat operation at 0.3 μ_{max} under carbon-limiting conditions (CH_3COONa) and incubated for 170 h at 37°C in a walk-in incubator.

2.2.2 Consortium biofilm

The consortium biofilm on quartz sand grains consisted of *Arthrobacter* sp. BP7/26, *Alcaligenes eutrophus* ER121, *Methylobacillus* sp. MB127, *Pseudomonas mendocina* AS302, *Pseudomonas stutzeri* EM77, *Ralstonia metallidurans* CH34, *Rhodococcus* sp. EM80, *Serratia* sp. N14, and an unclassified strain K1/8a. The biofilm was cultured in a manner identical to *Serratia* sp. N14 biofilm (see above).

2.3 Biofilm on PVC cylinders

A biofilm of *Pseudomonas aeruginosa* strain CMG156 was produced aerobically in 1 litre stirred-tank chemostat (Fig. 2) in a medium identical in all respects to the acetate wastewater medium (described above) with the exception of using synthetic wastewater instead of real (as described in Qureshi et al. [11]). The biofilm support (i.e. PVC-cylinders) was contained in a wire-mesh cage (Fig. 2) and the agitation system was designed to move the fluid through the biofilm support-bed. The agitation system also aerated the medium at the bottom of the agitation system just above the magnet, not shown in Fig. 2-right. The chemostat was operated in batch mode for the initial 24 h and then switched over to the continuous mode during continuous culture at 0.3 μ_{max} under carbon-limiting conditions (CH_3COONa) for 170 h at room temperature (30°C). PVC (polyvinylchloride) biofilm supports were fashioned out of flexible PVC conduit pipe used commercially in electrical wiring in accordance with Qureshi et al. [11].

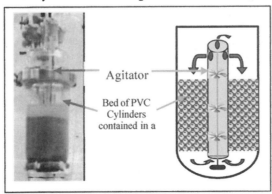

Agitator

Bed of PVC
Cylinders
contained in a

Fig. 2. Stirred-tank chemostat arrangement used for biofilm production on PVC-cylinders. The schematic on right shows fluid dynamics as blue arrows

2.4 Scanning Electron Microscopy

The Scanning Electron Microscopy (SEM) was carried out in accordance with Qureshi et al. [11]. At least 5 randomly selected samples from each biofilm were analyzed and the results averaged. Biofilm thickness was estimated from counting cell-layers per sample in all biofilms (Table 1). Though in the case of PVC biofilms, biofilm thickness was also measured directly from cross-sectional scanning electron micrographs and about 200 measurements were obtained and averaged (data not shown).

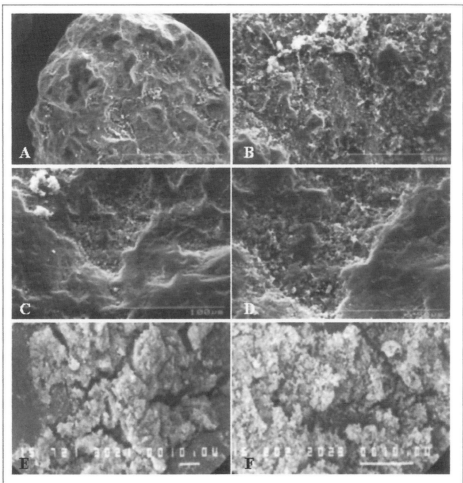

Fig. 3. Scanning electron microscopy of biofilms. A, biofilm of *Serratia* sp. N14 on quartz sand (scale bar = 500 μm); B, biofilm of *Serratia* sp. N14 on quartz sand (scale bar = 50 μm); C biofilm of consortium on quartz sand (scale bar = 100 μm); D, biofilm of consortium on quartz sand (scale bar = 50 μm); E, biofilm of *Pseudomonas* sp. CMG156 on PVC cylinders (scale bar = 10 μm); and F, biofilm of *Pseudomonas* sp. CMG156 on PVC cylinders (scale bar = 10 μm).

It is apparent in panels A and C that only topologically deeper areas were colonized by the biofilm while panels B and D demonstrate that the biofilm was only mono/oligo-layered. Panels E and F clearly demonstrate that the biofilm is multi-layered and existed as a thick crust over the support surface.

3 Results

The biofilm of *Serratia* sp. N14 grown on quartz sand grains was found generally to be tenuous (Fig. 3A, 3B). Topologically higher areas of the sand grains were found to be uncolonized without exception. Biofilm was observed only in topologically low-lying areas on the quartz sand grains. The consortium of *Arthrobacter* sp. BP7/26, *Alcaligenes eutrophus* ER121, *Serratia* sp. N14, unidentified strain K1/8a, *Methylobacillus* sp. MB127, *Pseudomonas mendocina* AS302, *Pseudomonas stutzeri* EM77, *Ralstonia metallidurans* CH34, *Rhodococcus* sp. EM80 was found to produce a biofilm that was marginally better than in terms of proliferation (considering greater number multi-cell foci) but still mono/oligo layered (Fig. 3C, 3D; Table 1). Similar to the monoculture, topologically high areas of the sand grain surface were uncolonized. *Pseudomonas* sp. CMG156 was used to produce a biofilm on the PVC cylinders custom fashioned from the commercial flexible conduit used in electrical fixtures [11]. This biofilm was found to be visible to the unaided eye at points of high proliferation (Fig. 3E, 3F; Table 1). Under the scanning electron microscope, topologically both high and low areas were found to be equally well colonized and the biofilm appeared as a thick crust with visible three-dimensional structure.

Table 1. Biofilm thickness measured from SEMs

Biofilm	Support	Sample size (N)	Average biofilm thickness	Error
Serratia sp. N14	Quartz sand	5	1	±1
Consortium	Quartz sand	5	1	±1
Pseudomonas sp. CMG156	PVC cylinders	5	31.5	±1
Pseudomonas sp. CMG156	Wire-mesh cage	Not determined		

4　Discussion

Biofilms are an important method of cell entrapment in many different kinds of bioreactors. In the context of application of contained-biofilms (such as in bioreactors) in the field of environmental bioremediation, the greater is the number of cells in the bioreactor the better would be the pollutant-treatment potential. Similarly, greater is the surface area of the active biomass in a reactor, greater would be the pollutant-biomass contact and hence efficiency of treatment. In addition to this, the economics of the operation are important to the feasibility of the any operation, therefore, a delicate balance of all of these factors is required. Several options are available with respect to the suitable support materials both naturally occurring and designed specifically for this purpose. Materials designed for the purpose of supporting biofilms are available commercially but can add to the cost of the operation.

Therefore, it was proposed to use quartz sand grains, available abundantly in the nature in deserts and beaches all over the world, as an economical biofilm support. A biofilm of *Serratia* sp. N14 strain was cultured on quartz sand grains in a fluidized bed reactor in an aerated tower configuration (Fig. 1). The resultant biofilm (Fig. 3A, 3B) was found to be colonizing only the topologically low areas and such areas under high magnification were found to be generally mono-/oligo-layers of bacterial cells (Table 1).

Due to dissatisfaction with the biofilm, it was hypothesized that a consortium of bacteria might perform better than a pure culture [2] and hence a biofilm of a consortium was cultured on the quartz sand grains under identical conditions. The resultant biofilm

(Fig. 3C, 3D) was found to be slightly better in terms of cell proliferation and density but it was also largely monolayer and the hypothesized effect was not observed.

Upon analysis of the entire process it was concluded that the inadequacy of the biofilm was due to the high shearing forces being experienced by the sand grains due to the fluidity of bed of sand in the aerated tower chemostat. Fig. 1 presents a schematic depicting the movement of the sand grains in the bed, due to the rising of the air-bubbles through the sand-bed for the purpose of maintaining aerobic conditions and for allowing the medium to reach inner regions of the sand-bed. This convective movement of sand grains (which is a characteristic of a fluidized-bed) was thought to be the cause for prevention of biofilm formation on the topologically higher areas on the sand grains or perhaps a cause for erosion of the biofilm from the surface of the sand grains. Although it has not been measured, but this suggests that the rate of cell detachment was higher than that of cell adhesion to the biofilm support.

In order to avoid friction among the support particles and thus avoiding the shearing forces, hydrodynamics in the chemostat were required to be altered. Thus an aerated stirred-tank configuration (Fig. 2) was adopted along with the designing and manufacturing of a unique agitation system (Fig. 2 right) that pumped fluid from the bottom of the vessel and released it above the bed of biofilm-support. This vertical fluid movement resulted in a gentle flow from top-to-bottom direction across the bed of support material thus conducting nutrition to the inner regions of the biofilm support bed (Fig. 2 right), which in this case was PVC-cylinders. The resultant biofilm (Fig. 3E, 3F) was found to be high in cell proliferation (Table 1). The biofilm was thick enough to be visible to the unaided eye as biomass-aggregates adhering to the PVC support. Under the scanning electron microscope both topologically high and low areas were found to be colonized. These observations suggested that shearing forces such as those experienced in the sand biofilms were indeed reduced due to immobilization of the biofilm-support bed. In other words elimination of bed-fluidity helped to alleviate shearing forces, though some shearing would still remain due to the chemostat hydrodynamics.

It was observed that a biofilm had co-formed on the PVC cylinder containment-cage during biofilm culture in the chemostat. This biofilm formation was unintentional and although it was not analyzed under the microscope, it was thought to be the best in terms of attached biomass and cell proliferation, since it could be easily recognized with the unaided eye as a macroscopic object covering the stainless-steel-cage in its entirety, as shown in Fig. 4.

Although a thorough statistical analysis remains to be carried out, comparing the quartz-sand and PVC-cylinder biofilms, observations in this study demonstrated that the improvement in biofilm could be attributed to the altered hydrodynamics and the resultant minimization of the shearing forces. Biofilms, natural or grown in laboratory, exist as a dynamic equilibrium between the rate of cell adhesion and the rate of cell detachment. Either increasing the rate of cell adhesion or decreasing the rate of cell detachment can create conditions conducive or favourable for biofilm production. In the light of current observations, the improvement in biofilm can be explained as a decrease in cell detachment rate resulting from minimization of shearing in biofilm support-bed due to altering hydrodynamics.

The stainless-steel cage biofilm, i.e. the macroscopic biofilm, demonstrated a pattern of growth (Fig. 4) that suggested that the fluid moved spirally in a top-to-bottom direction, instead of vertical top-to-bottom flow (depicted schematically in Fig. 2) that was anticipated during the designing stage. This biofilm experienced hydrodynamics in a slightly different manner than the PVC cylinders that also experienced spirally downward moving currents of fluid. The flow across the cage walls was observed to be directed centrifugally outwards (in addition to downwards) and transverse through the cage and the bed. This resulted in a centrifugally directed flow through the sieve like structure of the cage, which resulted in formation of thread-like biomass that was entangled in the wire-mesh cage-walls. This colonization pattern gave rise to heavy growth covering the entire surface area of the cage.

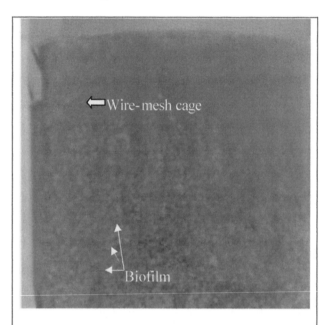

Fig. 4. The biofilm of *Pseudomonas aeruginosa* CMG156 co-formed on the stainless steel wire-mesh cage is visible to the unaided eye.

Hydrodynamics of natural biofilms have been documented by Stoodley et al. [20–24], by Lewandowski et al. [17], by Hall-Stoodley and Stoodley [24], and by Puredorj-Gage & Stoodley [25] etc. In addition to other factors, these studies show that a steady stream promotes good and fast-holding biofilm but the effects of hydrodynamics on the *laboratory-cultured-biofilms* beyond nutrient distribution and in-vessel homogenization especially with respect to maximizing biofilm (or biomass) yield has escaped thorough scientific investigation. Present study lacks in providing adequate biofilm quantification but the evidence presented in here emphasizes the need for thorough investigation of the effects of chemostat hydrodynamics on biofilm formation supported by quantification data.

Acknowledgements

Work on biofilms on the PVC cylinders was carried out by Fouad M. Qureshi at the Centre for Molecular Genetics, University of Karachi and was funded by the Ministry of Science and Technology, Government of Pakistan; Sind Environmental Protection Agency, Karachi. While work on biofilms on the quartz sand was carried out jointly by Fouad M.

Qureshi and John A. Finlay at the School of Biosciences, University of Birmingham and was funded by the Department for International Development, UK and the British Council.

References

[1] Costerton, J.W., Stewart, P.S., and Greenberg, E.P. Bacterial biofilms: a common cause of persistent infections. Science **284**, 1318.(1999)

[2] Costerton, J.W. A Short History of the Development of the Biofilm Concept. In: Microbial Biofilms. (Ghannoum, Mahmoud and O'Toole, George A, Eds.), 1. (ASM Press, Washington, DC). (2004)

[3] Stickler, D.J., Morris, N.S., McLean, R.J.C., and Fuqua, C. Biofilms on indwelling urethral catheters produce quorum-sensing signal molecules in situ and in vitro. Applied and Environmental Microbiology **64**, 3486.(1998)

[4] Walker, J.T., Bradshaw, D.J., Bennet, A.M., Fulford, M.R., Martin, M.V., and Marsh, P.D. Microbial biofilm formation and contamination of dental-unit water systems in general dental practice. Applied and Environmental Microbiology **66**, 3363.(2000)

[5] Ahmed, N., Badar, U., Qureshi, F.M., and Fasim, F. Biosorption and bioaccumulation of heavy metals by bacteria isolated from contaminated sites of Karachi, Pakistan. In: Biohydrometallurgy: A Sustainable Technology in Evolution (Tsezos, M., Hatzikioseyian, A., and Remoundaki, E., Eds.), 771. (National Technical University of Athens, Athens). (2004)

[6] Pumpel, T., Macaskie, L.E., Finlay, J.A., Diels, L., and Tsezos, M. Nickel removal from plating waste water using a biologically active moving-bed sand filter. Biometals **16**, 567.(2003)

[7] Ahmed, N., Badar, U., Fasim, F., and Qureshi, F.M. Recycling of metal contaminated wastewater: A case study. In: Environmental and Groundwater Pollution (Sherif, M. M., Singh, V. P., and Rashed, M. A., Eds.), 175. (Swets and Zeitlinger, Lisse). (2002)

[8] Ahmed, N., Qureshi, F.M., and Badar, U. Biotechnological approaches for the control of environmental pollution: Bioremediation. In: Biotechnology Applications for the Arid Regions (Al-Shayji, Y. A., Sidhu, J. S., Saleem, M, and Guerinik, K., Eds.), 281. (Kuwait Institute for Scientific Research, Kuwait). (2002)

[9] Macaskie, L.E., Thomas, R.A.P., and Lloyd, J.R. Application of microorganisms to the decontamination of heavy metal-bearing wastes. In: Industrial and Environmental Biotechnology. (Ahmed, Nuzhat, Qureshi, Fouad M., and Khan, Obaid Y., Eds.), (Horizon Scientific Press, Norfok). (2001)

[10] Silver, S., Phung, L.T., Lo, J.-F., and Gupta, A. Toxic metal resistances: molecular biology and the potential for bioremediation. In: Industrial and Environmental Biotechnology. (Ahmed, Nuzhat, Qureshi, Fouad M., and Khan, Obaid Y., Eds.), 33. (Horizon Scientific Press, Norfolk). (2001)

[11] Qureshi, F.M., Badar, U., and Ahmed, N. Biosorption of copper by a bacterial biofilm on a flexible polyvinyl chloride conduit. Applied and Environmental Microbiology **67**, 4349.(2001)

[12] Ahmed, N, Qureshi, F.M., and Khan, O.Y., Eds. Industrial and Environmental Biotechnology. (Horizon Scientific Press, Norfolk, UK). (2001)

[13]   Badar, U., Ahmed, N., Beswick, A.J., Pattanapipitpaisal, P., and Macaskies, L.E. Reduction of chromate by microorganisms isolated from metal contaminated sites of Karachi, Pakistan. Biotechnology Letters **22**, 829.(2000)

[14]   Nicolella, C., van Loosdrecht, M.C.M., and Heijnen, J.J. Wastewater treatment with particulate biofilm reactors. J. Biotech. **80**, 1.(2000)

[15]   White, C. and Gadd, G.M. An internal sedimentation bioreactor for laboratory-scale removal of toxic metals from soil leachates using biogenic sulphide precipitation. Journal of Indutrial Microbiology **18**, 414.(1997)

[16]   Branda, S.S. and Kolter, R. Multicellularity and Biofilms. In: Microbial Biofilms. (Ghannoum, Mahmoud and O'Toole, George A, Eds.),   20. (ASM Press, Washington, DC). (2004)

[17]   Lewandowski, Z., Stoodley, P., and Altobelli, S. Experimental and conceptual studies on mass transport in biofilms. Water Sci. Technol. **31**, 153.(1995)

[18]   Tortora, G.J., Funke, B.R., and Case, C.E. Environmental Microbiology. In: Microbiology: An Introduction **6th**, 714. (Benjamin/Cummings Publishing, Menlo Park, California). (1998)

[19]   Hentzer, M., Givskov, M., and Eberl, L. Quorum Sensing in Biofilms: Gossip in Slime City. In: Microbial Biofilms. (Ghannoum, Mahmoud and O'Toole, George A, Eds.),   119. (ASM Press, Washington, DC). (2004)

[20]   Stoodley, P.S., deBeer, D., and Lewandowski, Z. Liquid flow in biofilm systems. Applied and Environmental Microbiology **60**, 2711.(1994)

[21]   Stoodley, P., Lewandowski, Z., Boyle, J.D., and Lappin-Scot, H.M. The formation of migratory ripples in a mixed species bacterial biofilm growing in turbulent flow. Environmental Mic robiology **1**, 447.(1999)

[22]   Stoodley, P., Dodds, I., Lewandowski, Z., Cunningham, A.B., Boyle, J.D., and Lappin-Scot, H.M. Influence of hydrodynamics and nutrients on biofilm structure. J. App. Microbiol. **85**, 19S.(1999)

[23]   Stoodley, P., Lewandowski, Z., Boyle, J.D., and Lappin-Scot, H.M. Structural deformation of bacterial biofilm caused y short-term fluctuations in lquid shear: an in situ investigation of biofilm rheology. Biotechnol. Bioeng. **65**, 83.(1999)

[24]   Stoodley, P., Wilson, S., Hall-Stoodley, L., Boyle, J.D., Lappin-Scot, H.M., and Costerton, J.W. Growth and detatchment of cells from mature mixed spiecies biofilms. Applied and Environmental Microbiology **67**, 5613.(2001)

[25]   Puredorj-Gage, L.B. and Stoodley, P. Biofilm Structure, Behavior, and Hydrodynamics. In: Microbial Biofilms. (Ghannoum, Mahmoud and O'Toole, George A, Eds.), 160. (ASM Press, Washington, DC). (2004)

Effect of organic management on soil chemical and biochemical properties of a Xerofluvent of the Guadalquivir River Valley (SW Spain)

S. Melero[*1]**, J.C Ruíz**[1]**, J.F. Herencia**[1] **and E. Madejón**[2]

[1] Instituto de Investigación y Formación Agroalimentaria y Pesquera (IFAPA) "Las Torres-Tomejil" Sevilla, SPAIN
[2] Instituto de Recursos Naturales y Agrobiología (CSIC), Av. De Reina Mercedes 10, 41012 Sevilla, SPAIN

*Corresponding author: sebastiana.melero.ext@juntadeandalucia.es, Phone: +34955045517, Fax: +34 955045625

Organic farming is an alternative agriculture that reduces the negative effects caused by conventional management. The aim of this work was to study the effect of organic residues on the chemical and biochemical properties of a soil after five years of organic management. Two managements were compared: conventional management (use of inorganic fertilisers, no pesticides in that study and mechanical tillage for weed control) and organic management (use of manure and mechanical tillage for weed control) in a silty loam soil (Xerofluvent). Other aim was to evaluate seasonal effects on above properties along three different sampling periods (winter, early spring and early summer).
Soil chemical and biological status were evaluated by measuring the total organic carbon (TOC), available-P, Kjeldahl-N, microbial biomass carbon (Cmic), soil respiration, enzymatic activities (urease, alkaline phosphatase), metabolic quotient (qCO_2) and Cmic/TOC ratio.
Organic plots showed higher values of TOC, available-P, Kjeldahl-N; enzymatic activities, microbial biomass, and soil respiration than conventional plots. Therefore an improvement of soil fertility was achieved in organic plots. A progressive increase of biochemical properties, above all in summer, was observed as outcome seasonal changes along the study. Conventional soils showed a higher qCO_2 than organic soils, sign of a stress situation due to their lower C availability.

Keywords organic farming; soil enzymes; soil microbial biomass; total organic carbon

1   Introduction

The quantification of the activity of the microbial biomass can be fundamental to evaluate the effects of organic management on soil, as this management rely on an organic input of residues, which are broken down by microorganisms. Although the microbial biomass is a small and labile fraction (1 to 3%) of organic matter, the principal organic matter transformations are carried out by soil microorganisms. Microorganisms decompose organic substrates to obtain energy producing carbon dioxide (CO_2). The measurement of CO_2 has been used to determine biological activity in soils in relation to changes in chemical soil properties and in agricultural practices [1]. Furthermore, the microbial biomass is one of the main agents of the chemical transformations releasing essential nutrients to the plant. The microbial biomass has been suggested as a valuable index of the changes of soil organic matter, due to the clear relation between biomass and soil organic carbon content [2].
Moreover the study of the simple ratios between microbial biomass-C (Cmic) and total organic carbon (TOC), and the metabolic quotient (qCO_2), which is the rate of CO_2 per unit

of biomass and time, can give a better understanding of changes that are promoted by agricultural practices on soil microbial communities [1].

Soil enzymes have been suggested as potential indicators of soil quality because of their essential role in soil biology and their rapid response to changes in soil management. Soil enzyme activities also respond to agronomic practices such as fertilizers, organic amendments, vegetation cover, and pesticides [3].

Our main objective was to evaluate the soil status, during a crop cycle after five years of organic management, through the study of biochemical and chemical properties of a soil. The seasonal variability on biochemical properties through of three different sampling periods (winter, early spring and early summer) along the study was also evaluated.

2   Materials and methods

The field study was carried out in a silty loam soil classified as a Xerofluvent (Soil Survey Staff, 1996). The study site (latitude: 37° 8′33′′ N and longitude: 5° 16′4′′ W) was located in the Guadalquivir River Valley (SW Spain), at the "Las Torres-Tomejil" farm in Alcalá del Río (Seville). A Carrot crop (*Daucus carota L*) was conducted under organic and conventional soil management in plots of 6 x 12.5m. Four replicates per management were established randomly. The conventional system has been managed with chemical fertilizers and the organic system has been managed organically (Regulation (EEC) No 2092/91) since 1995. No pesticides were used in either management system and the weed control was performed by mechanical tillage. Before the sowing, mature compost was added to organic plots at a dose of 30 t ha^{-1} as deep fertilization. Some of the chemical properties of the manure compost are: moisture= 420 g kg^{-1}; pH= 7.6; EC=6.4 dSm^{-1}; organic matter = 268 g kg^{-1};C/N=11.4; available phosphorous=4.3 g kg^{-1}; available-K= 12 g kg^{-1}. A mineral fertilizer, (N-8 P$_2$O$_5$-15- K$_2$O-15), at a dose of 0.8 t ha^{-1} as deep fertilization and a top dressing (ammonium nitrate 33.5%) 0.1 t ha^{-1} more potassium nitrate (N-13 K$_2$O-46) 0.17 t ha^{-1} was added to the conventional plots.

From November 1999 to June 2000 three soil samplings were accomplished: the first two weeks after sowing (November 1999), the second at the flowering period (April 2000) and the third after harvesting and after tillage to prepare the soil for the following crop (June 2000). The soil samples were collected from the upper 15 cm of soil

Microbial biomass C (Cmic) content (determined by the chloroform fumigation-extraction), soil respiration and enzyme activities were determined as described in [4]. Results were expressed based on oven-dry weight of soil.

The statistical analyses were carried out using the program SPPS 11.0 for Windows and the results were expressed as mean values. Significant statistical differences of all variables between the different treatments were established by the students-t test at p<0.05.

3   Results and discussion

Results of chemical properties are showed in Table 1. The decrease of soil pH in organic plots can be due to their higher soil microbial activity. In soils with high carbonate content, as in this case, the variation of partial pressure of CO$_2$ constitutes a fundamental parameter of variation of pH [5].

The organically managed plots had a higher EC because of the addition of nutrients and salts through the compost. However, values of soil EC recorded in the following year (data not shown) decreased in organic plots to 0.55 (dSm^{-1}) and in conventional plots to 0.45

(dSm^{-1}). It is known that continuos application of compost can cause significant increases of EC, so precautions should be taken into account when composts are repeatedly applied to soils with restricted drainage or on unirrigated lands.

Organic management and the use of organic residues maintain soil organic matter at higher levels than inorganic fertilization [6]. The application of compost may account for the highest levels of the Kjeldahl-N and available-P in the organically managed plots. Excessive accumulation of some nutrients, and particularly P and N, can arise from the long-term use of manures, relative to the use of fertilisers. Under these conditions greater runoff of P, and leaching of N may result [6].

Table 1. Soil (0-15 cm) mean values[a] of pH and electrical conductivity (EC$_{1:2.5}$); total organic carbon (TOC); Kjeldahl-N; C/N ratio; available-P. n=24

	Treatment	pH	EC (dS m^{-1})	TOC (g kg^{-1})	Kjeldahl-N (g kg^{-1})	Available-P (mg kg^{-1})
November 1999	Inorganic	8.27 a	0.52 a	9.20 a	1.10 a	19.0 a
	Organic	8.13 a	0.72 a	24.0 b	2.60 b	97.0 b
April 2000	Inorganic	8.38 b	0.46 a	9.00 a	1.00 a	22.0 a
	Organic	8.10 a	0.68 b	22.0 b	2.30 b	72.0 b
June 2000	Inorganic	8.10 b	0.50 a	9.40 a	1.10 a	25.0 a
	Organic	7.80 a	0.75 b	22.4 b	2.00 b	75.0 b

EC electrical conductivity, TOC total organic carbon.
[a] Values of inorganic and organic treatments at the same sampling followed by the same letter do not differ significant

In general, soil respiration and Cmic was significantly higher in the organic treatments than in the inorganic treatment (Figure 1a, b). That can be explained to both the addition of microorganisms with the amendments and the incorporation of easily degradable organic matter and other nutrients, which stimulate the growth of the microorganisms of the soil. Furthermore the higher organic matter content holds a higher protective capacity to microbial biomass against cell death or predation.

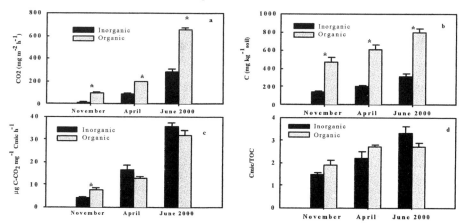

Figure 1. Mean values of a) soil respiration, b) carbon microbial biomass (Cmic), c) qCO$_2$, d) Cmic/TOC. Vertical bars are standard deviation. Sigificant differences between treatments are indicated in the figures (*).

76

The increase of soil respiration and Cmic along the study in both treatments could be also explained by seasonal changes [7]. Seasonal fluctuations in microbial biomass are outcome probably of variations in moisture, increase of temperature of the soil.

The qCO_2 values in conventional plots were higher than in organic plots (Figure 1c). These results indicated that in conventional plots, a lower microbial community respired at a greater rate, which can be due to the lower carbon availability.

The Cmic/TOC ratio is an indicator of the availability of soil organic matter to microorganisms. In the first and second sampling of the crop cycle, an increase of Cmic/TOC ratio was observed in organic and conventional plots (Figure 1d). This could be due to a higher microbial biomass content, as a result of the input of carbon from the fertilizers and exudates and sloughed-off cell of rhizosphere and crop residues during crop growing. In last sampling the Cmic/TOC ratio was higher in conventional plots due to the increase in microbial biomass and the lower TOC content in these plots.

Organic management plots had significantly higher levels of urease and phosphatase alkaline activities than those obtained in the conventional plots at the three different samplings (Figure 2). Therefore addition of organic residues to soil activates microbial growth and consequently increases enzyme activities, in order to breakdown the organic matter. In the last sampling the increase in enzymatic activities could also be due to rising of activity and microbial biomass by increasing of soil temperature.

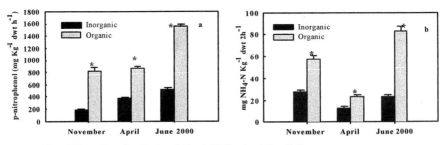

Figure 2. Mean values of enzimatic activities a) Alkaline phosphatase, b) Urease.
Vertical bars are standard deviation. Significant differences between treatments are indicated in the figure (*).

4   Conclusions

A general improvement of soil chemical properties was observed in plots under organic management. Organic residues added to the soil promote microbial and enzyme activities improving biochemical properties. All above underlined is positively translated to an improvement of soil quality. In summer higher activity and microbial biomass was observed.

References

[1]   P. Nannipieri, S. Grego,B. Ceccanti. Ecological significance of the biological activity in soil. In: Bollag J, Stotzky G. (Eds.), Soil Biochemistry 6. Marcel Dekker, New York,1990, pp 293-355.

[2]   D.S.Jenkinson, J.N Ladd,. Microbial biomass in soil: Measurement and turnover. In: Paul, E.A., Ladd, J.N (Eds), Soil Biochemistry, 5. Marcel Dekker, New York,1981, pp.415-471.

[3]   L.Gianfreda, J.M Bollag. Influence of natural and anthropogenic factors on enzyme activity in soil. In: Stotzky,G., Bollag, J.M. (Eds), Soil Biochemistry, vol.**9**. Marcel Dekker, New York,1996, pp 123-193.

[4]   K. Alef, P. Nannipieri (eds.), Methods in Applied Soil Microbiology and Biochemistry, Academic Press, London, 1995, chap. 7, pp. 311-373.

[5]   S.Bruckert, J. Rouiller. Mecanismos de regulación del pH de los suelos. In Bonneau, M., Souchier, B. (Eds), Edafologia. 2Constituyentes y propiedades del suelo. Publicaciones Masson, S.A. Barcelona,1987, pp 356-367.

[6]   D.C.Edmeades. Nutr. Cycl. Agroecosyst **66**(2), 165-180 (2003).

[7]   S.U.Sarathchandra, K.W. Perrot, M.R. Boase,J.E Waller. Biol. Fertil. Soils **6**, 328-335 (1988).

Environmental virological monitoring for the epidemiological surveillance and risk assessment

A. Carducci, M. Verani, F. Pizzi, E. Rovini, E. Andreoli, and B. Casini

[1] Department of Experimental Pathology, Medical Biotecnologies, Infectivology and Epidemiology – University of Pisa, via S. Zeno 35, 56127 Pisa, ITALY

Keywords: Environment, Risk Assessment, Virus

1   Introduction

Waters represent the main vehicle of spreading for the human enteric viruses that can survive longer in the environment than the majority of enteric non sporogenic bacteria, and have a lower infective dose (1,2). The different uses of water, like drinking, irrigating, bathing, and growing food (i.e. shellfishes) can frequently expose people to enteric viral infections (3,4). Therefore, water virological monitoring could constitute an important instrument both for the epidemiological surveillance and for the risk assessment (5). At present, knowledge on virological contamination of waters is still incomplete owing to the technical difficulties in virus detection, that limit the possibility of monitoring to few specialized laboratories and then obstacle a very large spread of this kind of investigation. Besides that, the number of possible viral species to be detected is very large and will became even larger in the future as knowledge about viral environmental spread will increase. The advent of molecular biology techniques (mostly PCR) has opened new possibilities for virological environmental monitoring, allowing to detect non culturable agents and to perform molecular epidemiology studies. However many questions remain still unsolved: the standardization of detection methods, the choice of the most significant agents for risk assessment, the selection of a reliable indicator for viral contamination, and so on. For this reasons we planned a study aimed to analyse the human enteric viruses environmental spread and its relations with virological diagnosis of gastroenteritis, to identify the most frequent viral pathogens in different types of water and to evaluate the possible correlations between pathogenic enteric viruses and commonly used faecal indicators.

2   Materials and methods

Since May 2004 an epidemiological surveillance of viral gastroenteritis diagnosed on faecal samples in the Clinical Virology Laboratory of the Department of Experimental Pathology of the University of Pisa was carried out, parallely with an environmental monthly monitoring of raw and treated wastewaters by a treatment plant, of waters from the river receiving this effluent and from the sea at this river outfall.

The monitoring concerns enteric viruses (enterovirus, Adenovirus 40 and 41, Rotavirus, Norovirus genotypes I and II, HAV) besides to bacterial (E. coli), and viral (somatic coliphages and TTVirus) indicators.

This last virus for its large diffusion, for the lack of pathology's association and for the faecal elimination, was studied as a potential indicator of water virological quality (6).

The volume of water samples was different according to the level of microbial pollution and the measured parameter: for *E. coli* always 100 ml; for coliphages and viruses 1 liter for raw sewage and 10 liters for the others materials. *E.coli* was isolated on TBX agar and counted by membrane filtration methods. For virological examination water samples were concentrated using two stage tangential-flow ultrafiltration. For coliphages analysis the concentrated samples were tested by plaque assay according to the double agar layer method (using *E. coli C*, ATTC 13706, as host strain) (7). For enteric viruses the concentrated samples, decontaminated with chloroform, were assayed with bimolecular tests (PCR, RT-PCR) and identified with genic sequencing (8). Data concerning *E. coli* and coliphages concentrations were statistically elaborated to search a correlation between these two parameters and between them and the virus presence (Excel, Microsoft).

3   Results

The surveillance of cases has revealed in the period from May 2004 to February 2005, 29 positive on 110 faecal samples, for one of investigated viruses at least, of which 8/109 (7,3%) for rotavirus, 2/110 (1,8%) for adenovirus, 3/111 (2,7%) for astrovirus, 7/102 (6,8%) for norovirus genotype I and 10/101 (9,9%) for norovirus genotype II, without to characterize particular epidemic peaks, but only a small cluster of rotavirus infections in May.

The water monitoring is at present completed for *E. coli*, while for phage counts it has been performed till January 2005 and for viruses till August 2004. The study of *E.coli* and somatic coliphages counts allowed to show a softly decrease due to the treatment plant (on average from $8,74 \times 10^6$ to $2,6 \times 10^5$ CFU/100ml for *E.coli* and from $3,1 \times 10^6$ to $2,4 \times 10^5$ PFU/100ml for the somatic coliphages), followed by a new increase in the river ($7,8 \times 10^5$ CFU/100ml and $2,4 \times 10^5$ PFU/100ml) probably due to further pollutions (Fig. 1–4).

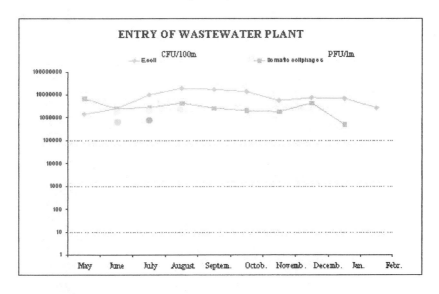

Figure 1: Results of microbiological analysis in sewage at entry of plant.

At the river mouth *E. coli* was found only once and coliphage counts were highly variable, probably due to the seawater dilution. No correlation was found between *E. coli* and somatic coliphages concentrations.

Enteric viruses were mostly detected in raw sewage: three of the four of analysed samples were positive, one of these for norovirus genotype I and adenovirus, one for adenovirus and norovirus genotype II, and one only for adenovirus (uncultured adenovirus AY747675).

In the plant effluent the viral presence was lower than at the entrance showing the relative efficacy of the wastewater plant in removing these agents; two of four samples were positive for adenovirus, in May and June 2004 (uncultured adenovirus AY747675, uncultured adenovirus AY747672.1).

River water resulted contaminated only in three cases, with a single virus (rotavirus) in May, June and August 2004. The rotavirus found in river water were different strains (human rotavirus A AF373896.1, human rotavirus A AF260950.1, rotavirus AY660563.1). No virus were found in seawater. No significant association was found between the enteric viral presence and *E. coli* or coliphages counts.

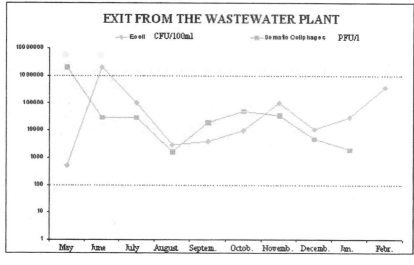

Figure 2: Results of microbiological analysis in sewage at the exit of plant.

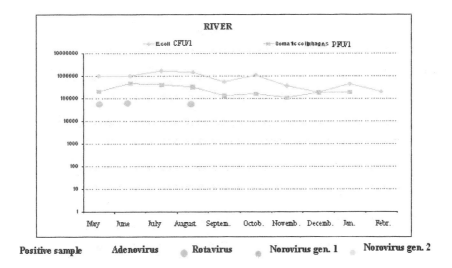

Figure 3: Results of microbiological analysis of the river water.

4   Conclusions

The data on viral gastroenteritis diagnosis did not pointed out any epidemic peak, but a continuous circulation of enteric viruses, mainly rotavirus and norovirus. According to these results, virological analysis of different matrices for the first four months of the monitoring, showed the frequent presence of enteric viruses in the environment, even if sewage were treated. The *E.coli* and somatic coliphages counts indicated a poor reduction

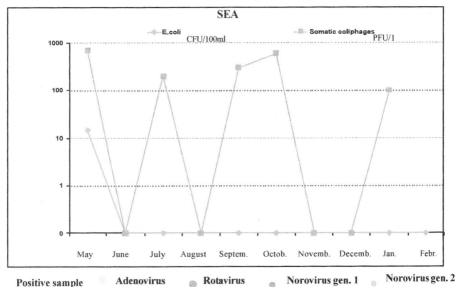

Figure 4: Results of microbiological analysis of seawater

82

(less than two logs on average) from wastewater treatment plant entry to exit, in association with the uncomplete removal of enteric viruses. In the river the microbiologic contamination newly increased, probably due to other pollution sources, and three samples tested positive for viruses, while the data on seawater showed a drastic reduction, without virus detection. Although the results obtained till now are still insufficient to clarify the dynamic of viral spread, at present we cannot identify a predominant virus that could be used in the risk assessment, nor an indicator representing the viral pollution. TTvirus, tested as possible indicator was never found, and then its usefulness to this aim seems very poor.

Acknowledgements

Work funded by the Italian Ministery of University and Research (Cofin 2004-2005)

References

[1]   Leclerc H., Schwartzbrod L., Dei-Cas E. Microbial agents associated with waterborne disease. Crit. Rev. Microbiol., 28 (4): 371-409 (2002).
[2]   OMS. Guideliness for drinking water quality, Third Edition, 2003 (2003).
[3]   Barwick R.S., Levy D.A., Crawn G.F., Beach M.J., Calderon R.L. Surveillance for waterborne-disease outbreaks, United States, 1997-1998. CDC Surveillance Summary May 26, 2000. MMWR 49 (SS04); 1-35 (2000).
[4]   Jonathan S. Yoder, M.P.H., M.S.W., Brian G. Blackburn, M.D., et al. Surveillance for Waterborne – Disease Outbreaks Associated with Recreational Water - United States, 2001—2002. CDC Surveillance Summary October 22, 2004. MMWR 53 (SS08); 1-22 (2004).
[5]   P. Payment and P.R Hunter. Endemic and epidemic infectious intestinal disease and its relationship to drinking water in *Water Quality: Guidelines, Standards and Health Ed. World Health Organization Water Series,* 61-88 (2002).
[6]   M. Verani, B. Casini, R. Battistini, A. Carducci. Un nuovo possibile indicatore del rischio virale occupazionale in ambienti sanitari.- *Riv. It. Ig. printing* (2004).
[7]   ISO/DIS 10705-2: Water quality-detection and enumeration of bacteriophages. Part 2: Enumeration of somatic coliphages. 1999.
[8]   A. Carducci, B. Casini, A. Bani et al. Virological control of groundwater quality using biomolecular test. *Wat. Sci. Tech.,* 47, 261-26 (2003).

Heavy metal effects on Extremophiles and on enzyme biosynthesis in a new *Bacillus* strain from Mount Rittmann, Antarctica

A. Poli [*,1] **, E. Esposito**[1]**, L. Lama**[1]**, S. Pelliccione**[1]**, A. Gambacorta**[1] **and B. Nicolaus**[1]

[1] Istituto Chimica Biomolecolare (ICB), CNR, Via Campi Flegrei 34, 80078, Pozzuoli, (Na) ITALY

[*]Corresponding author: e-mail: apoli@icmib.na.cnr.it,Phone: +39 081 8675190, Fax: +39 081 8041770

Several extremophiles, in particular thermophilic microorganisms, coming from different and various environmental sites, were studied for their heavy metal resistance. These strains were exposed to heavy metals up to 60 ppm in the growth media. We focused our attention on a new acidothermophilic *Bacillus*, designated strain MR3C^T, isolated from geothermal soil samples collected from Mount Rittmann in Antarctica. MR3C^T synthesized an extracellular constitutive amylolytic activity. This activity was inhibited by exposure of cells, during the exponential or early stationary growth phases, with various concentration of heavy metals (Ni $^{2+}$, Cu $^{2+}$, Zn $^{2+}$, Hg $^{2+}$). It was shown that MR3C^T α-amylase activity was very sensitive to nature and quantity of heavy metals used. Hg $^{2+}$ was by far the strongest inhibitor of α-amylase activity among the heavy metals tested. This study suggests the potential role of MR3C^T strain as toxicological indicator of heavy metals and it evaluates the use of a microbial assay to detect heavy metals trace in industrial wastes.

Keywords Extremophiles; heavy metals; amylase; enzyme inhibition

1 Introduction

Heavy metals are natural components of the Earth's crust. They derive also from anthropogenic sources, such as industrial processes, transport and agricultural processes by fertilizer application. As trace elements, some heavy metals (e.g. copper, selenium, zinc) are essential to maintain the metabolism of the human safe. However, at higher concentrations they can lead to poisoning because they tend to bioaccumulate in living organisms through the food chain causing damage overall to lung, kidney, liver and nervous system [1]. The bioassays using growth inhibition of microorganisms and the inhibition of enzyme activity and enzyme biosynthesis are simple, rapid, cost-effective, and require only small sample volumes [2].

Extremophiles are microorganisms that live in prohibitive environmental conditions for the other forms of life. In fact they ask extreme parameters of pH, pressure, temperature and salinity for their survival. Studies on these microorganisms, belonging to the *Bacteria* and *Archaea* domains, are very useful to understand the evolution of the life. In fact, the unusual environmental conditions in which they live could be very similar to those present in the primordial biosphere. Besides, having been confined for million of years, and having preserved many native aspects, the extremophiles could represent the descending of ancestral forms of life [3].

There are many extremophilic enzymes and also endogenous compounds that are used with success in the food industry, in the preparation of the detergents, in the pharmacological applications and also in genetic studies.

In particular enzymes that derive from *thermophiles*, and for this reason called *thermozymes*, represent a good sources of novel catalysts of industrial interest.

In this paper we report the effects of heavy metals on some *thermophiles*. *Bacillus thermantarcticus* collected near the crater of Mount Melbourne in Antarctica [4]; *Geobacillus thermoleovorans* subspecies *stromboliensis* isolated from ash and geothermal vulcanic soils near the crater of Stromboli [5]; a new *Anoxybacillus* sp. nov. named MR3C^T collected from the Mount Rittmann in Antarctica (paper submitted); a new strain of *Geobacillus toebii* named L1^T isolated from compost. The objectives of our studies were to determine the effects of heavy metals on the inhibition of amylase biosynthesis in *Anoxybacillus* MR3C^T, and thus to evaluate this bioassay for use in microbial toxicity testing.

2   Materials and methods

2.1 Reagents

The chemicals assessed for toxicity were $NiSO_4 \cdot 6H_2O$ (Carlo Erba), $ZnSO_4 \cdot 7H_2O$ (Aldrich), $Co(NO_3)_2 \cdot 6H_2O$ (J.T.Baker), $HgCl_2$ (J.T. Baker), $MnCl_2 \cdot 4H_2O$ (J.T. Baker), $Cr(NO_3)_3 \cdot 9H_2O$ (J.T. Baker), $CuSO_4 \cdot 5H_2O$ (J.T. Baker), $FeCl_3$ (Carlo Erba), $CdSO_4$ (Aldrich). Solution were prepared by dissolving in distilled water and subsequently filtered. The chosen metal concentrations were calculated as mg/l (ppm part per million) of each metal.

2.2 Cell cultures

Bacillus thermantarcticus (DSM 9572^T) was isolated from Antarctic geothermal soil near the crater of Mount Melbourne [4]. The strain was grown at 65°C on a medium containing 0.6 % yeast extract and 0.3 % NaCl (YN medium) at pH 6.0.

Geobacillus thermoleovorans *subspecies* stromboliensis, *(DSM 15392^T=ATCC BAA-979^T) collected from "Pizzo sopra la Fossa" (Italy, Eolian Island, site Stromboli) was grown at 70°C on YN medium at pH 7.0 [5].*

Anoxybacillus sp. nov. MR3C^T (ATCC BAA-872^T=DSM 15939^T=CIP 108338^T) was isolated from geothermal soil located on Mount Rittmann in Antarctica. Cultures were grown at 61°C on a medium containing 0.6% yeast extract and 0.6% NaCl at pH 5.6. To test amylolytic activity MR3C strain was grown on solid YN medium plus 0.2% starch according to Nicolaus et al. 2002 [3].

Geobacillus toebii named L1^T (DSM 17041^T) was isolated from compost. Cells were grown at 65°C on media containing 0.4 % yeast extract, 0.8 % peptone and 0.2 % NaCl at pH 7.0.

Cells were grown by inoculating 90 ml media in a 250-ml Erlenmeyer flask. The bacteria was incubated at an optimal temperature for 12 h, after which the cells were transferred to fresh media to get an optical density of 0.1 (A540 nm). Cells were harvested both in late exponential growth phase and in a stationary growth phase by centrifugation at 9000*g* for 30 min and only in the case of MR3C^T the supernatant obtained was utilised for α-amylase assay.

Each heavy metal was added to fresh media at appropriate concentrations at To (together with inocula). Bacterial growth in controls and other samples was checked by measuring spectrophotometrically at 540nm, using UV Spectrophotometer (Varian DMS 90).

2.3 Enzymatic assay

α-amylase activity was assayed as reported by Lama et al. [6]. Enzymatic assays were conducted by adding 0.4 ml of PPE or PPEm to 0.6 ml of 50 mM acetate buffer pH 6.0. To start the reaction 0.5 ml of 1% soluble starch (boiled) were added. One enzyme unit is defined as the amount of enzyme which causes the loss of 100 optical density units in 30 min under our experimental conditions.

$MR3C^T$ was grown in the above medium at 61°C, pH 5.6 for 24 h. Ammonium sulphate was added to the cells-free broth (200 ml) to 80 % saturation. The precipitated was recovered by precipitation (10000g, 1h, 4°C), dissolved in 50 mM sodium acetate buffer pH 6.0, and then dialyzed over night against the same buffer (named Partial Purified Enzyme, PPE).

Hg^{++} (1, 3, 5 ppm); Zn^{++} (5, 15, 30, 50, 70 ppm); Ni^{++} (12, 20, 30, 40, 50 ppm); Cu^{++} (10, 20, 30 ppm); 0.5 ppm Hg^{2+} plus 2 ppm Ni^{2+}; 0.5 ppm Hg^{2+} plus 2 ppm Zn^{2+} were added to the bacterial growth at T 8h and incubated at 61°C for 16 h. Following exposure to heavy metals the samples were centrifuged at 9000g for 30 min, after which the supernatants were used for the enzyme activity. The same procedure of purification were performed and the Partial Purified Enzymes were named PPEm.

Analysis of reaction products were performed according to Lama et al. [6]. SDS-PAGE (10 %) gel was performed as described by Lama et al. [7] using the Pharmacia low molecular weight electrophoretic standards. Staining of protein bands was done with Coomassie Brilliant blue R-250 (Bio-Rad).

3   Results

The microorganisms have been grown on media containing various concentration of heavy metals. All microorganisms tested showed resistance to the presence of heavy metals during the growth (Fig. 1). Only Cd^{2+}, Zn^{2+} and in some cases Hg^{2+} caused a decrease of growth when added at high concentration. In particular Cd^{2+} affected the growth (up to 80 %) of strain L1 at 30 ppm (Fig. 1d), up to 87 % of MR3C strain (Fig. 1c) and *Bacillus thermantarcticus* (Fig. 1a) at 40 ppm, while the metal inhibited the growth (up to 85 %) of *Geobacillus stromboliensis* (Fig. 1b) at low concentration (5 ppm). No significant difference was noted for all strains grown in the presence of Mn^{2+}, Co^{2+}, Fe^{2+}, Ni^{2+}, Cr^{3+} and Cu^{2+} up to 60 ppm of concentration (Fig.1).

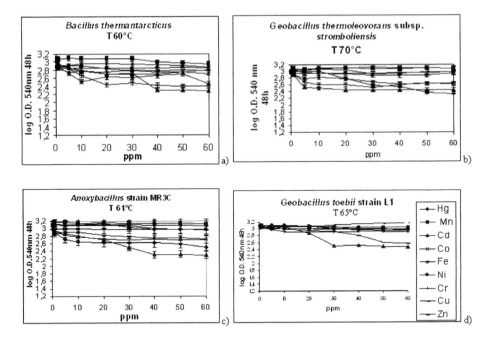

Fig. 1 Heavy metal effects on microorganism growth. The microorganisms have been grown on media containing various concentrations of heavy metals.

To study starch hydrolysis reaction MR3C strain was selected as a model system. This microorganism possessed a strong ability to hydrolyze starch due to the presence of an extracellular constitutive amylase (Fig. 2).

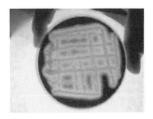

Fig. 2 Starch hydrolysis test on agar plate. After 24 h of growth the plate are loaded with Lugol Iodine solution (5.0 g I_2, 10.0 g KI in 100 ml distilled water, dil. 1:5). The medium zones containing non-hydrolysed starch appeared as blue night colour, while the medium zone around the colonies appeared transparent due to starch hydrolysis.

The enzymes were partially purified through out the ammonium sulfate precipitation of cell-free broth as describe in Materials and Methods and denominated Partial Purified Enzyme (PPEs), obtained in standard condition of growth, and Partial Purified Enzyme Metal (PPEm), obtained in presence of heavy metal added during the growth.

The PPEs worked at an optimal temperature of 61°C and at pH of 5.6. In optimal conditions of temperature and pH the PPEs presented a specific activity of 170 U/mg of proteins.

In Figure 3 was reported the effect of different heavy metals on PPEm activity.Among the heavy metals used, Hg^{2+} caused a rapidly inhibition of amylase activity also at low concentration. In fact, 5 ppm of Hg^{2+} were enough to reach the complete inhibition of the enzyme while the presence of Zn^{2+} caused 100 % enzyme inhibition at 70 ppm concentration. The presence of 30 ppm of Cu^{2+} and 50 ppm of Ni^{2+} caused 70 % enzyme inhibition.

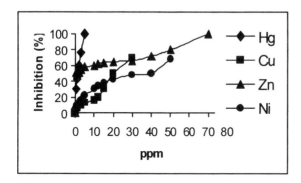

Fig. 3 Effect of different heavy metals on PPEm activity.

The products of α-amylase reaction, analysed by HPAE-PAD, were glucose, maltose, maltotriose and maltotetraose in a relative proportion of 20 %, 32 %, 38 % and 10 %, respectively for PPEs and 10 %, 40 %, 42 %, 8 % respectively for PPEm (Ni^{2+} at 50 ppm) as main reaction products after 24 h of incubation.

Partial purified enzymes obtained with or without heavy metals during the growth have been analysed by SDS-PAGE in denaturing condition (PPEs, Fig. 4, lane 8 and PPEm Fig. 4, lane 1, 2, 4, 5, 6). SDS-PAGE analysis showed a similar protein pattern both in PPEs and PPEm. However, the pattern referred to the presence of 5 ppm Hg (Fig. 4, lane 4), that caused 100 % enzyme inhibition, showed the decrease of alpha amylase protein stain.

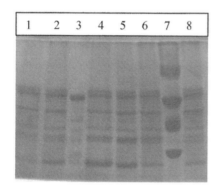

Fig. 4 SDS-PAGE 10%.
1) 0.5 ppm Hg^{2+}+ 2 ppm Ni^{2+}; 2) 0.5 ppm Hg^{2+}+ 2 ppm Zn^{2+};
3) α-amylase standard; 4) 5 ppm Hg^{2+}; 5) 3 ppm Hg^{2+};
6) 1 ppm Hg^{2+}; 7) Low Molecular Weight Standards ; 8) PPEs.

Several investigations have been carried out on the effects of heavy metals on various organisms [2]. This study demonstrates the potential of Extremophiles as toxicological indicators of heavy metals and it shows that the effect of heavy metals is detectable at very low concentration in MR3C *Anoxybacillus* strain α-amylase test system. Heavy metal resistance biotechnological process may be used for bioremediation of metal-contaminated environments and for bio-mining of expensive metals from effluents of any industrial process.

Acknowledgements

This work was financed by PNRA project and L5 Regione Campania

References

[1]   G. Ozdemir, T. Ozturk, N. Ceyan, R. Isler, and T. Cosar, bioresour technol. 90 (1), 71 (2003).

[2]   K. Guven , S. Togrul, F. Uyar, S. Ozant, and D. I. De Pomerai, enzyme microb. tech. 32, 658 (2003).

[3]   B. Nicolaus, L. Lama, and A. Gambacorta, Applications and Systematics of *Bacillus* and Relatives: Thermophilic *Bacillus* isolates from Antarctic Enviroments, (Blackwell Publishing, Berlin, 2002), chap. 5.

[4]   B. Nicolaus, L. Lama, E. Esposito, M.C. Manca, G. di Prisco, and A. Gambacorta, polar biol. 16, 101 (1996).

[5]   I. Romano, A. Poli, L. Lama, A. Gambacorta, and B. Nicolaus, j. gen. appl. microb. 51, 183 (2005).

[6]   L. Lama, B. Nicolaus, A. Trincone, P. Morzillo, V. Calandrelli, and A. Gambacorta, biotech. forum europe 8, 201 (1991).

[7]   L. Lama, V. Calandrelli, A. Gambacorta, and B. Nicolaus, res. microb. 155, 283 (2004).

Heavy metal toxicity in *Rhizobium leguminosarum* biovar viciae isolated from soils neighbouring metalomechanics industries

S.I.A. Pereira[*1], **A. I. G. Lima**[1] **and E.M.A.P. Figueira**[1]

[1]Centro de Biologia Celular, Departamento Biologia, Universidade de Aveiro, 3810-196 Aveiro, PORTUGAL

*Corresponding author: e-mail: siapereira@portugalmail.com, Phone: 00351 234 370 782, Fax: 00351 234 865 008

Heavy metals adversely influence microorganisms, affecting their growth, activities and survival. Metals exert a selective pressure on the organisms, resulting in microbial populations with higher tolerance to metals. In this context, *Rhizobium leguminosarum* biovar *viciae* was isolated from locations that have been influenced by metalomechanics industries. Physicochemical parameters were determined and heavy metal concentrations in soils were analysed by ICP-AES. Isolates were screened for their tolerance in YEM media supplemented with different heavy metals (Zn^{2+}, Pb^{2+}, Cd^{2+}, Ni^{2+}, Cr^{2+}). Proteins were extracted and separated by SDS-PAGE. MI_1 and MI_2 soils are the most polluted soils. Isolates showed different growth responses to heavy metals. MI_1 and MI_2 isolates were more tolerant than C isolates. Alterations on protein pool of *Rhizobium* populations were also analysed. Metals influenced their protein profiles, with most of the alterations corresponding to decreases in polypeptide expression. This work suggests that there is a relationship between Rhizobium's tolerance, heavy metal soil contamination and alterations in protein pool.

Keywords metal tolerance; protein alterations

1 Introduction

Soil is a key component of terrestrial ecosystems being essential for growth of plants and degradation and recycling of dead biomass [1]. Furthermore, it is important to recognize that soil is both a source of metals and also a sink of metal contaminants [2]. The discharge of heavy metals into the rivers and soils due to agricultural, industrial and domestic activities, and the effect of this pollution on the ecosystem and human health are growing concerns [3-4]. By acting as a sink for metals the soil function as a filter protecting the groundwater from inputs of potentially harmful metals, however soil microorganisms became exposed to these contaminants [3]. Soil is a complex dynamic system subjected to short-term fluctuations, such as variation in moisture status, organic matter content, pH and redox conditions. These changes in soil properties affect the form and bioavailability of these metals, additionally the amounts of metals accumulated in soils are also dependent on the emission levels, the transport of the metal from the source to the accumulation site and the retention of the metal once it has reached the soil [1].
Several studies have shown that metals adversely influence microorganisms, affecting their growth, morphology and activities [5, 6]. Metals exert a selective pressure on the organisms, resulting in microbial populations with higher tolerance to metals, but with lower diversity, when compared to unpolluted neighbouring areas [5, 7].
The present study aimed to investigate the effects of increase levels of heavy metals in indigenous *Rhizobium* populations isolated from soils subjected to metalomechanics

contamination. Maximum resistance level was determined and tolerances were compared. Since protein synthesis is a very sensitive cellular process to alterations in cell homeostasis [8] we examined alterations in protein expression in *Rhizobium* isolates from contaminated and non-contaminated locations, with the purpose to evaluate the level of stress imposed to each isolate.

2   Material and methods

2.1 Site selection and soil sampling procedures

Two agricultural soils were collected from contaminated locations: MI_1 and MI_2 soils have been influenced by metalomechanics industries since 30 years ago; a non-heavy metal contaminated agricultural soil was used as control (C). Soils were taken from the plough layer at 0-20 cm deep

2.2 Isolation of *Rhizobium*

Rhizobium was isolated from the root nodules of 30 days old *Pisum sativum* L. plants grown in contaminated (MI_1, MI_2) soils and non-contaminated soil (C). Nodules were sterilized in 2,5 % sodium hypochlorite for 2 min following a rinse in 95% ethanol and washed thoroughly in six changes of sterile water. Surface sterilized nodules were streaked on to the surface of yeast extract-mannitol (YEM) agar containing Congo Red [9].

2.3 Physicochemical analysis of soil

Physicochemical analysis was undertaken for soil water content, pH, redox potential and organic matter [10].

2.4 Determination of soils heavy metal content

For heavy metal concentration analysis representative soil sub-samples were air dried and sieved (< 2mm). Concentrations of Zn^{2+}, Pb^{2+}, Cd^{2+}, Ni^{2+} and Cr^{2+} were determined, after extraction with aqua-regia [11], by Inductively Coupled Plasma Atomic Emission spectrometry (ICP-AES).

2.5 Screening for heavy metal tolerance

Heavy metal tolerance of *Rhizobium* isolates was screened by plating in YEM media supplemented with metal (Zn^{2+}, Pb^{2+}, Cd^{2+}, Ni^{2+} and Cr^{2+}) increasing concentrations: 0, 0.065, 0.125, 0.165, 0.210, 0.250, 0.500, 0.750, 1.000, 2.000 and 3.000 mM. For growth measurements colonies were harvested, suspended in double-distilled sterile water (ddH_2O) and optical density (620 nm) was determined. The highest concentration of heavy metals supporting the growth was defined as the maximum resistance level (MRL).

2.6 Protein extraction and SDS-PAGE electrophoresis

Four isolates were selected from each location and were grown in YEM media, supplemented with different metal concentrations according to their tolerance. Cells were harvested by centrifugation for 15 min at 4000g and 4°C, and ressuspended in 200 μl of treatment buffer [12]. Samples were then sonicated for 1 min with an ultrasonic probe and submitted 5 min to 95 °C. Lysates were centrifuged to remove cell debris and the supernatant was collected. Proteins were separated by SDS-PAGE, carried out in 12.5% and 18% acrylamide slab gels [13]. Gels were stained with Coomassie brilliant blue R-250 (Bio-Rad). Densitometric readings were performed with a Bio-Rad – Model GS 710. The molecular weight and relative amount of proteins corresponding to each band were calculated using Quantity One Program Software (Bio- Rad).

2.7 Statistical analysis

To assess the overall differences in the soils properties from the various sites, statistically significant differences in the mean values of water content, pH, redox potential, organic matter and heavy metal concentrations were determined by one-way analysis of variance (ANOVA). Values are mean of three replicates.

3   Results

3.1 Physicochemical analysis and heavy metal contamination of soils

The physicochemical properties of soils and heavy metal concentrations are shown in Table 1 and 2.
Soils presented different physicochemical properties. pH and organic matter content varied significantly ($P < 0.05$) between locations.
Heavy metal concentration of soils also varied significantly ($P < 0.05$) has shown in Table 2. MI_1 and MI_2 soils presented higher concentrations of heavy metals than the control soil (C). Zinc was the only metal being near the limit or exceeding EC limits [14] in MI_1 and MI_2 soils.

Table 1 Physicochemical analysis of soils from different locations: C – Control isolates; MI_1 and MI_2 - Metalomechanics isolates. Data are the mean ± SE from three replicate experiments. Different letters indicate means that are significantly different ($P < 0.05$) from each other.

Soil	Physicochemical analysis			
	% water	pH	Eh	Organic matter (%)
C	19.1 ± 0.67 [a]	5.5 ± 0.07 [a]	376.3 ± 42.25 [a]	5.87 ± 0.06 [a]
MI_1	21.3 ± 0.90 [a]	6.1 ± 0.08 [b]	376.3 ± 28.75 [a]	6.24 ± 0.40 [b]
MI_2	25.1 ± 0.19 [b]	5.8 ± 0.06 [c]	392.7 ± 3.77 [a]	8.39 ± 0.25 [c]

Table 2 Metal concentration (mg kg^{-1}) and pH of soils from different locations: C – Control isolates; MI$_1$ and MI$_2$ - Metalomechanics isolates. Data are the mean ± SE from three replicate experiments. Different letters indicate means that are significantly different (P< 0.05) from each other.

Soil	Metal concentration (mg Kg^{-1} dry soil)					pH
	Zn^{2+}	Pb^{2+}	Cd^{2+}	Ni^{2+}	Cr^{2+}	
C	37.85 ± 17.98[a]	7.92 ± 3.25[a]	0.04 ± 0.01[a]	4.93 ± 0.14[a]	4.10 ± 3.75[a]	5.5 ± 0.07
MI$_1$	464.97 ± 7.10[b]	49.33 ± 1.06[b]	0.17 ± 0.01[b]	43.56 ± 0.26[b]	62.54 ± 0.93[b]	6.1 ± 0.08
MI$_2$	299.49 ± 15.27[c]	59.4 ± 2,29[c]	0.40 ± 0,02[c]	37.17 ± 1.42[c]	35.53 ± 2.06[c]	5.8 ± 0.06
EC limits*	300	300	3	75	200	5.5 - 7

* EC limits for sewage sludge treated soils (86/278/CEC) (CEC, 1986).

3.2 Isolates heavy metal tolerance

Rhizobium leguminosarum bv. *viciae* isolates showed distinct responses to metals. According to their MRL, isolates were divided in three groups: sensitive (MRL ≤ 0.210 mM), tolerant (0.250 ≤ MRL<1 mM) and extremely tolerant (MRL ≥ 2 mM). The percentage of sensitive, tolerant and extremely tolerant isolates from C, MI$_1$ and MI$_2$ soils are reported in Fig. 1.
MI$_1$ and MI$_2$ isolates showed similar responses to Ni^{2+} and Cr^{2+}, however to Zn^{2+} and Cd^{2+} the response was quite different. MI$_1$ isolates were more tolerant than MI$_2$ isolates to both metals. Some MI$_2$ isolates were sensitive to Pb^{2+}. Isolates from C soil were in general more sensitive than MI$_2$ e MI$_2$ to all metals.

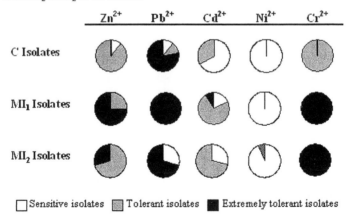

Fig. 1 Percentage of sensitive, tolerant and extremely tolerant isolates of *Rhizobium leguminosarum* bv. *viciae* from C, MI$_1$ and MI$_2$ soils. Isolates were obtained from root nodules of *Pisum sativum* L. plants as described. Data are the mean from three replicate experiments.

3.3 Alterations in protein expression

Alterations in protein expression of isolates grown under heavy metal stress are shown in Fig. 2. Although *Rhizobium leguminosarum* biovar *viciae* isolates displayed different protein profiles (data not shown) the quantitative alterations were similar when exposed to different metals, however isolates origin influenced the alterations in protein expression. In C and MI₁ isolates most alterations were decreases, especially in the last ones, MI₂ isolates showed distinct trend with increases surpassing decreases.

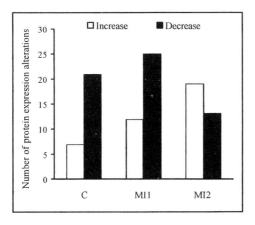

Fig. 2 Protein expression alterations induced by heavy metal stress in isolates from different locations: C - Control; MI₁ and MI₂ – Metalomechanics isolates. White – increase protein expression; Black – decrease protein expression. Data are the mean from three replicate experiments.

3   Discussion

Metal deposition into soil over long periods of time results in high metal concentrations, which therefore affects negatively soil microflora [15]. Soil is a complex environment where bacteria growth and development can be influenced by different edaphic factors, such as pH or organic matter content [16]. Ibekwe et al. [17] showed that under conditions of lower soil pH, the numbers of rhizobia were significantly reduced and the number of nodules observed on plants was also significantly lower. In this study, the pH values of most soils were below the optimal pH growth for rhizobia (6-7), however *Rhizobium* was able to withstand this pH values (Table 1).

Hirsch et al. [18] and Ibekwe et al. [17] reported that the distribution of *Rhizobium* isolates in agricultural soils is affected by the presence of the host plant, which generally leads to an increase in population effectives that will persist in soil for some years. Rhizobia was isolated from soils where no host plants were grown on the last few years, which lead us to conclude that rhizobia was far more vulnerable to the direct influence of heavy metal contamination. Although majority of metal concentrations determined in soils were below the EC limits [14] they could be enough to reduce bacteria effectives [6]. We isolated rhizobia from both contaminated soils indicating that these bacteria were able to survive under these metal concentrations. According to Ibekwe et al. [17] and Giller et al. [19] survival can be related with the physical protection of clay minerals and organic matter or with the existence of microsites where metal contamination may be minimal. These "niches" may harbour rhizobia that are not resistant to heavy metals.

Metalomechanics industries contributed to the increase of Zn^{2+}, Ni^{2+}, Cr^{2+} and Pb^{2+} concentrations in the MI₁ e MI₂ soils. Strong positive correlations between the amounts of heavy metals in these soils and the levels of tolerance to these elements were found, since MI₁ and MI₂ isolates were in general tolerant to all metals studied. These findings are consistent with the results of studies performed by Bååth [5] and Bååth et al. [7] who considered the possibility of heavy metals exert a selective pressure in bacteria of contaminated soils, resulting in the presence of more tolerant organisms in these areas.

Furthermore, Díaz-Raviña et al. [20] showed that the metal resistance patterns of bacterial populations were related to the total concentrations of metals in the soils.

It is also important to refer that the percentage of tolerant an extremely tolerant isolates to Zn^{2+}, Pb^{2+} and Cr^{2+} were similar in both contaminated soils. This lead us to conclude that *Rhizobium* isolates revealed multiple heavy metal tolerance, which is corroborated by Chander and Brookes [21] and Díaz-Raviña et al. [20].

Different heavy metals promoted similar alterations in the protein pool of *Rhizobium*. The stress imposed by heavy metals induced increases/decreases of polypeptides (Fig.1), which indicates an attempt of cells to adjust to adverse conditions. C isolates were less tolerant to heavy metals and most of the alterations were decreases of protein expression, suggesting a deleterious effect of metals on basic cell metabolism. Otherwise, MI_1 and MI_2 isolates were more tolerant and presented a high number of polypeptide increments. Saxena et al. [22] reported that these increments are usually related to metal tolerance mechanisms. These findings lead us to conclude that there is a relationship between *Rhizobium*'s tolerance and the alterations in protein pool. Thus, the analysis of protein expression alterations seems to be a good indicator to estimate the level of stress imposed to *Rhizobium* populations.

Acknowledgements

This work was supported by a grant from the Centre for Cell Biology.

References

[1] B.J. Alloway, B.J. Alloway (ed), Heavy Metals in Soils (Blackie Academic & Professional, New York, 1995a), p. 11-25.
[2] B. Robinson, C. Russell, M. Hedley and B. Clothier, agric. eco. environ. **87**, 315, (2001).
[3] B.J. Alloway, B.J. Alloway (ed), Heavy Metals in Soils (Blackie Academic & Professional, New York, 1995b), p. 3-9.
[4] S.P. McGrath, A.M. Chaudri and K.E. Giller, j. ind. microbial. **14**, 94, (1995).
[5] E. Bååth, M. Díaz-Ravina, A. Frostegård and C.D. Campbell, appl. environ. microbio. **64**, 238, (1998).
[6] A. Lakzian, P. Murphy, A. Turner, J.L. Beynon and K.E. Giller, soil boil. biochem. **34**, 519, (2002).
[7] E. Bååth, soil biol. biochem. **24**, 1167, (1992).
[8] A. Yeo, j. exp. bot, **49**, 915, (1998).
[9] P. Somasegaran, and H.J. Hoben., Handbook for Rhizobia (Springer-verlag, Berlin, 1994).
[10] I.V. Castro, Tese Doutoramento. Instituto Superior de Agronomia. Universidade Técnica de Lisboa. (1999).
[11] S.P. McGrath and C.H Cunliffe, j. sci. food agric. **36**, 794, (1985).
[12] B.D. Hames, in: B.D. Hames and D. Rickwood (eds.), Gel electrophoresis of proteins: a practical approach (IRL Press, Oxford, 1981), p.1-91
[13] U.K. Laemmli, nature, **227**, 680, (1970).
[14] CEC, off. j. europ. com. L181, Annex 1A, 10. (1986).
[15] A. Matsuda, F.M.S. Moreira and J.O. Siqueira, pesq. agro. bras. **37**, 343, (2002).
[16] K. Saeki, T. Kunito, H. Oyaizu and S. Matsumoto, j. environ. qual. **31**, 1570, (2002).
[17] A.M. Ibekwe, J.S Angle, R.L Chaney and P. van Berkum, soil sci. soc. am. j. **61**, 1679, (1997).
[18] P.R. Hirsch, M.J. Jones, S.P. McGrath and K.E. Giller, soil boil. biochem. **25**, 1485, (1993).
[19] K.E Giller, E. Witter and S.P. McGrath, soil biol. biochem. **30**, 1389, (1998).
[20] M. Díaz-Raviña, M. Bååth and A. Frostegård, appl. environ. microbial. **60**, 2238, (1994).
[21] Chander and P.C. Brookes, soil boil. biochem. **25**, 1231, (1993).
[22] D. Saxena, M. Amin and S. Khanna, can. j. microbiol. **42**, 617, (1996).

Isolation and screening of potential fungi for decolourization of distillery wastewaters

Deepak Pant and Alok Adholeya[*]

Centre for Mycorrhizal Research, Bioresources and Biotechnology Division, TERI School of Advanced Studies, DS Block, India Habitat Centre, Lodhi Road, New Delhi -110003, INDIA

[*]Corresponding author: e-mail: aloka@teri.res.in; Tel.: +91-11-24682100/24682111
Fax: +91 11 24682144/24682145.

The wastewater from molasses based distilleries is highly coloured with a high oxygen demand. Microbial decolourization is an environment-friendly and cost-competitive approach to treat this effluent. A broad screening of 32 fungi isolated from different environments on solid media and in broth resulted in the selection of 8 fungi, which were active in decolourizing the poly R-478 model system and capable of growing on and decolourizing post anaerobically treated distillery wastewaters. *Phanerochaete chrysosporium* was used as control organism for comparison. Effluent decolourization was carried out in static conditions with and without any additional carbon source. Wheat straw, bagasse and wood chips were used as immobilizing substrate offering the possibility of scaling-up the process. Tests on distillery effluent diluted at different COD concentrations revealed that all isolates performed well even at COD of 25 g l^1 (50% diluted effluent). Six isolates obtained from effluent contaminated soils performed better than the commercial strain, *Phanerochaete chrysosporium* MTCC 787 and displayed decolourization ranging from 68% to 90%. Removal in colour was accompanied by marked reduction in COD (88%) and BOD (94%) of the effluent. There was a substantial reduction in pH of the effluent to a range of 3.5 to 7.0 from a maximum of 9.0.

Keywords: distillery effluent; decolourization; fungi; COD; BOD

1   Introduction

Effluent of distillery industries known as spent wash is a major cause of soil as well as water pollution. Colour problem from distillery effluent is becoming increasingly important from environmental and aesthetic point of view. Due to the large volumes of effluent involved, their management is particularly difficult for the industry. The performance of conventional effluent treatment plant is such that the secondary treated effluent still contains high strength effluent quality (1). The physico-chemical methods alone for colour removal are expensive and environmentally unsatisfactory. The colour is hardly degraded by the conventional treatments and can even be increased during treatments, due to repolymerization of compounds (2). Current treatment processes such as chemical precipitation, chemical adsorption being used for removal of colour from treated wastewater have disadvantages due to high operation cost, high consumption of chemical agents, and fluctuation of the color removal efficiency and high volume of solid waste produced (3). Generally primary treated effluent (alkaline pH with strong objectionable odour, high BOD and COD values) is being released either for solar drying or loading in lagoons on soils for evaporation. There are 319 distilleries in India producing 3.25×10^9 L of alcohol and generating 40.4×10^{10} L of wastewater annually (4). The role of the added carbon to provide a nutrient source and the need to dilute the digested molasses spent wash before any biological treatment, are two major shortcomings as they exert both additional costs and greater waste effluent volumes (5). Fungi, especially white rot fungi have been traditionally used in studies involving decolourization of wastewaters such as distillery, pulp and paper and textiles (6-10). A detailed account of white-rot fungi and their enzymes for treatment of dyes and industrial effluents have given (11). Recently the isolation of a metal immobilizing fungus from polluted water in Egypt has been described (12). Quite a number of white-rots and other fungi have been used in previous studies for decolourization of distillery effluent (5, 13).

Keeping in view the applicability of biotreatment process, which becomes feasible only when it uses indigenous microorganisms with features to be used in the filed conditions, this study was carried out to isolate fungi from the polluted water and soils for distillery effluent treatment. The series of Poly B anthraquinone and Poly R anthraquinone dyes have been used as model substrate for the screening of fungi potentially degrading lignin and related compounds (14). These dyes have been used efficiently to differentiate the decolourizing activity of white-rot basidiomycota from other brown rot basidiomycota, ascomycota and deuteromycota (15). All the previous works involving immobilization of fungus have utilized an inert carrier such as polyurethane foam (16, 17). Most previous studies have been carried out using liquid culture conditions or solid cultures on agar plates, which however, do not reflect the natural living conditions (i.e. in wood and other lignocellulosic substrates) of white-rot fungi (18). Also, it is well established that the fungus needs a readily available carbon source for production of extracellular enzymes and decolourization activity (15, 19). The nutrient source is important because digested spent wash contain little, if any, readily available carbon in spite of its high total sugar content (5). So it would be a nice idea if the substrate being used for immobilization could also serve as a nutrient source for the microorganism. Based on this premise, we attempted different organic waste sources such as wheat straw, sugarcane bagasse, and wood chips etc., which themselves are difficult to degrade if left alone, for the establishment and growth of the microorganisms and subsequent application for effluent decolourization. These substrates apart from serving as a nutrient source also provide a buoyant support for the fungus to float when applied in wastewater bodies. Thus, this will be an attempt to develop low cost material that may be applied to industrial effluent in order to reduce the pollution load.

2   Materials and methods

2.1 Sampling sites and Fungi isolation

Wastewater and soil samples that were loaded with distillery effluents for six years were collected from Associated Alcohols and Breweries Limited, Barwaha, Madhya Pradesh located in central India. The distillery occupies 200 hectares of land. The Samples were collected from the 8 sites namely, a) Hot spent wash (without dilution from Fermenters); b) Cool spent wash (From cooling towers); c) Effluent after anaerobic treatment; d) Activated anaerobic sludge sample; e) Effluent from storage lagoons; f) Wet soil contaminated with effluent; g) Soil from nearby farmer's field, and h) Effluent from Effluent Treatment Plant (ETP) located within the industry premises.

The wastewater and soil samples collected in presterilized bottles and plastic bags respectively were kept at 4°C till further analysis was done. Both effluent and soil samples were serially diluted, and 0.1 ml of dilutions was spread onto potato dextrose agar (Hi media) plate (PDA) and nutrient agar (Hi media) plates. The plates were incubated at 28°C for 7 days after which, fungal colonies appearing on PDA plates were isolated into pure culture and routinely sub cultured on fresh PDA plates.

A commercial strain of *Phanerochaete chrysosporium* (MTCC 787) was used as control organism for comparison as it has been reported in decolourization studies previously (5, 20).

2.2 Solid-plate decolourization

Solid-plate decolourization was done according to method described previously (21). Solid medium in Petri plates were prepared using PDA and an aliquot of dye Poly R-478 (Sigma) to a final concentration of 10 ppm. Each of the isolated fungal strain was inoculated in triplicate on these plates. Uninoculated plats served as a control for abiotic decolourization.

2.3 Monitoring of Decolourization

Three samples were collected on 3rd, 7th and 14th days after inoculation and checked for changes in pH, colour, COD and BOD. Decolourization in solid medium was observed by visual disappearance of colour from the plates inoculated with Poly R 478 dye. Decolourization of effluent was measured after 100-fold dilution on a scanning UV-Visible Hitachi spectrophotometer (U-2000) by checking absorbance at 475 nm, which is the maximum visible wavelength for distillery effluents (16). COD and BOD were checked according to standard methods for examination of water and wastewaters (22).

2.4 Decolourization in aqueous medium

Post anaerobically digested distillery effluent at a final concentration of 50% (v/v) was inoculated with four mycelial agar discs cut with a (7 mm) borer from edges of an actively growing 3-5 week old colony growing on PDA plates. A detailed physico-chemical characterization of the effluent used in the study is given in table 1. Uninoculated flasks served as controls for abiotic decolourization.

2.5 Decolourization studies with substrates

Screened isolates were first grown on three substrates viz., wheat straw, bagasse and wood chips. Equal amount of each substrate (40 g) was soaked to saturation with water in a wide mouth glass jar. These were then sterilized by autoclaving and after cooling inoculated with eight discs from actively growing PDA plates cut with a 7 cm cork borer. These were kept at ambient temperature with loose caps until whole surface of substrate was covered with growing mycelia. The growth and penetration of fungi on different substrates was visually assessed (figure 1b). This substrate covered with fungal mat was evenly homogenized under sterile conditions and equal amounts (5 g of immobilized substrate) were used to inoculate 100 ml of distillery effluent in a 500 ml Erlenmeyer flask. Controls with no fungal inoculations were also used to account for decolourization due to natural microbial action. The flasks were prepared in triplicates for all the treatments.

Figure1 a. Poly R 478 plate decolourized by *Phanerochaete chrysosporium* MTCC 787
b. *Pleurotus florida* immobilized on wheat straw after 10 days of incubation

Figure 2 a. *Aspergillus* sp. growing on wheat straw along with control.
b. Effluent decolourized by *Pleurotus florida* along with control.

3   Results and discussion

3.1 Fungal strains used in the study

A total of 32 fungi were screened for their ability to decolorize the distillery effluent. Out of these, 23 fungi were isolated on potato dextrose agar plates from different contaminated environments by serial dilution. These strains have been named DB1 to DB23 in the text. Three isolates namely *Phanerochaete chrysosporium*, *Trametes hirsuta* and *Coriolus versicolor* were obtained from Institute of Microbial Technology (IMTECH), Chandigarh. Six isolates of oyster mushroom *Pleurotus sp*. were procured from Centre for Mycorrhizal Culture Collection (CMCC, TERI), New Delhi. Some of the best performing isolates were identified at Indian Type Culture Collection (ITCC) at Indian Agriculture Research Institute (IARI), New Delhi India (Table 2).

Table 1 Physico-chemical characteristics of distillery effluent

Parameter	Anaerobically treated Effluent (Released in the field)
Electrical Conductivity	33.16
pH	9.40
BOD5 (ppm)	5,000.00
COD (ppm)	50,000.00
Colour Pt-Co Unit	205×10^3
Total Organic Carbon	11400 mgl^{-1}
% nitrogen	3.92
Sodium (ppm)	500.00
Potassium (ppm)	2,500.00
Manganese (ppm)	259.44
Magnesium (ppm)	98.00
Zinc (ppm)	272.97
Copper (ppm)	395.51
Total Solids (%)	4.60
Total Dissolved Solids (TDS)	21,255.56
Total Sugar	2.8%
Reducing Sugar	0.227%

3.2 Agar plate screening

The decolourization of a type model dye, Poly R 478, is a simple method to assess the decolourization capability of fungi *in vitro*. The preliminary screening of 32 fungal isolates on poly R 478 plates led to eight isolates capable of decolourizing the dye in a period of 15 days. They were compared to *Phanerochaete chrysosporium* (figure 1a) as it has been previously reported in decolourization studies (5, 20). Isolate *Pleurotus florida* showed maximum decolourization on solid plate as shown in table 2.

3.3 Decolourization in aqueous medium

Increase in fungal biomass was observed during colour removal over a period of 14 days. All the fungi reduced the pH of the effluent (figure 3). Maximum reduction was by isolate DB1 where the pH was reduced to 3.56 from the initial pH of 9.42. This reduction in the pH of the effluent might be attributed to the release of organic acids by the fungi as most fungi are acidogenic by nature. Maximum decolourization achieved was 85% in case of isolate *Pleurotus florida* followed by isolate DB15 identified as *Fusarium monoliforme* where colour was removed by 76% after 14 days treatment. This is consistent with our findings earlier where, this particular isolate, *Pleurotus florida* grew best on substrate amended with distillery effluent (23).

Table 2 Morphological characteristics of screened fungi and decolourization of Poly R 478 inoculated plate in a 15-day period

Fungal isolate and Code	Source	Morphological characteristics	Decolourization
Phanerochaete chrysosporium	IMTECH, Chandigarh (MTCC 787)	White surface	+
PF (*Pleurotus florida*)	CMCC culture collection	White cottony surface	+++++
DB1 (not identified)	Effluent from ETP	White cottony surface	+++
DB6 (not identified)	Spent wash	White cottony surface	+++
DB11 (*Aspergillus flavus*)	Spent wash	Green surface	++
DB14 (not identified)	Effluent contaminated soil	Brown surface	+++
DB15 (*Fusarium monoliforme*)	Effluent contaminated soil	White surface	++++
DB18 (*Aspergillus niger*)	Effluent from storage lagoons	Black cottony surface	++
DB20 (*Aspergillus niger*)	Sludge of anaerobic treatment plant	Black surface	++++

+ indicates the degree of decolourization on solid plate inoculated with Poly R 478

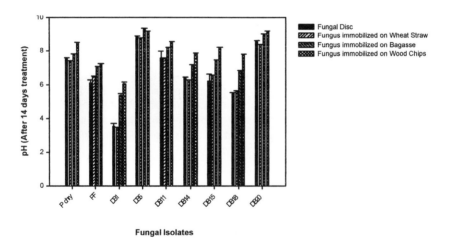

Fig 3 Change in pH of the effluent after 14 days treatment with fungal isolates

3.4 Immobilization of fungi on agro residue

Three agro residues viz. wheat straw, bagasse and wood chips were used for immobilization of the fungi. Fungi immobilized on wheat straw showed better decolourization and reduction in COD and BOD than the flasks inoculated with fungal disc alone. This was followed by immobilization on bagasse and wood chips where the reduction was least for all parameters, as shown in fig 3, 4 and 5. Immobilized fungal isolate of *Aspergillus niger* UM2 resulted in a 72 % decolourization of synthetic melanoidin and 80% decolourization of biogas effluent (24). Similar results were reported earlier (25), where stainless steel sponge was used as a carrier for immobilization of *Trametes hirsuta* and used for decolourization of textile dyes. Recently (26) it was demonstrated that lignocellulosic substrates such as hemp woody core and wheat straw, which on its own do not support the production of LiP and MnP, led to the production of these enzymes in submerged cultures of *P. chrysosporium*. Since these enzymes have previously been reported to play a role in decolourization, the use of such substrates gets emphasized for carrying out large-scale decolourization. Solid state cultures of *Lentinula edodes* were used (17) for decolourization of synthetic dyes when corncob was used as substrate.

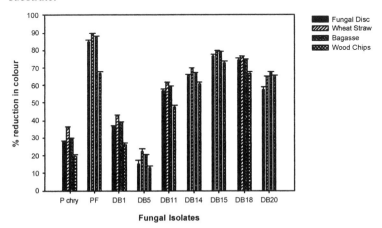

Fig 4 % reduction in colour of effluent after 14 days treatment with fungal isolates

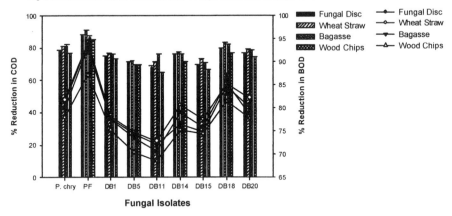

Fig 5 % Reduction in COD and BOD of the effluent after 14 days treatment with fungal isolates

All previous studies carried out for the decolourization of distillery effluent can be linked by the fact that almost all of them used effluent at very low to low dilutions ranging from 6.5% (5) to 20% (27). This is the first report of decolourization of highly recalcitrant distillery effluent using indigenously isolated fungi at such a high concentration (50% v/v).

Also, no readily available nutrient source like sucrose or glucose was added to the effluent. Further work is in progress to upscale the process by creating a consortium for large-scale treatment of effluent.

Acknowledgements

The authors wish to thank Dr R.K Pachauri, Director-General, TERI, India for providing the infrastructure for the study. The first author acknowledges University Grants Commission (UGC) for junior research fellowship. Thanks are also due to Mr. H. C. Bhandari, Manager, AABL, Barwaha for providing the effluent.

References

[1]   T. Nandy, S. Shastry, and S. N. Kaul, Journal of Environ. Manag. **65,** 25 (2002).

[2]   P. M. Miranda, G. G. Benito, N. S. Cristobal, and C. H. Nieto, Biores. Technol. **57,** 229 (1996)

[3]   S. Sirianuntapiboon, P. Somchai, S. Ohmomo, and P. Atthasampunna, Agric. Biol. Chem. **52,** 387 (1988)

[4]   J. Uppal, Water utilization and effluent treatment in the Indian alcohol industry – an overview In: Indo-EU workshop on Promoting Efficient Water Use in Agro-based Industries, TERI, New Delhi 15–16 January 2004, India (2004)

[5]   V. Kumar, L. Wati, P. Nigam, I. M. Banat, B. S. Yadav, D. Singh, and R. Marchant, Process Biochem. **33,** 83 (1998)

[6]   F. Zhang, J. S. Knapp, and K. N. Tapley, Enzyme Microb. Tech. **24,** 48 (1999)

[7]   D. Y. Kim, and Y. H. Rhee, Appl. Microbiol. Biotechnol. **61,** 300 (2003)

[8]   S. Ohmomo, Y. Kaneko, S. Sirianuntapiboon, P. Somchai, P. Atthasumpunna, and I. Nakamura, Agric. Biol. Chem. **51,** 3339 (1987)

[9]   V. L. Papinutti, and F. Forchiassin, FEMS Microbiol. Lett. **231,** 205 (2004)

[10]   R. Nagarathnamma, B. Bajpai, and P. K. Bajpai, Process Biochem. **34,** 939 (1999)

[11]   D. Wesenberg, I. M. Kyriakides, and S. N. Agathos, Biotechnol. Adv. **22,** 161 (2003)

[12]   E. M. El-Morsy, Mycologia. **96,** 1183 (2004)

[13]   A. M. Jimnez, R. Borja, and A. Martin, Process Biochem. **38,** 1275 (2003)

[14]   A. Jaouani, S. Sayadi, M. Vanthournhout, and M. J. Penninckx, Enzyme Microb. Tech. **33,** 802 (2003)

[15]   L. J. Cookson, Appl. Environ. Microbiol. **61,** 801(1995)

[16]   S. J. Kim, and M. Shoda, Journal of Biosci. Bioeng. **88,** 586 (1999)

[17]   C. Raghukumar, C. Mohandass, S. Kamat, and M. S. Shailaja, Enzyme Microb. Tech. **35,** 197 (2004)

[18]   C. G. Boer, L. Obici, C. G. M. Souza, and R. Peralta, Biores. Technol. **94,** 107 (2004)

[19]   J. Angayarkanni, M. Palaniswamy, and K. Swaminathan, Bull. Environ. Contam. Toxicol. **70,** 268 (2003)

[20]   J. Dahiya, D. Singh, and P. Nigam, Biores. Technol. **78,** 95 (2001)

[21]   J. Swamy, and J. A. Ramsay, Enzyme Microb. Tech. **24,** 130 (1999)

[22]   APHA, Standard Methods for the examination of water and wastewater. 17th ed. American Public Health Association, Washington, D.C. (1995)

[23]   D. Pant, U. G. Reddy, and A. Adholeya, World J. Microbiol. Biotchnol. DOI 10.1007/s11274-005-9031-2 (2005) (in press)

[24]   P. U. Patil, B. P. Kapadnis, and V. S. Dhamankar, AIDA News Letter. 53 (2003)

[25]   S. R. Couto, M. A. Sanroman, D. Hofer, and G. M. Gubitz, Biores. Technol. **95,** 67 (2004)

[26]   A. N. Kapich, B. A. Prior, A. Botha, S. Galkin, T. Lundell, and A. Hatakka, Enzyme Microb. Tech. **34,** 187 (2004)

[27]   T. Gonzalez, M. C. Terron., S. Yague, E. Zapico, G. C. Galletti, and A. E Gonzalez, Rap. Comm. Mass Spec. 14, 1417 (2000)

Microbial corrosion in zinc surface layer of galvanized steel by biofilms of sulfate reducer bacteria

Khanafari Anita[*1], Tahernia Alireza[1], Montasser Koohsari Shiedeh[3], Mazaheri Assadi Mahnaz[2], Noohi Ashrafulsadat[1]

1- Islamic Azad University- Microbiology Department, North of Tehran-Iran
2- Biotechnology Center, Iranian Research Organization for Science and Technology, Tehran, Iran
3-Islamic Azad University Science and Research Campus of Iran-Tehran

[*]Corresponding author: e-mail: khanafari_a@yahoo.com, Tel: +98 21 208 21 41, Fax: +98 21 284 4282.

Sulfate reducing bacteria (SRB) isolated from sewage, produces a thick black precipitate on galvanized steel coupons. The isolated strains produced dissolved H_2S up to 12m moles l^{-1} in 72h. The increase in corrosion rate produced in batch and semi-continuous cultures was studied by scanning electron microscopy and XRF methods. The results showed that the specific mass loss in semi–continuous culture was 89.82% more than in batch culture. Pitting was observed uniformly over the galvanized steel surfaces after exposure to Baar's medium with SRB for 30 days. The results of SEM indicated that, there was 0.01% Zn and 95% Fe on the galvanized steel surface. X-ray fluorescent spectroscopy (XRF) and Atomic absorption methods showed that, there was 83.459% zinc in the pellet but no zinc element was detected in the supernatant. The maximum content of zinc element was searched in biofilm on galvanized steel coupon by SEM analyses.

Keywords: Galvanized steel, SRB, Biofilm, and Microbial Corrosion.

1   Introduction

Galvanized steel is commonly used in small cooling towers associated with commercial air conditioning systems. In order to influence either the initiation or the rate of corrosion, microorganisms usually must become intimately associated with the corroding surface. In most cases, they become attached to the metal surface either in a form of a thin, distributed film or a discrete biodeposit. The thin film, or biofilm, is most prevalent in open systems exposed to flowing seawater, although the same can also occur in open fresh water systems (R.E.Tatnall). These films will usually be spotty rather than continuous in nature, but will nevertheless cover a large proportion of the exposed metal surface. In contrast to the distributed films are discrete biodeposits. These biodeposits may be up to several centimeters in diameter, but will usually cover only a small percentage of the total exposed metal surface (R.E.Tatnall).

The organisms in the films and deposits will generally have a large effect on the chemistry of the environment with in the deposit- that is, at the metal/film or the metal/deposit interface- without having any measurable effect on the bulk electrolyte properties. Occasionally, however, the organisms will be concentrated enough in the environment to influence corrosion by changing the bulk chemistry (R.E.Tatnall).

Sulfate Reducer Bacteria (SRB) is a group of bacteria that are commonly associated with MIC. Since the sulfate-reducing bacteria (SRB) has been considered the major bacterial species in causing metal corrosion under anaerobic environments (Crombie et al., 1980; Pankhania, 1988; Ford and Mitchell, 1990; Hao et al., 1996). Most of the previous studies concentrated on the corrosion problems caused by SRB encountered in those environments. Little is known about the potential of this type of bacteria to corrode metals under aerobic situations. Recently, a few researchers have studied SRB corrosion problems in aerobic environments. Hardy and Bown (1984) suggested that the most aggressive conditions associated with SRB were those, which were not entirely anaerobic, but where small quantities of oxygen might be present from time to time. Zinc is a well known, common and relatively inexpensive coating material for iron and steel. Zinc acts as a sacrificial anode, i.e. it corrodes in a corrosive environment and lets the steel play the role of cathode(J. Zhu, et al). In this study, the colonization, growth, and decay of SRB on galvanized steel coupons surfaces was investigated. An analysis of the corrosion products on corroded metal surfaces using X-Ray Fluorescent spectroscopy (XRF) and Scanning Electron Microscopy (SEM) was undertaken to determine the corrosion and zinc surface damage.

2 Materials and methods

2.1 Sampling

Sulfate reducing bacteria was isolated from sewage and cultured periodically in the presence of 0.1% ferrous ammonium sulfate (Postgate B medium). All subsequent cultures were grown inocula from at 30 °C after 4–7 days without shaking with gas Pak. The presence of sulfate reducers was confirmed by the detection of black iron sulfide in the culture tubes. Desulfoviridin can be produced by *Desulfovibrio* sp (Postgate, 1984) and was also routinely performed with the detection of a pink color under UV light.

2.2 Metal coupon preparation

Galvanized steel (2.93× 2.55 × 0.025mm) coupons were cut from sheet stock, polished with 240-grit paper and stored as described by Jayaraman et al. 1997c.

2.3 Batch culture corrosion experiments

Batch culture corrosion experiments were performed in 250ml conical flasks with Baar's medium at 30 °C without shaking as described by Jayaraman et al. 1997a. Six galvanized steel coupons (triplicates) exposed to SRB, were cleaned by wiping the surfaces with 0.01% chromic acid followed by repeated washes in warm water, as described earlier (Jayaraman et al. 1997a). The specific mass loss (in mg cm^{-2} for the total surface area of the coupon) was used as an indicator of the extent of corrosion.
The growth medium was replenished every 7 days by gentle addition along the walls of the flask.

2.4 Semi-continuous culture corrosion experiment

Biofilms on galvanized steel surfaces were developed in a semi-continuous reactor with modified Baar's medium. This medium supports the growth of SRB. A 3% SRB inoculums (A$_{600}$= 0.8–1, growth in Postgate C medium) was used for all semi-continuous experiments.

2.5 Zinc analysis of galvanized steel coupons

Three of the six corrosion coupons were released from the reactor after 45 days. The samples were centrifuged at 10000 rpm; 15 min at room temperature and the supernatant was examined under spectra AA.200, Varian by the atomic absorption method for zinc. The pellet was dried in an oven, 24 h a 50 °C, and examined for zinc using the X-Ray Fluorescent Spectroscopy (XRF) method.

2.6 Scanning electronic microscopy

The remaining three coupons, were fixed with 2% glutaraldehyde at 4°C for 24h, then rinsed with a phosphate buffer of 0.2 M pH 7.2, and fixed for 2h in osmium tetroxide. After washing three times in the phosphate buffer, the biofilms were dehydrated by passing through a graded series of alcohol washes (30-100%) and frozen rapidly in liquid nitrogen prior to freeze-drying. The dried coupons were then mounted on an aluminum stub, sputter coated with carbon and examined under a LEOi 5900 LV scanning electron microscope at an accelerating voltage of 13 kV, as described previously (Lopez et al, 2001).

3   Results and discussion

The depicted results of electron microscopy showed that the bacterial strains were able to form biofilms in corrosion coupons and the tests indicated that the bacterial strains isolated from sewage deposits, were sulfate-reducing bacteria. The strains exhibited rapid growth in mineral media with a lactate carbon source, and produced H_2S when sulfate was used as the electron acceptor. The bacteria also showed an ability to adhere to the metal substrates (Figure 1).

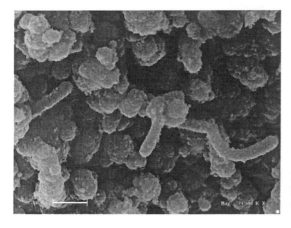

Fig. 1 Scanning electron micrograph of SRB in biofilm on galvanized steel coupon. Scale bar is 3μm.

The mass loss from the galvanized steel coupon in modified Baar's medium using SRB was evaluated every 7 days for 60 days in stationary batch cultures at 30°C under aerobic and anaerobic conditions. The presence of Bacteria in the biofilm has been investigated. The coupons were covered with a thick, black deposit, and difficult to clean (Figure 2).

106

Fig. 2 Scanning electron micrograph of thick, black surface deposits of SRB from corrosion products on galvanized steel coupon. Scale bar is 10μm.

The mass loss from the galvanized steel coupons in modified Baars medium with SRB and control medium were evaluated after 14 days in stationary batch cultures at 30°C under aerobic and anaerobic conditions. The results showed that the specific mass loss in semi – continuous culture was 89.82 % more than in batch culture. These results are similar to the studies of Pederson et al., 1998.

Pitting was observed uniformly over the galvanized steel surfaces after exposure to Barr's medium with SRB for 30 days. Similar observations were obtained by the studies of Ornek et al., 2002, and Mansfeld et al, 2000.

Element analysis by SEM demonstrated that, there was 80.78% Zn and 10.76% Fe on the galvanized steel surface. The galvanized steel coupons were exposed to SRB, cleaning were performed using 0.01% chromic acid followed by repeated washes in warm water; these results were changed to 0.01% zn and 95% Fe.

The results showed that SRB could damage the zinc surface layer on galvanized steel coupons. It would appear the zinc surface layer was somehow important to the mechanisms, perhaps acting as a source of energy for these particular bacteria. The results, from Figure 3 were in accordance with Schults (1981).

X-ray fluorescent spectroscopy (XRF) showed that, there was 83.459% of zinc element in the pellet whereas no zinc element was detected in supernatant by using the atomic absorption method.

SEM method was employed to analyzed the zinc rate on bacterial biofilm upon metal samples surfaces after preparation.

Zinc is a well known, common and relatively inexpensive coating material for iron and steel. Zinc acts as a sacrificial anode, i.e. it corrodes in a corrosive environment and lets the steel play the role of cathode. The zinc surface layer was somehow important to the mechanisms, perhaps acting as a source of energy for these particular bacteria. According to the results, which obtained from Figure 3, the maximum content of Zinc element was searched in biofilm on galvanized steel coupon. These findings were in accordance with Schults (1981).

Fig. 3 Scanning electron micrograph of SRB from corrosion products on galvanized steel coupon (on the left) and Zinc element was searched in biofilm by SEM analyzed as light spots (on the right). Scale bar is 3µm.

Acknowledgements

This project is supported by the Islamic Azad University- Microbiology Department, North of Tehran, Biotechnology Center, Iranian Research Organization for Science and Technology, Tehran and Islamic Azad University Science and Research Campus of Iran.

References

[1]   Angell, P., White, D. C. Is metabolic activity by biofilms with SRB consortia essential for long-term propagation of pitting corrosion of stainless steel?, J. Ind. Microbial. **15**: 329-33(1995).

[2]   Bozzola, J. J., & Russell, L. D. Electron microscopy; Jones and Bart let Publishers.U.S.A (1991).

[3]   Cord Ruwish R. A quick method for the determination of dissolved and precipitated sulfides in cultures of sulfate-reducing bacteria; J. Microbiol. Meth. **4**: 33-36(1987).

[4]   Crombie, D.J., Moody G.J, and Thomas J.D.R. Corrosion of Iron by Sulfate-Reducing Bacteria, Chem. Ind. **21(6)**: 500-505(1980).

[5]   Ford, T., and Mitchell R. Advances in Microbial Ecology, New York, N.Y.: Plenum Press (1990).

[6]   Geesey, G. G., What is Biocorrosion? In: International workshop on industrial biofouling and biocorrosion, Stuttgart, Germany, Springer, New York Berlin Heidelberg (1990).

[7]   Hao, O.J., J.M. Chen, L. Huang, and Buglass R.L. Sulfate-Reducing Bacteria, Critical Reviews in Environ, Sci. & Technol. **26(1)**: 155-187(1996).

108

[9]   Iverson, W. P. Microbial corrosion of metals, Adv. Appl. Microbiol. **32**: 1-36 (1987).

[10]   Jayaraman, A., et al. Axenic aerobic biofilms inhibit corrosion of SAE 1018 steel through oxygen depletion, Appl. Microbiol. Biotechnol. **48**:11-17(1997).

[11]   Kobrin, G. A Practical Manual on Microbiologically Influenced Corrosion, Published by NACE International (1993).

[12]   Mack, R. D., Wilhelm, S. M., Steinberg, B. G. Laboratory corrosion testing of metals and alloys in environments containing hydrogen sulfide, 246-259 in Laboratory Corrosion Tests and Standards, ASTM Special Technical Publication 866, Eds. Gardner S. Haynes and Robert Baboain, USA (1983).

[13]   Ladd, T. I., Costerton, J. W. Methods for studying biofilms bacteria, Meth. Microbiol. **22**, 285-307(1990).

[14]   Lopez, J, et al. Biofilms and Adhesion Protein in Anaerobic Bacteria Isolated from Mexican Gas Pipelines (2001).

[15]   Ornek, D., et al. Pitting corrosion inhibition of aluminum 2024 by Bacillus biofilms secreting polyaspartate or $\wp$-polyglutamate, Appl. Microbiol. Biotechnol., **58**:651-657(2002)

[16]   Pankhania, I.P. Hydrogen Metabolism in Sulfate-Reducing Bacteria and its Role in Anaerobic Corrosion, Biofouling **1**:27-29(1988).

[17]   Postgate, J. R. The sulfate-reducing bacteria, Cambridge University Press, New York (1984).

[18]   Schultz, R.A. Pitting of Galvanized Steel in a Cooling Tower Basin, Materials Performance**20**, **8**:44(1981).

Microbial diversity and metabolic pathway analyses of a mesophilic acetate-degrading methanogenic community in a chemostat cultivation

Toru Shigematsu[*], **Yueqin Tang, Shigeru Morimura,** and **Kenji Kida**

Department of Materials and Life Science, Graduate School of Science and Technology, Kumamoto University. 2-39-1Kurokami, Kumamoto 860-8555, JAPAN

[*]Corresponding author: e-mail: shige@kumamoto-u.ac.jp, Phone: +81 96 342 3668, Fax: +81 96 342 3679

The community structure and acetate conversion pathways of methanogenic consortia in a mesophilic acetate-fed chemostats were investigated. Microbial community analysis based on 16S rRNA genes on the chemostat at dilution rates of 0.025 and 0.6 d^{-1} revealed that a significant number of the genus *Methanoculleus*, a hydrogenotrophic methanogen, was detected only at the low dilution rate, although larger populations of aceticlastic methanogens, *Methanosaeta* and *Methanosarcina*, were detected at both dilution rates. *Bacteria* belonging to the phylum *Firmicutes* dominated at the low dilution rate. Acetate conversion pathways at the two dilution rates were analyzed using ^{13}C-labeled acetates. Syntrophic oxidation by acetate oxidizing syntrophs and hydrogenotrophic methanogens was suggested to occupy a primary pathway in total methanogenesis at the low dilution rate. In contrast, aceticlastic cleavage of acetate by aceticlastic methanogens was suggested to occupy a primary pathway at the high dilution rate.

Keywords methanogenesis, acetate conversion, stable-isotope, methyl-coenzyme M reductase, chemostat

1   Introduction

Acetate is quantitatively the most major intermediate of anaerobic degradation of organic matters under methanogenic conditions. It is estimated that approximately 70-80% of methane is derived from acetate in anoxic environments. Two processes by which acetate is converted to methane have been described. The acetate-utilizing methanogens, the genera *Methanosaeta* and *Methanosarcina*, use the aceticlastic cleavage pathway in which the methyl group of acetate is converted to CH_4, while the carboxyl group is converted to CO_2 [1]. The second process includes the syntrophic oxidation of acetate to CO_2 and hydrogen by one organism, and the subsequent reduction of CO_2 to CH_4 by hydrogenotrophic methanogen. *Bacteria*, such as *Thermacetogenium phaeum* and *Clostiridum ultunense*, were demonstrated to be capable of acetate oxidation in co-culture with hydrogenotrophic methanogens [2, 3]. The quantitative information of these two acetate conversion pathways in total methanogenic microbial communities has been limited to date.

In this study, a chemostat cultivation of a mesophilic acetate-degrading methanogenic consortium was constructed. We analyzed the concentrations of coenzymes related to methanogenesis in the chemostat to estimate the activities of methanogens. The community structures at two dilution rates of 0.025 and 0.6 d^{-1} were analyzed. Moreover, the acetate conversion pathways of the methanogenic consortia was analyzed using ^{13}C-labeled substrates followed by GC-MS analysis of the CH_4 and CO_2 produced. We also analyzed the transcripts of the *mcrA* gene, which encodes the methyl-coenzyme M reductase, at the both dilution rates.

2   Materials and methods

2.1 Operation of the acetate-fed chemostat

The continuous cultivation was carried out using a continuous stirred tank reactors (CSTRs), with a working volume of 1.7 *l*. The reactor was fed with a synthetic wastewater containing acetate as a sole carbon source at 37°C. Acclimatized mesophilic digested sludge from a domestic sewage treatment plant was used as the seeding sludge. The concentration of Ni^{2+} and Co^{2+} in the synthetic wastewater was adjusted to 52.6 μg/*l* and 61.2 μg/*l*, respectively. Detailed methods were described previously [4].

Methanogenic activity from acetaet of the culture broth was determined based on specific gas evolution rate. Coenzyme F_{420} was extracted from culture broth and relative concentration of coenzyme F_{420} was determined based on fluorescent strength at a wavelength of 460 nm after excitation at 420 nm. Coenzymes F_{430} and corrinoids were extracted from the culture broth, partially purified using an Amberlite XAD-2 column and quantified based on nickel and cobalt concentrations, respectively, as described previously [4].

2.2 Microbial community analyses

Fluorescence *in situ* hybridization (FISH) experiment was carried out as described previously [5]. Two domain-specific probes, EUB338 and ARC915 for *Bacteria* and *Archaea*, respectively, were used.

DNA from the microbial community was extracted according to our previous description [5]. Amplification of 16S rRNA gene from the extracted DNA was performed by PCR using the universal primer 530F and prokaryote-specific primer 1490R. The amplified 16S rRNA gene fragments were cloned and sequenced. Phylogenetic analysis based on determined nucleotide sequence was analysed using the BLASTN programm, Clustal X program and MEGA program. For detection and quantification of methanogens, a real-time quantitative PCR experiment was carried out using the TaqMan fluorogenic PCR system as described previously [5].

2.3 Acetate conversion pathway analysis

A 10-ml culture broth was taken from the chemostat and centrifuged at 15,000 rpm for 15 min. The precipitate was washed and transferred into vials and supplemented with $[1-^{13}C]$, $[2-^{13}C]$ or $[1,2-^{13}C]$ sodium acetate to give final concentrations of 100 mM. The vials were incubated at 37°C for 12 h. The CH_4 and CO_2 in the headspace were analyzed using a GC-MS as described previously [6].

Total RNA from the culture broth in the chemostat was extracted and purified. Reverse transcription (RT)-PCR reaction was performed for amplification of *mcrA* transcripts. Real-time quantitative RT-PCR experiments were conducted to quantify *mcrA* transcripts of different taxonomic groups using the TaqMan fluorogenic PCR system as described previously [6].

3 Results

3.1 The methanogenic actibvity and concentrations of coenzyme related to methanogenesis in the acetate-fed chemostat

A continuous cultivation of acetate-degrading methanogenic consortia was constructed. The chemostat cultivation was stable at dilution rates between 0.025 and 0.7 d^{-1}. Acetate concentrations in the chemostats were approximately 0.2 and 4 mM at dilution rates of 0.025 and 0.6 d^{1}, respectively, which suggest that the acetate fed into the reactor was almost completely degraded to CH_4 and CO_2 at a dilution rate between 0.025 and 0.6 d^{-1}.

Under steady state conditions at each dilution rate, methanogenic activity from acetate, based on specific gas evolution rate, and concentration of coenzyme F_{420}, F_{430} and corrinoids were analyzed. The relative concentration of F_{420}, which is involves methane production from H_2-CO_2, was approximately seven times higher at a low dilution rate of 0.025 d^{-1} than that at a high dilution rate over 0.6 d^{-1}. The methane concentration in biogas evolved at a dilution rate of 0.6 d^{-1} was approximately 51%, which was lower than the 65% obtained at a dilution rate of 0.025 d^{-1}. These results suggested that aceticlastic methanogens were dominant at the high dilution rate and that a significant number of hydrogenotrophic methanogens, possibly having a role in the syntrophic oxidation of acetate, were present at the low dilution rate. We therefore analyzed and compared the community structures of the acetate-fed chemostat at the dilution rates of 0.025 and 0.6 d^{-1}.

3.2 The microbial community structures of the chemostat at dilution rates of 0.025 and 0.6 d^{-1}

The community structures of the mesophilic acetate-degrading methanogenic consortia at dilution rates of 0.025 and 0.6 d^{-1} were analyzed by FISH and phylogenetic analyses based on 16S rRNA gene clonal sequences and quantitative real-time PCR. FISH experiments with archaeal and bacterial domain-specific probes showed that archaeal cells were predominant and only a small number of bacterial cells were detected at both dilution rates. In the domain *Archaea*, the number of cells closely related to the genus *Methanosarcina* was shown to be greater at the high dilution rate using FISH with species-specific probes. Taxonomic analyses based on rRNA gene clonal sequences obtained at the low and high dilution rates showed that 43% of 100 clones and 72% of 92 clones, respectively, were affiliated with the domain *Archaea* and the remainders at each dilution rate were affiliated with the domain *Bacteria*. Within the domain *Archaea*, all rRNA gene clones at both dilution rates were affiliated with the genera *Methanosaeta* or *Methanosarcina* of the aceticlastic methanogens. Within the domain *Bacteria*, the rRNA gene clones obtained at the low dilution rate were affiliated with four phyla, *Firmicutes* (Low G+C gram positive bacteria) (36%), *Bacteroidetes* (9%), *Chloroflexi* (6%) and candidate division OP12 (5%). The rRNA gene clones obtained at the high dilution rate were affiliated with four phyla, *Firmicutes* (16%), *Bacteroidetes* (8%), *Proteobacteria* (1%) and candidate division OP12 (3%).

Real-time quantitative PCR experiments showed that the number of rRNA gene sequences affiliated with the genus *Methanosarcina* was greater at the high dilution rate (Table 1). In addition, a significant number of rRNA gene sequences affiliated with the genus *Methanoculleus* were detected only at the low dilution rate. Detection of a hydrogenotrophic methanogen at the low dilution rate suggests that the syntrophic acetate

112

oxidation by hydrogenotrophic methanogens and acetate-oxidizing bacteria could occur at the low dilution rate.

Table 1 Quantification of 16S rRNA gene of methanogens in culture broths[a]

Primer / probe set	Target organism	Dilution rate (d^{-1})	
		0.025	0.6
MS1b / SAE835R / SAE761TAQ	*Methanosaeta*	9.27×10^6 (5.37×10^5)	1.99×10^7 (2.04×10^6)
MB1b / SAR835R / SAR761TAQ	*Methanosarcina*	6.45×10^5 (1.87×10^5)	7.43×10^7 (1.36×10^7)
AR934F / MG1200b / MCU1023TAQ	*Methanoculleus*	2.70×10^5 (6.74×10^4)	N.D.

[a] Unit: 16S rDNA copies / 50 ng DNA. All values are means of four independent experiments. SDs are shown in parentheses. N.D., Not detected.

3.3 Acetate conversion pathway analysis

Acetate conversion pathways of methanogenic consortia in acetate-fed chemostats at dilution rates of 0.025 d^{-1} and 0.6 d^{-1} were investigated using ^{13}C-labeled acetates followed by GC-MS analysis of CH_4 and CO_2 produced (Table 2). Non-aceticlastic syntrophic oxidation by acetate oxidizing syntrophs and hydrogenotrophic methanogens was suggested to occupy a primary pathway (approximately 62-90%) in total methanogenesis at the low dilution rate. In contrast, aceticlastic cleavage of acetate by aceticlastic methanogens was suggested to occupy a primary pathway (approximately 95-99%) in total methanogenesis at the high dilution rate. Phylogenetic analyses of transcripts of the methyl coenzyme M reductase gene (*mcrA*) confirmed a significant number of transcripts of the genera *Methanoculleus*, hydrogenotrophic methanogen, and *Methanosarcina*, aceticlastic metha-nogen, were present in the chemostats at the low and high dilution rates, respectively. The *mcrA* transcripts of the genus *Methanosaeta*, aceticlastic methanogen, which dominated the population, were poorly detected at the both dilution rate due to the limited diversity coverage of the primers used. These results demonstrated that dilution rate could cause a shift in the primary pathway of acetate conversion to methane in acetate-fed chemostats.

Table 2 GC-MS analysis of CH_4 produced from ^{13}C-labeled acetate.

Dilution rate (d^{-1})	Substrate	Peak intensities		m/z 17(^{13}CH$_4$)	CH_4 from carboxyl base / total CH_4	CH_4 from methyl base / total CH_4
		m/z 15 (^{12}CH$_4$)				
		Actual	Background subtracted[a]	(actual)		
0.025	^{13}CH$_3$^{12}COONa	2,695[b]	2,695	5,474	0.33	0.67
	^{12}CH$_3$^{13}COONa	1,139	1,139	928	0.45	0.55
	^{13}CH$_3$^{13}COONa	0		1,361		
0.6	^{13}CH$_3$^{12}COONa	81,973	14,066	614,114	0.022	0.98
	^{12}CH$_3$^{13}COONa	636,401	562,089	13,860	0.024	0.98
	^{13}CH$_3$^{13}COONa	74,312		588,433		

[a] Peak intensities at m/z values of 15 from ^{13}CH$_3$^{13}COONa were regarded as the background.
[b] All values are the average of duplicated experiments.

4   Discussion

The specific growth rates of the mesophilic acetate-oxdizing syntroph *C. ultunense* cocultured with a hydrogenotrophic methanogen, *Methanosaeta* and *Methanosarcina* using acetate as a substrate were reported to be 0.027–0.035 d^{-1}, 0.24–0.28 d^{-1} and 0.98 d^{-1}, respectively. At the high dilution rate of 0.6 d^{-1}, *Methanosarcina* would be predominant in the chemostat and engaged in aceticlastic cleavage of acetate to methane as the primary pathway. On the other hand, the low dilution rate of 0.025 d^{-1} is sufficiently low for growth of the three acetate-utilizing members. In this case, competition between the substrate affinities of the three members would be decisive for dominance. The apparent K_m value for acetate of an acetate-oxidizing syntroph, *Methanosaeta* and *Methanosarcina* were reported to be 0.65 mM, 0.8–0.9 mM and 3–5 mM, respectively. The acetate-oxidizing syntroph associated with *Methanoculleus* is better adapted to convert acetate primarily in the chemostat at a low dilution rate. The genus *Methanosaeta*, which showed the highest population among the three members, would play a secondary role for acetate conversion by aceticlastic cleavage in the chemostat at the low dilution rate. The dominance of bacteria belonging to the phylum *Firmicutes*, with which *C. ultunense* and *T. phaeum* are affiliated, were shown by 16S rRNA gene clonal sequence analysis. Some members of this phylum may contribute to the syntrophic acetate oxidation in the chemostat at the low dilution rate.

References

[1]   J. G. Ferry, in: Methanogenesis-ecology, physiology, biochemistry & genetics, (Chapman & Hall, New York, 1993), pp. 304-334.

[2]   S. Hattori, Y. Kamagata, S. Hanada, and H. Shoun, Int. J. Syst. Evol. Microbiol. **50,** 1601 (2000).

[3]   A. Schnürer, B. Schink, and B. H. Svensson, Int. J. Syst. Bacteriol. **46,** 1145 (1996).

[4]   K. Kida, T. Shigematsu, J. Kijima, M. Numaguchi, Y. Mochinaga, N. Abe, and S. Morimura, J. Biosci. Bioeng. **91,** 590 (2001).

[5]   T. Shigematsu, Y. Tang, H. Kawaguchi, K. Ninomiya, J. Kijima, T. Kobayashi, S. Morimura, and K. Kida, J. Biosci. Bioeng. **96,** 547 (2003).

[6]   T. Shigematsu, Y. Tang, T. Kobayashi, H. Kawaguchi, S. Morimura, and K. Kida, Appl. Environ. Microbiol. **70,** 4048 (2004).

Microbiota associated with *Posidonia oceanica* in Western Mediterranean sea

Ester Marco-Noales[*,1], Mónica Ordax[1], Armando Delgado[1], María M. López[1], María José Saavedra[2], Antonio Martínez-Murcia[2], Neus Garcias[3], Núria Marbà[3], Carlos M. Duarte[3]

[1] Departamento de Protección Vegetal y Biotecnología, Instituto Valenciano de Investigaciones Agrarias (IVIA), Carretera Moncada-Náquera km 5, 46113 Moncada (Valencia), SPAIN
[2] Molecular Diagnostics Center –MDC–, Apdo. 169, Orihuela (Alicante), SPAIN
[3] Grupo de Oceanografía Interdisciplinar, Instituto Mediterráneo de Estudios Avanzados, IMEDEA (CSIC-UIB), C/ Miquel Marqués, 21, 07190 Esporles, Islas Baleares, SPAIN

*Corresponding autor: e-mail: ester.marco@ivia.es, Phone: +34 96 3424000, Fax: +34 96 3424001

Culturable microbiota associated with the seagrass *Posidonia oceanica* from the Western Mediterranean coast was identified and characterized, in order to determine if certain bacterial species can be related and/or contribute to health status of the meadows. The most abundant genera recovered from rhizome, roots and leaves, according to 16S rDNA sequencing, were *Vibrio* spp., *Pseudoalteromonas* spp., and *Marinomonas* spp. A preliminary screening of selected strains for a phytopathogenicity trait, the hypersensitive response on tobbaco leaves, was not successful. The most remarkable finding was that isolates of *Pseudoalteromonas* spp. were the most abundant ones in the meadows with the highest seagrass mortality rate, which suggests a putative contribution of these bacteria to the meadow decline.

Keywords *Posidonia oceanica* meadows; seagrass decline; seagrass mortality rate; microbiota; *Pseudoalteromonas* spp.; *Vibrio* spp.; *Marinomonas* spp.

1   Introduction

Posidonia oceanica (L.) Delile is a clonal, slow-growing marine angiosperm, developing extensive, highly productive and millenary submarine meadows across the Mediterranean coast [1, 2]. *P. oceanica* plays an important ecological role in the marine environment through the high primary production of the ecosystems it forms, where a varied fauna and microbiota have their habitat. Moreover, *P. oceanica* meadows constitute a barrier for preventing coastal erosion. There is evidence that *P. oceanica* meadows are declining throughout the Mediterranean [3, 4], which is believed to derive from a complex combination of both physical and chemical factors, consequence from human activities and natural events, not all of them fully understood. In addition, different pathogens could also contribute to the observed seagrass decline. Despite ecological aspects related to bacterial activity and production in the meadows are receiving increasing attention [5, 6], very little attempts have been made to identify and characterize the bacterial microbiota associated with *P. oceanica*.

In order to investigate the possible relationships between bacterial communities and the health status of this seagrass species, here we report a preliminar characterization of bacterial microbiota associated with *P. oceanica* and the search of possible bacterial pathogens.

2   Materials and methods

2.1 Study site

P.oceanica samples were collected in March 2004 at depths of 49 m in four seagrass meadows in Mallorca, Balearic Islands (Spain): Illetes, Porto Colom, Magalluf and Pollença. Shoot demographic estimates (i.e. shoot mortality rate, shoot recruitment rate, net population growth rate) were assessed through annual direct shoot census in 3 permanent plots, of an area varying between 0.12 m^2 and 0.25 m^2 as to include at least 100 shoots plot^{-1}. The plots were installed between summer and fall 2001 and all shoots present inside them were counted and tagged. One or 1.5 years later, depending on the meadow, the plots were re-visited, and surviving shoots (i.e. those with a tag) and young (i.e. those not tagged) were counted. Shoot mortality rate (M; in yr^{-1}) was calculated as,

$$M = \frac{(\ln(NT_0 / NS_1)) \cdot 365}{t_1 - t_0}$$

being NT_0 the number of marked shoots at the start of the study (t_0, days) at each plot and NS_1 the number of marked shoots that survived at t_1 (days).

2.2 Laboratory analyses

2.2.1 Sampling

Leaves (with anomalous traits or not), roots and internal part of rhizome were sampled by comminuting them in small pieces in Artificial Sea Water (ASW) [7]. Then, extracts were incubated approximately 2 h with shaking at 25°C.

2.2.2 Count, isolation, and purification of bacterial strains

After incubation, culturable cells were counted on Marine Agar (Difco, USA) (MA) plates. Colonies with different morphology were isolated, purified on MA, and conserved at − 80°C, generating a broad collection of isolates.

2.2.3 Hypersensitive reaction

Tobacco plants (*Nicotiana tabacum* L.) were inoculated on the leaves with bacterial suspensions (10^9 cfu/ml) of selected isolates according to Klement and Goodman [9] to observe the induction of the hypersensitive reaction (HR), as a trait indicative of the potential pathogenicity of the strains. The hypersensitive response was observed daily for one week.

2.2.4 Identification

A selection of isolates was tentatively identified by three methods: Fatty Acid Methyl Ester (FAME) profiles, Biolog carbon substrate utilization patterns and ribosomal 16S DNA (rDNA) sequencing. For FAME analysis a HP-5890 chromatograph was used; the carrier gas was hydrogen, and the oven temperature was ramped from 170 to 270°C at a rate of 5°C per min. The software employed for comparison of peaks was Sherlock System 3.10 (MIDI, Inc.). Amplification of the rDNA 16S of the whole collection was performed as described by Martínez-Murcia *et al.* [10]. The DNA sequence was determined by direct sequencing of the PCR product on an ABI 3100 Avant sequencer (Applied Biosystems).

3 Results

3.1 Study site

Demographic data from the four meadows sampled are presented in Table 1. The meadows with the highest mortality rates were Pollença and Porto Colom.

Table 1 Demographic data of the sampled *Posidonia oceanica* meadows from Mallorca, Balearic Islands (Spain).

Meadow	Depth (m)	Average mortality rate (y^{-1})
Porto Colom	6,4	0,17
Pollença	4	0,19
Illetes	9	0,11
Magalluf	6	0,12

3.2 Bacterial populations

Counts of culturable bacteria ranged approximately between 10^3 and 10^6 cfu/g of rhizome, 10^6-10^7 cfu/g of root, 10^5-10^6 cfu/g of leaf, and 10^5-10^8 cfu/g of spotted leaf. These values were similar for all the meadows. Regardless of the sampled plant material and meadow, 3 to 6 different colonial morphotypes were observed. A collection of 104 isolates was generated, sixteen of them agarolytic.

3.3 Identification of bacterial isolates

3.3.1 Comparison of techniques

No coincidence at species level was found between FAME or Biolog and 16S rDNA sequencing, and only 50% of matchs were reached at genus level. Consequently, 16S sequencing was chosen to identify the whole collection of isolates.

3.3.2 Identification by 16S rDNA sequencing

Isolates belonged mostly to the genera *Vibrio, Pseudoalteromonas, Marinomonas*, and *Halomonas*, the percentage of each one depending on the meadow sampled (Fig. 1). *Marinomonas* spp. and *Halomonas* spp. appeared mainly associated with rhizome and roots, whereas *Vibrio* spp. and *Pseudoalteromonas* spp. were equally distributed in the belowground and aboveground plant parts. Isolates corresponding to other genera came from rhizome. The abundance of *Pseudoalteromonas* spp. was greater in the meadows with the highest mortality rate, Porto Colom and Pollença. It is remarkable that the sixteen agarolytic isolates belonged to this genus.

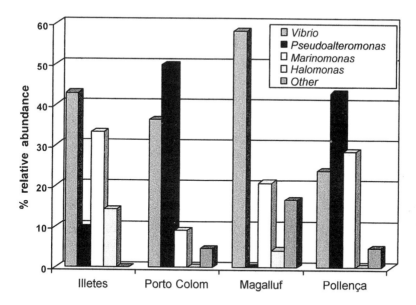

Fig. 1 Relative abundance of the main bacterial genera associated with *Posidonia oceanica* in four meadows from Balearic Islands (Spain).

3.4 Hypersensitive reaction

A total of 35 isolates, belonging to different genera, were assayed for hypersensitivity in tobacco leaves and all of them were negative. Negative control of ASW was absolutely harmless to the plant.

4 Discussion

This study provides information on the microbiota associated with *P. oceanica* in four Western Mediterranean meadows with different mortality rates. The previously studied abiotic stresses do not seem to explain completely the observed decline in the meadows.
The identification of isolates recovered from rhizome, roots, and leaves of *P. oceanica* samples was performed by 16S rDNA sequencing, as FAME and Biolog techniques were not suitable for marine bacteria. Two handicaps make difficult the identification of this kind of bacteria: firstly, marine bacteria do not grow well on the standard media stated by the manufacturers for a reliable identification; and secondly, databases are made mainly from clinical isolates, lacking most of the marine bacterial species. The most abundant genera recovered in the four sampled meadows were *Vibrio* spp., *Pseudoalteromonas* spp., and *Marinomonas* spp. The relative percentage of each genus was dependent on the meadow. The abundance of isolates belonging to *Pseudoalteromonas* spp. in the more damaged meadows suggests a putative role of this bacterial species in the decline of the meadows.
A widespread distribution of the genus *Pseudoalteromonas* across several marine habitats (seawater, rocks, macroalgae, and animals) has been hypothesized recently [12].This genus

includes some species pathogenic for fish and algae and others less virulent and even beneficial to their natural hosts [13, 14]. Thus, it has been associated with many hosts, such as fish [14], algae [13], corals [15], molluscs [13, 16], sponges [17] and larvae of several invertebrate [18]. *Pseudoalteromonas* spp. are also known to produce a variety of extracellular compounds with biological activity [14, 19]. Although in the present work it has not been possible to associate any of the anomalous traits observed in leaves of some *P. oceanica* samples with microbial infection, the mixture of microorganisms present inside chlorotic regions resembled biofouling communities common on surfaces in marine environment [20, 21]. Maybe some enzymes excreted by *Pseudoalteromonas* spp. cells could cause lesions on *P. oceanica* leaves, followed by their colonization by different bacteria, diatoms, invertebrate. Moreover, the effect of *Pseudoalteromonas* spp. cells could be enhanced in *P. oceanica* meadows as these ecosystems are subjected to different physical and chemical stress factors (e.g water temperature, organic matter and sulfide sediment concentrations, sediment anoxia, [22].

The hypersensitive response on tobacco leaves was evaluated as a trait to test phytopathogenicity. All the assayed isolates resulted negative, which does not demonstrate that strains are non pathogenic, because not all the phytopathogenic bacteria cause an hypersensitive response [23]. In addition, it is possible that the tobacco plant used routinely is not adequate for marine bacteria assays or that HR test is not the most appropriate one in this case. Consequently, approachs to test pathogenic traits of the bacterial isolates on *P. oceanica* seagrasses are under study.

In conclusion, this is the first report on characterization of bacterial microbiota associated with *P. oceanica* meadows in Western Mediterranean sea. Further studies are needed to know if some bacterial species have a role in the decline of these important marine ecosystems, which are currently under work in our laboratory.

Acnowledgements

This work has been partially supported by the projects Cabrera Project and Life-Posidonia (LIFE 2000/NAT/E/7303) funded by the Spanish Ministery of Environment and the European Commission, respectively. Authors thanks F. García from Laboratorio de Sanidad Vegetal de Barcelona (Generalitat de Cataluña); C.I. Salcedo, C. Morente, and J. Peñalver from IVIA; V. Perea from MDC; and Servicio de Microscopía de la Universidad de Valencia, for technical assistance.

References

[1]   M. Canals, and E. Ballesteros, Deep-Sea Res. Part II-Top. Stud. Oceanogr. **44**, 611 (1997).

[2]   C.M. Duarte, and C.L. Chiscano, Aquat. Bot. **65**, 159 (1999).

[3]   N. Marbá, C.M. Duarte, J. Cebrián, S. Enríquez, M.E. Gallegos, B. Olesen et al., Mar. Ecol. Prog. Ser. **137**, 203 (1996).

[4]   N. Marbà, C.M. Duarte, E. Díaz-Almela, J. Terrados, E. Álvarez, R. Martínez et al., Estuaries **28**, 51 (2005).

[5]   R. Danovaro, and M. Fabiano, Aquat. Microb. Ecol. **9**, 17 (1995).

[6]   N. López, and C.M. Duarte, J. Sea Res. **51**, 11 (2004).

[7]   P.W. Wolf, and J.D. Oliver, FEMS Microb. Ecol. **101**, 33 (1992).

[8]   E. Marco-Noales, E.G. Biosca, and C. Amaro, Appl. Environ. Microbiol. **65**, 1117 (1999).

[9]   Z. Klement, and R.N. Goodman, Annu. Rev. Phytopath. **5**, 17 (1967).

[10]   *A. J. Martinez-Murcia, A.I. Antón, and F. Rodríguez-Valera, Int. J. Syst. Bacteriol.* **49**,*1409*

(1999).

[11]   S. Kumar, K. Tamura, I.B. Jakobsen, and M. Nei, Bioinformatics **50**, 602 (2001).

[12]   T.L. Skovhus, N.B. Ramsing, C. Hölstrom, S. Kjelleberg, and I. Dahllöf, Appl. Environ. Microbiol. **70**, 2373 (2004).

[13]   T. Sawabe, H. Makino, M. Tatsumi, K. Nakano, K. Tajima, M.M. Iqba et al., Int. J. Syst. Bacteriol. **48**, 769 (1998).

[14]   C. Hölstrom, and S. Kjelleberg, FEMS Microbiol. Ecol. **30**, 285 (1999).

[15]   A.P. Negri, N.S. Webster, R.T. Hill, and A.J. Heyward, Mar. Ecol. Prog. Ser. **223**, 121 (2001).

[16]   E.P. Ivanova, T. Sawabe, A.M. Lysenko, N.M. Gorshkova, V.I. Svetashev, D.V. Nicolau et al., Int. J. Syst. Bacteriol. Evol. Microbiol. **52**, 235 (2002).

[17]   E.P. Ivanova, L.S. Shevchenko, T. Sawabe, A.M. Lysenko, V.I. Svetashev, N.M. Gorshkova, Int. J. Syst. Bacteriol. Evol. Microbiol. **52**, 263 (2002).

[18]   C. Hölstrom, D. Rittschoff, and S. Kjelleberg, Appl. Environ. Microbiol. **58**, 2111 (1992).

[19]   S.O. Lee, J. Kato, N. Takiguchi, A. Kuroda, T. Ikeda, A. Mitsutani, and H. Ohtake, Appl. Environ. Microbiol. **66**, 4334 (2000).

[20]   J.R. Henschel, and P.A. Cook, Biofouling **2**, 1 (1990).

[21]   M. Wahl, Mar. Ecol. Prog. Ser. **58**, 175 (1989).

[22]   M. Holmer, C M. Duarte, and N. Marbà . Mar. Biogeochem. **66**, 223 (2003).

[23]   C. Noval, Manual de Laboratorio. Diagnóstico de hongos, bacterias y nemátodos fitopatógenos, chapter **10**, 137 (1991).

Novel methodologies for the detection and classification of cultured and uncultured microorganisms from cultural heritage samples

J.M. Gonzalez[*,1], and C. Saiz-Jimenez[1]

[1]   Instituto de Recursos Naturales, CSIC, Apartado 1052, 41080 Sevilla, Spain

[*]Corresponding author: e-mail: jmgrau@irnase.csic.es, Phone: +34 954 624711, Fax: +34 954 624002

Microorganisms are important biodeterioration agents of cultural heritage and need to be studied in order to better preserve valuable art work. The detection and classification of microorganisms represent a preliminary step towards understanding their development and control as well as for the monitoring of these microbial communities. Methods based on cultures of microbes were the first ones available. Recently, the use of culture-independent, molecular methods is providing with unexpected information on the existence of a high microbial diversity on cultural assets. At present, method development is improving our knowledge, and facilitate the analysis, of microbial communities thriving on cultural objects, their potential biodeteriorating effect and possibilities of control. This report presents recent novel methodologies from our laboratory for the detection and identification of uncultured microbial communities and for the classification of cultured microorganisms.

Keywords , PCR, DNA amplification, MDA-PCR, GC mol%, DNA-DNA relatedness, library screening, chimera, microbial community, biodiversity, DNA, RNA, cultural heritage

1   Introduction

Microorganisms are known for their negative effects on cultural heritage and the society urge protective measurements to be arranged for the conservation of these art objects. As a consequence, the microorganisms present in colonizing microbial communities need to be identified. Classic methods for the study of microorganisms required the culture of microorganisms on supossely adequated media and conditions. Today, it is known that cultured microorganisms from any given sample represent a minimal fraction of the total microbial community (at most 1% [1, 2]). Recently, the introduction of molecular techniques to the analysis of microbial communities is providing with information on the presence of a high microbial diversity on cultural heritage.

Molecular techniques allow the detection and unambiguous identification of microorganisms and analysis of complex microbial communities without the need for any culturing or incubation. Thus, bias involved in using specific media compositions and conditions is not required any more. However, culturing methods provide with essential information on the physiology and metabolism of the microorganisms that escapes the possibilities of simple molecular survey of microbial communities. The complementarity of culturing and molecular methods must be emphasized in order to gain a completely understand the potential role of microorganisms on the conservation of cultural heritage.

The objective of this report is to show the availability of recent improvements of molecular protocols originally designed for cultural heritage studies although they can be applied to any microbiological field. Some of these methods have been designed for the analysis of uncultured microbial communities and their diversity, and some others planned to facilitate the identification and description of cultured microorganisms.

2 Molecular surveys of microbial communities

Molecular surveys of complex microbial communities aim to gather information on the diversity of microorganisms present in a given sample including the identification based on molecular methods of the microorganisms constituting the studied community. The procedure involves the extraction of DNA from the sample, amplification of gene-specific sequences by PCR, DNA library construction and screening of the obtained clones, followed by their sequencing. The last steps of the procedure involve the bioinformatic analysis of these sequences to identify the detected microorganisms from their sequences by comparison to DNA databases. Although the protocol is well established, novel improvements help to perform more efficient processing and to obtain new information on unknown questions.

In the following paragraphs, we will present a novel amplification procedure, an original screening strategy for complex DNA libraries, an initiative to analyze the metabolically active microbial communities, and some sequence analysis tools.

2.1 MDA-PCR

Cultural heritage samples often present relatively high amount of PCR inhibitors, such as humic acids. As well, the tyical minute samples (below 1 mg) collected during cultural heritage studies contain poor quantities of DNA. In this scenario, PCR amplification reactions frequently do not lead to successful PCR products. In these cases, alternative procedures are required. We proposed a novel, two steps amplification procedure to obtain successful amplification from those samples were standard PCR reaction do not provide results. This procedure has been described in detail [3].

The first step consists on a Multiple Displacement Amplification (MDA) reaction. This is a whole-genome amplification (a non specific amplification) carried out with random hexamers as primers, the F29 DNA polymerase (a thermolabile enzyme), and an isothermal reaction (30°C for 16h). The reagents and enzyme for this procedure is commertialized (GenomiPhi, Amersham). This reaction amplifies any DNA present in the sample. The objective of this step is to increase the amount of DNA in the sample or increase the relative amount of DNA with respect to the concentration of PCR inhibitors. The DNA product from this first step is to be used as DNA template in the second step. The second step of the proposed procedure is just a standard PCR reaction using gene-specific primers (in our case, 16S rRNA gene specific primers), a thermostable DNA polymerase, and a thermal cycling conditions composed of denaturing, annealing and extension steps.

The result of the procedure is that successful DNA amplifications can be obtained from those samples when standard PCR did not work. We have proved that the proposed MDA-PCR protocol can provide with successful amplifications at concentrations of DNA template above 10-fold the concentration of DNA template required by standard PCR reactions. As well, the MDA-PCR procedure can lead to successful amplifications in samples with a concentration of PCR inhibitors (i.e., humic acids) even above 10-fold the concentration of these substances tolerated by standard PCR. These satisfactory results increase the window of possibilities available for the successful complexion of our experimental molecular surveys of microbial communities in complex samples. Cultural heritage studies usually involve the process of difficult samples and generally they are so tiny samples that established procedures might not function properly on many occasions.

2.2 Original DNA library screening strategy

PCR products are generally cloned into DNA libraries in order to separate the different sequences amplified from distict microorganisms and to select specific clones of interest. Let´s imagine that by comparing several microbial community fingerprints from different samples taken over time, we conclude that one of the bands visualized during the analysis, for instance by DGGE (Denaturing Gradient Gel Electrophoresis), is characteristic of the studied biodeteriorating process. After preparing DNA libraries from a sample generating that band, we will have to select the clone containing the DNA fragment of interest. The procedure to follow in order to pick up the desired clone implies the screening through every clone in the library, generally well above 100 clones [4].

The procedure we propose is to perform the screening process in two steps [4]. First, we would group the clones into sets. Each set could consist of about ten clones. We analyze these sets or groups for the presence of the band of interest by PCR amplification and DGGE analysis. This will let us detect groups of clones containing the desired clone. So, the second step of the procedure will consist on the processing individually of the clones included in the selected group and detected which clone in that group contains the desired DNA fragment. This strategy resolves the objective of selecting a clone from a complex DNA library saving as much as 90% of time, costs and labor, allowing a more efficient process of cultural heritage samples, as well as samples from any other origin.

2.3 Sequence analysis and chimera detection

Once selected clones have been sequenced, the raw sequences need to be edited, analyzed for the possible existence of chimeras, and carry out homology searches for the identification of the microorganism corresponding to each sequence.

We perform chimera evaluations using our own algorithm [5] which in our opinion is more efficient in evaluating chimeras than previously published procedures. The computer program has been implemented , Ccode, and is freely available from http://www.irnase.csic.es/users/jmgrau/index.html. Those sequences that pass the test go on to a homology search analysis carried out by the Blast algorithm [6] online at the NCBI (National Center of Biotechnology Information; http://www.ncbi.nlm.nih.gov/Blast). We have also developed batch submission programs at our bioinformatics laboratory than can be used to automatically retrieve homology search results from any number of sequences.

2.4 DNA- vs. RNA-based molecular surveys of microbial communities

Recently, we have started the comparison of DNA- and RNA-based molecular surveys of microbial communities. The major advantage of this procedure is to be able to obtain information on the total (DNA-based) and the metabolically active (RNA-based) microbial communities. The amount of RNA in the cells is proportional to their metabolic activity [7], thus RNA-based analysis permit the detection of those microorganisms metabolically active in the samples under study. DNA-based analyses are very useful since they detect the total microbial community present in the samples. The comparison is very useful in cultural heritage and environmental studies since we can study which microorganisms are present and which ones are also metabolically active in the studied environment. The presence of a microorganism does not necessarily means it is actively participating in a biodeteriorating process. However, an inactive (showing undetectable activity) microorganism could represent a potential risk if the environmental conditions change becoming adequate for the development of that specific microorganism.

3   Characterization of cultured microorganisms

The characterization of novel isolates or cultured microorganisms is a requirement for the understanding their function and potential relationships with other species. As well, validly describing a microbial species allow future studies on that microorganism and increase our database of knowledge on the capabilities, habilities or risks for that specific organism. In order to fully characterize a microorganism, one must perform a series of analysis including both phenotypical and genotypical tests. Besides determining the use of carbon and energy sources and growth conditions, the 16S ribosomal RNA sequence for that microorganism might be obtained, as well as two estimates: the G+C mol% content and DNA-DNA relatedness between the microorganism under analysis and its closest relatives. These tests will allow to fit the studied microorganism within a microbial species or might define if the tested microorganism represents a novel, previously undescribed species. We have recently proposed two novel methods to estimate the G+C mol% content of a microorganism [8] and easily perform DNA-DNA relatedness analysis [9].

The proposed methods are both based on melting curves or denaturating curves and the use of a fluorescent dye, SYBR Green I. This fluorochrome attaches to double-stranded DNA emitting fluorescence. This fluorescence is practically null in the absence of double-stranded DNA. Consequently, at increasing temperatures the DNA will start denaturing and fluorescence emission will sharply decrease. Fluorescence measurements and temperature increases over the experiment cn be simultaneously carried out with an optical PCR thermocycler which, today, is available in most molecular biology and microbiology laboratories.

G+C mol% content determinations require the construction of a preliminary calibrarion curve. The equation for this line is obtained by calculating with the proposed protocol the melting temperatures for the DNA of several species of known G+C mol% content. We recommend the use of microbial species whose genome is completly sequenced [8]. This will allow a precise determination of the G+C mol% using bioinformatic methods and we will be free of estimating errors depending on the method used for the calibration.

DNA-DNA relatedness analyses are a required procedure for species delineation and classification. The difference in melting temperatures between DNA from a single cultured microorganism or species (homologous DNA) and the mixed DNA from that microorganism and a closely related species (heterologous DNA) will be used to differentiate microorganisms. A difference of 5°C or more will mean that we the compared DNAs belong to different species while differences below that level will imply that the microorganisms belong to the same species [9]. The method could be compared to other methods for the group of microorganisms under study providing a warranty for unambiguous results.

4   Remarks

The design of novel procedures, protocols and strategies are welcome since they serve to improve the analysis of microorganisms from cultural heritage samples and/or provide with additional knowledge about their role in the microbial communities under study. Cultural heritage interests are the understanding of microbial communities developing, colonizing, or threatening art objects. Any initiative facilitating or collaborating to better understand these communities is of great importance in the field. The ultimate goal is to be able to know enough about these microorganisms and microbial communities so that efficient and solid conservation programs can be designed. Microbiology and molecular biology can provide with the means, methodology and expertise required for that objective.

Acknowledgements

J.M.G. acknowledges support from the "Ramon y Cajal" programme, Spanish Ministry of Education and Science (MEC). This work was funded through a MEC project BTE2002-04492-C02-01.

References

[1]   D. M. Ward, R. Weller, M. M. Bateson, *Nature* **345**: 63 (1990).
[2]   J. M. Gonzalez and C. Saiz-Jimenez, *J. Sep. Science* **27**: 174 (2004).
[3]   J. M. Gonzalez, M.C. Portillo, and C. Saiz-Jimenez, *Environ. Microbiol.* (in press) (2005).
[4]   J. M. Gonzalez et al., *J. Microbiol. Methods* **55**: 459 (2003).
[5]   J. M. Gonzalez, J. Zimmermann, and C. Saiz-Jimenez, *Bioinformatics* **21**: 333 (2005).
[6]   S. F. Altschul et al., *J. Biol. Mol.* **215**: 403 (1990).
[7]   S. Molin, and M. Givskov, *Environ. Microbiol.* **1**: 383 (1999).
[8]   J. M. Gonzalez, and C. Saiz-Jimenez, *Environ. Microbiol.* **4**: 770 (2002).
[9]   J. M. Gonzalez, C. Saiz-Jimenez, *Extremophiles* (2004).

Perylene toxicity in the estuarine environment of Ria de Aveiro (Portugal)

Ângela Cunha[1], Adelaide Almeida[1], Ana Ré[1], Aida Martins[2] and Fernanda Alcântara[*,1]

1 Departamento de Biologia, Universidade de Aveiro, 3810-196 Aveiro, PORTUGAL
2 Instituto Nacional Engenharia Tecnologia Industrial, Estrada Lumiar 1649-038 Lisboa, PORTUGAL

*Corresponding author: e-mail: falcantara@bio.ua.pt

This article is focused on the toxic effects of perylene in the estuarine benthos. We pursued three objectives: (1) to register the effects of a high load of perylene on the bacteriobenthos and on the amphipod *Corophium multisetosum*, (2) to determine the effects of a small load of perylene ($2~\mu gL^{-1}$) on the activity of the suspended bacteriobenthos as well as (3) the response of the particle-free bacterioplankton. No acute effects of perylene were detected in the amphipod. However, the chronic toxicity assay revealed statistically significant negative effects on survival, growth and number of pregnant females. No apparent effect on the evolution of the bacteriobenthos was registered under a heavy perylene load but it is speculated that toxic effects may have been veiled. However, the doped suspension of the bacteriobenthos and the bacterioplankton showed altered profiles of activity in relation to the control, suggesting the evolution of a different bacterial community under the pressure of perylene.

Keywords Toxicity, perylene, *Corophium multisetosum*, bacteriobenthos, estuary

1 Introduction

Perylene is considered the predominant PAH in estuarine sediments [1] where microorganisms have prevalent roles on their degradation. PAH in sediments or in the water column have been found to elicit clear and negative effects on aquatic life [2]. It is, however, still unclear the impact of PAH on natural bacterial communities. Low PAH solubility and high sorption onto particulate organic matter was reported [2] to affect negatively PAH biodegradability and bacterial toxicity in contaminated areas. We focused on the effects of perylene in the soft sediment of a shallow estuary historically exposed to modest perylene loads. The extent of the responses induced by sudden inputs of perylene on the metabolic activity of benthic bacteria and the effects of tidal resuspension are studied in this context. The toxic effects of perylene at a higher level of the benthic community were tested through the acute and chronic toxicity responses of the locally common amphipod *Corophium multisetosum*.

2 Methods

Microcosm experiments: Two sets of 20 microcosms were prepared with structured sediment of a bare intertidal silty station in Ria de Aveiro and covered with a 3-cm layer of estuarine water. One set was kept as the control and the other was amended with 122mg perylene dissolved in 2mL acetone to give the average final concentration of 110 µg perylene.mL^{-1}. All microcosms were incubated at 20°C with aeration and 12-hour illumination per day. PAH in structured sediments were determined spectrofluorimetrically. Sediment suspension experiments: 1:500 (V/V) sediment suspensions were prepared with 1µm-filtered estuarine water and incubated at 20°C under agitation. A cetonic perylene solution was previously dispersed on the walls of one of the incubation vessels to reach the perylene saturation value in water of 2 µg L^{-1}. PAH analysis. The analysis of perylene in sediment suspensions followed the EPA methods [3, 4].

Toxicity tests: Acute and chronic toxicity assays were performed on doped sediments keeping non-doped microcosms as the control. Acute toxicity was determined according to the 10-day survival bioassay [5]. The biological endpoint for these sediments was survival (%). The results obtained were statistically compared with those of the control. Chronic toxicity was tested using the full life cycle (28 days).

Bacterial variables: Total bacterial number was determined by epifluorescence microscopy [6]. Bacterial biomass productivity (BBP) was calculated through incorporation of tritiated leucine [7]. Ectoenzymatic activity (EEA), including leucine-aminopeptidase (Leu-MCA) and β-glucosidase (GLU), was studied fluorimetrically [8]. The rate of O_2 consumption in sediment suspensions was determined with readings of dissolved oxygen against time (Strathkelvin 6-channel O_2-Measurement System) in 4 replicate respiration vials.

3   Results

Structured sediments: The collected sediments were anoxic silts with 4.1% organic matter. This sediment showed two peaks in the range of 400-450 nm, corresponding to 5-ring PAHs. Perylene-equivalent levels at collection time averaged 0.28±0.21 and 0.49±0.40 mgkg^{-1}sed in the surface- and near-surface layers, respectively. The PE-doped microcosms showed, at day-1, 38 and 10 mgkg^{-1} dwsed µg perylene-equivalents at the surface and sub-surface layers, respectively. After 9 days, the perylene levels were close to the initial range of values. During incubation of these sediments bacterial density increased by 53% in the control and by 55% in the test-sediment.

Sediment suspensions: Extraction recovered 94.6% of the added perylene in the PE-suspension. Perylene concentration in the control and in the PE-suspension increased with time from 0.127 to 0.387 µgL^{-1} and from 1.310 to 2.026 µgL^{-1}, respectively. The added perylene migrated rapidly to the suspension and was mostly (92%) adsorbed to the sediment particles. The relative fluctuation of bacterial activity is shown in Fig.1. The particle-free water used in the preparation of suspensions, when incubated in parallel showed to contain 60-70 % of the original bacterioplankton. BBP was negligible in this population (1–13 µgC L^{-1}h^{-1}). BBP in the control-suspension varied from 24 to 184 µgC L^{-1}h^{-1} with a peak at day-1. The PE-suspension appeared quite stable with BBP ranging from 134 to 210 µgC L^{-1}h^{-1} exhibiting the higher values at day-0 and day-14.

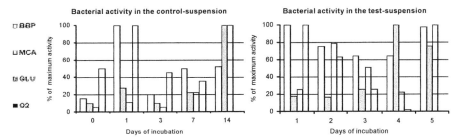

Fig. 1 Profiles of bacterial activity in sediment suspensions represented as percentages of the maximum values determined during the two-week incubation.

In the filtered bacterioplankton Leu-AMP varied from 924±0 to 1578±16 nmol $L^{-1}h^{-1}$ in the control and from 731±15 to 3046±78 nmol $L^{-1}h^{-1}$ in the test-filtrate. It increased by factors of 2.6 and 3.1 from day-0 to day-3 in the control and in the perylene-doped filtrate, respectively, but after that it dropped by factors of 3 to 4 with a considerable recover only at day-14. In the doped and control particle-free water, GLU oscillated from 6±2 to 94±4 nmol $L^{-1}h^{-1}$. This corresponded to 6-83% of the total suspension activity in the control, and to 7-25% of the total activity in the test-suspension. The sediment suspensions exhibited Leu-AMP activities in the range of 1464±303 to 15488±938 nmol $L^{-1}h^{-1}$ (control) and from 1824±541 to 11127±124 nmol $L^{-1}h^{-1}$ (PE-suspension) while GLU ranged, respectively, from 57±4 to 1278±62 nmol $L^{-1}h^{-1}$ and from 87±34 to 400±7 nmol $L^{-1}h^{-1}$.

The PE-suspension showed at day-0 the highest (1.46 µmol $O_2.L^{-1}h^{-1}$) of all rates of O_2 consumption. The parallel reading in the control was 3.5 times lower (0.41µmol $O_2.L^{-1}h^{-1}$). At day-1 the consumption increased to 0.82 µmol $O_2.L^{-1}h^{-1}$ in the control, a value similar to that in the test-suspension. Since then the profile decreased steadily. New readings at day-7 showed that O_2 consumption had almost ceased in the PE-suspension (0.02 µmol $O_2.L^{-1}h^{-1}$) while the control maintained the rate of 0.29 µmol $O_2.L^{-1}h^{-1}$, similar to those determined at day-0 and day-3.

Toxicity tests: Growth of the amphipod juveniles was similar in the non-doped test sediments and in the ordinary growth sediment (negative control). Survival of juveniles showed to be affected in one of the samples but not in the other two. The number of pregnant females per replicate showed significant differences between each of the test sediments and the control. After

Survival (%)

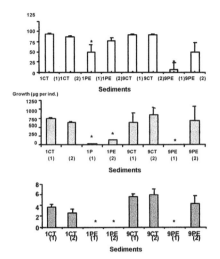

Fig. 2 *C. multisetosum* mean survival (%), mean growth, and mean pregnant females in 21-days chronic toxicity assay to test the effects of added perylene. (*) Statistical differences between the negative control and treatments (p<0,05). CT-sediment control; 1CT(1) and 9CT(1)-sediment surface (1) collected after 1 and 9 day incubation; 1CT(2) and 9CT(2) - sediment sub-surface (2) collected after 1 and 9 day incubation; 1PE(1) and 9PE(1)-sediment surface (1) doped with perylene and collected after 1 and 9 day incubation; 1PE(2) and 9PE(2)- sediment sub-surface (2) doped with perylene and collected after 1 and 9 day incubation. Results analysed by ANOVA and Tukey tests.

1 day incubation survival in acute tests was around 90% in the control sediment and 82% in PE1 (0-1 cm) but the results were not statistically different. No effects were observed after 9 days incubation. The number of pregnant females per replicate showed significant differences between each of the non-doped sediments and the negative control. Survival was mostly affected at the surface layer but growth and reproduction were impaired across the two test-sediment layers.

After 9 days, the added perylene affected negatively and significantly survival, growth and reproduction in the surface sediment (Fig. 2). Survival decreased in both sediment layers but growth and reproduction were more affected at the surface.

4 Discussion

Perylene showed to affect negatively, both at low or high concentrations, the reproduction of *C. multisetosum*. The low perylene concentration present in the native test sediments did not show to affect survival and growth of the juveniles but the added dose of perylene elicited, after 9 days, drastic chronic toxicity effects including the negative impact of this PAH in survival and growth of juveniles.

With respect to the bacteriobenthos, the high dose of perylene added to the microcosms did not change the profile of bacterial abundance at the sediment surface suggesting the absence of direct negative PAH effects. However, the doubling-time of surface sediment bacteria, determined as 1 to 7.8 days [9], was estimated in our sediments as 16-17 days in the control- and PE-sediment. On this basis we may hypothesise that the apparent growth was attenuated by bacterial predation in the control. Unless perylene was completely ineffective, predation may have declined in the PE-sediment due to toxic effects on bacterial predators. In this case we should expect a higher bacterial growth rate in the PE-suspension unless bacteria were under pressure in both microcosms, in one case due to predation and in the other through direct or indirect effects of perylene on bacterial activity. We mimicked the immediate and medium-term effects of tidal erosion on the mud flat bacteria with the sediment suspensions. Most of the added perylene was immediately sorbed to the fine sediment particles. In the first three days of incubation Leu-MCA and GLU were high in the bacterioplankton when compared to the suspensions. Later on, particle-attached activity predominated and the suspended benthic cells took over exhibiting very high levels of EEA, particularly GLU. However, the temporal profiles of MCA and GLU were different in the control and PE-suspension. The first showed two combined peaks at days 1 and 14. The doped suspension exhibited two peaks but with uncoupling of the two EEA. In this case, Leu-MCA joined BBP and both preceded GLU peaks. The addition of perylene to the suspension diminished, in relation to the control, the overall Max/Min ratios of activity. In the case of BBP the reduction was from 8 to 0.2, GLU from 22 to 5, and AMP from 11 to 6. It was not the case of O_2 consumption which increased from 3 to 73. This denotes a greater uniformity of bacterial activity in the PE-suspension regardless of the ext remely variable respiration rate.

Acknowledgements

The support of Project EICOS, Praxis XXI 2/2.1/MAR/1750/95 is gratefully acknowledged.

References

[1] H Budzinski, I. Jones, J Bellocq, C Piérard & P Garrigues. Evaluation of sediment contamination by polyaromatic hydrocarbons in the Gironde estuary. Mar. Chem. **58**:85. (1997).

[2] J O'Riordan. Ambient Water Quality Criteria for Polycyclic Aromatic Hydrocarbons (PAHs). Ministry of Water, Land and Parks, Province of British Columbia, PO Box 9340 STN Prov Govt Victoria BC V8W 9M1. (1993).

[3] EPA Method 610, Polynuclear Aromatic Hydrocarbons, U. S. Environmental Protection Agency, Washington, D.C., (1984).

[4] EPA Method 3550, Sonication Extraction, U. S. Environmental Protection Agency, Washington, D.C., (1986).

[5] ASTM. American Society for Testing and Materials. Standard guide for conducting 10-days static sediment toxicity tests with marine and estuarine amphipods. In Annual Book of ASTM Standards, Water and Environmental Technology, Vol.**11.04**, E1367-90 (1992). ASTM, Philadelphia.

[6] J Hobbie, R Daley & S Jasper Use of Nuclepore filters for counting bacteria by fluorescence microscopy. Appl. Environ. Microbiol. **33**:1225 (1977).

[7] M Simon, and F. Azam. Protein content and protein synthesis rates of planktonic bacteria. Mar. Ecol. Prog. Ser., **51**:201 (1989).

[8] H-G Hoppe. Use of fluorogenic model substrates for extracellular enzyme activity (EEA) measurement of bacteria. – In Handbook of methods in aquatic microbial ecology, P. F Kemp, P F Sherr, E B Sherr & J J Cole, (eds.) – Lewis Publishers, Boca Raton, p. 423-431 (1993).

[9] J A Novitsky. Microbial growth rates and biomass production in a marine sediment; evidence for a very active but mostly nongrowing community. Appl. Environ. Microbiol. **53(10)**:2368 (1987).

[10] L-A Meyer-Reil. Measurement of hydrolytic activity and incorporation of dissolved organic substrates by microorganisms in marine sediments. Mar. Ecol. Prog. Ser. **31**:143. (1986).

[11] J Leahy, & R R Colwell. Microbial degradation of hydrocarbons in the environment. Microbiol. Rev. **54(3)**:305 (1990).

130

Phenotypic and genotypic characterization of an outbreak of *Pseudomonas aeruginosa* from wild birds

Jiménez Gómez, P.A.[1*]

1 Sección de Microbiología. Facultad de Ciencias Experimentales y de la Salud. Universidad San Pablo C.E.U. Ctra. de Boadilla del Monte Km, 5,600. Madrid. Spain. Phone: 91 3724754. Fax: 91 3510496

*Corresponding author: e-mail: pgimgom@ceu.es

Several wild animals affected with sepsis were treated in BRINZAL (a wild life hospital). Samples were obtained from several organs of diseased animals and *Pseudomonas aeruginosa* strains were identified in such isolates. The aim of this study is to decide whether the infection was an outbreak or not. Therefore, we establish a typing method for tracing the epidemic relationship of *Pseudomonas aeruginosa* strains isolated from wild birds with sepsis.
Antimicrobial susceptibility testing, serotyping, phage-typing and pulsed field gel electrophoresis (PFGE) methodologies were applied to study the isolated strains. Another strain of *Pseudomonas aeruginosa* from a reference collection, was included as a control group to provide a basis of for the efficiency of the different markers used. All outbreak-related strains were susceptible to cotrimoxazol and resistant to phosphomycin and shared the same serotype, phage-type and PFGE pattern.

Here we describe for the first time that PFGE analysis are a valuable and useful epidemiological tool to discriminate unrelated and outbreak-related *Pseudomonas aeruginosa* strains from wild birds.

Keywords/Index Terms: PFGE, Wild birds, *Pseudomonas aeruginosa*, Phage-typing, Serotyping.

1 Significance and impact of the study

Preventing contamination and dissemination of phatogenic strains in wild animal hospitals is essential to avoid out-breaks. When outbreaks of these kind occur, it is essential to have powerful discriminative tools to identify possible out-breaks. This will enable to treat rapidly affected animals. Microbiological analyses of the hospitalary tools are essential to avoid cases like the one reported here.

2 Introduction

Pseudomonas aeruginosa is a well characterized gram negative bacteria responsible for sporadic infections in birds (**Walker et al. 2002**) and human. These bacteria produce an extra-cellular and hemolytic toxic which is the critical agent for tissue and organ damages.
These bacteria are divided in different serotypes according to antigen thermo-stability criteria and as reported before the invasiveness of these bacteria is reduced.
Pseudomonas aeruginosa is quite common an avian pathogen. Then, the bacteria are frequently associated to cold aquatic environments. It shows β-hemolytic activities in blood containing agar. Therefore, the specific tissue damage is related to hemorrhagic and necrosis.

On the other hand, *Pseudomonas aeruginosa* is highly resistant to wide spectrum antibiotics (Markarian, 1975; Schidger *et al.,* 1989) causing additional secondary infections in patients previously treated against other unrelated bacterial infections.

Regarding the prevalence of specific strains and its presence in certain animal groups, the O6 serotype is mostly identified in birds, although not exclusively.

Due to the high resistance capabilities, the bacteria are able to develop in disinfectant solutions (Walker et al., 2002). Moreover, its resistance to antibiotics (Wissing et al., 2001), allows *P. aeruginosa* to behave as a saprophyte in the epithelial mucous and in secretions of healthy animals. Occasionally, those infections are localized at the respiratory tract and as reported before, in this case the infection can lead to a 100% of mortality (Lin *et al.,* 1993).

3   Materials and methods

Strains: Six Pseudomonas aerugionsa isolates from wild birds used in this study are listed in Table 1. The control groups for the analyses of epidemiological markers are: three bacterial strains isolated from animal serum of the "BRINZAL" WILD BIRD hospital, and one strain from Pseudomonas aeruginosa collection of the Spanish National Reference Laboratory of Salmonella (SNRLS). Samples were collected in the period from December 2003 to February 2004.

3.1 Antimicrobial Susceptibility Determination

Pseudomonas strains were tested for their susceptibility to 12 antimicrobial agents by disk-diffusion assay, following the NCCLS guidelines (1997).

3.2 Serotyping

Identification of O antigens was carried out by a microtechnique with all available O antiserum. All antiserum were adsorbed with the corresponding cross-reacting antigens to remove non-specific agglutinins.

3.3 DNA extraction

Genomic DNA was extracted from cultures of Pseudomonas by a Cetylmethylammonium Bromide (CTAB) miniprep procedure as described by Murray and Thompson (19?0).

3.4 Genotyping by PFGE procedure

PFGE was performed as described by Baquar et al. (1994). Agarose-embedded genomic ONA was digested with 15 U of restriction enzyme Xba 1 overnight at 37 °C, and DNA fragments were separated by PFGE in a 1.2% 0-5 agarose gel (Conda, Spain) using the CHEF OR- 111 electrophoresis apparatus (Bio-Rad, Spain). Electrophoresis was done for 22 h with a voltage of 150 V and a linearly ramped pulse time of 1-50 s. A lambda ladder comprising 48'5 kpb concatemers (Roche, Spain) served as the MW markers for PFGE. The bands were visualized by staining with ethidium bromide.

4   Results

Six different animal species were collected in the Wild Life Hospital during December 2003 and February 2004. So far, the animals were unrelated as they were found in different locations and they did not share any common characteristic that could connect them.
After physical examinations, the animals were fed using glucose-serum via parenteral. All of them suddenly suffered acute death. This made impossible neither to follow the symptoms nor to establish a robust diagnosis of the death. Data taken from these animals are summarized in table I.

Table I. Description de los animales implicados en el brote septicémico por *P. aeruginosa*

CASE	SPECIES	LOCATION	REF. BRINZAI	CHECK-IN	DATE DEATH	OBSERVATION	AGE	WEIGHT
Br2/00	*Bubo bubo*	Fresnedillas	2921	8/12/99	12/12/99	Skin injuries, dehydratation.	Adulto	1600 g
Br3/00	*Columba palumbus*	Pozuelo	2943	31/12/99	1/1/00	Broken bones, mouth injuries.	Joven	297 g
Br5/00	*Bubo bubo*	Brunete	2958	15/01/00	16/01/00	Shooted, broken bones, inability to fly.	Joven	1622 g
Br4/00	*Tyto alba*	Extremera	2967	23/01/00	24/01/00	Depression.	Adulto	209 g
Br8/00	*Bubo bubo*	Aranjuez	2974	29/01/00	30/01/00	Weakness.	Sub-adulto	1091 g
Br9/00	*Asio flameus*	Las Azores	2999	23/02/00	6/3/00	Normal	Adulto	454 g
1/Br11	Serum 1	BRINZAL	-	-	-	-	-	-
2/Br11	Serum 2	BRINZAL	-	-	-	-	-	-
3/Br11	Serum 3	BRINZAL	-	-	-	-	-	-

The post-mortem analyses showed a massive presence of *Pseudomonas aeruginosa*. Samples from the following organs were taken from each animal: heart (all compartments), trachea, siringe, bronchial tubes, lungs, liver, gall bladder, pancreas, spleen, kidney, adrenal, sexual organs, oesophagus, gizzard, small and large intestines, cloaca, muscles, joints, thymus and tiroids.

Although the remaining organs and tissues did not show any remarkable injuries, all the observations pointed to a possible septicemia.

The phenotypic profile of each strain of *Pseudomonas aeruginosa* was obtained and further compared. All the profiles were coincident. An interesting observation was the absence of the natural resistance to cotrimoxazol in every strain tested. As this feature suggested the existence of an out-break, the cells were serotyped and also lysotyped. Again, all of them showed the same profile (table II). Finally, PFGE analyses were conducted. As shown in figure 1 all of them showed the same profile.

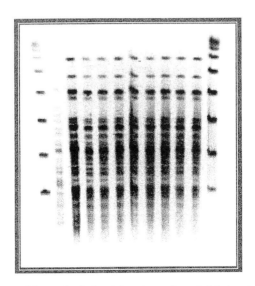

Figure 1. PFGE comparison of the strains isolated from heart. Lane 1: Marker, lane 2: *P. aeruginosa* control (not related), lane 3: Br2/00; lane4: Br3/00; lane5: Br5/00; lane6: Br4/00; lane7: Br8/00; lane8: Br9/00; lane9: 1/Br11; lane10: 2/Br11; lane11: 3/Br11; lane12: Molecular marker.

Table II. Resistance profiles of serotyping markers.

SPECIES	STRAIN	TICARCILINA	TICARCILINA/ÁCIDO CLAV	PIC/UREIDOPENICILI	CEFSULODINA	PIPER.+TAZOBACTAM	IMIPEMEN	AZTREOANM	CEFTAZIDIMA	AMIKACINA	GENTAMICINA	NETILMICINA	TOBRAMICINA	COLISTINA	CIPROFLOXACINA	FOSFOMICINA	COTRIMOXAZOL	SEROTYPE	LYSOTIPE
Ps. aeruginosa	Br2/00	S	S	S	S	S	S	S	S	S	S	S	S	S	S	R	S	O:2	21/44/F8/1214
Ps. aeruginosa	Br3/00	S	S	S	S	S	S	S	S	S	S	S	S	S	S	R	S	O:2	21/44/F8/1214
Ps. aeruginosa	Br5/00	S	S	S	S	S	S	S	S	S	S	S	S	S	S	R	S	O:2	21/44/F8/1214
Ps. aeruginosa	Br4/00	S	S	S	S	S	S	S	S	S	S	S	S	S	S	R	S	O:2	21/44/F8/1214
Ps. aeruginosa	Br8/00	S	S	S	S	S	S	S	S	S	S	S	S	S	S	R	S	O:2	21/44/F8/1214
Ps. aeruginosa	Br9/00	S	S	S	S	S	S	S	S	S	S	S	S	S	S	R	S	O:2	21/44/F8/1214
Ps. aeruginosa	1/Br11	S	S	S	S	S	S	S	S	S	S	S	S	S	S	R	S	O:2	21/44/F8/1214
Ps. aeruginosa	2/Br11	S	S	S	S	S	S	S	S	S	S	S	S	S	S	R	S	O:2	21/44/F8/1214
Ps. aeruginosa	3/Br11	S	S	S	S	S	S	S	S	S	S	S	S	S	S	R	S	O:2	21/44/F8/1214

5 Discussion

P. aeruginosa shows a high resistance to most of the widely used antibiotics. Therefore, this agent might cause secondary infections on patients treated for different bacterial infections. Analyses conducted in avian internal organs indicate the presence of *Pseudomonas* as the primary agent of the death of animals (Markarian, 1975; Schildger *et al.*, 1989). The same authors describe the high antibiotic resistance showed by those strains.

The infection can induce diarrhea, dehydration, disnea, septicemia, and finally death (Hernández, 1997). However, none of these symptomps were clearly identified in the animals affected by this out-break, only showing a slightly dehydration. This observation guided the standard applied treatment consisting in applying serum and kept the animals resting. As in previous cases, this treatment was successful for other affected animals showing the same vague symptomps, allowing a further release into nature. The sudden death of these animals made impossible to provide any antibiotherapy to treat them.

On the other hand, the infections by this bacteria often take place in young animals, which are more susceptible to be affected by other bacterial infections. *P. aeruginosa* is usually associated as secondary invader with *Corynebacterium pyogenes* in open wunds and necrotic injuries, and cardiac damage. These symptoms have been widely described in animals, however, very little information regarding birds has been provided. In the case shown here, the bacteria did not find any barrier to get introduced inside the animals which were very actually very weak. This obviously aided in the bacterial dispersion. According to the results obtained here, the isolates are from an outbreak strain originated by a catheter contamination.

The pathological injuries are always necrotic. The localized infections can be found in the respiratory tract causing local inflammation in larynx, etc. The hemorrhagic lesions are usually found in the postmortem analyses. In our isolates, the results were unclear as the fast development of the disease did not yield strong injuries identified after the death of the animals.

Histological symptoms associated to these infections include severe inflammations, sometimes localized in veins and arterial walls. In fact this bacterium usually produces hemorrhagic and necrosis wunds in the vascular system. These observations have been found in wild animals and in laboratory animals. By contrast to other septicemias, the process did not start in an individual animal. This was simultaneously detected in other independent animals without any common characteristic and always in different time frames. All together indicate that this case is an out-break (Eisenstein, 1990; Orskov y Orskov, 1983).

This is supported by the results obtained in serotype and phagotype, where all the isolates showed the same profile. Moreover, the PDGE confirmed the observations (Grothues et al., 1988; Tenover et al., 1995) as this technique is a powerful epidemiological tool to discriminate different isolates (Grothues et al., 1988; Chetouli et al., 1997).

References

[1] Baquar, N., Burnens, A. and Stanley, J. 1994b. "Comparative evaluation of molecular typing of strains fron a national epidemic de to *Salmonella brandenburg* by rRNA gene and IS200 probes and Pulsed-field Gel Electrophoresis". J. Clin. Microbiol. 32: 1876-1880.

[2] Eisenstein, B.I. 1990. "New molecular techniques for microbial epidemiology and the diagnosis of infectious diseases". J. Infect Dis. 161: 595-602.

[3]   Grothues, D., Koopman, V., Van der Horst, H. and Tümler, B. 1988. "Genome fingerpriting of *Pseudomonas aeruginosa* indicates colonization of cystic fibrosis siblings with closely related strains". J. Clin. Microbiol. 26: 1973-1977.

[4]   Hernandez, 1997 "Exploración ytécnicas laboratoriales de dignóstico usadas en aves salvajes"En Curso Teórico práctico de medicina y cirugía de aves salvajes" GREFA.

[5]   Lin, M.Y., Cheng, M.C., Huang, K.J. and Tsai, W.C. 1993. "Classification, pathogenicity, and drug susceptibility of hemolytic gram-negative bacteria isolated from sick or dead chickens". Avian Dis. 37 (1): 6-9.

[6]   Markarian, M. 1975. "*Pseudomonas aeruginosa*, the causative agent of infection in birds". Vet Med Nauki. 12 (8): 33-9.

[7]   Murray, M.G. and Thompson, W.F. 1980. "Preparation of genomic from bacteria. In: Current Protocols in Molecular Biology". Vol. 1. F.M., Ausubel, R. Brent, R.E. Kingston, D.D., Moore, J.G.

[8]   Seidman, J.A., Smith and K. Struhl. Eds. Greene Publishing Associates and Wiley-Interscience, New York.

[9]   Orskov, F. Y. and Orskov, I. 1983. "Summary of a workshop on the clone concept in the epidemiology, taxonomy, and evolution of the *Enterobacteriaceae* and other bacteria". J. Infect Dis. 148: 346-357.

[10]   Schildger, B.J., Zschock, M. and Gobel, T. 1989. "O-serovar distribution and antibiotic sensitivity of *Pseudomonas aeruginosa* strains from birds and reptiles". Zentralbl. Veterinarmed. 36 (4): 292-6.

[11]   Tenover, F.C., Arbeit, R.D., Goering, R.V., Mickelsen, P.A., Murray, B.E., Persing, D.H. and Swaminathan, B. 1995. "Interpreting chromosomal DNA restriction patterns produced by pulsed-field gel electrophoresis: Criteria for bacterial strain typing". J. Clin. Microbiol. 33: 2233-2239.

[12]   Walker SE, Sander JE, Cline JL, Helton JS. 2002. Characterization of Pseudomonas aeruginosa isolates associated with mortality in broiler chicks. Avian Dis. 46(4):1045-50.

[13]   Wissing A, Nicolet J, Boerlin P. 2001. The current antimicrobial resistance situation in Swiss veterinary medicine. Schweiz Arch Tierheilkd. Oct;143(10):503-10.

Single and multi-metal removal by an environmental mixed bacterial isolate

P. Sannasi[*,1], J. Kader, O[2]. Othman[2], S. Salmijah[2]

[1] School of Science & Mathematics, INTI College Malaysia, 71800 Bandar Baru Nilai, N. Sembilan, MALAYSIA
[2] Faculty of Science and Technology, Universiti Kebangsaan Malaysia, 43600 UKM Bangi, Selangor, MALAYSIA

[*] Corresponding author: email: palsan@intimal.edu.my, Phone: (6)06-798 2000, Fax: (6)06-799 7531

This paper reports the removal of metals from aqueous solution containing Cd(II), Cr(VI), Cu(II), Ni(II) and Pb(II) by a consortia culture (CC) comprising of environmental mixed bacterial culture. The metal removal capability of growing (active) and non-growing (inactive) cell populations at initial total metal concentrations of 100 mg/L in single and multi-metal systems were examined by determining percentage metal removal and specific metal uptakes (q_p). The removal and uptake performance of consortia culture in the multi-metal system was 23-58% lower compared to the single metal system ($p < 0.05$). The efficiency of metal removal in the single metal systems were in the order of : Pb > Cu > Ni > Cd > Cr. Growing cells displayed higher metal removal and uptake capability compared to non-growing cells for all metals tested except for Ni. For both growing and non-growing populations, the order of metal removal was Pb > Cu > Ni > Cd > Cr. TEM studies showed that metals were deposited both intracellularly and extracellularly for the two cell populations. XRF analysis confirmed the identities of the sorbed metals.

Keywords bacteria; biomass; heavy metal; metal removal; biosorption; bioaccumulation

1   Introduction

Sorption of multi-metallic ions by microbial cells is not only influenced by the physico-chemical properties of the solution but also by the surface specific properties of the biomass and types of micro organisms involved. Types of metal species and their various concentrations could lead to interference and competition at the bacterial sorption sites. Furthermore, the aspect of metal removal ability as a function of cell viability needs considerable additional study especially with respect to uptake by growing (active) and non-growing (inactive) cells. The uptake of single metal ions by various mono bacterial cultures has been much documented; relatively less is known for multi-metal removal by bacterial consortia. Heavy metals rarely exist singly in natural and wastewater streams. In nature, bacteria too, seldom exist as a single species.

2   Materials and methods

2.1   Source of bacterial biomass

The bacterial biomass was prepared from a mixed culture collectively known as consortia culture (CC) comprising of six Gram negative (*Pseudomonas* sp, *Serratia* sp, *Flavo-bacterium* sp, *Chryseomonas* sp, *Xanthomonas* sp and *Agrobacterium* sp) and three Gram positive (*Bacillus* sp, *Arthrobacter* sp and *Micrococcus* sp) environmental isolates. The isolates have been screened intensively and developed from a pool of mixed culture strains isolated from point and non-point sources in areas related to metal-based activities [1].

2.2 Media & growth conditions

Consortia culture was maintained at room temperature (28-30 °C) in basal medium containing yeast extract (0.5 g/L), peptone (0.5 g/L) and NaCl (8.5 g/L) during growth and acclimatisation periods, with fortnightly media refreshments. Growth culture ensures maximum biomass production whilst acclimatisation culture (enriched initially with 1 mg/L of each Pb(II), Cu(II) and Cr(VI) and subsequently the concentrations increased to 10 mg/L each), generates metal acclimated bacterial communities. Microbial growth was monitored by optical density (OD) at 600 nm.

2.3 Cell preparation

Aliquots from both growth and acclimatisation cultures (0.5%, v/v) were inoculated into 10 mL nutrient broth (NB), initial pH of 6.8 ± 0.2 and incubated under static conditions at room temperature for 48 h. Cell biomass was separated by centrifugation (4000 rpm, 15 min); pellets were washed, centrifuged and rinsed twice before re-suspended in either NB or deionised distilled water (ddH$_2$O). For biomass dry weight (DW) determinations, pellets were dried in an oven (45-50 °C) for 24-48 h and then heated further to 110-120 °C for 10-15 min or until they attained constant weights.

Growing cells. Tests were done in NB with an inocula size of 1%, v/v; for each inoculum, cell density was standardised at OD$_{600}$ of ≈ 0.500 (approx. 10^7 cells/mL).

Non-growing cells. Tests were done in sterile ddH$_2$O suspended with approximately 0.1g/L DW biomass.

2.4 Metals

Metals solutions were prepared from the following salts with NANOpure ddH$_2$O; Cd(II) from Cd(NO$_3$)$_2$.4H$_2$O, Cr(VI) from K$_2$CrO$_4$, Cu(II) from Cu(NO$_3$)$_2$ 2.5H$_2$O, Ni(II) from Ni(NO$_3$)$_2$.6H$_2$O and Pb(II) from Pb(NO$_3$)$_2$. Test solutions were prepared by diluting stock solutions (1000 mg/L) to the desired concentrations in ddH$_2$O.

2.5 Metal removal in single metal system

For assessment of percentage removal and metal uptake with individual ions (Cd, Cr, Cu, Ni, Pb) by consortia culture, each metal concentration was fixed at 100 mg/L for uniformity purposes [2]. All tests were conducted at an initial pH of 5.0–5.5, which was shown not to interfere with solution chemistry, and caused no precipitation of metals. They were incubated statically at room temperature for 24 h. Cells were removed by centrifugation and the DW measured before cells were digested with conc. HNO$_3$ at 60–65 °C for 15–16 h. Concentrations of Cd, total Cr, Cu, Ni and Pb in the cells and the supernatant were determined by atomic absorption spectrophotometer (Perkin Elmer Model 1100B).

Uptake capability of CC were expressed as metal uptake (q$_p$), calculated using the general definition q$_p$ = C$_p$ (V)/w, where q$_p$ is the metal uptake (mg metal/g DW biomass) in solution, V (mL). C$_p$ is the concentration of metal contained (mg/L) in the biomass of known dry weight, w (mg). Percent removal was calculated as C$_p$/C$_i$ × 100% where C$_i$ is the initial metal concentration (mg/L) [3, 4].

2.6 Metal removal in multi-metal systems

In the multi-metal system comprising of Cd(II), Cr(VI), Cu(II), Ni(II) and Pb(II), total metal ions tested for both growing and non-growing cells were kept consistent at 100 mg/L, comprising 20 mg/L of each metal [2]. Percentage metal removal and metal uptake determinations were as described for the single metal system.

2.7 Microbial-metal interactions

In order to view the localisation of the metal ions within cells, cells were subjected to transmission electron microscopy (TEM; Hitachi H-7100 EM). Samples were placed on Formvar coated 200-mesh copper grids. Cellular outline of control cells (unexposed to metals) were observed stained with 2% methylamine tungstate; whilst cells exposed to heavy metals (test cells) were observed as dark/dense areas produced by sorbed metals acting as contrasting agent. To confirm the identity of sorbed metals, harvested cells treated with 20 mg/L each of Pb(II), Cu(II), Cr(VI), Cd(II) and Ni(II) were subjected to lyophilisation (Labconco) and then analysed by X-ray fluorescence spectroscopy (XRF; Philips PW1480 X-ray spectrometer) in the pressed pellet form.

3 Results & discussion

Metal uptake and percentage metal removal by growing and non-growing cells in single and multi-metal systems are shown in Table 1.

Table 1 Percentage of metal removal from solution and metal uptake by consortia culture in single and multi-metal systems.

Ions	Metal system	Concentration (mg/L)	Growing cells		Non-growing cells	
			% removal	Metal uptake (mg/g DW biomass)	% removal	Metal uptake (mg/g DW biomass)
Pb	Single	100	*23.53 ± 0.02	*198.44 ± 0.19	20.64 ± 0.10	176.11 ± 0.87
	Multi^	20	15.10 ± 0.98	30.73 ± 2.00	14.58 ± 0.66	30.03 ± 1.36
Cu	Single	100	18.85 ± 0.26	166.39 ± 6.35	18.99 ± 0.12	156.78 ± 1.01
	Multi^	20	*13.88 ± 0.70	*28.17 ± 1.43	8.82 ± 0.17	18.63 ± 0.35
Ni	Single	100	13.34 ± 0.53	121.64 ± 1.26	*15.43 ± 0.23	*134.61 ± 2.04
	Multi^	20	6.41 ± 0.05	12.90 ± 0.10	6.39 ± 0.23	12.90 ± 0.46
Cd	Single	100	*10.62 ± 0.16	*101.88 ± 1.54	9.42 ± 0.15	89.88 ± 1.47
	Multi^	20	5.92 ± 0.41	11.90 ± 0.82	5.04 ± 0.39	10.37 ± 0.81
Cr	Single	100	*6.68 ± 0.08	*69.57 ± 0.86	5.98 ± 0.05	55.88 ± 0.46
	Multi^	20	*5.16 ± 0.42	*10.10 ± 0.82	4.01 ± 0.25	8.07 ± 0.50

Values are mean ± s.d. of triplicates.
^ : % metal removal is significantly different in comparison to single metal system at $\alpha = 0.05$
* : % removal and uptake is significantly different between growing and non-growing cell population at $\alpha = 0.05$

In the single metal system, percentage removal by consortia culture was highest with Pb (20.64-23.53%), followed by Cu (18.85-18.99%), Ni (13.34-15.43%), Cd (9.42-10.62%) and lastly Cr (5.98-6.68%). For both growing and non-growing cells, the highest metal uptake (q_p) was obtained with Pb (176.11-198.44 mg/g). Similar trends in removal were observed for the multi-metal systems but removal was significantly reduced ($p < 0.05$) as indicated by the decrease (23-58%) in both % metal removal and metal uptake. The removal of metals by CC in the multi-metal system was in the order : Pb > Cu > Ni > Cd > Cr. It is interesting to note that here, Pb which is relatively large ion (1.19 Å) was sorbed the most, followed by the relatively smaller ions Cu (0.73 Å), Ni (0.69 Å), Cd (0.97 Å) and Cr (0.52 Å). Metals with smaller ionic radius, for example Zn (0.74 Å) have been reported to be more easily sorbed onto a biosorbent compared to those with bigger radius such as Cd (0.97 Å) [4]. It is however also been shown that the more electronegative and heavier ions have greater affinity towards cell surfaces [2]. This could explain the higher percentage removal observed with Pb (2.33 Pauling; 207.2 amu) and Cu (1.90 Pauling; 63.55 amu) when compared to Ni (1.91 Pauling; 58.69 amu), Cd (1.69 Pauling; 112.41 amu) and Cr (1.66 Pauling; 51.99 amu). Furthermore, various inter-related factors such as ionic properties of the metal ion (ion affinity, electronegativity, ionisation potential, ionic radius and redox potential), the solution chemistry and multiple ions interaction and compatible biomass properties may also collectively play their part in determining the strength of metal sorption capacity [3–5].

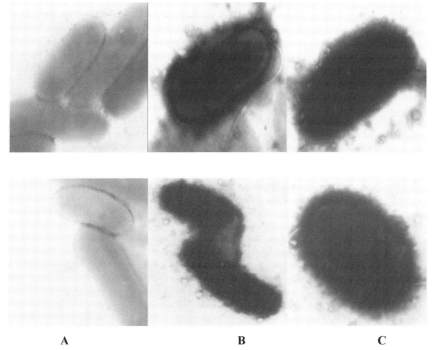

| A | B | C |

Fig. 1   Metal sorption by CC biomass. **I**: Growing cells. **II**: Non-growing cells. Plates **A**- control; **B**- cells in 100 mg/L Pb, Cu, Cr; **C**- cells in 100 mg/L Pb, Cu, Cr, Cd, Ni. Depositions of metals are observed as electron scattering/dense areas on the biomass.

Figure 1 shows metal ions were sorbed onto the outer cell surface as well as in the interior, in both growing and non-growing cells. Non-growing (inactive biomass) have been shown to behave in the same way as advanced ion-exchangers due to the presence of functional groups that facilitates passive sorption on the outer cell surface [2, 3]. Inactive biomass can also act as a gel matrix, permitting passive metallic ion translocation (accumulation) into the cell thus improving and increasing metal sorption capacity. This suggests that

140

intracellular accumulation, probably assisted by changes in membrane permeability can also occur in non-metabolically active cells [9]. XRF analysis confirmed the identities of the metals sorbed as Cd, Cr, Cu, Ni and Pb (Figure 2).

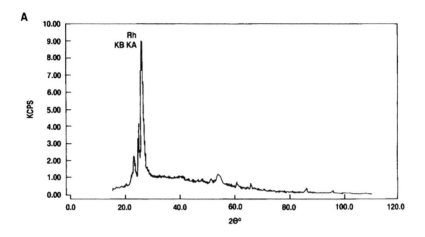

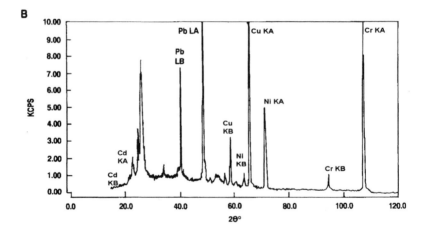

Fig. 2 XRF spectrum of CC biomass pellet. (**A**)- Control (**B**)- Test (in 100 mg/L Cd, Cr, Cu, Ni, Pb). XRF spectroscopy of bacterial pellet confirmed the identities of metals sorbed. The pellet gave major characteristic peaks at 21.72 $2\theta°$, 107.1 $2\theta°$, 65.56 $2\theta°$, 71.27 $2\theta°$ and 48.73 $2\theta°$ that were attributable to Cd(II), Cr(VI), Cu(II), Ni(II) and Pb(II), respectively. These peaks were absent in control biomass (unexposed to metals).

The use of a mixed culture biomass provides an abundance of functional groups acting as ligands facilitating metal sorption. Infra-red (IR) spectrum studies of raw CC biomass revealed major characteristic peaks belonging to the hydroxyl (OH) groups, amines (NH$_2$), carbonyls/carboxyl's (C-O stretching), amides ($CONH_2$), sulfates (SO_4^{2-}), phosphates (PO_3^{-}, PO_4^{3-}) and or sulphydryl (SH) groups (data not shown). The presence of such groups could be expected to cause precipitation of metals as metal carbonates, sulfides and phosphates [2, 4]. Most of these functional groups are however, non-specific towards metal binding and the various ions in multi-metal systems would compete for the same sites, which results in lower metal removal.

Growing cells were shown to remove significantly higher amounts of metals from solutions as well as increased metal uptake except for Ni (Table 1). This is likely due to their ability to participate in non-metabolic interactions as well as other interactions involving active metabolism such as reduction of the toxic Cr(VI) to Cr(III), active precipitation of the metal on the cell surface as well as modification on cellular uptake by altered pathway or specific translocation of the metals within the cytoplasm [3, 6–8].

4 Conclusion

From our study it can be concluded that the % removal of metal from a multi metal system is generally reduced as compared to that from single metal system in the following pattern: Pb > Cu > Ni > Cd > Cr. Electron microscopic studies showed that sorbed metals were deposited internally and externally, whilst XRF analysis confirmed the ability of CC to immobilise and retain metal ions within the biomass. Therefore, the results from this work have considerable importance in developing a bacterium-based biosorbent for practical applications. Further in-depth studies concerning system up scaling by column and bioreactor design will be carried out.

Acknowledgements

The study was funded under IRPA program 08-02-02-006 from the Ministry of Science, Technology & Innovation, Malaysia.

References

[1] P. Sannasi, S. Salmijah, and J. Kader, in: Proceedings of the 12th National Biotechnology Seminar, Damai Laut   Country Resort, Lumut, Malaysia, 2000, pp.
[2] K. C. Sekhar, S. Subramaniam, J. M. Modak, and K. A. Natarajan, Int. J. Mineral Processing **53,** 107 (1998).
[3] F. Veglio and F. Beolchini, Hydrometallurgy **44,** 310 (1997).
[4] S. Singh, S. Pradhan, L. C. Rai, Process Biochem **36,** 175 (2000).
[5] A. Lodi, C. Solisio, A. Converti, M. D. Borghi, Bioprocess Engineering **19,** 197 (1998).
[6] R. F. Unz and K. L. Shuttleworth, Current Opinion in Biotechnology **7,** 307 (1996).
[7] J. S. Chang, R. Law, C. C. Chang, Water Research **31,** 1651 (1997).
[8] V. D. Appanna, L. G. Gazso, M. S. Pierre, J. Biotechnology **52,** 75 (1996).
[9] B. Volesky, Chemical Eng. Dept., McGill University, Montreal, Canada. Personal communication (2002).

Study of *Helicobacter pylori* viability in aquatic microcosms by epifluorescence stain and *in situ* hybridization

Jiménez, A., Moreno, Y. Piqueres, P., González, A. and Ferrus, M.A. [*]

Dep. Biotecnología, Universidad Politécnica, Cno. Vera, s.n., 46022 Valencia, SPAIN.

[*]Corresponding author: e-mail: mferrus@btc.upv.es

H. pylori is one of the most common infective agents worldwide. Some studies have suggested a waterborne transmission pathway. However, under stress conditions cellular changes are observed and culturability is lost. The aim of this study was to determine the *in vitro* survival and morphological changes of *Helicobacter pylori* in aquatic microcosms, by measurement of membrane integrity, RNAr content and culturability changes. When stored in PBS, *H. pylori* maintained culturability for 15 to 21 days, while viable rod and coccoid cells were detected until 3 months. In water, *H. pylori* was culturable for 5 days and viable all along the assay. This results show that aquatic environments could be a potential infection source for *H. pylori*.

Keywords: Helicobacter pylori, Survival, VNC

1   Introduction

H. pylori is one of the most common infective agents worldwide. It has been established as an etiologic factor in the development of peptic ulcer and gastric neoplasia [1]. Design of prevention measures is difficult due to our limited knowledge of transmission pathways. Studies from developing countries have shown an association between the prevalence of *H. pylori* and the source of drinking water [2]. Presence of *H. pylori* in water systems has been the subject of previous studies. [3,4], and recently, *H. pylori* has been isolated from wastewater [5]. These studies show that *H. pylori* may survive in water for extended periods of time, which strongly supports a waterborne route of transmission.

However, under stress environmental conditions cellular changes are observed from bacillar to coccoid forms. It has been suggested that these could be Viable-Non-Culturable (VNC) cells, responsible for transmission in the environment but, as in this form *H. pylori* is non-culturable by ordinary techniques, its ability to survive, viability and virulence is still a matter of controversy [6,7].

A number of viability markers of different cellular functions have been developed. Many of these techniques are indirect staining analysis, monitoring membrane potential or integrity [8]. The LIVE/DEAD® Baclight™ (BL) viable staining method was developed to differentiate live and dead bacteria based on membrane integrity. This kit comprises two fluorescent nucleic acid stains, SYTO 9 and Propidium Iodide (PI). SYTO 9 penetrates both viable and nonviable bacteria, while PI penetrates bacteria with damaged plasma membranes only quenching the SYTO 9 fluorescence. Thus, bacterial cells with compromised membranes fluoresce red and those with intact membranes fluoresce green [9].

The integrity of nucleic acids have been also proposed to assess the viability of cells [9]. In this sense, fluorescent in situ hybridization (FISH) with rRNA probes has been reported to allow for the detection of viable but non-culturable forms [11]. Recently, a FISH method to detect *Helicobacter pylori* has been used to evidence its presence in water and wastewater [12].

Therefore, the aim of this study was to determine the in vitro survival and morphological changes of *Helicobacter pylori* in aquatic microcosms, by measurement of membrane integrity, RNAr content and culturability changes.

2 Materials and methods

2.1 Bacterial strains and culture conditions

H. pylori HBTC6 strain (belonging to our collection) was used for all the assays. Routine cultures were grown on Columbia Agar Base (DIFCO) with 10% defibrinated horse blood, and incubated under microaerobic conditions at 37 °C for 48 h.

2.2 Survival assay

Cells from a 48 h Nutrient Broth culture of *H. pylori* were harvested by centrifugation at 9000 g for 5 min, washed with sterile PBS and suspended in flasks containing 500 ml of PBS (130 mM sodium chloride, 10 mM sodium phosphate, pH 7.2) or sterilized declorinated fresh water. Flasks containing PBS and water were stored at 4°C and 10°C respectively, both in the dark. Samples were removed aseptically immediately after inoculation and then periodically during the following 3 months. Assays were performed in duplicate.

2.3 Culturable cells counts

Samples were serially diluted in PBS buffer and 100 µl of each dilution was plated onto Columbia Blood Agar plates, incubated under microaerobic conditions at 37 °C and examined for the presence of characteristic colonies at 48 h, 3, 7 and 10 days. Detected colonies were confirmed to be *H. pylori* by Gram stain and rapid urease production test. Colony counts were made in duplicate.

2.4 Fluorescent staining

Viability and morphology were determined by staining cells with the LIVE/DEAD BacLight kit (Molecular Probes, Eugene, Oreg.). Samples (1 ml) were mixed with 3 µl of a stain mixture of Syto-9 and PI (1:1) and incubated in the dark for 15 min at room temperature. *H. pylori* stained cells were immobilized on glass surfaces [13]. Fluorescent cells were counted using an epifluorescence Olympus microscope BX50 using a double band-filter cube (XF 53, Omega). A minimum of 10 fields were counted. Data were expressed as means at different interval times

2.5 Fluorescent in situ hybridization analysis

Cells from one ml of each sample were harvested by centrifugation and fixed for 3 h in 4% paraformaldehyde at 4°C. A 3 µl portion of fixed sample was placed on a gelatine-coated slide, air-dried, dehydrated (50, 80, 100% ethanol) and hybridized with a HPY probe (-CTGGAGAGACTAAGCCCTCC-) according to Moreno *et al* [12].

The EUB 338 probe, complementary to a region of 16S rRNA of the domain Bacteria was used as a positive control to assure that hybridization procedure had been properly performed. A fluorescent oligonucleotide sequence not complementary to Eubacteria rRNA was used as a negative control, to check for non-specific binding of HPY probe to sample components.

After hybridisation, slides were rinsed with distilled water, air-dried, and mounted in FluoroGuard™ (Bio-Rad Laboratories, USA). Samples were examined under an epifluorescence Olympus BX50 microscope with a set of U-MWB, U-MWIB and U-MWIG filters. Micrographs were taken with a digital Olympus DP 10 camera. Fluorescent green intensity signal was measured with Olympus DP Soft program.

3   Results and discussion

3.1 Culturable and Viable cell counts

Fig. 1 shows the viable, dead and culturable cell counts in PBS at 4°C: Culturability of *H. pylori* cells decreased progressively from 5.8 x 10^5 u.f.c./mL (initial count) to 2.5 x 10^3 u.f.c./mL at 21 days, and was definitively lost at 30 days. The number of cells with an intact membrane (viable cells) maintained stable until 72 h, and then decreased slightly up to three months (7 x 10^5). The number of cells that showed membrane damage (dead cells) remained stable until the end of the assay (1.3 x 10^5 at 105 days).

Fig. 2 shows viable, dead and culturable cell counts after starvation in fresh non-clorinated water at 10°C. Counts decreased from 2.8 x 10^5 (initial) to 1 x 10^1 at 5 days. No growth was observed after 5 days of storage in fresh water. Viable cell counts showed a slight decrease at 48h, while total counts remained constant during all the assay.

3.2 Fluorescent in situ hybridization analysis

The levels of 16S rRNA in bacillar forms, as measured by the intensity of fluorescent signal after hybridisation, decreased during the assay. Fluorescent signal did not change in coccoid forms.

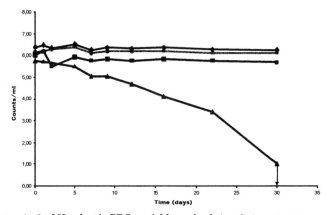

Fig. 1. Survival of *H.pylori* in PBS: o viable, □ dead, ▲, ufc, ◊ total cells (in cells/mL)

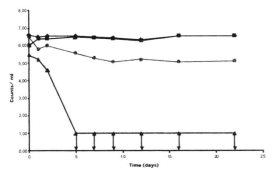

Fig. 2. Survival of *H.pylori* in water: o viable, □ dead, ▲, ufc,◊ total cells (in cells/mL)

3.3 Morphologic changes

For PBS experience, fluorescence microscopy showed the presence of viable spiral-shaped bacillar and coccoid cells until 30 days, with a significant decrease of bacillar morphology from the first 7 days. At 3 months, no bacillary forms were observed. Coccoid and degenerative forms began to appear after 5 days of inoculation. However, a significative amount of coccoid forms maintained its cellular integrity up to 3 months.

After inoculation in fresh water, red (non-viable) bacillar forms were observed from the first 24 hours. Viable rods were observed until 12 days, when culturability was already been lost. Bacillar forms disappeared at 16 days. Coccoid and degenerative forms began to appear after 48 h of inoculation, most of them being viable. A population of these coccoid forms maintained its viability during all the assay period.

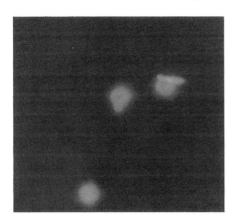

Fig.3. Bacillar, coccoid and transition cellular forms of *H. pylori* in PBS

4   Discussion

Membrane permeability is a key element in cell viability, as its loss involves the degradation of several vital processes linked to the cytoplasmic membrane [14]. Our results strongly supports other authors data, about the maintenance of viability and metabolic activity of *H. pylori* non-culturable cells [6, 15]. Survival studies of *H. pylori* in PBS stored

146

at 4°C indicated that these organisms can be culturally detected for up to 21 days. When assays where performed in fresh water, culturability was maintained up to 5 days. Viability was maintained after 3 months of starvation in PBS and 1 month in water. Detection of total rRNA by intensity of fluorescence after hybridization, or of DNA by PCR, are techniques that have been used to investigate cells viability [11, 16]. The intensity of the fluorescence signal by FISH, and therefore the levels of 16S rRNA of cells, decreased in bacillar forms during storage, so probably loss of culturability involved ribosome degradation.

Results showed that the transition to non-culturable cells was not always associated with coccoid forms. Spiral unculturable *Helicobacter* cells were present after 16 days in fresh water (Fig. 3) and 60 days in PBS. These results agree with that of other authors [17] who report that cells may exist in all morphologies in culturable and non-culturable states.

Our work seem to confirm too the ability of *H. pylori* to survive in water in the viable but non-culturable form, what suggests that aquatic environments could be a potential infection source for *H. pylori*.

Acknowledgements

This work was supported by by Ministerio de Ciencia y Tecnología, Spain and FEDER fundings (Research Project AGL2002-04480-C03-03). P. P. has a fellowship from the Universidad Politécnica de Valencia. A. G. has a F.P.U. fellowship from Ministerio de Educación, Spain.

References

[1]   Dunn B. E., Cohen H. and Blaser, M. J. Clin. Microbiol. Rev. **10**, 720 (1997).
[2]   Percival, S. L., Chalmers, R. M., Embrey, M., Hunter, P. R., Sellwood, J., and Wyn-Jones (eds). Microbiology of Waterborne Pathogens (Elsevier, London, 2004).
[3]   Hegarty, J.P., Dowd, M and Baker, K.H. J. Appl. Microbiol. **87**, 697 (1999).
[4]   Sasaki, K., Tajiri, Y., Sata, M., Fujii, Y., Matsubara, F., Zhao, M., Shimizu, S., Toyonaga, A. and Tanikawa, K. Scan. J. Infec. Dis. **31**, 275 (1999).
[5]   Lu, Y., Redlinger, T.E., Avitia, R., Galindo, A. and Goodman, K. Appl. Environ. Microbiol. **68**, 1436 (2002).
[6]   Nilsson, H. O., Blom, J., Al-Soud, W. A., Ljungh, A., Andersen, l. P. and Wadström, T. Appl. Environ. Microbiol. **68**, 11 (2002).
[7]   Engstrand L. J. Appl. Microbiol. **90**, 80S (2001).
[8]   J.T. Lisle, B.H. Pyle, G.A. McFeters. Lett. Appl. Microbiol. **29**, 42 (1999).
[9]   Haugland, R. P., in: Handbook of Fluorescent Probes and Research Chemicals (Molecular Probes Inc., Eugene, Oregon, 1996), 365.
[10]   Keer, J.T. and Birch, L. J. Microbiol. Methods **57**, 175 (2003).
[11]   Amann, R.I., W. Ludwing, and K.H. Schleifer. Microbiol. Rev. **59**: 143 (1995).
[12]   Moreno, Y., M.A. Ferrús, J.L. Alonso, A. Jiménez, and J. Hernández. Wat. Res. **37**: 2251(2003).
[13]   Alonso, J. L., Mascellaro, S., Moreno, Y., Ferrús, M. A., Hernández, J. Appl. Environ. Microbiol. **68**, 5151 (2002).
[14]   R.L. Kepner, J.R. Pratt. Microbiol. Rev. **58**, 603 (1994).
[15]   West, A. P., Millar, M. R., Tompkins, D. S. J. Clin. Pathol. **45**, 228 (1992).
[16]   L. Birch, C.E. Dawson, J.H. Cornett, J.T. Keer. Lett. Appl. Microbiol. **33**, 296 (2001).
[17]   Adams, B.L., Bates, T. C., Oliver, J. D. Appl. Environ. Microbiol. **69**, 7462 (2003).

Survival of *Ralstonia solanacearum* biovar 2 in river water: influence of water microbiota

B. Álvarez[1] , E.G. Biosca[1,2], and M.M. López[*1]

[1] Instituto Valenciano de Investigaciones Agrarias (IVIA), 46113-Moncada, Valencia, SPAIN
[2] Dpto. de Microbiología y Ecología, Universidad de Valencia, 46100-Burjasot, Valencia, SPAIN

*Corresponding author: e-mail: mlopez@ivia.es, Phone: +34 96 3424000, Fax: +34 96 3424001

Ralstonia solanacearum is a plant pathogenic bacterium which causes wilt disease producing severe economic losses in crops world-wide. Waterways are major dissemination routes, where the pathogen is able to survive for variable periods. In this study, the survival of this pathogen in river water microcosms with different biotic fractions was investigated at 25 and 14ºC. At both temperatures, in control sterile river water microcosms *R. solanacearum* cell numbers remained constant while in non-sterile microcosms a rapid decline was observed, being stronger at 25ºC. Lytic phages, present in this river water, were the main responsible agents of the reduced persistence of this pathogen in natural water, but indigenous bacteria and protozoa also contributed. Overall, the present results show a strong influence of water microbiota on the survival of *R. solanacearum* in natural river water.

Keywords bacterial wilt; potato brown rot; aquatic microbiota; microcosms; lytic phages

1   Introduction

Ralstonia solanacearum is a devastating plant pathogenic bacterium which causes diseases affecting over 200 plant species worlwide, many of them of agricultural interest. Symptoms are usually wilting of the plant as a result of vascular disorders, which progress to death. The biovar (bv) 2 of this species is the causal agent of potato brown rot and bacterial wilt in many solanaceous plants. During the last decade, the introduction of *R. solanacearum* bv 2 in Europe has been responsible for severe potato brown rot outbreaks and, in many cases, contaminated irrigation water was the main dissemination route of this pathogen. *R. solanacearum* is able to survive for long periods in sterile water and the influence of several abiotic factors has been studied [1]. However, a strong effect on its survival has been reported in unsterilised drainage water which has been related to water microbiota [1] although the nature and the influence of diverse biotic agents (protozoa, other bacteria and/or bacteriophages) on the fate of *R. solanacearum* bv 2 were not investigated. To address these issues, the population dynamics of a Spanish strain of *R. solanacearum* bv 2 was monitored in natural water microcosms from a Spanish river where the bacterium had previously been isolated[2, 3].

2   Materials and methods

2.1 Bacterial strain and growth conditions

A Spanish strain of *R. solanacearum* by 2 isolated from potato (IVIA 1602.1) was used in this study. The strain was grown on Yeast Peptone Glucose Agar (YPGA) [4] at 29 ºC for 72 h for detection of R. solanacearum lytic phages and for survival experiments in river water, as described below.

2.2 Water analysis

Prior to survival experiments, water samples were collected from September to November 2001 from three sites in the Tormes river (Northwest of Spain) where *R. solanacearum* bv 2 had previously been detected [2, 3] and analysed for *R. solanacearum* detection as previously described [2, 3]. Culturable cells of the pathogen were determined in duplicate on modified SMSA agar [5] after 72 h at 29°C and confirmed as described before [3]. River water culturable bacteria on Wilbrink medium [6] after 72 h at 29°C and the presence of lytic phages to *R. solanacearum* bv 2 were also determined. Lytic phages were isolated from river water samples using an enrichment method as described below and the Spanish strain of *R. solanacearum*. Briefly, phage enrichment was done by the addition of 1 ml aliquots of the filtered river water to log-phase cultures of host strain in 5 ml of modified Wilbrink broth [7]. Clearance of cultures after overnight incubation with shaking at 29°C was recorded as positive detection and they were assayed for lytic phages by a direct phage enumeration procedure on YPGA medium.

2.3 Water microcosm experiments

Survival experiments in river water microcosms were performed similarly to [8] by using seven water samples taken in September, October and November 2001. Four type of microcosms were prepared in duplicate from each water sample: untreated river water, containing the whole microbiota; 0.8 μm filtered water without most of protozoa; 0.2 μm filtered water without protozoa and bacteria, and filtered and autoclaved water without active phages, used as a negative control. Each microcosm was inoculated with the *R. solanacearum* strain IVIA 1602.1 at a final concentration of 5×10^6 cfu/ml as previously described [2, 3]. Microcosms from September and October water samples were kept in the static state at 25°C during one month while those of water sample from November were maintaned at 14°C.

Microscopic counts for total and viable *R. solanacearum* cells from water microcosms were done at time zero, after 24 and 48 h and 4, 8 and 28 days by a modified Kogure method [1] combined with acridine orange staining or by indirect immunofluorescence (IIF) using polyclonal antibodies specific to *R. solanacearum* bv 2 according to [3]. Culturability was monitored by plate counts on SMSA semiselective medium, as described before [2, 3].

3   Results

3.1 Water analysis

Total culturable indigenous bacteria and *R. solanacearum* counts on Wilbrink and SMSA media respectively, and phage detection results of the analysis of Tormes river water samples taken in 2001 are shown in Table 1. In September and October 2001 the temperature of surface water ranged between 18 to 14° C and *R. solanacearum* was isolated on SMSA agar from all the water samples (from 65 to 5 cfu/ml) while indigenous bacteria counts on Wilbrink medium varied between 2×10^4 to 1×10^3 cfu/ml. In November, when water temperature declined to 7-8°C the pathogen was not detected by direct isolation but river water bacteria counts remained at about 10^3 cfu/ml. Using the enrichment method, lytic phage to *R. solanacearum* were detected and isolated from water samples only when the pathogen was recovered on SMSA agar (at 14°C or higher).

Table 1 Characteristics of Tormes river water samples analysed.

N° of water samples analysed	Month of sampling	T^a (°C)	pH	Total culturable bacteria on Wilbrink agar (cfu/ml)	Culturable *R.solanaceareum* on SMSA agar (cfu/ml)	Phage detection
3	Sep	18	6.9-7.2	1.6×10^3-2.0×10^4	5.5-6.5×10^1	+
3	Oct	14	7.3-7.4	1.1-2.2×10^3	5.0×10^0-1.5×10^1	+
1	Nov	7-8	6.5	2.5×10^3	-	-

3.2 Effect of indigenous water microbiota

One typical result of the survival of *R. solanacearum* strain IVIA 1602.1 inoculated in untreated and treated river water microcosms incubated at 25°C is shown in Fig.1. This experiment was performed with all September and October water samples in duplicate with similar results. In filtered and autoclaved river water microcosms, used as control, total and viable cell counts of *R. solanacearum* stabilized at around 10^7 cells/ml (Fig 1 a). In contrast, in the presence of indigenous aquatic microorganims in the untreated and 0.8 µm filtered river water microcosms a 3 log decrease (10^4 cells/ml) was observed at days 2 and 4 post-inoculation respectively, total cell counts remaining slightly higher than viable counts (Fig. 1 b,c); for 0.2 µm filtered river water microcosms containing only the river water viral fraction a shorter decline of total and viable cell counts was observed, followed by an ulterior stabilization at about 10^6 cells/ml and 10^5-10^6 cells/ml, respectively (Fig. 1 d). Regarding culturable counts of *R. solanacearum* cells on SMSA agar in control river water microcosms, they slightly increased to 10^7 cfu/ml upon inoculation and remained constant to the end of the experiment (Fig. 1 a). A strong reduction was observed for the untreated, 0.8 µm filtered and 0.2 µm filtered river water microcosms (Fig. 1 b,c,d). For untreated and 0.8 µm filtered river water microcosms, the number of culturable cells of *R. solanacearum* on modified SMSA agar dropped below the detection limit ($<10^1$ cfu/ml) at day 2 post-inoculation. Afterwards, only indigenous bacteria were observed by plate counts on SMSA agar (data not shown). In comparison, in the 0.2 µm filtered river water microcosms an initial decline with similar slope was observed after 24 h, although culturable counts were maintained at about 10^3 cfu/ml until day 2 and then increased and stabilized at 10^4-10^5 cfu/ml until the end of the experiment (Fig. 1 d). This decrease in culturability along with *R. solanacearum* phage detection assays suggested lytic phages as the cause of the rapid drop in the population of the inoculated strain in 0.2 µm filtered river water microcosms. A putative mutant of the strain IVIA-1602.1 resistant to the phages showing smaller colonies with irregular borders was the responsible for the subsequent increase in culturability after day 2 in these microcosms (Fig. 1 d). This mutant was not observed in untreated and 0.8 µm filtered river water microcosms where the decline in *R. solanacearum* populations was stronger than in 0.2 µm filtered river water microcosms. The comparison of untreated and 0.8 µm filtered river water microcosms showed an added effect of indigenous protozoa and bacteria on the survival of *R. solanacearum* bv 2 in this river water (Fig. 1 b,c). At 14°C, cell counts for control sterile microcosms were similar to those at 25°C, but the declines in total, viable and culturable cell numbers for untreated, 0.8 µm filtered and 0.2 µm filtered river water microcosms were softer with respect to those at 25°C (data not shown). At both temperatures, the decline in culturability (about 10^{4-3} cfu/ml) of the inoculated strain in 0.2 μm filtered river water microcosms was followed by the appearance of the small colonies with irregular borders described above.

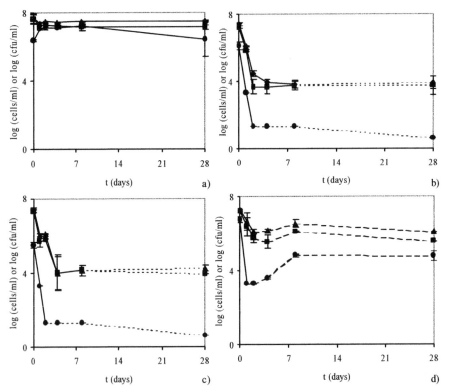

Fig. 1 Dynamics of total, viable and culturable cells (triangle, square and circle symbols respectively) on SMSA agar of strain IVIA-1602.1 of *R. solanacearum* in river water microcosms at 25 °C in: a) filtered and autoclaved control microcosms; b) untreated water microcosms; c) 0.8 μm filtered water microcosms and d) 0.2 μm filtered water microcosms. Error bars indicate standard deviation and dotted lines values below detection limit of the techniques used. Broken lines represent values for a putative phage-resistant variant.

4   Discussion

This paper reports the presence and survival of *R. solanacearum* in natural river water. Our data on the culturable populations of *R. solanacearum* in river water samples used for survival experiments agree with other studies in the same river, showing similar levels of *R. solanacearum* for water temperatures above 14°C, but becoming nondetectable by plating at lower temperatures [2, 3]. Biotic factors also influence the fate of *R. solanacearum* bv 2 in aquatic environments [1] but the available information is scarce. We have previously reported the detection of R. solanacearum phages from this river [9] and, in this study, the isolation of lytic phages from all the river water samples from which this bacterium was recovered, suggests an association between *R. solanacearum* and its river water phage(s).

Their characterisation is now under study.

In the survival experiments in control sterile river water microcosms, at 25 and 14°C, total, viable and culturable cell numbers of *R. solanacearum* remained at similar levels to those at inoculation time. This confirmed the survival of this pathogen under the nutrient limitation conditions of aquatic ecosystems, according to previous studies [1, 2, 3], the two temperatures assayed being favourable for its persistence [1]. In contrast, in the non-sterile microcosms containing different biotic fractions its survival was severely affected as reported in natural agricultural drainage water [1] but without investigation of the biotic agents involved. In our study at 25°C, a decrease in total and viable counts was observed from 1-2 days post-inoculation which was stronger for culturable counts. At 14°C the slope of decline in these microcosms was slighter, which implied that biotic interactions were slower at lower temperature. Lytic phages were considered to be the main responsible agents for the reduced persistence of *R. solanacearum* in river water, as all the non-sterile microcosms kept the viral fraction of the natural water and earlier phage detection in this river had been positive. However, such drop in culturability in the microcosms containing only the viral fraction was followed by a slight increase in culturable counts a few days post-inoculation along with the appearance on plates of small colonies with irregular borders. They correspond to a putative phage-resistant variant of the inoculated strain since it was identified as *R. solanacearum* bv 2. Phage-resistant LPS-defective mutants of *R. solanacearum* showing smaller and less fluidal colonies have been isolated by other authors [10]. Further research is in progress to know if our phage-resistant strain is LPS-defective.

Additional biotic effects were observed in the untreated river water and 0.8 µm filtered river water microcosms, suggesting that competitive and/or antagonistic bacteria and/or predatory protozoa also contributed to the reduced persistance of *R. solanacearum*. In untreated river water microcosms the survival of *R. solanacearum* cells, present in larger numbers than indigenous bacteria upon inoculation, was lower than that of aquatic bacteria. When predatory protozoa were removed, bacterial predation and/or competition for nutrients would occur. However, all these events were simultaneous to phage activity in both type of microcosms and affected the phage-resistant variant as well, which was not recovered from them, probably because of predation or bacterial competition for nutrients. Overall, our results show a strong impact of water microbiota on the survival of *R. solanacearum* bv 2 in natural water, with lytic phages being the main cause of its reduced persistence in the river water assayed.

Acknowledgements

This work was supported by projects QLK3-CT-2000-01598 and OCYT GV04B-313.

References

[1]   J. D. van Elsas, P. Kastelein, P. M. de Vries, and L. S. van Overbeek, Can. J. Microbiol. **47**, 842 (2001).

[2]   E.G. Biosca, P. Caruso, E. Bertolini, B. Alvarez, J.L. Palomo, M.T. Gorris, and M.M. Lopez, in: Bacterial wilt disease and the *Ralstonia solanacearum* species complex, C. Allen, P. Prior and C. Hayward (eds.) (APS Press, 2005).

[3]   P. Caruso, J.L. Palomo, E. Bertolini, B. Alvarez, M.M. Lopez, and E.G. Biosca, Appl Environ Microbiol. **71**, 140 (2005).

[4]   R. A. Lelliot, and D. E. Stead, in: Methods in Plant Pathology, T. F. Preece, (ed.) Vol. 2

152

(1987), p. 216.

[5]   J. G. Elphinstone, J. Hennessy, J. K. Wilson, and D. E. Stead, EPPO Bulletin, **26**, 663 (1996).

[6]   H. Koike, Phytopathology, **55**, 1317 (1965).

[7]   P. Caruso, M.T. Gorris, M. Cambra, J.L. Palomo, J. Collar, and M.M. Lopez, Appl Environ Microbiol. **68**, 3634 (2002).

[8]   E. Marco-Noales, E.G. Biosca, C. Rojo, and C. Amaro, Environ Microbiol. **6**, 364 (2004).

[9]   M. M. Lopez, and E.G. Biosca, in: Bacterial wilt disease and the *Ralstonia solanacearum* species complex, C. Allen, P. Prior and C. Hayward (eds.) (APS Press, 2005).

[10]   C. A. Hendrick, and L. Sequeira, Appl. Environ. Microbiol. **48**, 94 (1984).

The control of biological corrosion in cooling water system of a power plant

C.-M. Lee[*,1], L.-F. Wu[1], S.-D. Chyou[2], S.-T. Kuo[2], S.-J. Chen[2], H.-E. Wu[3], H.-Y. Chen[3], W.-L. Shih[3]

[1]   Department of Environmental Engineering, National Chung-Hsing University, Taiwan, R.O.C.
[2]   Taiwan Power Research Institute, Taiwan Power Company, Taiwan, R.O.C.
[3]   Taichung Thermal Power Plant, Taiwan Power Company, Taiwan, R.O.C.

[*]Corresponding author: e-mail: cmlee@dragon.nchu.edu.tw, Phone: +886-4 22850024, Fax: +886-4 22862587

There were many different physiological types of bacteria which interacted in complex ways within the slimes of the cooling water system, inclusive of mesophiles and thermophiles. Strain TPH-13-3 produced a large amount of extracellular polymeric substances (EPS), about four to fifteen times of the amount of others. Strain TPM-8 · TPH-2 · TPH-5-2 · TPH-7 were identified as iron oxidizing bacteria by methods based on 16S rDNA gene sequence. Results of the long-term experiments of inhibitory effects of antimicrobial agents indicated that most strains could be inhibited by antimicrobial agent N; some strains could only be inhibited by antimicrobial agent T; some strains could be inhibited by both antimicrobial agents; and strain TPH-13-3 could be inhibited neither by antimicrobial agent N nor by antimicrobial agent T, the possible reason was that EPS protected the cells from inhibition by antimicrobial agents. The presence of molybdate might decrease the inhibitory effect of antimicrobial agents to some strains.

Keywords microbiological influenced corrosion; antimicrobial agent; cooling water system; extracellular polymeric substance; iron-oxidizing bacteria

1   Introduction

Microbiologically influenced corrosion (MIC) is a general problem of industry. It often takes place in neutral, still water environment (pH 4-9, $10-50^{o}C$), induces the corrosion of carbon steel, stainless steel, aluminum alloys and copper alloys in industrial instruments, cause the serious loss of the economics. This way of corrosion takes place in the presence of microbial consortia in which many different physiological types of bacteria interact in complex ways within the structure of biofilm.

In the presence, molybdate is added as corrosion inhibitor to the cooling water system of a power plant in Taiwan [1]. Monitoring the corrosion of this cooling water system by potentiodynamic polarization, ZRA, Rp/Ec trends, and AC impedance methods, it is shown that, in the presence of 50 ppm corrosion inhibitor, general corrosion can be inhibited, but localized corrosion can not be inhibited [2]. It is thus necessary to examine whether the concentration of corrosion inhibitor is enough, and to find out the causes and solutions of MIC, in order to determine the best dose of corrosion inhibitors and to find suitable antimicrobial agent.

The main purpose of this research is to isolate the microorganisms from slimes of the cooling water system of this power plant. The long-term inhibitory effects of antimicrobial agent alone as well as in the presence of corrosion inhibitor to each strain of bacteria are examined, in order to find appropriate antimicrobial agent and to determine the optimal dosage of corrosion inhibitor.

2   Materials and methods

2.1 Isolation and identification

Bacterial strains were isolated from the slimes of the cooling water system of the power plant. The cooling water temperature ranged between 30 and 50°C. Pure cultures were isolated by using a spread plate method in Leptothrix2xPYG medium [3]. The cultures were incubated at 30°C and 55°C respectively in order to isolate mesophilic and thermophilic bacteria.
The DNA of isolated bacteria was extracted with commercial kit (Genomic DNA Mini Kit, Geneaid). After eluting the total genomic DNA, the 16S ribosomal genes were amplified from the bacterial total genomic DNA by using the polymerase chain reaction (PCR). The PCR reaction was performed by a method described as previously[4].The 16S rDNA was purified using Viogene's Gel Extraction System (gel extraction miniprep kit) and sequencing was performed at Mission Biotechnology company (Taichung, Taiwan). The 16S rDNA sequence was compared against the GenBank database using the NCBI Blast program.

2.2 The optimal growth temperature experiments of thermophilic bacteria

The optimal growth temperature of the thermophilic bacteria was examined by measuring the optical density (600nm) of cultures incubated in L-type glass tubes with cotton-stopper, containing 15 ml of Leptothrix2xPYG liquid medium in a temperature gradient incubator (TN-12,TOYO®). The temperature gradient ranged from 20 to 60°C.

2.3 The extracellular polymeric substance (EPS) determination

The extracellular polymeric substance (EPS) of bacteria was extracted by using the steam extraction method [5]. Hexose sugar concentrations were used as measures of extracellular slime polymer. Hexose sugar was measured by phenol-sulfuric acid method [6]. The concentration of EPS was expressed as mg of glucose per 1g of cell dry weight.

2.4 The long-term experiments of inhibitory effects of antimicrobial agents

The long-term experiments of inhibitory effects of antimicrobial agents were examined by measuring the optical density (600nm) of culture incubated in screw-cap tube containing 10 ml of Leptothrix2xPYG liquid medium into which various dosages of antimicrobial agents were added. The initial optical density of culture in tubes was 0.1. The tubes were shaken at 100rpm in the dark at 30°C or 55°C for about 1400 hrs. In order to understand whether presence of molybdate might influence the inhibitory effect of antimicrobial agents or not, the long-term experiments have been repeated by the addition of 107ppm molybdate.

3   Results and discussion

3.1 Characteristic of the isolated strains

Using the spread plate method, thirteen mesophilic and ten thermophilic bacterial strains capable of growth on Leptothrix2xPYG medium were isolated from the slime of the cooling water system. Among them one strain was coccus-shaped, the others were rod-shaped. Table 1 showed the identifications of the strains based on 16S rDNA gene sequence, strain TPM-8 · TPH-2 · TPH-5-2 · TPH-7 were iron oxidizing bacteria; strain TPH-3 was *Brevibacillus thermoruber*, endospore-forming bacteria; strain TPH-13-3 was *Nocardia farcinica*. According to the Results, there were many different physiological types of bacteria which interacted in complex ways within the slimes of the cooling water system, including iron oxidizing bacteria. The strain TPM-8 and TPH-7 were denitrifying iron oxidizing bacteria which could carry out iron oxidization utilizing nitrate as electron acceptor in an anoxic condition.

According to the specific growth rate calculating from the results of the experiments of temperature gradient incubator, optimal growth temperature of each strain was obtained. As shown in table1 the optimal growth temperature of all test strains were higher than $40^{\circ}C$, strain TPH-3 were at $53^{\circ}C$.

In order to explore the reasons of MIC in the cooling water system, the ability of the strains produced extracellular polymeric substance (EPS) was determined. The results of the EPS extraction of bacteria were shown in table1. The concentration of the EPS were ranged from 7.4~107.3 (mg glucose/g cell dry weight), strain TPH-13-3 could produce the most amount of EPS which was 4~15 times than other strains.

Table 1   Characteristic of some strains isolated from slime of the cooling water system.

Strains	Identification of the strains	Optimal growth temp.($^{\circ}$C)	EPS(mg glucose/ g cell dry weight)
TPM-3	*Acidovorax* sp.	-	22.52
TPM-8	Denitrifying Fe-oxidizing	-	19.97
TPH-2	*Leptothrix* sp.	44	7.4
TPH-3	*Brevibacillus thermoruber*	53	15.4
TPH-5-2	*Caldimonas manganoxidans*	45	12.3
TPH-7	Denitrifying Fe-oxidizing	49	27.3
TPH-13-3	*Nocardia farcinica*	44	107.3

3.2 The long-term experiments of inhibitory effects of antimicrobial agents

As shown in Fig1, results of the long-term (1400 hrs) experiments of inhibitory effects of antimicrobial agents indicated that strain TPM-3 could be inhibited by more than 150mg/L antimicrobial agent N but couldn't be inhibited after about 650 hr by antimicrobial agent T up to 200mg/L. As shown in Fig2, strain TPH-2 could be inhibited by more than 200mg/L antimicrobial agent N but couldn't be inhibited after about 800 hr by antimicrobial agent T up to 200mg/L. Strain TPM-10 could be inhibited by more than 150mg/l antimicrobial agent N ; strain TPM-26 could be inhibited by more than 200 mg/l antimicrobial agent N ; strain TPH-5-2 could be inhibited by antimicrobial agent N and antimicrobial agent T when their concentration are more than 100mg/L; and strain TPH-13-3 could be inhibited neither

156

by antimicrobial agent N nor by antimicrobial agent T(data not shown), the possible reason was that extracellular polymeric substances protected the cells from inhibition by antimicrobial agents. The presence of molybdate might decrease the inhibitory effect of antimicrobial agents to strainTPM-26 but it's not effect on the others.

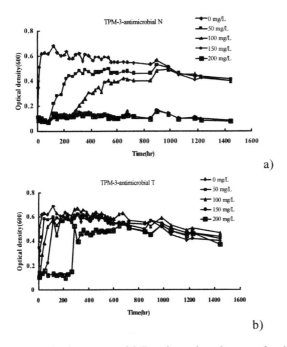

a)

b)

Fig. 1 The time course of O.D under various dosages of antimicrobial agent N (a) and antimicrobial agent T (b) for strain TPM -3

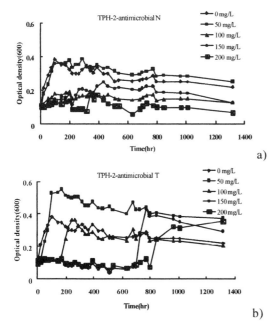

Fig. 2 The time course of O.D under various dosages of antimicrobial agent N (a) and antimicrobial agent T (b) for strain TPH-2

4   Conclusions

The water used in the cooling water system has been pretreated, and there wasn't any organic substance provided as carbon and energy sources of bacteria. It has been verified in this study that the presence of molybdate couldn't promote the growth of bacteria. However, many different genuses of bacteria have been isolated from the slimes of cooling water system. According to the results of identification, we supposed that the autotrophic iron oxidizing bacteria might colonize the steel first, and most of them were thermophilic was due to there were high temperature regions in the system. Iron-oxidizing bacteria oxidized soluble Fe^{2+} to less soluble Fe^{3+} ions. The lower Fe^{2+}activity increased the rate of the anodic reaction (Fe? $Fe^{2+}+2e^-$), and the iron oxidizers converted significant amount of Fe^{2+} to Fe^{3+} at the anode side. As a result, insoluble tubercles, consisting of hydrated ferric oxides and biological slimes, grew on the surface. Under the sheltered area beneath deposits oxygen was reduced ($O_2 + 2 H_2O + 4 e^-$? $4 OH^-$). Increased surface concentration of OH^- promoted further precipitation of $Fe(OH)_3$ or $Fe_2(CO_3)_3$, and then the corrosion rates were accelerated. After the death of the iron oxidizing bacteria, the degradation of the bacterial biomass could serve as carbon and energy sources of the other heterotrophic bacteria, so there were many heterotrophic bacteria existed in the system. Furthermore, the denitrifying iron oxidizing bacteria could carry out iron oxidization utilizing nitrate as electron acceptor in an anoxic condition, so it could reproduce in this condition.

158

Acknowledgements

This work was founded through Project TPC-546-91-2410-1-1 from the Taiwan Power Company.

References

[1]   C.M. Mustafa, S.M. Shahinoor Islam Dulal. Corrosion Science. **52.1**,16 (1996).
[2]   S.-D. Chyou, H.- N. Hon, M. -S. Tsai, C.-Y. Wu . W.-L. Shih, K.-L.-Hsieh, H.-Y. Chen and H.-E. Wu . Taiwan Power Plant Report, no. **92**, (2000)
[3]   The Web page of environmental analysis laboratory, EPA, executive yuan, R.O.C - Environmental biological testing method ° 2001
http://www.niea.gov.tw/niea/LIVE/E20951C.htm
[4]   C. F. Yang, C. M. Lee, C. C. Wang. Proceedings of the IWA-Asia Pacific Regional Conference, Asian Waterqual 2003, Bangkok, Thailand, 19–22 October 2003, Session F 1.; Industrial Wastewater Treatment and Management, **106** (2003)
[5]   M. J. Brown and J. N. Lester, Applied and Environmental Microbiology. **40.2**,179 (1980).
[6]   M. Dubois, K. A. Gilles, J. K. Hamilton, P. A. Rebers , and F. Smith, Analytical Chemistry. **28.3**,350 (1956).

The effects of various antibiotics on marine bacteria

B.Sabbaghzadeh [*] , M.Mazaheri Assadi[1] , T.Zahraie Salehi[2]

* Science and research campuse , Azad University ,Tehran,Iran
[1]Biotechnology Center, Iranian Research Organization for Science and Technology.
[2]Faculty of Veterinary medicine,Tehran University.

In this research we studied the symbiosis bacteria with single cell algae *(Chlorella)*.After the recognition of bacteria , the effect of various antibiotics such as chloramphenicol , erithromycin, vancomycin ,streptomycin and etc, tested on bacteria .It is noteworthy that algae and bacteria have mutualism relation with each other. This research shows that choloramphenicol , erithromycin and streptomycin have best result for elimination of bacteria. But streptomycin antibiotic is suggested as the best antibiotics for elimination of bacteria because of its less effect on algae to the two mentioned antibiotics.

1   Introduction

Marine ecosystems are appropriate environment for different kinds of microorganisms because of having almost fix condition of environmental factors such as temperature , salinity and pH. Most of these microorganisms have environmental relations (symbiosis or parasitism). In nature, marine algae and bacteria are coexist (Wang et al.,2003).We can refer to single cell algae *(Chlorella)* that have symbiosis relation with bacteria .We should eliminate bacteria from algae culture   medium because of various reasons such as investigation on capability of production surface active compounds(surfactants) by algae alone. In past, it was used various antibiotics for elimination of bacteria from algae culture medium *(Chlorella)*,that because of disrecognition of bacteria , in some cases not only applicable antibiotics were not effective but also cause damage to structure of algae. Therefore at first, bacteria coexist microscopic algae *(Chlorella),* were recognized and then the effect of various antibiotics on this group of marine bacteria were studied in this research. Finally, appropriate antibiotic and its presize effective amount was suggested for elimination of bacteria from algae culture medium.

2   Materials and methods

2.1 Microorganisms and cultures

The *Chlorella* sp. employed in this study was supplied by the Iran of shrimp research center (Boshehr, Certol station). The microalgal cells were grown in natural seawater supplemented with the nutrient-enriched CH-medium (Sorokin and Krauss, 1958) which was autoclaved in 250 ml flasks. The autoclaved medium was used in all experiments. These flasks were plated on shaker with 150 rpm and under 14h/10h light/dark cyle, with a light intensity of 4200 lux provided by 4 cool-white fluorescent tubes at 24–27°C.The maximum growth of our culture was obtained in 7–10 days.

2.2 Bacteria and culture medium

Each 100 μl volume of an appropriate dilution of algal culture was plated on agar plates prepared with seawater agar (S.W.A ; Himedia).The S.W.A plates were incubated at 37 °C for 48-72 hours. In order to study bacteria , smears were prepared and stained by gram method. Finally the slides were observed by light microscope.Also we took photographes for more exact observation of algae culture medium by electron microscope.

2.3 Biochemical characteristic

Several methods were used such as cultivation in Blood agar and MacConkey media and also gram staining for recognition of bacteria. For final recognition ,the isolated bacterium was cultured in Simmon-Citrate-Agar ,TSI, SS-Agar, Glucose Broth, Lcotose Broth, EMB, SIM, MR-VP, TCBS and also other required tests such as Oxidas (OT), Catalase (CAT) , OF, Amino acids like Ornithin, Cystein and lysine and nitrat test. Meanwhile we used 75% seawater and 25% distilled water instead of 100% distilled water in making all medium except MR-VP medium. Also bacteria cultured in differents pH (4-7-9) and salt (0.1,0.5 and 1gr in 100 ml) in order to determine appropriate pH and salinity for bacteria growth.

2.4 Anthony methods

We employed Anthony method to observe bacteria capsule. Non stained capsule was observed in a purple field and bacteria will take extreme ly stain in this method.

2.5 Study resistance of bacteria to antibiotics

After recognition of bacteria antibiogram test were carried out for investigation resistance of bacteria to different kinds of antibiotics. We completely spread sample of bacteria on Muller Hinton plate by sterile swab and plated different antibiogram disks like tetracycline, ampicillin, vancomycin, chloramphenicol, kanamycin, gentamycin, cefalexin, erythromycin and streptomycin in appropriate distances.
It is noteworthy that dl of the experiments were done in sterile conditions. Then we incubated the plates in 37°C for 48-72 hours .Then the plates were observed for judgment. It should be mentioned that the Muller Hinton medium prepared by seawater.

2.6 The resistance of microalgal cells to antibiotics

Antibiogram tests were also used to determine resistance of microalgal cells to various antibiotics. We spread wholly microalgal cells with sterile swab on seawater plates and then we plated various antibiogram discs in appropriate distances in aseptic conditions. Then we transferred the plates in growth of algal conditions (appropriate light and temp) for 48-72 hours. In the meantime all of the experiment have done for triplet.

2.7 Supplementation of appropriate liquid medium for bacteria growth

The solutions were used in supplementing appropriate growth liquid medium of bacteria are as follows;

Solution 1 contains: Beef extract 10 gr Pepton 10 gr Water 250 cc
Solution 2 contains:
NaCl 28.13 gr
KCl   0.77 gr
Cacl2(2H2O) 1.6 gr
Mgcl2(6H2O) 4.8 gr
NaHCO3 0.11 gr
MgSO4 (7H2O) 3.5 gr
Water 100 cc

For preparation of working solution we used 250ml of solution 1 and 750ml of solution 2. Of course we could use from seawater sampling instead of solution 2. Then final solution was autoclaved for 15 Min. This medium was employed as appropriate liquid medium for growth of bacteria during the experiments.

2.8 Determination of presize effective amount of antibiotic

Serial dilution was used in this regard. Different dilutions of streptomycin antibiotic prepared as follows:
0.1,0.2,0.3,0.4,0.5,0.6,0.7,0.8,0.9,1,2,3,4,5mg/ml.
Then 1ml of every dilutions of antibiotic were pipeted into another tubes and also 1ml of suspension of bacteria that compared with control 1 macfarland, added to the said tubes. Then the tubes were incubated in 37°C for 24h.
Meanwhile we had 2 tubes as control:
Control 1: Culture medium contains bacteria without antibiotic.
Control 2: Culture medium without bacteria contains antibiotic.

3   Results

3.1 Microoroganisms

Microscopic green algae *(Chlorella)* is coexist with bacteria. In study of morphological characteristics of this bacteria, as it is specified in picture, their colonies have characteristics as follows: Tiny, transparent, smooth, convex, aerobic with butter consistency. Gram-negative rod were observed in study of smears under light microscope. They were pleumorphism. In the meantime we observed dark-violet spots in some parts of rod that can show existence of pigment in bacteria. Mean while pictures showed that bacteria have capsule. It have stated some characteristics of bacteria in tables1 and 2. Also we observed the best growth of bacteria in pH=7 and in 1gr salt in 100 ml of medium. Thus bacteria are halophile as we expected.

Table1. Characteristics of bacteria

Characteristics	bacteria
Oxidase reaction	(+)
Nitrate reduce to nitrite	(+)
Glucose	(-)
Lactose	(-)
Indole	(-)
Catalase	(-)
OF	(+)
SIM	(-)
TCBS	(+)
TST	(-)
Growth at 37°C	(+)

Table2. Nutritional characteristics

Characteristics	bacteria
L- Ornithine	(-)
L- Lysine	(-)
L- Lysteine	(-)
MacCoonkey- Agar	(+)
Blood Agar	(+)
EMB	(+)
SS-Agar	(+)
MR-VP	(-)

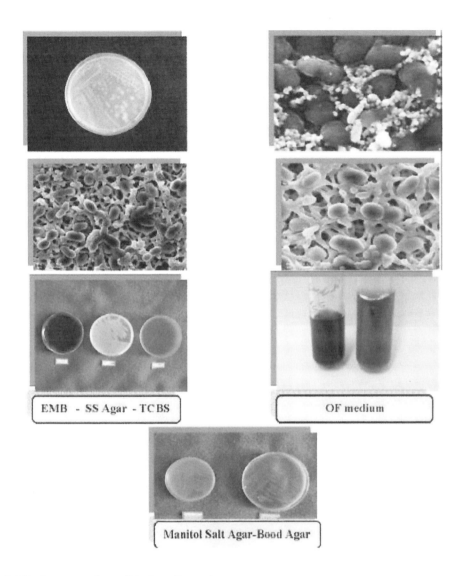

EMB - SS Agar - TCBS

OF medium

Manitol Salt Agar-Bood Agar

3.2 Resistance to the antibiotics of bacteria

Differences in the level of antibiotic resistance of bacteria were determined (Table3).Bacteria was resistant to clinically used antibiotics such as penicillin, vancomynic, cefalexin and erythromycin.Also bacteria was sensitive to tetracycline, ampicillin, chloramphenicol, kanamycin, gentamycin and streptomycin.

164

Table3. Resistance to the antibiotics of bacteria

Antibiotics	Layers(cm)
Ampicillin	23
Cefalexin	12
chloramphenicol	30
Erithromycin	22
Gentamycin	23
Kanamycin	20
Penicillin	12
Streptomycin	21
Tetracycline	20

3.3 Resistance to the antibiotics of microalgal cells

Differences in the level of antibiotic resistance of microalgal cells were determined (Table4). Microalgal cells were resistant to cefalexin,penicillin, streptomycin and vancomycin antibiotics. Also they were sensitive to ampicillin, chloramphenicol, erithyromycin, gentamycin, kanamycin and tetracycline.

Table4. The resistance of microalgal cells to antibiotics

Antibiotics	Layers(cm)
Ampicillin	32
Cefalexin	10
chloramphenicol	36
Erithromycin	30
Gentamycin	36
Kanamycin	33
Penicillin	14
Streptomycin	17
Tetracycline	30

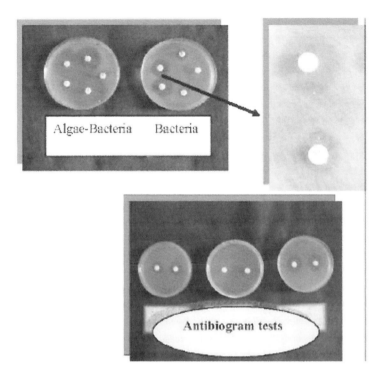

Algae-Bacteria   Bacteria

Antibiogram tests

3.4 Presize effective amount of streptomycin antibiotic

After 24 hours because of becoming turbid, we exit controll of tube from incubator and studied it. We obtained following result in study of tubes test with various dilutions of antibiotic. All tubes (1 to 9) tubes with 0.1 to 0.9 mg/ml dilutions, became turbid. On the other hand, said delutions were not effective in order to prevent bacteria growth but we did not notice growth in next tubes test with 1 to 5 mg/ml dilutions of antibiotic. And tubes were completely transparent. In this manner, we suggest that 1mg/ml is minimum inhibitor concentration (MIC).

4   Discussion

Marine ecosystems with fix condition are appropriate ecosystems for living of different kinds of microorganisms that have various kind of symbiosis relation with each other such as green microalgae*(Chlorella)* that have symbiosis with bacteria.
For different kinds of research like capability of microalge cells for production of surfactants we should eliminate bacteria from algae culture medium.For recognition bacteria coexist with *Chlorella* we done diagnostic and biochemical tests at the first in this research.
This research shows that the bacteria are belong to *Pseudomonas* genus (Bergey et al.,2001). We used different kinds of antibiotics for elimination of bacteria at the second research. It is suggested streptomycin antibiotic for elimination of bacteria from algae culture medium because of less damage to bacteria.

Finally minimum inhibiting concentration of streptomycin (MIC) is 1 mg/ml.

References

[1]   Arvanitodou et al., 1997 M.Arvanitodou, A.Tsakris, T.C. Constantindis and V.C. Katsouyannopulus. Transferable antibiotic resistance among *Slmonalla* strains isolated from surface water ,*water research* 37(1997),pp.1112-1116.

[2]   Ash et al.,2002 R.J.Ash, B.Mauck and M.Morgan. Antibiotic resistance of Gram-negative bacteria in rivers, Unites States, Emerging infections Diseases 8 (2002), pp. 713-716.

[3]   Krieg et al., 2001 N.R.Krieg, J.G.E.Murray,D.J.Brenner,M.P.Bryant, J.G.Holt, J.W.Moulder, N.Pfennig,P.H.A.Sneath,J.T.Staley. Bergey's manual of Systematic bacteriology (2001), pp.196

[4]   Chandrasekarn et al.,1998 S.Chandrasekarn, B.Venkatesh and D.Laithakumari. Transfer and expression of a multiple antibiotic resistance plasmid in marine bacteria, *Current Microbiology* 37 (1998),pp. 347-351

[5]   Davis, 1992 J. Davis. Another look at antibiotic resistance, *Journal of General Microbiology* 138 (1992), pp. 1553-1559.

[6]   De Vincente et al., 1990 A. De Vincente, M. Aviles, J.C.Codina, J.J. Borrego and P. Romero. Resistance to Antibiotics and heavy metals of *Pseudomonas aeruginosa* isolated from natural water, *Journal of Applied Bacteriology* 68 (1990), pp. 625-6320

[7]   Mudryk 2004, Z.J. Mudryk. Occurrence and distribution antibiotic resistance of heterotrophic bacteria isolated from a marine beach.

[8]   Hirsch et al., 1999 R. Hirsch, T. Ternes, K. Haberer and kL.Kratz. Occurrence of antibiotics in the aquatic environment, The Science of the total Environment 225 (1999), pp. 109-118.

[9]   Wang et al., 2003 C.H.Wang, P.Qian, P.Wong and P.H.Hsieh. Antibiotic treatment enhances C2 toxin production by *Alexandrium tamarense* in batch cultures.

[10]   Jan mudryk 2004. Occurrence and distribution antibiotic resistance of heterotrophic bacteria isolated from a marine beach.

The influence of temperature and type of illumination on the biochemical composition and EPA production of *Nannochloropsis gaditana*

M. H. F. Henriques[*], and **J. M. S. Rocha**

University of Coimbra, Department of Chemical Engineering, Pólo II - Pinhal de Marrocos, 3030-290 Coimbra, PORTUGAL

[*]Corresponding author: e-mail: marthe3@eq.uc.pt, Phone: +351-239-798700, Fax: +351-239-798703

Nannochloropsis gaditana is a marine microalga well appreciated in aquaculture due to its nutritional value and valuable chemical compounds, like EPA. Cultures of *Nannochloropsis gaditana* were carried out in tubular bubble reactors, sparged with a flow rate of 120 mL min^{-1} of air and submitted to a 12:12 h light/dark regimen. Two factors have been tested: temperature and type of illumination (fluorescent and halogen lamps). Different light intensities were also used, respectively 15 µE for fluorescent lamps and 33 µE for halogen lamps. The effect of light intensity by fluorescent lamps on the cell growth rate and cell biochemical composition was studied at 7, 11 and 16 µE. Low temperature and the type of light source were found to be determinant either on the growth rate and on pigments and fatty acids production. Fluorescent lamp illumination is more favourable. Both the use of fluorescent lamps and the increase of light intensity (low range) led to higher cell growth rates, as well as an increase of the total fatty acids content. The last increase is mainly due to the accumulation of storage lipids. However, increasing the light intensity does not favour the EPA content of the cells.

Keywords *Nannochloropsis gaditana*, light intensity, EPA

1   Introduction

New sources of biologically active compounds with potential in pharmaceutical, nutraceutical and cosmetic applications are of great economical interest and have been intensively researched. Microalgae are very efficient photosynthetic microorganisms that convert solar energy into different metabolites. Their use as a source of pigments acting as antioxidants (*e.g.* carotenoids) and polyunsaturated fatty acids (PUFA), *e.g.* eicosapentaenoic acid (EPA, C20:5n3), with recognised benefits to human health, such as preventing cardiovascular diseases and alleviating inflammatory symptoms, is therefore of great interest [1]. However, microalgae are very sensitive microorganisms that grow very slowly. Economic production of high value compounds with microalgae will only be possible after optimization of cell growth and accumulation of those metabolites. The best culture conditions must be found for each species, to achieve that goal and to allow their commercial exploitation [2].

2   Material and methods

2.1 Microorganism and growth conditions

Cultures of *Nannochloropsis gaditana* were obtained from Necton [3]. Microalgae were grown on synthetic seawater medium in Erlenmeyer flasks with cotton covers before inoculation in the tubular bubble reactors. The composition of the medium used for both cultures systems was described by Henriques [4]. The culture systems used in this work were 0.7 L tubular bubble reactors (glass tubes with 70 cm length and 5 cm diameter) sparged with air at 120 mL/min flow rate, under two temperatures (18 °C and 25 °C) and two light regimes (15 µE and 33 µE). The light regimes were promoted by different types of light sources, the lowest by fluorescent lamps and the highest by halogen lamps.

2.2 Analytical methods

The cell density concentration was evaluated by measuring the optical density of the culture at 540 nm in a Beckman Du Series 600 spectrophotometer. The absorption at 540 nm was correlated with the dry weight and the cell number per millilitre of culture. The cells were counted using a Neubauer hemocytometer.

The biochemical composition of *Nannochloropsis gaditana* was evaluated during the growth. The content of chlorophyll a was evaluated using the Mackinney method [5]; total proteins were determined by the Lowry method modified by Peterson [6], and carbohydrates were measured with the colorimetric phenol-sulfuric acid method [7]. The fatty acids methyl esters (FAME) were prepared by transesterification of lyophilized cells, according to the Sato and Murata method [8], using nonadecanoic acid (C19:0) as an internal standard. FAME samples were analyzed using a TREMETRICS 9001 chromatograph with a FID detector using a DB225 J&W SCIENTIFIC capillary column (30 m length, 0.25 µm internal diameter and 0.15 µm film thickness). The oven time-temperature program was the following: 70 °C (1 min), 20 °C/min until 180 °C, 5 °C/min until 220 °C (5 min) and 4 °C/min until 240 °C (2 min), giving a total heating time of 27 min. Fatty acids were identified by comparison with retention times of known standards obtained from Sigma.

3   Results and discussion

3.1 Influence of temperature and type of illumination

The type of light was found to be determinant on the autotrophic growth of this microalga. The cultures exposed to fluorescent light exhibited faster growth than the ones that received radiation from halogen lamps, even with superior light intensity (Fig. 1). The range of intensities used was lower than the values mentioned in literature as inducing photoinibition [9, 10, 11]. A stationary phase was reached after 15 days for the cultures under 15 µE (15 µmol photons m^2 s^{-1}). Under these conditions significant growth is observed, and the temperature becomes a determinant factor for the total amount of biomass obtained. Low temperature (18 °C) was more favourable.

3.1.1 Pigments / Chlorophyll

Under fluorescent light, the content of chlorophyll a in the culture increases, until the stationary phase is reached. A decrease is observed afterwards, probably due to the shadow effect caused by neighbour cells when their density becomes high. This phenomenon was not observed with halogen lamps, despite the use of more intense radiation. This is related to the quality of light and the presence of specific radiation wavelengths. Some authors defend that the pigment content in the cells changes according to the availability of light [12]. Cells submitted to intense radiation do not need high concentration of pigments for efficient photosynthesis; on the other hand, low radiation promotes the production of chlorophyll a. The type of light was not considered by these authors. The electromagnetic spectrum of fluorescent lamps is more similar to the sun light and contains more radiation with adequate wavelengths (680 nm) for the absorption by chlorophyll a, while halogen lamps are relatively poor on that specific radiation [13].

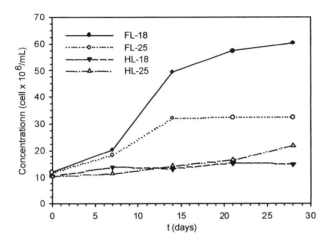

Fig. 1 Cell concentration. FL-18: 18 °C and 15 μE (fluorescent lamp - FL). FL-25: 25 °C and 15 μE (FL). HL-18: 18 °C and 33 μE (halogen lamp - IIL). IIL-25: 25 °C and 33 μE (HL).

3.1.2 Total proteins and carbohydrates

The evolution of total proteins during the cellular growth followed a similar pattern to the content of chlorophyll a. This can also be used as an indicator of the growth, since it is proportional to the cellular concentration during the exponential phase.

Carbohydrates are consumed during the growth phase because these compounds are the most accessible energy sources that can be used by the cells during growth. Older cells have lower proportions of proteins and carbohydrates per unit of dry weight, but their lipidic content is higher [14].

3.1.3 Fatty acids

Nannochloropsis gaditana is referred as a species with a high content of polyunsaturated fatty acids (PUFAs) acting as structural lipids (galactolipids); the most abundant are the eicosapentaenoic acid (EPA) and the arachidonic acid (AA). However, their relative proportion with the saturated and monounsaturated fatty acids, called storage lipids (triacilglycerols), is still quite low [14]. When the growth is intense (fluorescent light), the difference between the contents of these classes of lipids becomes even more evident. A duplication of the storage lipids was observed with a temperature of 18 °C. Using halogen lamps, with higher light intensity, the contents of EPA and AA in the stationary phase were only half of the ones achieved with fluorescent lamps [4].

3.2 Influence of light intensity on growth rate and biochemical composition

As the radiation from fluorescent lamps favoured the cell growth, the effect of its light intensity was evaluated (Fig. 2a). The results obtained with the light intensities tested (7, 11 and 16 µE) show this variable as a limiting factor on growth, because its increase intensifies cell division. In the range of light intensities used, no photoinhibition was found. The biochemical composition was only evaluated at the stationary phase of growth. The content of chlorophyll a decreased with the increase of lightning; a greater availability of radiation caused a reduction in the production of pigments. The protein and carbohydrates contents also decreased under conditions where growth was favoured. The total lipids content was approximately doubled when the light intensity was 16 µE. In Fig. 2b it is possible to observe the fatty acids distribution. A high content of storage lipids (palmitic acid - C16:0 and palmitoleic acid - C16:1) was responsible for the strong increase in the total content of lipids at the highest light intensity tested. If the main purpose is the maximization of structural lipids content - like EPA (C20:5n3) - smaller light intensities are preferable. However, a balance must be achieved between the content of structural lipids and the cell density obtained.

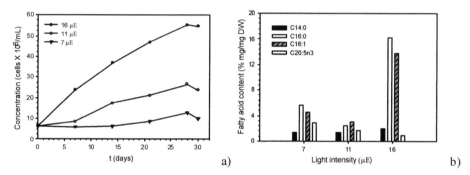

Fig. 2 (a) Cells concentration for different light intensities. (b) Fatty acids content *versus* light intensity (fluorescent lamps).

4 Conclusions

The type of light source was found to be determinant both on the growth rate and on the production of pigments and fatty acids. Fluorescent lamps illumination and lower temperatures were more favourable in all cases. An increase of light intensity with fluorescent lamps induced higher cell growth rates and total fatty acids content. The last increase is mainly due to the accumulation of storage lipids, particularly palmitic and palmitoleic acids. However, higher light intensities do not favour the EPA content of the cultures. The total contents of proteins and carbohydrates, as well as their evolution during cell growth, seem to be not influenced by the environmental factors tested.

References

[1]   M. R. Brown, S. W. Jeffrey, C. D. Garland, CSIRO *Mar. Lab. Rep.*, **205**, 44, (1989).
[2]   A. Richmond, *Hydrobiologia*, **512**, 33-37, (2004).
[3]   Necton, http://www.necton.pt/, (2004).
[4]   M. H. F. Henriques, Valorização tecnológica da microalga marinha *Nannochloropsis gaditana*, MSc. Dissertation, Department of Chemical Engineering, University of Coimbra, Portugal, (2004).
[5]   G. Mackinney, *J. Biol. Chem.*, **140**, 315-322, (1941).

[6]   G. L. Peterson, *Anal. Biochem*., **83**, 346-356, (1977).

[7]   J. A. Hellebust, J. S. Craigie (eds.), Handbook of Phycological Methods. Physiological and Biochemical Methods (Cambridge University Press, Cambridge, 1978), pp. 95-97.

[8]   N. Sato, N. Murata, *Biochim. Biophys. Acta*, **619**, 353-366, (1980).

[9]   S. M. Renaud, D. L Parry, L. V. Thinh, C. Kuo, A. Padovan, N. Sammy, *J. Appl. Phycol*, **3**, 43-45 (1991).

[10]   L. M. Lubián, O. Montero, I. Moreno-Garrido, I. E. Huertas, C. Sobrino, M. González-del Valle, G. Parés, *J. Appl. Phycol*., **12**, 249-255 (2000).

[11]   A. A. Gitelson, Y. A. Grits, D. Etzion, Z. Ning, A. Richmond, *Biotech. Bioeng*., **69**, 516-525 (2000).

[12]   T. Fisher, T. Berner, D. Iluz, Z. Dubinsky, *J. Phycol*., **34**, 818-824, (1998).

[13]   A. Kommareddy, G. Anderson, Study of light requirements of a photobioreactor, *In Proceedings of the Annual ASAE International Meeting*, Nevada, EUA, (2003).

[14]   J. Fábregas, A. Maseda, A. Domínguez M. Ferreira, A. Otero, *Biotech. Letters*, **24**, 1699-1703, (2002).

Two-phases olive mill solid waste treatment in an anaerobic continuous stirred tank reactor at diluted influent concentrations: evaluation and implicated microorganisms

B. Rincón[1], R. Borja[*1], F. Raposo[1], J. M. Gonzalez[2], M. C. Portillo[2] and C. Saiz-Jimenez[2]

[1] Instituto de la Grasa, CSIC, Avda. Padre García Tejero 4, 41012 Sevilla, SPAIN.
[2] Instituto de Recursos Naturales y Agrobiología, CSIC, Apartado 1052, 41080 Sevilla, SPAIN.

* Corresponding author: e-mail: rborja@cica.es, Phone: +34 95 4689654, Fax: +34 95 4691262

Olive oil manufacturing, like most industrial processes, has undergone evolutionary changes. Technological transformation of the olive oil industry has caused the massive change of the traditional three-phases manufacturing systems by the new two-phases processes. This work focused on the new aqueous solid residue from the primary centrifugation that contains all the vegetation water of the olive, and produce considerable pollution problems. Among the different alternatives for the treatment of the two-phases olive mill solid waste (OMSW), anaerobic digestion has a number of important advantages. Herein, the performance and microbial communities of a continuous stirred tank reactor (CSTR) treating two-phases OMSW was analysed. A molecular characterization of the microbial communities involved in the process was also accomplished. Molecular identification of microbial species was performed by PCR amplification of 16S ribosomal RNA genes, denaturing gradient gel electrophoresis (DGGE), cloning and sequencing.

Keywords: Two-phases olive mill solid waste; anaerobic digestion; performance; microbial community

1 Introduction

Olive oil manufacturing has a great importance in the Mediterranean basin. This industry has undergone important changes in the last years. Lately, a two-phases centrifugation process, for separating the olive oil fraction from the vegetable solid material and vegetation water, has been introduced [1]. This improvement reduces water consumption [2]. Currently, more than 90 % of the Spanish olive oil is extracted with the two-phases system. The aqueous solid waste from a primary centrifugation, also called olive mill solid waste (OMSW), has an average composition of: 60-70% water, 13-15% lignin, 18-20% cellulose and hemicellulose, 2.5-3% olive oil retained in the pulp, and about 2.5% mineral solids. Among the organic components, the following components can be highlighted: sugars (3%), volatile fatty acids (C2-C7) (1%), polyalcohols (0.2%), proteins (1.5%), polyphenols (0.2%) and other pigments (0.5%) [3].
The anaerobic digestion is a biological process in which a complex community of microorganisms work in a stable, self-regulating steady-state converting waste organic matter into a mixture of carbon dioxide and methane gases [4]. Anaerobic digestion has a great number of advantages: low nutrient requirements, energy savings, generation of low quantities of sludge, excellent waste stabilization, production of biogas (methane) without the requirement of pre-treatments of the residues [5, 6].
The aim of this work was to carry out an evaluation of the substrate utilization and methane production rates in the anaerobic digestion of two-phases OMSW using low organic

loading rates (OLR). Simultaneously a molecular characterization of the microbial communities involved in this anaerobic digestion process of this substrate was also carried out by using a stirred tank reactor operating at mesophilic temperature (35°C).

2 Material and methods

2.1 Equipment

Experiments were carried out in an anaerobic continuous stirred tank reactor (CSTR)(2 L working volume) with a thermostatic jacket set at 35 °C. The reactor was fed daily by means of an external feeder and liquid effluent removed daily through a hydraulic seal, comprising 25 cm liquid column, designed to prevent air from entering the reactor and biogas from leaving it. The volume of CH_4 produced in the process was measured using a 5 litre Mariotte reservoir [7] fitted to the reactor. CO_2 produced in the process was removed by bubbling the gas mixture through a NaOH solution (3 M).

2.2 Two-phases olive mill solid waste (OMSW)

Two-phases OMSW was collected from the Experimental Olive Oil Factory located in the "Instituto de la Grasa" (CSIC) of Sevilla, Spain. The characteristics of the OMSW are summarised in Table 1.

Table 1 Composition and features of the two-phases OMSW used as influent

pH	5.3
Moisture	86.7
Total COD	162
Soluble COD	57.5
TVFA	1.4
Alkalinity	1.1
Total Solids	143
Mineral Solids	17
Volatile Solids	126
Total suspended solids	106
Mineral suspended solids	11
Volatile suspended solids	95

COD, chemical oxygen demand; TVFA, total volatile fatty acids (as acetic acid); alkalinity (as $CaCO_3$). All amounts, except pH and moisture, are expressed in g/L, and the moisture is expressed in %.

2.3 Experimental Procedure

During the experiments the organic loading rate (OLR) was gradually increased from 0.75 to 3 g COD/L·d, using feed flow-rates of 0.0093, 0.0185, 0.0278 and 0.0370 L/d of the OMSW studied. Once steady-state conditions were achieved at each feed flow-rate (Table 2), the daily volume of CH_4 produced, total and soluble COD (COD_t and COD_s, respectively), pH, total volatile fatty acids (TVFA), alkalinity and volatile solids (VS) of the different effluents were determined. Samples were collected for at least six consecutive days. Steady-state values for the analyzed parameters were the average of these consecutive measurements. Each experiment had a duration of 1-2 times the corresponding HRT.

2.4 Chemical Analyses

All analyses were carried out according to the recommendations of the Standard Methods of APHA [8].
The pH and CH_4 production were determined daily, whilst the remaining parameters were the averages of six determinations taken over six days after the steady-state conditions were reached.

2.5 Molecular characterization of microbial communities

Diversity of microbial communities was analyzed by molecular techniques. DNA was extracted from steady-state samples using the Nucleospin Food DNA extraction kit (Macherey-Nagel, Düren, Germany). The 16S ribosomal RNA sequences were amplified by PCR using the primer pairs 27F and 1522R for bacterial sequences and 20bF and 1492bR for archeal sequences [9]. Microbial community fingerprints were obtained for each sample by Denaturing Gradient Gel Electrophoresis (DGGE) using the PCR amplification products obtained from the primer pairs 341F-GC and 518R for bacterial sequences, and 344F-GC and 518R for archaeal sequences [10].

3   Results and discussion

3.1 Anaerobic reactor performance.

Table 2 shows the experimental parameters obtained for the different organic loading rates (OLR) studied during this work.

Table 2 Steady-state parameters for different operational conditions (Influent total COD=162 g/L)

OLR (g COD/L·d)	0.75	1.5	2.25	3.00
q (flow-rate) (L/d)	0.0093	0.0185	0.0278	0.0370
pH	7.7	7.3	7.3	7.4
*r_{CH4} (L $_{CH4}$/L·d)	0.164	0.326	0.479	0.659
Total COD (g/L)	4.75	5.25	6.24	7.04
Soluble COD (g/L)	2.67	2.91	3.56	4.43
VS (g/L)	3.7	4.2	5.1	7.0
TVFA (mg acetic acid/L)	95	180	270	440
Alkalinity (mg CaCO₃/L)	4490	4060	2900	3000
TVFA/Alkalinity	0.02	0.04	0.08	0.12

Values are the averages of six determinations taken over six days after the steady-state conditions were reached. Deviations between determinations were lower than 5 % in all cases. *r_{CH4}: methane production rate.

Total and soluble COD of the effluents increased at increasing OLR, as illustred in Figure 1. These COD increases were related to similar increases of TVFA in the effluent, indicating that a stable fraction of total COD was constituted by TVFA.

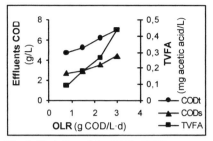

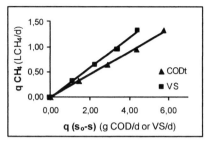

Fig. 1 Variation of the COD_t, COD_s and TVFA with the OLR.

Fig. 2 Variation of the q_{CH4} as a function of the product of $q(S_0-S)$ for the two-phases OMSW used.

As can be seen in Figure 2, the volume of CH_4 produced per day is related to the product of $q(S_0-S)$ according to equation:

$$q_{CH4} = Y_p q (S_0 - S)$$

where S_0 and S are the substrate concentrations (expressed as g COD/L or g VS/L) at the digester inlet and effluent respectively, and q is the feed flow-rate. These data result in CH_4 yield coefficients (Y_p) over the OLR range studied of: 0.225 (95% confidence range: ± 0.004) L methane STP/g COD removed and 0.290 (95% confidence range: ± 0.006) L methane STP/g VS removed. These values are higher than those obtained in previous studies of anaerobic digestion of other types of two-phases OMSW with similar influent substrate concentration [3].

3.2 Microbial communities

In this study, the microbial communities of a continuous stirred tank anaerobic reactor treating two-phases OMSW at low OLRs were surveyed using molecular techniques. Figures 3 and 4 show the diversity of Bacteria and Archaea, respectively, found in the analyzed samples at four different OLRs (A:0.75, B:1.50, C:2.25, D: 3.00 g COD/L·d).

Results showed differences in the microbial, both bacterial and archaeal, communities when varying OLRs as suggested by comparative DGGE analysis of microbial community fingerprints.

During this experimental study the most frequently encountered microbial group were the **Firmicutes** (53.3 % of analyzed sequences), represented mostly by members of the Clostridiales, as Figure 3 shows. The **Chloroflexi** represents an important bacterial group under the studied conditions (23.4 %) and has been reported as a major constituent in anaerobic systems. The Chloroflexi group represented percentages >15 % in previous studies [11, 12] which is in agreement with our results for the anaerobic processing of two-phases OMSW. The **Gamma-Proteobacteria** (8.5% of sequences; represented mainly by the genus *Pseudomonas*), **Actinobacteria** (6.4%), and **Bacteroidetes** (4.3%) are significant components of the microbial communities during the anaerobic decomposition of OMSW.

176

The methanogenic community showed low diversity at the lowest OLR studied (0.75 g COD/L·d) and the community became increasingly diverse for increasing OLRs within the studied OLR range (Figure 4).

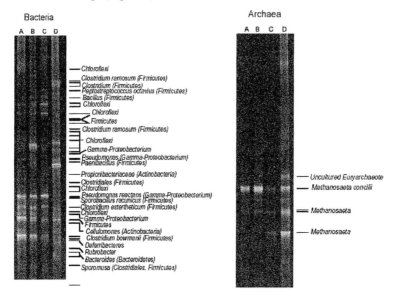

Fig. 3 (left) DGGE analysis of the diversity of bacterial communities at diferents OLRs.

Fig. 4 (right) DGGE analysis of the diversity of archaeal communities at diferents OLRs.

The major component detected within the Archaea was ***Methanosaeta concilii*** (formerly ***Methanothrix soehngenii***). Results showed the existence of molecular diversity within the genus ***Methanosaeta*** in the anaerobic process under study, as can be seen in Figure 4, stable synthrophic associations have been reported between ***Methanosaeta* spp.** and bacteria [13]; for instance, the genus ***Syntrophobacter* (Firmicutes)** has been detected in our study and could be playing a role in this type of interations.

Acknowledgements

The authors wish to express their gratitude to the "Comisión Interministerial de Ciencia y Tecnología" (project REN2001-0472/TECNO) and "Junta de Andalucía" for providing financial support.

References

[1]   R. Borja, J. Alba and C.J. Banks, Process Biochem. **31**, 219 (1996)
[2]   J Alba, Proceedings of the Simposium Científico-Técnico Expoliva 93, Jaén, Spain, May 1993.
[3]   R. Borja, B. Rincón, F. Raposo, J. Alba and A. Martín, Process Biochem. **38,** 733 (2002).

[4]   Jr. M. C. Sterling, R. E. Lacey, C. R. Engler, S. C. Ricke, Bioresour. Technol. **77**, 9 (2001).

[5]   H. Kang, P. Weiland, Bioresour. Technol. **40**, 245 (1992).

[6]   P. Weiland, Water Sci. Technol. **27**, 145 (1993).

[7]   A. Martín, R. Borja, I. García and J. A. Fiestas, Process Biochem. **26**,101 (1991).

[8]   APHA (American Public Health Association). Standard Methods for the Examination of Water and Wastewater, 17th ed., APHA, Washington DC, USA (1989)

[9]   V. J. Orphan, K. U. Hinrichs, W. Ussler III, C. K. Paul, L. T. Taylor, S. P. Sylva, J.M. Hayes, E. F. DeLong, Appl. Environ. Microbiol. **67**, 1922 (2001).

[10]   J.M. Gonzalez, C. Saiz-Jimenez, J. Separat. Sci. **27**, 174 (2004).

[11]   M. S. Elshahed, J.M. Senko, F. Z. Najar, S. M. Kenton, B. A. Roe, T. A. Dewers, J. R. Spear, L. R. Kumholz, Appl. Environ. Microbiol. **69**, 5609 (2003)

[12]   S. Rossetti, L. L. Blackall, M. Majone, P. Hugenholtz, J. J. Plumb and V. Tandoi, Microbiol. **149**, 459 (2003)

[13]   W. M. Wu, R. F. Hickey and J. G. Zeikus, Appl. Environ. Microbiol. **57**, 3438 (1991).

Use of rRNA gene restriction patterns in *Colibacillosis* by *E. coli* O78, O88 and non typable strains in wild birds

Jiménez, P.A.[*][1]

1 Sección de Microbiología. Facultad de Ciencias Experimentales y de la Salud. Universidad San Pablo C.E.U. Ctra. de Boadilla del Monte Km. 5,600. Madrid. Spain. Phone: 91 3724754. Fax: 913510496.

[*]Corresponding author: e-mail: pgimgom@ceu.es

Since 1999 we have isolated and selected the most relevant colibacillosis cases. To prove these diagnoses it is very important to carry out the anatomopathological and histopathological study. Moreover, it is essential to obtain a microbiological culture of *E. coli* in all injured organs.

To properly finish this diagnosis, all the isolates must have a clonal origin. We did phenotypic characterization, serotyping and rRNA gene restriction patterns.

In each case, all the strains showed the same phenotypic profile. Serotyping method couldn't always be used because new, yet undescribed serotypes and non typable APEC strains exist, as found in case 2. For this reason, we needed yet another discrimination method. We employ rRNA gene restriction patterns, never used before in wild life birds to resolving this problem. The result was satisfactory. We concluded that the rRNA gene restriction pattern is a good molecular method to confirm a colibacillosis diagnosis, even in new and non-described strains.

This is the first report of septicemia in the Little owl *(Athene noctua),* White stork *(Oconia ciconia)* and Osprey *(Pandion haliaetus)* in Spain.

1   Introduction

E. coli is found in natural environments such as soil, sewage, and water. It is an important agent of avian colibacillosis and has been responsible for causing disease in a variety of wild and domestic birds. Although avian pathogenic *E.coli* (APEC) (Dho-Moulin & Fairbrother, 1999) have been well documented in both domestic poultry and wild waterfowl, there has been little information concerning this disease in wild birds.

APEC are found in the intestinal microbiota of healthy birds and most of the diseases associated with them are secondary to environmental and host predisposing factors. The type of anatomopathologic and histopathologic lesions associated with avian colibacillosis depends on the duration of the disease and the virulence of the strains (Baliarsingh *et al.*, 1993). These studies are vital to determine a correct diagnosis but they need microbiological, genetic and molecular explanations (Dho-Moulin & Fairbrother, 1999).

One of the most important substances involved in a colibacillosis process is aerobactin. This iron sequestering system is an important virulence factor identified in APEC (Ngeleka et al., 1996; Blanco et al., 1997; Dho-Moulin & Fairbrother, 1999).

The emergence of bacterial pathogens resistant to antimicrobial agents has compromised the ability of veterinarians to control many of these pathogens with antimicrobial agents (CES, 1998; Fedorka-Cray *et al.*, 1998; Tollefson et al., 1998). These analyses could reveal important signs of these animals I health and they are necessary to apply an effective treatment. For this reason, it is necessary to determine the resistances of our isolates in wild birds.

A classification system in which E *coli* is subdivided on the basis of O (Lipopolysaccharide), H (Flagellar) and K (surface) antigens has been described. APEC isolates commonly belong to certain serogroups, 01/ 02 and 078 (Dho-Moulin & Fairbrother, 1999; Gomis *et al.*, 2001) and to a restricted number of clones (Dho-Moulin & Fairbrother, 1999). More recently, molecular biology techniques directly detecting nucleotide sequence variations in chromosomal DNA of bacterial isolates have been applied to clinical and epidemiological investigations. Moreover to diagnose a colibacillosis, it is essential to isolate E *coli* in different organs and to prove that these strains have a donal origin, but serotyping is not always a definitive method to proving the donal origin of the isolates.

Recently, Machado, 1998 has suggested that computer-interpreted ribotyping may replace serotyping when the latter is not available or when the strains are not typable, such as in this study. rRNA gene restriction patterns is a good method to distinguish if the strains have a donal origin or not in *Salmonella* (Gruner *et al.*, 1994; Guerra *et al.*, 1997; Landeras *et al.*, 1997). Yet, until now, nobody has used this method in APEC in wild birds. The same as in humans (Machado *et al.*, 1998), this kind of analysis can be useful in proving a definitive colibacillosis djagnosis in wild birds.

The aim of this study is to establish a reliable molecular method for the diagnostic of colibacillosis in wild birds.

2   Materials and methods

2.1 Clinical and Necropsy data

Cases included in this study involved the Little owl (*Athene noctua*) case 1, White stork (*Ciconia ciconia*) case2 and Osprey (*Pandion haliaetus*) case 3, which had died and were submitted for necropsy to the Pathology Service of the Veterinary Medical Hospital GREFA (Grupo Para la Recuperación de la Fauna Autóctona) according to the recommendations of the Proceedings of the Association of Vetinarians (Montali, 1988). Swabs for microbiologic isolation were collected during necropsy from selected tissues. A diagnosis of avian colibacillosis was made based on gross and caracterization by phenotypic probes in APY Expression Plus Vitec System (BioMèrieux).

2.2 In vitro studies

Serotyping Identificaction of O antigens was carried out by a microtechnique described by Guinee *et al*, 1972 and modified by Blanco *et al.*,1998 with all available O (O1 to O171) antisera. All antisera were adsorbed with the corresponding cross-reacting antigens to remove non-specific agglutinins.

Aerobactine virulence gene. PCR for amplification of aerobactine-specific sequences was performed as previously described by Yamamoto *et al* (1995) using the primers AER 1 (5′-TAC CGG ATT GTC ATA TGC AGA CCG T-3′) and AER 2 (5′-AAT ATC TTC CTC CAG TCC GGA GAA G-3′).

rRNA gene restriction patterns. *Mlu*I-ribotypificaction and *Cla*I-ribotypificaction were processed according to Machado *et al.*, 1998 protocol.

Antibiotic resistance profiles. Antimicrobial resistance profiles of the test organisms were determined by the ATB SYSTEM Expression Plus (BioMèrieux) in a semiautomatic gallery

ATB G-5. The antimicrobics tested were amoxicillin (8 and 16 mg/l, amoxicillin-clavulanic acid (8/4 and 16/8 mg/l), piperacillin (16 and 64 mg/l), piperacillin – tazobactam (16/4 and 64/4), Ticarcillin (16 mg/l), cefalotin (8 mg/l), cefotaxim (8 and 32 mg/l), ceftriaxon (8 and 32mg/l), ceftazidim (8 and 32 mg/l), aztreonam (8mg/l), imipenem (4mg/l), ceftazidime 1mg/l, cotrimoxazol (2 and 38 mg/l), tobramycin (4mg/l), Amikacin (16 mg/l), gentamycin (4mg/l), netilmicin (8mg/l), pefloxacin (2 and 4 mg/l), ciprofloxacin (1 and 2 mg/l).

3 Results

Necropsy results: Three animals presented for necropsy were diagnosed as having avian colibacillosis. All birds had died during the course of their illness. The more relevant lesions observed in the postmortem study were: Case 1.- Polyarthritis Polyserositis, with presence of fibrinous exudate on pericardial and air sacs surface. Hepatomegaly and splenomegaly with fibrinous exudates on serosal surface. Case 2.- Generalized caquexia, Hidropericardium, enteritis of small and large bowell. Case 3.- Focal areas of pneumonia and aerosaculitis

Isolate characterization: The antimicrobial susceptibilities are shown in Table 1. In each case all the isolates have the same antibiotic resistance profiles, but not the same among each other.

Table 1. Comparison of serotype and antimicrobial susceptibilities of avian isolates.

	Aerob	Serotype	Antibiotics																		
			AMO	AMC	PIC	TZP	TIC	CFT	CTX	CRO	CAZ	AZM	IMI	CA1	TSU	TOB	AKN	GEN	NET	PEF	CIP
CASE 1	+	O 78	R		I		R	R							R					R	R
CASE 2	+	NON TYPABLE	R	I	R		R	R												R	R
CASE 3	+	O 88	R		R		R	R							R					R	R

AMO: amoxicillin; AMC: amoxicillin-clavulanic acid; PIC: piperacillin; TZP: Piperacillin – tazobactam TIC: Ticarcillin; CFT: cefalotin; CTX: cefotaxim; CRO: ceftriaxon; CAZ: ceftazidim; AZM: aztreonam, IMI: Imipenem; CA1: ceftazidim (1mg/l); TSU: cotrimoxazol; TOB: tobramycin; AKN: Amikacin, GEN: gentamycin; NET: netilmycin; PEF: pefloxacin; CIP: ciprofloxacin.

Aerobactin production. All 18 isolates examined were aerobactin-positive (Table 1).
Serotyping. Among the isolates"11 (78%) were typable with specific antisera (Table 1). The serogroups observed were 078 (7 isolates) and 088 (4 isolates), which belonged to case 1 and case 3, respectively. On the other hand, every isolates in case 2 were non-typable. Isolates of serogroup 078 (case 1) were found on pure culture in hearts, lungs, abdominal air sacks, guts, pancreas, spleen and kidneys. Non-tjpable serogroups (case 2) were isolated on pure culture in heart, lungs, guts, pancreas and spleen. Finally, isolates of serogroup 088 (case 3) were found on pure culture in heart, lungs, guts and spleen.
*rRNA gene restriction patterns. Mlu*J-ribotyping is recommended as a primary epidemiological marker. Strains with similar *Mlu*I-ribotype should then be submitted to *Cla*I- ribotyping. In our cases all the isolate seemed to have a clonal result similar to a clonal origin but with no epidemiological relationship among the 3 cases. The results are

shown in figure 1a and figure 1b for case 1 (*Mlu*I-ribotype and *Cla*I-ribotype, respectively); figure 2a and 2b for case 2 *(Mlu*I-ribotype and *Cla*I-ribotype, respectively) and figure 3a and 3b for case 3 *(Mlu*I ribotype and *Cla*I-ribotype, respectively).

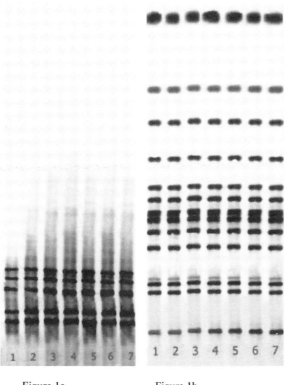

Figure 1a   Figure 1b   Jiménez, P.A.

Figure 1. rRNA gene restriction patterns of case 1 from hearts, lungs, abdominal air sacks, guts, pancreas, spleen and kidney isolates (lanes 1 to 7) after digestion with *Mlu*J (1a) and *Cla*I (1b).

4   Discussion

Wild birds only show symptomatology when they are really ill. This is an adaptation mechanism for survival in the environment, since the symptomatology makes them less competent in their environment. For this reason, in the absence of a veterinary symptom, it is important for the avian clinician to be capable to conduct a systematic necropsy. Proper preparation of tissues and other specimens for the laboratory ensures the quality of the necropsy results. Keeping records of necropsies assists in understanding disease trends in an avian collection, thus giving the practitioner the ability to carry out prophylactic programs (Montali, 1988).

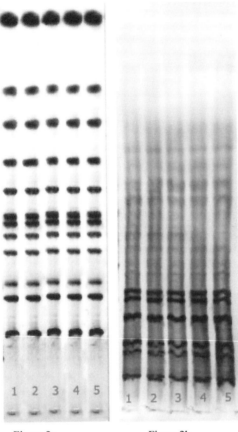

Figure 2a   Figure 2b   Jiménez, P.A.

Figure 2. rRNA gene restriction patterns of case 2 from hearts, lungs, guts, pancreas and spleen isolates (lanes 1 to 5) after digestion with *Mlu*J (2a) and / *Cla*I (2b). Migration from bottom to top

There are many different anamopathological and histopathological lesions associated with colibacillosis (Dho-Moulin & Fairbrother, 1999). Normally, these lesions depend on two factors: the duration of the disease and the virulence of the APEC strains (Baliarsingh *et al.,* 1993). According to Dho-Moulin & Fairbrother, 1999 our animals showed a colibacillosis process. Unfortunately, anamopathological and histopathological analyses are not sufficient to determine a correct colibacillosis diagnosis. Molecular, genetic and microbiological studies are essential in these cases (Dho-Moulin & Fairbrother, 1999).

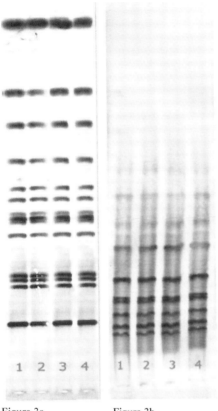

Figura 3a   Figura 3b   Jiménez, P.A.

Figure 3. rRNA gene restriction patterns of case 3 from hearts, lungs, guts and spleen isolates (lanes 1 to 4) after digestion with *Mlu*J (3a) and *Cla*I (3b). Migration from bottom to top.

Iron is the best known and perhaps the most important inorganic cation micronutrient. Essentially all bacteria cells contain Fe2+ (incorporated in heme -containing proteins such as cytochromes and in nonheme iron proteins). In *E coli*, six Fe $^{3+}$ transport systems can operate in aerobic conditions (Braun & Hantke, 1991). These systems are considered in depth. The major potential virulence factor associated with replication of systemic *E coli* is the production of siderophores such as aerobactin. Much of the relationship between siderophores and virulence is based on epidemiological correlations between the production of siderophores by systemic *E coli* isolates (Gulig, 1996) in vivo in a diffusible form, which may be an important step in the production of disease by intestinal *E coli* (Dho-Moulin & Fairbrother, 1999). Aerobactin is an important virulence factor, particulary in invasive diseases in birds (Blanco *etal.*, 1997). We detected aerobactin production in all the strains. It is very important virulence factor in producing colibacillosis by APEC, but there was not sufficient evidence to confirm a clonal origin of the isolates.

Recently, antibiotic resistance in *E coli* has been reported in wild birds world-wide (Bass *et al.*, 1999). The possibility of resistance organisms developing with no antimicrobial agent is

184

a reality. However, when choosing a treatment, it is wise to choose an antimicrobial agent that has the least predisposition for creating resistance organisms. The of antibiotics in both humans and animals contributes to selection pressure resulting in this resistance (Fedorka-Cray *et al.*, 1998). There is also a concern that resistance in animal pathogens may be transferred to wild birds via food-borne pathogens (Tollefson *et al.*, 1998). This argument deserves deeper reflection. As in Bass *et al.* (1999) results, our strains showed a multi-drug-resistance to quinolones, β-lactamics and sulfamides.

On the other hand, in each case, all the isolates showed the same phenotipic profile of antibiotic resistance. This is another indication of the clonal origin of the isolates, but it is not definitive.

In the colibacillosis diseases, serogroups 01, 02, 071, and 078 are the most prevalent (Allan *et al.*, 1992; Dho-Moulin & Fairbrother, 1999) and 088 is also contemplated by Blanco *et al*, 1998. Cases 1 and 3 are included in this group. But new, yet undescribed serotypes exist (Allan *et al.*, 1992). These strains, currently considered non-typable, are important in clinical examinations of wild birds and should be the object of further study, due to their singularity as well as their virulence. Another important aspect of this analysis is that in a colibacillosis diagnosis, all the isolates must belong to the same serotype. In this sense, our strains fall inside this premise.

Finally it was necessary to find a technique capable of proving the donal origin of the isolates in each case. rRNA gene restriction patterns have emerged over the last few years as one of the best strain-typing methods we have for bacterial infections, both environmentally-acquired and wild life hospital-acquired. Almost all bacteria are typable by this method. Thus, rRNA gene restriction patterns with one restriction endonuclease could not separate some serovars, but the ribotype pattern combinations with the two endonudeases *MluI*, and *ClaI* showed good differentiation in *E coli* serovars (Machado *et al.*1998) and in our strains. The same strains showed the same common bands.

In conclusion, all results suggest that the diversity of phenotypes detected partially explains the multifactorial nature of avian colisepticemia. But the essential proof in determining a colibacillosis diagnosis was rRNA gene restriction patterns. This is the first time that rRNA gene restriction patterns have been used to determine colibacillosis in wild birds.

This is the first report of septicemia in the Little owl *(Athene noctua)*, White stork *(Oconia ciconia)* and Osprey *(Pandion haliaetus)* in Spain.

Acknowledgements

This work was supported by the Universidad San Pablo C.E.U. "Estudio de marcadores epidemiológicos y resistencia a antibióticos en Microbiota normal y patógena de origen Humano y animal 13/99 proyect". We wish to thank Linda Hamalainen for help in preparing the manuscript. We wish to thank also Dr Jorge Blanco to the Universidad de Santiago de Compostela (facultad de Veterinaria, laboratorio de referencia de *E. coli*. ,Lugo, Spain) for aerobactin determination and serotiping.

References

[1]   Allan B.J., Van-den-Hurk J.V. & Potter A.A. (1993). Characterization of *Escherichia coli* isolated from cases of avian colibacillosis. *Can. J. Vet. Research*. 57: 3, 146-151, 33 ref.

[2]   Baliarsingh-S.K., Rao A.G. & Mishra P.R. (1993). Pathology of experimental colibacillosis in chicks. *Indian Vet. J*. 1993, 70: 9, 808-812.

[3]   Bass L., Liebert C.A., Lee M.D., Summers A.O., White D.G., Thayer S.G. & Maurer J.J. (1999): Incidence and characterization of integrons, genetic elements mediating multiple-drug resistance, in avian *Escherichia coli*. *Antimicrob Agents Chemother* ,43(12):2925-9

[4]   Blanco, J.E., Blanco, M., Mora, A. and Blanco, J. 1997. "Production of toxins (enterotoxins, verotoxins, and necrotoxins) and colicins by *Escherichia coli* strains isolated from septicemic and healthy chickens. Relationship with in vivo pathogenicity". J Clin. Microbiol. 35 (11): 2953-2957.

[5]   Blanco J.E., Blanco M, Mora A, Jansen WH, García V, Vazquez ML, Blanco J.1998. Serotypes of *Escherichia coli* isolated from septicaemic chickens in Galicia (northwest Spain). Vet Microbiol. 61(3):229-35.

[6]   Braun V. & Hantke, K. (1991). Genetics of bacteria iron transport, (pp. 107-138). In G. Winklemann (Ed.) *Handbook of microbial Iron Chelates*. CRC Press, Inc., Boca Raton, Fla.

[7]   CES: Comité Económico y Social (1998): 1118/98 EN-MAI/FGA/ORT/sf/mf. *"Dictamen sobre la resistencia a los antibióticos como amenaza para la salud pública"* (Iniciativa) Bruselas, 9/10.

[8]   Dho-Moulin, M. & Fairbrother, J.M. (1999). "Avian pathogenic *Escherichia coli* (APEC)". *Veto Res*. 30 (2-3): 299-316.

[9]   Fedorka Cray P.J., Dargatz D.A., Wells S.J., Wineland N.E., Miller M.A., Tollefson L. & Petersen K.E. (1998). Impact of antimicrobic use in veterinary medicine. *J Am Vet Med Assoc*. 15,213(12):1739-41.

[10]   Gomis, S.M., Riddell, C., Potter, A.A. & Allan, B.J. (2001). Phenotypic and genotypic characterization of virulence factors of Escherichia coli isolated from broiler chickens with simultaneous occurrence of cellulitis and other colibacillosis lesions. *Can. J. Vet. Research*. 65: 1, 1-6, 37.

[11]   Guerra, B., Landeras, E., González-Hevia, M.A. & Mendoza, M.C. (1997). "A three-way ribotyping scheme for *Salmonella* serotype Typhimurium and its usefulness for phylogenetic and epidemiological purposes". *J. Med. Microbiol*. 46: 307-313.

[12]   Guinee PAM, Agterberg CM, Jansen WH. 1972. *Escherichia coli* O antigen typing by means of a mechanized microtechnique. Appl Microbiol. 24:127-31.

[13]   Gulig, P. 1996. Pathogenesis of systemic disease. 2774-2787 p. In Neidhardt F.C. *et al* (Eds.) *Escherichia coli* and *Salmonella*. Celular and molecular biology. ASM Press.

[14]   Gruner, E., Lucchini, G.M., Hoop, R.K. & Altwegg, M. (1994). Molecular epidemiology of *Salmonella enteritidis*. Eur. J. Epidemiol. 10: 1, 85-89.

[15]   Landeras, E., Usera, M.A., Calderón, C. & Mendoza, M.C. (1997). "Usefulness of phage typing and "two-way ribotyping" to differentiate *Salmonella enteritidis strains"*. *Microbiol. SEM*. 13: 471-480.

[16]   Machado, J., Grimont, F. & Grimont, P.A.D. (1998). "Computer identification of *Escherichia coli* rRNA gene restriction patterns". *Res. Microbiol*. 149: 119-135.

[17]   Montali, R.J. (1988). Necropsy procedure for birds. *Proceedings Association of Avian Veterinarians*.

[18]   Ngeleka, M., Kwaga, J.K.P., White, D.G., Whittam, T.S., Riddell, C., Goodhope, R., Potter, A.A. & Allan, B. (1996). *Escherichia coli* cellulitis in broiler chickens: clonal relationships among strains and analysis of virulence- associated factors of isolates from diseased birds. *Infect. Immun*. 64: 8, 3118-3126, 44 ref.

[19]   Tollefson L., Angulo F.J. & Fedorka-Cray P.J. (1998). National surveillance for antibiotic resistance in zoonotic enteric pathogens. Vet Clin North Am Food Anim Pract. 14(1):141-50.

[20]   Yamamoto S, Terai A, Yuri K, Kurozono H, Takeda Y, Yoshida O. 1995. Detection of urovirulence factors in Escherichia coli by multiplex polymerase chain reaction. FEMS Immunol Med Microbiol 12:85-90

Industrial Microbiology - Future Bioindustries

Acrylic acid removal from synthetic wastewater and actual industrial wastewater by pure culture of bacteria isolated from the acrylonitrile-butadiene-styrene resin manufactured wastewater treatment system

C.C. WANG[1*], C.M. LEE[2]

[1] Department of Environmental Engineering, Hungkuang University, 34, Chung-Chie Rd. Shalu, Taichung 433, Taiwan, ROC.
[2] Department of Environmental Engineering, National Chung Hsing University, 250, Kuokuang Rd. Taichung 402, Taiwan, ROC.

*Corresponding author: e-mail: chunchin@sunrise.hk.edu.tw, Phone: +886 4 26318652 4110, Fax: +886 4 26525245

The aim of this study is to understand the performance of strain DNA-7 and DNA-8 for treating different initial acrylic acid concentrations from synthetic wastewater and actual industrial wastewater. Besides, the aim is also to understand the ability of strain DNA-7 and DNA-8 for tolerating different initial acrylonitrile, acrylamide and ε-caprolactam concentrations from synthetic wastewater. The results are: strain DNA-7 and DNA-8 could utilize acrylic acid from synthetic wastewater for growth, when the initial acrylic acid concentration was below 1746.1 mg/l and 1690.4 mg/l, respectively. Both strains could tolerate 198.4 mg/l acrylonitrile toxicity and strain DNA-7 could remove it completely. But the growth of both strain were inhibited by 293.7 mg/l ε-caprolactam. Strain DNA-8 could tolerate 295.7 mg/l acrylamide but strain DNA-7 couldn't. Besides, strain DNA-7 and DNA-8 could utilize acrylic acid from actual industrial wastewater for growth, when the initial acrylic acid concentration was below 1741.1 mg/l. Strain DNA-7 degraded acrylic acid faster than strain DNA-8, whether the acrylic acid existed in synthetic wastewater or actual industrial wastewater.

Keywords Acrylamide; Acrylic acid; Acrylonitrile; Biodegradation; ε-Caprolactam

1   Introduction

Acrylic acid is the intermediate product from the acrylonitrile and acrylamide metabolism. Besides, its effective concentration for fish and invertebrates ranged from 27 to 236 mg/l [1]. It may lead to an adverse environmental impact on water quality and thus endanger public health and welfare when the large amount of acrylic acid accumulates with the acrylonitrile and acrylamide removal. Strain DNA-7 and DNA-8 were isolated from the acrylonitrile-butadiene-styrene (ABS) resin manufactured wastewater treatment system and they could utilize 600 mg/l acrylic acid for growth in the preliminary test. The aim of this study is to understand the performance of strain DNA-7 and DNA-8 for treating different initial acrylic acid concentrations from synthetic wastewater and actual industrial wastewater. Besides, the aim is also to understand the ability of strain DNA-7 and DNA-8 for tolerating different initial acrylonitrile, acrylamide and ε-caprolactam concentrations from synthetic wastewater.

2   Methods

2.1 Acrylic acid removal in a batch reactor with synthetic wastewater

These experiments were conducted with a series of batch reactors. Each reactor contained microorganisms at an initial concentration of 10^7 cfu/ml and 40 ml of PBM, different concentrations of acrylic acid were added and sealed with teflon/silicon stoppers. The reactors were shaken at 120 rpm in the dark at 30°C. Acrylic acid concentration, pH and O.D. value at 600 nm were followed as a measurement. The compositions of the PBM for the experiment contained the following (in grams per liter): $MgSO_4 \cdot 7H_2O$, 0.2; $CaCl_2 \cdot 2H_2O$, 0.02; NH_4Cl, 0.1; K_2HPO_4, 1.0; KH_2PO_4, 1.0. The trace element solution in the PBM was 10 ml/l and the composition of the trace element solution was as follows (in milligrams per liter): $FeSO_4 \cdot 7H_2O$, 300; $MgCl_2 \cdot 4H_2O$, 180; $CoCl_2 \cdot 6H_2O$, 106; $Na_2MoO_4 \cdot 2H_2O$, 34; $ZnSO_4 \cdot 7H_2O$, 40. The final pH value was 7.5 at PBM.

2.2 Acrylic acid removal in a batch reactor with actual industrial wastewater

Actual industrial wastewater (ABS resin manufacturing wastewater) was obtained from an ABS wastewater treatment plant and centrifuged at 8000 rpm for 20 minutes under 4°C. The supernatant was filtered using a strainer to remove the white suspended solid. A 0.8 μm membrane filter followed by 0.45 μm, 0.2 μm and 0.1 μm membrane filters were used to filter the wastewater and sterilize it.

Each reactor contained 40 ml sterilized actual industrial wastewater; different concentrations of acrylic acid were added and sealed with teflon/silicon stoppers. Finally 10^7 cfu/ml cell suspension concentration was added and the reactors were shaken at 120 rpm in the dark at 30°C. Acrylic acid concentration, pH and O.D. value at 600 nm were followed as a measurement.

3   Results and discussion

3.1 Acrylic acid removal by acrylic acid utilizing bacteria in a batch reactor with synthetic wastewater

3.1.1 Strain DNA-7

Figure 1 shows acrylic acid removal at 315.5 mg/l by strain DNA-7 in a batch reactor with synthetic wastewater. At 4 hours and 7 hours, 28.3% and 84.2% of 315.5 mg/l acrylic acid was removed by strain DNA-7 and its residual concentration was 5.9 mg/l at 19.3 hours. The $O.D._{600}$ value increased with the acrylic acid removal and reached 0.574 at 12.2 hours. The pH values decreased with the reaction progress. In order to know if the strain DNA-7 could tolerate the acrylamide toxicity and utilize acrylamide as a substrate for growth, 295.7 mg/l of acrylamide was added to the reactor bottles at 24.9 hours. The added acrylamide was remained relatively constant until the experiment was terminated at 69.3 hours. It indicated that strain DNA-7 could not utilize acrylamide for growth.

Because of the higher initial acrylic acid concentration, a longer reaction time was needed for its removal. During the initial 9 hours, 12.8 hours and 19.1 hours, 22.4%, 42.9% and 61.5% of 1746.1 mg/l acrylic acid were removed by strain DNA-7. Its residual concentration was 493.4 mg/l at 69.3 hours. The $O.D._{600}$ value increased with the acrylic acid removal and reached 0.678 at 27.7 hours. (data not shown).

In the series of experiments, it showed that strain DNA-7 could utilize acrylonitrile for growth but could not remove ε-caprolactam(data not shown).

3.1.2 Strain DNA-8

Figure 2 shows acrylic acid removal at 305.5 mg/l by strain DNA-8 in a batch reactor with synthetic wastewater. At 51.2 hours, 15.4% of 305.5 mg/l acrylic acid was removed by strain DNA-8 and its residual concentration was 14.8 mg/l at 92 hours. The $O.D._{600}$ value showed an apparent lag period of 37.5 hours. It then increased with increasing removal of acrylic acid and reaches 0.291 at 92.5 hours. The pH values decreased with the reaction progress. In order to know if the strain DNA-8 could tolerate the acrylamide toxicity and utilize acrylamide as a substrate for growth, 295.7 mg/l of acrylamide was added to the reactor bottles at 141.5 hours. The added acrylamide was remained 277.9 mg/l at the end of the study period of 289.2 hours. It showed that strain DNA-8 could tolerate the acrylamide toxicity but do not like to utilize acrylamide for growth.

Because of the higher initial acrylic acid concentration, a longer reaction time was needed for its removal. At 91.6 hours and 141.6 hours, 21.8% and 59.0% of 1690.4 mg/l acrylic acid were removed by strain DNA-8. Its residual concentration was 677.9 mg/l at 288.8 hours. The $O.D._{600}$ value still showed an apparent lag period of 50.8 hours. It increased with the acrylic acid removal and reached 0.543 at289.4 hours. (data not shown).

In the series of experiments, it showed that strain DNA-8 could utilize acrylonitrile for growth but

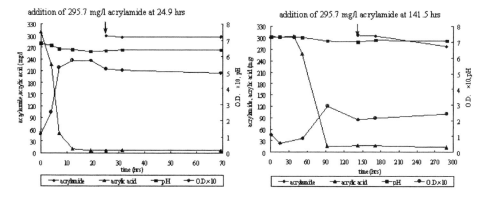

Figure 1. Time course of the removal of 315.5 mg/l acrylic acid by strain DNA-7 in a batch reactor with synthetic wastewater

Figure 2. Time course of the removal of 305.5 mg/l acrylic acid by strain DNA-8 in a batch reactor with synthetic wastewater

3.2 Acrylic acid removal by acrylic acid utilizing bacteria in a batch reactor with actual industrial wastewater

3.2.1 Strain DNA-7

Figure 3 shows acrylic acid removal at 420.3 mg/l by strain DNA-7 in a batch reactor with actual industrial wastewater. At 6.5 hours and 14.4 hours, 36.4% and 92.3% of 420.3 mg/l acrylic acid was removed by strain DNA-7 and its residual concentration was 13.8 mg/l at

75.9 hours. The O.D.$_{600}$ value increased with the acrylic acid removal and reached 0.698 at 14.4 hours. The pH values decreased with the reaction progress.

During the initial 6.5 hours, 25.9 hours and 75.3 hours, 6.6%, 25.1% and 42.2% of 1741.1 mg/l acrylic acid were removed by strain DNA-7. Its residual concentration was 696.5 mg/l at 187.1 hours. The O.D.$_{600}$ value increased with the acrylic acid removal and reached 0.876 at 187.1 hours. (data not shown).

3.2.2   Strain DNA-8

Figure 4 shows acrylic acid removal at 420.3 mg/l by strain DNA-8 in a batch reactor with actual industrial wastewater. At 31.7 hours and 53.6 hours, 36.9% and 77.4% of 420.3 mg/l acrylic acid was removed by strain DNA-8 and its residual concentration was 84.9 mg/l at 214.6 hours. The O.D.$_{600}$ value showed an apparent lag period of 11.3 hours. It then increased with the acrylic acid removal and reached 0.541 at 53.6 hours. The pH values decreased with the reaction progress.

At 47.4 hours, 53.6 hours and 71.1 hours, 22.2%, 26.8% and 36.9% of 1741.1 mg/l acrylic acid was removed by strain DNA-8 and its residual concentration was 1062.4 mg/l at 214.7 hours. The O.D.$_{600}$ value showed an apparent lag period of 11.4 hours and it reached 0.811 at 103.5 hours. (data not shown).

From Figure 1~Figure 4 and the contents, it showed that the strain DNA-7 and the strain DNA-8 could utilize acrylic acid for growth in a batch reactor with synthetic wastewater and actual industrial wastewater, when the initial acrylic acid concentration was below 1690.4 mg/l. Strain DNA-8 had an apparent lag period but strain DNA-7 did not. It might indicate that the acrylic acid degrading enzyme produced by strain DNA-8 need to induce. Strain DNA-8 could tolerate 295.7 mg/l acrylamide toxicity but the strain DNA-7 growth was inhibited by the acrylamide. Both strains could utilize acrylonitrile for growth but could not remove ε-caprolactam. From Figure 1~Figure 4, it also showed that the higher initial acrylic acid concentration, a longer reaction time was needed for its removal. Besides, strain DNA-7 degraded acrylic acid faster than strain DNA-8, whether the acrylic acid existed in the synthetic wastewater or actual industrial wastewater.

The microbial degradation of nitriles proceeds through two enzymatic pathways; nitrile hydratase catalyzes the hydration of a nitrile to an amide, followed by its conversion to an acid and an ammonium by amidase [2, 3], whereas nitrilase catalyzes the hydrolysis of a nitrile to an acid and ammonium [4]. Strain DNA-8 could utilize acrylonitrile, acrylamide and acrylic acid for growth. It indicated that strain DNA-8 include nitrile hydratase and amidase. Strain DNA-7 could utilize acrylonitrile and acrylic acid for growth but not remove acrylamide. It indicated that strain DNA-7 include nitrilase but not include nitrile hydratase and amidase. Because of strain DNA-7 and strain DNA-8 had different enzymatic system; it is expected to play a different for treating ABS resin manufacturing wastewater.

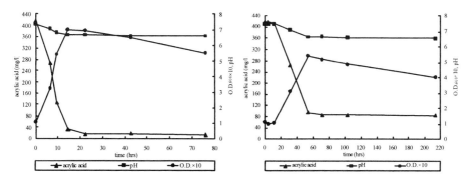

Figure 3. Time course of the removal of 420.3 mg/l acrylic acid by strain DNA-7 in a batch reactor with actual industrial wastewater

Figure 4. Time course of the removal of 420.3 mg/l acrylic acid by strain DNA-8 in a batch reactor with actual industrial wastewater

4   Conclusions

Strain DNA-7 could utilize acrylic acid for growth in a batch reactor with synthetic wastewater and actual industrial wastewater, when the initial acrylic acid concentration was below 1741.1 mg/l. Strain DNA-7 could remove acrylonitrile but could not degrade acrylamide and ε-caprolactam.

Strain DNA-8 could utilize acrylic acid for growth in a batch reactor with synthetic wastewater or actual industrial wastewater, when the initial acrylic acid concentration was below 1690.4 mg/l. Strain DNA-8 could remove acrylonitrile and acrylamide but could not degrade ε-caprolactam.

Whether the acrylic acid existed in the synthetic wastewater or actual industrial wastewater, Strain DNA-7 degraded acrylic acid faster than strain DNA-8.

Acknowledgements

The research was supported by the National Science Council, Taiwan, ROC (NSC 92-2211- E241-003).

References

[1]   C. A. Staples, S. R. Murphy, J. E. McLaughlin, H. W. Leung, T. C. Cascieri and C. H. Farr, Determination of selected fate and aquatic toxicity characteristics of acrylic acid and a series of acrylic esters. Chemosphere. 40, 29-38 (2000).

[2]   H. Yamada and M. Kobayashi, Nitrile hydratase and its application to industrial production of acrylamide. Biosci. Biotech. Biochem. 60, 1391-1400 (1996).

[3]   M. Kobayashi and S. Shimizu, Metalloenzyme nitrile hydratase: Structure, regulation, and application to biotechnology. Nat. Biotechnol. 16, 733-736 (1998).

[4]   M. Kobayashi and S. Shimizu, Versatile nitrilases: Nitrile-hydrolysing enzymes. FEM Microbiol. Lett. 120, 217-224 (1994).

Biodegradation of hydrolyzed polyacrylamides in aqueous solution

Wei-Dong Chen[*], Hong-Yan Liu

Department of Biochemical Engineering, Faculty of Chemical Science and Engineering, University of Petroleum (Beijing), Beijing 102249, China

[*] Corresponding author: E-mail: wdchen@bjpeu.edu.cn, Tel & Fax: +86-10-89733852

The biodegradation of hydrolyzed polyacrylamides (HPAM) in aqueous solution with the consortium of pure strains isolated from oilfield sediment was studied in batch experiments after enrichment culture. Factors that influence the biodegradation of HPAM were studied. Degradation rate was estimated by the biochemical oxygen demand (BOD) and chemical oxygen demand (COD). The biomineralization rate of HPAM with molecular weight of 15,000,000 was as high as 45.5% at an initial concentration of 0.3 g/l after being incubated for 5 days at 20°C, being about three times greater than that with sulfate reducing bacteria after being incubated for 7 days (about 20%). Compared with the other ways of HPAM degradation, there is no acrylamide monomer released in biodegradation process. Due to the presence of suitable enzymes, the consortium of pure strains can efficiently degrade HPAM by decomposing the main chain carbon backbone. The results suggest that the HPAM degradation by the enzymes extracted from these strains is promising in practical applications.

Keywords Hydrolyzed polyacrylamides, Biodegradation, biomineralization rate, consortium of pure strains

1   Introduction

Hydrolyzed polyacrylamide is an important chemical promoter of oil-field development by polymer waterflooding in recent years in china [1]. With the ever-increasing usage of these chemicals in the field of enhanced oil recovery, more and more wastewater with increasing content of polyacrylamides is being produced, 1.0 g/l in Daqing oil-field, for example [2], wastewater with concentrated polyacrylamides must be treated before being discharged into environment to keep operating costs low (come up to emission or reinjection standard). It is important to depolymerize polyacrylamide effectively. Biodegradation is the most advisable way because of its mild reactive conditions, convenient operations and economic attractiveness. It is affirmable that the biodegradation products will not include acrylamide [3] but the other degradation ways possibly will.

However, microorganisms that can effectively degrade HPAM have not been found so far. In this paper, the degradation of polyacrylamide by consortium of the pure strains *Plesiomonas shigelloides*, *Micrococcus roseus* and *Micrococcus varians*, respectively was investigated and the factors influencing the degradation process were also studied.

2   Experimental and results

2.1 Materials

All reagents used were AR grade and obtained from the VAS Lab Supplies Inc. (Tianjin, China). The polyacrylamide used is a commercial grade polymer produced by Tianjin Kermel Chemical Reagents Development Center (Tianjin, China).

2.2 Glassware and PH measurement

All glassware used in the experiments was washed with tap water and rinsed with distilled water, and then sterilized at 120°C for 30min in an electric heater portable pressure steam sterilizer (model YX-280).
The *pH* values of polyacrylamide and culture medium solutions were measured with a digital *pH* meter (Analytical Sensors Inc.).

2.3 Culture medium

The culture medium is composed of 0.2 g of $MgSO_4$, 0.02 g of $CaCl_2 \cdot 2H_2O$, 1.0 g of KH_2PO_4, 1.0 g of K_2HPO_4, 1.0 g of NH_4NO_3 and 1.0 g of yeast extract powder per liter. The *pH* of the medium was adjusted to a value between 6.5 and 7.0 with HCl and NaOH solutions.

2.4 Purification of polyacrylamides

To remove the residues of acrylamide monomer and impurities from the commercial polyacrylamide, polyacrylamides solution was dialyzed three times with dialysis bag, and then dried at 60°C to a constant weight. Purified polyacrylamides were dissolved in culture medium solution and the solution was heated at 111°C for 30min to reduce substantially the unchanged residual acrylamide monomer [4] and to sterilize the polymer solution. All the experiments performed after sterilizing was carried out in Super Clean Bench (Harbin Donglian electronic and technology development corporation of China) to prevent contamination.

2.5 Degradation of polyacrylamide

The degradation of HPAM in aqueous solution with the consortium of the pure strains was estimated by the biochemical oxygen demand (BOD) and chemical oxygen demand (COD). To 1 ml polymer solution (0.15 g Polyacrylamides in 500 ml culture medium solution) 19 ml distilled water and 10 ml potassium dichromate (0.25 mol/l) were added to make up a sample solution. The oxidation-reduction reaction in the solution was continued for 2 hours at the boiling point of the solution. An ammonium ferrous sulfate solution was used to titrate the excess potassium dichromate. The color of the solution is changed from yellow to cyan and then to mahogany, which is the end-point of the titration. The same experiments were also carried out with distilled water for comparison. The volume of the ammonium ferrous solution at the critical point (see Table 1) was used to calculate the COD value of the samples using Eq. (1) below

$$COD(O_2, mg/l) = \frac{(V_0 - V_1) \times C \times 8 \times 1000}{V} \qquad (1)$$

Where C denotes the concentration of ammonium ferrous sulfate solution (mol/l); V_0 is the volume of the ammonium ferrous sulfate solution used to titrate distilled water (ml); V_1 is the volume of ammonium ferrous sulfate solution used to titrate the sample solution (ml); V is the volume of polymer solution (ml); 8 is the mole weight of 1/2O [5].

Table 1 The titration volume of ammonium ferrous sulfate solution and COD values of samples

Sample	V ($\times 10^{-3}$ l)	COD (g/l)
Distilled water	32.8	—
Sample solution	28.8	2.3667
Culture medium solution	30.1	1.5975
Polyacrylamides solution		0.7692

The concentration of the ammonium ferrous sulfate solution was determined by titration with datum potassium dichromate solution. 10 ml datum potassium dichromate solution was added into a 500 ml conical flask, into which 100 ml distilled water was added to get a total volume of 110 ml. 30 ml sulfuric acid (>99%) was then added into the flask and the final solution was titrated with ammonium ferrous sulfate solution (0.2500 mol/l), the color of the solution is changed from yellow to cyan and then to mahogany at the end-point of the titration (see Table 2). The concentration of the ammonium ferrous sulfate solution is determined by Eq. (2) below

$$C[(NH_4)_2 Fe(SO_4)_2] = \frac{0.2500 \times 10.00}{V} \qquad (2)$$

Where V is the volume of the ammonium ferrous sulfate solution (ml) [5]. The COD value of the HPAM at an initial concentration of 0.3 g/l was measured (see Table 1).

Table 2 The titration volume and concentration of ammonium ferrous sulfate solution at the critical point

V ($\times 10^{-3}$ l)	Concentration (mol/l)
16.9	0.07396

The degradation of polyacrylamide with the consortium of the pure strains was determined in batch assays at a constant temperature of 20°C under aerobic conditions with stir. Determinate amount (80 percent of corresponding COD value) of sterilized polyacrylamides solution was added into a BOD bottle and inoculated with 3 ml adapted consortium of the pure strains and then incubated for 5 days. Values of BOD were autosaved by the apparatus (see Table 3).

Table 3 BOD value of HPAM (0.3 g/l; 15,000,000) at 20°C (g/l)

Sample	1d	2d	3d	4d	5d
Culture medium solution	—	—	—	—	—
Polymer solution	0.200	0.250	0.300	0.350	0.350

The biomineralization rate of polyacrylamides solution is determined by Eq (3) below

$$BMR = COD(HPAM) / BOD_5 \times 100\% \qquad (3)$$

Where BMR is the biomineralization rate. The biomineralization rate of the polyacrylamides at an initial concentration of 0.3 g/l is 45.5%.

3   Discussion

The objective of this study was to examine the degradation of polyacrylamide by a consortium of the pure strains (identified as *Plesiomonas shigelloides*, *Micrococcus roseus* and *Micrococcus varians*, respectively). The results indicated that it was feasible to degrade polyacrylamide macromolecules by the mixed culture. The biomineralization rate of the HPAM with molecular weight of 15,000,000 at an initial concentration of 0.30 g/l reached 45.5% regardless of stir after being incubated for 5 days at 20°C, being about three times greater than that with sulfate reducing bacteria after being incubated for 7 days (about 20%) [18]. The experiments showed that the biomineralization rate of polyacrylamides decreased as the concentration of corresponding polyacrylamides solution increased (as shown in Fig.1). This may indicate that the metabolism of 3 ml microorganism can mineralize a determinate amount of polyacrylamides. The inhibition of concentrated solutions on the extraction of necessary enzymes responsible for the decomposition of the main carbon chain may serve as another explanation. It was found that the biomineralization rate of polyacrylamides decreased with increasing of molecular weight of polyacrylamides, this could be attributable to the increasing difficulties for the bacteria to attack the main carbon backbone with increasing in the length of polymer chain. The limitation of necessary enzymes from the cells may be another explanation.

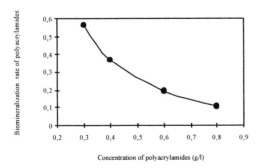

Fig.1. Biodegradation rate as a function of polyacrylamide concentration (molecular weight 15,000,000)

Polyacrylamides may also undergo photo degradation during the experiments [6]. The photolysis of a polymer depends on the chemical structure of the material, the type of chemical bonds themselves. Polyacrylamides are composed primarily of C- C, C-H, and C- N bonds. These bonds have energies of approximately 340, 420 and 414 kJ/mol. Therefore, to break a C- C, C- H, and C-N bond with electromagnetic energy, wavelengths of 325, 250 and 288 nm are required, respectively [7]. During the experiments of this study, the polyacrylamide samples were exposed to visible light with wavelength in the range of 400-760 nm, which has not the capability of breaking any of the above chemical bonds. Further more, during the measurement of BOD, the apparatus was covered by cloth to avoid illumination.

The effects of *pH* on the biodegradation of polyacrylamides could also be neglected, since the *pH* values of the samples used were all between 6.5-7.0. It was reported that *pH* within the range of 5 to 9 did not lead to depolymerization of polyacrylamide [8].

Below temperatures of about 200°C, polyacrylamides are generally thermally stable and undergo very little physical change apart from a slight weight loss. This slight weight loss is probably due to surface and matrix-bound water of the polymer from the environment and other volatile impurities [9-11]. If employed a drying step before analysis, there is no weight loss before approximately 220°C from pretreated polyacrylamide samples during thermal analysis [12]. During the experiments in this study, the polyacrylamide samples were either exposed to approximately 110°C for 30 minutes during purification or to room temperature during formation. Therefore, thermal degradation during biodegradation is negligible.

The biomineralization rate of polymer solutions containing yeast extract powder was found to be higher than that of solutions without it, which suggests that the addition of yeast extract powder promotes or assists in the degradation of polyacrylamides. To examine this issue, solutions that contain HPAM and yeast extract powder as sole and mixture of carbon sources were formulated for comparison. Results showed that the oxygen demand for microorganism metabolism (produce CO_2) of a solution formulated with HPAM (0.3 g/l; 15,000,000) and yeast extract powder was 350 mg/l, being much higher than 200 mg/l of a solution formulated with the HPAM alone, and also higher than that of solutions formulated with the yeast extract powder alone, which was zero. Therefore, yeast extract powder does not serve as a carbon source. How does yeast extract powder promote the degradation of polyacrylamide? The possible reason is that it changed the metabolic route of the microorganism to break C-C bond within the polyacrylamide.

Literatures contain numerous reports on chemical degradation of polyacrylamide. A work by Yu-ming Nan et al. (1997) showed that $S_2O_8^{2-}$ and Fe^{2+} could depolymerize polyacrylamide effectively. The viscosity loss of HPAM at initial concentration from 0.2 g/l to 2.0 g/l with molecular weight of 10,500,000 could be up to 90% in a hour when $FeSO_4 \cdot 7H_2O$ (0.01 g/l) and $(NH_4)_2S_2O_8$ (0.07 g/l) were present [13]. A recent work by Bao-hui Wang et al. (2004) showed that the degradation rate of polyacrylamide with molecular weight of 18,000,000 was up to 90% in 90 minutes by K_2FeO_4, and the viscosity of polyacrylamides solution (0.02%, w/w) dropped to that of distilled water in 15 minutes [14].

A major concern associated with the degradation of polyacrylamide is the possible formation of acrylamide by some degradation mechanisms [8, 15]. Acrylamide is an acute eye, skin, and respiratory tract irritant and is absorbed by all routes of exposure, including intravenous, intraperitoneal, subcutaneous, intramuscular, oral, and dermal. In higher animals, once absorbed, acrylamide causes damage to the central nervous system, producing an ascending central/peripheral axonopathy [16]. But earlier work by MacWilliams noted that biodegradation of polyacrylamide did not result in the formation of acrylamide [17]. As it is well recognized that acrylamide is completely biodegradable and acrylamide is not formed during the biodegradation of the polyacrylamides. Thus, it is a promising technique to combine the biodegradation and chemical degradation as a more effective way for the mineralization of polyacrylamides macromolecule.

4   Conclusion

The results showed that it is feasible to degrade polyacrylamides in aqueous solution under aerobic conditions with the mixed culture *Plesiomon shigelloides*, *Micrococcus roseus* and *Micrococcus varians*, respectively. The biomineralization rate of HPAM at an initial concentration of 0.3 g/l reached 45.5% after being incubated for 5 days at 20°C, which was three times greater than that with sulfate reducing bacteria after being incubated for 7 days (about 20%) [18].

Acknowledgements

This work is supported by the Natural Science Foundation of China National Petroleum Corporation. (Grant No. 2004012)

References

[1]   Y.L. Zhan, S.H. Guo and G.X. Yan, Polymer Bull (China). **17**, 70 (2004).

[2]   L.B. Chen and H.T. Zhang, Engineering and technology. **1**, 20 (2004).

[3]   M.J. Caulfield, G.G. Qiao and D.H. Solomon. Chem. Rev. **102**, 3067 (2002).

[4]   M.W.C. Coville, U.S.Patent 4,132,844. (1979).

[5]   Methods of monitoring and analyzing water and sewage, edited by S.F. Wei, the third edition, China Environmental Science Press, Beijing, (1998), p354.

[6]   C. Decker, Handbook of Polymer Science and Technology. (1989), p.541.

[7]   D.G. Crosby, Herbicide: Chemistry, Degradation and Mode of Action, p.836.

[8]   E.A. Smith, S.L. Prues, and F.W. Oehme, Ecotoxicology and environmental safety. **35**, 121 (1996).

[9]   M.E.S. R e.Silva, E.R Dutra, V. Mano and J.C. Machado, Polym.Degrad. Stab. **67**, 491 (2000).

[10]   N. Grassie, I. C. McNeill, R., S. J. N. Eur. Polym. J. **14**, 931 (1978).

[11]   R. Vilcu, F. Irinei, J. Ionescu-Bujor, M. Olteanu and I. Demetrescu, J. Therm. Anal. **30**, 495 (1985).

[12]   J. D. Van Dyke and K.L. Kasperski, J. Polym. Sci., Part A: Polym.Chem. **31**, 1807 (1993).

[13]   Y.M. Nan, J. Daqing Petoleum institute. **21**, 49 (1997).

[14]   B.H. Wang, F.G. Kong, T.K. Zhang, J.H. Yang and Y. Chen, Industrial water treatment. **24**, 21(2004).

[15]   V. Molak, US Department of Health and Human Services. (1991).

[16]   J.J. Collins, G.M.H. Swaen, G.M. Marsh, H.M.D. Utidijian, J.C. Caporossi and L.J. Jucas, J. Occup. Med. **31**, 614 (1989).

[17]   D.C. MacWilliams, I. Kirk and D.F. Othmer, Encyclopedia of Chemical Technology (New York). 1978, p.298.

[18]   F. Huang, H.X. Fan, Z.H. Dong and L.M. Xu, Petroleum processing and petrochemicals. (China). **30**, 33 (1999).

Biomethanation of sugar industry wastewater by using down-flow anaerobic filter

Amit kumar[#,1], Bram Sercu[#], Vinay kumar[**], D.B. Sapkal[*], B.B. Gunjal[*], Herman Van Langenhove[#]

*Department of Environmental Science and Alcohol Technology, V.S.I. Manjari (B.K) Pune- 412307 MH- India.
**Department of Environmental Science, University of Pune, 411007, MH, India.
[#]Research Group of EnVOC, Faculty of Bioscience Engineering, University of Ghent, Coupure Links 653, B- 9000 Ghent, Belgium.

[1]Corresponding author: Email: amitmalikes@yahoo.co.in; Tel + 32-9264-5998 (o); +32-9264-5948(o); fax: +32-9264-6243.

The feasibility of a laboratory-scale down-flow anaerobic filter (DAF) reactor for the treatment of sugar industry wastewater was investigated. Physical-chemical pre-treatment was used to recover the oil and grease and to remove suspended solids from wastewater. The pre-treated wastewater was used to feed the DAF reactor, at pH 4.6- 5.9, biochemical oxygen demand (BOD) 1500-2000 mg l^{-1} and chemical oxygen demand (COD) 3200-4500 mg l^{-1}. The effect of hydraulic retention time (HRT) on the extent of the degradation of this wastewater was examined. Varying the HRT from 2.5 d to 0.8 d at 27 $\pm$ 1°C, had no significant effect on COD removal efficiency and the pH remained in the range of 6.6-7.4. Under these conditions the COD removal efficiency was 85-90% after steady state at a organic loading rate (OLR) between 2.06-8.54 kg m^{-3} d^{-1}. At HRT < 0.5 d the COD removal efficiency was decreased and a decreased pH and methane yield was also observed. The study proves that DAF can be used for the treatment of sugar industry wastewater under low mesophilic condition at a optimal HRT of 0.8 d.

Keywords: Sugar industry wastewater treatment; Down flow anaerobic filter; Hydraulic retention time.

1   Introduction

In the last years, anaerobic treatment technology has developed remarkably and gained wider acceptance for the treatment of specific industrial wastewaters (Garcia et al., 1998; Lettinga et al., 1999; Frankin, 2001; Plumb et al., 2001). There are about 393 large and medium scale sugar industries (mills) in India. In 1997 about 26% of these units, still had to attain satisfactory performance level with regard to installation and operation of effluent treatment plants (CPCB, 1997a). Since sugar industry wastewater normally has a COD of higher then 4000 mg l^{-1}. It has the potential to supply carbon in a form that anaerobic microorganisms can convert into methane (Frankin, 2001). Lata et al (2002) made an assessment of biomethanation potential of selected industrial organic effluents in India, and reported that sugar effluent has a bioenergy potential. There have been several studies on the treatment of sugar industry wastewater using anaerobic reactors (Sastry et al., 1990; Pathe et al., 1995; Radwan and Ramanujam, 1995; Khusheed et al., 1997; Reddy and Shivalingaiah., 1997). Since growth rates of anaerobic bacteria are low, long start-up periods are usually required. In addition, long HRT are needed for sufficient degradation of organic matter. Therefore, much research has been directed to the development of techniques for maintaining a high biomass concentration in anaerobic reactors (Jewell et al., 1981; Kuba et al., 1990; Jianlong et al., 2000). Anaerobic filters (AF) offers important advantages over other type of anaerobic reactors. There is a little sludge production and effluents are substantially free of suspended solids (Veiga et al., 1994). Successful operation of an anaerobic filter treating a simulated starch wastewater has been reported (Ahn and Forster, 2000). To our knowledge no published information is available for the treatment of sugar industry wastewater by using a DAF reactor. On this basis, a study was conducted to assess the feasibility of a DAF reactor for the treatment of pre-treated sugar industry wastewater. The main objective of this study was to investigate the effect of loading rate and hydraulic retention time (HRT) on the reactor performance under low mesophilic condition ($27 \pm 1°C$).

2 Materials and methods

2.1 Down- flow anaerobic filter (DAF) design

The anaerobic down-flow filter used in the study was fabricated with Perspex materials and has an internal diameter of 0.105 m, column height of 1.35 m and effective reactor volume of 10 l. The reactor was packed with polyvinyl chloride (PVC) cylindrical perforated rings with a diameter of 1.2 cm, as a support for the growth of microorganisms. The feed thank was made of high density poly ethylene (HDPE), has a diameter of 0.39 m and height of 0.42 m ($0.05 m^3$). The schematic diagram of the experimental digester is shown in fig. 1.

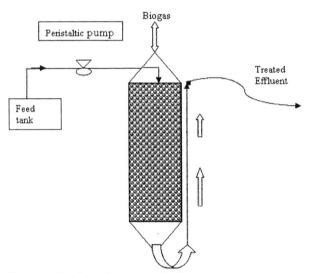

Fig. 1. Schematic diagram of the down-flow anaerobic filter.

2.2 Substrate characteristics

The wastewater was obtained from Yaswant Shakari Sakar Karkhana, Theure (Maharastra-India). The wastewater was free from oil and grease; it was recovered by the industry itself.

The sedimentation in feeding tank eliminates 15% of COD and 30% of total suspended solids (TSS). The total solids (TS), total suspended solids (TSS), biological oxygen demand (BOD), chemical oxygen demand (COD) and pH were determined according to standard methods (APHA, 1992). Biogas samples were analysed using an ORSAT apparatus. The physical-chemical characteristics of the wastewater entering the DAF reactor are given in table 1. Daily effluent samples were taken from the digester effluent for the analyses.

Table 1. Physical-chemical characteristics of sugar industry wastewater entering the DAF reactor.

Parameter	Value
pH	$4.60 - 5.85$
BOD$_{(5d\ at\ 20°C)}$	$1500 - 2000$ mg l^{-1}
COD	$3200 - 4200$ mg l^{-1}
TS	$5000 - 5500$ mg l^{-1}
TSS	$800 - 1000$ mg l^{-1}

3 Results and discussion

3.1 Inoculation and reactor start-up

A mixed consortium of sludge from an active biogas digester of distillery wastewater and cow dung slurry was used as inoculum. The inoculum was recirculated in the packed filter for a period of 5 d for acclimatisation of the microorganisms. After 5 d operation, 80% COD was removed. Then the DAF was started in continuous feeding with sugar industry wastewater at a OLR of 2.06 (HRT 2.5 d). After the system had stabilized the OLR was increased by shortening the HRT gradually from 2.5 to 0.5 d (by changing initial feed flow rate) to give an OLR increase from 2.06 to 10.15 kg m^3 d^{-1}. The loading changes were made to the reactor after the COD removal was stable at the applied HRT or OLR.

3.1.2 Influence of loading rate on COD removal and pH variation

The COD removal and pH variation at different OLR are shown in Fig. 2. At a OLR between 2.06 to 8.54 kg m^3 d^{-1} the COD removal efficiency was > 86% after stabilisation. After increasing the loading rate steady-state was always reached after 45 d. The pH remained neutral (6.6-7.4) under these conditions. However, at 10.15 kg m^3 d^{-1} loading rate, the COD removal efficiency and the pH decreased rapidly to a value between 5.0-5.6. The pH decreased was most likely caused by the accumulation of volatile fatty acids (VFA) at a high OLR.

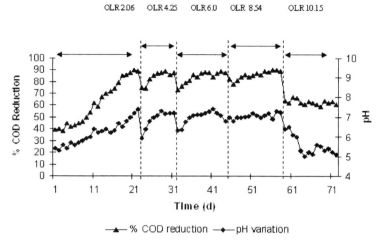

Fig. 2. COD removal and pH variation at different loading rates.

3.1.3 Biogas production

The effect of varying the OLR on the DAF reactor on biogas production and methane yield is shown in Fig. 3. Biogas production rate (l d^{-1}) from anaerobic digestion of sugar industry wastewater was improved by increase of OLR from 2.06 to 8.54 kg m^3 d^{-1}. However, there was a significant decrease in conversion of substrate to biogas when the OLR increased from 8.54 to 10.15 kg m^3 d^{-1}. The methanogenesis was inhibited due to pH decrease from 7.2 to 4.8 because rapid degradation of substrate to volatile fatty acids (VFA). The highest biogas production rate of 2.2 l d^{-1} was obtained with OLR of 8.54 kg m^3 d^{-1}.

The methane yield (l kg^{-1} removal of COD) was decreased by increasing OLR and highest conversion of wastewater to methane was obtained at low OLR (2.06 kg m^3 d^{-1}). Methane yield decreased by decreasing the HRT from 0.8 d to 0.5 d. The failure of reactor performance at low HRT is due to washout of substrate without submitting efficient biodegradation.

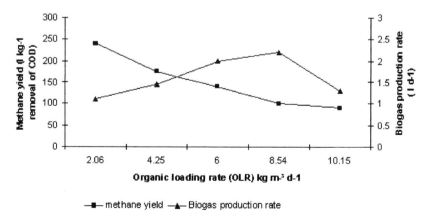

Fig.3. Influence of OLR on biogas production and methane yield.

3.1.4 Comparison to other anaerobic reactors

The results of this study were compared with those reported by other investigators, as mentioned in table 2. It was found that the DAF reactor shows high COD removal efficiency 85-90% at applied loading rate with HRT 19 h under low mesophilic condition (27 ± 1°C). However, some investigator observed that UASB reactor also gives high COD removal efficiency at 34 °C, but this study proves that DAF can also perform well at 27 ± 1°C.

Table 2. Comparison to other anaerobic reactors.

Types of reactor	Methane yield $m^3 kg^{-1} cod^{-1}$	COD loading rate $(kg\, m^{-3}\, d^{-1})$	HRT (h)	Temp (°C)	COD reduction (%)	Reference
RBC	NA	NA	NA		90-97	Rawan and Ramanujam., 1995
UASB	0.22	13	5.5	34	80	Khusheed et al., 1997
UASB	0.34 – 0.28	13	4 – 24		> 90	Pathe et al., 1995
UASB	0.22	1.8 – 7.2	NA		95	Reddy and Shivalingaiah, 1997
UAF	NA	0.073 – 0.52	NA		NA	Sastry et al., 1990
DAF	0.22	2.06 - 8.54	19	27 ± 1	85-90	this work

RBC: rotating biological contactor, UASB: up flow anaerobic sludge blanket reactor, UAF: up flow anaerobic filter, DAF: down flow anaerobic filter, NA: not available.

4   Conclusions

The study established the feasibility of a down-flow anaerobic filter for treating sugar industry wastewater. A COD removal efficiency of 90% was observed at a relatively short retention time (0.8 d), at 8. 54 kg m^3 d^{-1} OLR and 27 ± 1°C. Varying the organic loading rate from 2.06 to 8. 54 kg m^3 d^{-1} did not affect the COD removal efficiency, always 4-5 d adaptation time was needed for steady state COD removal. However, at OLR > 8. 54 kg m^{-3} d^{-1}, the DAF reactor shows 25-30% decreased COD removal efficiency.

Acknowledgements

The authors wish to thank Yaswant Shakari Sakar Karkhana, Theure, Maharastra, for providing wastewater throughout the study.

References

[1]   Ahn, J.H., Forster, C.F., 2000. Kinetic analyses of the operation of mesophilic and thermophilic anaerobic filters treating a simulated starch wastewater. Process Biochemistry 36, 19-23.

[2]   APHA, 1992. Standard methods for the examination of water and wastewater (APHA, AWWA and WEF), (18th ed.).

[3]   CPCB, (Eds.), 1997a. Central pollution control board. National inventory of large and medium industry and status of effluent treatment and emission control system. Programme objective series, PROBES/68, Akashdeep printers, New Delhi, India.

[4]   Frankin, R. J., 2001. Full-scale experiences with anaerobic treatment of industrial wastewater. Water Science and Technology 8, 1-6.

[5]   Frankin, R.J., 2001. Full-scale experiences with anaerobic treatment of industrial wastewater. Water Science and Technology 8, 1-6.

[6]   Garcia, C. D., Buffiere, P., Moleetta, R., Elmaleh, S., 1998. Anaerobic digestion of wine distillery wastewater in down-flow fluidized bed. Water Research 32, 3593-3600.

[7]   Jewell, W.J., Switzznboum, M.S., Morris, J.W., 1981. Municipal wastewater treatment with the anaerobic attached microbial film expanded bed process. Journal of Water Pollution 53, 482- 490.

[8]   Jianlong, W., Hanchang, S., Yi. Q., 2000. Wastewater treatment in a hybrid biological reactor: effect of organic loading rates. Process Biochemistry 36, 297-303.

[9]   Khusheed, A., Farooqi, I.H., Siddiqui, R.H., 1997. Development of granular sludge on cane sugar mill waste treatment using a pilot scale UASB reactor. Indian Journal of Environmental Health 39, 315-325.

[10]   Kuba, T., Furumai, H., Kusuda, T., 1990. A kinetic study on methanogenesis by attached biomass in a fludized bed. Water Research 24, 1365-1372.

[11]   Lata K., Kansal A., Balakrisnan M., Rajeshwari K. V., Kishore V.V.N. 2002. Assessment of biomethanation potential of selected industrial organic effluents in India. Resources, Conservation and Recycling 35, 147-161.

[12]   Lettinga, G., Rebac, S., Parshina, S., Nozhevnikova, A. N., Vanlier, J. B., Stam, A. J. M., 1999. High rate anaerobic treatment of wastewater at low temperature. Applied and Environmental Microbiology 65, 1969- 1702.

[13]   Pathe, P.P., Nandy, T., Kaul, S.N., 1995. UASB reactor for the treatment of sugar effluents. Indian Journal of Environment Protection 15, 174- 180.

[14]   Plumb, J. J., Bell, J., Stukey, D.C., 2001. Microbial populations associated with temperature of an Industrial dye effluent in an anaerobic baffled reactor. Applied and Environmental Microbiology 67, 3226- 3235.

[15]   Radwan, K.H., Ramanujam, T.K., 1995. Treatment of sugarcane wastewater using modified biological contactor. Indian Journal of Environmental Health 37, 77- 83.

[16]   Reddy, U., Shivalingaiah, 1997. Studies on the treatment of sugar Industry wastewater. Proceeding of International Conference on Industrial pollution and Control Technologies Hyderabad, India, pp. 177-180.

[17]   Sastry, C.A., Srinivas, V., Subrahmanyam, Y.V., Lolla, V.L.P.S., 1990. Treatment of sugar mill waste water using anaerobic filter. Indian Journal of Environmental Protection 10, 570- 582.

[18]   Veiga, M.C., Mendez R., lema, J.M., 1994. Anaerobic filter and SEFF reactors in anaerobic treatment of tuna processing wastewater. Water Science and Technology 30, 425- 432.

Biosynthesis of calcium gluconate by *Aspergillus niger* in shake flask

Irfana Mariam, Sikander Ali and **Saeed Ahmad Nagra**

Institute of Chemistry,University of the Punjab, Lahore, Pakistan

The present study describes the production of calcium gluconate by locally isolated strains of *Aspergillus niger* in shake flasks using glucose salt, $CaCO_3$ medium. Thirty cultures of *Aspergillus niger* were isolated from different soil samples and were examined for acid production. Among all the cultures tested, *Aspergillus niger* IC-15 gave better production of calcium gluconate (85.50 g/l), 72 hours after spore inoculation. The cultural conditions optimised for maximum calcium gluconate production were, glucose concentration (150 g/l), pH (6.5) and $(NH_4)_2SO_4$ (2.5 g/l).

Keywords: Calcium gluconate, *Aspergillus niger*, Production, Gluconic acid, Glucose oxidase, production.

1   Introduction

Calcium gluconate i.e., D-gluconic acid calcium salt ($C_{12}H_{22}CaO_{14}$) is one of the most important salts of calcium, which occurs as a white crystalline or granular powder, without taste or odour. It is widely used in food, textile, and leather, pharmaceutical, chemical and concrete industries and this has stimulated various scientists to undertake intensive investigation for increased production of the salt to meet its commercial demand (Buzzini *et al.,* 1993; Pedrosa *et al.,* 2000). It is stable in air and its solution is neutral to litmus paper. It is insoluble in alcohol and many other organic solvents (Trager and Qazi, 1991).
Calcium gluconate decomposes by mineral acids and other acids, which are stronger than the gluconic acid. It is incompatible with soluble sulphates, carbonates, bicarbonates, citrates, tart rates, salicylates and benzoates. Calcium gluconate fills the need for a soluble; non-toxic well tolerated form of calcium. Calcium therapy is indicated in conditions such as parathyroid deficiency (tetany), general calcium deficiency (during pregnancy, growth, lactation, decreased dietary calcium intake, menopause, old age etc.) and when calcium is the limiting factor in increased clotting time of the blood. It can be used orally, intramuscularly and intravenously (Delgado and Remers, 1991; Ray and Banik, 1999).
Gluconic acid its salts are commercially produced by three different methods; i- Chemical oxidation of glucose with a hypo chlorite solution (Kundu and Das, 1984;), ii- Electrolytic oxidation of glucose solution containing a measured amount of bromide (Ambekar *et al.,* 1965) and, iii- Fermentation process where specific micro-organisms are grown in medium containing glucose and other ingredients (Hill and Robinson, 1988; Shah and Kothari, 1993; Lee *et al.,* 1998). Oxidation of glucose to gluconic acid presents incomplete utilization of the sugar, this acid is known to be produced by several microorganisms. Gluconic acid was considered to be the product of incomplete oxidation of glucose to gluconic acid. However, later studies revealed that enzyme activity of some oxidases and dehydrogenases is responsible for oxidizing glucose to gluconic acid (Rose 1961; Pons *et al.,* 2000).
The main object of the present work is to isolate and select a potent strain of *Aspergillus niger* from local habitats capable of producing large amounts of calcium gluconate.

## 2	Materials and methods

Isolation of organism

Thirty isolates of *Aspergillus niger* were obtained from soils of different selected areas of Lahore by "Standard pour plate method". The mould cultures were screened by submerged fermentation method, using shake flasks each containing 25 ml of fermentation medium. All the experiments were performed in triplicates. The soil samples were taken from different areas of Lahore in polythene bags. This work is a part of M.Phil thesis (2001) at University of the Punjab, Lahore. The method of Johnson *et al.* (1959) was used for the isolation of *Aspergillus niger* strains. The young colonies of *Aspergillus niger* were picked up and grown on potato dextrose agar slants (BDH Germany) for culture maintenance. Conidia from a young colony of *Aspergillus niger* were then inoculated on solidified PDA slants.

Inoculum preparation

The spore suspension from an agar slant of 3-5 days old culture of *Aspergillus niger* was prepared by adding 10 ml sterile Monoxal O. T. (Dioctyl ester of sodium sulpho succinic acid). The conidial suspension was aseptically transferred to a sterile empty MacConkey bottle containing glass beads. The mixed suspension was shaken vigorously to break the clumps of conidia. The number of conidia per ml of inoculum was determined with the help of Thomas Counting Chamber.

Fermentation technique

Calcium gluconate fermentation was carried out by submerged fermentation in 250 ml cotton wool plugged Erlenmeyer flasks containing 25 ml of fermentation medium containing (g/l); glucose 150, $(NH_4)_2SO_4$ 2.5, $MgSO_4.7H_2O$ 0.5, KH_2PO_4 1.0, $CaCO_3$ 32.0 having pH 6.0. The medium was inoculated by transferring 1.0 ml of conidial suspension (3 × 10^6 conidia/ml). The flasks were placed on a rotary shaker (model: GFL 544) for incubation at 30±1°C with an agitation speed of 200 rpm for 72 hours. All the experiments were carried out in duplicate. The nitrogen sources and their concentration have marked influence on the metabolic system of the mould culture. Thus, the effect of adding various nitrogen sources on the production of mycelial dry weight, glucose consumption and Calcium gluconate formation was investigated (Table 5). The nitrogen sources added were NH_4NO_3, $(NH_4)_2SO_4$, NH_4Cl, $NaNO_3$ or peptone.

Analytical methods

The samples for analysis were taken out at the end of fermentation and the contents of flasks were filtered through Whatman filter paper No. 1. Then the filtrate was centrifuged at 5000 rpm for 10 minutes. The supernatant liquid was used for estimation of glucose as well as for calcium gluconate.

208

Estimation of glucose

DNS (Dinitro salicylic acid) method was used for glucose estimation (Tasun *et al.,* 1970). The transmittance was measured at 530 nm by spectrophotometer (Hitachi model: U-2000, Japan).

Estimation of Calcium gluconate

The fermented broth was centrifuged and the supernatant liquid was used for analysis of calcium gluconate by the method of Pharmacopoeia (1990).

Mycelial Dry Weight

The mycelial dry weight was determined according to Chaturvedi *et al.,* (1978).

3   Results

Screening of A*spergillus niger* strain

The amount of calcium gluconate produced by different isolated mould cultures, ranged from 42.35 to 85.50 g/l. The mycelial dry weight was ranging from 6.50 to 15.50 g/l (Table 1).

Rate of Calcium gluconate production

The data of Table 2 shows the rate of Ca-gluconate production by *Aspergillus niger* IC-15. After 12 hours inoculation, the production of Ca-gluconate was negligible. However its production was increased 24 hours after spore incubation & at 48 hours, Ca-gluconate production was about 55.25 g/l. The production of Ca-gluconate was 65.25 g/l at 60 hours after inoculation. The production was reached maximum i.e., 96.35 g/l at 72 hours after the incubation. Further increase in the incubation period, resulted in lowering the yield of Cal-gluconate. The mycelial weight and glucose consumption was also increased during fermentation. The amount of mycelial dry weight was 8.00 g/l, 72 hours after spore inoculation.

Table 1: Screening of *Aspergillus niger* for the production of Calcium gluconate in shake flask

Strain	Glucose (g/l)		Calcium Gluconate produced (g/l)	% age yield of Calcium Gluconate	Mycelial dry weight (g/l)
	Residual	Used			
IC-1	65.23	84.77	71.49	47.66	7.98
IC-2	70.25	79.75	66.48	44.32	8.58
IC-3	68.10	81.90	77.50	51.66	8.15
IC-4	69.30	80.70	75.80	50.30	8.45
IC-5	72.40	77.60	68.63	45.75	8.80
IC-6	63.50	86.50	60.70	40.76	10.00
IC-7	66.10	83.90	80.20	53.46	7.20
IC-8	74.25	75.75	70.10	46.73	7.10
IC-9	80.50	70.50	66.00	44.00	9.90
IC-10	86.50	63.50	55.25	36.83	11.20
IC-11	68.75	81.25	75.15	50.10	9.40
IC-12	98.00	52.00	45.25	30.16	8.75
IC-13	65.75	84.25	78.90	52.60	7.50
IC-14	95.20	54.80	45.50	30.33	10.60
IC-15	71.60	78.40	85.50	57.00	6.76
IC-16	80.00	70.00	63.80	42.53	9.25
IC-17	73.50	76.50	72.25	48.16	8.90
IC-18	81.50	68.50	63.20	42.13	9.75
IC-19	67.20	82.80	78.00	52.00	7.50
IC-20	100.25	49.85	42.35	28.23	14.75
IC-21	62.50	87.50	82.45	54.96	6.75
IC-22	66.00	84.00	80.12	53.41	6.85
IC-23	83.40	66.60	58.64	39.09	10.50
IC-24	71.50	78.50	73.50	49.00	8.00
IC-25	96.80	53.20	56.75	37.83	12.50
IC-26	69.40	80.60	83.10	55.40	6.50
IC-27	84.25	65.75	66.80	43.20	15.00
IC-28	100.15	49.85	48.20	32.13	15.50
IC-29	77.60	72.40	78.80	52.53	7.90
IC-30	82.50	67.50s	65.50	43.66	10.50

Glucose added = 150 g/l
pH = 6.0
Temperature = 30°C
Incubation period = 72 hours
Agitation = 200 rpm

Table 2: Rate of Calcium gluconate production by *Aspergillus niger* IC-15

Incubation period (hrs)	Glucose (g/l)		Calcium Gluconate produced (g/l)	%age yield of Calcium Gluconate	Mycelial dry weight (g/l)
	Residual	Used			
12	127.75	22.25	0	0	3.0
24	100.50	49.50	38.60	25.73	4.50
36	90.50	59.50	46.75	31.16	5.80
48	81.60	68.40	55.25	36.83	6.50
60	75.40	74.60	65.25	43.50	7.10
72	63.66	86.44	96.35	64.23	8.00
84	55.10	94.90	89.15	59.43	7.80
96	37.10	112.90	79.75	53.16	8.50
108	28.32	121.68	75.82	50.54	9.50
120	15.15	134.85	70.33	46.88	10.00

Glucose residual = Glucose added - Glucose used
Glucose added = 150 g/l
pH = 6.0
Temperature = 30°C

Effect of pH

Table 3 shows the effect of pH (4.0–7.5) on Cal-gluconate production. The pH of fermentation medium was adjusted with 1N HCl solution or 1N NaOH solution. The production of Ca-gluconate was minimum at pH 4.0 (i.e., 29.50 g/l) and mycelial dry weight was also poor at this pH. Raising the pH from 4.0–6.0 increased the rate of calcium gluconate production. It was maximum (i.e., 112.77 g/l) at pH 6.5. Thus it was used in further research. Rate of Cal-gluconate production was again decreased above this pH i.e., about 72.75 g/l at pH 8.0.

Table 3: Effect of pH on the production of Calcium gluconate by *Aspergillus niger* IC-15

Initial pH	Glucose (g/l)		Calcium Gluconate produced (g/l)	%age yield of Calcium Gluconate	Mycelial dry weight (g/l)
	Residual	Used			
4.0	105.60	44.40	29.50	19.66	3.88
4.5	94.25	55.75	49.75	33.16	4.70
5.0	70.10	79.90	69.35	46.23	5.75
5.5	66.45	83.55	86.76	57.84	7.98
6.0	50.25	99.75	98.36	65.57	8.40
6.5	40.20	109.80	112.77	75.18	8.96
7.0	62.70	87.30	86.37	57.58	8.75
7.5	69.25	80.75	79.36	52.90	9.50
8.0	60.30	89.70	72.75	48.50	9.65

Glucose residual = Glucose added – Glucose used
Glucose added = 150 g/l
Fermentation period = 72 hours
Rotation rate = 200 rpm

Effect of glucose concentration

The concentration of carbon sources plays an important role on the conversion of glucose into gluconic acid and its calcium salt. The data of Table 4 shows the effect of different glucose concentration (10–30 % w/v) on the consumption of glucose, mycelial dry weight and calcium gluconate formation. In all of these levels of sugar, the maximum conversion into gluconic acid was found when the glucose concentration was 15 % w/v i.e., 11.35 g/l (74.90 %). The production of calcium gluconate however was reduced with the increase in the concentration of sugar. The mycelial growth was increased at higher concentration of glucose added.

Table 4: Effect of glucose concentrations on the production of calcium gluconate by *Aspergillus niger* IC-15

Concentration of glucose (%)	Glucose (g/l)		Calcium Gluconate produced (g/l)	%age yield of Calcium Gluconate	Mycelial dry weight (g/l)
	Residual	Used			
10.0	40.00	60.00	49.00	49.00	6.20
12.5	41.65	83.35	80.00	64.00	7.10
15.0	39.50	110.50	112.35	74.90	8.70
17.5	66.25	108.75	109.00	62.28	9.80
20.0	87.50	112.50	115.00	57.50	10.30
22.5	113.15	111.85	110.33	49.03	12.75
25.0	140.25	109.75	112.20	44.88	13.25
27.5	170.45	104.55	107.33	39.02	13.80
30.0	210.15	89.85	85.44	28.48	14.20

Glucose residual = Glucose added – Glucose used
Glucose added = 150 g/l
Fermentation period = 72 hours
Rotation rate = 200 rpm

Effect of nitrogen sources

The glucose consumption, hence its conversion into calcium gluconate was maximum in the presence of ammonium sulphate i.e. 109.33 g/l and 120.10 g/l respectively. Thus ammonium sulphate was selected for further experiments. The mycelial dry weight produced by various nitrogen sources ranged from 7.31 to 12.99 g/l in the presence of ammonium sulphate, it was 8.23g/l.

Table 5: Effect of nitrogen sources on the production of calcium gluconate by *Aspergillus niger* IC-15

Nitrogen sources	Glucose (g/l)		Calcium Gluconate produced (g/l)	% age yield of Calcium Gluconate	Mycelial dry weight (g/l)
	Residual	Used			
NH_4NO_3	45.25	104.75	103.50	69.00	7.31
$(NH_4)_2SO_4$	40.67	109.33	120.10	80.06	8.23
NH_4Cl	63.67	86.33	82.18	54.78	10.78
Peptone	51.75	98.25	96.77	64.51	10.00
$NaNO_3$	77.00	73.00	61.75	41.16	12.99

Fermentation period = 72 hours
Nitrogen source added = 2.5 g/l

Effect of different concentrations of ammonium sulphate

Effect of different concentrations of ammonium sulphate (1.0-4.0 g/l) on the growth of *Aspergillus niger* IC-15 and calcium gluconate formation was investigated (Table 6). The optimum level of nitrogen source was 2.0 g/l. The glucose consumption, calcium gluconate formation and mycelial dry weight were 120.23, 125.33,and 8.25g/lrespectively. At a concentration of 1.0 and 1.5g/l of ammonium sulphate the yield of calcium gluconate were 88.25 and 107.00 g/l respectively. However the yield of calcium gluconate was also decreased with increase in the level of ammonium sulphate. The mycelial dry weight was ranged from 7.35 to 1.98 g/l and its growth was directly proportional to the amount of ammonium sulphate.

Table 6: Effect of different concentration of ammonium sulphate on the production of calcium gluconate by *Aspergillus niger* IC-15

Amount of ammonium sulphate (g/l)	Glucose Used (g/l)	Calcium Gluconate produced (g/l)	% age yield of Calcium Gluconate	Mycelial dry weight (g/l)
1.0	91.55	88.25	58.83	7.35
1.5	105.25	107.00	71.33	8.10
2.0	120.23	125.33	83.55	8.25
2.5	125.88	110.20	73.46	8.90
3.0	111.66	89.22	59.48	9.33
3.5	92.75	69.88	46.58	9.27
4.0	97.65	62.36	41.57	11.98

pH = 6.5
Glucose added = 150 g/l
Fermentation period = 72 hours

4 Discussion

The production of calcium gluconate by *Aspergillus niger* is a worth praising achievement in the field of Fermentation Technology. The optimisation of cultural conditions such as pH, temperature, incubation period & selection of the suitable substrate is very essential for maximum production of enzyme glucose oxidase and the bioconversion of glucose into calcium gluconate or gluconic acid (Prescott & Dunn's, 1987; Yoshie 1999). The present studies describe the production of calcium gluconate by *Aspergillus niger*. Thirty mould cultures of *Aspergillus niger* were isolated from different soil samples by "pour plate method". Of all the cultures examined, *Aspergillus niger* IC-15 was selected as it gave better results of calcium gluconate. The selected culture consumed 86.44 g glucose, 72 h after conidial inoculation and the amount of calcium gluconate produced was about 96.35 g/l. The mycelial dry weight was 8.00 g/l. The initial concentration of the glucose in the basal medium was kept at 150 g/l.

Optimum fermentation period is one of the most important factors in calcium gluconate fermentation (Pons *et al.*, 2000). The time course fermentation studies during fermentation showed that the calcium gluconate production was maximum at 72 hrs, after conidial inoculation. Further incubation did not increase the production of calcium gluconate which might be due to the over growth of mycelium. The results also revealed that actual biosynthesis of the product was started 24 h after inoculation as there was no production prior to this period. Buzzini *et al.* (1993) have reported maximum yield of calcium gluconate at 72 h of time. However, according to Yasin *et al.* (1975) the maximum amount of calcium gluconate was produced by *Aspergillus niger* strain was 96 h after spore inoculation. Hence, the present finding is more encouraging as compared to Yasin *et al.* (1975) because reduction in the time period reduced the cost of calcium gluconate production.

The maintenance of a favourable pH is very essential for the successful fermentation of calcium gluconate (Milikovic & Vukojevic, 1989; Lee *et al.* 1998). The influence of pH of fermentation medium on the growth of the mould culture and hence the concentration of gluconic acid was also investigated. At an initial pH of 5.0 and less than 5.0, both mycelial growth and calcium gluconate production was poor while at initial pH of 6.5, the consumption of glucose as well as calcium gluconate production was improved significantly. It may be due the fact that at pH 6.5, the mycelium contained maximum glucose oxidase. However, further increase in the initial pH reduced the rate of bioconversion of glucose into calcium gluconate. These results are actually in accordance with the work reported by Takao & Sasaki (1964). At a lower pH, glucose oxidase was not formed yet its formation started at pH value greater than 5.0. Therefore, the effect of initial pH on calcium gluconate production might reflect the effect on glucose oxidase activity. In a similar study, Ray and Banik (1999) found maximum amount of calcium gluconate by mutant *Aspergillus niger* at a pH of 6.5. However, the study is not in good agreement with Kundu and Das (1994) who got maximum yield of calcium gluconate at initial pH 5.5. This may be due to the type of strain and its physiological conditions. As the pH was increased up to 7.0 or 8.0, the production of calcium gluconate was remarkably decreased.

The initial glucose concentration has been found to determine the amount of calcium gluconate by *Aspergillus niger*. Effect of glucose concentration on calcium gluconate production and mycelial dry weight was investigated. When glucose concentration in the medium was 15% (w/v) the yield of calcium gluconate was maximum. However, further increase in the glucose concentration resulted in lowering both the glucose utilization and calcium gluconate formation. Increase in calcium gluconate yield at 150 g/l of glucose might be due to fact that at this concentration of glucose oxidase activity in *Aspergillus niger* was optimum which resulted in the high yield of calcium gluconate. The decrease in the production may be due to catabolic repression of *Aspergillus niger* (Doneva *et al.*, 1999). Qadeer *et al.* (1975) found that 10-15 % of glucose is more suitable and economical, for calcium gluconate production and when glucose concentration was increased up to 30%, the yield of calcium gluconate was markedly decreased. Moreover, at a higher

214

concentration of glucose (20 - 30 %), the occurrence of foam was noted which disturbed the proper mycelial distribution. Mycelium accumulated on the foam layer decreased the contact surface of mycelium with fermentation medium, and this resulted in slower bioconversion of glucose to calcium gluconate. A glucose concentration of 15% was found optimum for calcium gluconate production. Ray and Banik (1994) have also observed higher yield of calcium gluconate when the concentration of glucose was 150 g/l than any other concentration of glucose. They found 100% of yield from this substrate concentration. Our observation was similar to the above-mentioned workers.

Nitrogen constituent has a profound effect on the yield of gluconic acid because the type of nitrogen source and its concentration affect the performance of the fungus considerably. Effect of different nitrogen sources on both the glucose consumption as well as calcium gluconate production was examined. The most suitable nitrogen source for the production of calcium gluconate by *Aspergillus niger* IC-15 was ammonium sulphate. However, sodium nitrate resulted minimum yield of calcium gluconate production. The other nitrogen sources like ammonium nitrate, ammonium chloride, sodium nitrate and peptone were favourable for mycelial growth of *Aspergillus niger* but not suitable for calcium gluconate production. In a similar study, Elanghy and Megalla, (1995) have reported that the peptone was best sources for the production of calcium gluconate. But in present series of experiments, ammonium sulphate was found as best source of nitrogen for the production of calcium gluconate so its various concentrations (1.0-40 g/l) were further studied. The optimum level of $(NH_4)_2SO_4$ was 2.0 g/l. Further increase in its concentration did not enhance the production of calcium gluconate. However, at higher concentration of nitrogen, the growth rate of the fungus was increased but the amount of glucose provided in to the medium was consumed only for the growth of *Aspergillus niger* and not for the production of calcium gluconate. Thus only 2.0 g/l of ammonium sulphate was optimum and used in further studies. From the studies it is evident that the biosynthesis of calcium gluconate is strongly dependent on the selection of strain as well as cultural conditions employed.

References

[1]   Amberkar, G.R., S.B. Thadani, and V.M. Doctor. 1965. Production of calcium gluconate by *Penicillium chrysogenum* in submerged culture. App. Microbiol. 13: 713-719.

[2]   Buzzini, P., M. Gobbetti, J. Ross and S. Haznedari. 1993. Calcium gluconate *Aspergillus niger*. Microbiol. Technol, 43(2) 195-198.

[3]   Chaturvedi, S. K., G. N. Qazi, C. N. Gains, G. M. Gupta, C. L. Choptra and C. K. Atal. 1978. Proc: Indust ferment symp. 40.

[4]   Delgado, J. N. and W.A. Remers. Text Book of Organic Medical and Pharmaceutical Chemistry. 774, 1991.

[5]   Doneva, T., C. Vassilieff and R. Donev. 1999. Catalytic and biocatalytic oxidation of glucose to gluconic acid in a modified three phase reactor. Biotechnol Lett. 21(12): 1107-1111.

[6]   Elnaghy, M. A. and S.E. Megalla. 1975. Gluconic acid production by Penicillium puberculum. Folia. Microbiol. 20: 504-508.

[7]   Hill, G.A., and C.W. Robinson. 1988. Morphological behaviour of *S. cerevisiae* during continuous fermentation of calcium gluconate. Biotechnol Lett. 11(11), 805-810.

[8]   Johnson, L. F., E.A. Curl, J.H. Bond and H.A. 1959. Fribourg. Methods for studing soil micro flora-plant disease relationship. Burgess publ. Co. Minneapolis Minn., U. S. A.

[9]   Kundu, P. N. and A, Das. 1984. Utilization of cheap carbohydrate sources for production of calcium gluconate by *Penicillium funiculosum* mutant MN-238. Ind. J. Exp. Biol., 22:279-281.

[10]   Lee, H. W., J.G. Pan and J.M. Lebeault. 1998. Calcium gluconate from glucose substrate. Applied Microbial. Biotech. 49(1): 9-15.

[11]   Miljkovic, D. and N. Vukojevic. 1989. New method for the preparation of gluconic acid from d-glucose in neutral medium. Zb. Matice Srp. Pir. Nauke, 77: 89-93.

[12]   Pedrosa, A., and M.L. Serrano. 2000. Solubility's of ca- gluconate in water and in aq. solution of ethanol and methanol. J.Chem. Eng 45(3): 461-463.

[13]   Pharmacopoeia, U. S.1990. "Assay of calcium gluconate".

[14]   Pons, A.J.,F. Sagues, M.A.Bees and P.G. Sorenson. 2000. pattern formation of calcium gluconate in MB- glucose. J. Phys. Chem. 104(10), 2251-2259.

[15]   Prescott, S.C. and C.G. Dunn.1987. Industrial microbiology. 4th Edition. Mcgraw Hill Book Co. Inc. New York.

[16]   Qadeer, M.A. Baig, M. Afzal and O. Yunus. 1975. production of calcium gluconate by *Aspergillus niger* in 50-L fermentor. Pak. J. Sci. Ind. Res. 18 (5): 227-228.

[17]   Ray, S. and A.K. Banik.1994. Development of a mutant of *Aspergillus niger* and optimisation of some physical factors for improved calcium gluconate production. Ind. J. Exp. Biol. 32: 965-868.

[18]   Ray, S. and A.K. Banik.1999. Effect of ammonium and nitrate ratio on glucose. Ind. J. of Experimental Biology. 37(4): 391-395.

[19]   Rose, A. H.1961. Gluconic acid In: Industrial Microbiology, Butter Worths London

[20]   Shah, D.N. and R.M. Kothri.1993. Glucose oxidase rich *Aspergillus niger* strain an economical substrate for the preparation of tablet grade calcium gluconate. Biotechnol. Letters, 15(1): 35-40.

[21]   Takao, S. and Y. Sasaki.1964. gluconic acid fermentation by *Pullularia pullulans*. Screening of gluconic acid production strain and some conditions for its production. Agr. Biol. Chem. 28: 752.

[22]   Tasun, K., Chose, and Ghen. 1970. Sugar determination by DNS method. Biotech. Bioeng. 12: 921.

[23]   Trager, M. and G.N. Qazi. 1991. Contribution of endo and exo-cellular glucose oxidase to gluconic acid production at increased dissolved oxygen concentration. J. Chem. Technol. Biotechnol. 50(1): 1-11.

[24]   Yasin, M., A. Hameed. And M.A. Qadeer1975.. Selection of a hyper-producer strain of *Aspergillus*, for the production of calcium gluconate. J. Sci. Res., 43:67-72.

[25]   Yoshie, T. 1999. Carbon and nitrogen utilization and acid production by mycelia of *Aspergillus niger*. Mycoscience, 40(1): 51-56.

Combined effect of doxorubicin and metal on the yeast *Candida utilis*

J. Keyhani[*,1], **E. Keyhani**[1,2], **S. Khavari-Nejad**[2], **F. Attar**[2], and **F. Azzari**[2]

[1] Laboratory for Life Sciences, Saadat Abade, Sarve Sharghi 34, 19979 Tehran, IRAN
[2] Institute of Biochemistry and Biophysics, University of Tehran, 13145 Tehran, IRAN

[*]Corresponding author: e-mail: keyhanie@ibb.ut.ac.ir, Phone: +98-21-695-6974, Fax: +98-21-640-4680

The combined effect of doxorubicin and nickel on some vital parameters of the yeast *Candida utilis* was investigated and compared to the effect of the drug or the metal alone. Addition of nickel to doxorubicin enhanced the inhibitory effect of the drug on the parameters investigated, including cell growth, morphology and cytochromes content, suggesting a greater toxicity of doxorubicin when combined with nickel.

Keywords doxorubicin; nickel; anthracyclines-metal complexes; *Candida utilis*

1 Introduction

A majority of therapeutic agents are organic molecules that interact with cell components through weak linkages of hydrogen bonding and Van der Waals forces, allowing for nonspecific interactions with nontarget molecules. The formation of metal-drug complexes allows for stronger and more specific types of interactions through covalent and ionic bonds due to the coordination ability of metals [1].

Doxorubicin is an anthracycline antibiotic highly effective against solid and haematologic malignancies [2]. However its use is limited because it induces acute and chronic cardiotoxicity [3,4]. Anthracyclines are known to intercalate between DNA base pairs [5] and to act on protein kinase C [6], a family of serine/threonine kinases involved in signal transduction, cell growth/differentiation and hormone secretion. As interaction between drug and protein is favored by the presence of a metal, anthracyclines-metal complexes may become more effective anti-cancer drugs [4], but they may also become more toxic. Thus it is mandatory to first explore the effect of such complexes on normal cells. Yeast, a eukaryotic cell easy to culture and to obtain in large quantities, is a good model for such investigations.

In this study, the combined effect of doxorubicin and nickel on some vital parameters of the yeast *Candida utilis* was investigated and compared to the effect of either the drug or the metal alone.

2 Materials and methods

Yeast *Candida utilis* (strain ATCC 8205) was cultured at 28°C in enriched medium (liquid or solid) containing increasing concentrations of either doxorubicin (doxo) (1-200 µg/ml), $NiSO_4$ (Ni^{2+}) (1-5 mM) or both (1-5 mM Ni^{2+} with either 10 or 20 µg/ml doxo). The cells cytochrome content was deduced from reduced-minus-oxidized spectra of cell suspension, using extinction coefficients 16 for cytochrome *c* ozidase, 18 for cytochrome *b*, and 19 for cytochrome *c* [7]. The effect of Ni^{2+} on doxo was assessed by spectrophotometry: scanning repetition of a solution of the drug (16 µg/ml) to which Ni^{2+} was added in 0.4 mM increments, was performed with a Perkin-/Elmer spectrophotometer.

3   Results and discussion

Addition of Ni^{2+} to doxo resulted in the formation of a metal-drug complex as revealed by spectrophotometry (Fig. 1). In the presence of 1mM Ni^{2+}, colony count remained unaffected in up to 50 µg/ml doxo, and was inhibited by only 20% in 200 µg/ml doxo, while in the presence of 2 mM Ni^{2+}, cell count was diminished by 30% in as little as 20 µg/ml doxo and by 99% in 50 µg/ml doxo (Fig. 2).

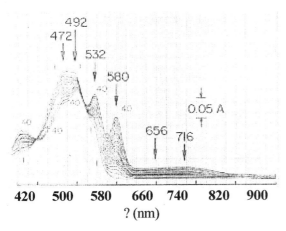

Fig. 1 Absorption spectra of doxorubicin (16 µg/ml) alone (spectrum 1) and in the presence of increasing amounts of Ni^{2+} added in 0.4 mM increments (spectra 2-40). Addition of Ni^{2+} produced a decrease in the extinction coefficients at 472 and 492 nm, a progressive enhancement of the shoulder at 532 nm and the formation of another absorption band at 580 nm. These observations suggested an interaction between Ni^{2+} and doxorubicin leading to the formation of a doxo-Ni^{2+} complex.

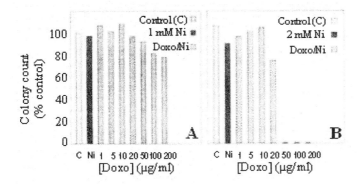

Fig. 2 Colony count (% control) after incubation at 28°C for 72 h on solid enriched medium supplemented with 1 mM Ni^{2+} (A), or with 2 mM Ni^{2+} (B) and various doxorubicin concentrations as indicated.

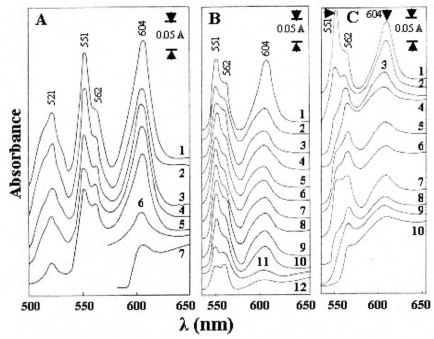

Fig. 3 Dithionite-reduced minus H2O2-oxidized difference spectra of yeast cells cutured in the presence of doxo (A), Ni2+ (B), and doxo-Ni2+ (C). (A): spectrum 1 is control; spectra 2-7: respectively 1, 5, 10, 20, 50 and 100 µg/ml doxo. (B): spectrum 1 is control; spectra 2-12: 1, 1.5, 2, 2.1, 2.2, 2.3, 2.4, 2.5, 3, 4, and 5mM Ni2+, respectively. (C): spectrum 1 is control; spectrum 2: doxo (20 µg/ml); spectra 3-10: doxo (20 µg/ml) + 1, 1.5, 2, 2.1, 2.2, 2.4, 2.5, and 3 mM Ni2+, respectively.

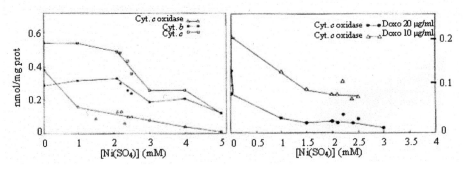

Fig. 4 Cytochromes content, expressed in nmol/mg prot, of cells grown in the presence of 1-5 mM Ni^{2+} (A) and of cells grown in the presence of 1-3 mM Ni^{2+} and either 10 or 20 µg/ml doxo (B).

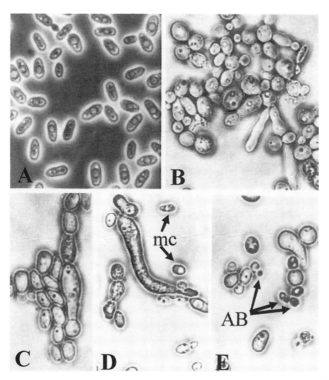

Fig. 5 Phase-contrast light microscopy of *Candida utilis:* control (A); + 2 mM Ni^{2+} (B); + 20 µg/ml doxo (C); + 20 µg/ml doxo and 1 mM Ni^{2+} (D,E). AB: apoptotic bodies; mc: microcells. Magnif.: × 930.

Table 1: Cell yield after 16 h culture at 28 °C in the presence of either doxo (A), Ni^{2+} (B), or doxo-Ni^{2+} (C)

A		B		C	
Doxo (µg/ml)	Cell yield (wet weight) (% control)	Ni^{2+} (mM)	Cell yield (wet weight) (% control)	Ni^{2+} (mM) (+ 20 µg/ml doxo)	Cell yield (wet weight) (% control)
0	100	0	100	0	98
3	107	1.0	82	1.0	47
5	112	1.5	62	1.5	36
10	105	2.0	47	2	28
20	98	2.5	43	2.5	28
50	63	3.0	42	3.0	27
100	63	4.0	30	4.0	20
200	23	5.0	23	5.0	18

220

The effect of doxo-Ni^{2+} on the cell yield in liquid medium was more drastic even in the presence of 1 mM Ni^{2+} (Table 1C), probably because of an increased uptake by the cells. In the presence of up to 20 μg/ml doxo, growth was not inhibited and even stimulated (Table 1A), but in the presence of 20 μg/ml doxo with 1 mM Ni^{2+}, the cell yield dropped by 50% (Table 1C). The dithionite-reduced minus H_2O_2-oxidized difference spectra in Fig. 3 showed that Ni^{2+} caused a decrease in the absorbance of cytochrome c oxidase together with a broadening of the absorbance band and the formation of a shoulder at 592 nm. The same alterations in the cytochrome c oxidase absorbance were caused by doxo-Ni^{2+} but not by doxo alone, and they probably reflect changes in the heme environment caused by the binding of Ni^{2+} to the cytochrome as was shown with another hemoprotein [8]. The amount of cytochrome c oxidase detectable in cells grown in doxo-Ni^{2+} was considerably reduced compared to the amount detectable in cells grown doxo alone (Fig. 4). As shown in Fig. 5, cells grown in the presence of Ni^{2+} or doxo alone appeared swollen, occasionally elongated and tended to agglutinate (Fig. 5B,C). Cells grown in the presence of the drug and the metal were often necrotic and sometimes rhizomorphe (Fig. 5D); apoptotic bodies and microcells were also seen (Fig. 5D,E). The number of necrotic cells seen after growth in the presence of doxo-Ni^{2+} was largely increased compared to that seen in doxo alone (see also reference 9). Results suggest a greater toxicity of doxo when combined with Ni^{2+}, probably due in part to the binding of Ni^{2+} to some vital proteins as seen with cytochrome c oxidase.

Acknowledgements

This work was supported in part by the University of Tehran, and in part by the J. and E. Research Foundation, Tehran, Iran.

References

[1]   A. Y. Louie, and T. J. Meade, Chem. Rev. **99**, 2711 (1999).
[2]   J. S. Sandberg, F. L. Howsder, A. Di Marco, et al., Cancer Chemother. Rep. **54**, 1 (1970).
[3]   R. L. Basser, and M. D. Green, Cancer Treat. Rev. **19**, 57 (1993).
[4]   R. D. Olson, and P. S. Mushlin, FASEB J. **4**, 3076 (1990).
[5]   A. Di Marco, Cancer Chemother. Rep. **6**, 91 (1975).
[6]   E. Monti, F. Monzini, F. Morazzoni, et al. , Inorg. Chim. Acta **205**, 181 (1993).
[7]   E. Keyhani, and J. Keyhani, Biochim. Biophys. Acta **717**, 355 (1982).
[8]   J. Keyhani, E. Keyhani, S. Zarchipour, et al., Biochim. Biophys. Acta **1722**, 312 (2005).
[9]   E. Keyhani, and J. Keyhani, Ann. N.Y. Acad. Sci. **1030**, 369 (2004).

Hydrodynamic analysis and inoculum effect on *Rhizopus nigricans* growth for cellulase production in bubbling column bioreactor

E. Botello-Álvarez, S. Pérez-Castrejón, H. Jiménez-Islas, R. Rico-Martínez, J.L. Navarrete-Bolaños[*]

Departamento de Ingeniería Química-Bioquímica, Instituto Tecnológico de Celaya, Av. Tecnológico s/n, C.P. 38010. Celaya, Gto., México., Phone: (+52) 461 61 1 75 75, Fax: (+52) 461 61 1 79 79.
[*]Corresponding author: e-mail: jlnb@itc.mx., Phone: (+52) 461 61 1 75 75, Fax: (+52) 461 61 1 79 79.

Image processing techniques were used to study the different forms of growth of *Rhizopus nigricans* to produce cellulases. Previous efforts had concentrated on characterizing the mechanisms involved in different forms of fungal growth, identifying the hydrodynamic conditions as the most important factor on determining the specific growth morphology in the culture of filamentous fungi. The tentative range of conditions of superficial air velocity, flow regime, and quantity and morphology of the initial inoculum for cellulase production, in a bubbling column bioreactor, associated to different growth morphologies were identified. Superficial air velocities below 0.965×10^{-3} m/s, where the laminar regime prevails, induce the formation of small, spongy pellets and favors cellulase production. Superficial air velocities with predominantly turbulent flow (11.4×10^{-3} m/s) induce fungal growth in both pellets and amorphous compact forms with restricted generation of new biomass and cellulase production. In addition, it was found that the maximum cellulase production yield was obtained from cultures with 2.5% of pelletized inoculum with laminar flow, conditions that lead to completely pelletized morphological forms with air mycelia. These conditions do not limit nutrients transport and do not propitiate microorganisms' autolysis.

Keywords: *Rhizopus nigricans*, cellulase production, bubbling column bioreactor.

1 Introduction

Aerobic submerged fermentation cultures of filamentous fungi are involved in a wide scale of industrial bioprocesses (e.g. antibiotics, enzymes, vitamins, colorants, etc.). One relevant aspect of these cultures is their morphological growth that is determinant of the process yield (1, 2). Previous studies have shown that fungi cultured on submerged fermentation systems exhibits different morphological growth phases ranging from free dispersed mycelia or mycelia grouping (clumps), up to interlaced mycelia mass formation (pellets) that are spherical or ellipsoidal and present hyphal fungal growth with internal structure that may be classified as spongy or compact (3–5). As a function of these growth patterns, the apparent viscosity of culture medium varies from that of a newtonian fluid up to one resembling a pseudoplastic fluid; these transitions limit the mass transfer mechanism among phases, altering the metabolic activities of the microorganism to produce a specified product (2, 6). The specific conditions leading to a particular morphology over another are complex, and include factors such as the genetics of the fungi, inoculum nature, medium culture composition, and physical factors as pH, temperature, agitation and mixing (2). However, special notice has been placed on the hydrodynamic conditions as the leading cause for morphologic transitions present in filamentous fungi cultures (4,6). Here, assisted with image processing techniques, the different forms of Rhizopus nigricans growth to produce cellulases are studied. This microorganism was previously isolated from endogenous microbiota of the marigold flower, and defined as the most significant in marigold flower ensilage due to its high cellulase activity (7, 8). The objective is to characterize the hydrodynamic variables effect in a bubbling column bioreactor, as well as the effect of quantity and initial morphology of starter inoculum, on the morphological growth and cellulase production. The friction factors, associated to each flow regimen in the bubbling column bioreactor, are used as the basis to establish the correlation among morphological development, cellulase yield and hydrodynamic regimen.

2   Materials and methods

Microorganism. *Rhizopus nigricans,* microorganisms associated as normal flora of the marigold flower, were used in all assays.

Culture preparation. *Rhizopus n,* was cultured on nutrient and potato dextrose (PDA) agars (Difco laboratories, Detroit Michigan) slants at 28 °C for 24 h. Biomass taken from the slants was transferred to 250 mL Erlenmeyer flasks containing 100 mL of potato dextrose broth (Difco laboratories, Detroit Michigan), incubated on a rotary shaker at 28 °C and 100 rpm (Forma Scientific, model 4520) for 48 hours to obtain pelletized inoculum, and 28 °C and 175 rpm (Forma Scientific, model 4520) for 48 hours to obtain dispersed mycelia inoculum.

Fermentation assays in a bubbling column. The products obtained from culture preparation were used as starter inoculum in the fermentation assays. In the study, the variables analyzed were the initial morphology of fungi (pellets or dispersed mycelia), the quantity (2.5–15 % (v/v)) and superficial air velocity ($0.0965 \times 10^3 – 11.377 \times 10^3$ m/s). The assays were performed in two stages: Firstly for each type of inoculum morphology (pellets or dispersed mycelia) proportional increases were applied to the ratio inoculum amount-superficial air velocity. In the second stage, again for each type of inoculum morphology, one of the variables (inoculum amount or superficial air velocity) was kept constant while the other was varied using regular increments. In all experiments, the friction factor was evaluated as a means to assess the flow regime.

Fermentation system. A bubbling column bioreactor (height 0.95 m, 0.095 m diameter, and diffuser of 200 ? m average) was used for the fermentation process. The column design includes a jacketed body for temperature control, six ports for sensors (pH, dissolved oxygen, foam/level control), and auxiliary equipment. The system includes a Bio Console (ADI 1035, Applikon Biotechnology), as well as peripheral equipment needed for any fermentation process (gas flow, pump section, foam/level controller, temperature control), and a Bio Controller (ADI 1030, Applikon Biotechnology) for accurate setting of parameters. In addition, the ADI 1030 Bio Controller is equipped with a bi-directional serial communication port for supervisory control and data acquisition (Bio expert software, Applikon Biotechnology).

Morphological studies. A digital camera (Sony model DCR-TRV730 NTSC) was connected to the fermentation system to obtain digital images (sigma scan pro V5). At the conclusion of the fermentation, microorganism samples were analyzed by microscopy techniques (stereomicroscopy, Leica model).

Enzymatic extract and biomass. Samples of known volume taken from the fermentation system were centrifuged at 6000 rpm (Hermle, model Z383) for phase separation. The liquid phase (supernatant) obtained was used as raw cellulase extract, and the solid phase (biomass) was used for microorganism growth kinetics.

Enzymatic activity for the raw enzymatic extract. For enzymatic activity studies, a 2 mL volume of raw extract was added to 100 mL of a mixture of carboxymethyl cellulose (CMC, high viscosity; Sigma Chemical Co., St. Louis, MO) of 3.73 g/L that had a viscosity of 2500 centipoises (cp) (Brookfield digital rheometer model DV-III+). The solutions obtained were kept on a rotary shaker at 28 °C at 175 rpm (Forma Scientific, model 4520) for 24 h. The enzymatic activity was measured indirectly as a function of viscosity reduction.

Microorganism growth kinetics. For determining cellular dry weight (cdw), the solid phase obtained from the samples previously centrifuged was dissolved, and filtered using Millipore membranes (0.45 ím pore size). The membranes and cell pellets were then dried at 90 °C for 12–16 h, until a constant weight was obtained.

3   Result and discussion

In general, the results show (Figure 1) that cellulase production is favored by pelletized starter inoculum and superficial air velocities below 2×10^{-3} m/s where a laminar regime prevails. Increasing the superficial air velocity up to 4×10^{-3} m/s leads to diminishing cellulase production independently of the starter inoculum morphology. Seemingly, this latter superficial air velocity value falls in the transition zone between laminar to turbulent flow regimen.

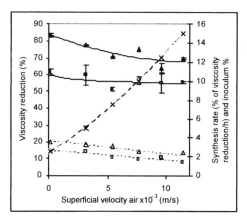

Figure 1. Cellulases production and synthesis velocity as a function of superficial air velocity and inoculum percentage (v/v): (▲) Cellulase production with pelletized inoculum, (■) Cellulase production with dispersed mycelia inoculum, (△) synthesis velocity with pellets, (□) synthesis velocity with dispersed mycelia, and (x) inoculum percentage (v/v).

The turbulence induces fungal growth in pelletized and amorphous compact forms, restricting new biomass generation and cellulase production. Analysis by microscopy of the amorphous compact form shows an exterior rigid layer and a hollowed interior; the hollowed region can be associates to microorganism autolysis due to limited diffusion of nutrients. Kinetic studies of cellulase synthesis confirm that the pelletized initial inoculum lead to larger cellulase production over the dispersed mycelia starter inoculum. For both types of initial morphology the production declines steadily when the superficial air velocity and inoculum amount are increased. In addition, kinetic biomass growth studies show (Figure 2) the same trend: a steady decrease of biomass production (normalized by the amount of starter inoculum) as the superficial air velocity and starter inoculum concentration are increased. Once again, such observation appears to be consistent with the onset of fungal morphological transition associated to the flow. For pelletized starter inoculum, this transition appears to be completed at superficial air velocities around 5.4×10^{-3} m/s for which the fungi, at the end of the fermentation cycle, are grouped in compact pellets devoid of dispersed mycelia with hollow interiors. Below this velocity, the pellets preserve a high level of porosity ("spongy structure") with significant exterior mycelia which are roughly half the size of the pellet diameter. The transition occurs at much lower superficial air velocities when the starter inoculum is in the form of dispersed mycelia.

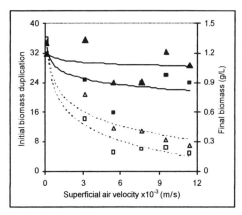

Figure 2. Normalized biomass growth: (Δ) with pelletized inoculum, (□) with dispersed mycelia inoculum, (▲) final biomass with palletized inoculum,(■) final biomass with dispersed mycelia inoculum.

Assays belonging to the second stage of the experimental strategy, described above, showed that cellulase production is independent on the starter inoculum concentration, and instead depends only on superficial air velocity. Once again, low superficial air velocities are associated to larger cellulase yields, thus confirming that laminar flow is determinant in the growth and cellulase production by *Rhizopus nigricans* fungi. Previous efforts have elucidated the influence of the hydrodynamic state on the final morphological form, via different measurements such as the turbulent intensity, for processes taking place on stirrer vessels and air lift bioreactors (9, 10). A similar characterization can be achieved for the bubbling column bioreactor, discussed here, by relating the flow regime to an estimated Reynolds number via experimental estimation of the friction factor. The experimental friction factor was defined as the ratio between the average pressure drop on the column (neglecting the hydrostatic pressure contribution) and the kinetic energy based on the superficial air velocity. By analogy with the regular plots of friction factor versus Reynolds number (see e.g. ref. 11), one can clearly define air velocities that belong to either laminar or turbulent regimes. The results show (Figure 3) that superficial air velocities below 3×10^{-3} m/s belong to a laminar regimen.

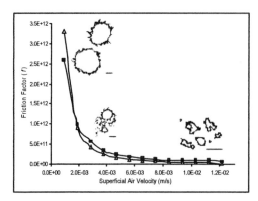

Figure 3. Superficial air velocity versus friction factor. The final morphology for the laminar flow (upper left corner) and turbulent flow (lower right) are exemplified using digitized experimental images (scale: all bars = 5 mm).

For this regime, digitized images exhibit spongy-looking pellets with exterior mycelia for pelletized starter inoculum. In addition, it is observed that the pellets move along the liquid streamlines with negligible radial velocity, thus there are little fluid-mycelia interactions leading to shear stresses. The assays at the high superficial velocities (turbulent regime) lead to the formation of compact, hollowed pellets, seemingly this is the result of the fungi response to the increasing shear stresses induced by the significant radial velocities. When the starter inoculum is in the form of dispersed mycelia, there are occurrences of growth characterized by interlacing among mycelia, forming clumps. This occurs at low superficial velocities (laminar regime) and can be interpreted as the result of small fluid-mycelia interactions (small, but significant, radial velocities induced by the mycelia presence). At high superficial velocities (turbulent flow) one still observed compact hollowed-pellets that are also interlaced by mycelia.

4 Conclusions

The morphology of the starter inoculum and the flow regime has significant effect on cellulase production by *Rhizopus nigricans* in a bubbling column bioreactor. Pelletized starter inoculum and laminar regime appear to favour the biomass growth and cellulase synthesis. While turbulent regime limit both biomass growth and cellulase synthesis. This study has shown that the starter inoculum morphology, and the interaction solid-liquid (biomass-culture medium), resulting from the hydrodynamic state, determine the growth pattern (amount and developmental morphology) of filamentous fungi, and can be considered as the main factors to be considered in fermentation process in bubbling column bioreactors. In general, the best yields can be obtained using pelletized starter inoculum; even if biomass growth is undesirable, the pelletized starter inoculum leads to the more favourable formation of compact hollowed-pellets under turbulence. A morphology that exhibits biomass growth limitations, such as these hollowed pellets, may be advantageous for production processes of metabolites not associated to the growth. Finally, in the study it was found that the maximum cellulase yield was obtained from cultures with 2.5% of inoculum and 0.965×10^{-3} m/s of superficial air velocity. In these conditions, the fungi show completely pelletized morphological growth. Image analysis reveals a sponge-like consistency with air mycelia. These characteristics do not limit nutrients transport, avoid microorganism autolysis, and do not affect the hydrodynamic regime. The kinetic studies show that the maximum yield is reached in 24 hours processing time, with a substrate uptake of the 82.7%, and pH decrease of the 54.7%. The latter condition enhances cellulase activity.

References

[1]   J. Nielsen, C.L. Johansen, M. Jacobsen, P. Krabben, and J. Villadsen. Pellet formation and fragmentation in submerged cultures of *Penicillium chrysogenum* and its relation to penicillin production. Biotechnol Prog. 11:93-98 (1995).

[2]   M. Papagianni. Review. Fungal morphology and metabolite production in submerged mycelial processes. Biotechnol. Adv. 22:189-259. (2004).

[3]   P. W. Cox, and C. R. Thomas. Classification and measurement of fungal pellets by automated image analysis. Biotechnol and Bioeng 39:945– 952 (1992).

226

[4]   Z. J. Li, V. Shukla, A. P. Fordyce, P. A. Gade, S. Kevin, S. K. Wenger, and R.M. Marten. Estimation of Hyphal Tensile Strength in Production-Scale *Aspergillus oryzae* Fungal Fermentation, Biotechnol. and Bioeng. **77**: 601-613, 2002. (2000).

[5]   P. Znidarsic and A. Pavko. The morphology of filamentous fungi in submerged cultivations as a bioprocess parameter. Food Technol. and biotechnol. **39**: 237-252. (2001).

[6]   Y. Q. Cui, R. G. J. M. van der Lans, and K. Ch. A. Luyben. Effects of Dissolved Oxygen Tension and Mechanical Forces on Fungal Morphology in Submerged Fermentation. Biotechnol. and Bioeng. **57**: 409-419. (1998).

[7]   J. L. Navarrete-Bolaños, H. Jiménez-Islas, E. Botello-Alvarez, and R. Rico-Martinez. Mixed Culture Optimization for Marigol Flower Ensilage via Experimental Design and Response Surface Methodology. J. Agric. Food Chem. **51**: 2206-2211. (2003).

[8]   J. L. Navarrete-Bolaños, H. Jiménez-Islas, E. Botello-Alvarez, R. Rico-Martinez and O. Paredes-Lopez. Improving Xanthophyll Extraction from Marigold Flower Using Cellulolytic Enzymes. . J. Agric. Food Chem. **52**: 3394-3398. (2004).

[9]   K. E. Morud, and B. H. Hjertager. LDA Measurements and CFD Modelling of Gas-Liquid Flow in a Stirred Vessel. Chem. Eng. Sci. **51**: 233-249, (1996)

[10]   V. V. Ranade. Modeling of Turbulent Flow in a Bubble Column Reactor. Trans Ichem E. **75**: 14-23, (1997).

[11]   R. B. Bird, W. E. Stewart, and E. N. Lightfoot. Transport Phenomena. 2^{ed}. John Wiley and Sons, Inc. 2002. USA.

Inactivation mechanisms of his-tagged D-amino acid oxidase from *Trigonopsis variabilis*

M. Arroyo[*1], **I. de la Mata**[1], **M. P. Castillón**[1], **C. Acebal**[1], **J. L. García**[2], and **M. Menéndez**[3]

[1] Departamento de Bioquímica y Biología Molecular I, Facultad de Ciencias Biológicas, Universidad Complutense de Madrid, c/ José Antonio Novais 2, 28040 Madrid, SPAIN
[2] Departamento de Microbiología Molecular, Centro de Investigaciones Biológicas, Consejo Superior de Investigaciones Científicas (CSIC), 28040 Madrid, SPAIN
[3] Departamento de Química-Física de Macromoléculas Biológicas, Instituto de Química-Física "Rocasolano", Consejo Superior de Investigaciones Científicas (CSIC), Madrid, SPAIN

[*]Corresponding author: e-mail: arroyo@bbm1.ucm.es, Phone: +34 913945120, Fax: +34 913944672

The activity of the chimeric D-amino acid oxidase from *Trigonopsis variabilis* (HIS-*Tv*DAAO) falls progressively with time after dilution into the reaction medium prior to the addition of the substrate. The cause of this inactivation is the dissociation of the holoenzyme into FAD and apoenzyme by dilution, a process which requires reaching several minutes reaching an equilibrium state in which most of the enzyme is dissociated and inactive. When exogenous FAD is added to the reaction medium, this kinetic constant is decreased and the inactivation of the holoenzyme is retarded. On other hand, holoenzyme shows higher thermal stability than apoenzyme at the temperature range between 25°C to 42°C, indicating that FAD may contribute to the stabilization of the protein.

Keywords D-amino acid oxidase; *Trigonopsis variabilis*; inactivation; FAD dissociation

1 Introduction

In recent years, a major biotechnological application of D-amino acid oxidase (EC 1.4.3.3, DAAO) has emerged for the industrial production of 7-aminocephalosporanic acid (7-ACA), a key starting material for the preparation of semisynthetic cephalosporin antibiotics [1,2]. D-amino acid oxidase is a flavoenzyme which can catalyze the conversion of cephalosporin C to glutaryl-7-aminocephalosporanic acid, the first intermediate in the two-step enzymatic route that leads to 7-ACA. Several DAAOs from microorganisms such as the yeasts *Rhodotorula gracilis* (*Rg*DAAO) and *Trigonopsis variabilis* (*Tv*DAAO) have demonstrated to work efficiently in the oxidative deamination of cephalosporin C, whereas DAAO from pig kidney (*pk*DAAO) has been considered a poor catalyst in the same reaction [3]. *Tv*DAAO is nowadays the most industrially used for the production of 7-ACA although no detailed investigations have been performed on this enzyme like those for *Rg*DAAO and *pk*DAAO. In this sense, we have studied the inactivation mechanisms of a chimeric *Tv*DAAO which contains a polyhistidine tag in its N-terminal end (HIS-*Tv*DAAO) [4].

2 Materials and methods

2.1 Enzyme Purification

Chimeric D-amino acid oxidase from *Trigonopsis variabilis* (HIS-*Tv*DAAO) was produced and purified as previously described [4]. HIS-*Tv*DAAO is largely expressed as apoenzyme (about 90%), and further enzyme purification using a tailor-made metal chelate support containing a very low density of cobalt ligands lead to a 100% pure apoenzyme preparation. The holoenzyme could be reconstituted from apoenzyme by addition of exogenous FAD to yield a fully active enzyme.

2.2 Protein determination

Protein concentration was measured according to Bradford method using bovine serum albumin as standard [5].

2.3 Enzyme activity assay

The reaction mixture contained 80 μl of 100 μM FAD dissolved in distilled water and 100 μl of 100 mM D-alanine dissolved in 100 mM potassium phosphate buffer pH 8.0. The mixture was preincubated for 5 minutes at 30°C, and the reaction was started by adding 20 μl of enzyme solution. After 10 minutes, the reaction was stopped with 20 μl of a saturated solution (approximately 10 mM) of 2,4-dinitrophenyl-hydrazine (DNPH) in 1N HCl. After 15 minutes of incubation with DNPH, the colour was developed by the addition of 180 μl of 2N NaOH. Pyruvic acid formation proceeded linearly with time during this period. The activity was calculated by spectrophotometric absorption measurement at 450 nm, using a pyruvic acid calibration curve. All samples were measured three times and the standard error was always below 5%. One activity unit (U) was defined as the amount of enzyme producing 1 μmol of pyruvic acid per minute under the conditions mentioned.

2.4 Kinetics of thermal inactivation of HIS-$T\nu$DAAO

One ml of HIS-$T\nu$DAAO at 1.4 μg/ml concentration was stored in 100 mM potassium phosphate buffer pH 8.0 at different temperatures in the range of 25°C-42°C. Apoenzyme samples were stored in the abscence of the flavin cofactor whereas holoenzyme samples were stored in the presence of 100 μM FAD. 20 μl aliquots of these solutions were withdrawn at periodic intervals and cooled in an ice bath prior to enzyme activity determination according to DNPH method. Residual activity was measured and expressed as percentage of initial activity. From the semilogarithmic plot of residual activity versus time, the inactivation rate constants (k_d) were calculated from the slopes, and the apparent half-lives ($t_{1/2}$) were estimated. The temperature dependence of k_d was analyzed from Arrhenius plot (natural logarithm of k_d versus reciprocal of the absolute temperature); the activation energy (E_a) was obtained from the slope of the plot. In addition, the thermodynamic activation parameters of the inactivation process such as Gibbs free energy (ΔG^*), enthalpy (ΔH^*) and entropy (ΔS^*) were calculated from the following equations: (1) $\Delta H^* = E_a - RT$; (2) $\Delta G^* = - RT\ln[(k_d.h)/(k_B T)]$ and (3) $\Delta S^* = (\Delta H^* - \Delta G^*)/T$, where k_d is the inactivation constant (min^{-1}), k_B is the Boltzmann´s constant (1.38×10^{-23} J K^{-1}), h is the Planck´s constant (1.84×10^{-37} J h), R is the gas constant (8.314 J mol^{-1} K^{-1}) and T is the temperature (K).

2. 5 Loss of enzymatic activity caused by dilution

At t=0 min, an holoenzyme stock of 10 μg/ml HIS-$T\nu$DAAO (0.025 μM) in 100 mM potassium phosphate buffer pH 8.0 was diluted to 0.05 μM in 100 μl of the same buffer at 30°C which contained different amounts of FAD. At various times after dilution, enzymatic activity was measured by the addition of 100 µl of 100 mM D-alanine dissolved in 100 mM potassium phosphate pH 8.0 as mentioned for enzyme activity determination by the DNPH method. According to Neet and Ainslie´s theory for hysteretic enzymes [6], the decrease of enzymatic activity upon dilution could be adjusted to the equation (4) $A_t = A_r + (A_0 - A_r)$. e^{-kt} in which A_t is the activity at time "t", A_r is the remaining activity at infinite time after dilution, A_0 is the activity at time zero following dilution and k is the rate constant for the inactivation. It was possible to describe the relationship between the remaining activity (A_r) and the enzyme concentration as a holoenzyme to apoenzyme equilibrium, assuming that all activity derived from the holoenzyme, and the apoenzyme form was inactive. To make the analysis we assumed that the enzyme sample contained only fully active holoenzyme at t=0 min, and it gradually distributed itself into inactive apoenzyme and active holoenzyme during the dissociation reaction. Thus, we could calculate the concentration of inactive apoenzyme by the expression [apoenzyme]=$[E]_{total}$ × $(A_0 - A_r)$ / A_0, and the concentration of holoenzyme by the expression [holoenzyme]=$[E]_{total}$ × (A_r / A_0), in which $[E]_{total}$ is the total enzyme concentration. The dissociation constant (K_D) for the equlibrium could be calculated from the decay curve after equilibrium was reached in the abscence of FAD by the equation (5) $K_D = $[apoenzyme][FAD]/[holoenzyme] under the assumption that [FAD] is equal to the [apoenzyme] as one molecule of holoenzyme dissociates into one molecule of apoenzyme and one molecule of FAD.

3 Results and discussion

3.1 Thermal inactivation kinetics of apoenzyme and holoenzyme of HIS-TvDAAO

Thermal inactivation kinetics of apoenzyme and holoenzyme of HIS-TvDAAO were studied in the range of 25°C–42°C. Thermal stability was followed by measuring the residual enzyme activity with time. Both apoenzyme (Figure 1a) and holoenzyme (Figure 1b) forms followed first-order deactivation kinetics since semilogarithmic plots of residual activity versus storage time could be adjusted to linear curves in all cases. The Arrenhius plots (Figure 1c) were linear in the temperature range studied. From this plot and making use of Equations (1-3), the thermodynamic parameters for the inactivation process, Gibbs´ free energy (ΔG^*), enthalpy (ΔH^*) and entropy (ΔS^*) of activation were calculated (Table 1).

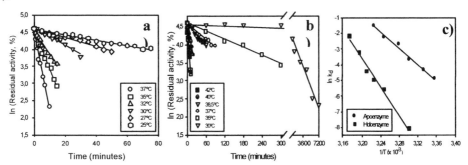

Fig 1. Kinetics of thermal inactivation of HIS-TvDAAO: a) apoenzyme, b) holoenzyme, c) Arrhenius plots.

Table 1. Activation parameters for the thermal deactivation process of HIS-*Tv*DAAO

Inactivation temperature °C	Half-life $t_{1/2}$ (min)	Inactivation rate constant k_d (min^{-1})	?G* KJ/mol	?H* KJ/mol	?S* KJ/mol.°K
APOENZYME (E$_a$ = 215 KJ/mol)					
25	91.2	7.6 ×10^{-3}	95.4	212.5	0.4
27	53.3	13 ×10^{-3}	94.6	212.5	0.4
30	26.6	26 ×10^{-3}	93.7	212.5	0.4
32	10.2	68 ×10^{-3}	92.0	212.5	0.4
35	6.4	109 ×10^{-3}	91.6	212.5	0.4
37	3.1	221 ×10^{-3}	90.4	212.5	0.4
HOLOENZYME (E$_a$ = 381 KJ/mol)					
30	2221.6	0.3 ×10^{-3}	104.9	378.4	0.9
35	185.9	3.7 ×10^{-3}	100.3	378.4	0.9
37	98.0	7.1×10^{-3}	99.3	378.4	0.9
38,5	67.8	10.2 ×10^{-3}	98.9	378.4	0.9
40	17.4	39.7 ×10^{-3}	95.8	378.4	0.9
42	6.1	114.5 ×10^{-3}	93.6	378.4	0.9

The half-life of apoenzyme was found to increase in the presence of FAD at all temperatures studied, indicating that holoenzyme was more stable than the apoprotein form of the enzyme. The magnitude of free energy of activation reflected the effectiveness of relative stabilization of holoenzyme compared to apoenzyme due to the presence of FAD. The significant change in the activation entropy (?S*) and the difference in the slopes of Arrhenius plots indicated that the stabilization of HIS-*Tv*DAAO was of conformational origin. On the other hand, the change in the activation entropy, ?S*, can be explained in terms of enhancement of the order and compactness of the structure, thus favoring intra-molecular stabilizing forces and consequently increasing the stability of the enzyme. Such stabilization may be related to the association of subunits which could lead to a dimeric state in the holoenzyme form. In this sense, contribution of the dimeric state to the thermal stability has been recently reported for *Rhodotorula gracilis* D-amino acid oxidase [7].

3.2 Effect of dilution on the holoenzyme of HIS-*Tv*DAAO

We have observed that holoenzyme suffered an inactivation with time after dilution into the reaction medium prior to the addition of the substrate. Such inactivation upon dilution was studied by allowing the holoenzyme to stand in the diluted state at 30°C and pH 8.0. After a fixed period of time, substrate was added to begin the reaction and the activity was measured. As observed in the bottom curve of Fig. 2, holoenzyme activity gradually decreased as a function of time and approached a constant value. Such behaviour is easily explained by assuming that holoenzyme undergoes a slow and reversible dissociation into inactive apoenzyme and FAD. Experimental plot of activity versus time after dilution could be adjusted to a single exponential decay, and we could calculate the pseudo-first kinetic constant for the inactivation process due to FAD dissociation. Such constant was $k = 0.104$ min^{-1}, indicating that there was a small difference between this value and that described for *pk*DAAO ($k = 0.45$ min^{-1}) [8]. Taking into account the initial and remaining activities of the holoenzyme in the abscence of FAD, we could calculate the dissociation constant of FAD-apoenzyme complex (K_D=2.6×10^{-7} M) as explained in the methods section. This value is very similar to that determined for pkDAAO (K_D=2.2×10^{-7} M) [9], and almost one order of magnitude higher than that described for RgDAAO (K_D=2.0×10^{-8} M)[10]. That means that flavin cofactor is tightly bound to RgDAAO whereas this binding is weaker in pkDAAO and HIS-TvDAAO. From the other curves in Fig. 2 we observed that addition of increasing amounts of FAD in the medium had two effects: (i) it reduced the net rate of inactivation, and (ii) it affected the final remaining activity when the steady state was reached.

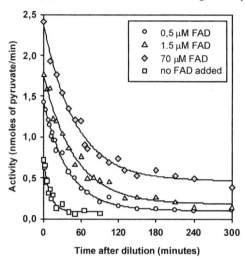

Fig. 2. Progress of inactivation of HIS-TvDAAO holoenzyme after dilution in presence of different concentrations of FAD. No cofactor was added in the bottom curve, being 0.05 μM the amount of added holoenzyme prior to activity determination.

Acknowledgements

Financial support from the Universidad Complutense (PR78/02-10961) is gratefully acknowledged.

References

[1]   L. Fischer. Recent Res. Devel. Microbiol. **2**, 295 (1998).
[2]   M. S. Pilone and L. Pollegioni. Biocatal. Biotrans. **20**, 145 (2002).
[3]   L. Pollegioni, L. Caldinelli, G. Molla, S. Sacchi and M. S. Pilone. Biotechnol. Prog. **20**, 467 (2004).
[4]   J. Alonso, J. L. Barredo, P. Armisén, B. Díez, F. Salto, J. M. Guisán, J. L. García and E. Cortés. Enzyme Microb. Technol. **25**, 88 (1999).
[5]   M. M. Bradford. Anal. Biochem. **72**, 248 (1976).
[6]   K.E. Neet and G.R. Ainslie. Meth. Enzymol. **64**, 192 (1980).
[7]   L. Pollegioni, S. Iametti, D. Fessas, L. Caldinelli, L. Piubelli, A. Barbirolli, M. S. Pilone and F. Bonomi. Prot. Sci. **12**, 1018 (2003).
[8]   M. Dixon and K. Kleppe. Biochim. Biophys. Acta **96**, 357 (1965).
[9]   V. Massey and H. Ganther. Biochemistry **4**, 1161 (1965).
[10]   P. Casalin, L. Pollegioni, B. Curti and M. Pilone. Eur. J. Biochem. **197**, 513 (1991).

Killer activity of yeasts isolated from Spanish dry-cured ham

F. Pérez-Nevado[*,1], M.G. Córdoba Ramos[1], E. Aranda Medina[1], A. Martín González[1], M.J. Andrade[2], J.J. Córdoba Ramos[2]

[1] Department of Zootecnia, University of Extremadura, Crtra. de Cáceres s/n, 06071 Badajoz, SPAIN
[2] Department of Zootecnia, University of Extremadura, Avda. Universidad s/n, 10071, Cáceres, SPAIN

[*]Corresponding author: e-mail: fpen@unex.es, Phone: +34 924286200, Fax: +34 924286201

Killer activity of 86 yeasts strains isolated from Spanish dry-cured ham from 4 Spanish Designation of Origin (SDO) was studied. Most of isolates were identified as *Debaryomyces spp.* (80.5%). An elevated percentage of these strains were able to produce killer toxins (61.6%). In general, killer activity was noticed at pH lower than 5, and only 10.5% of killer yeasts showed killer activity at pH 5.5. Killer yeast distribution was different between the four Specific Denominations of Origin studied. The higher percentages of killer yeasts were found at *Jamón de Huelva* and *Dehesa de Extremadura* SDO. Additionally, when killer activity was analyzed at similar conditions of the cured ham, i.e. pH 5.5 and salt concentrations between 3 and 5%, only 9.3% of strains showed killer activity.

Keywords Yeast; killer toxin; dry-cured ham; identification; characterization; starter culture.

1   Introduction

Killer yeasts have been isolated from a great variety of fermentation food processes [1] [2] [3]. These yeasts are able to produce toxic proteins or glycoproteins (the so-called killer toxins) that can cause death in other killer-sensitive yeast strains. Although the best killer system studied is that of *Saccharomyces cerevisiae*, this phenomenon is very frequent among many other yeasts genera [4]. Thanks to their ability to control spoilage in the preservation of food, killer yeasts could be adequate to be selected as starter cultures. In addition, the variability of the killer phenomenon in nature embodies an exceptional potential for the discrimination of yeasts at the strain level. In the production of Spanish dry-cured ham, yeasts are proved to be present at different levels until final stages of maturation. There are previous works about the ability of yeasts to generate volatile compounds in meat products [5]. However, other activities, like killer toxins production, have not been studied in detail. Thus, the aim of this work was to study the killer activity of 86 yeasts strains isolated from Spanish dry-cured ham from 4 Spanish Designation of Origin (SDO).

2   Material and methods

Yeast strains: 86 yeasts were isolated from several industries of 4 Specific Designation of Origin where this product is traditionally elaborated (**Table 1**). Yeasts isolated at different stages of dry-cured ham maturation were identified and characterized.

Plate assay for killer activity at different pH: Production of killer toxin by yeasts isolated from ham was analysed in MB medium at different pH (3.5, 4, 4.5, 5, 5.5 and 6). Plates were seeded with 48-hour cultures of sensitive *Saccharomyces cerevisiae* strains.

Strains to be tested for killer activity were loaded onto the seeded agar to produce patches of about 5 mm diameter, killer yeasts produced growth-free zones around themselves (**Figure 1**).

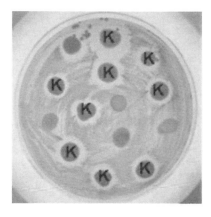

Fig.1 Growth-free zone around different killer yeasts (K) at pH 3.5.

Killer activity at different NaCl concentrations: The killer activity of ham yeast strains was examined in MB medium in the absence and presence of different salt concentrations (3, 4, 5% of NaCl) at similar pH conditions of the cured ham surface, i.e. at pH 5.5. Killer activity was determined by the same procedure used above.

3   Results and discussion

In this study, most of yeasts isolated from Spanish dry-cured ham were identified as *Debaryomyces spp.* (80.5%) and 13.8% were identified as *Candida spp.* When killer activity of these isolates was studied, an elevated percentage of killer yeasts were found (61.6%). In general, yeasts were able to produce active killer-toxins at pH between 3.5 and 4.5. Only 10.5% of yeasts were killer at pH 5.5 and no one killer yeast was found at pH so high than 6. Thus, low percentages of yeasts isolated were killer at typical pH values of dry-cured ham (5.5-6). Like us, other authors found that most killer yeasts analyzed showed an optimum of killer activity at pH lower than 5.5 [1].

Differences in number of killer yeasts between the four SDO studied were found (Table 1). The SDO *Jamón de Huelva* and *Dehesa de Extremadura* showed the highest percentages of killer yeasts (77.8% and 68.4% respectively). *Guijuelo* was the SDO with less percentage of killer yeasts (50%).

Table 1   Number of yeasts isolated and percentage of killer yeasts at the 4 SDO studied.

SDO	Dehesa de Extremadura	Guijuelo	Jamón de Huelva	Los Pedroches
Yeasts isolated	19	30	18	19
% killer yeasts	68.4	50	77.8	57.9

234

When activity of killer yeasts was analyzed at different pH, all the SDO, except *Guijuelo*, showed yeasts with killer activity at pH between 3.5 and 5.5 (**Figure 2 A-D**). In contrast, higher percentages of killer yeasts at pH 5.5 were found in isolates from *Los Pedroches*.

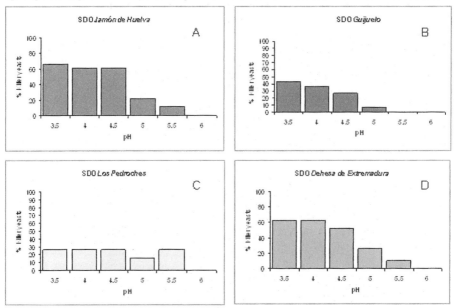

Fig. 2   Killer activity at different pH (percentage) of yeasts isolated from Spanish dry-cured ham of 4 SDO.

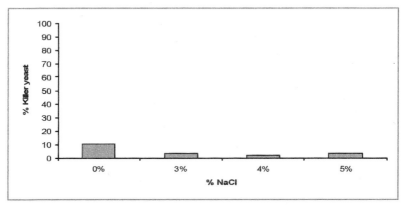

Fig. 3   Percentage of killer yeasts isolated from Spanish dry-cured ham at different NaCl concentrations and without salt added.

To eliminate other yeasts it is important that killer yeasts were able to produce an active killer toxin at similar conditions of the cured ham surface (i.e. pH 5.5 and salt concentrations between 3 and 5%). No enhance activity of killer action was found when killer activity was examined in the absence and presence of different salt concentrations (3, 4, 5% of NaCl) at pH 5.5 (Figure 3). Only 9.3 % of yeasts strains were able to produce killer effect in presence of different salt concentrations, 3.5% of yeasts were killer at 3% of NaCl, 2.3% at 4% of NaCl and 3.5% in medium with 5% of salt. These results are in contrast with those of Llorente et al. [2] and Aguiar and Lucas [6] who described that killer action was enhanced by salt, especially in strains that have been isolated from other salt environments. Our results can be explained by the fact that, in conditions of high pH (5.5 or upper), only 10.5 % of yeasts isolated were killer. On the other hand, most killer yeasts at higher salt concentrations were isolated from SDO *Guijuelo* (20%) and *Dehesa de Extremadura* (10.5%). In *Los Pedroches* and *Jamón de Huelva* there was not found any killer yeast in these conditions.

4   Conclusions

- Among the yeasts isolated from Spanish dry-cured ham, an elevated percentage of killer strains were found (61.6%). Most of them were killer at pH lower than 5, and only 10.5% of killer yeasts had killer activity at pH 5.5.
- Percentages of killer yeasts were different between the four Specific Denominations of Origin studied. *Jamón de Huelva* and *Dehesa de Extremadura* showed the higher percentages, whereas *Guijuelo* presented the lower killer yeasts percentage.
- No significant differences in killer activity of yeasts were noticed in the presence or not of salt concentrations studied (3, 4, 5% of NaCl) at pH 5.5.
- Only 4 strains were killer at 5% of NaCl and pH 5.5. Further studies would be necessary to clarify the effect of these killer yeasts or their toxic compounds against other yeasts present at dry-cured ham.

Acknowledgements

This work was supported by project AGL2001-0804 funded by Spanish Ministerio de Ciencia y Tecnología.

References

[1]   G. Gulbiniene, L. Kondratiene, T. Jokantaite, E. Serviene, V. Melvydas, and G. Petkuniene, Food Technol. Biotechnol. **42**, 159-163 (2004).

[2]   F. Pérez, M. Ramírez, and J.A. Regodón, Antonie van Leeuwenhoek, **79**, 393 (2001).

[3]   P. Llorente, D. Marquina, A. Santos, J.M. Peinado, and I. Spencer-Martins, Applied and Environmental Microbiology. Mar. 1997 (1997), p. 1165.

[4]   M.J. Schmitt and F. Breinig, FEMS Microbiology Reviews, **26**, 257 (2002).

[5]   A. Martín, J.J. Córdoba, M.J. Benito, E. Aranda, and M.A. Asensio, International Journal of Food Microbiology, **84**, 327 (2003).

[6]   C. Aguiar and C. Lucas, Food Technol. Biotechnol. **38**, 29 (2000).

Limiting factors of photo-hydrogen production by *Rhodopseudomonas palustris* WP3-5

Chi -Mei. Lee[*1],Kwei. -Min. Yu[1],and Pei-Chung Chen[2]

[1] Department of Environmental Engineering, National Chung-Hsing University, Taichung 402, TAIWAN, R.O.C.
[2] Department of Biotechnology, Hung-Kuang University, TAIWAN, R.O.C.

[*]Corresponding author: e-mail: cmlee@dragon.nchu.edu.tw, Phone:+886-4 22850024, Fax: +886-4 22862587

Hydrogen production from waste using photosynthetic bacteria is attractive because energy can be recovered from renewable resources. In a previous study, *Rhodopseudomonas palustris* WP3-5 was isolated from piggery wastewater and has proven to efficiently produce hydrogen from the effluent from the hydrogen fermentation reactors. In this study, the factors (NH_4^+ concentration, PO_4^{3-} concentration, EtOH concentration, Mo and Fe cofactor, vitamine B_{12}, yeast extract and Fe-citrate) that affect the hydrogen gas production ability of this isolated purple nonsulfur bacteria will be investigated using batch reactors. The hydrogen production efficiency will also be investigated by using continuous flow photosynthetic reactor.

Keywords photohydrogen production; ammonia; yeast extract; nitrogenase; Fe-citrate; purple nonsulfur photosynthetic bacteria

1   Introduction

World energy requirements and consumption have dramatically increased with the development of technology. It is therefore necessary to search for an alternative energy resources. Hydrogen gas, a clean energy source that does not cause the greenhouse effect, serves well for this purpose. It is known that different taxonomic groups including some photosynthetic bacteria species can produce hydrogen gas. Among these microorganisms, hydrogen gas production using purple nonsulfur bacteria can be characterized as: 1) High hydrogen gas production rate, 2) Hydrogen production is not inhibited by the produced oxygen because water is not used as the electron donor. 3) These bacteria can produce hydrogen gas in the dark or under light, 4) When used as the organic substrate source, wastewater can be treated because the wastes are used as the electron donors [1].

Hydrogen production from waste using photosynthetic bacteria is attractive because energy can be recovered from renewable resources. The effluent from anaerobic hydrogen fermentation reactors contains high concentrations of organic acids. The possibility of utilizing these initial dark hydrogen fermentation stage effluents for photohydrogen production using purple nonsulfur bacteria should be elucidated.

In a previous study, *Rhodopseudomonas palustris* WP3-5 was isolated from piggery wastewater and has proven to efficiently produce hydrogen from the effluent from the hydrogen fermentation reactors[2].In this study, the factors (NH_4^+ concentration, PO_4^{3-} concentration, EtOH concentration, Mo and Fe cofactor, Vit. B_{12}, yeast extract and Fe-citrate) that affect the hydrogen gas production ability of this purple nonsulfur bacteria will be investigated using batch reactors. The hydrogen production efficiency will also be investigated by using continuous flow photosynthetic reactors.

2   Materials and methods

2.1 Organisms

The photosynthetic bacteria *Rhodopseudomonas palustris* WP3-5 used in this study was isolated from piggery wastewater . This strain was identified using the 16S rRNA gene sequencing method.

2.2 Batch culture

The microorganisms obtained from the liguid culture were used for a batch study. The liquid medium was an enrichment medium for *Rhodospirillaceae* [3]. After centrifugation, the concentrated cells were suspended in 10 ml medium with mixed organic acid as the carbon source. The mixed organic acid consisted of 244 mg formic acid, 520 mg acetic acid and 800 mg butyric acid per liter. This mixed organic acid exhibited an organic profile identical to that from an anaerobic hydrogen fermentation reactor. After gassing with argon gas to retain 40 ml of medium and sealed with teflon/silicon stoppers, the serum bottles were placed into a climate room at 35°C providing 155 µmole/m^2/s of illumination.

2.3 Continuous culture

As shown in Fig 1, a lab-scale photobioreactor was used. The reactor had a 2.5 L volume and was equipped with a magnetic stirrer.

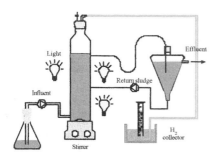

Fig 1. Schematic diagram of a lab-scale Photo-bioreactor

3   Results and discussion

3.1 NH_4Cl concentration effect on photohydrogen production

Different NH_4Cl concentrations , 100, 200 mg/L were used as the nitrogen source, after NH_4Cl exhaustion, 200mg/L NH_4Cl was added. As shown in Fig. 2, hydrogen production occurred only when the NH_4^+ concentration was below 17 mg-NH_4^+/L. In the experiment without NH_4Cl, hydrogen production was continued although the cell dry weight increased only slightly. In the presence of 80mg/L NH_4Cl, hydrogen production was not improved when glutamate and viamine B_{12} were added.

3.2   Phosphate concentration effect on photo-hydrogen production

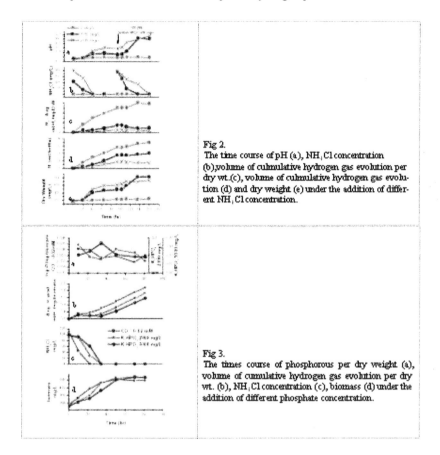

Fig 2.
The time course of pH (a), NH₄Cl concentration (b),volume of cumulative hydrogen gas evolution per dry wt.(c), volume of cumulative hydrogen gas evolution (d) and dry weight (e) under the addition of different NH₄Cl concentration.

Fig 3.
The times course of phosphorous per dry weight (a), volume of cumulative hydrogen gas evolution per dry wt. (b), NH₄Cl concentration (c), biomass (d) under the addition of different phosphate concentration.

Phosphate was used as a medium buffer, in this experiment 2000mg/L K_2HPO_4, 5000 mg/L K_2HPO_4 and 0.02 mM CO_3^{2-} were used as the buffer to investigate the phosphate concentration effect. When 0.02mM CO_3^{2-} was used, 100 mg/L K_2HPO_4 was supplemented as the phosphorous source. Fig 3 shows that the phosphate buffer concentration had no effect on growth and hydrogen production. However the lag period increased with the phosphate buffer concentration. When CO_3^{2-} was used as the buffer, significant microbial growth involving the NH_4Cl consumption and the least lag period was observed. This showed that the carbonate concentration was the factor limiting the *Rhodopseudomonas palustris* growth rate as reported by Takabatake, et. al. [4].

3.3 Ethanol concentration effect on photo-hydrogen production

To investigate the ethanol concentration effect , 5 sets of experiments were conducted with ethanol added at 0, 200, 400, 800 and 1000 mg/L concentration. The hydrogen production volume and cell dry weight after 70 hrs incubation were determined (data not shown).

There was no significant ethanol concentration influence on hydrogen production up to 1000 mg/L.

3.4 Yeast extract and Fe-citrate effect on photo-hydrogen production

To investigate the yeast extract and Fe- citrate effects, 200 mg/L yeast extract and 5 mg/L Fe-citrate were added respectively. As shown in Fig 4, the addition of Fe-citrate increased the hydrogen production without increasing the biomass. The addition of yeast extract efficiently increased the growth .

3.5 Cofactor concentration effect on photo-hydrogen production

Hydrogen production is associated mainly or completely with the action of nitrogenase. Mo and Fe are the metal cofactors for nitrogenase [5]. Four sets of experiments were conducted with different combinations of Mo and Fe concentrations. The results showed that the hydrogen production capability increased by increasing the $FeCl_3$ concentration but not the Mo concentration.

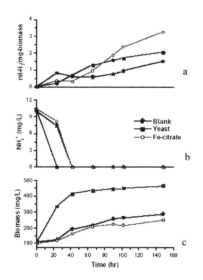

Fig 4.
The time course of cumulative volume of hydrogen gas evolution per dry wt. (a), NH_4^+ concentration (b),
Biomass (c) under the addition of yeast extract and Fe-citrate.

3.6 Continuous photo-hydrogen production

When the load was 2.11 Kg COD/m^3/day, the influent C/N ratio of 39:1 showed higher hydrogen percentage than that with a C/N ratio of 19:1 (data not shown). When the load was 2.11 Kg COD/m^3/day, there was little difference in hydrogen percentage between the influent C/N ratio of 41:1 and 78:1.When the C/N was 78:1, maximum hydrogen gas percentage reached 75%. After 18 days of operation, the hydrogen production decreased gradually. On the 28[th]day , the addition of 0.25 mM $FeCl_3$ caused the hydrogen production to recover and maintained for two weeks.

4 Conclusion

When effluent from the hydrogen fermentation reactors was used for photo-hydrogen production using *Rhodopseudomonas palustris* WP 3-5, the tolerable NH_4^+ concentration was 17 mg/L. In the presence of NH_4^+, the addition of glutamate or vitamine B_{12} could not improve hydrogen production. The carbonate concentration was the factor that limited the growth rate and hydrogen production. The phosphate buffer and ethanol concentration had no effect on growth and hydrogen production. The yeast extract was required for growth. In contrast to the yeast extract, iron citrate was not required for growth, but it improved hydrogen production. The hydrogen production capability increased with increasing Fe concentration but did not increase with an increase in the Mo concentration. The continuous experimental results indicated that the higher the C/N concentration in the influent, the higher the hydrogen gas percentage.

Acknowledgements

The support of the National Science Council, Taiwan, R. O. C. through project NSC 91-2211-E-005-030 is gratefully acknowledged.

References

[1]   H.Zürrer, R. Backhofen. Biomass **2**, 165 (1982)
[2]   C.-M. Lee, P.-C. Chen, C.-C. Wang and Y.-C. Tung, Int. J. Hydrogen Energy **27**, 1309 (2002)
[3]   N. Pfennig, Int. J. Syst. Bacteriol. **28**,283 (1978)
[4]   H. Takabatake, K. Suzuki, I.-B. Ko and T. Noike, Bioresource Technol. **95**,151 (2004)
[5]   E. Fascetti, E. D'addario, O.Todini and O. Robertiello, J. Hydrogen Energy **23**,753 (1998)

Mineral-phosphate solubilization activity of iron ore associated microflora

P. Delvasto[1], A. Ballester [*1], C. García[2], J.M. Igual[3], J. Muñoz[1],F. González[1], M. Blázquez[1]

[1] Department of Materials Science. Universidad Complutense. Ciudad Universitaria, 28040 Madrid, Spain.
[2] Department of Industrial Technology. Universidad Alfonso X "El Sabio". Avda. de la Universidad 1, 28691 Villanueva de la Cañada, Madrid, Spain.
[3] Natural Resources and Agrobiology Institute. CSIC. Apartado 257, 37071 Salamanca, Spain.

[*] Corresponding author: e-mail: ambape@quim.ucm.es, Phone: +(34) 91 394 4339, Fax: +(34) 91 394 4357

A high P iron ore was screened for mineral phosphate solubilizing activity within its associated microflora. Microorganisms belonging to genera *Burkholderia*, *Clavibacter* and *Aspergillus*, were isolated. A pH-sensitive dyed medium was used to approach the solubilization mechanism of selected isolates. Only isolate *Aspergillus niger* was able to acidify the medium while solubilizing phosphate. By way of shake flask experiments, P-solubilization of some isolates was assessed. Implications of these findings on biodephosphorization of iron ores were discussed.

Keywords: iron ore; phosphate solubilization; biobeneficiation; *Burkholdelia spp*; *Aspergillus niger*.

1   Introduction

Many current iron ore resources contain over 0.08 w% of P, making this material useless for the manufacture of metallic iron and steel. Geochemical conditions during deposition of iron ores permit the association of iron minerals with P either as accessory phosphatic inclusions or into the crystal structure of the iron oxide itself. Phosphatic phases found in iron ores include several Ca and Al phosphates like apatite, wavellite and turquoise [1]. Little attention has been paid to biodephosphorization of iron ores. Works by Buis [2] and Parks et al. [3] have dealt with the use of fungal strains to treat phosphatic iron ores. The main drawback of these investigations was that strains used were not related to the ores being treated. To the knowledge of the authors, no attempts to characterize or evaluate the phosphate solubilizing activity of the microorganisms naturally associated to iron ores have been reported in the literature. In the authors' opinion, this is a crucial step if a biobeneficiation process is to be chosen as a way to diminish phosphorus contents of iron ores, mainly because using indigenous microorganisms in biomining processes assures few ecological distortions and less time consumption in terms of adaptation of exogenous microorganisms to new environments. Based on the afore mentioned situation, the aims of this study are: a) Determine if there exist phosphate solubilization activity among the different microorganisms associated to a Brazilian high phosphorous iron ore, b) Approach the solubilization mechanisms of some of the isolates by plate assays and c) Evaluate the performance of these isolates in terms of their phosphate solubilization ability in broth. Implications of the results obtained on the possibilities of establishing a biodephosphorization treatment of iron ores are discussed.

2   Materials and methods

2.1 Screening for phosphate solubilizing microorganisms

A high-phosphorus iron ore (P content 0.18 w%) from the Jangada mine, located in the Brazilian region of Minas Gerais, was screened for phosphate solubilizing microorganisms (PSM). Cells and spores were detached from ore surface by shaking at 150 rpm a mixture of 5 g of fresh mineral and 100 ml of sterile distilled water for 24 hours. Supernatant was serial diluted in sterile distilled water and suitable aliquots spread over Petri dishes containing differential growth media for PSM. This PSM isolation technique is based on the formation of halos around the colonies of microorganisms capable to solubilize a Ca-phosphate insoluble compound [4, 5]. Inoculated plates were incubated at 30 °C for at least 12 days. PSM colonies were collected and subcultured. Culture media employed to differentially isolate PSM strains consisted on variations of a chemically defined medium proposed elsewhere [12]. All media contained per liter: 50 ml of a salts solution ($MgCl_2 \cdot 6H_2O$, 100 g/l; $MgSO_4 \cdot 7H_2O$, 5 g/l; KCl, 4 g/l; $(NH_4)SO_4$, 2 g/l), 10 g of a sugar as carbon source (glucose or sucrose), and 2.5 g of tricalcium phosphate, $Ca_3(PO4)_2$, as an insoluble source of phosphate. In some cases, 6 ml/l of a pH indicator dye (bromophenol blue, 0.4% ethanol solution) was added to the media [5]. For gelification, 20 g/l of agar was added. Prior to autoclaving, pH of all media was adjusted to 8 using diluted NaOH.

2.2 Identification of the isolates

Four phosphate solubilizing (PS) bacteria were isolated from ore microflora and identified by means of 16S rRNA sequential method. This identification was carried out in the Microbiology Laboratory of the Natural Resources and Agrobiology Institute CSIC of Salamanca, Spain. The fungal PS isolate found was identified in the Spanish Type Culture Collection Centre (CECT) in Valencia, Spain by morphological and molecular means (5.8S rRNA sequential method).

2.3 Shake flask solubilization experiments

Shake flask experiments were carried out in 150-ml Erlenmeyer flasks. Each flask received a quantity of an insoluble phosphate mineral, i.e. $Ca_3(PO4)_2$, $AlPO_4$ (berlinite, Fluka Chemika) or a natural siliceous rock containing up to 6 w% of P in the form of $CuAl_6(PO_4)_4(OH)_8 \cdot 5H_2O$ (turquoise). The flasks were autoclaved (20 min.,126 °C) and, in sterile conditions, 30 ml of glucose broth was added to each flask. Final P concentration in the system was intended to be 500 mg/l, however in some cases it could not be attained (see information on table 1). Inoculation was made by adding 0.05 ml of purified bacteria inoculum (10^8 cells/ml) or 0.05 ml of a pure spores suspension (10^8 spores/ml). Uninoculated flasks were left as control. Flasks were kept for 7 days in an orbital incubator (150 rpm, 30 °C). The pH of spent broth was measured immediately after the sampling using a pH meter. Total phosphorus in solution was determined by colorimetric methods, as described elsewhere [6]. All values of pH and P concentration shown are averages of three replicates.

3   Results and discussion

3.1 Isolation and identification of PSM strains

After incubation of inoculated plates with supernatant from detachment process, abundant bacterial and fungal activity was found, revealing that ore-associated microflora was diverse. Four halo-producing bacterial strains were isolated and purified. Bacterial strains were identified as *Burkholderia cepacia*, *Burkholderia caribiensis* and *Clavibacter xyli*. One Bacterial isolate could not be identified to species level but was found belonging to genus *Burkhordelia* (hereafter, *Burkhordelia* sp.). One halo-producing fungal strain was isolated and identified as *Aspergillus niger*. All isolates found were able to grow on glucose plates and sucrose plates as well. As only exception, *Burkhordelia cepacia* isolate was exclusively found in plates containing sucrose. Some screening routines included the addition to media formulations of bromophenol blue, a pH-indicator that turns from blue to yellow at pH ≤ 3. It was found that all isolates able to produce PS halos on glucose and sucrose plates, had the same behavior when bromophenol blue was added. However, only the fungal isolate *Aspergillus niger* produced at the same time strong yellow acidification halos and PS halos.

3.2 In-plate approach to solubilization mechanisms of selected isolates

Diverse PS mechanisms have been proposed, including acidification [7, 8], organic acid production [7, 8, 9] and metabolites secretion i.e. organic ligands or exopolysaccharides[7, 8, 9]. Acidification and organic acids production will be associated to a pH drop on surrounding medium, whilst secretion of other metabolites are not necessarily related to pH drops. As shown in table 1 only isolate HNA-1, *Aspergillus niger*, was able to produce a strong acidification halo, turning yellow its surrounding medium. It was not the case with other bacterial isolates as *Burkholderia sp.* and *B. caribiensis*. Both bacterial isolates produced strongly clear PS halos with no color change. This in-plate results suggest that PS activity of *A. niger* is correlated with a strong acidification. *A. niger* is known to be an excellent organic acids producer [8]. On the other hand, *Burkholderia sp.* and *B. caribiensis* isolates PS activity seems not to be directly related to a significant pH drop. *Burkhordelia* species are known to be good exopolisaccharides producers [10]. It has been reported that in some cases these metabolites are responsible for dissolution of P-rich minerals, by cationic-complexing mechanisms [9]. So, for bacterial isolates studied in this work, a non-acidification dissolution mechanism could be taking place.

3.3 Shake-flask solubilization performance of selected isolates

Al-phosphates, as well as Ca-phosphates, have been described as accessory P-bearing phases in iron ores [1]. So, from the stand point of biobeneficiation of P-rich iron ores, it is important to study how these PSM isolates obtained from Jangada iron ore behave when typical iron ore gangue phosphatic minerals are added to the medium. Quantitative PS performance of isolates *A. niger*, *B. caribiensis* and *Burkholderia sp.* is shown on table 1.

Table 1. Mineral phosphate solubilization studies carried out in glucose broth after 7 days of incubation.

Isolated species	Ca$_3$(PO$_4$)$_2$			AlPO$_4$			CuAl$_6$(PO$_4$)$_4$(OH)$_8$·5H$_2$O		
	Total P released to solution (mg/l)	Solubilization yield (%)	Final pH	Total P released to solution (mg/l)	Solubilization yield (%)	Final pH	Total P released to solution (mg/l)	Solubilization yield (%)	Final pH
Burkholderia sp.	157	31.4	4.8	21	4.2	3.2	-	-	-
Burkholderia caribiensis	126	25.2	4.0	-	-	-	-	-	-
Aspergillus niger	197	39.4	2.2	39	15.6*	2.2	9.7	4.0**	2.1

* Original P concentration was 250 mg/l instead of 500 mg/L. ** Original P concentration was 240 mg/L.

Considering the case of Ca-phosphate, it can be seen that all isolates are able to solubilize it to extents over 25 % in 7 days. The leading isolate is *A. niger,* with almost a 40% of solubilization attained. Considering that original pH of all formulated media was close to neutrality (6.5 – 7), the fungal isolate, as previously shown by plate assays, was the one that produced the maximum pH drop, around 5 units, in comparison with bacterial isolates on which pH drop while solubilizing Ca-phosphate was of around 3 pH units. In the case of berlinite, only *A. niger* and *Burkholderia sp*. isolates were tested. It was found that *A. niger* solubilized up to 15% of the original phosphate added, while solubilization yield of bacterial isolate *Burkhordelia sp*. was around 4%. It is interesting that *Burkholderia sp*. isolate had an acidification response when placed in contact with berlinite (final pH=3.2) in comparison when placed in contact with Ca-Phosphate (final pH=4.8). It has been described that bacterial PS activity based on production of cation-complexing polymers can be enhanced on slightly acidic media [9]. So, this acidification behavior of *Burkholderia sp*. can be interpreted as an adaptation reaction to a more recalcitrant P-source. On the other hand, *A. niger* acidification level remained unchanged, which indicates that acid production by *A. niger* is not dependant on P-source. In fact, when A. niger cultures were placed in contact with the other substratum tested, turquoise-containing rock, the final pH was almost the same, as seen on table 1. This results suggest differences in solubilization mechanisms between both microorganisms. Mechanism followed by fungi is indirect in nature (i.e. fungal cells secrete acids which in turn dissolve phosphates), while bacterial solubilization mechanisms seems to be more complex and dependant on surface attachment, generation of metabolites and, as a consequence, attack of mineral surface [7, 9], making the latter mechanism of a more adaptable nature. In the case of turquoise, only *A. niger* could grow and solubilization yield attained was the lowest for all studied P-sources, 4.0%.

4   Implications of these findings on biobeneficiation of iron ores

Existence of PSM among iron ore associated microflora is promising. It can be considered as a first step on the establishment of biological routes to reduce P-content of iron ores. Bioactivation of this microflora by addition of proper nutrients, in order to enhance its PS activity, can open the possibility to heap leach iron ore to reduce P. Indirect procedures as those proposed in the literature [2, 3], can be also used. Since isolate A. niger showed a strong acidification activity, production of leaching liquors based on organic acids generated by this isolate can be promoted and used to treat ground ores or fines during wet agglomeration procedures. Nevertheless, more research must be conducted on this direction and a good mineralogical characterization of the ore to be treated is of paramount importance [1]. As a first approach, biobeneficiation using these isolates can be performed on ores containing manly Ca-phosphates as impurities. Although Al-phosphate solubilization exhibited by these microorganisms was not relevant, variations on nutrients sources should be implemented to verify if obtained solubilization yields can be increased, opening in such a way the possibility to treat Al-phosphate-bearing iron ores.

Acknowledgments

The authors wish to thank the personnel of the biology laboratory of Universidad Alfonso X "El Sabio" for their cooperation. P. Delvasto was the recipient of a doctoral fellowship from the Venezuelan Foundation for Science, Technology and Innovation (FONACIT) and the Simón Bolívar University of Venezuela.

References

[1]   Kokal, H. "The origin of phosphorus in ironmaking raw materials and methods of removal. A review". 63rd annual meeting Minnesota section AIME. (1990). pp 225-257.

[2]   Buis, P. "Bioremediation techniques for the removal of phosphorus from iron ore". PhD Dissertation. Mining engineering. Michigan Technological University. U.S.A. (1995)

[3]   Parks, E., Olson, G., Brinckman, F., Baldi, F. "Characterization by high performance liquid chromatography (HPLC) of the solubilization of phosphorus in iron ore by a fungus". Journal of industrial microbiology, 5. (1990). pp 183-190

[4]   Nautiyal, C.S. "An efficient microbiological growth medium for screening phosphate solubilizing microorganisms" FEMS microbiology letters, 170. (1999). pp 265-270

[5]   Gupta, R., Singa l, R., Shankar, A., Kuhad, R.C., Saxena, R.K. "A modified plate assay for screening phosphate solubilizing microorganisms". J. Gen. Appl. Microbiol. 40. (1994). pp 255-260

[6]   AOAC. "973.55 Phosphorus in water, photometric method." AOAC Official Methods of Analysis. (1990). pp. 328-329

[7]   Illmer, H., Schinner, F. "Solubilization of inorganic calcium phosphates – Solubilization mechanisms" Soil. Biol. Biochem. , 27. No. 3. (1995). pp 257-263

[8]   Illmer, H., Barbato, A., Schinner, F. "Solubilization of hardly soluble $AlPO_4$ with P-solubilizing microorganisms." Soil. Biol. Biochem. , 27. No. 3. (1995). pp 265-270

[9]   Welch, S., Taunton, A., Banfield, J. "Effect of microorganisms and microbial metabolites on apatite dissolution". Geomicrobiology Journal, 19. (2002). pp 343-367

[10]   Cescutti, P., Lagatolla, C., Tonin, E., Sist, P., Urbani, R., Rizzo, R. "Exopolysaccharides produced by CF strains of Burkholderia cepacia". Minutes of the meeting of the international *Burkhordelia cepacia* working group held at April 5-7, 2002 in San Antonio, Texas. U.S.A. (2002). pp 10-11

Optimisation of two recombinant whole cell systems for the production of optically pure D-amino acids

A. I. Martínez-Gómez, S. Martínez-Rodríguez, J. Pozo-Dengra, J. M. Clemente-Jiménez, F. Rodríguez-Vico and F. J. Las Heras-Vázquez*

Departamento de Química-Física, Bioquímica y Química Inorgánica. Edificio C.I.T.E. I. Universidad de Almería. La Cañada de San Urbano, Spain. E-04120.

*Corresponding author: e-mail: fjheras@ual.es, Phone: + 34 950 015055, Fax: + 34 950 015008.

Two whole cell recombinant systems for the production of optically pure D-amino acids were optimised and compared. Each system contained three enzymes, 2 of which were D-hydantoinase and D-carbamoylase from *Agrobacterium tumefaciens* BQL9. The third enzyme was hydantoin racemase 1 for the first system, and hydantoin racemase 2 for the second one, both from *Agrobacterium tumefaciens* C58. The analysis of the induction parameters for the two whole cell systems, in *Escherichia coli* JM109, showed that the optimal induction conditions were the same as regards induction time and temperature, i.e. 8 hours and 34°C. However optimum concentration of the inducer, isopropyl-β-D-thiogalactosidase (IPTG), was different (0.1 mM of IPTG for system 1 and 0.2 mM for system 2). The production of D-amino acids was analysed after transforming four *E. coli* strains with the two constructions, and *E. coli* strain BL21 was found to be the best host strain for both systems. Comparison of the two systems in BL21 showed that system 1 produced 100% conversion from the substrate D,L-5-methylthioethyl-hydantoin (D,L-MTEH) to the product D-methionine in 180 minutes, while system 2 took twice as long.

Keywords Hydantoinase process; optimisation; D-amino acids

1 Introduction

Optically pure D-amino acids are valuable intermediates for the preparation of semisynthetic antibiotics, pesticides and other products of interest for the pharmaceutical, food and agrochemical industries [1]. Enzymatic production of optically pure D-amino acid from D,L-5-monosubstituted hydantoins has been proved to be cheaper and less contaminating than chemoenzymatic production [2]. In this enzymatic reaction, called "hydantoinase process", firstly the chemically synthesised D,L-5-monosubstituted hydantoin ring is hydrolysed by a stereoselective hydantoinase enzyme. Further hydrolysis of the resulting enantiospecific *N*-carbamoyl-D-amino acid to the free D-amino acid is catalysed by highly enantiospecific *N*-carbamoyl-D-amino acid amidohydrolase (D-carbamoylase). At the same time as D-hydantoinase hydrolyses the enantiospecific D-5-monosubstituted hydantoin, the chemical and/or enzymatic racemisation of L-5-monosubstituted hydantoin starts.

To satisfy the growing demand for mainly natural and non-natural D-amino acids as building blocks for pharmaceuticals it is necessary to develop a biocatalytic process. Unfortunately, industrial enzymes are often not available or show properties which are incompatible with industrial demands. However, the application of recombinant DNA technology is the natural evolution for successful development of biocatalytic processes, as it allows the rapid design of biocatalysts for new industrial applications.

The aim of this work was to design a biocatalyst for the production of optically pure D-amino acids. For this purpose two whole cell recombinant systems were developed. These systems contained three co-expressed enzymes, 2 of which were D-hydantoinase and D-carbamoylase from *Agrobacterium tumefaciens* BQL9. The third enzyme was hydantoin racemase 1 [3] for the first system, and hydantoin racemase 2 [4] for the second one, both from *Agrobacterium tumefaciens* C58. Optimal induction conditions of the resulting whole cell systems such as induction time and temperature and inducer concentration were determined. Likewise, *Escherichia coli* strains BL21, DH5a, JM109 and TOP10F were studied as suitable hosts for both recombinant constructions. Finally, the optimised systems were compared in order to decrease the reaction time and increase D-amino acid production.

2 Materials and methods

Both constructions were transformed in *E. coli* and grown in Luria-Bertani (LB) liquid medium supplemented with 100 μg mL^{-1} of ampicillin at 37°C. For induction of the expression of the three genes, isopropyl-ß-D-thiogalactosidase (IPTG) was added at final concentrations from 0 to 0.4 mM, with induction temperatures from 28 to 37°C for 0 to 10 additional hours. After induction the cells were collected by centrifugation and stored at –20°C.

Prior to the reaction the pellets were resuspended at a concentration of 1 mg mL^{-1} in potassium phosphate buffer (pH 7.5). For the enzyme reactions the cells were added to prewarmed D,L-5-(2-methylthioethyl)-hydantoin (D,L-MTEH) solution at a final concentration of 15 mM and incubated at 40°C. Aliquots were taken during the reaction and stopped by addition of twice the reaction volume of 1M HCl. After centrifuging the stopped samples, the supernatant was analysed by HPLC [5].

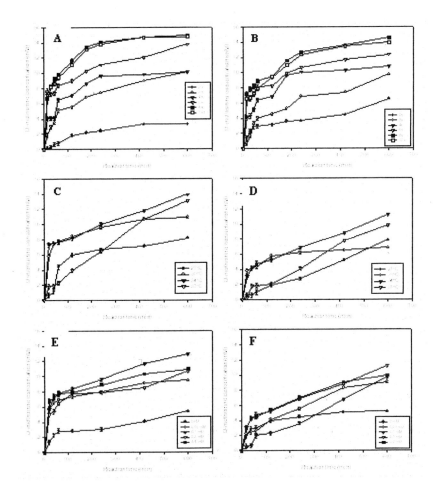

Fig. 1. Study of the induction parameters in the two recombinant whole cell systems, in JM109 *E. coli* strain, based on D-methionine production. A and B: induction time at 34°C, after adding IPTG at final concentration of 0.1 mM for the systems 1 and 2, respectively. C and D: induction temperature for 6 hours, after adding IPTG at final concentration of 0.2 mM for the systems 1 and 2, respectively. E and F: inductor (IPTG) concentration at 34°C for 6 hours for the systems 1 and 2, respectively. The reactions and measurements were carried out as described in Materials and Methods.

3 Results

Analysis of the induction parameters for the two whole cell recombinant systems, in JM109 *E. coli* strain, was performed in order to obtain maximum production of D-amino acid (D-methionine) in the shortest possible time. This was obtained after 8 hours (Fig. 1A and 1B). The optimal induction temperature was clearly 34°C (Fig. 1C and 1D), while the highest production was obtained with 0.1 mM of IPTG for system 1 and 0.2 mM for system 2 (Fig. 1E and 1F).

The two plasmids have been introduced in four *E. coli* strains in order to select the host strain in which D-amino acid production is fastest. After analysis of the reaction profile (Fig. 2A and 2B), strain BL21 was selected as the best host strain for both systems.

After ascertaining the best host strain, the final step was to compare both whole cell recombinant systems used as biocatalysts for the production of optically pure D-amino acids. As Figure 3 shows, system 1 including D-hydantoinase and D-carbamoylase from *Agrobacterium tumefaciens* BQL9 and hydantoin racemase 1 from *Agrobacterium tumefaciens* C58 produced 100% conversion from the substrate D,L-MTEH to the product D-methionine in 180 minutes, while system 2 took twice as long.

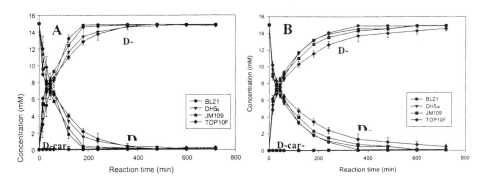

Fig. 2. Reaction profile of D-methionine production from D,L-MTEH using the different *E. coli* host strains BL21, DH5, JM109 and TOP10F transformed with the plasmids containing the enzymes for systems 1 (A) and 2 (B), respectively. Reactions and measurements were carried out as described in Materials and methods.

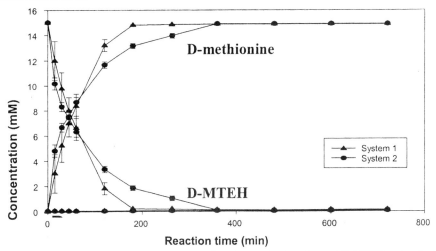

Fig. 3. Comparison of reaction profile of D-methionine production from D,L-MTEH using both whole cell systems in *E. coli* strain BL21. Reactions and measurements were carried out as described in Materials and methods.

4 Conclusions

To satisfy the growing demand for natural and non-natural D-amino acids as building blocks for pharmaceuticals it is necessary to develop a biocatalytic process. To this end we have designed a novel biocatalyst for the production of optically pure D-amino acids developing two whole cell recombinant systems that co-express D-hydantoinase, D-carbamoylase and hydantoin racemase enzymes. After optimising induction and selecting the best host strain for the recombinant systems, both multi-enzymatic systems were compared, showing that system 1 produced 100% conversion twice as fast as system 2.

Acknowledgements

This work was supported by project PTR-1995-0755-OP and BIO2004-02868 from Ministerio de Educación y Ciencia, Spain. The authors thank Andy Taylor for critical discussion of the manuscript and Pedro Madrid-Romero for technical assistance.

References

[1]   C. Syldatk, A. Läufer, R. Müller, and H. Höke. Adv. Biochem. Eng. Biotechnol. **41**, 29 (1990).

[2]   M. Pietzsch, and C. Syldatk, in: K. Drauz, H. Waldmann H. (Eds.), Enzyme Catalysis in Organic Synthesis (Wiley-VCH, Weinheim, 2002), pp.761-799.

[3]   F.J. Las Heras-Vázquez, S. Martínez-Rodríguez, L. Mingorance-Cazorla, J.M. Clemente-Jiménez, F. Rodríguez-Vico. Biochem. Biophys. Res. Commun. **303**, 541 (2003).

[4]   S. Martínez-Rodríguez, F.J. Las Heras-Vázquez, J.M. Clemente-Jiménez, F. Rodríguez-Vico. Biochemie **86**, 7 (2004).

[5]   S. Martínez-Rodríguez, F.J. Las Heras-Vázquez, J.M. Clemente-Jiménez, L. Mingorance-Cazorla, F. Rodríguez-Vico. Biotechnol. Prog. **18**, 1201 (2002).

Single cell protein production by *Saccharomyces* sp no12 by utilizing Lignocellulosic waste and its nutritional evaluation

Nivedita Sharma[1,*], **M.Chandel*** and ***Bhanu Neopaney*****

*Department of Basic Sciences, Dr. Y.S. Parmar University of Horticlture and Forestry, Nauni, Solan, HP, India,
**Department of Biotechnology, Government of Himachal Pradesh, Shimla, HP, India
E-mail: bhanuneopaney@rediffmail.com

[1]Corresponding author: e-mail: nivea_64@yahoo.co.in

Among 73 isolates grown for SCP production, *Saccharomyces* sp. 12 isolated from oilcake of *Prunus armeniaca* showed the maximum production of SCP (2.975%) and thus was selected for further coculture studies. The lignocellulosic waste used in the present study included a mixture of agricultural waste (cobs of *Zea mays* and husk of *Oryzae sativa*) and forestry waste (saw dust of *Cedrus deodara* , *Shorea robusta* and *Toona ciliata*).For the bioenrichment of *Saccharomyces* sp. 12 along with spores/ cells of *Phanerochaete chrysosporium* 1556, *Pleurotus sajor-caju* ,*P. ostreatus, Agaricus bisporus, Bacillus sublitis, B coagulans, B licheniformis* and *Pseudomonas aeroginosa* . The designed sequential culturing with *Azotobacter chroococcum* in the order of *S.* sp. 12 +*A.chroococcum* + *S.* sp. + *A. chroococcum* + (Coculture of *P. sajor-caju* + *P. chrysoporium*) resulted in remarkable increase in bioproteins i.e. 18.38 % . The enriched microbial biomass was evaluated for its nutritional status. The novel combination of the bioproteins with different microorganisms had enriched the lignocellulosic biomass with protein and fats thus recommending it as an attractive additive in various feed and also as an important food supplement after its refining from left over residue.

Keywords: Single cell protein (SCP), biodegradation, sequential culture, microbial biomass

1   Introduction

The alarming rate of population growth, world wide preference for the vegetarian food and adverse climatic conditions for the agricultural crops have led to the inadequacy of good quality food throughout the world. The exploitation of non-conventional sources for the production of food / food supplements can be a viable substitute in this direction. Single cell protein is a fine example of it (Puniya et al, 1995). The most promising way of producing microbial protein is through rapid and efficient growth of microorganisms by degrading inexpensive and locally available wastes. Lignocellulosic waste is generated in bulk in nature, thus posing a serious threat to the environment. The lignocellulosic waste is can be bioconverted to microbial biomass by employing the suitable degrading microorganisms. It will help not only to generate food supplement by applying microbial technology to utilize lignocellulosic waste as carbon source but also to lower the environmental pollution to an extent (Ani 2000, Cameron and Aust, 2001) .
Saccharomyces sp no12 isolated from oil cake of *Prunus armeniaca* showed the highest production of SCP of 2.98% (Chandel, 2002) hence it was selected for further study of biodegradation of unutilized forest and agriculture lignocellulosic waste.

2 Materials and methods

Lignocellulosic waste selected for degradation was;

Sawdust of *Pinus roxburghii, Cedrus deodara, Toona ciliata, Shorea robusta* and powder of *Zea mays* cob and rice husk of *Oryza sativa*. After collecting them locally these were washed and dried , ground to 2.0mm mesh size and were mixed in equal ratio.

Bioenrichment of SCP:

Standard cultures for coculture studies were *Pleurotus sajor-caju, Agaricus bisporus, Pleurotus ostreatus* collected from National Research Center for Mushroom, Solan, India, *Phanerochaete chrysosporium* from International Collection Center, Braunschweig, Germany, *Bacillus coagulans, Bacillus licheniformis, Bacillus subtilis and Pseudomonas aeroginosa* from Microbiology Research Lab, Deptt of Basic Sciences, UHF , Nauni, Solan, India .

1. *Coculture method*: The inoculum (1%) of 1 O.D. culture of *Saccharomyces* sp no 12 and standard cultures (1:1) were grown in the Modified medium of Sharma and Chandel (Chandel 2002) containg 2.0 g of lignocellulosic mixture in the following fashion:
These were incubated at $28^{\circ}C$ for 10days Biomass was collected as pellet after centrifugation at 5000rpm for 15 min. Crude protein and nitrogen in pellet of each coculture combination were evaluated by following the method of Ranganna (1986a).
2. *Coculture + Free Nirogen Fixing Azotobacter chroococcum.* *A.chroococcum* was procured from Indian Agriculture Institute, N. Delhi, India. *A.chroococcum* was added in each coculture in the ratio of 3:3:4 and high protein producing combinations were selected for further study. Incubation conditions were same as mentioned above in coculture method.
3. *SCP enrichment by sequential culture method:* Sequential culture method was designed in the following order:

Sequence S-1: *S*.sp no 12 + *A.. chroococcum* (added 3rd day) + coculture of {*P.sajor-caju* + *P. chrysosporium*}(added 6th day).
S-2: *S.* sp no12 + coculture of {*P.sajor-caju*+ *P.chrysosporium*}added 3rd day+ *A. chroococcum* (added 6th day).
S-3: *S*.sp no 12 +*P.chrysosporium* (added 3rd day) + *A. chroococcum* (added 6th day).
S-4: *S*.sp no12 + *P. sajor-caju* (added 3rd day) + *A.chroococcum* (added 6th day).

All above sequential cultures were incubated for 15 days in total at $28^{\circ}C$ and protein and nitrogen were estimated as given above.

Nutritional Evaluation of Microbial Proteins:

The sample of bioproteins so prepared was evaluated for its nutritional constraints on (dry weight basis) except moisture.

Nitrogen contents, Crude Protein, Soluble Protein, Total Sugars, Reducing Sugar, Fats, Moisture, Fibers and Nucleic Acid were analysed by standard methods of Micro- Kjeldahl Method of Ranganna(1986a) , N%x 6.25, Lowry method (1951), Dubois et al(1956), Miller (1959), Folch method (1957), Ranganna (1986b), Ranganna (1986c), and Cetyl-Trimethyl Ammonium Bromide method of Murray and Thompson (1980).

3   Results and discussion

The hyper protein producer isolate *S. sp.* no12 was grown in coculture conditions with other microbial standard strains procured from different national and International Collection Centers and Research Labs to enhance the protein contents when lignocellulosic waste was utilized as substrate. The different microorganisms used for coculture study were *P.ostreatus, P. chrysosporium, A.bisporus, P.sajor-caju, P.aeroginosa, B.licheniformis, B.sublitis* and *B.coagulans.* All the cocultures showed the increase in the total proteins though this increase varied from culture to culture(Table1). Similar studies had also been reported by Barinotto and Bendent (2000) for coculture of *Lactobacillus bulgaricus* and *Streptococcus thermophilus* on dairy waste and Rodriguez and Gallardo (1993)for *Cellulomonas* and *Pseudomonas* in agrocellulosic waste. Since cocultures of *S.sp* no12+ *P.crysosporium* (4.55%) and *S.sp* no12 +*P.sajor-caju* (4.025%) had shown the highest crude proteins therefore these combinations were selected for the next experiment.

Table1: Bioenrichment of *S.* sp No. 12 for SCP Production with other microbes in mixed cultures

Co-cultures	Crude protein (%)	N(%)
Saccharomyces sp. + *Pleurotus ostreatus*	3.341	0.535
Saccharomyces sp. + *Phanerocheaete chrysosporuum*	4.550	0.728
Saccharomyces sp. + *Agaricus bisporus*	3.675	0.588
Saccharomyces sp. + *Pleurotus sajor-caju*	4.025	0.645
Saccharomyces sp. + *Pseudomonas aeroginosa*	3.500	0.560
Saccharomyces sp. + *Bacillus licheniformis*	3.500	0.560
Saccharomyces sp. + *Bacillus subtilis*	2.275	0.364
Saccharomyces sp. + *Bacillus coagulans*	2.100	0.336
SEm	0.004	0.002
CD$_{0.0}$.	0.008	0.003

Each experiment was conducted in triplicate

When *A. chroococcum* was added in the selected cultures, there was a steady increase in their overall protein contents .It proved the capacity of *A. chroococcum* to enhance the microbial proteins probably by converting the soluble nitrogenous substances present in the culture medium into more complex proteins (Shawky et al, 1988).Maximum proteins, nitrogen, wet weight and dry weight i.e. 5.075%, 0.812%, 4.980% and 1.467% respectively. However during the cultivation of *A. chrococcum* with cocultures , crude proteins were statistically different from each other were formed in *S.* sp no. 12 + *P.chrysosporium* + *A.chroococcum* mixed culture (Table 2).

Table 2: Bioenrichment of selected coculture with *A. chroococcum* for SCP production

Cocultures	Crude Protein (%)	N (%)	Wet wt of Biomass (g/g)	Dry wt of Biomass (g/g)
S.sp+ +P. sajor-caju+ A chroococcum	4.550	0.728	4.855	1.280
S. sp.+(P. sajor-caju+P.chrysosporium)+A chroococcum	3.325	0.532	4.725	1.148
S.sp.+ P.chrysosporium+A chroococcum	5.075	0.812	4.980	1.467
S.sp.+ (P. sajor-caju+ P. chrysosporium)+ A chroococcum	3.679	0.580	4.729	1.148
SEm	0.004	0.001	0.005	0.012
CD$_{DM}$	0.008	0.003	0.012	0.027

These mixed cultures were taken for sequential culture study. Sequential culture is a method where inoculation of different strains was done in a specific order at a fixed interval of time . The different sequences, S-1, S-2, S-3 and S-4 (Table 3) followed in the present study had shown the encouraging results. The microbial proteins increased to 18.38% in sequence S-1 *S.sp no12 + A.chroococcum + coculture of{ P. sajor-caju + Phanerochaete chrysosporium}* after addition of methionine and glutamic acid (1.5%each) in it. Statistically crude proteins of all sequences were found to be significantly different from each other at 5% level of significance. The appreciable increase in proteins in a sequential culture suggested the usefulness of such type of culture methods. The reason for higher increase in sequential method was due to the metabolic products formed by one strain were utilized by next strain for its growth. Statistically crude protein of all sequences was found to be significantly different from each other at 5 % level of significance.

Table 3: Effect of Sequential Culture Method on Microbial Protein Production

Sequence No.	Sequence of Culture	Crude Protein (%)	N (%)	Wet wt of Biomass (g/g)	Dry wt of Biomass (g/g)
S-1	*S.Sp+A chroococcum +{P. sajor-caju+ P. chrysosporium}*	18.380	2.800	5.283	1.490
S-2	*S. sp.+(P. sajor-caju+P.chrysosporium)+A chroococcum*	7.700	1.232	5.00	1.400
S-3	*S.sp.+ P.chrysosporium+A chroococcum*	4.025	0.644	4.790	1.710
S-4	*S.sp.+ Pleurotus sajor-caju +A chroococcum*	2.975	0.476	4.755	1.156
SEm		0.005	0.114	0.009	0.013
CD$_{DM}$		0.012	0.263	0.022	0.302

Nutritional Evaluation Of Microbial Protein Biomass: The bioprotein sample containing different microorganisms and unutilized lignocellulosic residue was analyzed for various contents mentioned in Table 3 to explore its nutritional characteristics. Data presented in the table exhibited. The presence of crude proteins (18.380%), nitrogen (2.94 0%),soluble protein (0.040%), total sugar (0.360%), reducing sugar (0.013%),nucleic acid (0.042%),

255

fats (1.320%) and fibers (72.0%) in it.This showed that waste sample had been enriched nutritionally after its bioconversion with degrading microbes. Lignocellulosic waste, which otherwise contained negligible protein (Balagopalan, 1996) had become enriched in crude protein to 18.38 % along with the other elements mentioned in the Table 4.

Table4: Nutritional Evaluation of Microbial Protein Biomass#

Elements	Approximate Composition (%)
Crude Protein	18.380
Nitrogen	2.940
Soluble Protein	0.040
Total Sugar	0.360
Reducing Sugar	0.013
Nucleic acid	0.042
Fats	1.320
Fibers	72.000

*On dry weight basis
#Loss in moisture content= 68.12%(on wet weight.basis) at 65°C

4   Conclusion

The prepared sample of microbial proteins became highly rich in protein and fats and thus could serve as an attractive supplement for animal feed and could serve as an important food additive after refining the sample.

References

[1]   Al-Ani,F.A. 2000. Dirasat Agr Sci **27**(1),p58
[2]   Balagopalan,C. 1996 . J Sci Ind Research **55** , p479
[3]   Barinotto, M.E.P. and Bendent H.D. 2000 Anais do XV11 Congress National de Laticinos **54,** p126
[4]   CameronM.D. and Aust S.D. 2001. Enz Microbial Technol **28** ,p129
[5]   Chandel M. 2002. M.Sc. Thesis , UHF Nauni , Solan, India
[6]   Dubois, M.; Gilles, G.A.; Hamilton, J.K. and Rebers, P.A.1956. Anal. Chem.**28** ,p350
[7]   Folch, J.; Lees M. and Sloane-Stanely, G.1957. J.Biol. Chem. **226** ,p 497
[8]   Lowry O.H.; Rosebrough, N.J.; Farr, A.L. and Randall, R.J. 1951. J . Biol. Chem.**193** , p 265
[9]   Miller, G.L. 1959. Annal. Chem. **31**,p 426
[10]   Murray, M.G. and Thompson, W.F.1980. Nucleic Acid Research **8**, p 4321
[11]   Puniya A.K.; Singh, S.; Kumar C.G. and Singh, K. 1995. Indian J exp Biol **33**, p545
[12]   Ranganna,s. 1986a.,1986b, 1986c. In: Handbook of analysis and quality control for fruit and vegetable products. Tata M c Graw Hill Publishers,N. Delhi, India pp4-26
[13]   Rodriguez, H. and Gallardo ,R. 1993. Acta Biotechnol.**13**,p141
[14]   Shawky, B.T. ;Ghali, Y.; Ahmed, F.A. and Kahil, T. 1988. Egyptian J Microbiol **23**, p159

Study on the hemoprotein at 503 nm in *Salmonella typhimurium*

Hossein Tayefi-Nasrabadi and **Ezzatolah Keyhani**[*]

Institute of Biochemistry and Biophysics, University of Tehran, 13145 Tehran, IRAN

[*]Corresponding author: e-mail: keyhanie@ibb.ut.ac.ir, Phone: +98-21-695-6974, Fax: +980210640-4680

Besides its respiratory chain (cytochrome *bd*) *Salmonella typhimurium* (S. *typhimurium)* also contains a hemoprotein exhibiting a single absorption band at 503 nm (hemoprotein-503). This hemoprotein-503 was detectable when S. *typhimurium* was grown aerobically in glucose minimal medium, but was not found in enriched medium. High amounts of hemoprotein-503 were obtained when cells were grown in the presence of lactate, methanol, mannitol and excess of metals. Cyanide, azide, nitrate and nitrite bound to hemoprotein-503, forming a complex that exhibited a trough at 503 nm. Spheroplasts and membrane fragment preparations showed that hemoprotein-503 was bound to the inner plasma membrane of S. *typhimurium.* Substrate oxidation showed that hemoprotein-503 was reducible in vitro by glucose, succinate, gluconate, glutamate, ethanol, mannitol, ascorbate, NADH, lactate, citrate and pyruvate. Data showed that hemoprotein-503 is a redox center in S. *typhimurium.*

Keywords hemoprotein-503; substrate oxidation; redox center; Salmonella typhimurium.

1 Introduction

Salmonella typhimurium belongs to the Enterobacteriacea genera and is a common cause of food poisoning worldwide [1]. It is a major threat to human and to animal health, especially damaging for the poultry industry. This facultative anaerobic bacterium can extract energy from substrate both aerobically and anaerobically, but without oxygen and other electron acceptors [2]. Energy production depends upon fermentation which is less efficient than the aerobic metabolism. The respiratory chain of S. *typhimurium* has now been well studied by others [3–5]. It includes two terminal oxidases, the cytochrome *bo* complex and the cytochrome *bd* complex, which are quinol oxidases [6]. Expression of the cytochrome *bo* complex, which functions as a coupling site in the respiratory chain, is elevated during aerobiosis [7]. Cytochrome *o* belongs to the family of oxidases which contain a heme/copper binuclear oxygen reducing center and functions as a proton pump [8]. When the oxygen concentration drops in the culture medium, the cytochrome *bo* complex is replaced by the cytochrome *bd* complex which is then predominantly synthesized in the cells [9].

Several investigators have reported the occurrence of a pigment absorbing at 503 nm in microorganisms, e.g. *Azotobacter vinelandii, Saccharomyces cerevisiae, Escherichia coli* B, *Bacillus megaterium* and in yeast [10-14]. Lindenmayer and Smith [14] suggested that the 503-nm pigment in yeast was the semiquinone form of ubiquinone, based on oxidation-reduction rate, spectral characteristics, and the observation that this pigment disappeared upon rupturing the cell. Pierre et al. [13], as well as Olden and Walter [11], concluded that the 503-nm pigment was a tetrahydroporphyrin intermediate arising during the interconversion of porphyrinogens and porphyrins. Because of several fold increases of hemoprotein-503 under stress conditions (e.g. growth in the presence of lactate, methanol, mannitol, metals), this protein might be a redox stress center in *S. typhimurium*. Thus the purpose of this research was to investigate 1) hemoprotein-503 biosynthesis under a wide range of substrates such as glucose, lactate, methanol, mannitol, gluconate, etc.; 2) the effects of inhibitors on this hemoprotein in vitro. The increase of this protein under stress conditions showed that hemoprotein-503 acted as a redox center, presumably as a signalling molecule in response to stress condition (such as: presence of metals, lactate, methanol and mannitol).

2   Materials and methods

S. typhimurium (strain 3507 - Johns Hopkins University Collection) was grown in enriched and minimal media containing lactate, methanol, mannitol, glucose as substrate, or glucose in the presence of metal such as various cadmium concentrations. Cells were harvested by centrifugation (5,500 ×g); the cell pellet was washed twice in phosphate buffer 0.1 M, pH 7.0 (centrifugation at 8,500×g). Spheroplasts were prepared according to [15]. For membrane fragment preparation, spheroplasts were suspended in Tris buffer 0.05 M, pH 9.0. The suspension was homogenized in a Waring blender, then centrifuged at 10,000×g for 40 min.The yellow cloudy supernatant consisting of membrane fragments and soluble proteins was centrifuged at 40,000×g for 40 min. The pellet consisting of membrane fragments was washed once with phosphate buffer 0.1 M, pH 7.0 and recentrifuged at 40,000×g for 40 min. Hemoprotein-503 was assayed spectrophotometrically by obtaining difference spectra with a DW2 UV/Vis Aminco spectrophotometer (USA). The amount of hemoprotein-503 was reported as the difference between the optical density at 503 and 485 nm. A millimolar extinction coefficient of 56 was used to calculate the concentrations of hemoprotein-503 [12]. Substrates used in vitro are shown in table 1. The inhibitors were KCN, KNO_3, KNO_2 and azide.

3   Results and discussion

Dithionite reduced-minus-H_2O_2 oxidized difference spectra of suspensions of *S. typhimurium* cells grown aerobically in enriched or glucose minimal medium, exhibited an absorption peak at 630 nm and a trough at 649 nm due to cytochrome *d*, a small absorption peak at 595 nm due to cytochrome b_{595}, a peak at 560 nm due to cytochrome b_{560} and a peak at 431 nm (Soret region) due to the γ band of *b*-type cytochromes (Fig. 1) [3]. Because peptone and dithionite rendered hemoprotein-503 colorless [13, 14], it was not detected in difference spectra of cells grown in enriched medium but rather in glucose reduced-minus-air oxidized spectra of cells grown in glucose minimal medium (Fig. 2). As seen in that figure, the 503 nm band would appear, reach a maximum, and then gradually decrease with time while a new band at 444 nm appeared that would grow as the 503 nm band diminished. No cytochrome *d* was detected in these spectra. However, the respiratory chain of *S. typhimurium* is a *bd* terminal oxidase and has no cytochrome *c*; therefore, the 444-nm band was a new peak, probably related to *b*-type cytochrome or to a new branch of the respiratory chain. The highest amount of hemoprotein-503 was seen in cells grown in the presence of lactate, methanol, mannitol, or glucose + cadmium (stress conditions). Several in vitro substrates including succinate, NADH, ethanol, glucose, glutamate, ascorbate, mannitol, gluconate, lactate, citrate and pyruvate were used to reduce the respiratory chain of *S. typhimurium* (Fig. 3, 4, 5). As seen in these figures, all used substrates could reduce hemoprotein-503, but were unable to reduce cytochrome d, therefore no cytochrome d was detected in these spectra. This suggested that *S. typhimurium* was using a new respiratory chain or a new branched respiratory chain with hemoprotein-503 and the hemoprotein at 444 nm as essential components. To evaluate the effect of nitrogen-containing ligands on hemoprotein-503, KCN, KNO_3, KNO_2 and azide were used; results showed that in enriched and minimal media, KCN significantly bound to both reduced and oxidized forms of hemoprotein-503 and induced a trough at 503 nm. Furthermore, KCN partially bound to cytochrome d in reduced form, resulting in trough at 630-633 nm, and significantly bound to the oxidized form of cytochrome d, resulting in a trough at 649 nm (Fig. 6). KNO_3 as well as KCN significantly bound to both reduced and oxidized forms of hemoprotein-503 resulting in a trough at 503 nm. KNO_2 partially bound to hemoprotein-503 in reduced form and significantly bound to the oxidized form of hemoprotein-503 resulting in a trough at 503 nm. Sodium azide, partially bound to both reduced and oxidized forms of hemoprotein-503. (data not shown). To determine whether hemoprotein-503 was located in *S. typhimurium* plasma membrane or cytoplasm, spheroplasts were prepared from cells grown aerobically in glucose minimal medium (late logarithmic phase), and the KCN+ dithionite reduced-minus-dithionite reduced difference spectrum from the supernatant was recorded. Results showed that the hemoprotein-503 was in the spheroplast. Moreover when the spheroplasts were disrupted and membrane fragments separated, results showed that the hemoprotein-503 was detected only in the inner plasma membrane. When pyridine hemochrome spectra were prepared from membrane fragments, results showed that hemoprotein-503 was localized in the inner membrane of *S. typhimurium*. Pyridine binds to the heme of cytochromes or is as a solvent for porphyrin and induces a blue shift in the absorption spectra of cytochromes and porphyrins. Therefore, a peak at 610 nm due to cytochrome d and peaks at 557 nm and 525 nm, respectively due to the a and ß bands of cytochrome b_{560}, were detectable, along with a peak at 495 nm related to hemoprotein-503 and one at 422 nm due to the ? band of b-type cytochromes.

The amount of hemoprotein-503 was obtained from substrate reduced-minus-air oxidized spectra of cell suspensions that were grown aerobically in glucose minimal medium (Table 1). As shown in this table, gluconate is the best substrate for the reduction of hemoprotein-503 giving a maximum value of 9.50×10^{-3} nmole/mg prot, while glutamate and ethanol gave the lowest value (1.7×10^{-3} nmole/mg prot). In minimal medium, at late logarithmic phase of growth, the amount of hemoprotein-503 was four times higher than at the stationary phase. Because gluconate (for which $NADP^+$ is the first electron acceptor) and succinate (for which flavin is the first electron acceptor) can reduce hemoprotein-503, this hemoprotein, as a component of the inner plasma membrane of *S. typhimurium*, acts as a redox center which is an electron acceptor, preferably from NADPH.

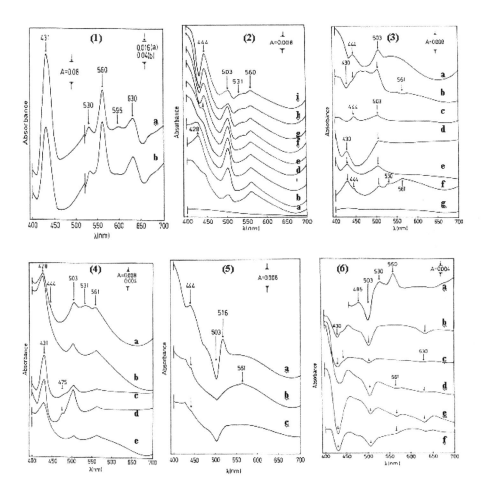

Fig. 1: Dithionite reduced-minus-H_2O_2 oxidized difference spectra of *S. typhimurium* cell suspensions grown aerobically in (a) enriched medium and (b) glucose minimal medium, at stationary phase.

Fig. 2: In vitro glucose-reduced-minus-H_2O_2 oxidized difference spectra of cells grown in glucose minimal medium as substrate, recorded after 2 (a), 3 (b), 5 (c), 10 (d), 12 (e), 20 (f), 22 (g), 30 (h) and 33 min (i).

Fig. 3: In vitro substrate-reduced-minus-H_2O_2 oxidized difference spectra of cells suspensions. Substrate was: (a) glucose, (b) rotenone + glucose, (c) ethanol, (d) NADH, (e) rotenone + NADH, (f) succinate.

Fig. 4: In vitro substrate reduced-minus-H_2O_2 oxidized difference spectra of cells. Substrate was: (a) gluconate, (b) mannitol, (c) ascorbate + TMPD, (d) ascorbate, (e) glutamate.

Fig. 5: In vitro substrate reduced-minus-H_2O_2 oxidized difference spectra of cells. Substrate was: (a) pyruvate, (b) citrate, (c) lactate.

Fig.6: In vitro glucose + inhibitor-reduced-minus- glucose-reduced difference spectra of cells. Inhibitor was: KCN (b), KNO_2 (c), KNO_3 (e). H_2O_2 + inhibitor -oxidized-minus- H_2O_2 -oxidized difference spectra of cells. Inhibitor was: KCN (a), KNO_2 (d), KNO_3 (f).

260

Table 1: Amount of hemoprotein-503 obtained from substrate reduced-minus-air oxidized spectra of cell suspensions that were grown aerobically in glucose minimal medium.

Reducing substrate	Hemoprotein-503 (nmole/mg prot) $\times$ 10^{-3}	Reducing substrate	Hemoprotein-503 (nmole/mg prot) $\times$ 10^{-3}
Late logarithmic phase		Late logarithmic phase	
gluconate	9.5	ascorbate	2.83
glucose	8.6	NADH	1.82
pyruvate	7.0	glutamate	1.7
lactate	6.6	ethanol	1.7
succinate	6.35	Stationary phase	
citrate	6.0	glucose	2.3
mannitol	5.4	mannitol	3.6

References

[1] J. M. Miller, Clin. Microb. Newsletter 9 (1987) 173.
[2] S. Iuchi, L. Weinder, J. Biochem. 120 (1996) 1055.
[3] S. Junemann. Biochim. Biophys. Acta 1321 (1997), p. 107.
[4] E. Keyhani, D. Minai-Tehrani. In: BioThermoKinetics in the Post Genomic Era C. Larsson, I.L. Pahlman. L. Gustafsson. eds), Chalmers Reproservice, Goteborg (1998), p. 236.
[5] E. Keyhani, D. Minai-Tehrani. Biochim. Biophys. Acta 1506 (2001), p. 1.
[6] Y. Anraku, Annu. Rev. Biochem. 57 (1988) 101.
[7] P.A. Cotter, V. Chepuri, R.B. Gennis, R.P. Gunsalus, J. Bacteriol. 172 (1990) 6333.
[8] M. Saraste, L. Holm, L. Lemieux, M. Lubben, J. van der Oost, Biochem. Soc. Trans. 19 (1991) 608.
[9] W.S. Shipp, Arch. Biochem. Biophys. 150 (1972) 459.
[10] Joyce R, et al, Biochem. J. 120 (1970) 771.
[11] Kenneth Olden, Walter P. Hempfling, J. Bacteriol. 113 (1973) 914.
[12] Rozanne Polsun, W. James Polglase, Biochim. Biophys. Acta, 329 (1973) 256.
[13] Pierre, L. C., C. Volland, and P. Chaix, Biochim.Biophys. Acta, 143 (1967) 70.
[14] Lindenmayer, A. and L. Smith, Biochim. Biophys. Acta, 93 (1964) 445.
[15] H. R. Kaback, Bacterial membrane, Methods Enzymol. 22 (1971) 99.

The knowledge flow and commercialisation along the continuous chain: education - research - implementation by innovative methods

U.Viesturs*[1], and A. Zilevica[2]

[1] Institute of Wood Chemistry 27 Dzerbenes Str., LV-1006 Riga, LATVIA
http://www.lza.lv/scientists/viestursu.htm
[2] University of Latvia, 1a Šarlotes Str., LV-1001 Riga, LATVIA
http://www.lza.lv/scientists/zilevica.htm

* Corresponding author: e-mail: lumbi@lanet.lv, koks@edi.lv, Phone: +371 7034884, +371 7553063, Mob. phone: 9284923, Fax: +371-7550635, +371 7034885

The Lisbon/Barcelona goals are unattainable without the relevant updating of study curricula. 10-15% of population must feel an aspiration for creation (inventions, innovations). We must teach students to become not only skilled employees, but also innovative employers. We must assure the continuity in the ERTDI(+I)P line, where: E - higher education; RTD - development of technologies / products / services; I - implementation; + I - special methods for commercialisation of RTD results; P - production. This work focuses attention on the continuity of ERTDI(+I)P processes starting at universities and ending in industry. To increase compatibility on the university level, we should start with the correction of the study curricula; correct the structure of public investments; change the attitude at universities, and declare that their concern is the full formula ERTDI(+I)P; activate the so-called innovation programs, which means state incentives for TDI steps – implementation of special methods to facilitate the commercialisation of RTDI results.

Keywords: education, research, innovations; biotechnology; teaching creativity; entrepreneurship; technology development; technology; ERTDI(+I)P system

1 Introduction

The well-known Lisbon / Barcelona goals [1] are unattainable without the relevant updating of study curricula.

A part of the population (10-15%) must feel an aspiration for creation (inventions, innovations). To foster this is the task of universities. A considerable role in establishing the above-mentioned approach is played by the educational system, first of all, higher education. We must teach students to become not only skilled employees, but also innovative employers. Actually, we must assure the continuity in the ERTDI(+I)P line [2, 3], where:

    E - higher education;
    RTD - development of technologies /products/services;
    I - implementation;
  + I - special methods for commercialisation of RTD results;
    P - production.

The commercialisation of RTD results (I, +I, P steps) could be facilitated mainly by the interactions between the driving forces of progress: technology performers, private / public investors, independent experts (technological / financing auditing), the public, policy-makers (state and municipal levels) [4–6] keeping in mind sustainable development, especially environmental issues [2].

262

2   Aims

The aims of the present paper were as follows:
1. analysis of the possibilities to improve the interactions along the whole chain: education – research – production (Fig. 1);
2. elaboration of recommendations for correction of study curricula, at least within professional programs.

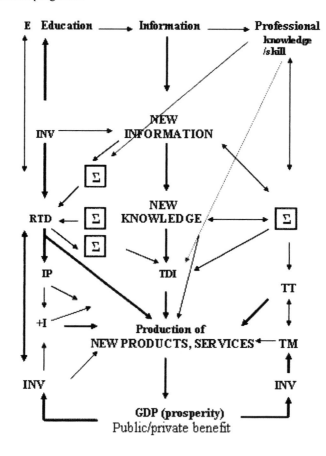

Fig. 1 Full scenario of the activated ERTDI+(I) P system:

S – EUREKA (I have found / created it: appearance of the initial idea);   TT – technology transfer;
TM – technology multiplication;   INV – investments - public / private;   IP – intellectual property.

3   Methods

The methods used were as follows:

• analysis of interviews with

-   students in biotechnology and industrial microbiology,

- spin-off companies in biotechnology and bioengineering,
- and entrepreneurs,
• analysis of the state of the art (success stories).

Traditionally, the general content of higher education includes acquiring of:

1) theoretical knowledge, which provides a basis for a profession during study courses;
2) skills, which are necessary for practical professional activities;
3) knowledge and skills, which set the stage for research and development within a professional field;
4) a knowledge of the conditions in society, which affect the economic situation;
5) an understanding of other countries' experience and international circumstances that are important for a student's professional job.

However, competitiveness and globalisation are currently hot key words and reality in the job market. Therefore, extending of expertise (A + B) is very desirable.

Typical (A) and suggested (B) functions of universities and activities realised by researchers are as follows.

Functions of universities:

development of knowledge,

accumulation of knowledge,

return of knowledge,

development of new products, services (not always recognised as the responsibility of classical universities).

Activities realised by teachers:

A – Professors and scientists well known in the corresponding scientific field; authors of monographs, scientific articles, etc.; leaders of scientific research projects.

B – Professors, high-quality specialists in a professional field;

Experts and advisers in a profession, heads of professional associations and other leaders of practical projects.

Criteria: patents, certificates, licensing of IP, the spin-off process, IPO in the stock market, expertise in relevant branches of industry or other practical applications, etc.

An optimum is a combination of A and B.

4   Results and discussion

In order to increase the compatibility on the university level, especially after graduating,

1. We have to start with the correction of the study curricula:
   - to teach creativeness
   - to teach also entrepreneurship
   - to suggest / advise / teach students to become employers
   - to teach the formation of target-oriented products, which includes the following steps:
     a) problem formulation (need, opportunity),
     b) problem restatement,

c) idea development (use brain-storming and other rules, get as many ideas as possible),

d) solution – find positive and negative aspects, develop any idea to find a solution,

e) planning,

f) control.

2. To correct the structure of public investments:
    - to increase investments in applied R and D
    - to invest in marketing of R and D products
    - to improve the tax policy in favour of new / better products in the market encouraging a regulatory environment.

3. To change the attitude (climate) in universities and to declare that their concern is the full formula ERTDI(+I)P, not only E and R, as it is, unfortunately, now in the majority of European universities.

4. To activate the so-called innovation programs, which means state incentives (taxes, joint financing, etc.) for TDI steps – implementation of special methods to facilitate the commercialisation of RTDI results (+I in the above-mentioned formula).

As a result, we could expect an increase in new high-tech products and simultaneously IPO and other innovative indicators.

5   Conclusion and recommendation

Focusing of attention on the continuity of ERTDI(+I)P processes that start in universities (even in pre-university schools) and end in industry is a very important issue for strengthening the EU25 competitiveness.

References

[1] Science and Technology: the Key to the Future of Europe – Guidelines for Future European Union Policy to Support Research. COM (2004) 353, 16 June 2004.

[2] U. Viesturs, and A. Zilevica, Creation of a knowledge-based innovative society: public – private partnership (2003). http://www.innovation.lv/irc/Publikacijas/6/KatowiceInternet.pdf

[3] K. Sedlenieks, Z. Ozola, A. Žilevica, and U. Viesturs, in: Public and Private Sector Partnerships: Sustainable Success, L. Montanheiro, F. Kužnik, and A. Olchojski (eds.) (Sheffield Halam University Press, UK, 2003), pp. 403-414.

[4] Irish Council for Science, Technology and Innovation, National Code of Practice for Managing Intellectual Property from Publicity Funded Research, January 2004, the National Policy and Advisory Board for Enterprise, Trade, Science, Technology and Innovation, 24.

[5] S. Serger, and E. Hansson, Competing in the Single Market. SMEs and Innovation in the Baltic Countries and Poland. IKED. Malmo: IKED (2004). http://www.iked.org http://www.iked.org/default.asp?id=27&mnu=27

[6] U. Viesturs, J. Sabulnieks, A. Silins, and J. E. Ekmanis, in: Technology Transfer for Economic Development: Experience for Countries in Transition, M. Cavlek, J. Švarc, and D. Hubner (eds.) (MOST Publication, 2001), pp. 261-278.

Xanthan gum production using whey for preculture preparation

M.R. Soudi [*], M. Ebrahimi, and S. Sharyat Panahi

Department of Microbiology, Faculty of Science, Alzahra University, Tehran 19938, Iran

[*]Corresponding author: e-mail: msoudi@alzahra.ac.ir or msoudi@yahoo.com , Phone: +98-21-8044058, Fax: +98-21-8047861

Xanthan gum, the unique exopolysaccharide of *Xanthomonas campestris* pv campestris, is produced commercially with an annual world wide amount of approximately 30 million kg, and mainly consumed as a food ingredient. Inexpensive inorganic and organic ingredients are usually used as production medium. However, in order to prepare inoculum, a commercially available and expensive culture medium (YM broth) is used for preculture preparation. In this study, we propagated a native strain of *Xanthomonas campestris* pathovar campestris using 2l bench top fermentor (Biostat B, B.Braun) and different substrates as preculture medium. In all cases, a 5% inoculum was used and all processes were terminated after 72 h of culture. Aeration with atmospheric air (1 v/v.m), regulation of pH at 7.0 $\pm$ 0.2, incubation at 28 °C, and agitation at 250 rev min^{-1} were taken as constant parameters of processes. Foaming was controlled automatically using technical silicon oil diluted 1:10 (v/v). The results showed that peptonized whey at 3% (W/V) concentration can successfully replace YM broth. The process yielded higher amounts of xanthan gum (16.5 gl^{-1}) and higher viscosity (1560$\pm$ 45 cp) than those of commercial YM medium.

Keywords peptonized whey; preculture; xanthan gum; *Xanthomonas campestris*

1   Introduction

Whey, a waste by-product of dairy industries, contains about 7% solids and is composed of about 12% protein and 70% lactose. Sewerage of whey is an expensive procedure due to the high biological oxygen demand which is imposed on sewer systems [1].

Production of high-quality materials such as xanthan gum from whey is preferable method of handling this waste. Xanthan gum, the unique exopolysaccharide of *Xanthomonas campestris* pv campestris, is produced commercially with an annual world wide amount of approximately 30 million Kg [2]. This valuable microbial polysaccharide has extensive uses in food, petroleum and other chemical and pharmaceutical industries. Inexpensive inorganic and organic ingredients have been usually reported as the substrate for production of xanthan. However, in order to prepare inoculum, a commercially available and expensive culture medium (YM broth) is apparently used for preculture preparation.

Production of xanthan gum by lactose-utilizing strains of *Xanthomonas campestris* has been studied by investigators [3–5]. Many researchers have focused their research on whey as the main substrate in composition of production medium. However, we have used whey as the sole component of preculture medium in gum production process. This helps lower the cost of culture medium in intermediary fermentor and it will be cost-effective in large scale processes.

2   Materials and methods

All pure chemicals and ingredients of culture media were purchased from Merck Company. Dry whey was obtained from domestic producers. All experiments were carried out in triplicate and average of data was shown here.

2.1   Microbial strain

A native isolate of *Xanthomonas campestris* pv *campestris* strain b82 was used [6].

2.2   Whey preparation

Dry whey powder was dissolved in tap water at different final concentrations from 10 g l^{-1} to 100 g l^{-1}. Intact solution and partially hydrolyzed (peptonized) solutions of whey were used as preculture media. The latter solution was prepared by alkaline treatment of reconstituted whey in presence of NaOH. A normal solution of Caustic soda was added to whey to obtain final concentrations from 0.04 to 2 g l^{-1}. Solutions were heated to boil for 15 minute and neutralized by normal solution of hydrochloric acid.

2.3   Production medium

Synthetic culture medium containing sucrose (as the main carbon source) and ammonium sulfate (as the main nitrogen source) plus a combination of nutrient minerals and trace elements were used as production medium. Basic composition of the synthetic culture medium for xanthan gum production has been previously reported [7].

2.4   Fermentation process

An overnight culture of *Xanthomonas campestris* strain b-82 in YM broth was used as primary culture. Inocula were prepared in 500-ml shake flasks containing 100 ml of whey solutions and incubated in shaking incubator at 28 °C and 150 rev min^{-1} for 24 h. A control preculture medium (YM broth) was also used. All cultures were incubated in shaking incubator at 28°C and 150 rev min^{-1}. *Growth* was measured using viable cell count method, as required. These precultures were used to inoculate final synthetic production medium in a 2 l bench top fermrntor (Biostat B Fermenter, B.Braun Company). In all cases, a 5% inoculum was used and all processes were terminated after 72 h of culture. Aeration with atmospheric air (1 v/v.m), regulation of pH at 7.0 $\pm$ 0.2, incubation at 28 °C, and agitation at 250 rev min^{-1} were taken as constant parameters of processes. Foaming was controlled automatically using technical silicon oil diluted 1:10 (v/v).

2.5   Biomass determination

For *biomass* determination, cultures were diluted in distilled water and cells were harvested by centrifugation (15000 $\times g$), washed and dried in 80 °C oven, and weighed.

2.6   Xanthan extraction

Exopolysaccharides were separated from the culture supernatant, subsequent to vigorous shaking in presence of 20% methanol and dehydration with three volumes of methanol. In order to obtain purified products, the process was repeated as many times as required.

2.7 Viscosity measurement

One gram of dried xanthan gum, in each case was dissolved in 1% KCl solution. *Viscosity* was determined on a Brookfield Viscometer system, having a 3° spindle at 60 rev min^{-1} and 25 °C.

3 Results and discussion

Preliminary studies were carried out with cultivation of *Xanthomonas campestris* strain b82 on different concentrations of reconstituted whey from 10–100 g l^{-1}, as preculture medium. Synthetic medium containing sucrose was constantly used as production medium and amount of separated *biomass* and extracted *xanthan* was calculated in each case. As shown in Figure 1, maximum concentration of *xanthan gum* is produced when preculture medium contained at least 30 g dry whey l^{1}. Higher concentrations of whey beyond this did not show remarkable increase in *xanthan* concentration. It is interesting that lower amount of *xanthan* was produced when YM broth used as preculture.

Onc main reason is that the cells grown on YM broth are surrounded by a halo of exopolysaccharide due to preexisting glucose in preculture medium which can easily be shown under light microscope. Preexisting exopolysaccharide plays the role of a barrier inhibiting superior absorptive properties of the cells. Thus, mass transfer is weakened when such cells are introduced to production medium. In contrast, bacterial cells grown on whey don't form the exopolysaccharide layer.

Although relatively less *biomass* is produced in whey compared with the use of YM broth, the cells are more active and are better gum producers.

When whey is used as preculture medium, the bacterial cells use pertinacious materials. Moreover our experiments on strain b82 showed to weak growth on lactose as sole carbon source in minimal synthetic medium.

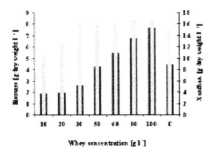

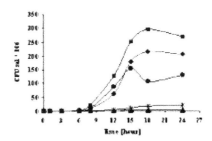

Fig. 1 Productivity of *X.campestris* strain b82 in the same synthetic production medium and different whey concentrations as preculture media. Columns indicate Yield of xanthan gum (gray) and biomass (black). C: YM broth as control.

Fig. 2 Growth curves of *X.campestris* in hydrolyzed whey (30 g l^{1}) treated with different alkalinities in terms of NaOH concentration (g l^{1}): 0.04 (■); 0.4(●); 1(×); 2(▲); not treated(◆).

To enhance the *growth* of *Xanthomonas campestris* strain b82 on whey, the substrate was treated with alkaline and heat. As shown in Figure 2, the results indicate that *growth rate* increases when culture medium is treated with 0.01 g l^{-1} of caustic soda. Exposing whey to higher alkalinity resulted in lowering of *growth rate*, probably due to harsh treatment conditions and production of undesirable by-products. In comparison with YM broth, more amount of *gum* was produced and higher *viscosity* obtained when hydrolyzed whey was used as preculture (Table 1).

These results are promising, when we consider that huge amounts of substrate must be used for preparation of inoculum and feeding of intermediary fermentors.

Table 1 Characteristics of xanthan gum production with different preculture media.

Characteristic	Preculture medium	
	Hydrolyzed whey	YM broth
Biomass (g l^{-1})	2.6	4.5
Xanthan gum (g l^{-1})	16.5	15.4
Viscosity (cP)	1560 (±45)	1434 (±37)
Inoculum: Maximum cell count (CFU ml^{-1})	2.9×10^8	3.1×10^8
Needed cultivation time (h)	18	12

Acknowledgements

This work was financed by vice-chancellor (Research), Alzahra University, Tehran, Iran. We gratefully acknowledge Mrs. Sara Gharavy for editing of the manuscript.

References

[1]   R.D. Schwartz and E.A. Bodie, Applied and Environmental Microbiology (a) **51**, 203 (1986).
[2]   S. Kalogiannis, G. Iakovidou, M. Liakopoulou_Kyriakides, D.A. Kyrikidis, G.N. Skaracis, Process Biochemistry (a) **39**, 249 (2003).
[3]   L.V. Ekateriniadou, S.V. Papoutsopoulou , and D.A. Kyrakidis, Biotechnology letters (a) **16**, 517 (1994).
[4]   J.-F. Fu , Y.-H. Tseng, Appl. Env. Microbiol. (a) **56**, 919 (1990).
[5]   S.V. Papoutsopoulou, L.V. Ekateriniadou and D.A. Kyriakidis, Biotechnology Letters (a) **16**, 1235 (1994).
[6]   M.R. Soudi, Alzahra University J of Science (a) **12**, 45 (1990).
[7]   F. García-Ochoa, V.E. Santos and A.P. Fritsch, Enzyme and Microbal Technology (a) **14**, 991 (1992).

Food Microbiology

An assessment of differential media for the recovery of histamine producing bacteria

E. Economou [1], C. Papadopoulou [1*], S. Levidiotou [1], M. Brett [2], K. Seferiadis [3]

[1] Food, Water and Environmental Microbiology Unit, Department of Microbiology, Medical School, University of Ioannina, 45110 Ioannina, Greece
[2] Marine Biotoxins Unit, Food Hygiene Laboratory, Central Public Health Laboratory Service, Colindale, London, United Kingdom
[3] Department of Clinical Chemistry, Medical School, University of Ioannina, 45110 Ioannina, Greece

*Corresponding author: e:mail: cpapadop@cc.uoi.gr Tel.: ++30 2651097592, cel: ++30 6944690511, Fax: ++302651097855 and ++32651093563

1   Introduction

Histamine fish poisoning (HFP) is one of the most frequently reported intoxications associated with the consumption of fish worldwide [1]. HFP is associated with the ingestion of foods with a high content of histamine produced by decarboxylation of L-histidine by bacterial decaboxylases. HFP has been reported in many countries and is the most prevalent seafood-borne disease in the United States [1]. Symptoms include skin rashes, urticaria, oedema, localized inflammation, nausea, vomiting, diarrhoea, cramping, hypotension, headaches, palpitations and an oral burning and blistering sensation [2]. Histamine is not inactivated by heat during cooking or processing. The formation of histamine in fish is prevented by storage at low temperatures.
Histamine is produced in fish by the bacterial decarboxylation of histidine. Histamine production is favoured by the presence of bacterial species that are prolific histamine producers, like *Enterobacteriaceae* (*Morganella morganii*, *Enterobacter cloacae*, *Citrobacter freundii*, *Escherichia coli*, *Hafnia alvei*, *Klebsiella pneumoniae*, *Proteus* spp., *Serratia macrescens*), lactic acid bacteria and *Photobacterium* spp.. Other bacterial genera such as *Staphylococcus* spp., *Pseudomonas* spp., *Aeromonas* spp., *Plesiomonas shigelloides* and *Clostridium perfringens* have also been implicated in HFP [1]. The histamine producing bacteria produce histidine decarboxylases that are responsible for the production of histamine in foods. The bacterial histidine decarboxylases are heat stable enzymes that can decarboxylate histidine even in the absence of bacteria.
Various histidine decarboxylase media have been developed to detect the presence of histamine producing bacteria in fish. The use of a reliable method for detection and enumeration of histamine producing bacteria is important for the prediction of histamine build up in fish flesh. The objective of this study was to evaluate 3 liquid (broths) and 3 solid (agars) media for the differentiation of histamine decarboxylase bacteria isolated from fish.

2   Materials and methods

2.1 Bacterial strains

The evaluation of the media was carried out employing 76 strains, which were isolated from temperature abused tuna [3]. The strains belonged to the families of *Entero-*

bacteriaceae (39 isolates including 4 *Morganella morganii* and 6 *Klebsiella* spp. strains), *Staphylococci* (15 strains), *Pseudomonaceae* (16 strains) and other families (6 strains).

2.2 Histamine detection

The isolates were subcultured in trypticase soy broth (Scharlau) supplemented with 1% histidine (pH=6.5) for 48 hr at 30°C. The supernatant was collected and centrifuged twice at 3500 rpm for 10 min using a Sorvall GLC-4 desktop centrifuge (Kendro Laboratories, Stortford Hall Park, Hertfordshire GM32 5GZ, UK). The histamine concentration of the supernatants was measured with the ELISA kit Histamine ELISA (IBL-Hamburg GmbH Flughafenstr. 52A, D-22335 Hamburg, Germany). The ELISA plate was read in an ELISA DAS type A1 photometer (DAS s.r.l., Viali Tivoli Km 18642, Palombara Sabina, Roma, Italy). The isolates were characterised as strong, weak and non – histamine producers according to the classification of Behling and Taylor [4].

2.3 Media evaluated

The agar media evaluated were the Niven medium [5], modified Niven medium [6] and a solid medium based on Møeller decarboxylase broth [7] with the addition of 1.5% agar. Histidine containing broths were prepared using the same formulation without the addition of agar. Gas production in histidine decarboxylase broths was detected by the insertion of an inverted Durham tube. The materials used were tryptone (EZMix Tryptone, Sigma Chemical Co, St. Louis, USA), beef extract (Sigma Chemical Co, St. Louis, USA), yeast extract (Scharlau Chemie S.A., Sentmenat, Barcelona, Spain), NaCl (Merck KgaA, Darmstadt, Germany), CaCO$_3$ (Aldrich Chemical Co, Milwaukee, USA), glucose (SERVA Electrophoresis GmbH, Heidelberg, Germany), agar (purififed agar, Oxoid Ltd, Hampshire, England), bromocresol purple (Merck KgaA, Darmstadt, Germany), cresol red (Aldrich Chemical Co, Milwaukee, USA) and pyridoxal phosphate (Sigma Chemical Co, St. Louis, USA). Solid and liquid media with the same composition but without the addition of histidine (DL – histidine monohydrochloride monohydrate, Sigma Chemical Co, St. Louis, USA) were used as controls.

Table 1. Composition of tested media

	Niven medium *	Modified Niven medium *	Møeller decarboxylase medium *
Tryptone	0,5%	0,5%	0,5%
Beef extract			0,5%
Yeast extract	0,5%	0,5%	
L-histidine	2,7%	2%	1%
NaCl	0,5%	0,5%	
CaCO$_3$	0,1%		
Glucose			0,05%
Agar	1,5%	1,5%	1,5%
Bromocresol purple	0,006%		0,001%
Cresol Red		0,02%	0,0005%
Pyridoxal phosphate			0,0005%
pH	5,3	6,5	6,0

* Solid and liquid media with the same composition but without the addition of histidine were used as controls.

All strains were inoculated into trypticase soy broth (Scharlau Chemie S.A., Sentmenat, Barcelona, Spain) supplemented with 1% histidine (pH=6.5), they were incubated overnight at 30°C and they were subcultured onto the 3 solid media (agar plates) and into the 3 different liquid media described above. The inoculated plates were incubated aerobically at 30°C for 72 hr. The same procedure was followed for the media not containing histidine. The whole experimental procedure was performed in duplicate. The cultures were monitored every 24 hr for bacterial growth, change of color and gas production. Growth of histamine producers was signaled in solid media by formation of colored colonies surrounded by colored halo and in broths by change of color and gas production. No similar changes were observed in the media without histidine.

3   Results

Among the strong histamine producers, Niven agar, Niven broth and Møeller decarboxylase agar differentiated 71%, 86% and 86% of the isolates respectively. Modified Niven agar, modified Niven broth and Møeller decarboxylase agar differentiated 56%, 29% and 56% of the isolates respectively.

Among the weak histamine producers, Niven agar differentiated 71% of the isolates respectively. Modified Niven agar, Møeller decarboxylase agar, Niven broth, modified Niven broth and Møeller decarboxylase agar differentiated 38%, 42%, 29%, 17% and 56% of the isolates respectively.

All the solid media used, showed false positive results, varying from 8 to 14% for Niven agar, 29 to 43% for modified Niven agar and 4 to 14% for Møeller decarboxylase agar. No false positive results were recorded in the broths tested.

All isolates have grown well in the media used with the exception of a *Staphylococcus auricularis* isolate. One *Proteus mirabilis* isolate produced a yellow colour in Niven agar and three isolates (one *Morganella morganii* and two *Proteus mirabilis* isolates) have produced a yellow colour in Møeller decarboxylase agar

The results of the evaluation of media are shown in the following.

Table 2. Results of the inoculation of isolates in selective media for the detection of histamine producing bacteria

	Agar			Broth		
	Niven (%)	Modified Niven (%)	Møeller (%)	Niven (%)	Modified Niven (%)	Møeller (%)
Strong histamine producers (7 isolates)						
Positive	71	57	57	86	29	86
Negative	14	14	29	14	71	14
False Positive	14	29	14	0	0	0
Weak histamine producers (23 isolates)						
Positive	71	38	42	29	17	21
Negative	17	25	46	71	83	79
False Positive	8	33	4	0	0	0
Bacteria that do not produce histamine (36 isolates)						
Positive	33	20	28	11	0	7
Negative	54	35	63	85	96	91

4   Discussion

Compared to the rest tested media, the Niven agar and broth had the best performance in detecting histamine producing bacteria, exhibiting adequate detection rates for strong and weak histamine producing bacteria. However we agree with other researchers [8–11] that the main problem in utilizing Niven medium are the false positive results. Ababouch et al [8] have reported that 48% of the isolates tested in Niven agar have given false positive results. Ben – Gigirey et al [11] also reported that among 19 isolates, 10 of them have given false positive result (53%). Also, Hernández – Herrero et al [10] reported that, among 2192 isolates mainly Gram positive (83%), the 37% of the isolates exhibited false positive result. The high number of false positive results is attributed to the production of alkaline compounds like ammonia or related compounds [12].

Another problem with the Niven agar was that 33% of the non – histamine producing bacteria gave a positive result after inoculation in Niven medium supplemented with histidine, but not in Niven medium without histidine which was used as control. This observation does not coincide with the theory that alkaline products from the degradation of substances other than histidine are responsible for the false positive results encountered. Perhaps a different metabolic pathway involving histidine degradation but not histamine production is responsible for this phenomenon. This observation comes into terms with the description of a metabolic pathway for histidine by *Klebsiella aerogenes* [13] and *Pseudomonas* spp. [14].

Modified Niven medium was introduced into food microbiology as a medium suitable for detecting acid sensitive histamine producing bacteria, like *Clostridia*. According to our results, modified Niven agar and modified Niven broth are rather unsuitable as screening mediums for the detection of histamine producing bacteria, regardless of their acid sensitivity. Actis et al [12] have tested modified Niven medium with the inoculation of five known histamine and non – histamine producing bacteria. They reported that modified Niven medium showed false positive results and they agree that modified Niven medium is unsuitable for the detection histamine producing bacteria.

References

[1]   L. Lehane, and J. Olley. *Int J Food Microbiol* **58** (2000), p. 1.
[2]   S.L. Taylor. *Crit. Rev. Toxicol.* **17** (1986), p. 91.
[3]   V. Economou, M. Brett. C. Papadopoulou, T. Nichols. Proceedings of the International Conference on Veterinary Public Health and Food Safety, Rome, Italy, 22 – 23 October 2004, pp. 63 – 64 (2004).
[4]   A.R. Behling, S.L. Taylor. *J. Food Sci.* **59** (1982), p. 517.
[5]   J.C.F. Niven, M.B. Jeffrey, and J.D.A Corlett. *Appl. Env. Microbiol.* **41** (1981), p. 321.
[6]   D.H. Yoshinaga, and H.A. Frank. *Appl. Environ. Microbiol.* **44** (1982), p. 447.
[7]   V. Moeller. *Acta Pathol. Microbiol. Scand.* **36** (1955), p.158.
[8]   L. Ababouch, M.E. Afilal, S. Rhafiri, and F.F. Busta. *Food Microbiol.* **8** (1991), p. 127.
[9]   G.C. Fletcher, G. Summers, and P.W.C. van Veghel. *J. Food Prot.* **61** (1998), p. 1064.
[10]   M. Hernández – Herrero, A. Roig – Sagués, J. Rodríguez – Jerez, and T. Mora – Ventura. *J. Food Prot.* **62** (1999), p. 509.
[11]   B.J.M. Ben-Gigirey, V.B. de Sousa, T.G. Villa, and J. Barros-Velasquez. *J. Food Prot.*, **62** (1999), p. 933.
[12]   L. Actis, J. Smoot, C. Barancin, and R. Findlay. *J. Microbiol Meth.* **39** (1999), p. 79.
[13]   B. Magasanik, H.R. Bowser. *J. Biol. Chem.* **213** (1955), p. 571.
[14]   T.G. Lessie, F.C. Niedhardt. *J. Bacteriol.* **93** (1967), p. 1800

Aeromonas growth under low temperatures

T. Zuccolotto[1], A.P.L. Delamare[1,2], S.O.P. Costa[1,2], S. Echeverrigaray[1]

[1] Instituto de Biotecnologia, Universidade de Caxias do Sul, R. Francisco G. Vargas 1130, 95001-970 Caxias do Sul, RS, BRAZIL.
[2] Instituto de Biociências, Universidade de São Paulo, São Paulo, BRAZIL.

Considering the importance of *Aeromonas* as food contaminants and pathogenic bacteria, the present work aimed to analyze *Aeromonas* growth and pathogenicity factors production (haemolysins and proteases) under low temperatures. Eleven representatives of different *Aeromonas* species, and 11 *A. hydrophila*, including clinical and environmental isolates, were evaluated. Screening and kinetic experiments were conducted at 4, 10 and 30°C on M9 minimal medium. Results showed that 86% of the *Aeromonas* isolates included in this study were able to grow at 10°C after 10 days. However, only 31.8% of the strains exhibited relevant growth after 15 days at 4°C. Most *A. hydrophila* strains (54%) exhibited just weak growth or not grow at 4°C. Kinetic results showed that *A. hydrophyla* growth at low temperatures occurs after a long lag phase that varied between less than 24h to 264h at 10°C, and between 144h to 336h at 4°C. The increment of lag phase was accompanied by a drastic reduction of maximum growth yield (μ_m). The tolerant strains attained stationary phase after 96h at 10°C and initiated growth at 4°C after 144h to 216h. Kinetic profiles obtained at 4°C with *Aeromonas* pre-grew at 10°C and 30°C showed that these bacteria pre-adapted to low temperature reducing the lag phase, but not the generation time. Low growing temperatures negatively affect both haemolytic and proteolytic extracellular activities. However, the cold tolerant strain M4 maintained relatively high activities even growing at 4°C.

Keywords: *A. hydrophila*, cold tolerance, virulence factors.

1   Introduction

Aeromonas are a ubiquous group of aquatic bacteria currently associated with animal and human diseases. Epidemiological studies have shown that *Aeromonas* are mainly transmitted by food and water, causing gastric infections on childrens and adults [1, 2]. Moreover, several studies also considered the possibility of intoxication by the consumption of pre-formed *Aeromonas* toxins present in contaminated food [3]. Their pathological potential is associated with the production of toxins, invasins (proteases, lipases, DNAses) and biofilms [4].

Although they are sometimes considered as opportunistic human pathogens and reported to be susceptible to food processing procedures, the Food and Drug Administration now considers *Aeromonas* hydrophila as an emerging foodborne pathogen of concern [5].

Their high frequency in processed and non-processed foods may be associated, in part, with their ability to tolerate low temperatures, high salt concentrations, and low pH, conditions usually used for food preservation. This is supported by the detection *Aeromonas* growth and toxin production in contaminated food maintained at low temperatures [6, 7]. This characteristic posses a serious problem for food preservation enhancing the risk of contamination and intoxication, and reducing the market validity of the products.

Low temperature leads to a decrease of membrane fluidity that affects membrane-associated cellular functions. The organisms overcome this problem by decreasing the degree of saturation in the membrane phospholipids to attain more flexibility. Furthermore, cold shock also causes stabilization of the secondary structure of RNA and DNA resulting in reduced efficiency of translation, transcription and DNA replication [8]. Bacterial cold tolerance involves to distinct steps, a first shock response characterized by the synthesis of cold shock proteins like cspA of *A. hydrophila* [9], and general stress response, and an adaptation step that involves specific proteins like.

In this context, the present work aimed to analyze *Aeromonas* growth and patogenicity factors production (haemolysins and proteases) under low temperatures. Eleven representatives of different *Aeromonas* species, and 11 *A. hydrophila*, including clinical and environmental isolates, were evaluated. Screening and kinetic experiments were conducted at 4, 10 and 30°C on M9 minimal medium until the bacteria reached the stationary phase.

2   Material and methods

Eleven epresentatives of different *Aeromonas* species, and 11 *Aeromonas hydrophila*, including clinical and environmental isolates, were evaluated. Experiments were conducted on M9 minima medium to avoid the effect of potential protectors on the culture media. *Aeromonas* were grown at 4, 10 and 30°C with no shaking. Growth was monitored by the increase on D.O. at 560nm. Cell viability after temperature downshift (4°C- 6h) was evaluated by the appropriate dilution of bacterial suspensions and plating on LB medium. *A. hydrophila* strain OC was used just in the first experiments due to its poor growing on minimal medium.

Selected bacterial strains (M3, M4, ATCC7966 and NCIB9233) were grown on M9 medium at 4, 15, and 30°C until stationary phase. Supernatant was collected by centrifugation and filtration (0.22 μm). Crude extracts were used to evaluate the enzymatic activities. Extracellular haemolytic and proteolitic activity were assayed using rabbit blood, and azocasein, respectively, as described by Handfield et al [10] and Shift et al. [11].

Total proteins were evaluated by the Bradford method [12].

3   Results and discussion

A first screening (Figure 1) showed that 86% of the *Aeromonas* isolates included in this study were able to grow at 10°C after 10 days. However, only 31,8% of the strains exhibited relevant growth after 15 days at 4°C.

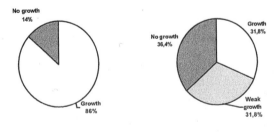

Fig. 1. Growth of *Aeromonas* strains on M9 medium at low temperatures. A- 10°C (10 days), and B- 4°C (15 days). No growth − less than 0.1 OD_{560nm} units, Weak growth − 0.1 to 0.2 OD_{560nm} units, Growth->0.2 OD_{560nm} units

The cold tolerant strains were the representatives of *A. salmonicida*, *A. media*, *A. ichtiosmia*, *A. encheleia*, *A. allosaccharophila*, and just two of the 11 strains of *A.*

hydrophila (M4 and NCIB 9233). Most *A. hydrophila* strains (54%) exhibited just weak growth at 4°C, and three of them (M3, M1 and OC) did not grow at all.

Table 1. Kinetic parameters of 10 isolates of *A. hydrophila* grown on M9 medium at 37°C, 10°C e 4°C.

	Temperature	Lag (h)	μ_m*		Temperature	Lag (h)	μ_m
CECT191	37°C	1	0,1750	FC	37°C	>1	0,1665
	10°C	>24	0,0052		10°C	24	0,0041
	4°C	216	0,0008		4°C	216	0,0005
M3	37°C	1	0,1842	ATCC7966	37°C	>1	0,1748
	10°C	264	0,0005		10°C	48	0,0033
	4°C	---	0,0000		4°C	264	0,0003
ATCC14468	37°C	1	0,1562	NCIB9233	37°C	1	0,1706
	10°C	24	0,0039		10°C	>24	0,0049
	4°C	168	0,0005		4°C	144	0,0015
CECT839	37°C	>1	0,1773	M2	37°C	>1	0,1663
	10°C	168	0,0014		10°C	>24	0,0022
	4°C	336	0,0007		4°C	192	0,0004
M4	37°C	1	0,1705	M1	37°C	1	0,1486
	10°C	>24	0,0042		10°C	192	0,0012
	4°C	192	0,0012		4°C	---	0,0000

* Maximum growth yield – D.O. units x h^{-1}

The kinetic results (Table 2) showed that *A. hydrophyla* growth at low temperatures occurs after a long lag phase that varied between less than 24h to 264h at 10°C, and between 144h to 336h at 4°C. The increment of lag phase was accompanied by a drastic reduction of maximum growth yield (μ_m). The tolerant strains NCIB9233 and M4 showed faster growth attaining stationary phase after 96h at 10°C and initiating growth at 4°C after 144h and 216h, respectively.

The high bacterial viability (>95%) observed after cold shock (4°Cm 6h) indicates that *Aeromonas* growth at low temperatures depends on the capacity to adapt metabolic machinery to the stress condition, rather than the capacity to resist the temperature downshift. However, longer periods at low temperature (4°C, 15 days) lead to a drastic reduction of the viability of non-growing strains, indicating that, although not 100% efficient, freezing temperatures may be considered as an important control system against most *Aeromonas* food contaminants.

Kinetic profiles obtained at 4°C with *Aeromonas* pre-grew at 10°C and 30°C showed that these bacteria pre-adapted to low temperature reducing the lag phase, but not the generation time. These results are expected as low temperatures drastically affect not only cell metabolism by reducing enzymatic activities, but also the uptake of organic and mineral nutrients by the reduction of substrate affinity [13].

Kinetic data differs from those obtained by Palumbo et al. [14] on Heart Infusion Broth (HIB) indicating that culture medium highly influence *Aeromonas* response to low temperatures, specially in terms of generation time and maximum growth yield.

Figure 2 shows the growth behavior of four *Aeromonas* strains at 15 and 4°C. As can be observed, M3 strain growth at 15°C after a lag fase of 48h, and did not growth at 4°C. Conversely, strains M4, ATCC7966 and NCIB9233 exhibited good growth at 15°C reaching an OD_{560} of 0.6 to 0.7 units (8-9 x 10^7 cells/ml) after 48 to 96 h. Growth at 4°C

278

was preceded by an adaptation period of 72 to 96h after which the three strains grow exponentially but with a drastic reduction of the generation time.

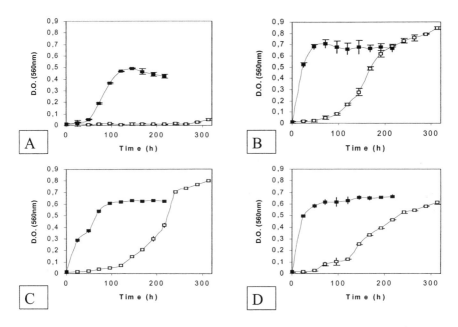

Figure 2. Growth curves at 4°C (open symbols) and at 15°C (full symbols) *A. hydrophila* strains (A) M3, (B) M4, (C) NCIB9233 e (D) ATCC7966.

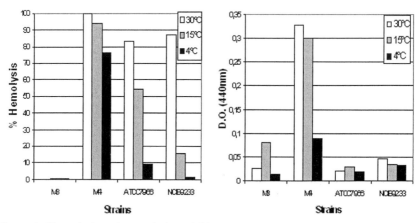

Figure 3. Haemolytic and proteolytic activities exhibited by stationary phase cultures of four *A. hydrophila* strains grown at 4, 15 and 30°C.

To investigate the effect of temperature downshift and bacterial growth under low temperatures on the production of extracellular enzymes involved in *Aeromonas* pathogenicity, extracellular haemolytic and proteolytic activities of crude extracts were evaluated. Results present in Figure 3 shows that low growing temperatures negatively affect both haemolytic and proteolytic extracellular activities.

The reduction on extracellular enzymatic activity varied among the strains independent of their growing ability at low temperatures. In this sense, NCIB9233 considered as a cold tolerant strain exhibited a drastic reduction on its haemolytic activity at 15 and 4°C, whereas M4 showed a moderate reduction on the same conditions. Cold tolerant strain M4 maintained relatively high activities (>70% of haemolytic activity) even growing at 4°C, indicating that depending on the strain cold tolerance *Aeromonas* can maintain their virulence potential, and can produce extracellular proteins like haemolysin-aerolysin, one of the responsible for human intoxication by contaminated food.

The evaluation of specific activity (enzyme activity per mg protein) of the crude extracts showed that growing temperature does not influence the specific activity, and that the reduction observed in overall activity is due to a decrease in enzyme production or secretion.

Acknowledgments

This work was founded by FAPERGS and UCS. Authors thanks Dr. Naharo for the representatives of *Aeromonas species*.

References

[1]   B.J.Freij. Pediatr. Infect. Dis. *3 (1984), p.164*
[2]   S.M. Kirov. Int. J. Food Microbiol. *20 (1993), 179.*
[3]   K.N. Majeed, A.F. Egan, I.C. Mac Rae. *J. Appl. Bacteriol.* **69** (1990), 332
[4]   S.L. Abbott, L.S. Seli, J.M. Catino, M.A. Hartley, J.M. Janda. *J. Clin. Microbiol.* **36** (1998), 1103.
[5]   J.H. Isonhood, M. Drake. *J. Food Protect.* **65** (2002), 575.
[6]   L.R. Beuchat. *Int. J. Food Microbiol.* **13** *(1991), p.217*
[7]   K. Krovacek, A. Faris, I. Mansson. *Int. J. Food Microbiol.* **13** *(1991), 165*
[8]   S. Phadtare, J. Alsina,M. Inouye. *Curr. Op. Microbiol.* **2** *(1999), 175.*
[9]   K.P. Francis, G.S. Stewart. *J. Ind. Microbiol. Biotechnol.* **19** *(1997) 286.*
[10]   M. Handfield, P. Simard, M. Couillard, R. Letarte. *Appl. Environ. Microbiol.* **62** (1996), 3459.
[11]   S. Swift, M.J. Lynch, L. Fish, D.F. Kirke, J.M. Thomas, G.S.A.B. Stewart, P. Williams. *Infect. Immun.* **67** (1999), 5192.
[12]   M. Bradford. *Anal. Biochem.* **72** (1976), 248.
[13]   D.B. Netwell. *FEMS Microbiol. Ecol.* **30** (1999), 101.
[14]   S.M. Palumbo, A.C. Williams, R.L. Buchanan, J.G. Phillips. *J. Food Protec.* **54** (1991), 429.

Comparison of two processes for isolation of exopolysaccharide produced by *Lactobacillus acidophilus*

Ana. I. E. Pintado*,[1], K. Truszkowska[2], José A. Ferreira[3], Manuela M. E. Pintado[1], Ana. M. P. Gomes[1], Manuel A. Coimbra[3] and F. Xavier Malcata[1]

[1] Escola Superior de Biotecnologia, Universidade Católica Portuguesa, Rua Dr. António Bernardino de Almeida, P-4200-072 Porto, Portugal
[2] Warsaw Agricultural University, Poland
[3] Universidade de Aveiro, P-3810-193 Aveiro, Portugal

*Corresponding author: e-mail: apintado@esb.ucp.pt, Phone: +351225580045, Fax: +351225090351

Exopolysaccharides (EPS) are currently employed as additives in a wide variety of food products, in which they serve as thickening, stabilizing, emulsifying or gelling agents. On the other hand, probiotic microorganisms that belong to a group of lactic acid bacteria — which nowadays are recognized to exhibit beneficial effects upon health, produce EPS. In this work, the strain *Lactobacillus acidophilus* LAC has been examined for its EPS production capacity; two process of EPS isolation were also assessed, via testing of two alternative routes for precipitation of medium proteins, using xanthan gum as control. The treatment with sulfosalicylic acid to remove proteins during EPS isolation revealed a decrease of mannoprotein co-precipitation, whereas treatment with trichloroacetic acid decreased the co-precipitation of lactose and produced less variability in EPS quantification.

Keywords: Carbohydrates, lactic acid bacteria, polymer

1   Introduction

Many bacteria are known to produce exopolysaccharides (EPS) – which are either excreted in the growth medium, or remain attached to the bacterial cell wall thus forming capsular EPS [1]. A large variety of EPS can be produced by lactic acid bacteria (LAB); in fact, EPS-producing LAB play an important role in the dairy industry, because of their contribution to attain specific rheological and textural profiles in dairy products. Furthermore, LAB are recognized as safe for human consumption, and EPS produced thereby has strongly been suggested to provide health benefits. It has indeed been speculated that the increased viscosity of EPS – containing foods may extend the residence time of ingested fermented milk in the gastrointestinal tract, and therefore be beneficial to transient colonization by probiotic bacteria [2]. The total yield of EPS produced by LAB depends on medium composition and growth conditions. The different broth media may be a source of contaminants – viz. polysaccharides and proteins [3], that may induce false results; hence, attempts to isolate and characterise EPS from broth media using different isolation procedures are in order, in attempts to eliminate interferences caused by the medium. In this work, the ability of *Lb. acidophilus* to produce EPS in MRS medium, and two different processes for EPS isolation were investigated.

2   Materials and methods

2.1 Bacterial strains and culture conditions

Lactobacillus acidophilus LAC was obtained as a freeze-dried concentrated starter culture, marketed by DSM Food Specialties (Australia). The strains were maintained frozen (-80 °C), were sub-cultured in 20 ml of MRS broth, and incubated anaerobically at 37 °C for 48 h. Before experimental use, the culture was propagated twice in the same medium and under similar environmental conditions.

2.2 Fermentation performance

Fermentations were carried out in a Braun Biostat B 2-L fermentor, filled with 1.5 L of MRS broth containing 2 %(w/v) lactose. The yield of EPS and the two isolation methods were tested by adding xanthan gum to the medium. The experiments were carried out at 37 °C, at 150 rpm, at pH 5.5 and under N_2. A 5 %(v/v) standard inoculum was prepared from a subculture of *Lb. acidophilus*, previously grown in the corresponding medium for 20 h at 37 °C, and was used to start-up every fermentation batch. Fermentation batches lasted up to 48 h.

2.3 Sampling, growth determination and chemical composition

Samples were aseptically withdrawn from the fermentor at (0, 2, 4, 6, 8, 10 12, 24, 26, 28, 30, 32, 34 and 48 h). Growth was monitored spectrophotometrically (OD at 650 nm) and as viable counts (after plating on MRS agar). The weight of isolated and dried polymers was measured, and the total amount of carbohydrates was determined by the phenol-sulphuric acid colorimetric test [10], using glucose as standard. The total residual sugars (viz. lactose and glucose) and lactic acid were determined by high pressure liquid chromatography (HPLC) (Merck), with detection by refractive index at 30 °C for the disaccharides and U.V. absorbance for the organic acid. The colunmused was an Aminex HPX-87x, the eluant was 5 mM H_2SO_4 at a flow rate 0.6 mL/min, the sample injection volume was 50 μL, and the separation temperature was 40 °C. Prior to analysis, all samples were pretreated in order to eliminate protein interference; 1 mL of sample was accordingly precipitated with 200 μL of 35 %(v/v) perchloric acid (PCA), allowed to stand for 10 min in ice, added with 55 μL of 7.0 M KOH (to neutralize PCA) and centrifuged for 10 min at 4000 rpm, and the supernatant was filtered through a 0.2 μm membrane filter [4].

For monosaccharides analysis, all samples taken at 48 h of fermentation were dyalised using a membrane with a cutt-off of 12-14 kDa. Neutral sugars were released by Saeman hydrolysis [5], and analyzed as their alditol acetates by GLC [6, 7] using a Carlo Erba 6000 apparatus with a split injector (split ratio 1:60) and a FID detector. A 30 m — column DB-225 (J&W), with 0.25 mm inner diameter and 0.15 μm film thickness, was used. The injector and detector temperatures were 220 and 230 °C, respectively. The oven temperature program used was: 220 °C for 4 min, followed by 230 °C for 6.5 min at a rate of 25 °C min^{-1}. The flow rate of carrier gas (H_2) was set at 1 mL min^{-1}, at 220 °C. Hexuronic acids (HexA) were determined colorimetrically according, to a modification [8] of the method of Blumenkrantz and Asboe-Hansen [9].

2.4 Isolation of exopolysaccharide

To optimise EPS isolation, cells and residual polypeptides were removed by centrifugation (at 4000 rpm for 20 min). Aliquots of the cultures (10 mL) were aseptically withdrawn, and precipitation of proteins was via one of two different processes: addition of pronase E solution and one volume of 20 %(w/v) trichloroacetic acid (TCA), or precipitation with 2 %(w/w) 5-sulfosalicylic acid, followed by heating and stirring for 1 h. Cells and precipitated proteins were removed by centrifugation (at 4000 rpm, for 20 min, at 4 °C). The clear supernatant was collected, and the EPS was precipitated overnight of with three volumes of cold ethanol, followed by centrifugation of the precipitate (at 4000 rpm, for 20 min, at 4 °C). Afterwards, the pellet was recovered, and EPS were duly lyophilized and weighed.
All determinations were done in duplicate, and expressed as mean values and associated standard deviations.

3   Results and discussion

3.1 Bacterial growth and exopolysaccharide production

The fermentation profiles of the various experiments — depicted in Fig. 1, revealed that the stationary growth phase occurred between 12 and 24 h for all fermentation. However, slight variations were observed in the growth rates: when one considered fermentations with addition of xanthan, the growth rates were lower whereas when one compared fermentations without addition of xanthan, it became clear that xanthan inhibits growth. The growth rate was 0.29 h^{-1}, with a final cell number of 3.65×10^8 cfu/ml, for medium without adding xanthan gum and precipitated via 20 %(w/v) TCA; 0.16 h^{-1}, with a final cell number of 2.94×10^9 cfu/ml, for medium without xanthan gum and precipitated via 2 %(w/w) 5-sulfosalysilic acid; 0.08 h^{-1}, with a final cell number of 3.17×10^8 cfu/ml, for medium with xanthan gum and precipitated via 20 %(w/v) TCA; and 0.04 h^{-1}, with a final cell number 3.47×10^8 cfu/ml, for medium with xanthan gum and precipitated via 2 % (w/w) 5-sulfosalysilic acid.

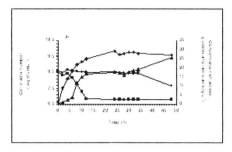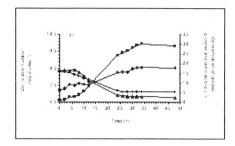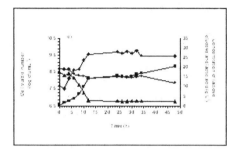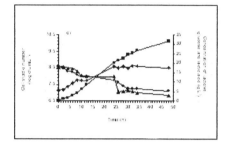

Fig. 1 Fermentation profile of Lb. acidophilus, in terms of cell viable numbers and concentrations of lactose, glucose and lactic acid, at 37° C and constant pH 5.5 – ◆ cell number (log cfu mL-1); ●, lactose; ■, glucose, and ▲, lactic acid (g L-1). a) MRS medium without xanthan gum and with precipitation via 20 %(w/v) TCA; b) MRS medium with xanthan gum and with precipitation via 2 %(w/w) 5-sulfosalicylic acid; c) MRS medium without xanthan gum and with precipitation via 2 %(w/w) 5-sulfosalicylic acid; d) MRS medium with xanthan gum and with precipitation via 20 % (w/v) TCA.

Both isolation methods yielded similar results, in terms of final EPS yield – but variations in EPS production were noticed (see Fig .2). EPS quantifications in the initial experiments indicate that yeast extract, beef extract and proteose peptone can interfere on EPS quantification – as also suggested by *Kimmel* and *Roberts* [11]. These interferents are more problematic at the beginning of fermentation, however, if the strain uses carbohydrate-reacting material during growth, the concentration thereof will be lower.

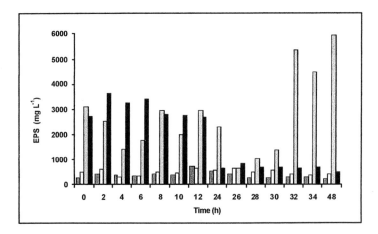

Fig. 2 Crude EPS production, at 37° C and constant pH 5.5, in MRS medium: ▨, EPS production (mg/l) in MRS medium without xanthan gum and with precipitation via 20 %(w/v) TCA gum; □ , EPS production (mg/l) in MRS medium without xanthan gum and with precipitation via 2 %(w/w) 5-sulfosalicylic acid; ▨ , EPS production (mg/l) in MRS medium with xanthan gum and with precipitation via 20 %(w/v) TCA; ■ , EPS production (mg/l) in MRS medium without xanthan gum and with precipitation via 2 %(w/w) 5-sulfosalicylic acid.

Sugar and lactic acid analysis

Monitoring of residual sugars and lactic acid (see Figs. 1, 2, 3 and 4), one observed that in all batch fermentations the main sugar converted into lactic acid was glucose – which is converted homofermentatively to lactic acid, via glycolytic degradation; in all experiments by 48 h, glucose was essentially exhausted. Lactose started to be consumed later than glucose in the absence of xanthan; however, in presence of xanthan, it was consumed earlier and at higher rates, thus leading to low residual lactose by 48 h. The contents of lactic acid measured in all experiments were very high, and lied between 21 and 32 g L^{-1} by 48 h.

Table 1 Monosaccharide composition of EPS isolated by 48 h of fermentation, following dialysis.

Samples	Mass	Sugar composition (mol%)								Total sugars
	(mg)	Rha	Fuc	Ara	Xyl	Man	Gal	Glc	HexA	(mg g-1)
1	303			11		70	5	8	6	446
2	399			5		44	23	23	5	346
3	1036					60	9	19	12	520
4	956			1		36	31	42	13	644

1- MRS medium without xanthan gum and with precipitation via 20 %(w/v) TCA.
2- MRS medium without xanthan gum and precipitation via 2 %(w/w) 5-sulfosalicylic acid.
3- MRS medium with xanthan gum and with precipitation via 20 %(w/v)TCA.
4- MRS medium with xanthan gum and with precipitation via 2 %(w/w) 5-sulfosalicylic acid.

Acid hydrolysis and sugar analysis of the isolates revealed mannose as the major sugar component (36-70 % of sugars); however, glucose, galactose, uronic acid and arabinose were also present (Tab. 1). This realisations showed that both techniques probably are ineffective to remove mannoproteins from the MRS medium. The samples precipitated with sulfosalicylic acid showed lower percentage of sugars (below 10%) than in samples precipitated with TCA; nevertheless, the fraction of mannoproteins present in samples precipitated with sulfosalicylic acid was lower than in TCA — precipitated samples, and richer in glucose and galactose (probably derived from lactose hydrolysis). The increase in polymeric material in samples 3 and 4 was due to addition of xanthan to the fermentatiom medium.

Acknowledgements

Financial support for A.I.E.P was through a PhD fellowship by FCT (Fundação para a Ciência e Tecnologia, Portugal) (SFRH/BD/5212/2001). Financial support for the research effort presented herein was partially provided support by PROCHEESE: PRObiotic CHEese from whEy with in Situ-generated Exopolysaccharide (POCTI/1999/AGR/36163) administered by FCT, and PROBIOSORO — Utilização de matrizes obtidas a Partir do lactossoRo para incorpOração de estirpes proBIÓticas, como procesSo de valORizaçao daquele subprodutO administered by AdI.

References

[1]   I. W. Sutherland, Advances in Microbiology and Physiology, 8: 143-213 (1972).
[2]   B. German, E. Schiffrin, R. Reniero, B. Mollet, A. Pfeifer, and J.R. Neeser, Trends in Biotechnology, 17, 492-499 (1999).
[3]   A. Laws, Y. Gu, and V. Marshall, Biotechnology Advances, 19: 597-625 (2001).
[4]   M. E. Pintado, A. I. E. Pintado, J. A. Lopes da Silva, and F.X. Malcata, Carbohydrate Polymers, 37: 1-6 (1998).
[5]   R. R. Selvendran, J. F. Marchand, and S. G. Ring, Analytical Biochemistry, 96: 282-292 (1979).
[6]   A. B. Blakeney, P. J Harris, R. J Henry, B. A. Stone, Carbohydrate Research, 113, 291–299 (1983).
[7]   P. J. Harris, A. B. Blakeney, R. J. Henry, and B. A. Stone, Journal American Official Analytical Chemistry International, 71, 272–275 (1988).
[8]   M. A. Coimbra, I. Delgadillo, K.W. Waldrom and R. R. Selvendran. In H. F. Linskens and J.F. Jackson (eds.), Modern Methods of Plant Analysis Vol. 17, pp. 19-44. (Berlin, Springer-Verlag.) (1996).
[9]   N. Blumenkrantz, and G. Asboe-Hansem, Analytical Biochemistry, 54: 484-489 (1973)
[10]   M. Dubois, K. A., E. Gilles, J. K. Hamilton, P. A. Rebers, and F. Smith, Analytical Chemistry, 28: 350-356 (1956).
[11]   S. A. Kimmel, and R. F. Roberts, International Journal of Food Microbiology, 40: 87-92 (1998)

Detection of *Listeria monocytogenes* from fresh and marine water fish using Real Time PCR, PCR and standard ISO methods

C. Salamoura[1], C. Papadopoulou[1], G. Filioussis[1], C. Dontorou[1], G. Zakas[1], S. Levidiotou[1], M. Malamas[2]

[1]*Food, Water, Environmental Microbiology Unit, Microbiology Department, Medical School, University of Ioannina, Greece*
[2]*Department of Pharmacology, Medical School, University of Ioannina, Ioannina, Greece*

The objectives of the present study were the assessment of *Listeria monocytogenes* contamination of fish and the evaluation of standard ISO methods and modern molecular techniques. Two hundred and fifty five samples of fresh and marine water edible fish were examined for *L. monocytogenes* contamination using the ISO 10560 cultural method and two different molecular techniques (PCR, RT PCR). The samples were collected from local fish markets and local aquacultures. Half-Fraser broth and Fraser broth were used for pre-enrichment and enrichment reciprocally, while Oxford and PALCAM agar (ISO 10560) were used as selective agars. Two PCR automated systems, were employed for the molecular detection. Seven strains of *L. monocytogenes* isolated using the cultural techniques, were also detected by both PCR systems. One strain of *L. monocytogenes* (serotype 4) was isolated from a fresh fish (cephalus) of local capture, one strain *L. monocytogenes* (serotype 4) was isolated form imported fresh salmon filet and four strains *L. monocytogenes* (serotype 1) were isolated from smoked rainbow trout. All isolates were susceptible to routinely used antibiotics. The results indicate the possible risk of *L. monocytogenes* infection transmitted from fish to humans through inadequately cooked seafood meals and/or through contamination of culinary tools. The results also suggest that molecular techniques accelerate considerably the recovery time of *L. monocytogenes* from food, enabling a more efficient safeguarding of public health with regard to foodborne infections.

Keywords: *Listeria monocytogenes*, ISO methods, PCR, Real Time PCR, fish

1   Introduction

Listeria monocytogenes is a Gram positive, aerobic and/or microaerophilic rod, isolated from humans, animals, environmental samples (soil, water) and a variety of foods (Gray et al, 1966). Epidemiological investigations following major epidemics (Schlech et al, 1983, Fleming et al, 1985) have shown that epidemic listeriosis is a food borne disease. The more commonly implicated foods are milk, cheese, meat, seafood, fruits and vegetables (Dalton et al, 1997). Also *L. monocytogenes* has been isolated from production lines of fresh and cold-smoked fish (Farber 1991, Dillon et al. 1992, Rorvik et al. 1995, Vaz-Velho et al. 1998, 2000) and thus seafood has been included as potential source of *Listeria* infection in humans. The numbers of *L. monocytogenes* cells in foods, environmental and clinical samples are very small and the selection of appropriate methodology for the detection of the bacterium is very important. Two selective agar media, Oxford and PALCAM, recommended by ISO-10560 (ISO 10560, 1999), are suggested for detection of *L. monocytogenes* in foods. Other methods for the detection of *L. monocytogenes* from foods include immunofluorescence, immunoenzymic methods (ELISA, mini-VIDAS), immunomagnetic separation (IMS) and molecular methods such as PCR, RT PCR and LCR (Candrian, 1995, Simon et al, 1996) The objectives of the present investigation were the comparative study of the standard cultural techniques vice the modern molecular methods used for the detection of *L. mococytogenes* and surveillance of *L. monocytogenes* contamination of fish in Greece.

2   Material and methods

The fish samples

Two hundreds and fifty five samples of fresh and marine water edible fish were collected from local fish markets and local aquacultures. Specifically there were collected 30 samples from each of the following fish species *Oncorhynchus mykiss* (Rainbow trout), *Salmo salar* (Salmon), *Sparus aurata* (Sea Bream), *Disentrarchus labrax* (Bass), *Mugil cephalus* (Cephalus), *Diplodus puntazzo,* plus 30 samples of smoked salmon and 45 samples of smoked rainbow trout. All samples were transported to the laboratory at 4°C inside portable insulated boxes and the fresh fish samples were analyzed upon arrival at the laboratory, while the processed fish (smoked) samples were maintained at 4°C until they were further processed for *L. monocytogenes* detection.

The microbiological examination procedure

I. ISO cultural techniques
From each sample 25g of fish flesh were placed into 225ml half-Fraser broth (Merck, Cat. No. 1.10398) and were homogenised in a stomacher for 1–2min according to the following procedure:

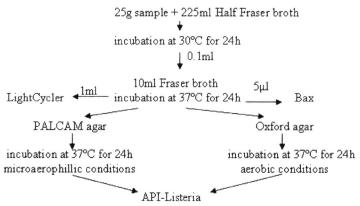

II. The molecular techniques
1. Equipment
Bax® system
The Bax® pathogen detection system is an automated microbial screening method for food and environmental samples, that eliminates the need for gel electrophoresis and photo documentation. The PCR tablet contains fluorescent dye, which binds with DNA and emits

a fluorescent signal in response to light. After amplification, the Bax® system begins to measure the fluorescent signal. During detection, the temperature of the samples is raised to the point, where the DNA strands denature, releasing the dye and lowering the signal. This change in fluorescence can be plotted against temperature to generate a melting curve, which is interpreted by the Bax® system software.

LightCycler
The LightCycler instrument consists of a cycler and a fluorimeter component. The cycler component has been optimized for rapid PCR applications. Unlike conventional PCR, in which cycling programs take several hours, PCR analyses conducted with the LightCycler take only 40min. The Real Time Fluorimeter (RTF) allows continuous monitoring of fluorescence without running an Experimental Protocol. One sample at a time can be monitored. The graph at the screen shows temperature and fluorescence output of a sample over time. The use of different samples and standards gives the opportunity to determine the initial concentrations of the microbial of the unknown samples
2. DNA extraction
The extraction of DNA for each molecular system was carried out following the instructions of the producer company.

3   Results

The results are shown in Tables 1 and 2.

Table 1. Number of isolated *L. monocytogenes* strains determined from raw and processed fish samples analyzed by cultural and PCR methods.

Raw and processed Fish samples analyzed	No. of samples	Number of isolated *L. monocytogenes* strains	Positive samples (%)
Oncorhynchus mykiss	30	0	0
Salmo salar	30	2	6,7
Sparus aurata	30	0	0
Disentrarchus labrax	30	0	0
Mugil cephalus	30	1	3,3
Diplodus puntazzo	30	0	0
Smoked rainbow trout	45	4	8,9
Smoked salmon	30	0	0
Total	**255**	**7**	**18.9**

Table 2. Serotype of isolated *L. monocytogenes* strains and their sensitivity to antibiotics

Raw and processed Fish	Number of isolated *L. monocytogenes* strains	Serotype	Antibiotics	
			Ampicillin	Penicillin
Mugil cephalus	1	4	S	S
Salmo salar	2	1	S	S
Smoked rainbow trout	4	1	S	S

S: Sensitive

4   Discussion

Standard culture methods ISO 10560 are currently used to detect *L. monocytogenes* in food, although they are far from being 100% effective (Donnelly, 2001). The isolated strains of *L. monocytogenes* were confirmed by the two molecular methods employed. The results demonstrate that molecular methods surpass cultural techniques in speed and accuracy. The time required for the LightCycler to finish the PCR runs, was 40min, the time for the Bax® system was 4hrs, while the cultural methods required at least 24hrs for the appearance of suspect colonies. The combination of automated PCR systems and standard cultural methods enables the detection of viable pathogens necessary for the food risk assessment. The potential risk of *L. monocytogenes* infection transmitted to humans through the consumption of inadequately cooked or mal-processed fish has to be emphasized. The detection of *L. monocytogenes* in processed fish (smoked fish) may be attributed either to defective processing or to post-processing contamination due to environmental infestation. The cold-smoking process does not generate sufficient heat to inactivate *Listeria* organisms which may be present on fish (Guyer and Jemmi, 1991, Eklund et al, 1995). In addition the psychotropic nature of this bacterium, allows survival and even multiplication during refrigeration storage.

References

[1]   Candrian U., 1995. Polymerase chain reaction in food microbiology. J. Microbiol. Methods 23:89-103.
[2]   Dalton, C.B., Austin, C.C., Sobel, J., Hayes, P.S., Bibb, W.F., Graves, L.M., Swaminathan, B., Proctor, M.E., Griffin, P.M., 1997. Listeriosis from chocolate milk: linking of an outbreak of febrile gastroenteritis and "sporadic" invasive disease. N. Engl. J. Med. 336:100-105.
[3]   Dillon, R., Patel, T., Ratman, S., 1992. Prevalence of *Listeria* in smoked fish. J. Food Prot. 55:866-870.
[4]   Donnelly, C.W., 2001. *Listeria monocytogenes*: a continuing challenge. Nutr. Reviews. 59: 183-194.
[5]   Eklund M.W., Poysky F.T., Paranijpye R.N., Lashbrook L.C., Peterson M.E., Pelroy G.A. (1995) Incidence and sources of *Listeria monocytogenes* in cold-smoked fishery products and processing plants. J. Food Prot. 58:502-508.

[6]   Farber, J., 1991. *Listeria monocytogenes* in fish products. J. Food Prot. 54:922-924, 934.

[7]   Fleming, D.W., Cochi, S.L., MacDonald, K.L., 1985. Pasteurized milk as a vehicle of infection is an outbreak of listeriosis. N Engl J Med 312: 404-407.

[8]   Gray, M.L., Killinger, A.H., 1966. *Listeria monocytogenes* and listeric infections. Bacterio Rev 30: 309-382.

[9]   Guyer, S., Jemmi, T., 1991. Behavior of *Listeria monocytogenes* during fabrication and storage of experimentally smoked salmon. Appl. Environ. Microbiol. 57:1523-1527.

[10]   Herman L. (1997) Detection of viable and dead *Listeria monocytogenes* by PCR. Food Microbiol. 14:103-110.

[11]   ISO 10560 (1999) Milk and milk products – detection of *Listeria monocytogenes*. Geneva: International Organization for Standardization.

[12]   Masters C.I., Shallcross J.A., Mackey B.M. (1994) Effect of stress treatments on the detection of *Listeria monocytogenes* and enterotoxigenic *Escherichia coli* by the polymerase chain reaction. J. Appl. Bacteriol. 77:73-79.

[13]   Rorvik, L.M., Caugant, D.A., Yndestad, M., 1995. Contamination pattern of *Listeria monocytogenes* and other *Listeria* spp. in a salmon slaughterhouse and smoked salmon processing plant. Int. J. Food Microbiol. 25:19-27.

[14]   Simon, M.C., Gray, D.I., Cook, N., 1996. DNA extraction and PCR methods for the detection of *Listeria monocytogenes* in cold-smoked salmon. Appl. Environ. Microbiol. 62:822-824.

[15]   Schlech, W.F., Lavigne, P.M., Bortolussi, R.A., Allen, A.C., Haldane, E.V., Wort, A.J., Hightower, A.W., Johnson, S.E., King, S.H., Broome, C.V., 1983. Epidemic listeriosis – Evidence for transmission by food. N Engl J Med. 308: 203-206.

[16]   Vaz-Velho, M., Duarte, G., Gibbs, P., 1998. Occurrence of *Listeria* spp. in salmon trout (*Oncorhynchus mykiss*) and salmon (*Salmo salar*). Food Science and Technology International 4:121-125.

[17]   Vaz-Velho, M., Duarte, G., Gibbs, P., 2000. Evaluation of mini-VIDAS rapid test for detection of *Listeria monocytogenes* from production lines of fresh to cold-smoked fish. J. Microbiol. Methods 40:147-151.

Direct identification of *Campylobacter* species in poultry samples by multiplex PCR

A. González, R. Cervera, M.A. Ferrús*, **J. Hernández**

Dpto. Biotecnología. Universidad Politécnica de Valencia, Cno. Vera s.n., 46022 Valencia, SPAIN.

*Corresponding author: e-mail: mferrus@btc.upv.es

Campylobacter jejuni and *C. coli* are recognized as the most common cause of food-borne bacterial gastroenteritis in human, and poultry consumption has been implicated as a primary source of transmission. The detection and identification of these microorganisms by conventional tests is problematic and time consuming. Therefore, the objective of this study was the use of multiplex PCR for simultaneous distinction of the two species in poultry samples collected from different local markets. A total of 10 strains of *Campylobacter* were isolated from the 14 poultry samples processed. Multiplex PCR assay was applied to preenrichment and enrichment broths and to isolated strains. There was a perfect correlation between results obtained when the PCR reaction was performed directly onto the samples and when the technique was carried out on the isolates. The species-specific multiplex PCR assay is a rapid method for the detection and identification of thermotolerant *Campylobacter* species in poultry samples, and it can be a good alternative to identify isolated strains instead to biochemical identification.

Keywords: *Campylobacter*, Detection, multiplex PCR, poultry

1   Introduction

Thermotolerant campylobacters, particularly C. *jejuni* and C. *coli* are recognized as the most common human enteric pathogens causing acute bacterial diarrhoea worldwide [1]. A variety of contaminated foods and water have been implicated as vehicles for the transmission of campylobacteriosis to man. Animal origin foods are widely regarded as the main source of food-borne infection due to the presence of the organisms as part of the intestinal flora of many animals, and poultry consumption has been implicated as a primary source of transmission [2].

Detection and identification of these microorganisms by conventional biochemical tests is problematic and time consuming because they are relatively inert biochemically and have fastidious growth requirements. Isolation of campylobacters may require about 4-5 days due to slow growth and lack of a really suitable selective medium [3]. Moreover, campylobacters, like other food-borne pathogens are often stressed by non-favourable conditions such as nutrient starvation, pH in food or temperature variation and they can be transformed into non-culturable coccoid forms [4]. Accurate identification of C. *jejuni* and C. *coli* is difficult too, as they are closely related, and a large number of strains show atypical biochemical responses [5].

Over the last decade, molecular techniques such as PCR-based systems have been applied to develop improved detection methods for campylobacters in stool and food samples [6]. Due to its high sensitivity, specificity and rapid results, PCR is presented as an alternative to conventional methods. However, food samples may contain inhibitory substances with a significant effect on the activity of the *Taq* polymerase enzyme [7]. Direct PCR amplification of campylobacteria from food samples has proved to be difficult, as levels of bacterial contamination in food products are lower than in clinical samples [8]. Therefore, a short enrichment step and purification of the bacterial DNA prior PCR reaction is required for food analyses [9].

The ability of PCR to amplify specific regions of DNA has been used to identify certain campylobacteria too. A prerequisite for designing primers in any diagnostic assay is the availability of genomic sequence information, and 16S and 23S rRNA gene sequence data are widely used as a basic tool for the development of PCR assays for identifying bacteria [10].

Therefore the objective of this study was the development of a simple PCR method for simultaneous detection and identification of the two main pathogen *Campylobacter species*. A multiplex PCR assay was optimised to analyse poultry samples collected from different local markets. mPCR was compared with conventional isolation methods and direct simple PCR.

2   Material and methods

Fourteen chicken liver samples were taken from several supermarkets with intervals of at least one week. Portions of 10 gr from each sample were homogenized with Nutrient Broth (1:4 w/v) and incubated in aerobic conditions at 37°C for 2 h to stimulate bacterial growth. A 5 ml portion of this preenrichment broth was then enriched in Preston broth for 24 and 48 hours at 37°C under microaerophilic conditions.

For cultural detection, portions of 100 μl of each broth (preenrichment, enrichment at 24 h and at 48 h) were platted onto Columbia agar Base (Oxoid) supplemented with 5% defibrinated sheep blood (ASO) with cephoperazone (16mg/L) and Campylobacter Growth Supplement (Oxoid) added, and incubated at 37°C in microaerobic atmosphere. Presumptive colonies were identified by conventional phenotypic and biochemical tests (Gram stain, aerobic growth, oxidase, catalase, hippurate test and API-*Campy* system).

For PCR detection, aliquots of 1 ml of each broth were collected, harvested by centrifugation at 9000 g for 5 min, suspended in 500 μl TE buffer and processed for DNA extraction, following the CTAB method [11].

Campylobacter PCR detection assay was performed by using primers THERM1 (5'-TATTCCAATACCAACATTAGT-3') and THERM4 (5'-CTTCGCTAATGCTAACCC-3'), which amplify a 491 bp 23S rRNA fragment from thermotolerant campylobacters [12]. A final reaction volume of 50 μl was made by addition of 5 μl of each sample, 200 ng of each primer, 0.2 mM of each deoxynucleotide, 1.5 mM $MgCl_2$ and 2 U of *Taq* polymerase (Ecogen, Spain). The amplification consisted of an initial DNA denaturing step at 94°C for 12 min, followed by 45-cycle reaction (94°C for 1 min; 56°C, 1 min; 72°C, 1min). The cycling included a final extension step at 72°C for 5 min to ensure full extension of the product.

Multiplex PCR was carried out according to Denis et al. [13]. As shown in Table 1, primers amplified a 857 bp fragment of 16S rRNA gene for *Campylobacter* genus, a 589 bp fragment of *map*A gene for *C. jejuni* species and 462 bp fragment of *ceu*E gene for *C.coli* species.

Table 1: Primers used for the multiplex PCR assays [13]

Primer	Sequence	Product
MD16S1	5'-ATC TAA TGG CTT AAC CAT TAA AC-3'	857 bp 16S rRNA
MD16S2	5'-GGA CGG TAA CTA GTT TAG TAT T-3'	
MdmapA1	5'-CTA TTT TAT TTT TGA GTG CTT GTG-3'	589 bp *mapA*
MdmapA2	5'-GCT TTA TTT GCC ATT TGT TTT ATT A-3'	
COL3	5'-AAT GAA AAT TGC TCC AAC TA TG-3'	462 bp *ceuE*
MDCOL2	5'-TGA TTT TAT TAT TTG TAG CAG CG-3'	

Multiplex PCR was applied directly on the preenrichment and the enrichment broth after 24 and 48 h of incubation. DNA was extracted from 1ml of the broths as described above. The mPCr assay was also performed on bacterial colonies isolated from the agar plates: cells were harvested, suspended in in 500 µl TE buffer and processed for DNA extraction.

Optimal PCR conditions were established by testing different sample sizes, primers and MgCl$_2$ concentrations. A final reaction volume of 30 µl was made by addition of 5 µl of each sample. The PCR reaction mixture contained 0.6U Taq polymerase (Ecogen, Spain), 100 µM of each dNTP, 1.5mM of MgCl$_2$, 0.5 µM MD16S1 and MD16S2 primers, 0.42 µM of MDmapA1, MDmapA2, COL3 and MDCOL2. The amplification consisted of an initial denaturation step at 95°C for 10 min followed by 35 cycles (denaturation at 95°C for 30 s, annealing at 59°C for 1 min 30 s and extension at 72°C for 1 min), ending with a final extension step at 72°C for 10 min.

All PCR reactions were performed with an automatic thermal cycler (PHC-3 Thermal Cycler, Techne Corporation, Cambridge, UK). A negative control in which DNA was replaced with sterile distilled water was also included. DNA templates from Reference strains *C. jejuni* NCTC 11828 and *C. coli* NCTC 11366 were used as positive controls.

PCR products were analyzed by electrophoresis at 100 V for 1 h through 1% (w/v) SeaKem LE agarose (FMC Bioproducts, Denmark) gels. mPCR products were analyzed by electrophoresis at 90 V for 2 h through 1.5% (w/v) SeaKem LE agarose. Amplimers were visualized by staining with ethidium bromide under UV light. A 100 bp DNA ladder was used as a molecular weight marker.

3   Results and discussion

A total of 10 strains of *Campylobacter* were isolated from 9 out the 14 poultry samples processed. mPCR identified 3 isolates as *C. jejuni* and 7 as *C. coli* . The best rates of isolation were obtained after 24h enrichment.

A total of 13 out of the 14 chicken samples, yielded the *Campylobacter* PCR 491 bp band (Fig.1a). Direct detection by PCR was only possible in 3 samples, in the other cases a 24h enrichment was neccessary. This result agrees with that of other authors[14], and is probably due to low contamination levels in food samples. Longer enrichment times did not improve the results. Four *Campylobacter* PCR positive samples were negative by cultural analysis after 48 h of incubation. This may indicate that either *Campylobacter* was stressed,

294

remaining viable without capacity to grow on media culture, or only bacterial DNA was present in the samples. An enrichment step was included to avoid this last possibility, as dead cells-DNA is diluted during the process [6].

Multiplex PCR assay was applied to preenrichment and enrichment broths. Identification at level specie was achieved by mPCR after 48h enrichment. Simultaneous detection of *C. jejuni* and *C. coli* in mPCR assays is shown in Figure 1b. Of the 13 positive samples, 4 showed to contain *C. jejuni*, in 8 samples *C. coli* was detected and 1 sample was both *C. jejuni* and *C. coli* positive.

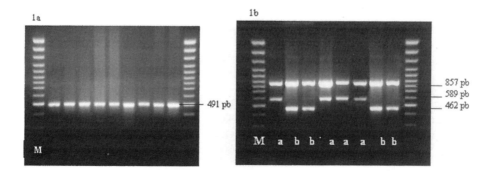

Figure 1a. Direct PCR detection of *Campylobacter* spp. in poultry samples.
1b.Direct mPCR detection of *C. jejuni (a)* and *C. coli (b)* in poultry samples. M: Molecular weight marker.

Four samples were PCR-positive although no *Campylobacter* colonies were visible on ASO-cephoperazone (16mg/l) plates, confirming the higher sensitivity of PCR versus culture [15], but 1 sample was PCR negative and positive with the traditional method. PCR inhibition has been reported in food samples [6, 8] and is, probably, the cause of this fail in detection.

There was a perfect correlation between the results obtained when the PCR reaction was performed directly onto the liver samples and when the technique was carried out on the isolates, as the specie assigned to each strain and the species detected in the sample from which the strain has been isolated were exactly the same.

In conclusion, the species-specific multiplex PCR assay is a rapid method for the detection and identification of thermotolerant campylobacters species in poultry samples, and it can be a good alternative to identify isolated strains instead to biochemical identification.

Acknowledgements

This work was supported by the GV04B-413 Research Project from Generalitat Valenciana, Consellería de Sanitat, Spain. A. G. is the recipient of a F. P. U. grant from the Ministerio de Educación y Ciencia, Spain.

References

[1]   Frost, J.A. J. Appl. Microbiol., **90**, 85S (2001).
[2]   Kramer, J. M., Frost, J. A., Bolton, F. J., Wareing, D. R. J. Food. Prot. **63**,1654 (2000).

[3]   On, S.L.W. Identification methods for Campylobacters, Helicobacters and related organisms. Clin. Microbiol. Rev. **9**, 405 (1996).

[4]   Jones, D.M., E.F. Sutcliffe, and A. Curry. J. Gen. Microbiol. **137**:2477 (1991).

[5]   Harmon, K. M., Ransom, G. M., Wesley, I. V. Mol. Cel. Probes, **11**, 195 (1997).

[6]   Denis, M., Refrégier-Petton, J., Laisney, M.-J., Ermel, G., Salvat, G. J. Appl. Microbiol. **91**, 255 (2001)

[7]   Lawson, A.J., Linton, D., Stanley, J. and Owen, R.J. J. Appl. Microbiol. **83**, 375 (1997).

[8]   Moreno, Y., Hernández, M., Ferrús, M. A., Alonso, J. L., Botella, S., Montes, R., Hernández, J. Res. Microbiol. **152**:577 (2001).

[9]   Giesendorf, B., Quint, W., Kenkens, M., Stegman, H., Huf, F., Niesters, H. Appl. Environ. Microbiol **58**, 3804 (1992)

[10]   Van Camp, G., Fierens, H., Vandamme, P., Goossens, H., Huyghebaert, A., De Wachter, R. Sys. Appl. Microbiol. **16,** 30 (1993).

[11]   Wilson, K. In: Ausubel, F.M., Brent, R., Kingston, R.E., Moore, D.D., Smith, J.A., Seidman, J.G. and Struhl, K. (eds). *Current Protocols in Molecular Biology* (John Wiley & Sons, New York, 1987)

[12]   Fermér, c., Engvall, E. O. J. Clin. Microbiol. **37**, 3370 (1999).

[13]   M. Denis, C. Soumet, K. Rivoal, G. Ermel, D. Blivet, G. Salvat and P. Colin. Let. Appl. Microbiol. **29**, 406 (1999).

[14]   Houf, K., Tutenel, A., De Zutter, L., Van Hoof, J., Vandamme, P. FEMS Microbiol. Let. **193, 89** (2000).

[15]   Lawson, A.J., Linton, D., Stanley, J., Owen, R.J. J. Appl. Microbiol. **83**, 375 (1997).

Ecology and biodiversity of microbial populations in spontaneous grape must fermentations

M. Maqueda Gil[1], M. L. Álvarez Franco[2], E. Zamora-de-Alba[2] and M. Ramírez Fernández[*],[1]

[1] University of Extremadura, Department of Microbiology, Avda. Elvas s/n, 06071 Badajoz, SPAIN
[2] Estación Enológica de Almendralejo, Conserjeria de Agricultura y Medio Ambiente, Junta de Extremadura, Carretera de Sevilla 114, 06200, SPAIN

[*]Corresponding author: e-mail: mramirez@unex.es, Phone: +34 924289426, Fax: +34 924289427

The biodiversity and population dynamics of 53 spontaneous vinifications were analyzed. The fermentation speed increased with increasing initial amounts of *Saccharomyces* yeasts. The population dynamics depended on grape sanitary quality. The amount of apiculate yeasts was significantly higher in must from deteriorated grapes. Red wine quality was significantly improved by the presence of apiculated yeasts at the beginning of the spontaneous fermentation only. This result shows that wine quality may be improved by using selected apiculated yeasts at the beginning of fermentation followed by the addition of a massive inoculum of a selected yeast to replace the apiculated yeast population.

Keywords: yeast, wine, biodiversity, must fermentation, mitDNA.

1   Introduction

Spontaneous must fermentation involves the sequential participation of different microbial populations from the vineyard and the winery [1]. Generally, it begins with the involvement of many non-*Saccharomyces* yeasts (*Hanseniaspora/Kloeckera, Pichia, Candida*, etc.). The proportion of these yeasts decreases with the increase of alcohol concentration through fermentation, and *S. cerevisiae* becomes the main yeast of must fermentation [2]. Different proportions of yeast species at each fermentation stage may affect wine quality [3].

2   Materials and methods

2.1 Materials

Fifty-three 53 spontaneous vinifications were performed using 14 grape varieties during six consecutive vintages in several wineries of south-western Spain. 27 fermentations were made at commercial wineries, and 26 at the laboratory. Monitoring of fermentation: density, °Brix, and temperature were measured each day.

2.2 Microbiological analysis

Samples from the beginning stage of fermentation (BS), tumultuous stage (TS) and final stage (FS) were analyzed. Total viable yeasts (CFU/ml) [4], *Saccharomyces* sp., non-*Saccharomyces*, apiculate yeasts, and bacteria were determined. Different *Saccharomyces* sp. strains were differentiated by RFLP of mitDNA [5], and the fragments were separated by electrophoresis in 0.7% agarose gel. The results were analyzed with the Diversity Database software package (BioRad).

2.3 Organoleptic test

Organoleptic test were performed by 12 experts. The scores were normaliced assigning 100% of preference to the wine with the highest raw score.

2.4 Statistical analysis

The data were analyzed for statistical significance with the Mann-Whitney and Spearman non-parametric tests (SPSS software package, version 11.5 for Windows, Chicago, IL).

3   Results and discussion

The amount of microorganisms in must from deteriorated grapes was significantly (p=0.034) higher than in must from healthy grapes (Table 1), mostly in the amount of apiculate yeasts (p=0.004).

Table 1. Number of viable microorganisms/ml ($\cdot 10^6$) of each microorganism type at the beginning of fermentation. Mann-Whitney test to analyze the effect of grape sanitary quality on the microbial population of spontaneous fermentations. Only the types of microorganisms with significant differences are shown.

	Grape sanitary quality		p
	Deteriorated	Good	
Total microorganisms	12.4±35.9	5.95±18.4	*
Apiculate yeasts	0.88±2.46	0.13±0.38	**

** p=0.01, * p=0.05. The data are the means of 53 independent experiments and the standard deviation.

Table 2. Number of viable microorganisms/ml ($\cdot 10^6$) of each microorganism type at each fermentation stage. Test of Mann-Whitney to analyze the effect of the place of fermentation on the ecology of spontaneous fermentations. Only the types of microorganisms with significant differences are shown.

	Installation		p
	Winery	Laboratory	
Total-BS	15.19±6.50	0.36±0.14	**
Total-TS	26 782±11 189	89.24±12.1	**
Saccharomyces-BS	14.49±31.82	0.05±0.19	***
Saccharomyces-TS	25 155±52 392	60±66.02	***
Apiculate yeasts-TS	42.6±222	31.4±42.4	***
Apiculate yeasts-FS	0±0	2.1±8.6	*
Bacteria-BS	0.00014±0.0005	0.07±0.22	***
Bacteria-TS	0.00003±0.0001	0.36±0.7	**
Bacteria-FS	0.0002±0.0012	0.43±1.1	*

*** p=0.001, ** p=0.01, * p=0.05. The data are the means of 53 independent experiments, and the standard deviation.

The population of *Saccharomyces* sp. and the total population of microorganisms in winery fermentations were significantly higher than in the laboratory ones (Table 2) at the

298

beginning and the tumultuous stages. This is probably due to an extra inoculation by resident winery yeasts from previous wine fermentations [6]. There were many apiculate yeasts at the final stage of some laboratory fermentations, probably due to the better temperature control and the lower ethanol content (11%) than at the winery, conditions that generally favour the survival of non-*Saccharomyces* yeasts [1, 6].

We observed three major general types of spontaneous fermentation dynamics (Fig.1): (a) full domination of *S. cerevisiae* from the beginning to the final stage (Fig. 1A), (b) several microorganism subpopulations at the beginning stage followed by *S. cerevisiae* domination at the final stage (Fig. 3B) (the most frequent fermentations [6]), and(4) (c) complete absence of *S. cerevisiae* (Fig. 3C) (the case of rotten grapes).

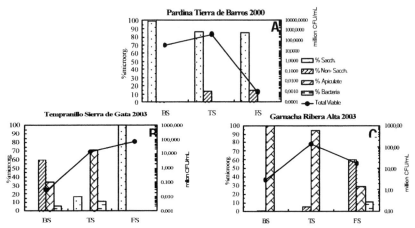

Fig. 1A, B, C. *Three major types of microorganism population dynamics among the 53 spontaneous fermentations.*

The Spearman correlations (Tables 3-5) indicate that the presence of high amounts of *Saccharomyces* prevented the growth of apiculate yeasts and bacteria (Table 3), and sped the onset of fermentation. The presence of apiculate yeasts at the tumultuous fermentation stage increased with the amount of bacteria at the beginning of fermentation (Table 4). This happened when fermentation started slowly and halted before the sugars were fully consumed, so that T100 was significantly higher (Table 5).

Table 3. Spearman correlation between the number of *Saccharomyces* (CFU/mL) and other microorganism subpopulations or T15. Only the significant correlations are shown.

	Saccharomyces- BS	*Saccharomyces*-TS	*Saccharomyces*-FS
Apiculate yeasts-TS	-0.598**	-0.443**	n.s.
Apiculate yeasts-FS	-0.437**	-0.461**	-0.385**
Bacteria-SF	-0.637**	-0.504**	Ns
Bacteria-TS	ns	-0.533**	Ns
Bacteria-FS	ns	-0.388**	Ns
T15 (days)	-0.351*	ns	Ns

** Significant correlation at 0.01 level (bilateral). * Significant correlation at 0.05 level (bilateral). ns not significant (bilateral). T15= days to consume 15% of initial sugar.

Table 4. Spearman correlation between the CFU/mL of apiculate yeasts and other microorganism subpopulations. Only the significant correlations are shown.

	Apiculate yeasts - BS	Apiculate yeasts -TS	Apiculate yeasts -FS
Apiculate yeasts -TS	0.422**	1	0.556**
Bacteria-BS	ns	0.522**	0.283*

** Significant correlation at 0.01 level (bilateral). * Significant correlation at 0.05 level (bilateral). ns, not significant (bilateral).

Table 5. Spearman correlation between the CFU/mL of bacteria and the fermentation kinetics parameters. Only the significant correlations are shown.

	Bacteria-BS	Bacteria-FS
T15 (days)	0.303*	ns
T100 (days)	ns	-0.376**

** Significant correlation at 0.01 level (bilateral).* Significant correlation at 0.05 level (bilateral). ns not significant (bilateral). T15= days to consume 15% of initial sugar. T100= days to consume 100% of initial sugar.

The presence of high amount of total microorganisms in white must fermentations reduced wine quality (Table 6). However, the presence of apiculate yeasts at the beginning of fermentation improved red wine quality, while the presence of no-*Saccharomyces* yeasts at the same stage reduced it.

Table 6. Spearman correlation between the CFU/mL of each microorganism subpopulations at each fermentation stage and the wine quality. Only the significant correlations are shown.

	White wines % Preference	Red wines % Preference
Total microorganisms - TS	-0.482*	n.s.
Non-*Saccharomyces*- BS	n.s.	-0.614*
Apiculate yeasts -BS	n.s.	0.664*

* Significant correlation at 0.05 level (bilateral). ns, not significant (bilateral).

4 Conclusions

The presence of high amounts of total microorganisms in white must fermentation reduced wine quality. Although the absolute dominance of *S. cerevisiae* improved must fermentation kinetics, red wine quality was significantly improved by the presence of apiculate yeasts at the beginning stage of fermentation. Therefore, red wine quality could be improved by using selected apiculate yeasts at the beginning of fermentation followed by the addition of a massive inoculum of selected yeast to replace the apiculate yeast population.

300

Acknowledgements

This work was funded by grant 2PR04B003 from the Extremadura Local Government, Spain. Matilde Maqueda is the recipient of a grant from the Spanish Ministerio de Educación y Ciencia.

References

[1]   Fleet, H.F., Heard, G.M. (1993) Yeast growth during fermentation. In *Wine Microbiology and Biotechnology* (G. H. Fleet, ed) Harwood Academic Publishers, Newark, N.J. 27-54.

[2]   Heard, G.M., Fleet, G.H. (1985) Growth of natural yeast flora during the fermentation of inoculated wines. *Appl. Environ. Microbiol.* **38**, 22-25.

[3]   Vianna, E., Ebeler, S.E. (2001) Monitoring ester formation in grape juice fermentations using solid phase microextraction coupled with gas chromatography-mass spectrometry. *Journal of Agricultural and Food Chemistry* **49**, 589 - 595.

[4]   Regodón, J.A., Pérez, F., Valdés, M.E., De Miguel, C., Ramírez, M. (1997) A simple and effective procedure for selection of wine yeast strains. *Food Microbiol.* **14**, 247-254.

[5]   Querol, A., Barrio, E., Ramón, D. (1992) A comparative study of different methods of the yeast strain characterization. *System. Appl. Microbiol.* **15**, 439-446.

[6]   Ribéreau-Gayon, P., Dubourdieu, D., Donèche, B., Lonvaud, A. (2000) Handbook of Enology: The Microbiology of wine and vinifications. Volume 1. (1998, ed) John Wiley & Sons Ltd, Chichester, England.

Effects of NH_4VO_3 on mitochondrial electron-transport chain and glutathione peroxidase activity of *Saccharomyces cerevisiae* UE-ME₃

Susana Magriço[1], Isabel Alves-Pereira[1,2] and Rui Ferreira [*1,2]

[1]Departamento de Química, Universidade de Évora, Apartado 94, 7002-554 Évora, Portugal
[2]Instituto de Ciências Agrárias Mediterrânicas (ICAM-UATSA), Apartado 94, 7002-554 Évora, Portugal

[*]Corresponding author: Tel.: +351 266745313; fax: +351 266745394, E-mail address: raf@uevora.pt (R. Ferreira).

The effects of two concentrations of NH_4VO_3 (25 and 75 mM) on complex I and complex III of mitochondrial electron-transport chain and glutathione peroxidase (GpX, EC 1.11.1.9) were studied in wild-type wine yeast *Saccharomyces cerevisiae*, strain UE-ME₃ grown in agar plate medium . A significantly decrease of NADH-Q_2 reductase, and GpX enzymatic activities for the 25 and 75 mM concentration and a significantly decrease of NADH-cit c reductase for the 25 mM concentration were observed in treated cells. The result show that NH_4VO_3, a pentavalent salt of vanadium (V), caused oxidative damages at mitochondrial respiratory chain of *Saccharomyces cerevisiae*, strain UE-ME₃.

Keywords: Vanadate toxicity; Heavy metal stress; NADH-Q_2 reductase; NADH-cit c reductase; glutathione peroxidase; *Saccharomyces cerevisiae*

1   Introduction

Vanadium a heavy metal from natural occurrence in the hearth crust had been described as having an important role in biological processes as component of some enzymes and other biological molecules, particularly those involved in nitrogen fixation and NADH oxidation [1–3]. Moreover, Human activities such as fuel oil burning, steel empowerment and other industrial processes contribute sometimes, to the increment of vanadium environmental levels in the air, water and food, which turns it a pollutant. Consequently, vanadium (V) in the +5 oxidation state like ammonium metavanadate (NH_4VO_3), and others heavy metals had been implicated in environmental hazards to the living beings [4].

Yeasts has been used by the humans since milliards years ago which manipulation caused a great impact on food production and the socio-economical development. Bread, beer and wine symbolize the more representative products of this manipulation along the years [5–7] in which, the yeast *Saccharomyces cerevisiae* had a great role. Beside it has been considered one of the most important microorganism for the man, *Saccharomyces cerevisiae* is one of the best known eukaryotic systems. The wine fermentation is a complex microbiologic process in which are evolved several fermentative yeasts and bacteria. Traditionally the wine is produced by natural fermentation, achieved by the yeast although being in slighter concentration in the grape surface than other microorganisms, because during the fermentation this one multiplies itself, avoiding other microorganisms proliferation [8]. The origin of these yeast strains could be the grape itself or even the wine factory environment [8]. Moreover, the capacity of wine fermentation is greatly influence by the yeast resistance to the stress conditions, including the oxidative stress [9]. In the yeast like in other eukaryotic organisms, the mitochondrial respiratory chain and by products of the cellular metabolism are the biggest source of reactive oxygen species (ROS) and the main O_2 consumer [10]. It is known that ROS are implicated in number of biological processes including apoptosis [11, 21]. Because ROS are commonplace in aerobic organisms, they have enzymatic as well as non-enzymatic defence systems against them. For example, superoxide dismutase (SOD) catalyses disproportions of O_2^- to O_2 and H_2O_2, and H_2O_2 formed as well as LOOH could be reduced to H_2O and corresponding alcohol by glutathione peroxidase (GpX). Though GpX has been evolutionary acquired by mammals, it has been demonstrated that GpX is expressed in yeast cells and plays a crucial role in the defence line against ROS [12] under *S. cerevisiae* [13].

Several studies has showed that mitochondrial-mediated reactive oxygen species generation occur mainly at complex I and complex III, a major source of oxidative stress in the cell [11, 14]. Thus, the main purpose of this work was to study the effects of NH_4VO_3, on mitochondrial electron chain transport and GpX antioxidant enzyme of wild-type wine yeast *Saccharomyces cerevisiae*, strain UE-ME$_3$.

2   Materials and methods

2.1. Microorganisms and growth conditions

The eukaryotic model used was the yeast *Saccharomyces cerevisiae* vinic UE-ME$_3$, a strain isolated from regional wine (Alentejo-Portugal) belonging to the Enology laboratory collection of Évora University. The isolated colonies of strain were stored in glycerol (30%, w/v) at -80°C. The cells were grown to mid-exponential phase in a water bath, with orbital stirring, at 28°C, in 250-ml flasks containing 100 ml of mineral medium [15] with vitamins, oligo-elements and 2% (w/v) of glucose. Exponential-phase cells were inoculated in agar plate enriched mineral medium and incubated during 72 hours at 28°C in the absence or presence of 25 and 75 mM NH_4VO_3,.

2.2. Enzymatic assays

Cells growing in agar plate enriched mineral medium, containing 0; 25 and 75 mM NH_4VO_3 were harvest, disrupted with glass beads and cell homogenates were sequential centrifuged at 3000 g and 12000 g. The sediments obtained post-centrifugation at 12000 g were used for NADH-Q2 reductase (for complex I monitoring) and NADH-cit c reductase (for complex III monitoring) determination according to [16]. The supernatant was used to determine GpX activity according to Tran [17] and Inoue [12]. All enzymatic measurements were carried out with a double beam spectrophotometer, Hitachi-U2001.

NADH-Q$_2$ reductase activity was assayed in a mixture containing 60 mM phosphate buffer (pH 7.5); 0.5 mM EDTA; 1.3 mM BSA; cell extracts in adequate concentration and 0.033 mM NADH which oxidation was monitored at 340 nm after reaction initiated with 0.063 mM Q$_2$, after 1 min incubation period at 30°C.

NADH- cit c reductase activity was assayed in a mixture containing 10 mM phosphate buffer (pH 7.5) with 100 µM KCN; 22.5 µM oxidized cytochrome c and cell extracts in adequate concentration. After an incubation period of 2 min at 30°C, the reaction was started by the addition of 0.5 mM NADH and the reduction of cytochrome c was monitored at 550 nm at timed intervals [16].

GpX activity was measured in a reaction mixture of 50 mM potassium phosphate buffer (pH 7.0) containing 5.0 mM glutathione, 1.0 mM *tert*-butyl hydroperoxide (*t*-BHP), 0.16 mM NADPH, 0.24 unit/ml glutathione reductase and cell extracts in adequate concentration at 25°C. The reaction was started by the addition of *t*-BHP. The reaction was measured by following the decrease in absorbance of NADPH at 340 nm with spectrophotometer [12, 17].

Protein determination was realised according to Lowry [18].

2.3. Statistical analysis

All the experiments were repeated at least five times independently and the data presented are mean values performed in five experiments ± S.D. The statistical analysis of results were realized by ANOVA I and Tukey tests were carried out to determine significant differences ($p < 0.01$) between the enzymatic activities of cells growing in the absence or presence of 25 and 75 mM NH_4VO_3 [19].

3 Results

In order to study the effects of NH_4VO_3 on mitochondrial electron chain transport and antioxidant defences of wild-type wine yeast *Saccharomyces cerevisiae* UE-ME_3, exponential-phase cells were harvest, suspended and grown in agar plate enriched mineral medium, containing NH_4VO_3. Enzymatic activities were assayed using cellular extracts prepared from cells growing in agar plate enriched mineral medium in the presence of 0; 25 and 75 mM of NH_4VO_3 in the culture medium. In each graphic are represented the mean values of five replicates of the enzymatic activities.

In this study it was determined the mitochondrial NADH-Q_2 reductase (complex I) and NADH-Cit c reductase (complex III) activities to evaluate the effects of ammonium metavanadate on mitochondrial electron-transport chain viability [20].

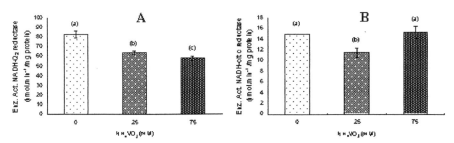

Fig 1: Effects on mitochondrial electron transport chain enzymes of *Saccharomyces cerevisiae* UE-ME_3 induced by 25 and 75 mM NH_4VO_3 on the culture medium: A) NADH-Q_2 reductase activity and B) NADH-cit c reductase activity. The data shown are means from five independent experiments ± S.D. Bars with no common letter are significantly different ($p<0.01$).

The results obtained shows a significantly decreased of NADH-Q_2 reductase activity ($p<0.01$), which directly depend on vanadium concentration in agar plates (r=0,874) (Fig.

1A). What concerns to NADH-Cit c reductase, we only observed a significantly decrease (p<0.01), around 23% of control, in the cells grown in the presence of 25 mM NH_4VO_3 (Fig. 1B). These effects strongly suggest that ammonium metavanadate inhibit mitochondrial electron transport chain at complex I level in the concentration of 25 and 75 mM, and at complex III in the concentration of 25 mM. Considering mitochondrial-derived ROS could increase when the mitochondrial respiratory chain was interrupted by respiratory chain inhibitors, like rotenone [21, 22, 23], it is possible that ammonium metavadate increase mitochondrial ROS of *S. cerevisiae* by mitochondrial electron transport chain inhibition.

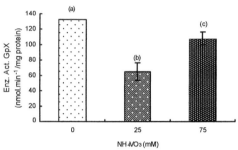

Fig 2: Effect of NH_4VO_3 on GpX antioxidant enzyme of *Saccharomyces cerevisiae* UE-ME$_3$ on cells growing in agar plate enriched mineral medium containing 0, 25 and 75 mM NH_4VO_3. The data shown are means from five independent experiments ± S.D. Bars with no common letter are significantly different (p<0.01).

In this study we also observed a significantly decrease (p<0.01) of GpX, an antioxidant enzymatic activity, approximately 50% of control, in the case of the cells grown in the presence of 25 mM, and 20% of control in treated cells with 75 mM NH_4VO_3 (Fig. 2). These results suggest that NH_4VO_3 a pentavalent salt of vanadium could disturb the antioxidant defences of *Saccharomyces cerevisiae* UE-ME$_3$, particularly in the culture medium concentration of 25 mM.

4   Discussion

The usual source of vanadium in the +5 oxidation state is ammonium metavanadate. This metal is widely used in industry and consequently it is present in atmosphere, water and soil in several regions of planet as contaminant. Thus, the aim of this work was to study the effects of NH_4VO_3, on mitochondrial electron chain transport and GpX antioxidant enzymes of the wild-type wine yeast *Saccharomyces cerevisiae*, strain UE-ME$_3$. Beside it has been considered one of the most important microorganism for the man, there are scarce studies to evaluate the effects of heavy metals on the yeast mitochondrial electron transport and GpX antioxidant enzymes. When reactive oxygen species (ROS) are produced in yeast cells as normal by products of cellular metabolism, the molecular defence mechanisms, such as the glutathione peroxidase (GpX) enzymes, under physiological conditions, are able to avoid oxidative damages [12, 13]. Moreover, this balance could be disturbed when yeast cells are exposed to diverse environmental stress conditions, such as metal ions as pentavalent vanadium. In these study we observed that ammonium metavanadate in concentration of 25 mM induce a significantly decrease of mitochondrial NADH-Q$_2$ reductase activity (23% of control), mitochondrial NADH-cit c reductase activity (23% of control) and a significantly decrease of cytosolic GpX activity (50% of control). The cytotoxic effects observed at mitochondrial respiratory chain (complexes I and III), with eventually increase of mitochondrial ROS [11, 14] and antioxidant defences inhibition caused by ammonium metavanadate in *Saccharomyces cerevisiae*, strain UE-ME$_3$ could leads to cellular apoptosis [21]. Moreover the ammonium metavanadate in the concentration of 75 mM only cause a significantly decrease (29.5% of control) of mitochondrial NADH-Q$_2$ reductase activity.

Our results show that NH$_4$VO$_3$, a pentavalent salt of vanadium (V), caused oxidative damages at mitochondrial respiratory chain of *Saccharomyces cerevisiae*, strain UE-ME$_3$.

References

[1]   Robson, C. N., Alexander, J., Harris, A. L. and Hickson, I. D. (1986). Cancer Research. 46, 6290-6294.
[2]   Liochev, S. and Friedovich, I. (1986) Arch Biochem Biophys. 250, 139-145
[3]   Coulombe, R. A., Briskin, D. P., Keller, R. J., Thornley, W. R. and Sharma, R. P. (1987) J. Pharmacol. Exp. Ther. 242, 354-363
[4]   Vido, K., Spector, D., Lagniel, G., Lopez, S., Toledano, M. B. and Labarre, J. (2001) J. Biol. Chem.. 376, 8469-8474
[5]   Prescot, L. M, Harley, J. P. and Klein, D. A. (1996) Microbiology. W. C. Brown Publishers, London
[6]   Davis, Dulbecco, Eisen, Ginsberg and Wood. (1967) Microbiologia- Infecções Bacterianas e Micóticas. Edart-São Paulo-Livraria Editora, S. Paulo.
[7]   Deacon, J. W. (1980) Introduction to Modern Mycology. John Wiley and Sons, New York.
[8]   Ciani, M., Mannazzu, I., Marinangeli, P., Clementi, F. and Martini, A. (2004) Antonie van Leeuwenhoek. 85 (2), 159-164.
[9]   Carrasco, P., Querol A.and del Olmo M. (2001) Arch Microbiool,. 175(6), 450-457
[10]   Costa, V. and Moradas-Ferreira, P. (2001) Molecular Aspects of Medicine. 22, 217-246
[11]   Chen, Q., Vazquez, E. J., Moghaddas, S., Hoppel, C. L. and Lesnefsky, E. J. (2003) J. Biol. Chem. 278, 36027–36031
[12]   Inoue, Y., Matsuda, T., Sugiyama, H-I., Izawa, S. and Kimura, A. (1999) J. Biol. Chem. 274, 27002-27009.
[13]   Dormer, U. H., Westwater, J., McLarent, N. F., Kent, N. A., Mellor, J. and Jamieson, D. J. (2000) J. Biol. Chem. 275, 32611-32616.
[14]   Petrosillo, G., Ruggiero, F. M.; and Paradies, G. (2003) The FASEB J. 17, 2202-2208
[15]   Van Uden, N. (1967) Arch. Microbiol. 58, 155-168.
[16]   Tzagoloff, A., Akai, A. and Needleman, R. B. (1975) J. Biol. Chem.. 250, 8228-8235
[17]   Tran, L.-T., Miki,T., Kamakura, M.,Izawa, S., Tsujimoto, Y., Maybe, S., Inoue, Y. and Kimura, A.(1995) J. Fermentation and Bioengineering. 80, 606-609.
[18]   Lowry, O. H., Rosenbrough, N. J., Farr, L. and Randall, R.J. (1951) J. Biol. Chem. 193, 265-275
[19]   Sokal, R. R. and Rohlf, F. J. (1997) Biometry. W. H. Freeman, New York.
[20]   Ludovico, P.; Sousa, M. J.; Silva, M. T.; Leão, C. and Côrte-Real, M. (2001) Microbiology. 147, 2409-2415
[21]   Li, N.; Ragheb, K.; Lawler, G.; Sturgis, J.; Rajwa, B.; Melendez, J. A. and Robinson, J. P. (2003) J. Biol. Chem. 278, 8516-8526
[22]   Turrens, J. F. and Boveris, A. (1980) Biochem. J. 191, 421-427
[23]   Turrens, J. F. (1997) Biosci. Resp. 17, 3-8

Identification by PCR of trichothecene-producing *Fusarium*

E. D. Gomez, G. Cuesta, R. M. Montes[*]**, E. Hernandez**

Department of Biotechnology, Polithecnic University of Valencia 46022 VALENCIA, SPAIN

[*]Corresponding autor: e-mail: rmontes@btc.upv.es

Traditional methods for isolation and identification of *Fusarium* sp. are tedious, expensive and take time. Moreover, to check if they are mycotoxin-producing strains, long time incubation in producing-conditions on adequate substrates and purify the mycotoxin chromatographic methods are needed. Identification based on PCR is an effective alternative that can be used in identification of fungal species as well as in the detection of genes involved in mycotoxin production. The aim of this work is the PCR identification of *Fusarium* strains isolated from corn and the detection of gene sequences that code for trichothecene mycotoxins.

Keyword: trichothecene; mycotoxins; *Fusarium.*

1 Introduction

Species of the genus *Fusarium* are known to cause diseases of economic importance in many field crops resulting in significant losses to agriculture. This genus produce additionally detrimental mycotoxins, mainly trichothecenes and zearalenols. Among all the trichothecenes produced, the most abundant and potentially toxic are deoxynivalenol (DON), nivalenol (NIV), and acetylated derivates. These toxins can be present in crop products [1–4]. The consumption of mycotoxin contaminated food and feed products suppose an risk to human and animal health, as these mycotoxins are carcinogenic and can potentially impair the immune system [2, 5].

Traditionally, chemotyping of *Fusarium* isolates has been carried out by using gas chromatography/mass spectroscopy, but this method is time consuming and expensive. Moreover, to check if they are mycotoxin-producing strains, long time incubation in producing-conditions on adequate substrates and purify the mycotoxin chromatographic methods are needed [6].

The use of molecular techniques based on the polymerase chain reaction (PCR) for species identification and as diagnostic tool have become very popular during the last decade. PCR identification is an effective alternative that can be used in fungal identification as well as in the detection of genes involved in mycotoxin production [7].

In the present work, we have identified by PCR *Fusarium* strains isolated from maize and we have amplified the gene sequences that code for trichothecene mycotoxins production.

2 Material and methods

2.1 Isolation of *Fusarium* strains

Strains were first isolated from maize in Valencia, Spain. First of all, corn grains were decontaminated in order to reduce the external contamination. Fifty grains were immerse in 1% sodium hypochloride for 1 min following by two washed for 1 min each with sterile water. Decontaminated grains were placed in potato dextrose agar (PDA) and malachite green agar [14]. Both culture mediums were supplemented with chloranphenicol (0,05%). Each isolated strain was grown on PDA plates. Cultures were incubated at 28° C for 9 days. The morphological identification was performed according to Nelson *et al* [8]. The reference *Fusarium* strains used in this study were: *Fusarium culmorum* CECT 2148, *F. oxysporum fsp. lycopersici* CECT 2715, *F. poae* CECT 20165, *F. roseum* CETC 2218, *F. solani* CECT 20232, *F. sporotrichioides* CECT 20166, *F. tricinctum* CECT 20150, *F. verticillioides* CECT 2982, *Gibberella intricans* CECT 2149, *G. zeae* CECT2150. All these strains were supplied by the Spanish Collection of Cultures Type (CECT).

2.2 DNA extraction and PCR amplification

The frozen mycelia were ground into fine powder in liquid nitrogen using a sterile pestle and mortar, and total genomic fungal DNA was extracted using CTAB method [9], re-suspended in TE and stored at -20°C.

The primers ITS-Fu-f and ITS-Fu-r (Table 1) are specific for *Fusarium* genus [10]. The amplification reactions were carried out in a 50 µl volume containing 5µl of buffer (100 mM Tris- Cl, pH 7.5; 100mM NaCl; 0.1 mM EDTA; 2 mM DTT; 50% Glicerol; 0.1% Tween-20) , 3µl $MgCl_2$, 0.4µl DNTP's (Ecogen, Spain), 1µl of primers , 0.5 µl Taq polymerase (Ecogen, Spain) and 3µl of fungal DNA. These primers amplify specifically a 398-bp fragment within the ITS region of *Fusarium* spp. PCR conditions were: one cycle at 95° C for 2 min, followed by 30 cycles of 94°C for 1 min, 61° C for 30 s, and 72° C for 1 min, and finally, one cycle of elongation at 72°C for 5 min.

Primers FF2 and FR1 (Table 2) were used to amplify a 425-bp of the 18S rRNA gene. This fragment has been related with trichothecene production [11]. The amplification reactions were carried out in a 50 µl volume containing 5µl of buffer (100 mM Tris- Cl, pH 7.5; 100mM NaCl; 0.1 mM EDTA; 2 mM DTT; 50% Glicerol; 0.1% Tween-20), 1.5µl $MgCl_2$, 0.4µl DNTP's (Ecogen, Spain), 1µl of random primers, 1.5µl formamide, 0.25 µl Taq polymerase (Ecogen, Spain) and 3µl of fungal DNA. The conditions for PCR were: initial denaturation of DNA at 94° C for 4 min followed by 35 cycles of 94° C for 1 min, 58° C for 30 s, and extension at 72° C for 45s, and finally, one cycle of elongation at 72° C for 5 min.

3 Results and discussion

The *Fusarium* strains studied in this work were isolated from commercial grains from different areas in Valencia. Morphological identification of strains was confirmed by PCR using ITS-Fu-f and ITS-Fu-r primers. These primers (Table 1) showed good specificity for the genus *Fusarium* and a 389 pb product was amplified in all references strains. No fragment was amplified in *Acremonium* sp.. In 180 isolated strains from maize grains were amplified the same product with this primers.

Primers FF2 and FF1 (Table 1) amplified a 425-pb fragment related with trichothecene production. With these primers we have amplified the same fragment in 61 of the 180 *Fusarium* strains.

The contamination of corn and derivates with *Fusarium* strains and its toxins is a worldwide problem and sensitive detection methods of micotoxins-producing strains are necessary. A diagnostic by PCR is a rapid protocol to identify *Fusarium* at genus level and trichothecene-producing strains was used.

Fungal identification is still achieved through traditional phenotypic typing that requires an expert taxonomist, takes a long time, and has often led to miss-identification of species due to paucity and plasticity of the characters used, loss of cultural viability and degeneration of the cultures [2, 12]. The two internal transcribed spacer (ITS) regions are variable and are frequently used to distinguish at the species level. Primers ITS-Fu-f and ITS-Fu-r are useful to identify all *Fusarium* strain at genus level because amplify a 398 bp fragment from all *Fusarium* reference strains. With this primers we have identify *Fusarium* strains isolated from corn.

Differences in the sequence of the tri5 gene regions were utilised to set up PCR-based assays for the detection of trichothecene-producing *Fusarium sp.* Trichothecene biosynthesis is regulated by environmental conditions such a water activity, temperature and nutrients [13]. For this reason to test if any strain is mycotoxin producer, optimal conditions of growth must be optimized and this is tedious and time consuming. A rapid alternative is direct detection of the gene that codified these toxins. Primers FF2 and FR1 amplify a 425-bp fragment in almost all the *Fusarium* reference strains. The PCR assay is useful for the rapid identification of *Fusarium* strains by it specificity and sensitivity, it is essential that the parameters used in the PCR are optimal. This technique is a good choice for *Fusarium* identification since it avoids long time incubation.

The results obtained in this work shown that PCR assays are highly selective and sensitive in detecting the *Fusarium* genus and the trichothecene-producing strains. Nevertheless, this method is culture dependent in the sense that is necessary to isolate and culture the strain for DNA extraction. To consider usage of this method for detection of *Fusarium* strains in infected corn grains the PCR will have to be further evaluated and optimized.

Table 1. Oligonucleotide primers used in the present study

Primers	DNA sequence
ITS–Fu-f	CAACTCCCAAACCCCTGTGA
ITS-Fu-r	GCGACGATTACCAGTAACGA
FF2	GGTTCTATTTTGTTGGTTTCTA
FR1	CTCTCAATCTGTCAATCCTTATT

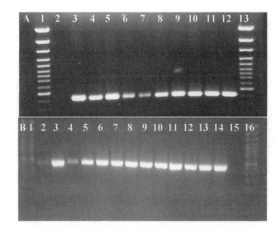

Figure 1. A. PCR assay (ITS-Fu-f and ITS-Fu-r primer pair) with genomic DNA from different fungal species of *Fusarium* strains from CECT. 1and 13 Molecular markers, 2 negative control, lanes

3-10 *Fusarium culmorum* CECT 2148, *F. oxysporum fsp. lycopersici* CECT 2715, *F. poae* CECT 20165, *F. roseum* CETC 2218, *F. solani* CECT 20232, *F. sporotrichioides* CECT 20166, *F. tricinctum* CECT 20150, *f. verticillioides* CECT 2982, *Gibberella intricans* CECT 2149, *G. zeae* CECT2150.
B. Isolates of maize 1 and 16. Molecular markers, lanes 2-14. v2.6.2, p2.7.5, p4.1.1, v2.6.5, v2.7.3, v2.3.3, v2.3.1, p2.7.4, v10.7.4a, p4.9.2, v5.3.2, v8.6.4, v2.11,p9.1.1,v9.6.1, v8.1.1, p8.6.4, p9.10.1, 15 negative control.

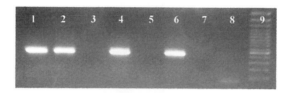

Figure 2. Analysis of PCR amplification (primers FF2 and FF1) of strains used for specificity test. Lane 1. Positive Control (*Fusarium culmorum CECT 2148)*, Lanes 2-7 p 10.1.3, v2.73, v 1.8.3, v 2.6.2, p9.10.1, p8.6.4, Lane 8 control negative, Lane 9 . Molecular weight marker.

References

[1]   Morrison, E., Rundberget, T., Kosiak, B., Aastveit, A. H. and Bernhoft, A. 2002. Cytotoxicity of trichothecenes and fusarochromanone produced by Fusarium equiseti strains isolated from Norwegian cereals. Mycopathologia. 153 (1): 49 - 56

[2]   Edel, V., Steinberg, C., Gautheron, N. and Alabouvette, C. 1996. Evaluation of restriction analysis of polymerase chain reaction (PCR) amplified ribosomal DNA for the identification of *Fusarium* species. Mycological Research. 101 (2):179-187

[3]   Mishra, P. K., Fox, R. T. V. and Culham, A. 2002. Restriction analysis of PCR amplified nrDNA regions revealed intraspecific variation within populations of *Fusarium culmorum.* FEMS Microbiology Letters. 215(2): 291-296

[4]   Knoll, S., Mulfinger, S, Niessen, L. and Vogel, R.F. 2002. Identification of *Fusarium graminearum* in cereal samples by DNA detection strips. Letters in applied microbiology. 34:144-148

[5]   Kim, H., Lee, T., Dawlatana, M., Yun, S. and Lee, Y. 2003. Polymorphism of trichothecene biosynthesis genes in deoxynivalenol and nivalenol-producing *Fusarium graminearum* isolates. Mycol. Res. 107 (2):190-197.

[6]   Jennings, P., Coates, M. E., Turner, J. A., Chandler E. A., and Nicholson, P. 2004. Determination of deoxynivalenol and nivalenol chemotypes of *Fusarium culmorum* isolates from England and Wales by PCR assay. Plant Pathology. 53 (2):182-190

[7]   Doohan, F. M, Weston, G., Rezanoor, H. N., Parry D.W. and Nicholson, P. 1999. Fusarium ear blight of wheat: the use of quantitative PCR and visual disease assessment in studies of disease. Plant pathology. 48: 209-217

[8]   Nelson, P. E., Toussoun, T.A. and Marasas, W.F.O. 1983. Fusarium Species: An Illustrated Manual for Identification. The Pennsylvania State University, Press: University Park., pp.193

[9]   Wilson, K. 1987. Preparation of genomic DNA from bacteria. In: Current protocols in molecular biology. (Ausbel, F. M., Brent, R., Kingston, R. E., Moore, D. D., Smith, J. A., Seidman, J. G. and Struhl, K., Eds.), unit 2.4.1.

[10]   Abd-Elsalam, K. A., Aly, I. N., Abdel-Satar, M. A., Khalil, M. S., and Verreet, J. A. 2003. African Journal of Biotechnology. 2(4): 82-85

310

[11]   Niessen, L., Schmidt, H. and Vogel, R. F. 2004, International Journal of Food Microbiology. 95(3): 305-319

[12]   O`Donnell, K. 2000. Molecular phylogeny of the *Nectria haematococca- Fusarium* solani species complex. Mycologia. 92: 919-938

[13]   Bakan, B., Giraud.delville, C., Pinson, L., Richard-Molard, D., Fourier, E. and Brygoo, Y. 2002. Identification by PCR of Fusarium culmorum strains producing large and small amounts of deoxinivalenol. Appl. Environ. Microbiol. 68(11): 5472-5479.

[14]   Castella, G., Bragulat, M. R., Rubiales, M. V. and Cabañes F. J. 1997. Malachite green agar, a new selective medium for *Fusarium* spp. Mycopathologia. 137: 173-178.

Isolation of *Arcobacter butzleri* from meats and its susceptibility to various acids and humectants

L. Cervenka[*,1], **I. Zachová**[2], **J. Vytrasová**[2]

[1] University of Pardubice, Department of Analytical Chemistry, nám. Cs. legií 565, 532 10 Pardubice, CZECH REPUBLIC
[2] University of Pardubice, Department of Biological and Biochemical Sciences, Štrossova 239, 530 03 Pardubice, CZECH REPUBLIC

[*]Corresponding author: e-mail: libor.cervenka@upce.cz, Phone: +420 466 037 703, Fax: +420 466 037 068

The aim of this study was to isolate and identified *Arcobacter butzleri* from meat and meat products using specific enrichment media and PCR. *A. butzleri* strains reacted positively with the primers in expected size amplicons of 1223 pb and 686 pb. The effect of weak organic acid addition and water activity (a_w) on the growth of *A. butzleri* CCUG 30484 in culture media at 30°C was studied. In total, 32 *Arcobacter* strains were isolated, of which 22 were *A. butzleri*. Chicken samples were found to be the main reservoir of *A. butzleri* (40.7% in retail level and 23.5% in processing plants). Generally, the growth of *A. butzleri* CCUG 30484 was inhibited at medium pH and the strain tested was extremely sensitive to broth environment with a_w values of < 0.980 using NaCl, glycerol or sucrose as humectants.

Keywords *Arcobacter butzleri*; meat; water activity, weak organic acid preservatives

1   Introduction

The genus *Arcobacter* was proposed to describe the organism previously known as "aerotolerant campylobacter" [1]. *Arcobacter butzleri* is an emerging food-borne pathogen and become widespread around the world. *A. butzleri* was recently isolated from cattle and swine faeces as well as from cloacal chicken swabs in Japan [2], from beef and dairy cattle in Texas [3], from broiler flocks slaughtered in poultry slaughterhouses in Belgium [4], and also from chicken carcasses sold in retail markets in Turkey [5].

Arcobacter and *Camplobacter* have similar properties. The low resistance of *Campylobacter* to conditions of reduced pH and a_w is well-described [6] but little is known about efficacy of these treatments for successful reduction of *A. butzleri*. Lactic and citric acids were reported to be effective in reducing number of *A. butzleri* [7]. The minimal water activity for growth of *Arcobacter* sp. was not determined but the ability to grow on media containing various amounts of sodium chloride was reported [8].

The aim of this study was to isolate and identified *Arcobacter butzleri* from meat and meat products using specific enrichment media and PCR. *A. butzleri* strains reacted positively with the primers in expected size amplicons of 1223 pb and 686 pb. The effect of weak organic acid addition and water activity (a_w) on the growth of *A. butzleri* CCUG 30484 in culture media at 30°C was studied.

312

2   Materials and methods

2.1 Isolation of *Arcobacter* spp. and *Arcobacter butzleri* from meat samples.

A swab sample taken from a sampling site was transferred to 10 ml of JM enrichment broth [9] and subsequently incubated for 48 h at 30°C under aerobic conditions. A 25 g portion of aseptically taken meat samp le was homogenized with 225 ml of JM broth to be incubated for 48 h at 30°C under aerobic conditions. The inoculation onto particular selective agars followed, namely JM, Formula 6 [10], together with Mueller-Hinton agar (Imuna, Slovakia). The plates were incubated aerobically for 2-3 days at 30°C. Suspect colonies were subcultivated with the use of a nonselective CASO agar (Merck, Germany), after an aerobic incubation for 24 h at 30°C, the colonies were evaluated according to morphology, physiology and biochemical tests (appearance of the colony on blood agar, colony size, Gram stain, oxidase and/or catalyse activity, growth on McConkey agar, growth on a nonselective agar at 25 and 30°C, growth at 37°C both under aerobic and anaerobic conditions, nitrate reduction, fermentation of lactose, and utilization of citrate). Only such colonies, for which tests corresponded to the genus *Arcobacter*, were taken for subsequent PCR analysis according to Harmon and Wesley [11] with some modification [12]. The presence of an *Arcoabcter* sp. was indicated by the presence of a PCR product of 1223 bp, whereas for *A. butzleri* two products of 1223 and 686 pb were detected.

2.2 pH and water activity testing

Arcobacter butzleri CCUG 30484 (University of Götteborg, Sweden) was used in this study. To prepare inocula for the test media, cultures were activated by transfer in brain heart infusion broth (BHI, HiMedia Lab., India) at 30°C for 48 h.
Weak organic acid used: formic, citric, tartaric, propionic, DL-lactic, DL-malic and ascorbic acid at different pH levels (6.5, 6.0, 5.5, 5.0 and 4.5). All the experiment was carried out in duplicate using 5ml BHI broth, initial count 2×10^6 CFU/ml, 30°C and 48h incubation period.
BHI agar plates were adjusted to a_w values ranging from 0.995 to 0.970 by NaCl, glycerol, sucrose, lactose and maltose. The water activity of BHI agar was determined at 25°C using Thermoconstanter Aw Sprint Novasina TH 500 (Axair Ltd., Switzerland). Growth/no growth pattern was determined after 48h incubation at 30°C.

3   Results and discussion

3.1 Isolation of *Arcobacter* spp. and *Arcobacter butzleri* from meat samples

In total, *Arcobacter* strains were isolated, of which 22 were *A. butzleri* as seen in Table 1. Chicken samples were found to be the main reservoir of *A. butzleri* (40.7% in retail level and 23.5% in processing plants). When only biochemical tests were applied to suspect colonies, false positive results were obtained. Results very similar to those for arcobacters were also obtained in the presence of other microorganisms which yielded similar biochemical tests and which were often found in the sample (e.g. *Alcaligenes*). For beef/pork production line, two samples were found to be positive for *A. butzleri* using PCR technique. Occurrence of arcobacters is evidently influenced by the sanitary conditions in meat-processing plants. For example, the collective of authors have reported 90% of positive samples only in one plant of all examined, for which only 5% of positive samples were found [13]. In the case of poultry meat and swabs, 19 samples were evaluated as positive for *A. butzleri*. Arcobacters have not been detected at all samples of pork, rabbit and poultry meat from domestic farming. The both isolation procedure and PCR technique are described in our recent study [12].

3.2 pH and water activity testing

The results of *Arcobacter* sp. growth under various pH conditions are summarized in Table 1. *Arcobacter butzleri* grow at pH 5.5 but no viable cells were detected in BHI broth acidified to pH 5.0 with ascorbic, malic, lactic and propionic acids after 48 incubation at 30°C. Formic, citric and tartaric acid were more inhibitory for *A. butzleri* with no growth observed at pH < 5.5 whereas at pH 6.0 the growth was observed. It is evident, in general, from the results of present study that tested strain of arcobacter was mildly acid tolerant microorganism capable to grow at pH > 5.5, as was concluded by Hilton *et al.* [14] using inorganic acid. The similar sensitivity of *Campylobacter jejuni* to moderately acid environment was determined in pasteurized milk whose pH has been adjusted by lactic or propionic acid to pH range 4.2-5.3 [15]

Table 1 Growth of *Arcobacter butzleri* CCUG 30484 in BHI broth adjusted to pH using weak organic acids after 48 h incubation at 30°C.

pH	Growth in BHI broth with given acid						
	Formic	Propionic	Lactic	Ascorbic	Malic	Tartaric	Citric
4.5	-	-	-	-	-	-	-
5.0	-	-	-	-	-	-	-
5.5	-	+	+	+	+	-	-
6.0	+	+	+	+	+	+	+
6.5	+	+	+	+	+	+	+

(+) growth was interpreted as development of turbidity after 48 h incubation at 30°C

Citric acid inhibited the growth of *Arcobacter butzleri* in *Arcobacter* selective broth acidified to pH 4.5 with no viable counts detected after 6 h incubation at 30°C but with viable cell numbers presented at low pH using lactic acid as preservative [7]. Because of differences in dissociation properties of the tested acids, different amounts of each acid were required to achieve the target pH values, in which the growth was inhibited. Thus the concentration at this pH values were calculated and found differs among tested acids [16]. On this basis, tested concentration of citric (10.2 mmol/l), tartaric (20.8 mmol/l) and malic (29.7 mmol/l) acids were the lowest observed, while the concentration of formic (73.3 mmol/l) and ascorbic (53.4 mmol/l) acids were the highest. Weak organic acids have optimal inhibitory activity at low pH because this favours undissociated state of the molecule which is chiefly responsible for acid's antimicrobial activity [17]. At pH 5.5, citric acid (6.4×10^{-4} mmol/l) and tartaric acid (4.4×10^{-4}) should be considered the most effective due to lowest undissociated concentration acid used to achieve the target pH in comparison of formic acids. Among acids which inhibited the growth of *A. butzleri* at lower pH of 5.0, concentration of undissociated malic acid of 0.58 mmol/l appeared to be the most effective.

Table 2   Water activity requirements for growth of *Arcobacter butzleri* CCUG 30484 on BHI agar.

Solute	a_w	Growth/ no growth	Solute	a_w	Growth/ no growth	Solute	a_w	Growth/ no growth
	0.995	+		0.995	+		0.995	+
	0.990	+		0.991	+		0.991	+
NaCl	0.986	+	Glycerol	0.985	+	Sucrose	0.985	+
	0.981	-		0.979	-		0.978	-
	0.970	-		0.974	-		0.973	-

(+) growth was interpreted as colony forming units observed on BHI agar after 48 h incubation at 30°C.

At concentration below 1mmol/l, ascorbic acid stimulated the growth of *Campylobacter jejuni* in nutrient broth whereas at 5 mmol/l the culture of *C. jejuni* (10^4 cells/ml) was destroyed [18]. The similar results were obtained in our work, in which undissociated ascorbic acid concentration of 4.9 mmol/l (pH 5.0) inhibited the growth of *A. butzleri* after 48 h incubation at 30^0C.

In this study, it was confirmed that *A. butzleri* was susceptible to decrease of water activity with no colony forming detected on agar plates at a_w < 0.980 adjusted with NaCl as well as with glycerol or sucrose as seen in Table 2. It was reported that in media containing 2%, 3.5% and 4.0% NaCl, 100%, 70% and 15% of *A. butzleri* isolates with different sources can grow, respectively, while growth of *A. cryaerophilus* isolates were only observed in media containing 2% NaCl [19]. On the contrary, none of 49 *A. butzleri* chicken isolates were able to grow at presence of 4% NaCl [20]. Additionally, 2.0%, 3.5% and 4.0% NaCl solutions corresponded with water activity values of 0.989, 0.980 and 0.977, respectively [21].

Acknowledgements

The support by Ministry Education, Youths and Sports (no. 0021627502) and by Grant Agency of Czech Republic (no. 203/05/2106) is gratefully acknowledged.

References

[1]   P. Vandamme, M. Vancanneytm, B. Pot., L. Mels., B. Hoste, D. Dewettinck, L. Vlaes, C. van der Borre, R. Higgins, J. Hommez, et al., Int. J. Syst. Bacteriol. 42, 344 (1992).
[2]   H. Kabeya, S. Maruyama, Y. Morita, M. Kubo, K. Yamamoto, et al., Vet. Microbiol. 19, 153 (2003).
[3]   S. C. Golla, E. A. Murano, L. G. Johnson, N. C. Tipton, E. A. Cureington, J. V. Savell, J. Food Protect., 65, 1849 (2002).
[4]   K. Houf, L. de Zutter, J. van Hoof, P. Vandamme, J. Food Protect. 66, 364 (2003).
[5]   H. I. Atabay, F. Aidyn, K. Houf, M. Sahin, P. Vandamme, J. Food Microbiol. 81, 21 (2003).
[6]   J. E. L. Corry and H. I. Atabay, J. Appl. Microbiol. 90, 96 (2001).
[7]   C. A. Phillips., Lett. Appl. Microbiol. 29, 424 (1999).
[8]   T. L. Maugeri, C. Gugliandolo, M. Carbone, D. Caccamo, M.T. Ferra, Microbiologica 23, 143 (2000).
[9]   L. G. Johnson, E. A. Murano, J. Food Protect. 62, 610 (1999).
[10] L. G. Johnson, E. A. Murano, J. Food Protect. 62, 456 (1999).
[11] K. M. Harmon, I.V. Wesley, Vet. Microbiol. 58, 215 (1997)
[12] J. Vytrasová, M. Pejchalová, K. Harsová, Š. Bínová, Folia Microbiol. 48, 2, 227 (2003).

[13]   C. I, Collins, I. V. Wesley, E. A. Murano, J. Food Protect., **59**, 448 (1996).
[14]   C. L. Hilton, B. M. Meckey, A. J. Hargreaves, S. J. Forsythe, J. Appl. Microbiol. **91**, 929 (2001)
[15]   Z. Cuk, A. Anna-Prah, M. Janc, J. Zajc-Salter, J. Appl. Bacteriol. **63**, 201 (1987).
[16]   L. Cervenka, I. Zachová, P. Minaríková, J. Vytrasová, Czech J. Food Sci. **21**, **6**, 203 (2003).
[17]   M. Stradford, in: R. K. Robinson, C. A. Batt, P. D. Pattel (eds) Encyclopedia of Food Mikrobiology (Academic Press, San Diego, 2000), 1729.
[18]   B. J. Juven, J. Kanner, J. Appl. Bacteriol. **61**, 339 (1986).
[19]   H. I. Atabay, J. E. L. Corry, S. L. W. On, J. Appl. Microbiol. **84**, 1007 (1998).
[20]   H. I. Atabay, F. Aydin, K. Houf, M. Sahin, P. Vandamme Int. J. Food Microbiol. **81**, 21 (2003).
[21]   S. L. Resnik, J. Chirife, J. Food Protect. **51**, 419 (1988).

PCR assay for identification of *Aspergillus* section *Nigri* using internal transcribed spacer regions 1 and 2

González-Salgado, A.[1], Patiño, B.[2], Jurado, M.[1], López-Errasquín, E.[2], Vázquez C.[2], González-Jaén, M.T.[1,*]

[1] Department of Genetics,
[2] Department of Microbiology III, University Complutense of Madrid, Jose Antonio Novais,2. Madrid. SPAIN.

[*]Corresponding author: e-mail: tegonja@bio.ucm.es, Phone: 0034913944830, Fax: 0034913944844

Species of *Aspergillus* belonging to section *Nigri* are commonly known as black *Aspergilli* and have a significant impact on modern society through their ability to produce food spoilage and mycotoxins, among which ochratoxin A (OTA) is one of the most important. The taxonomy of *Aspergillus* section *Nigri* has been studied by means of morphological and cultural criteria, but it is still controversial and problematic. The accurate identification of *Aspergillus* species in the section *Nigri* is of great importance because the toxic profile of the single species is different and could expose the contaminated food commodities to different toxicological risks. In this work we report the results of the analysis of the internal transcribed spacer of rDNA (ITS) in *Aspergillus* spp. strains in order to obtain diagnostic sequences of each species of the section *Nigri*.

Keyworlds: *Aspergillus* section *Nigri*; Ochratoxin A; PCR; ITS.

1   Introduction

Species of *Aspergillus* belonging to section *Nigri* [1] are commonly known as black *Aspergilli* and have a significant impact on modern society through their ability to produce enzymes and organic acids [2]. These species also cause food spoilage and produces mycotoxins [3], among which aflatoxins and ochratoxin A is the most important.

Identification up to now is based on morphological characteristics, such as conidial shape, color and size. Thom and Raper [4] and Raper and Fennell [5] divided the black *Aspergilli* into 15 and 12 species respectively. In 1980, Al- Musallam [6] revised this classification and recognized five distinguishable species and the *A. niger* aggregate, subdivided into seven varieties. Using molecular methods (nuclear and mithochondrial DNA polymorphism) separated the black *Aspergilli* into *A. japonicus, A. heteromorphus, A. ellipticus, A. carbonarius* and the *A. niger* aggregate. The *A. niger* aggregate would include according to Kusters-van Someren [7] two morphologically indistinguishable species, *A. niger* and *A. tubingensis* based in RFLP analysis which was corroborated by other studies [8, 9]

Ochratoxins are cyclic pentaketics, dihydroisocoumarin derivates linked to an L-phenylalanine moiety. The most potent ochratoxin derivate, ochratoxin A (OTA) is a secondary metabolite produced by *Aspergillus* and *Penicillium* species. OTA was proved to exhibit nephrotoxic, immunosuppressive, teratogenic and carcinogenic properties and has been classified by International Agency for Research on Cancer as a possible carcinogen to humans (group 2B) [10]. OTA occurs in various foodstuffs and beverages [11, 12]. Only two *Aspergillus* belonging section *Nigri* are known to produce OTA: *A. carbonarius* and *A. niger* [13–15], although OTA production by some *A. japonicus* strains has been also reported [16].

Early detection of ochratoxigenic fungi on raw materials, feed and foodstuffs, is vital if OTA is to be eliminated from the food chain. To achieve this, application of molecular diagnosis methods is required. PCR-based methods that target DNA are considered a good alternative for rapid diagnosis because of their high specificity and sensitivity, particularly enhanced, when multi-copy sequences are used to develop species specific primers [17].

2   Materials and methods

2.1 Fungal Isolates and culture conditions.

All the isolates used in this study, along with their sources, are given in Table 1.

Cultures were maintained on potato dextrose-agar (PDA, Scharlau Chemie, Barcelona, Spain) at 4°C and stored as spore suspension in 15% glycerol at –80°C. The isolates were cultured in 100 mL Erlenmeyer flasks containing 20 mL liquid medium Sabouraud (Scharlau Chemie, Barcelona, Spain). Cultures were inoculated with mycelial disks cut from the plates and incubated at 25°C, 150 rpm. Mycelia from 2day-old cultures were harvested by filtration through Whatman paper n° 1 and kept at –80°C for DNA isolation.

2.2 DNA extraction and PCR amplification.

Genomic DNA of the strains was obtained using either the genomic DNA Extraction Kit (Genomix, Talent, Trieste, Italy), according to the manufacturer's instructions.

All genomic DNAs used in this work were tested for suitability for PCR amplification using primers ITS1 and ITS4 [18], which amplify the ITS region in *Aspergillus*. The PCR reaction was performed in an Eppendorf Mastercycler Gradient (Eppendorf, Hamburg, Germany) using between 10 pg to 10 ng of genomic DNA. The amplification program used was described by Henry *et al.*[19] The amplification products were isolated using the High Pure PCR Product purification Kit (Roche, Germany) and were sequenced using the ABI PRISM DNA Sequencer (Applied Biosystems, Foster City, USA) according to the manufacturer's instructions in the Genomic Unit of the University Complutense of Madrid (Spain). All the strains were sequenced in both directions. Sequences were analysed and aligned by Clustal method using the program DNAstar (Lasergene, Wisconsin, USA).

PCR assays were carried out using a set of primers CAR1/CAR2 [20] for *A. carbonarius*. The rest of PCR assays were carried out using primer ITS1 [18] in all cases combined with a species specific primer: NIG for *A. niger*, JAP for *A. japonicus*, HET for *A. heteromorphus* and ELL for *A. ellipticus* located in ITS2 region. The PCR amplification protocol used for *A. carbonarius* was described in Patiño *et al* [20] In the case of *A. niger, A. japonicus, A. heteromorphus* and *A. ellipticus* the PCR program was the same that for *A. carbonarius* except for the annealing temperatures which were 66°C, 62 °C, 65°C and 65,5 °C, respectively. PCR products were detected in 2% agarose ethidium bromide gels in TAE 1X buffer (Tris-acetate 40 mM and EDTA 1.0 mM). The DNA ladder "Real escala n°2" (Durviz, Valencia, Spain) was used as molecular size marker.

2.3 RFPL analysis.

Aliquots of 10 μl of PCR amplification reactions obtained with ITS1/NIG primers were digested overnight with *Rsa* I (Amersham Pharmacia Biotech, Buckinghamshire, England) at 37° C in a total volume of 40 μl. The restriction fragments were separated by electrophoresis in 2.5% agarose gels. The DNA ladder "Real escala n°2" (Durviz, Valencia, Spain) was used as molecular size marker. At least two independent experiments were performed.

3   Results

The ITS1-5.8S-ITS2 sequences of several isolates of *A. niger, A. tubingensis A. japonicus, A. ellipticus, A. heteromorphus, A. carbonarius* and other related *Aspergillus* species were obtained and aligned together with other sequences of *Aspergillus* species available in the GenBank. The accession number of a representative sequence obtained in our laboratory of each specie are   *A. carbonarius* : **AJ876878**, *A. niger:* **AJ876876**, *A. tubingensis:* **AJ876877**, *A. heteromorphus* : **AJ876879**, *A. japonicus:* **AJ876880** and *A. ellipticus:* **AJ876881**. Six specific primers, CAR1/CAR2, NIG, JAP, HET and ELL, were designed on the basis of the sequence alignment above mentioned.
PCR amplifications of genomic DNA from all the strains with the different combination of primers are indicated in table 1.

Table1. Fungal strains analysed indicating, origin, specie, host and the occurrence of PCR amplification product with primers: ITS1-NIG, ITS1-HET, ITS1-JAP and ITS1-ELL. Strains supplied by Dr. Sanchís (University of Lleida, Spain)(*). Strains supplied by Dr. Moretti *(*CNR, Bari, Italy)(**). Strains supplied by Dr. Venancio (University of Minho, Portugal (+).

STRAIN	ORIGIN	SPECIE	HOST	ITS1-NIG	CAR1-CAR2	ITS1-HET	ITS1-JAP	ITS1-ELL
CECT 2091	Canada	*A. niger*		+	-	-	-	-
CECT 20157	Canada	*A. niger*		+	-	-	-	-
CECT 2574		*A. niger*		+	-	-	-	-
CECT 2775		*A. niger*		+	-	-	-	-
CECT 20156		*A. niger*		+	-	-	-	-
T.TT.A5	Spain	*A. tubingensis*	Grapes	+	-	-	-	-
ZD.MF.ZD. A9	Spain	*A. tubingensis*	Grapes	+	-	-	-	-
T.TT.A11	Spain	*A. tubingensis*	Grapes	+	-	-	-	-
T.TT.A2	Spain	*A. tubingensis*	Grapes	+	-	-	-	-
T.TT.A1	Spain	*A. tubingensis*	Grapes	+	-	-	-	-
T.TT.A7	Spain	A. tubingensis	Grapes	+	-	-	-	-
CBS 117.55	Brazil	*A. heteromorphus*	Culture contaminant	-	-	+	-	-
MUCL 13578		*A. japonicus*		-	-	-	+	-
ITEM 4158**	Italy	*A. japonicus*	Grapes	-	-	-	+	-
ITEM 4685**	Portugal	*A. japonicus*	Grapes	-	-	-	+	-
ITEM 4687**	Spain	*A. japonicus*	Grapes	-	-	-	+	-
MUCL 31303	Costa Rica	*A. ellipticus*	Soil	-	-	-	-	+

CBS 482.62	Costa Rica	*A. ellipticus*	Soil	-	-	-	-	+
CBS 707.79	Costa Rica	*A. ellipticus*	Soil	-	-	-	-	+
168*	Spain	*A. carbonarius*	Grapes	-	+	-	-	-
229*	Spain	*A. carbonarius*	Grapes	-	+	-	-	-
242*	Spain	*A. carbonarius*	Grapes	-	+	-	-	-
207*	Spain	*A. carbonarius*	Grapes	-	+	-	-	-
171*	Spain	*A. carbonarius*	Grapes	-	+	-	-	-
MUM 04.01+	Portugal	*A. carbonarius*	Grapes	-	+	-	-	-
MUM 04.02+	Portugal	*A. carbonarius*	Grapes	-	+	-	-	-
MUM 04.03+	Portugal	*A. carbonarius*	Grapes	-	+	-	-	-
CBS 589.68	USA	*A. ochraceus*		-	-	-	-	-
CL.1	Spain	*Cladosporium sp*	Grapes	-	-	-	-	-
UCO.1	Spain	*Alternaria sp.*	Grapes	-	-	-	-	-
BO.1	Spain	*Botrytis sp.*	Grapes	-	-	-	-	-
CECT 2906		*P. verrucosum*		-	-	-	-	-

In the case of the primers CAR1/CAR2 a single fragment of about 420 bp was only obtained when genomic DNA from *A. carbonarius* strains was used (Figure 1a). With ITS1/NIG A single fragment of about 420 bp was amplified only when was used genomic DNA from the *A. niger* aggregate (Figure 1a). ITS1/JAP amplified a fragment of 520 bp only in *A. japonicus*, ITS1/ELL a fragment of 420 bp was only obtained *A. ellipticus* and ITS1/HET a fragment of 540 bp only in *A. heteromorphus* (Figure 1b). Control amplifications of the genomic DNA with primers ITS1 and ITS2 were positive for all the strains analysed.

The product obtained with ITS1/NIG included a target sequence for *Rsa* I endonuclease which allows the differentiation between *A. niger* and *A. tubingensis* by digestion of the PCR amplification product, generating two bands in the case of *A. niger* (345 bp and 76 bp) and one in *A. tubingensis* (420 bp).

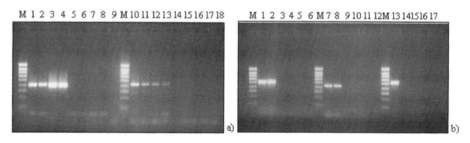

Fig. 1 PCR amplifications using primers for *Aspergillus* section *Nigri*. **a)** CAR1/CAR2 and ITS1/NIG. Lanes 1-4: *A. carbonarius* 168, 229, 242 and 207. Lanes 10-13: *A. niger* aggregate CECT 2091, CECT 20157, T.TT.A5 and T.TT.A11. Lanes 5-6 and 14-15: *A. japonicus* MUCL 13578 and *A. ellipticus* MUCL 31303. Lanes 7 and 16: *A. ochraceus* CBS 589.68. Lanes 8 and 17: *Botrytis* sp. BO.1. Lanes 9 and 18: Non-template control. **b)** ITS1/JAP, ELL and HET. Lanes 1-2: *A. japonicus* MUCL 13578 and ITEM 4158. Lanes 7-8: *A. ellipticus* MUCL 31303 and CBS 707.79. Lane 13: *A. heteromorphus* CBS 117.55. Lanes 3, 9 and 14: *A. niger* (CECT 2574). Lanes 4, 10 and 15: *A. ochraceus* (CBS 589.68). Lanes 5, 11 and 16: *Cladosporium* sp. (CL.1). Lanes 6, 12 and 17: Non-template control. M: DNA ladder "Real escala n°2".

4 Discussion

Specific PCR assays have been developed in this study for detection of both *A. carbonarius* and *A. niger* species, the main OTA producers in Section *Nigri*, as well as for the rest of species included in this Section.

The set of specific primers described in this work have been designed on the basis of ITS sequence comparisons of several strains of *Aspergillus* species and taking into account the phylogenetic and taxonomic analyses reported previously by other authors [6, 7]. The specificity of the assays was tested on a number strains of *Aspergillus* species as well as on other fungi commonly associated with grapes, cereals or coffee, such as *Penicillium*, *Cladosporium* or *Alternaria* species.

The PCR assays described in this work represent an advantage in terms of time of analysis and specificity in comparison with the conventional identification methods and the more laborious molecular methods based on AFLP profiles[21], SSCP of the PCR-IGS [22] or the secondary metabolite profiles [23] reported so far for *Aspergillus*. PCR assays reported for species of the group (*A. japonicus* and *A. carbonarius*) [24, 20] are useful to provide complementary or confirmation tests to assist correct identification.

The PCR assay with primers ITS1/NIG allowed discrimination of the two species of the *A. niger* aggregate (*A. niger* and *A. tubingensis)* from the rest of the species included in the Section though digestion with *Rsa* I. Our assay would provide a more accurate identification of *A. niger* isolates since the amplification fragment used for digestion was specific to *A. niger* aggregate.

Detection limit of ITS amplification product, defined as the clearly visible product on agarose gels containing ethidium bromide, has been estimated between 1 to 10 pg of DNA template in *Fusarium* [17,25]. We found similar detection levels (10 pg) with all set of primers when serial dilutions of genomic DNA of the different *Aspergillus* strains were used as template for PCR amplification (data not shown). The sensitivity of our PCR assay based on ITS sequences was, therefore, more sensitive than primers based on single copy gene, estimated between 0.1 and 1 ng of DNA template per reaction [17].

The specificity and high degree of sensitivity of the PCR detection assays developed for *A. carbonarius* and *A. niger* provide a good, rapid, sensitive and accurate tool for early detection of the main OTA-producing *Aspergillus* included in section *Nigri* in order to prevent OTA entering the food chain.

Acknowledgments

This work was supported by the Spanish McyT (AGL 2004-07549-C05-05/ALI).

References

[1]   W. Gams, M. Christensen, A.H. Onionsm, J.I. Pitt and R.A Samson, Intrageneric taxa of *Aspergillus*. In: Advances in *Penicillium* and *Aspergillus* systematics (Samson, R.A.and Pitt, J.I. Eds), (Plenum Press, New York, 1985), pp 55-62.
[2]   J.W. Bennet and M.A. Klich *Aspergillus*: Biology and industrial applications. (Butterworths-Heinemann, Boston, USA.,1992).
[3]   Z. Kozakiewicz, Mycol. Pap. **161**,1-188 (1989).
[4]   C. Thom and K.B Raper. In: A manual of the *Aspergilli*. (Williams and Wilkins Company, Baltimore, 1945), pp:1-373.

[5]   K.B Raper and D.I. Fennell. The genus *Aspergillus*. (Williams and Wilkins, Baltimore, 1965).

[6]   A. Al-Musallam. Revision of the black *Aspergillus* species. Ph.D. Thesis, Utrecht, The Netherlands, (1980).

[7]   M.A. Kusters-van Someren, R.A. Samson and J. Visser, Curr. Genet. **19**, 21-26 (1991)

[8]   F. Accensi, J. Cano, L. Figuera, M.L. Abarca and F.J Cabañes, FEMS Microbiol. Lett. **180**, 191-196 (1999).

[9]   J. Varga, E. Kevei, C. Fekete, A. Coenen, Z. Kozakiewicz and J.H. Croft, Mycol. Res. **97**,1207-1212 (1993).

[10]   International Agency for Research on Cancer (IARC) **56**, Lyon, France: World health organization (1993).

[11]   E. Petzinger and A. Weidenbach, Livest. Prod. Sci. **76**, 245-250 (2002).

[12]   J. Varga, K. Rigó, J. Téren and A. Mesterházy, Cereal Res. Comm. **29**, 85-92 (2001).

[13]   M.L. Abarca, M.R. Bragulat, G. Castellá and F.J. Cabañes, Appl. Environ. Microbiol. **60**, 2650-2652 (1994).

[14]   C.N. Heenan, K.J. Shaw and J.I. Pitt, J. Food Mycol.**1**, 67-72 (1998).

[15]   J. Téren, J. Varga, Z. Hamari, E. Rinyu and F. Kevei, Mycopathologia **134**, 171-176 (1996).

[16]   P. Battilani, A. Pietri, A.T. Bertuzzi, L. Languasco, P. Giorni and Z. Kozakiewicz, J. Food. Protec. **66**, 633-636 (2003).

[17]   B.H. Bluhm, J.E. Flaherty, M.A. Cousin and C.P. Woloshuk, J. Food Protec. **65**, 1955-1961 (2002).

[18]   T.J. White, T. Burns, S. Lee and J.W. Taylor, Amplification and direct sequencing of fungal ribosomal RNA genes for phylogenetics. In: PCR protocols: a guide to methods and applications (Innis, M.A., Gelgard, D.H., Sninsky, J.J. and White, T.J. Eds), (Academic Press, New York, 1990), pp: 315-322.

[19]   T. Henry, P.C. Iwen and S.H. Hinrichs, J. Clin. Microbiol. **38**, 510-1515 (2000).

[20]   B. Patiño, A. González-Salgado, M.T. González-Jaén and C. Vázquez, Int. J. Food Microbiol. (in press) (2005).

[21]   H. Schmidt, M. Ehrmann, R.F. Vogel, M.H. Taniwaki and L. Niessen, Syst. Appl. Microbiol. **26**, 138-146 (2003).

[22]   P.M. Rath and R. Ansorg, Mycoses **43**, 381-386 (2000).

[23]   L. Parenicová , P. Skouboe, J. Frisvad, R. Ramson, L. Rossen, M.T. Hoor-Suykerbyuk and J. Visser, Appl. Environ. Microbiol. **67**, 521-527 (2001).

[24]   G. Perrone, A. Susca, G. Stea and G. Mulé, Eur. J. Plant Pathol. **110**, 641-649 (2004).

[25]   M. Jurado, C. Vázquez, B. Patiño and M.T. González-Jaén, Syst. Appl. Microbiol. (in press) (2005).

Progress in prevention of aflatoxin contamination in food by preharvest application of a yeast strain, *Pichia anomala* WRL-076

Sui Sheng T. Hua

USDA-ARS, Western Regional Research Center, Albany, California 94710, USA

*Corresponding author: e-mail: ssth@pw.usda.gov, Phone: 510-559-5905, Fax: 510-559-5777

The major aflatoxin-producing fungus, *Aspergillus flavus* has a broad ecological niche and reproduces copiously. Contamination of aflatoxin in food such as tree nuts, corn and peanuts has been recognized as a significant problem in food safety and quality worldwide. A yeast strain, *Pichia anomala* WRL-076 has been demonstrated to reduce the spore production of *A. flavus* in both lab conditions and a few field experiments. Thus, *P. anomala* is likely to provide an economical means of managing aflatoxin contamination in food chain. Furthermore, biological control may also reduce problems of pest resistance, pest resurgence and environmental pollution due to the use of chemical pesticides in recent decades.

Keywords aflatoxin; *Aspergillus flavus*; *Pichio anomala*; biological control; food safety.

1 Introduction

The potential danger of aflatoxins to human health was initially observed in 1960 following their association with acute hepatoxicity in poultry (Turkey X disease) and subsequently with fatal toxicoses in India and West Africa. Aflatoxins were first identified in the 1960s as a family of toxic compounds. Aflatoxin B_1 is and the most potent carcinogen known and extremely toxic to the livers of mammals [1, 2]. Furthermore, aflatoxins are very heat-stable and are not eliminated by heat treatments. The toxins have also been implicated as a contributory epidemiological factor along with hepatitis B virus in areas of Africa, China, and Southeast Asia where there is an extremely high incidence of liver cancer. The major aflatoxin-producing fungus, *A. flavus* has a broad ecological niche and can reproduce copiously. The spores of *A. flavus* are airborne and can infect wounded plant tissues [3, 4]. Infection of corn, peanuts, cotton seeds, almonds, pistachios and other tree nuts by *A. flavus* before harvest and during storage results in the biosynthesis and accumulation of aflatoxin B_1. This mycotoxin poses a significant problem in food quality and safety worldwide.

The U. S. Food and Drug Administration has set a maximum level of 20 ppm of aflatoxin in food intended for human consumption. Recently the Commission of the European Community has set a more restrictive level of 2 ppm for aflatoxin B_1. Aflatoxin contamination of peanuts, corn, tree-nuts continues to be a regulatory deterrent to the growers and processors. Billions of dollars are spent each year in testing, sorting and paying expenses to deal with rejection of commodities contaminated with aflatoxin.

There is no conventional fungicide to control *A. flavus*. Numerous aflatoxin elimination and control studies are conducted, including breeding, genetic engineering. Managing aflatoxin contamination via biological control is a promising approach currently available. Furthermore extensive use of chemical pesticides has caused the development of pest resistance and resurgence as well as environmental pollution and risks to human health. Therefore, the use of antagonistic microorganisms to control plant-pathogenic fungi is receiving increasing attention as an environmentally - compatible approach.

Yeast species are promising biocontrol agents for postharvest protection of fruits and vegetables and offers alternatives to chemical fungicides. Several properties of yeasts make them useful for biocontrol purposes. Biological activity of antagonistic yeasts may involve nutrient competition, site exclusion, direct parasitism, and perhaps induced resistance of host plant [5, 6]. Yeast species do not produce allergenic spores or mycotoxins. They are usually non-pathogenic and many species are able to grow at low oxygen and water activity (aw) levels. In this paper we report the use of a saprophytic yeast strain, *P. anomala* WRL-076 [7] to reduce the population of *A. flavus*. The interaction of yeast and *A. flavus* was visualized using FUN-1 fluorescent stain.

2   Materials and methods

2.1 Yeast and fungal cultures

P. anomala was maintained on potato dextrose agar (PDA, Difco, Detroit, MI, USA). The strain was grown in liquid medium such as nutrient-yeast extract-dextrose broth in flask at 28^0C on a shaker at about 200 rpm. The yeast cells are harvested by centrifugation. Harvested yeast cells are stored at 4^0C until use. *A. flavus* was maintaineded on PDA. Spores were suspended in Tween-80 solution and counted by using a Beckman-Coulter Mutisizer.

2.2 Fluorescent staining

The fungal hyphae of *A. flavus* collected on the sieve screen were gently rinsed with sterile deionized water. FUN-1[2-chloro-4-(2,3 dihydro-3-methyl (benzo-1,3-thiozol-2-yl)-methylidene)-1-phenylquinolinium iodide] stain (Molecular probes, Eugene, OR, USA) in HEEPS (N-2-Hydroxyethylpiperazine-N'-2-ethanesulfonic acid) buffer was added to the hyphae at an optimum concentration determined. The hyphae were stained by incubating at 28°C in the dark for 30 minutes. The stained fungal hyphae were spotted on a slide and viewed through a fluoroscein filter in a Leica DMRB epifluorescence microscope (Leica Microsystems, Wetzlar, Germany). The fluorescein filter enabled the visualization of both green fluorescent hyphae and red fluorescent CIVS. Images were captured with a Sony DKC-5000 digital color camera (Sony Electronics Inc., Tokyo, Japan).

2.3 Field evaluation of the efficacy of the biocontrol agent, *P. anomala* WRL-076

Pistachio nut-fruits on the tree were individually wounded with a dental needle and sprayed with aqueous suspension of yeasts at 3×10^7 cells/ml in August. The wounded nut-fruits without yeast-spray were used as controls. Five weeks after the yeast spray, wounded nut-fruit was hand picked from the tree and immediately placed to a special agar medium and incubated at 28^0C for eight days. Microorganisms were eluted from the experimental samples in Tween 80 solution by shaking and sonicating in flasks. Viable counts of yeast and *A. flavus* were determined by spreading the eluted samples on dichloran rose bengal chloramphenicol (DRBC) agar plate using an Autoplate 4000 Spiral Plate (Spiral Biotech, MA, USA).

3   Results

3.1 Interactions of *Pichia anomala* and *A. flavus*

FUN-1 stain is a membrane-permeant, halogenated cyanine compound that binds nucleic acids and has an unique and useful property. FUN-1 stain was used to assess damage of hyphae by antifungal agents in several species of *Aspergilli* [8, 9]. In this paper FUN-I was used to visualize the metabolic status of *A. flavus* grown in the presence or absence of the biocontrol yeast, *P. anomala* in potato dextrose broth. Metabolically active *A. flavus* hyphae accumulated red fluorescence in vacuoles, while hyphae that were inhibited by *P. anomala* WRL-076 and were stained green. Red fluorescence accumulation in the vacuoles was greatly reduced in *A. flavus* hyphae when the yeast to fungus ratio was 50 in PD broth.

3.2 Effect of yeast on colonization of wounded pistachio nut-fruits

Experiments were conducted in a California pistachio orchard in 2003. To examine the variation among the nuts, every single nut collected was analysed for colonization of *A. flavus* and viable spores production. The percent of colonization by *A. flavus* on nut-fruits was 27.1% for the control and 5.1% for the yeast treated nut-fruits. The results are summarized in Fig. 1.

3.3 Inhibition of spore production of *A. flavus* by *P. anomala*

The colony forming unit (CFU) of *A. flavus* spores from each single nut was enumerated. Average spore production in *A. flavus* infected nuts was 5.6×10^6 and 1.3×10^6 respectively for the control and yeast sprayed pistachio nut-fruits The experiments demonstrated that the yeast, *P. anomala* can modulate spore production of *A. flavus* in wounded the pistachio nut-fruits. A reduction of spore number was observed in the range of 77% (Fig. 2). The total number of *A. flavus* spores produced on all wounded pistachios in the control is 1.28×10^8 and for yeast sprayed pistachios is 5.2×10^6 respectively. Therefore the *A. flavus* spore number present in the orchard environment was reduced by 96%.

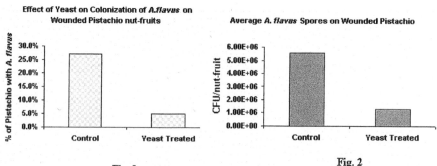

Fig. 1   Fig. 2

4   Discussion

The biocontrol efficacy of a strain of yeast, *P. anomala* WRL-076, was evaluated previously on pistachio flowers, leaves, nut-fruits and almond leaves in lab experiments. Spore production of *A. flavus* was reduced by about 80% in pistachio flowers sprayed with the yeasts. It was also effective in decreasing spore production by about 60% on almond and pistachio leaves sprayed with *P. anomala* WRL-076. Wounded pistachio nut fruits inoculated with yeast decrease spore production of *A. flavus* significantly. Almond nuts sprayed with the yeast showed much less fungal growth [10, 11].

Aflatoxin contamination is associated with wounding in corn, peant, cotton seed and tree nut [2–4, 10, 11]. Assessment of the efficacy of *P. anomala* has been achieved by mechanically wounding pistachio nuts on the tree in the orchard to increase the number of wounded nuts. The wounding experiments in pistachio orchard demonstrated that the yeast, *P. anomala* can modulate spore production of *A. flavus* on wounded pistachio nut-fruits and is very powerful in reducing the spore production of *A. flavus*. This unique biocontrol activity of *P. anomala* WRL-076 can prevent further infection of pistachio nut-fruits by *A. flavus* during the growing season in orchards. Overall the results clearly demonstrate that the production of *A. flavus* spores on wounded pistachio nuts was inhibited by spraying yeast to the tree. One can anticipate that field applicationof this effective yeast in pistachio will decrease the population of *A. flavus* in the orchards. The outcome will be a reduction of aflatoxin contamination in the edible nuts. Similar results of reducing fungal population by the yeast were observed in wounded almond fruits in orchard. Field test of *P. anomala* on corn are now in progress.

Future research directions should be: 1. Define effective ways for utilizing *P. anomala* WRL-076 for protecting tree-nuts, corn and fruits from infection by *A. flavus*; 2. Apply *P. anomala* as part of the integrated pest management (IPM); 3. Commercialize a cost-effective biopesticide through large-scale inexpensive production of *P. anomala* WRL-076.

Acknowledgements

The technical support of S. B. Ly and advice of Dr. M. Brandl on fluorescent staining is gratefully acknowledged.

References

[1]   W. O. Ellis, J. P. Smith, Simpson, and J. H. Oldham. Crit. Rev. Food Sci. Nutr. **30**, pp. 403-439 (1991).

[2]   G. A. Payne. Crit. Rev. Plant Sci. **10**, pp.423-440 (1992).

[3]   U. L. Diener, R. J. Cole, T. H. Sanders, G. A. Payne, L. S. Lee, and M. A. Klich. Annu. Rev. Phytopathol. **25**, pp 249-270 (1987).

[4]   P. J. Cotty, P. Bayman, D. S. Egel, and K. S. Elias. The genus *Aspergillus* (Plenm Press, New York, 1994), pp. 1-27.

[5]   C. L. Wilson, and M. E. Wisniewski. Ann. Rev. Phytopathol. **27**, 425 –451 (1989).

[6]   W. Janisiewicz, and L. Korsten. *Ann. Rev. Phytopathol.* **40**, pp. 411-441 (2002).

[7]   S. S. T. Hua, J. L. Baker, and M. Flores-Espiritu. *Appl. Environ. Microbiol.* **65**, pp. 2738-2740 (1999).

[8]   P. J. Millard, B. L. Roth, H. P. Thai, S. T. Yu, and R. P. Haugland. Appl. Environ. Microbiol. 63, pp. 2897-2905 (1997).

[9]   C. Lass-Flörl, M. Nagl, C. Speth, H. Ulmer, M. P. Dierich, and R. Würzner. Antimicrob. Agents Chemther. **45**, pp.124-128 (2001)

326

[10]   S. S. T. Hua, J. L. Baker, and O. K. Grosjean.   Proceedings of the Second International Symposium of Pistachios and Almonds, Davis, California, USA, August 24-29, 1997 pp.576-581(1998).

[11]   S. S. T. Hua. Proceedings of the Third International Symposium of Pistachio and Almond, Zaragoza, Spain, May 20-24, 2001 pp. 527-530 (2002).

Quantitative models application to fulfil microbiological criteria in foods

E. Carrasco[*,1], A. Valero[1], F. Pérez[1], E. Todd[2], R. M. García-Gimeno[1], and G. Zurera[1]

[1] Departamento de Bromatología y Tecnología de los Alimentos, Universidad de Córdoba, Campus Rabanales, Edif. Darwin-Anexo, 1014 Córdoba, SPAIN
[2] National Food Safety & Toxicology Center, 165 Food Safety &Toxicology Building, Michigan State University, East Lansing, MI 48824-1314, U.S.A.

[*]Corresponding author: e-mail: bt2cajie@uco.es, Phone: +34 957 212057, Fax: +34 957 212000

A recent Draft regulation adopted by the Commission of the European Communities establishes microbiological criteria in foods. For the pathogen *Listeria monocytogenes* in a specific food category, two different microbiological criteria are proposed, and the application of either depends on the demonstration that *L. monocytogenes* will not exceed a certain level in the product throughout its shelf-life. This demonstration should be based on predictive microbiology and quantitative models. Two types of quantitative modes were applied. From 10 cfu/g starting level of the microorganism, only 9 out of 126 conditions led to levels below the target value established by the Draft regulation. These conditions were different combinations of values of the factors CO_2 and shelf-life. This work is intended to illustrate that it is not compulsory to fulfil the microbiological criterion *absence in 25 g* as long as manufacturers establish proper values of these factors so that the target value is not exceeded.

Keywords food safety; *Listeria monocytogenes*; microbiological criteria; quantitative models

1   Introduction

Research on food microbiology is continuously promoting revisions in regulations, such as microbiological criteria in foodstuff, which are intended to be incorporated in national legislation.

In 1997, Codex alimentarius [1] stated that "a microbiological criterion defines the acceptability of a product or a food lot, based on the absence or presence, or number of microorganisms including parasites, and/or quantity of their toxins/metabolites, per unit(s) of mass, volume, area or lot".

Listeria monocytogenes has been recognized as a pathogen in animals for more than 70 years [2], but only in the last 25 years it has been considered as a microorganism of concern especially for food industries, because of being a zoonotic agent which causes a food-borne listeriosis.

In 1999, the Scientific Committee on Food (SCF) and the Scientific Committee on Veterinary relating to Public Health (SSCVPH), from the European Commission, were asked to give scientific advice on *L. monocytogenes* as a food-borne pathogen, as no microbiological criteria had been established with the exception of those just mentioned in the Council Directive 92/46/CE [3], and due to the substantial controversy existing in the intra-Community trade as result of the absence of agreed reference values. Finally, two documents were presented, by both SCF [4] and SCVPH [5]. Specifically, SCVPH discusses *L. monocytogenes* levels which could be relevant in a possible further effort to lower the incidence of food-borne listeriosis, and distinguishes between 2 categories of foods; foods supporting the growth of *L. monocytogenes*, and foods which not. Regarding the former, *L. monocytogenes* should not be detected in 25 gram at the time of production, while in the latter, *L. monocytogenes* levels should be < 100 cfu per gram at the time of consumption, and therefore throughout the shelf life of the commodity. SCF supported the recommendations of SCVPH.

Recently, a Draft Regulation on microbiological criteria adopted by the Commission of the European Communities [6] included the above microbiological considerations given by SCVPH, and took into account other aspects, such as the type of population (high risk population), sampling plan or actions in case of unsatisfactory results. In this Draft, for the food category *ready-to-eat foods able to support growth of L. monocytogenes other than those intended for infants and young children and ready-to eat-foods for special medical purposes*, two possible microbiological criteria are proposed, and the application of either depends on the demonstration or not by the manufacturer that the food product will not exceed the limit 100 cfu/g throughout the shelf-life; a non demonstration will lead to the microbiological criterion of absence in 25 g at the stage products ready to be placed on the market.

Given the prevalence of *L. monocytogenes* in different foods reported by SCVPH [5], reaching 36 % in minced meat, 52 % in meat products, 60 % in fish products or 12 % in salads, food industries may be interested in demonstrating that despite the presence of *L. monocytogenes* in the product (frequently at low numbers), its level will not exceed 100 cfu/g at the end of the shelf-life.

The Draft Regulation on microbiological criteria mentioned above states in its article 3(2) that, when necessary, the food business operators res ponsible for the manufacture of the product shall conduct studies in accordance with Annex II in order to investigate the compliance with the criteria throughout the shelf-life. This Annex establishes that these studies may include predictive mathematical modelling and tests to investigate the growth or survival under reasonably foreseeable conditions.

In this work, the application of quantitative models by means of computering skills together with the use of internet skills is proposed in order to demonstrate that criteria established for *L. monocytogenes* throughout the shelf-life of ready-to-eat foods are accomplished.

2   Material and methods

2.1 Growth kinetics of *L. monocytogenes*

Levels of *L. monocytogenes* were calculated through the Eq. (1), or equation of population dynamics.

$$LN(N) = LN(N_o) + \mu\, t \tag{1}$$

N = number of microorganisms at the end of the exponential phase (cfu/g); N_o = number of microorganisms at the beginning of the exponential phase (cfu/g); μ = maximum specific growth rate (h^{-1}); t = time, or shelf-life (h). The values of shelf-life ranged from 24 to 336 h (24 h-step).

The kinetic parameter *maximum specific growth rate* (μ) described by Baranyi and Roberts [7] was obtained through the use of a predictive model taken from Pin et al. [8]. Two factors were included in the model: temperature and carbon dioxide concentration. For temperature, a probability distribution in domestic refrigerators was used [9], and for carbon dioxide concentration, values ranging from 0 to 80 % (10 %-step) were considered.

In order to obtain probability distribution of levels of *L. monocytogenes* at both beginning (N_o) and end of the shelf-life (N), simulations were carried out with Palisade @Risk Professional© software (Newfield, NY USA). The simulation included 10,000 iterations.

The lag phase of the microorganism was not included in the growth model, as it was assumed that in some scenarios, such as when *L. monocytogenes* is present in salads, there would be none.

2.2 Growth data in food

Growth data in food was taken from the database ComBase [10]. The food selected was endives [11], in which only temperature and carbon dioxide concentration were monitored. In order to obtain the kinetic parameter μ, the model proposed by Baranyi and Roberts [7] was fitted to the growth data through the DMFit v.2 Excel add-in (Norwich, UK) [7].
The accuracy factor (A_f) and bias factor (B_f) proposed by Ross [12] enabled the assessment of the reliability of the predictive model of Pin et al. [8] when compared to growth data in endives. The kinetic parameter assessed was μ.

3   Results and discussion

3.1 Calculation of the level of *L. monocytogenes* just after the manufacture process

By considering the level of 100 cfu/g at the end of the shelf-life (N), a probability distribution of the level of *L. monocytogenes* at the beginning of the shelf-life (N_o), i.e. just after the manufacture, was obtained for every condition. Approximately, 18 % of conditions presented levels higher than 1 cfu/g in the 95 % percentile. For example, at 70 % CO_2 and 96 h-shelf-life, 95 % percentile was 13 cfu/g.
When trying to analyze the output (N_o), we came up with two drawbacks of this kind of study: (1) approximately, 82 % of conditions presented values between 0 and 1, which are unrealistic, even though they could be modelled; for instance, it is not possible that 0.02 cfu/g can grow exponentially until reaching 100 cfu/g at the end of the shelf-life; and (2) the 95 % percentile value corresponded to a very low temperature (3.49 °C, i.e. 5 % percentile of temperature distribution); this approach involves a potentially hazardous situation, as the temperature of the majority of domestic refrigerators is above 3.49 °C. In the example above, at 70 % CO_2 and 96 h-shelf-life, 13 cfu/g. will lead to > 100 cfu/g at the end of the shelf-life when the storage temperature is 3.49 °C or higher.
For this reason, it is proposed a study taking as a starting point several levels of *L. monocytogenes* just after the manufacture process (N_o) and assessing if the target level 100 cfu/g (N) will be reached or not.

3. 2 Calculation of the level of *L. monocytogenes* at the end of the shelf-life

3. 2. 1 Predictive model
Taking as a starting point several levels of *L. monocytogenes* just after the manufacture (N_o), the 95 % percentile of the level N of *L. monocytogenes* in all conditions was above 100 cfu/g. What is more, with a starting level of 10 cfu/g (minimum level detected) the target value 100 cfu/g corresponded to percentiles within the range 0-5 % percentile in 78.5 % conditions (99 out of 126 conditions).
With these results, it seems very difficult to accomplish with the microbiological criterion 100 cfu/g at the end of shelf-life. However, due to the use of a predictive model performed in laboratory with challenge tests in broth culture, it is possible that the model could be conservative.

3. 2. 2 Endives growth data

The indices B_f (4.36) and A_f (4.74) deviate very much from the ideal value 1 (perfect concordance between observations and predictions), assessing that the model is very conservative for the type of food considered, i.e. a fail-safe model. Such values of the indices could be in part due to the better growth of *L. monocytogenes* at high CO_2 concentrations in endives as a consequence of a reduction in levels of competing aerobic bacteria, as stated by Carlin et al. [11]. This fact highlights the importance in selecting properly the factors that influence the growth of the pathogen.

Therefore, the kinetic parameter μ obtained from endives at 10 °C was directly entered in Eq. (1) to assess if the level 100 cfu/g was reached at the end of the shelf-life or not. Results are shown in Fig. 1 for a starting level of 10 cfu/g.   Only 9 out of 126 conditions presented a level of *L. monocytogenes* below 100 cfu/g at 10 °C. However, this fact should not discourage manufacturers, as the advantage of this study is based on the no obligation of performing absence of *L. monocytogenes*; its presence is allowed, as long as 100 cfu/g is not reached at the end of shelf-life by controlling the factors CO_2 and shelf-life, although the quantification of the initial level is crucial for this study.

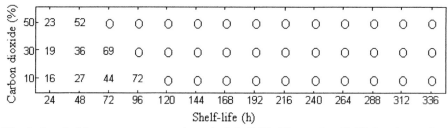

Fig. 1 Level of *L. monocytogenes* at the end of shelf-life (*N*) stored at 10 °C. Levels are given in cfu/g. Circles represent those conditions in which the level was undesirable, i.e. higher than 100 cfu/g.

4   Conclusions

Food industries which produce endives with certain prevalence of *L. monocytogenes* can still place the product on the market as long as they apply values of factors which lead finally to levels less than 100 cfu/g.

By means of predictive modelling, manufacturers could demonstrate, to the satisfaction of the competent authority, that the product will not exceed the microbiological criterion throughout the shelf-life.

Acknowledgements

This work has been performed in the framework of the collaboration between Michigan State University and University of Córdoba. The support of the projects CAL01-032 from INIA and AGL 2001-2435 from MYCT, are gratefully acknowledged.

References

[1]   Codex alimentarius (http://www.codexalimentarius.net/web/index_en.jsp) Ref. CAC/GL 21 (1997).

[2]   E.G.D. Murray, R.A. Webb and SM.B.R. Swann, Journal of Pathology and Bacteriology, **29**, 407 (1926).

[3]   Council Directive 92/46/EC, Official Journal N° L **268**, 1 (1992).

[4]   Scientific Committee on Food (SCF) (http://www.europa.eu.int/comm/food/fs/sc/scf/out63_en.pdf) (2000).

[5]   Scientific Committee on Veterinary relating to Public Health (SSCVPH) (http://www.europa.eu.int/comm/food/fs/sc/scv/out25_en.pdf). (1999).

[6]   Draft Commission regulation on microbiological criteria for foodstuff. SANCO/4198/2001 Rev. 12 (PLSPV/2002/3665/3665R6-EN.doc) (2004).

[7]   J. Baranyi, and T.A. Roberts, Int. J. Food Microbiol., **23**, 277 (1994).

[8]   C. Pin, G.D. García de Fernando, J.A. Ordóñez, and J. Baranyi, Food Microbiol., **18**, 539 (2001).

[9]   I. Azevedo, M. Regalo, C. Mena, G. Almeida, L. Carnerio, P. Teixeira, T. Hogg, and P.A. Gibbs P. A., Food Control, **16**, 121 (2005).

[10]   J. Baranyi, and M.L. Tamplin, J. Food Prot., **67**, 1967 (2004).

[11]   F. Carlin, C. Nguyen-the, A. Abreu Da Silva, and C. Cochet, Int. J. Food Microbiol., **32**, 159 (1996).

[12]   T. Ross, J. Appl. Bacteriol., **81**, 501 (1996).

Rapid biased evolution of genetically unstable wine yeast hybrids under non-selective conditions

Jesús Ambrona[1], **Antonia Vinagre**[1]and **Manuel Ramírez**[1]

[1] Departamento de Microbiología, Facultad de Ciencias, Universidad de Extremadura, 06071 Badajoz, Spain

Genetic instability causes very rapid asymmetric loss of heterozygosity (LOH) at the *cyh2* locus in some wild wine yeasts and their hybrids under non-selective conditions. The $cyh2^R/cyh2^S$ hybrids generally become $cyh2^R/cyh2^R$ homozygotes and lose a killer phenotype. Low rates of cell viability are associated with high rates of LOH at the *cyh2* locus and at other undetermined loci bearing possible deleterious alleles in heterozygosity. The sense of the biased evolution is determined by the mechanism of asymmetric loss of heterozygosity. But the speed of the processes seems strongly affected by the differences in cell viability between the new homozygous yeasts and the original heterozygous hybrid cells. The presence of active killer K2 toxin abolishes the asymmetric loss of heterozygosity of the hybrid populations. This phenomenon is interesting because it may cause important, sudden phenotype changes in industrial and pathogenic yeasts.

Keywords yeast; wine; genetic instability; LOH; adaptive mutation; killer toxin

1   Introduction

Most commercial wine yeasts are naturally occurring strains of *S. cerevisiae* isolated from wines and spontaneously fermenting musts. The phenotype of these wild yeasts usually is more variable than that of domesticated laboratory yeast strains, which have necessarily been selected for genetic stability to obtain reproducible research results. In wine yeasts, karyotype changes have been reported during vegetative growth, as have genetic changes that modify the yeast's metabolic properties. High genetic instability in natural wine yeasts, but not in the common laboratory strains of *S. cerevisiae*, has been described recently [1]. Spontaneous $cyh2^R/cyh2^R$ mutants, resistant to high cycloheximide concentrations, frequently arise from these unstable $cyh2^S/cyh2^S$ wine yeasts. Genetic instability may alter useful properties of industrial yeasts, resulting in problems in biotechnological processes or lower quality of products such as bread, pastry, beer, or wine. For example, loss of the killer phenotype may result in protracted wine fermentation. Thus, genetically stable industrial and natural yeasts guarantee the good functioning of the biotechnological processes they are involved in, which is of evident economic importance.

2   Materials and methods

Yeast strains. *S. cerevisiae* wine yeast strains were isolated from several Spanish wineries. EX33 is a virus-free killer-sensitive yeast. EX73, EX85, and EX88 are killer K2.
Liquid culture seeding on YEPD+CYH and colony analysis. The hybrids were inoculated from a YEPD plate culture in sterile tubes containing 2 ml of liquid YEPD and incubated at 30°C to saturation (two days, with shaking, about 6 doublings). A suitably diluted sample of each culture was spread onto a YEPD+CYH plate to obtain isolated colonies. The number of colonies of different sizes and the proportion of sectored colonies were

determined. Ten to twenty of each type of colony from each type of hybrid were isolated and subjected to killer phenotype and tetrad analyses.

Colony isolation and YEPD+CYH replica plating. Colonies isolated on YEPD plates were harvested, replicated on another YEPD plate, incubated at 30°C for two days (roughly 6 doublings), replica plated to YEPD+CYH and YEPD, and incubated 1-8 days at 30°C to determine cyh resistance. Ten to twenty colonies of each phenotype from each type of hybrid were harvested from the original YEPD isolation plate and subjected to killer phenotype and tetrad analysis.

Cell viability (via 100). One hundred (50 unbudded and 50 with a small bud) cells from each hybrid were micromanipulated onto YEPD plates and incubated at 30°C for four to eight days. The number of viable cells, colony size, and sectored colonies were determined.

3   Results and discussion

The genetic constitution of the unstable Ln hybrid populations changed only a little after three months storage, probably because of the low number of population doublings during freeze and thaw (around 12). Also, they still maintained the high genetic instability; i.e. they became $cyh2^R/cyh2^R$ homozygote and lost the killer phenotype under non-selective conditions (see changes of H7-8* and H6-7* hybrids after 100 doublings on YEPD, **Table 1**). However, after three years storage, heterozygous Ln hybrids became fully $cyh2^R/cyh2^R$ homozygote and lost the killer phenotype (**Table 1**). On the contrary, the genetically stable Pa hybrids did not change during freeze storage

The high proportion of sectored colonies, due to the low cell viability of newly obtained hybrids, indicates genome instability in both types of hybrids, Ln and Pa (**Fig. 1** and **Table 2**). However, the Pa hybrids did not behave as the Ln hybrid populations in evolving to $cyh2^R/cyh2^R$ homozygosis.

334

Table 1. Genetic analysis of isogenic hybrids (from the cross 85R6B×88P1A) newly obtained, and after 100 doublings.

| Hybrid | Phenotype | Newly obtained (32-49 doublings) | | | | | | Phenotype | After 100 doublings (total: 132-149) | | | | | |
| | | Liquid culture seeding in YEPD+CYH (+6 doublings) Colony size | | | Colony isolation and YEPD+CYH replica (+26 doublings) Clone phenotype | | | | Liquid culture seeding in YEPD+CYH (+6 doublings) Colony size | | | Colony isolation and YEPD+CYH replica (+26 doublings) Clone phenotype | | |
		G	M	P	Cs	Pa	Se		G	M	P	Cs	Pa	Se
H7-8	Ln, K	29	0	71	50	50	0	Ln, S	100	0	0	99	0	0.4
H7-8*	Ln, K	27	0	73	46	54	0	Ln, S	100	0	0	100	0	0
H7-8**	Ln, S	100	0	0	100	0	0	na	na	na	na	na	na	na
H6-7	Ln, K	22	0	78	48	52	0	Ln, S	100	0	0	100	0	0
H6-7*	Ln, K	29	0	71	47	53	0	Ln, S	100	0	0	100	0	0
H6-7**	Ln, S	100	0	0	100	0	0	na	na	na	na	na	na	na
H1-1	Pa, K	9.4	8.3	82	0	100	0	Pa, K	0.9	1.9	97	0	99	0.3
H1-1*	Pa, K	5.1	0	95	0	96	4.3	Pa, K	2.2	3.3	95	0	95	5.3
H1-1**	Pa, K	3	1	96	0	99	1	na	na	na	na	na	na	na
H2-1	Pa, K	0.3	1.8	98	0	100	0	Pa, K	0.9	3.4	96	0.8	99	0
H2-1*	Pa, K	0.2	4.4	95	0	100	0	Pa, K	1.6	1.9	96	0	99	1.1
H2-1**	Pa, K	1.4	1.4	97	0	100	0	na	na	na	na	na	na	na

Phenotype: Ln = lawn, Pa = papillae in YEPD+CYH (after replica plating); K= killer, S= sensitive to K2 killer toxin. Colony size (single-cell colony size in YEPD+CYH): La = large, Me = medium, Sm = small. Clone phenotype (phenotype after replica plating on YEPD+CYH of single-cell clones previously isolated on YEPD plate): Ln = lawn, Pa = papillae, Se = sensitive to cyh (no growth). The small colonies on YEPD+CYH and their spore-clones are killer, the large colonies and their spore-clones are killer-sensitive and have no ScV-M2 virus. Only two hybrids of each type are shown. Very similar results were obtained with the rest of the Pa or Ln hybrids analyzed from the 85R6B×88P1A and 85R6B×85P4D crosses. The data are percentages. *, after three months freeze storage; **, after three years freeze storage.

Table 2. Viability and frequency of sectored colonies of some freeze stored (*) isogenic hybrids (from the cross 85R6B×88P1A) just thawed, and after 100 doublings.

| Hybrid | Phenotype | Via 100 | Sectored colonies in YEPD+CYH (%) | | | Phenotype | Via 100 | Sectored colonies in YEPD+CYH (%) | | |
			La	Me	Sm			La	Me	Sm
H7-8*	Ln, K	56	11	--	53	Ln, S	99	0	--	--
H6-7*	Ln, K	90	3	--	18	Ln, S	98	0	--	--
H1-1*	Pa, K	72	73	--	93	Pa, K	99	67	70	4
H2-1*	Pa, K	47	72	70	89	Pa, K	100	82	85	3

Phenotype: Ln = lawn, Pa = papillae in YEPD-CYH (after replica plating); K= killer, S= sensitive to K2 killer toxin. Via 100 = frequency of viable yeast in 100 cells micromanipulated on YEPD plate. Colony size (single-cell colony size in YEPD+CYH): La = large, Me = medium, Sm = small. Only two hybrids of each type are shown. Very similar results were obtained with the rest of the Pa or Ln hybrids analyzed from the 85R6B×88P1A and 85R6B×85P4D crosses. -- = not applicable. The data are percentages.

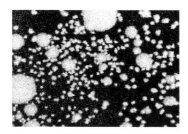

Fig. 1. Detail of sectored colonies (star shape in small colonies and involutions in large colonies) of the Pa-hybrid HA-5 in YEPD+CYH.

The proportion of $cyh2^R/cyh2^R$ cells in Sm colonies was higher in sectored colonies (52 %) than in non-sectored (28 %) (**Table 3**). These results suggest that he same genome instability that gives rise to sectored colonies in YEPD+CYH is the origin of the LOH at the *cyh2* locus.

Table 3. LOH in sectored and non-sectored colonies from Pa hybrids cultured in YEPD+CYH

Colony size	Sectored colonies				Non-sectored colonies			
	Via spo (%)	Segregation type (%)			Via spo (%)	Segregation type (%)		
		4R:0S	2R:2S	0R:4S		4R:0S	2R:2S	0R:4S
La	98±1.7	100±0	0±0	0±0	98±2.3	100±0	0±0	0±0
Me	100±0	100±0	0±0	0±0	81±16	64±32	36±32	0±0
Sm	98±1.9	52±25	40±21 4.8±4.8a 2.8±2.8b	0±0	83±14	28±28	69±31 2.6±2.6a	0±0
Each result is the mean (± standard error) of 72 analyzed tetrads from three single cell colonies (24 from each) of the same size from the Pa hybrids H1-1, H3-1 and H5-1. Colony size (in YEPD+CYH): La = large, Me = medium, Sm = small colonies. Via spo (%) = percentage of viable spores. a = 1R:3S, b =								

After serial passes of the hybrids (up to 30 doublings) in 4MB plates at 20 °C, the phenotype of Ln hybrids H6-7 and H7-8 changed from lawn to papillae (YEPD+CYH plate, **Fig. 2**), and they recovered a clear killer phenotype (4MB plate, **Fig. 2**) because the remaining $cyh2^R/cyh2^S$ killer yeasts killed the $cyh2^R/cyh2^R$ killer-sensitive yeasts. On the contrary, the Ln hybrids, already fully evolved to $cyh2^R/cyh2^R$ cells (such as H6-7 and H7-8 after 100 doublings), maintained the lawn phenotype, because not enough killer yeast remained to kill the $cyh2^R/cyh2^R$ killer-sensitive cells.

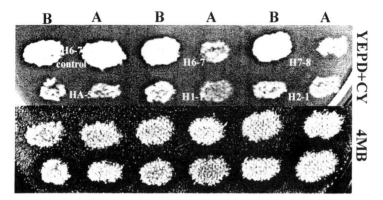

Fig. 2. Analysis of CYH^R phenotype (replica-plating in YEPD+CYH) and killer phenotype (replica-plating in 4MB seeded with the killer sensitive yeast EX33) of some stored (-80°C) hybrids before (B) and after (A) 30 doublings in the presence of active K2 toxin (4MB, 20°C). Ln hybrids: H6-7 and H7-8. Pa hybrids: H1-1, HA-5, and H2-1. Control: 6.7 after 100 doublings in YEPD at 30°C. The killer yeasts show an inhibition halo of sensitive yeast growth.

4   Conclusions

The sense of the biased evolution is determined by the mechanism involved in the LOH, but the speed of the processes is strongly affected by the differences in cell viability between the new homozygous yeasts and the original heterozygous hybrid cells. The loss of ScV-M2 virus may increase cell viability of $cyh2^R/cyh2^R$ yeasts and so favour biased evolution. However, the presence of active killer K2 toxin abolishes the asymmetric loss of heterozygosity of the hybrid populations.

Acknowledgements

This work was funded by grant 2PR01B002 from the Extremadura Regional Govemment, Spain.

References

[1]   Ramírez, M., A. Vinagre, J. Ambrona, F. Molina, M. Maqueda, and J. E. Rebollo. **2004**. Genetic instability of heterozygous hybrid populations of natural wine yeasts. Appl. Environ. Microbiol. 70:4686-4691.

Rodamine resistance as marker for monitoring yeasts in wine fermentations

Jesús Ambrona[1], Antonia Vinagre[1], Matilde Maqueda[1], Emiliano Zamora[2], María L. Álvarez[2] and Manuel Ramírez[1]

[1] Departamento de Microbiología, Facultad de Ciencias, Universidad de Extremadura, 06071 Badajoz, Spain
[2] Estación Enológica, Junta de Extremadura, Almendralejo, Badajoz, Spain

Winemaking with selected wine yeasts requires simple techniques to monitor the yeast population dynamics. This allows one to ensure the dominance of the inoculated yeast strain during must fermentation. We obtained new high concentration rodamine-resistant mutants and low concentration rodamine-white mutants, easy to detect by replica-plate assay, from selected wine yeasts. The rodamine-white mutants bear a mutation located at the *pdr5* locus. The mutants were genetically stable but most of them had lost the killer phenotype of the parent strain. They were genetically improved by elimination of recessive growth-retarding alleles followed by hybridization with selected killer wine yeasts. Several spore-clones were selected according to their must fermentation kinetics and the organoleptic quality of the wine. Some spore-clones were tested in industrial winemaking, and they were easily monitored during must fermentation using a colour-plate assay. They accounted for more than 96% of the total yeasts in the must, and the resulting wine had as good a quality as those made with reference commercial wine yeasts. The rodamine-white yeasts may also be detected by direct seeding on rodamine agar or by observation through a fluorescence microscope, which greatly reduces the time of analysis.

Keywords: yeast, wine, genetic marker, must fermentation, rodamine.

1   Introduction

Saccharomyces cerevisiae spontaneous mutants resistant to cycloheximide ($cyh2^R$) were isolated from industrial wine yeast [1]. The recessive mutations did not affect the fermentation kinetics, the quality of the wines, or the viability of active dry yeast made with the mutants. A fast, reliable, and economic method to monitor inoculated selected yeast through must fermentation by using homozygous $cyh2^R/cyh2^R$ mutants was developed [1]. The method can be recommended to the food industry because it is simple, and does not require sophisticated equipment or special personnel skills. Some $cyh2^R$ selected mutants have already been marketed and used to confirm their dominance during industrial grape juice fermentation for five years. The procedure has been working excellently, but recently we detected up to 10% of $cyh2^R$ yeasts in the spontaneous fermentations of some wineries. The profile of mitDNA restriction pattern of these $cyh2^R$ yeasts matches that of the marketed strain. We are beginning to be concerned about $cyh2^R$ yeasts becoming resident in the wineries that use $cyh2^R$ commercial yeasts. Moreover, if the presence of these $cyh2^R$ yeasts increases in many wineries, they could spread through the vineyard-winery ecosystem with unpredictable effects on autochthonous wild yeast populations.

To solve this problem, we obtained new wine yeasts resistant to rodamine (rod), easy to detect by replica-plate assay, direct seeding on rodamine agar, or by observation through a fluorescence microscope, what greatly reduces the time of analysis.

338

2   Materials and methods

Yeast strains: JP73, JP85, JP88, and JP33 are prototrophic and homothallic *S. cerevisiae* wine yeasts. JP73, JP85 and JP88 are K2-killer, and JP33 is virus-free killer-sensitive. E7AR1 is a K2-killer cycloheximide-resistant wine yeast from the hybrid 7AR, and is sold by BIOTEX (Talavera la Real, Spain). The genetic improvement of the mutants was done by hybridization with the 88P1A spore-clone from JP88.

Assay for killer activity was performed in low-pH (pH 4) blue plates (4MB) seeded with 100 μl of a 48-hour culture of the sensitive strain JP33. Strains being tested for killer activity were loaded (4 μl of a 48-hour culture to produce a patch approximately 5 mm in diameter), or replica plated onto the seeded 4MB plates and incubated for 4 days at 20°C. Killer strains produce a clear halo as a result of killing the seeded sensitive yeasts.

The percentage of wild parent yeasts was determined by analysing the mtDNA restriction pattern. Purified mtDNA was digested with *RsaI*. The fragments were separated in 0.5× TBE-0.8% agarose gel for 75-90 minutes and visualized on a UV transilluminator after ethidium bromide staining.

Analytical methods. Density, °Brix, pH, total acidity, volatile acid, reducing sugars, alcohol, and malic acid were determined according to the EC recommended methods. Lactic acid was determined using the EEC recommended method. Major volatile compounds and polyols were quantified by gas chromatography. T15 is the time needed to ferment 15% of the total sugars present in the must, and T100 is the time needed to ferment 100% of the total sugars.

Data were analysed for statistical significance by a one-way analysis of variance (ANOVA) with the software package SPSS version 11.5 for Windows (Chicago, IL).

3   Results and discussion

The frequency of spontaneous rodamine-white mutants (RodBC) obtained in YEPD+cyh 2μg/ml+rod 6G 5μg/ml was from 4×10^{-9} to 1.26×10^{-8} (**Table 1**).

Table 1.   Isolation and characterization of RodBC spontaneous mutants from wine yeasts.

Parent strain (MIC μg/ml of cyh, rod)	Frequency
JP73 (2, 300)	4×10^{-9}
JP85 (2, 100)	3.57×10^{-9}
JP88 (2, 100)	1.26×10^{-8}

The mutants were genetically improved by elimination of recessive growth-retarding alleles followed by spore-clone selection. Thereafter, a procedure for detection of the new improved spore-clones was designed and evaluated. This test allows one to easily differentiate the pink wild yeasts from the white RodBC mutants (**Table 2**).

Table 2. Results of the direct seeding test on YEPD+rod (5 µg/ml) to monitor RodBC mutants in mixtures of the spore-clone Rod1 and the parent yeast JP88.

Mix	Number of colonies detected		Number of colonies expected	
	White	Pink	White	Pink
50% Rod 1 + 50% JP85	137	126	131.5	131.5
90% Rod 1 + 10% JP85	135	21	140.4	15.6

The mutants were again genetically improved by hybridization with the selected wine yeast 88P1A. The spore-clones that showed the best fermentation kinetics, RodM2H3-1B and RodM2H5-6D, were pre-selected to perform micro-vinifications with fresh white must (Cayetana) and red grapes (Merlot). The RodBC spore-clones were monitored by the direct seeding test and by observation through a fluorescence microscope (**Fig. 1**). They accounted for 100% of the total yeasts in the must, and the fermentation kinetics parameters T15 and T100 were similar to those of the reference parent and commercial wine yeasts (**Table 3**).

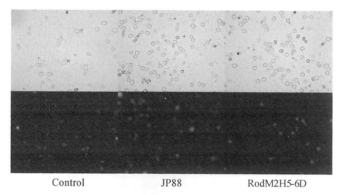

Control   JP88   RodM2H5-6D

Fig. 1. Nomarski (above) and fluorescence (below) microscopy of yeast samples from three Pardina juice vinifications: non-inoculated control, inoculated with JP88, and inoculated with the selected spore-clone RodM2H5-6D.

Table 3. Fermentation parameters of fresh white must (Cayetana) and red grape (Merlot) micro-vinifications inoculated with the pre-selected genetically-improved RodBC mutants.

Yeast strain	Cayetana		Merlot	
	T15	T100	T15	T100
Control	8	17	6	14.5
RodM2H3-1B	2	12.5	2.2	4
RodM2H5-6D	2.5	11	2.5	5
E7AR1	2.5	12.5	2.5	5
JP88	2	12.5	2.2	4

Six independent winery vinifications were carried out with fresh white must (Cigüentes and Macabeo) and red grapes (Cabernet-Sauvignon and Tempranillo, twice for both) using the two selected mutants RodM2H3-1B and RodM2H5-6D, the parent strain JP88, and a reference commercial wine yeast E7AR1. The mutants accounted for more than 96% of the total yeasts in the musts, and the resulting wine had as good a quality as those made with the parent and reference commercial wine yeasts. There were only significant differences between inoculated and control (non-inoculated) vinifications for T15 and the degree of dominance. In both cases, this was simply a reflection of the expected values (high T15 and low degree of dominance) of the control vinifications (**Table 4**).

Table 4. Must fermentation parameters and wine analysis results of six independent winery vinifications done with Cigüentes, Macabeo, Cabernet-Sauvignon, and Tempranillo grapes. ANOVA to study the effect of inoculation with RodM2H3-1B and RodM2H5-6D selected rodamine-white mutants.

Parameter	Yeast strain					p^a
	Control	E7AR1	JP88	RodM2H3-1B	RodM2H5-6D	
T15 (days)	3.8±0.39b	1.36±0.17a	1.6± 0.21a	1.7±0.26a	1.47±0.26a	0.000
T100 (days)	10.1±2.24a	5.9±1.45a	5.7±0.82a	6.01±1.49a	7.85±1.86a	0.304
Preference (%)	68±5.09a	66.9±4.8a	67±2.51a	61.3±6.33a	73±3.76a	0.559
Viable yeasts/ml ($\times 10^6$)	0.29±0.21a	1061±670a	1185±889a	423±209a	295±112a	0.426
Degree of dominance (%)	8.3±6.03a	100±0b	100±0b	98±1.62b	96.6±3.33b	0.000
Ethanol (%v/v)	13±0.36a	12.7±0.49a	12.8±0.41a	12.9±0.37a	12.7±0.49a	0.940
Alcohol yield	15.9±0.56a	16.6±0.66a	16.5±0.62a	16.3±0.6a	16.6±0.85a	0.948
pH	3.3±0.07a	3.3±0.08a	3.4±0.07a	3.3±0.07a	3.4±0.08a	0.999
Total acidity (g/l)	8.1±0.63a	8.3±0.5a	8.1±0.66a	8.7±0.55a	8.6±0.51a	0.899
Volatile acidity (g/l)	0.26±0.03ab	0.2±0.01a	0.18±0.01a	0.26±0.04ab	0.4±0.06b	0.017
Reducing sugars (g/l)	0.67±0.42b	0.05±0.03ab	0.01±0a	0.08±0.08ab	0.2±0.11ab	0.161
Malic acid (g/l)	1.3±0.13a	1.69±0.12b	1.56±0.09ab	1.8±0.06b	1.86±0.12b	0.016
Lactic acid (g/l)	0.16±0.1a	0.05±0.02a	0.08±0.04a	0.07±0.03a	0.05±0.02a	0.649
Acetaldehyde (mg/l)	18.5±4.04a	16.2±4.15a	16.8±3.97a	17±3.47a	19.5±3.6a	0.972
Ethyl acetate (mg/l)	17.3±6.18a	13.2±4.86a	15.8±5.99a	12±5a	10.5±4.21a	0.890
Methanol (mg/l)	132±32.5a	111±29.1a	136±41.2a	98±26.6a	125±30.8a	0.914
Fusel alcohols (mg/l)	342±65.5a	335±55.2a	348±47.4a	266±39.8a	285±51.2a	0.735
Butanol-1(mg/l)	0.83±0.54a	0.56±0.36a	1±0.63a	0.16±0.16a	0.41±0.41a	0.715
1-propanol (mg/l)	27.±4.19a	28.8±3.92a	34.3±8.04a	30.3±4.75a	30.6±5.46a	0.910
Isobutanol (mg/l)	55.7±10.7a	56.3±9.3a	53.2±14.1a	37.8±7.1a	42±5.4a	0.567
Amyl alcohols (mg/l)	259±54.3a	250±47.3a	260±34a	197±31.6a	228±43.1a	0.823
H_2S (U^b)	3.33±3.3a	3.33±2.1a	5±3.4a	11.6±4.7a	6.6±4.9a	0.535

The data are the means of six independent experiments and standard errors. [a] p values obtained by ANOVA for the wines made with each yeast. [b] U = arbitrary units. Different letters (a and b) mean significantly different groups found with a Duncan test at p<0.05.

4   Conclusions

The rodamine-white mutations of the mutants analysed do not significantly affect the yeast metabolism related to the production or elimination of the compounds responsible for the aroma and flavour of the wines. The statistical analysis showed that the RodM2H3-1B and RodM2H5-6D mutants are as good as their parent strains or as commercial winemaking yeasts, plus they can be monitored during fermentation by easy and inexpensive replica-plating or direct seeding on a YEPD+rod plate. Therefore they can be used in alternate years with other genetically marked yeasts, such as the cycloheximide-resistant strains, to avoid any inoculated yeast becoming resident in the wineries.

Acknowledgements

This work was funded by grant 2PR01B002 from the Extremadura Government, Spain. Matilde Maqueda is the recipient of a grant from the Spanish Ministerio de Educación y Ciencia.

References

[1]   Pérez, F.; Regodón, J. A.; Valdés, M. E.; De Miguel, C.; Ramírez, M. Cycloheximide resistance as marker for monitoring yeasts in wine fermentations. *Food Microbiol.* **2000**, *17*, 119-128.
[2]   Ramírez, M.; Vinagre, A.; Ambrona, J.; Molina, F.; Maqueda, M.; Rebollo, J. E. Genetic instability of heterozygous hybrid populations of natural wine yeasts. *Appl. Environ. Microbiol.* **2004**, *70*, 4686-4691.

Strength of attachment affects survival of *Salmonella* on inoculated cantaloupe treated with sanitizers

Dike O. Ukuku[*] and William F. Fett

USDA-ARS-ERRC, Wyndmoor, PA 19038

[*]Corresponding author: Tel: +1-215-233-6427; Fax: +1-215-233-6406; e-mail dukuku@errc.ars.usda.gov.

[†]Mention of trade names or commercial products in this article is solely for the purpose of providing specific information and does not imply recommendation or endorsement by the U.S. Department of Agriculture.

Difficulty in removing or inactivating bacterial human pathogens on whole cantaloupe surfaces is due both to the surface irregularities (crevices and pits in the netting) and to increasing time intervals between contamination and sanitizer treatments, presumably resulting in stronger attachment and/or biofilm formation.

The objective of this study was to determine the relationship between strength of attachment of *Salmonella* serovars to melon surfaces and the effectiveness of sanitizer treatments during a 24 h storage period.

Whole melons were inoculated at ~4.5 $\log_{10}$ CFU/cm^2 and stored at 25°C for up to 168 h before washing with water, chlorine (200 ppm) or hydrogen peroxide (2.5%). The ability of *Salmonella* serovars to resist removal by washing with water and killing by chlorine and hydrogen peroxide was compared at 20 min, 30 min, and 2, 6, 9 and 24 h post inoculation. Washing with water within 30 min after inoculation led to a significant ($p<0.05$) removal of a cocktail of *Salmonella* Poona strains (1.5 $\log_{10}$ unit reduction), but was totally ineffective thereafter. The efficacy of the sanitizer treatments in eliminating *Salmonella* from the melon surface was also dependent on the time interval between inoculation and treatment. Between 2 and 24 h storage, the strength of attachment for *Salmonella* on melon surfaces varied slightly among serovars, but generally increased from approximately 0.2 to 0.9. A 3 $\log_{10}$ unit reduction was achieved when chlorine and hydrogen peroxide treatments were applied up to 2 h after inoculation. Above 6 h, population reductions were approximately only 2 $\log_{10}$ units. Populations of *Salmonella* transferred to fresh-cut pieces during rind removal survived and grew during storage at 5, 10, 15 and 20°C for up to 10 days. However, storage at 5°C suppressed growth of *Salmonella* on fresh-cut pieces for up to day 8 when prepared within 2 h after inoculation and sanitizer treatment. A higher number of *Salmonella* was recovered in fresh-cut pieces prepared at day 7 immediately after sanitizing treatment. The results of this study indicate that *Salmonella* rapidly becomes strongly attached to the cantaloupe rind surface and that strongly attached bacteria are more difficult to remove by washing with water or inactivation with sanitizer treatments.

Keywords : *Salmonella*, strength of attachment, cantaloupe, sanitizer

1   Introduction

The ability of pathogenic bacteria to adhere to surfaces of fruits and vegetables continues to be a potential food safety problem of great concern to the produce industry. Surface structure and physiological characteristics of bacteria and surface structure and physical properties of the substratum, in this case the melon rind, play a major role on how and where bacteria may attach. The surface of cantaloupe is comprised of a meshwork of tissue commonly referred to as the "net" [1]. The presence of the raised net tissue gives the surface an inherent roughness which may favor microbial attachment and hinder detachment. Cantaloupe contaminated with various *Salmonella* serovars has been the cause of several outbreaks of foodborne illness in the U.S. The last three recorded U.S. cantaloupe-related outbreaks of salmonellosis have been due to *Salmonella* Poona.

Bacterial attachment to surfaces is influenced not only by cell surface charge [2] and hydrophobicity [3;4;5;6], but also by the presence of particular surface appendages such as flagella and fimbriae (pili) as well as extracellular polysaccharides [7;8]. Flagella, fimbriae, outer membrane proteins, and extracellular polysaccharide may influence bacterial attachment to plant surfaces [9]. Ukuku et al., [10] reported that bacteria attached on cantaloupe surfaces for more than 2 days were difficult to remove by washing treatment. Ukuku and Sapers [11] reported transfer of *Salmonella* Stanley inoculated on cantaloupe surface to fresh-cut pieces during rind removal. A better understanding of bacterial adhesion to cantaloupe is needed for the development of more effective washing treatments to control microorganisms on melon surfaces and their transfer to fresh-cut pieces. In this study, we investigated the effect of attachment strength on survival and removal of *Salmonella* populations on whole cantaloupe surfaces washed with water or sanitized with chlorine or hydrogen peroxide.

2   Materials and methods

2.1 Bacterial strains, growth conditions, and inoculum preparation.

Sixteen *Salmonella* strains were used in this study: *Salmonella* Stanley H0558, *Salmonella* Newport H1275, *Salmonella* Anatum F4317, *Salmonella* Infantis F4319, *Salmonella* Poona RM2350, *Salmonella* Hidalgo 02-517-2, *Salmonella* Typhimurium 45, *Salmonella* Gaminara 02-615, *Salmonella* Mbandaka 00-916, *Salmonella* Poona G-91-1595, *Salmonella* Poona 953, *Salmonella* Poona 348, *Salmonella* Poona 418, *Salmonella* Michigan, *Salmonella* Oranienburg 389, and *Salmonella* St. Paul 02-517. Bacteria were maintained on Brain Heart Infusion Agar (BHIA, BBL/Difco, Sparks, MD) slants held at 4°C. Prior to use, each culture was subjected to two successive transfers by loop inocula to 5 ml Brain Heart Infusion Broth (BHIB, BBL/Difco). A final transfer of 0.2 ml was made into 20 ml BHIB with incubation at 36°C for 18 h under static conditions. Bacterial cells were harvested by centrifugation ($10,000 \times g$, 10 min) at 4°C, and the cell pellets were washed in salt-peptone [0.85% NaCl, 0.05 % Bacto-peptone (BBL/Difco)]. The cell pellets were used to prepare two different types of inoculum as stated below. The first inoculum type consisted of the individual bacterial strains at 10^8 CFU/ml. The second inoculum type consisted of a mixture containing strains of all five strains of *Salmonella* Poona at approximately 1.4×10^8 CFU/ml per strain. Both types of inoculum were prepared in 3 L of 0.1 % (w/v) peptone-water.

2.2 Inoculation of cantaloupe.

Unwaxed whole cantaloupes (Western shippers) purchased from a local produce distributor were allowed to come to room temperature (~20°C) overnight before being inoculated. Individual cantaloupes were submerged in 3liters of each of the inocula (~18°C) and agitated by stirring with a glove covered hand for 10 min and then treated as stated below.

2.3 Washing treatments.

A commercial bleach containing 5.25% sodium hypochlorite (NaOCl, Clorox®, Clorox Company, Oakland, CA, USA), was diluted in sterile water to obtain a wash solution containing 200 ppm of chlorine. The pH was adjusted to 6.4 ± 0.1 by adding citric acid (Sigma Chemical Co., St Louis, MO, USA). Free chlorine in the solution was determined with a chlorine test kit (Hach Co., Ames, IA, USA). A second wash solution was prepared from a 30% stock solution of hydrogen peroxide (Fisher Scientific, Suwanee, GA, USA) that was diluted to 2.5% in sterile water. Washing treatments were performed by totally submerging the melons in 3 l of sterile tap water, or water containing 200 ppm chlorine, or 2.5% hydrogen peroxide. Melons were manually rotated for 5 min to assure complete contact of surfaces with the wash solution. Washed melons were placed on crystallizing dishes inside a biosafety cabinet to dry for 1 h.

2.4 Attachment experiments.

Cantaloupes inoculated with individual or cocktails of strains were washed as described above after intervals of 20, 40, 60 min or 1 to 7 days post inoculation with storage at 5 to 25°C. Bacterial cells in the wash water (loosely attached) and those remaining on the melon surfaces were enumerated as described below. The population remaining on the melon surface after washing (water) treatment was described as strongly attached bacteria. The strength of attachment (S_R) values were calculated as (strongly attached bacteria)/(loosely + strongly attached bacteria) as previously reported [6; 12]. The S_R value represents the percentage of the total bacterial population strongly attached to the cantaloupe.

2.5 Microbiological examination.

Plugs (n = 40) of cantaloupe rind (2.2 cm) weighing approximately 25 g total were cut with a sterile stainless steel cork borer and blended (Waring commercial blendor with 75 ml of 0.1% peptone-water at speed level 5 for 1 min.) Salmonella was enumerated on XLT4 agar (BBL/Difco, Sparks, MD) with incubation at 35°C for 48 h. For comparison, a pure culture of Salmonella was plated on XLT4 agar (BBL/Difco), incubated as above, and run parallel with the samples. Selected black or black-centered colonies from the agar plates were confirmed to be Salmonella according to the FDA Bacteriological Analytical Manual following conventional biochemical methods [13] as well as serological assays using latex agglutination (Oxoid, Ogdensburg, New York).

3   Results and discussion

3.1 Strength of attachment and removal by washing with water.

The relationship between S_R-values and log reduction for a cocktail of *Salmonella* Poona strains inoculated on cantaloupe surface and then washed with water following storage at 25°C, is shown in Figure 1.

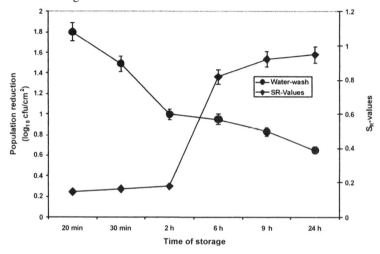

Fig. 1

Washing with water was most effective when applied within 20 min of inoculation resulting in a population reduction of 1.8 $\log_{10}$ units of *Salmonella* from the melon surface. Log reductions gradually decreased with increasing storage time before treatment to about 0.6 $\log_{10}$ at 24 h. After 2 h post inoculation, reduction of the *Salmonella* population by washing with water was not significant (p>0.05). The attachment strength (S_R-values) for *Salmonella* increased abruptly from 0.2 to 0.9 at 2 h to 6 h, followed by a slight increase between 6 and 24 h.

3.2 Strength of attachment and removal by sanitizer treatment.

The relationship between S_R-values and log reduction for a cocktail of *Salmonella* Poona strains inoculated on cantaloupe surface and then sanitized with chlorine or H_2O_2 following storage at 25°C is shown in Figure 2.

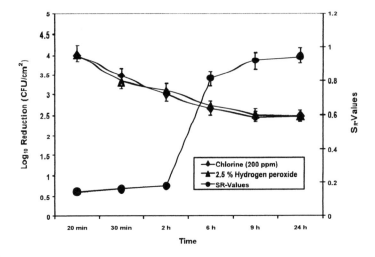

Fig. 2

Sanitizer treatments with chlorine or H₂O₂ were most effective in reducing populations of *Salmonella* when applied within 20 min after inoculation resulting in an ~ 4 log reduction for *Salmonella* on the cantaloupe surface. Population reductions decreased gradually with time of storage with a 2.5 log unit reduction at 24 h. Differences between sanitizers in the effect of post inoculation storage on population reductions obtained were not significantly ($p>0.05$) different. The extensive raised netting on the surface of cantaloupe melon no doubt provides numerous microbial attachments sites and helps to protect attached microbes from being washed from the surface, killed with sanitizers and possibly from environmental stresses such as UV radiation and desiccation. Surface irregularities such as roughness, crevices, and pits have been shown to increase bacterial adherence and reduce the ability of washing treatments to remove bacterial cells [14–16].

S_R-values of *Salmonella* populations on whole cantaloupe surfaces after storage of melons at 25°C for up to 2 h and log reductions due to chlorine treatments are shown in **Table 1**.

Salmonella Serovar/strain	S_R-value (30 min)	S_R-value (1 h)	S_R-value (2 h)
Poona G-91-1595	0.152 ± 0.012^a $(4.18\pm0.12)^b$	0.158 ± 0.021 (3.46 ± 0.15)	0.167 ± 0.012 (2.96 ± 0.20)
Stanley H0558	0.177 ± 0.023 (3.95 ± 0.15)	0.185 ± 0.013 (3.48 ± 0.16)	0.197 ± 0.016 (2.85 ± 0.14)
Poona RM2350	0.177 ± 0.021 (4.23 ± 0.16)	0.185 ± 0.020 (3.68 ± 0.15)	0.198 ± 0.022 (3.04 ± 0.18)
Newport H1275	0.184 ± 0.032 (3.60 ± 0.12)	0.189 ± 0.022 (3.09 ± 0.22)	0.199 ± 0.017 (2.69 ± 0.20)
Hidalgo 02-517-2	0.162 ± 0.012 (4.22 ± 0.18)	0.176 ± 0.013 (3.89 ± 0.13)	0.183 ± 0.015 (3.08 ± 0.15)
Typhimurium 045	0.179 ± 0.021 (3.89 ± 0.21)	0.181 ± 0.016 (3.14 ± 0.15)	0.189 ± 0.023 (2.79 ± 0.13)
Gaminara 02-615	0.168 ± 0.021 (3.96 ± 0.21)	0.178 ± 0.017 (2.88 ± 0.14)	0.188 ± 0.017 (2.58 ± 0.14)

Poona 953	0.177 ± 0.022 (3.67±0.14)	0.185 ± 0.022 (2.97±0.15)	0.193 ± 0.021 (2.72±0.21)
Poona 348	0.174 ± 0.020 (3.85±0.17)	0.188 ± 0.022 (3.19±0.16)	0.197 ± 0.016 (2.79±0.12)
Poona 418	0.173 ± 0.012 (3.89±0.15)	0.185 ± 0.016 (3.12±0.16)	0.198 ± 0.018 (2.79±0.15)
Michigan	0.178 ± 0.015 (3.49±0.22)	0.187 ± 0.021 (2.82±0.16)	0.208 ± 0.015 (2.58±0.14)
St. Paul 02-517	0.185 ± 0.022 (3.55±0.21)	0.197 ± 0.021 (3.08±0.21)	0.199 ± 0.014 (2.85±0.16)
Oranienburg 389	0.178 ± 0.018 (3.89±0.12)	0.185 ± 0.011 (2.87±0.13)	0.198 ± 0.011 (2.48±0.14)
Mbandaka 00-916	0.170 ± 0.013 (3.68±0.14)	0.183± 0.015 (2.89±0.12)	0.189 ± 0.015 (2.63±0.16)
Anatum F4317	0.172 ± 0.018 (3.85±0.15)	0.186 ± 0.038 (3.16±0.14)	0.194 ± 0.015 (2.86±0.15)
Infantis F4319	0.174 ± 0.020 (3.87±0.17)	0.185± 0.022 (2.99±0.14)	0.187 ± 0.020 (2.57±0.12)

[a]Values not in parentheses are means $\pm$ SD of three experiments with duplicate determinations per experiment. The average populations of individual strains of *Salmonella* recovered from the cantaloupe surface after inoculation was ~ 4.5 $\log_{10}$ CFU/cm^2.
[b]Values in parentheses represent log reductions after treatment with 200 ppm Cl_2 for 5 min.

The strength of attachment for all strains averaged 0.17 at 30 min and increased to an average of 0.19 after storage at 25°C for 2 h. *Salmonella* serovars Michigan and St. Paul had the highest S_R-values indicating stronger attachment on melon surfaces, and were slightly more resistant to the sanitizer treatment than the rest of the strains tested. Log reductions after treatment with Cl_2 and time of storage were negatively correlated. The shorter the storage time, the greater the log reduction.

S_R-values of *Salmonella* populations on whole cantaloupe surfaces after storage of melons at 25°C for 24, 72 and 168 h and log reductions due to chlorine treatments are shown in **Table 2**

Salmonella Serovar/strain	S_R-value (24 h)	S_R-value (72 h)	S_R-value (168 h)
Poona G-91-1595	0.822 ± 0.032[a] (2.88±0.12)[b]	0.856 ± 0.022 (2.46±0.12)	0.861 ± 0.012 (2.41±0.22)
Stanley H0558	0.815 ± 0.028 (2.85±0.18)	0.861 ± 0.023 (2.31±0.23)	0.885 ± 0.018 (2.35±0.18)
Poona RM2350	0.803 ± 0.021 (3.03±0.12)	0.862 ± 0.022 (2.72±0.21)	0.884 ± 0.020 (2.54±0.20)
Newport H1275	0.874± 0.052 (2.60±0.12)	0.881± 0.022 (2.51±0.22)	0.889± 0.012 (2.29±0.22)
Hidalgo 02-517-2	0.796 ± 0.018 (3.16±0.18)	0.806 ± 0.016 (2.76±0.16)	0.803 ± 0.015 (2.68±0.25)
Typhimurium 045	0.819 ± 0.022 (2.59±0.22)	0.834 ± 0.012 (2.34±0.12)	0.859 ± 0.026 (2.29±0.16)
Gaminara 02-615	0.798± 0.024 (2.76±0.24)	0.808± 0.027 (2.58±0.17)	0.818± 0.014 (2.48±0.14)
Poona 953	0.817 ± 0.027 (2.87±0.17)	0.847 ± 0.022 (2.57±0.16)	0.862 ± 0.020 (2.62±0.20)

Poona 348	0.871 ± 0.025 (2.85±0.15)	0.908 ± 0.022 (2.69±0.12)	0.901 ± 0.015 (2.49±0.15)
Poona 418	0.882 ± 0.032 (2.83±0.12)	0.922 ± 0.021 (2.62±0.21)	0.928 ± 0.018 (2.28±0.18)
Michigan	0.911 ± 0.020 (2.59±0.20)	0.942 ± 0.023 (2.22±0.13)	0.948 ± 0.015 (2.28±0.18)
St. Paul 02-517	0.905 ± 0.026 (2.75±0.26)	0.938 ± 0.023 (2.48±0.23)	0.955 ± 0.022 (2.35±0.12)
Oranienburg 389	0.783 ± 0.018 (2.83±0.18)	0.795 ± 0.013 (2.65±0.13)	0.798 ± 0.014 (2.68±0.14)
Mbandaka 00-916	0.800 ± 0.016 (2.70±0.16)	0.835 ± 0.012 (2.55±0.15)	0.840 ± 0.019 (2.40±0.19)
Anatum F4317	0.852 ± 0.028 (2.82±0.18)	0.866 ± 0.038 (2.66±0.18)	0.874 ± 0.018 (2.74±0.18)
Infantis F4319	0.842 ± 0.021 (2.84±0.21)	0.854 ± 0.026 (2.54±0.16)	0.872 ± 0.022 (2.72±0.12)

[a]Values not in parentheses are means ± SD of three experiments with duplicate determinations per experiment. The average populations of individual strains of *Salmonella* recovered from cantaloupe surface after inoculation was ~ 4.5 log CFU/cm^2.
[b]Values in parentheses represent log reductions after treatment with 200 ppm Cl$_2$ for 5 min.

As the strength of attachment is increased due to storage at 25°C, the efficacy of the sanitizer treatment is reduced. The population of *Salmonella* on cantaloupe surfaces was reduced by 2 log$_{10}$ unit at 168 h. Again, serovars Michigan and St. Paul that had higher S_R-values were slightly more resistant to the sanitizer treatment than the rest of the strains tested.

The population of a cocktail of *Salmonella* Poona strains on fresh-cut cantaloupe pieces prepared from inoculated melons sanitized with 200 ppm chlorine either 2 h or 168 h after inoculation and storage for various times and temperatures in shown in Figure 3.

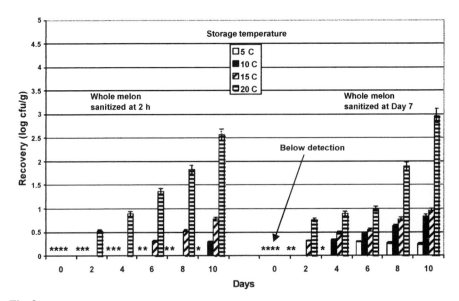

Fig. 3

No *Salmonella* were recovered on Day 0 of storage following fresh-cut processing prepared within 2 h and 168 h of inoculating the whole melon and sanitizer treatment. The population of *Salmonella* on fresh-cut pieces was greater with increasing storage temperature and time. Growth was detected earlier on fresh-cut pieces prepared from inoculated melons held 7 days prior to sanitizing and processing, compared to melons held only <2 h, suggesting greater transfer during fresh-cut preparation with the longer post-inoculation storage time before treatment with the sanitizer.

In conclusion, the strength of attachment of a cocktail of *Salmonella* Poona strains increased with time of storage at 25°C and was negatively correlated with population reductions after washing with water or sanitizer treatments (Figures 1 and 2).

The attachment strength for all individual *Salmonella* strains on the cantaloupe rind was low at 20 min to 2 h post-inoculation storage at 25°C, but increased greatly after storage for 24 to 168 h (Tables 1 and 2). As for the cocktail of *Salmonella* Poona strains, higher S_R-values due to longer storage times for the individual strains of *Salmonella* resulted in lowered efficacy of water washes or sanitizer treatments (Tables 1 and 2). Treatment of inoculated whole melons with 200 ppm Cl_2 was not effective in elimination of *Salmonella* Poona from the the surface and the pathogen was transferred to fresh-cut pieces. Continuous storage at 5°C is important to delay growth of the surviving pathogenic bacterial cells.

References

[1]   Webster, B. D. and M. E. Craig. 1976. Net morphogenesis and characteristics of the surface of muskmelon fruits. J. Amer. Soc. Hort. Sci 101:412-415.

[2]   Fletcher, M., and G. I. Loeb. 1979. Influence of substratum characteristics on the attachment of marine pseudomonad to solid surfaces. Appl. Environ. Microbiol. 37:67-72.

[3]   Van der Mei, H. C., M. Rosenberg, and H. J. Busscher. 1991. Assessment of microbial cell surface hydrophobicity, pp. 263-288. *In* N. Mozes, P. S. Handley, H. J. Busscher, and P.G. Rouxhet, (eds.), Microbial cell surface analysis. VCH, New York.

[4]   Van Loosdrecht, M. C. M., J. Lyklema, W. Norde, G. Scharaa, and A. J. B. Zehnder. 1987a. The role of bacterial cell wall hydrophobicity in adhesion. Appl Environ. Microbiol. 53:1893-1897.

[5]   Van Loosdrecht, M. C. M., J. Lyklema, W. Norde, G., Scharaa, and A. J. B. Zehnder. 1987b. Electrophoretic mobility and hydrophobicity as a measure to predict the initial step of bacterial adhesion. Appl. Environ. Microbiol. 53:1898- 1901.

[6]   Ukuku D. O. and Fett, W. F. 2002. Relationships of cell surface charge and hydrophobicity with strength of attachment of bacteria to cantaloupe rind. J. Food Prot. 65:1093-1099.

[7]   Fletcher, M., and G. D. Floodgate. 1973. An electron-microscopic demonstration of an acidic polysaccharide involved in the adhesion of a marine bacterium to solid surfaces. J. Gen. Microbiol. 74:325-334.

[8]   Frank, J. F. 2000. Microbial attachment to food and food contact surfaces. Adv. Food Nutr. Res. 43:320-370.

[9]   Romantschuk, M., E. Roine, K. Bjorklof, T. Ojanen, E-L. Nurmiaho-Lassila, and K. Haahtela. 1996. Microbial attachment to plant aerial surfaces, pp. 43-57. *In* C. E. Morris, P. C. Nicot and C. Nguyen-The (eds.), Aerial plant surface microbiology. Plenum Press, New York.

[10]   Ukuku, D. O, Pilizota, V., Sapers, G. M. 2001. Influence of washing treatment on native microflora and *Escherichia coli* population of inoculated cantaloupes. J. Food Safety 21:31-45

350

[11]   Ukuku, D. O, Sapers,   G. M. 2001. Effect of sanitizer treatments on *Salmonella*
         Stanley attached to the surface of cantaloupe and cell transfer to fresh-cut tissues
         during cutting practices. J. Food Prot. 64:1286-1292.

[12]   Dickson, J. S. and M. Koohmaraie. 1989. Cell surface charge characteristics and
         their relationship to bacterial attachment to meat surfaces. Appl. Environ.   Microbiol
         55:832-836.

[13]   Andrews, W.H., June, G., Sherrod, P., Hammack, T.S., and Amaguana, R.M. 1995.
         Salmonella. In FDA Bacteriological Analytical Manual, 8th ed., Chapter 5.

[14]   Austin, J.W., and Bergeron, G. 1995. Development of biofilms in dairy processing lines.
         J. Dairy Res. *62,* 509-519.

[15]   Frank,   J.F.   and Koffi,   R.A.   1990.   Surface   adherent   growth   of   *Listeria
         monocytogenes* is associated with increased resistance to surfactant sanitizer and
         heat. J. Food Prot. *53,* 550–554.

[16]   International commission on microbiological specifications for foods (ICMS).
         1980. Factors affecting life and death of microorganisms. In *Microbial Ecology of
         Foods (1)*. Academic Press, New York.

Sulfometuron methyl resistance as genetic marker for monitoring yeast populations in wine fermentation

Jesús Ambrona[1], Antonia Vinagre[1], Matilde Maqueda[1], Emiliano Zamora[2], María L. Álvarez[2] and Manuel Ramírez[1]

[1] Departamento de Microbiología, Facultad de Ciencias, Universidad de Extremadura, 06071 Badajoz, Spain
[2] Estación Enológica, Junta de Extremadura, Almendralejo, Badajoz, Spain

Winemaking with selected wine yeasts requires simple techniques to monitor the yeast population dynamics. This allows one to ensure the dominance of the inoculated yeast strain during must fermentation. We obtained new sulfometuron methyl resistant mutants, easy to detect by replica plate assay, from selected wine yeasts. The mutations were dominant and located at the *ilv2* locus that encodes for the enzyme acetolactate synthase. The mutants were genetically stable and maintained the killer phenotype of the parent yeast strain. They were genetically improved by elimination of recessive growth-retarding alleles followed by spore-clone selection, according to the must fermentation kinetics and the organoleptic quality of the wine. Some mutants were tested in industrial winemaking, and they were easily monitored during must fermentation using the plate assay. They accounted for more than 97% of the total yeasts in the must, and the resulting wine had as good a quality as those made with reference commercial wine yeasts.

Keywords: yeast, wine, genetic marker, must fermentation, sulfometuron methyl.

1   Introduction

Spontaneous *Saccharomyces cerevisiae* cycloheximide-resistant mutants ($cyh2^R$) were isolated from industrial wine yeast [1]. The recessive mutations did not affect the fermentation kinetics, the quality of the wines, or the viability of active dry yeast made with the mutants. A fast, reliable, and economic method to monitor inoculated selected yeast through must fermentation was developed by using homozygous $cyh2^R/cyh2^R$ mutants [1]. The method can be recommended to the food industry because it is simple, and does not require sophisticated equipment or special personnel skills. Some $cyh2^R$ selected mutants have already been marketed and used to confirm their dominance during industrial grape juice fermentation for five years. The procedure has been working excellently, but recently we detected up to 10% of $cyh2^R$ yeasts in the spontaneous fermentations of some wineries. The mitDNA restriction pattern of these $cyh2^R$ yeasts matches that of the marketed strain. Therefore, we are beginning to be concerned about $cyh2^R$ yeasts becoming resident in the wineries that use $cyh2^R$ commercial yeasts. Moreover, if the presence of these $cyh2^R$ yeasts increases in many wineries, they could spread through the vineyard-winery ecosystem with unpredictable effects on autochthonous wild yeast populations.

To solve this problem, we obtained new wine yeasts resistant to sulfometurom methyl (smr), easy to detect by replica plate assay, from selected diploid wine yeasts. The mutations were dominant and genetically stable, so that the mutants are even easier to obtain than the recessive $cyh2^R$ mutants [1; 2].

2 Materials and methods

Yeast strains: JP73, JP85, JP88 and JP33 are prototrophic homothallic *S. cerevisiae* wine yeasts. JP73, JP85 and JP88 are K2-killer, and JP33 is virus-free killer-sensitive. E7AR1 is a K2-killer cycloheximide-resistant wine yeast from the hybrid 7AR and sold by BIOTEX (Talavera la Real, Spain). The genetic improvement of the mutants was done by hybridization with 88P1A spore-clone from JP88.

Assay for killer activity was performed in low-pH (pH 4) blue plates (4MB) seeded with 100 μl of a 48-hour culture of the sensitive strain JP33. Strains being tested for killer activity were loaded (4 μl of a 48-hour culture to produce a patch approximately 5 mm in diameter) or replica plated onto the seeded 4MB plates and incubated for 4 days at 20°C. Killer strains produce a clear halo as a result of killing the seeded sensitive yeasts.

The percentage of wild parent yeasts was determined by analysing the mtDNA restriction pattern. Purified mtDNA was digested with *RsaI*. The fragments were separated in 0.5× TBE-0.8% agarose gel for 75-90 minutes and visualized on a UV transilluminator after ethidium bromide staining.

Analytical methods. Density, °Brix, pH, total acidity, volatile acid, reducing sugars, alcohol, and malic acid were determined according to the EC recommended methods. Lactic acid was determined using the EEC recommended method. Major volatile compounds and polyols were quantified by gas chromatography. T15 is the time needed to ferment 15% of the total sugars present in the must, and T100 is the time needed to ferment 100% of the total sugars.

The data were analysed for statistical significance by a one-way analysis of variance (ANOVA) with the software package SPSS version 11.5 for Windows (Chicago, IL).

3 Results and discussion

The frequency of spontaneous mutants obtained in SD+smr 20 μg/ml was from 2.22×10^{-8} to 1.26×10^{-7}. Around half of them resisted up to 100 μg/ml of smr (**Table 1**).

Table 1. Isolation and characterization of SMR^R spontaneous mutants from wine yeasts.

Parent strain (MIC, μg/ml)	Frequency of SMR^R spontaneous mutants isolated in SD+smr 20 μg/ml[a]	Number of mutants that grow at increasing smr concentration in μg/ml		
		20	50	100
JP73 (5)	$1.15 \times 10^{-7} \pm 7.10 \times 10^{-8}$	15	9	8
JP85 (5)	$2.22 \times 10^{-8} \pm 1.77 \times 10^{-8}$	15	10	10
JP88 (5)	$1.26 \times 10^{-7} \pm 9.21 \times 10^{-8}$	15	7	6
[a]The data are the means of three independent experiments and standard errors.				

The mutants were genetically improved by elimination of recessive growth-retarding alleles followed by spore-clone selection. The spore-clones that showed better fermentation kinetics were pre-selected to perform micro-vinifications with fresh white must (Cayetana) and red grapes (Merlot). They accounted for 100% of the total yeasts in the musts, and the fermentation kinetic parameters T15 and T100 were similar to those of the reference parent and commercial wine yeasts (**Table 2**).

Table 2. Fermentation parameters of fresh white must (Cayetana) and red grape (Merlot) micro-vinifications inoculated with the pre-selected SMR^R mutants.

Yeast	Cayetana		Merlot	
	T_{15}	T_{100}	T_{15}	T_{100}
Control	4.75	13	6.5	15
SMR10-11D	2.1	9	3.1	13
SMR3-2A	2.1	9	2.75	11
SMR12-1A	2.1	11	3.5	13
SMR16-5A	2.5	10	3.7	13
E7AR1	2.1	12	3.5	14
JP73	2.3	12	3.7	14
JP85	2.2	11	3.1	13
JP88	2.1	10	2.75	14

Six independent winery vinifications were carried out with fresh white must (Cigüentes and Macabeo) and red grapes (Cabernet-Sauvignon and Tempranillo, twice for both) using the two selected mutants SMR10-11D and SMR16-5A, the parent strain JP88, and a reference commercial wine yeast E7AR1 (**Table 3** and **Fig. 1**). The mutants accounted for more than 97% of the total yeasts in the musts, and the resulting wine had as good a quality as those made with parent and reference commercial wine yeasts. There were only significant differences between inoculated and control (non-inoculated) vinifications for T15 and the degree of dominance. In both cases, this was simply a reflection of the expected values (high T15 and low degree of dominance) of the control vinifications (**Table 3**).

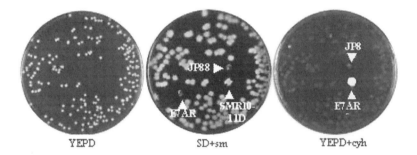

YEPD   SD+sm   YEPD+cyh

Fig. 1. Replica-plating on SD+smr and YEPD+cyh of colonies isolated in YEPD from a sample (day four) of the winery vinification made with Tempranillo grapes in 2002. The arrows indicate the patches of control yeasts added with sterile toothpicks onto a colony-free part of the plate.

Table 3. Must fermentation parameters and wine analysis results of six independent winery vinifications made with Cigüentes, Macabeo, Cabernet-Sauvignon, and Tempranillo grapes. ANOVA to study the effect of inoculation with SMR16-5A and SMR10-11D selected SMR^R mutants.

Parameter	Yeast					p^a
	Control	E7AR1	EX88	SMR16-5A	SMR10-11D	
T15 (days)	3.8±0.4b	1.36±0.17a	1.6±0.21a	1.95±0.28a	1.5±0.18a	0.000
T100 (days)	10.1±2.24a	5.9±1.45a	5.7±0.82a	6.8±0.83a	6.7±1.54a	0.258
Preference (%)	66±5.1a	67±4.8a	67±2.5a	66±4.2a	72±2.6a	0.877
Degree of dominance (%)	8.3±6a	100±0.0b	100±0.0b	97±1.43b	98±1.34b	0.000
Alcohol (% v/v)	13±0.36a	12.7±0.49a	12.8±0.41a	12.9±0.36a	12.8±0.38a	0.935
pH	3.3±0.07a	3.3±0.09a	3.4±0.07a	3.4±0.09a	3.3±0.07a	0.993
Total acidity (g/l)	8.1±0.63a	8.3±0.5a	8.1±0.66a	7.5±0.52a	8.1±0.36a	0.870
Volatile acidity (g/l)	0.26±0.03b	0.2±0.01ab	0.18±0.01a	0.20±0.02ab	0.18±0.02a	0.111
Reducing sugars (g/l)	0.67±0.42b	0.05±0.03a	0.01±0.00a	0.01±0.00a	0.00±0.00a	0.084
Malic acid (g/l)	1.3±0.13a	1.69±0.12a	1.56±0.09a	1.43±0.12a	1.52±0.16a	0.335
Lactic acid (g/l)	0.16±0.1a	0.05±0.02a	0.08±0.04a	0.08±0.04a	0.05±0.02a	0.706
Acetaldehyde (mg/l)	18.5±4.04a	16.2±4.15a	16.8±3.97a	18.2±4.39a	18.5±3.81a	0.991
Ethyl acetate (mg/l)	17.3±6.18a	13.2±4.86a	15.8±5.99a	15.3±6.15a	11.8±4.49a	0.960
Methanol (mg/l)	132±32.5a	111±29.1a	136±41.2a	130±20.6a	132±28.4a	0.981
Fusel alcohols (mg/l)	342±65.5a	335±55.2a	348±47.4a	368±30.1a	309±46.7a	0.947
Butanol-1 (mg/l)	0.83±0.54a	0.56±0.36a	1±0.63a	2.83±1.79a	1.66±1.66a	0.664
Butanol-2 (mg/l)	0.0±0.0	0.0±0.0	0.0±0.0	0.0±0.0	0.0±0.0	---
1-Propanol (mg/l)	27±4.19a	28.8±3.92a	34.3±8.04a	45.3±7.10a	38±6.34a	0.251
Isobutanol (mg/l)	55.7±10.7a	56.3±9.3a	53.2±14.1a	39.5±2.9a	32±3.8a	0.256
Amyl alcohols (mg/l)	259±54.3a	250±47.3a	260±34a	280±26.2a	237±38.7a	0.965
H_2S (U^b)	3.33±3.33a	3.33±2.1a	5±3.41a	3.33±3.33a	3.33±3.33a	0.994

The data are the mean values of six independent experiments and standard errors. [a] p values obtained by ANOVA for the wines made with each yeast. [b] U = arbitrary units. Different letters (a and b) means significantly different groups found with the Duncan test at p<0.05.

4   Conclusions

The smr^R mutations of the mutants analyzed do not affect significantly the yeast metabolism related to the production or elimination of the compounds responsible for the aroma and flavour of the wines. The statistical analysis showed that SMR10-11D and SMR16-5A mutants are as good as their parent or as commercial winemaking yeasts, plus they can be monitored by an easy and inexpensive replica-plating assay during fermentation. Therefore, they can be used in alternate years with other genetically marked yeasts, such as the cycloheximide resistant strains, to avoid any inoculated yeast becoming resident in the wineries.

Acknowledgements

This work was funded by grant 2PR01B002 from the Extremadura Government, Spain. Matilde Maqueda is the recipient of a grant from the Spanish Ministerio de Educación y Ciencia.

References

[1]   Pérez, F.; Regodón, J. A.; Valdés, M. E.; De Miguel, C.; Ramírez, M. Cycloheximide resistance as marker for monitoring yeasts in wine fermentations. *Food Microbiol.* **2000**, *17*, 119-128.

[2]   Ramírez, M.; Vinagre, A.; Ambrona, J.; Molina, F.; Maqueda, M.; Rebollo, J. E. Genetic instability of heterozygous hybrid populations of natural wine yeasts. *Appl. Environ. Microbiol.* **2004**, *70*, 4686-4691.

Use of carbon dioxide to control the microbial spoilage of bullfrog (*Rana catesbeiana*) meat

C. A. Conte-Júnior[1], M. Fernández[2*], S. B. Mano[1]

[1] Depto. de Tecnologia dos Alimentos, Faculdade de Veterinária, Universidade Federal Fluminense, Rua Vital Brazil Filho, n°64, 24.230-340, Niterói/RJ, Brasil.
[2] Dpto. de Nutrición, Bromatología y Tecnología de los Alimentos, Facultad de Veterinaria, Universidad Complutense de Madrid, 28040 Madrid, Spain.

*Corresponding author: email: manuela@vet.ucm.es, Phone: +34 91 394 3946, Fax: +34 91 394 3743

In the present work, the usefulness of modified atmosphere packaging (MAP) for preserving frog meat has been studied. Samples of bullfrog (*Rana catesbeiana)* meat were packaged in CO_2 enriched atmospheres (20/40 CO_2/N_2, 40/60 CO_2/N_2 and 80/20 CO_2/N_2, and 100% CO_2) and stored at 5±°C for 17 days. Two atmospheres containing 100% air and 100 N_2 were used as control. The atmospheres containing 80% and 100% CO_2 effectively extended the doubling time of the microbiota capable to grow in the storage conditions, increasing the shelf life 1.6-2.5-fold (10 and 15 days) in comparison with packaging in air (6 days). CO_2 concentrations lower than 80% appear not useful for MAP of frog meat, since they did not show any inhibitory effect on the natural microbiota, although other coexisting factors may influence this parameter.

Keywords: frog meat, *Rana catesbeiana*, MAP, CO_2

1   Introduction

Frog meat is not only appreciated for its exquisite flavor and texture but also as a source of protein of high biological value [1]. Brazil is a pioneer country in frog rearing, having developed and disseminated a frog farming system in a region where frogs were traditionally obtained by hunting [2]. In the international market, frog meat is usually commercialized as fresh or frozen legs, the former reaching higher prices than the unfrozen pieces [2, 3]. Furthermore, countries such as the United States, Canada and France import live animals, due to the consumers' preference for fresh meat [2]. In fact, many consumers prefer fresh meat despite its higher cost and lower shelf life [4].

Modified atmosphere packaging (MAP) technology provides a method of offering to consumers fresh products with a longer shelf life [5, 6]. This technology can be used by the food industry as an efficient tool to launch new products, providing convenience and practicability to them [7, 8]. At the present time, the food sector demands less severe technologies that can replace preservation methods which can alter food chemically and physically. This is the case of MAP technology [9,10].

The objective of the present study was to evaluate the growth/survival of the aerobic mesophilic population in leg fillets of bullfrog (*Rana catesbeiana*) packaged in MAP with different concentrations of CO_2 (100% CO_2, 100% N_2, 20/80 CO_2/N_2, 40/60 CO_2/N_2, 80/20 CO_2/N_2), by assessing the conditions and velocity of growth of those microorganisms in MAP environment and in aerobiosis, at 5±1°C.

2   Materials and methods

2.1 Sample preparation and modified atmosphere packaging

Fresh leg fillets of bullfrog (*Rana catesbeiana*) (supplied by a local slaughtering house in Rio de Janeiro) were used in this study. Seventy samples of approximately 70g were packaged in multilayer high barrier plastic bags ('Cryovac' BB4L) filled with approximately 1L of the following atmospheres: 100% air, 100% CO_2, 100% N_2 and three mixtures of CO_2/N_2 (20/80, 40/60 and 80/20). The atmospheres containing 100% air and 100% N_2 were used as controls. Once filled, the bags were immediately sealed and kept under refrigeration at $5\pm1°C$, during 17 days.

2.2 Analyses

Samples were analyzed for periodic counts of aerobic mesophiles and pH. A homogenate of 25g of frog meat was prepared in 225mL of a sterile physiological saline solution (0.85% NaCl). This homogenate was used for the microbiological analyses and to measure pH; this parameter was measured with a 'Horiba' mod M-13 automatic pH meter.
The growth of aerobic mesophiles was observed during 17 consecutive days. Counts were determined by the pour plate technique [11], in Plate Count Agar (PCA) (Merck), incubated at 35°C for 48h. Microbial analyses were performed in duplicate using two slices from different bags.
Bacterial growth parameters (lag phase and doubling time) were assessed using the Baranyi and Roberts's equation [12] and the shelf life of meat was defined as the number of days needed to reach 10^7 cfu g^{-1}.

3   Results and discussion

3.1 pH Changes

The changes in the pH of the samples are reported in Figure 1. The leg fillets presented an initial pH of 6.38, which showed an initial decrease in all the samples to 5.8-6.1 at day 2.
In the samples packaged in air, the pH stabilized around 5.8 until day 12, in which a remarkable increase was observed, coinciding with microbial counts above 10^7 cfu g^{-1}, and, therefore, spoilage. This increase may be attributed to the accumulation of basic compounds derived from the growth of *Pseudomonas* spp. and related microorganisms [13].
In the CO_2 enriched atmospheres, pH remained stable throughout the experience in values ranging from 5.5 to 6.1. Some authors have reported decreases in the pH of meat packaged with CO_2 as a consequence of the solubility of this gas in the food [14, 15], while other studies have shown no modifications in these values due to the effect of CO_2 [16–20].

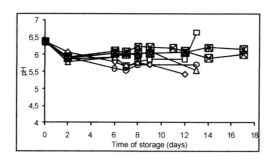

Fig 1. pH of the samples of frog meat packaged in: (─□─) 100% air, (─◇─) 100% N_2, (─△─) 20/80 CO_2/N_2, (─○─) 40/60 CO_2/N_2, (─✕─) 80/20 CO_2/N_2, and (─✳─) 100% CO_2, stored at 5±1°C.

3.2 Natural microbiota

The initial counts were low (10^4 cfu g^{-1}) (Figure 2), which reflects that the samples were obtained in good hygienical conditions, much better than the ones usually found in the slaughtering houses.

Counts remained below 10^6 cfu g^{-1} and 10^5 cfu g^{-1} until days 8 and 12 in the samples packaged in 80% and 100% CO_2, respectively, while by day 6, the other atmospheres studied showed counts ranging from $5 \times 10^6 - 5 \times 10^8$ cfu g^{-1}.

In the present work, the lag phase was calculated by the Baranyi and Roberts's equation. This parameter indicates the time needed for a microbial population to start to grow actively. Unfortunately, the lag phase is not constant, even in model systems, since it does not only depend on the species, but also on the origin of the microorganism or microbial group [21]. On the contrary, the doubling time is always constant in the same conditions (strain, substrate, temperature, atmosphere, etc).

In this study, as expected, the total microbiota needed more time to grow in the most selective atmospheres (Table 1), although noticeable differences were found among the CO_2 enriched atmospheres. The longest lag phase (8.6 days) was observed in the atmosphere containing 100% CO_2. The samples packaged in 80% and 100% CO_2 showed doubling times of 23.1 and 17.6 hours respectively, much higher than the doubling time observed in the atmospheres containing 20% and 40% CO_2, which was approximately 8 hours. The differences observed among the atmosphere containing CO_2, and also with the samples packaged in air, might be explained by the different dominant microbiota growing in each environment. As the purpose of this study was an initial approach to the shelf-life of frog meat packaged in MAP, no selective counts (lactic acid bacteria, Enterobacteriaceae, psychrotrophic bacteria, etc) were performed, although these studies are further needed to develop the most adequate MAP system for frog meat.

Finally, as it can be seen in Table 1, the values reached by the total microbiota during the lag phase were 0.7-1.1 logarithmic units lower in the CO_2 enriched atmospheres, in comparison with the samples packaged in air, which confirmed the selective effect of CO_2 on microbial growth, especially when 100% was used.

Table 1. Shelf life and growth parameters of micro-organisms of leg fillets of bullfrog (*Rana catesbeiana*) packaged in several gas mixtures and stored at 5°C for 17 days.

Atmospheres	SL	LP	DT	NC
Air	6	0.4	11.9	9.2
N_2 (100%)	5	0.6	8.0	8.1
20/80 CO_2/O_2	5	0.8	8.4	8.5
40/60 CO_2/O_2	5	0.6	8.0	8.4
80/20 CO_2/O_2	10	—	23.1	8.4
CO_2 (100%)	15	8.6	17.6	8.1

SL – Shelf life (days to reach 10^7 cfu g^{-1}).
LP – Lag phase (days).
DT – Doubling time (hours)
NC – Number of cells at the stationary phase (log cfu g^{-1})

3.3 Shelf life

As reported in Table 1, packaging in 80% and 100% CO_2 significantly increased the shelf life of frog meat (1.6 and 2.5-fold, respectively). In these samples, counts of 10^7 cfu g^{-1} were reached in approximately 10 days with 80% CO_2, and 15 days in 100% CO_2 (Figure 2). No favourable effect of the atmospheres containing a lower concentration of CO_2 and the one containing 100% N_2 was observed (Table 1, Figure 2). These results could be explained by the presence among the natural microbiota of frog meat of low numbers of psychrotrophic bacteria tolerant to medium levels of CO_2. These bacteria would have grown and have caused spoilage in a short period of time in those samples stored in 20% and 40 % CO_2, but they could have been inhibited by higher CO_2 concentrations.

The resulting shelf life of 10-15 days when packaging frog meat in 80% and 100% CO_2 atmospheres is solely based on a microbiological point of view. However, for establishing shelf life, the sensory properties of the product must also be taken into account. In our study a change in the colour (slightly paler than the normal colour) of the samples was noticed at day 10 in the 80/20 CO_2/N_2 atmosphere and at day 15 in the 100% CO_2 atmosphere, while odour did not significantly change.

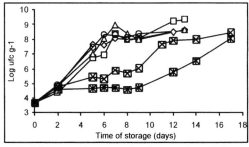

Fig 2. Aerobic mesophiles growth curves in the samples of frog meat packaged in: (—□—) 100% air, (—◇—) 100% N_2, (—△—) 20/80 CO_2/N_2, (—○—) 40/60 CO_2/N_2, (—✕—) 80/20 CO_2/N_2, and (—✱—) 100% CO_2, stored at 5±1°C.

4 Conclusions

In view of our results, it can be concluded that the atmospheres containing 80% and 100% CO_2 appear to be effective for extending the shelf life of frog meat stored under refrigeration. These atmospheres controlled the growth of aerobic mesophiles which, in those conditions, showed a low growth rate and moderate counts, although other coexisting factors may also have influence in these parameters. On the contrary, packaging in 20% and in 40% CO_2 did not appear to be useful for MAP of frog meat. Moreover, these atmospheres even enhanced the growth of the natural microbiota capable to grow in these conditions.

Further research is still necessary for the development of MAP of different animal origin products, such as frog meat (i.e. selective counting of individual/groups of microorganisms). It is likely that the effectiveness of the technique shall be optimized so that ideal concentrations of gases in MAP can be established for this product.

Acknowledgements

This work was supported by the ´Conselho Nacional de Desenvolvimento Científico e Tecnológico (CNPq)`, Brazil.

References

[1]   Vieira, M. I. (1993) *Ra-touro gigante: características e reproduçao* (4th ed.). Sao Paulo: INFOTEC.

[2]   Lima, S. L., Cruz, T. A., Moura, O. N. (1999) *Ranicultura: análise da cadeia produtiva.* Viçosa: Folha de Viçosa.

[3]   Palov, A., Garcia de Fernando, G. D. Ordoñez, J. A., Hoz, L. (1994) ß-hydroxyacyl-CoA-dehydrogenase (HADH) activity of unfrozen and frozen-thawed frog (*Rana esculenta*) legs. *Jounal of the Science of Food and Agriculture*, **64**, 141-143.

[4]   Ramos, E. M., Gomide, L. A. M., Ramos, A. L.S., Peternelli, L. A. (2004) Effect of stunning methods on the differentiation of frozen-thawed bullfrog meat based on the assay of ß-hydroxyacyl-CoA-dehydrogenase. *Food Chemistry*, **87**, 607-611.

[5]   Sarantópoulos, C. I. G. L. (1992) Novas tendências em embalagens de frango. In: Anais da Conferência Apinco de Ciência e Tecnologias Avícolas. pp. 67-77. Santos. Fundação Apinco de Ciência e Tecnologia Avícolas.

[6]   Mano, S. B., García de Fernando, G. D., López-Gálvez, D, Selgas, M. D., García, M. L., Cambero, M. I. and Ordoñez, J. A. (1995) Growth/survival of natural flora and *Listeria monocytogenes* on refrigerated uncooked pork and turkey packaged under modified atmosphere. *J. Food Safety.* **15**, 305-319.

[7]   Garcia de Fernando, G. D., Nychas, G. J. E., Peck, M. W. and Ordóñez J.A. (1995) Growth/survival of psychrotrophic pathogens on meat packaged under modified atmospheres. *International Journal of Food Microbiology*, **28**, 221-231.

[8]   Day, B. P. F. (1997) In *The Wiley Encyclopedia of Packaging Technology*, 2nd edition. A. L. Brody & K. S. Marsh (eds.). Wiley Interscience. New York.

[9]   Sarantópoulos, C. I. G. L., Alves, R. M. V., Contreras, C. J. C., Galvão, M. T. E. L. and Gomes, T. C. (1998) Use of a modified atmosphere masterpack for extending the shelf life of chicken cuts. *Packaging Technology and Science*, **11**, 217-229.

[10]   Mano, S.B. Ordónez, J.A. and Garcia de Fernando, G.D. (2000) Growth/survival of natural
        flora and *Aeromonas hydrophila* on refrigerated uncooked pork and turkey packaged in
        modified atmospheres. *Food Microbiology*, **17**, 657-669.
[11]   Vanderzant, C. and Splittstoesser, D. F. (1992) Compendium of methods for the
        microbiological examination of foods. 3nd edition. pp. 75-95. Washington. American Public
        Health Association.
[12]   Baranyi, J. and Roberts, T.A. (1994) A dynamic to predicting bacterial growth in food.
        International Journal of Food Microbiology, **23**, 277-294.
[13]   Mano, S. B., Ordóñez, J. A. and Garcia de Fernando G. D. (1999) Aumento de la vida útil y
        microbiología de la carne de pavo envasada en atmósferas modificadas. *Revista Brasileira
        de Ciência Veterinária*, **6**, 55-65.
[14]   Daniels, J. A., Krishnamurthi, R. and Rizvi, S. S. H. (1985) A review of effects of carbon
        dioxide on microbial growth and food quality. *J. Food Prot.*, **48**, 532-537.
[15]   McMullen, L. M. and Stiles, M.E. (1991) Changes in microbial parameters and gas
        composition during modifie atosfere storage of fresh pork loin chops. *J. Food Prot.*, **54**, 778-
        783.
[16]   López-Gálvez, D., Hoz, L., Ordóñez, J. A. (1995) Effect of carbon dioxide and oxygen
        enriched atmospheres on microbiological and chemical changes in refrigerated tuna
        (*Thunnus alalunga*) steaks. *J. Agric. Food Chem.*, **43**, 483-490.
[17]   Doherty, A., Sheridan, J. J., Allen, P., Mcdowell, D. A., Blair, Y. S. and Harrington, D. (1995)
        Growth of *Yersinia enterocolitica* O:3 on modified atmosphere packaged lamb. *Food
        Microbiology*, **12**, 251-257.
[18]   Sheridan, J. J., Doherty, A., Allen, P., Mcdowell, D. A., Blair, Y. S., Harrington, D. (1995)
        Investigations on the growth of *Listeria monocytogenes* on lamb packaged under modified
        atmospheres. *Food Microbiol.*, **12**, 259-266.
[19]   Doherty, A., Sheridan, J. J., Allen, P., Mcdowell, D. A., Blair, Y. S. and Harrington, D. (1996)
        Survial and growth of *Aeromonas hydrophyla* on modified atmosphere packaged normal and
        high pH lamb. *Int. J. Food Microbiol.*, **28**, 379-392.
[20]   Avery, S.M., Rogers, A.R. and Bell, G. (1995) Continued inhibitory effect of carbon dioxide
        packaging on *Listeria monocytogenes* and other microorgisms on normal pH beef during
        abusive retail display. *Int. J. Food Sci. Technol.*, **30**, 725-735.
[21]   Mano, S.B. Ordónez, J.A. and Garcia de Fernando, G.D. (2002) Aumento da vida útil e
        microbiologia da carne suína embalada em atmosfera moificada. *Ciênc. Tecnol. Aliment.*,
        22(1), 1-10.

Agriculture, Soil, Forest Microbiology

Agricultural non point sources control against microbiological contaminations of drinking water resources in dairy mountain areas

D. Trevisan[*], C. Renaers, J.M. Dorioz, G. Nicoud

INRA, UMR CARRTEL 75, avenue de Corzent 74203 Thonon les bains, France.

[*]Corresponding author: e-mail: trevisan@thonon.inra.fr, Phone :+ 33 (0)4 50 26 78 30, Fax: +33 (0)4 50 26 07 60

In mountain areas, microbiological contaminations of ground waters are frequent. Local authorities need appropriate tools for monitoring and preventing sanitary risks. Agriculture is a common source of microbiological contamination, due to livestock grazing and land spreading of manure. Protection against non-point agricultural pollution is necessary to prevent natural waters from the contamination before their sanitary treatment and their distribution. The aim of our work is to test an innovative approach for groundwater protection. We have developed a multi-criteria system for decision-making to define optimal practices for the prevention of microbe contamination. This system results from a multi-discipline analysis of the various factors governing water contamination. This multi-criteria system is tested on a representative sample of the groundwater supplies and production systems.

Keywords: fecal bacteria, non point source pollution, risk control, decision making.

1 Introduction

In most European countries the protection of drinking water supplies is guaranteed with the implementation of protection perimeters. Constraints on soil occupation or agricultural practices are defined on the basis of hydrogeologic criteria such as saturated zone permeability and microbiological decay rates [1]. Sets of regulated practices are defined and farmers are due to implement them, most often without financial means for effective controls nor precise evaluation of their efficiency.

In mountain areas, the microbiological quality of drinking water is frequently affected by breeding and dairy production. In such context, water supply protection does not allow, in many cases, an adequate control of risk: (i) constraints for agricultural management do not integrate the soil purification capacity on pathogens mortality and transfer [2]; (ii) regulated practices are not always adapted to agronomic constraints; neither do they refer to the farming systems functioning and to its ability to control the risks. Consequently the protection perimeters provide more often a juridical protection rather than a real prevention of sanitary risks.

Our objective is to set up a multidisciplinary framework connecting hydrology, pedology, agronomy, microbiology and social sciences in order to define agricultural technical projects for water conservation. Our goal is to promote a technical dynamic instead of the actual constraints implementation, relating this advance to the sustainability of mountain agriculture.

Models linking agricultural activities and water quality do not at present permit valuable tools for the evaluation of farming incidences on water resources. This is due to the numerous parameters for the estimation of stocks and pollutants loads and to the restricted and specific domain of models use [3]. As a consequence water management scientists have recently developed environmental indicators to link activities and water quality [4]. We have built up a similar approach to analyse and control microbiological contamination of drinking water. Our work is based on the analysis of two components: (i) the survival and transfer of fecal micro-organisms; (ii) the functioning of farming systems and its incidence on grazing and organic effluents management.

2 Method

We consider that water conservation has to be analysed considering two risk components: a bacteria transfer risk, depending on the environment purification capacity and a rejection risk of the regulation of the agricultural practices, depending on the farming systems. Transfer and rejection risks vary in opposite patterns with the increase of constraints implemented for water conservation. For example if we decide to adapt the fertilization levels to the purification capacity of the environment we will expect a decrease of transfer risk. This could bring up problems in the farming functioning system such as the necessity to find new spreading surfaces or to add new storage capacities. Consequently the probability of refusing the fertilization reduction increases and the conservation objective will likely not be achieved. We consider that the optimal changes or constraints to be implemented are those which minimize transfer and rejection risk (Fig. 1). The method we used to determine this optimum is based on ranking different changes or actions on agricultural practices regarding the two risk components, in order to identify the actions which simultaneously rank best on transfer and rejection risk hierarchy. The ranking process is founded on the development of indicators, which includes two work steps. The first one is concerned by an identification of elementary variables for the evaluation of phenomena (for example, temperature could be considered as an elementary measurable variable for the evaluation of bacterial decay). The second one corresponds to an aggregative process in which elementary variables are weighted and linked together to characterize the patterns of biological or environmental processes (for example one can note the temperature and the moisture and add or multiply these two elementary notes to evaluate the decay rates for fecal bacteria; if one considers that temperature is the most influent factor, the temperature note could previously be augmented by an appropriate weight factor). This aggregation between variables could be done by using models or by means of expert evaluation. Many works have been conducted for this purpose and have led to different tools that allow to classify, rank or establish typologies of various properties. The reader can refer to [5] for more precise details on available methods for indicators aggregation.

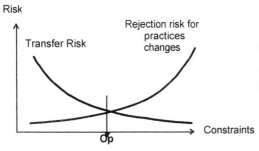

Figure 1 Evolution of transfer and rejection risk in relation to constraints for water conservation. Op represents the optimal constraint to attain to minimize the two risk components.

The ranking method ELECTRE III [6] has been chosen in the present work for the following reasons: (i) it is based on partial aggregation and fuzzy logic in order to rank different actions ; (ii) the method has previously been used to analyse various scenarios of farm and land management for water conservation [7].

We have used ELECTRE to compare water control actions over the two recognized risk components. The transfer evaluation includes indicators related to bacterial decay and fluxes which are recognized to be the main processes controlling the fate of microbiological water contamination (table I). Notes on survival or fluxes are deduced from previous studies carried out in various situations of soils and vegetations typical of dairy mountain areas [2]. The patterns for bacterial decay are : (i) increase of decay on soil surface with decrease of vegetation biomass ; (ii) increase with thinner spreads of organic effluent on soil surface ; (iii) increase in soils layers with enforcement of biological competition and protozoan predation. The fluxes notes take into account different properties of soils controlling velocity and amounts of water movements through the soil (leaching) or at the top surface (runoff).

The rejection risk indicators are built to account for changes feasibility in farming systems. They are identified by an analysis of the modifications impact on the different farming system components: grass production, organic matter management, herd displacement, additional works and supplementary costs for purchase of services, goods or materials (Fig. 2).

Surveys and indicators evaluations are carried out within 6 water supply protection perimeters. These study sites have been chosen to be representative of the diversity of hydrological conditions, soils and farming systems in mountain areas. Regarding hydrological conditions they are concerned by superficial water resources (from slope deposits), peri-glacial alluvial or karstic formations. Farms are involved in traditional or organic production mode. Soils and farm surveys are conducted in order to measure the different elementary variables introduced in the indicator aggregation. One site example will be treated, characterized by the following elements.

Hydrology and water quality. The site is concerned by a mid depth groundwater from peri-glacial deposits. Water is drawn on to supply a 3000 inhabitant municipality. Raw water is regularly contaminated by E. Coli or Streptococci, especially during high rain events from November to the end of March.

Pedology. Soils are diverse and represented by deep silt-clay slightly acid soils and shallow calcareous soils ; the first ones are characterized by a high purification capacity regarding the organo-mineral and hydrodynamic properties ; on the contrary the second must be considered as highly sensitive to bacterial transfer.

Soil occupation. Soil occupation is composed of perennial grass swards or annual cultures alternating wheat, maize and temporary clover production. Liquid or solid manure fertilization is applied in three major periods: (i) in early spring to empty effluent storage infrastructures and for agronomic purpose; (ii) during the late spring, after the first grass cut to favor regrowth;(iii) during late autumn after the last pastures or before ploughing and seeding of annual cultures, as well as to prevent exceeding of storage capacity.

3 Results

Considering the limited water storage of the calcareous soils, the farmers have introduced major limitations in the agricultural use of plots within this particular soil unit. In relation to the high sensitive risk for bacterial transfer we have enforced the limitations with additional constraints on stocking rate and grazing period and prohibition of solid or liquid manure fertilization. These protection measures have been set up without complementary analysis. The agricultural use of the plots concerned by the deepest silty clay soils has been subject to various proposals for water protection. Different solutions have been studied in order to avoid spreading during wet periods or to restrict the use of manure fertilization: (i) the delay of the spreads during spring, when soils moisture remains under field capacity; (ii) their advance during autumn, before the soils reach field capacity; (iii) the prohibition of manure fertilization and its substitution by mineral fertilizers.

Theses actions have been characterized by means of four indicators related to constraints on work, costs or manure transport. The related indicators for water risks have been evaluated on the basis of soils properties

Table I Notes obtained for fecal bacteria survival and fluxes indicators.

S1 : bacterial survival on soil surface	biomass < 1t.ha^{-1} for 1 cutting period	1 <biomass <1.75t ha^{-1}	biomass>1.75t ha^{-1}
note for S1	S1= 1	S1= 2	S1= 3
S2 : bacterial survival in organic effluent	fertilization level < 20 t. ha^{-1}	20 to 40 t. ha^{-1}	> 40 t. ha^{-1}
liquid manure	S2= 1	S2= 2	S2= 3
composted manure	S2= 1	S2= 1	S2= 2

S3 : bacterial survival in soil	Season*	Organic matter content < 2.5 %	2.5% < OM < 7.5%	MO > 7.5%
clay content < 25 %	spring - summer	S3= 1	S3= 2	S3= 3
	fall - winter	S3= 2	S3= 3	S3= 5
clay content 25 %<TA< 40%	spring - summer	S3= 2	S3= 2	S3= 3
	fall - winter	S3= 3	S3= 3	S3= 5
clay content > 40 %	spring - summer	S3= 3	S3= 3	S3= 3
	fall - winter	S3= 5	S3= 5	S3= 5

Td : Leaching intensity	Water Storage < 50 mm	50<WS< 100 mm	WS> 100 mm
spring - summer	Td=5	Td=2	Td=1
Fall - winter	Td=10	Td=5	Td=3

Tr : Runoff intensity	Water Storage < 50 mm	50<WS< 100 mm	WS> 100 mm
spring - summer	Tr=1	Tr=1	Tr=2
Fall - winter	Tr=2	Tr=5	Tr=10

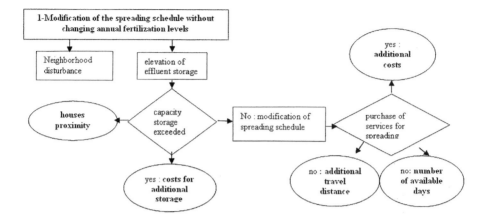

Fig. 2. Impact of changes on hay meadow fertilization. Squares identify modifications introduced by the change, diamonds choices for adaptation and ellipses elementary measurable variables to be taken into account for the evaluation of the change feasibility.

The ranking of the actions has been made independently for the two risk components after weights of the various indicators have been evaluated with an adapted Simo's procedure [5].

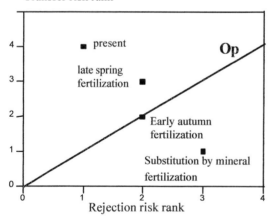

Fig. 3. Ranks of agricultural changes referring to water and refusal risk. The actions situated nearest to the bisector and the origin have to be considerer as optimum actions (Op)

Plotting together the two risk ranks enable us to evaluate the optimal changes to implement, i.e. the changes which minimize simultaneously the two component risks (fig. 3). The optimum corresponds to the nearest point to the bisector and along this line nearest to the origin. Early autumn fertilization corresponds to such an optimum; regarding the present situation it reduces the water by a two rank magnitude and increases the feasibility constraints by only one rank. This solution has been adopted and we are making longer observations on the evolution of bacteriological water quality.

4   Discussion

Our work deals with two main scientific stakes. The first concerns the necessity to resolve multidisciplinary problems, the second is to analyse a holistic system structured by various components and dimensions, from bacteria populations to groundwater dynamic and farming systems. The indicators we have developed give us some advances on these topics. They connect together thematic fields of knowledge and give the possibility to infer various phenomena for decision-making. Indicator development has been recently the object of innovative research in the field of environment management. To our knowledge there is no such equivalent work on microbiology risk for drinking water. The results we obtain have pointed out some possible changes to implement on farm system to avoid bacterial transfer. Some others have to be rejected due to their feasibility and the constraints they introduce on various aspects of the farm management. The validation of the sets of indicators is not directly possible and we are aware of their significance, but one can observe their functionality through water quality changes. Our work is still continuing for this purpose.

References

[1]   A. Lallemand-Barrès, J.C. Roux (1999). Périmètres de protection des captages d'eau souterraine destinée à la consommation humaine. Ed BRGM, Orléans, 334p.

[2]   D. Trevisan, J.Y. Vansteelant, J.M. Dorioz (2002). Survival and leaching of fecal bacteria after slurry spreading on moutain hay meadows: consequences for the management of water contamination risk. Water Research, 36, pp 275-283.

[3]   R. A. Dickinson (1991). Problems with using existing transport models to decribe microbial transport in porous media. In Modelling the fate of microorganisms, C.J. Hurst Ed. Am. Society for Microbiol. Washington, 21-47.

[4]   T.A. Wang, W.F. Mc Ternan (2002). The development and application of a multilevel decision nalysis model for the remediation of contaminated groundwater under uncertainty. J. of environmental management, 64, pp 221-235.

[5]   A. Schärlig, (1999). Décider à plusieurs critères : panorama de l'aide à la décision multicritère. Presses polytechniques et universitaires romandes, 304p.

[6]   B. Roy (1990). The outranking approach and the foundations of ELECTRE methods. In: Bana e Costa, C.A. (Ed.), Reading in multiple criteria decision aid. Springer-Verlag, Berlin, pp 155-183.

[7]   A. Arondel, P. Girardin (2000). Sorting croping systems on the basis of their impact on groundwater quality. European J. of operational research, 127 (3), pp 467-482.

Antagonistic activity of actinomycetes and fungi from horticultural compost against phytopathogen microorganisms

F. Suárez-Estrella[1], M.C. Vargas-García, M.J. López, and J. Moreno.

University of Almería, Department of Applied Biology, Area of Microbiology, CITE II B, La Cañada de San Urbano, 04120, Almería, SPAIN

[1]Corresponding author: e-mail: fsuarez@ual.es, Phone: +34 950 015891, Fax: +34 950 015476

In vitro studies on biological control of phytopathogen microorganisms by antagonistic actinomycetes and fungi from horticultural composts are reported. Two hundred and thirty-three microbial strains were isolated from compost samples obtained at different maturation stages. Selected phytopathogens from the Spanish Type Culture Collection (CECT) were *Erwinia carotovora* supsp. *carotovora CECT 225*, *Pseudomonas syringae* subsp. *syringae* CECT 127, *Xanthomonas campestris* CECT 95, *Fusarium oxysporum* f.sp. *melonis* CECT 20474, *Rhizoctonia solani* CECT 2824 and *Pythium ultimum* CECT 2365. *In vitro* experiments allowed the selection of isolates showing suppressing activity at least against four phytopathogens. Strains of actinomycetes showing the most important inhibitory effect were identified as *Streptomyces* spp. On the other hand, more effective fungal strains were identified as *Aspergillus ochraceus* and *Phialemonium obovatum*. Selected strains were, consequently, considered as potential biocontrol agents *in vitro* toward the most of phytopathogen microorganisms assayed.

Key words Biological control; compost

1 Introduction

Several widely distributed microorganisms (soils, composts and rhizosphere) have been tested for soilborne plant pathogen biocontrol over the last 20 years (1). However, few of these antagonistic microbes have been commercialized as biocontrol agents due to problems such as inconsistent performance in the field, lack of broad-spectrum disease suppression activity, or slower or less suppressive activity when compared with chemical pesticides (2).

Biocontrol fungi, such as *Trichoderma* and *Gliocladium spp.* or non-pathogenic strains of *Fusarium oxysporum*, have been used to reduce diseases caused by several phytopathogen microorganisms such as *Rhizoctonia, Pythium, Sclerotinia and Fusarium* (3). In addition, several less-known groups of biocontrol fungi and rhizobacteria, including *Penicillium spp.*, or *Pseudomonas* and *Bacillus spp.* have been used to control soilborne pathogens (4). Actinomycetes besides may inhibit or lyse soil specific fungi (5) or reduce soil fungal populations in general.

In this sense, compost obtained from heterogeneous vegetable wastes shows important suppressive effects against diseases caused by several plant pathogens such as *Pythium spp., Phytophthora spp.* (6), *Rhizoctonia spp.* and *Fusarium spp.* (7). The presence of beneficial microorganisms accounts, in part, for the suppressive effects of this product (1). *Bacillus spp., Enterobacter spp., Pseudomonas spp., Streptomyces spp.* and other bacterial genera, as well as *Penicillium spp., Aspergillus spp.*, several *Trichoderma spp.*, isolates of *Gliocladium virens*, and other fungi have been identified as biocontrol agents in compost-amended substrates (8).

Several soilborne microbes cause significant economic losses to most important cultures practiced at the South-East of Spain (9). The main goal of this work was the isolation and identification of effective biocontrol agents from several compost classes, for management of some of the most representative phytopathogens causing diseases at this zone.

2   Material and methods

Phytopathogen strains were supplied by CECT (Spanish Type Culture Collection). Bacterial cultures of *Erwinia carotovora* supsp. *carotovora* CECT 225 (E.c.c.), *Pseudomonas syringae* subsp. *syringae* CECT 127 (P.s.s.) and *Xanthomonas campestris* CECT 95 (X.c.), were kept in slant on nutrient agar (NA) while fungal cultures of *Fusarium oxysporum* f.sp. *melonis* CECT 20474 (F.o.m.), *Rhizoctonia solani* CECT 2824 (R.s.) and *Pythium ultimum* CECT 2365 (P.u.) were kept on potato dextrose agar (PDA) at 4°C.

Ten compost piles prepared from various materials were used. Raw materials in the mixtures and aeration conditions of composting processes are listed in Table 1. Piles were periodically turned and aerated while fluctuations in temperature were observed.

Table 1. Raw materials and aeration conditions of the composting process

	Raw Materials	**Aeration Conditions**
1	Pep[1]:Alm[2] (3:1)	Turned and aerated
2	Pep:Alm (3:1)	Turned and aerated
3	Pep:Alm (3:1)	Turned and aerated
4	Alm:Pep (3:1)	Turned and aerated
5	Alm:Pep (3:1)	Turned and aerated
6	Alm:Pep (3:1)	Turned and aerated
7	Pep:Alm (3:1)	Non aerated
8	Alm:Pep (3:1)	Non aerated
9	Pep:Alm (3:1)	Only turned
10	Alm:Pep (3:1)	Only turned

1: pepper plant wastes
2: almond peel wastes

Compost samples were collected at different stages of the composting process. Sampling times were 0, 14, 28, 42, 56, 70, 84, 120, 150, 180 and 240 days after starting the process, respectively. Most representative actinomycetes and fungi were isolated using SCA plates (sodium caseinate agar) and PDA plates (potato dextrose agar) respectively.

Finally, two hundred and thirty-two microbial strains were assayed against phytopathogenic microorganisms from CECT. *In vitro* antagonistic capacity of selected microorganisms was tested by the techniques of Marjan de Boer *et al.* [10], Kirby *et al.* [11], Landa *et al.* [12] and Monaco (13).

Some of the most interesting strains were identified by macro and microscopical observation and molecular methods. In this case, direct amplification of 16S rDNA gen, partial sequenciation and sequence analyses were used. NCBI Nucleotide Sequence Database was used to identify the antagonistic strains (http://www.ncbi.nlm.nih.gov/).

Statistical analyses were carried out to study the microbial antagonistic effect toward phytopathogen growth. Data were analysed by multifactorial analysis of variance (ANOVA) and compared using Fisher's protected least significant difference test (LSD) at 0.05 significance level.

3   Results

Seventy-one actinomycetes and one hundred and sixty-two fungal strains were isolated from compost samples obtained at different maturation phases.

Statistical analyses revealed the actinomycetes and fungi more antagonistic toward phytopathogen microorganisms. Factors considered were the compost pile from which the microorganisms were isolated (raw material), the time of composting and the aeration treatment applied (data not shown). Figures 1 and 2 show different examples of antagonistic effect between microorganisms assayed.

Fig. 1 *In vitro* antagonism test: antagonistic actinomycete toward phytopathogen bacteria. Observe the growth inhibition zones around the rolls.

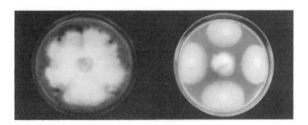

Fig. 2 *In vitro* antagonism test: antagonistic fungus toward phytopathogen fungus. Observe the growth inhibition (right) respect to control plate without antagonistic fungus (left).

In general, the factors related with composting process (raw material, aeration treatment and composting time) did not significantly affect the isolation of antagonistic microorganisms. However, when *in vitro* tests between antagonistic and phytopathogen fungi were performed, best inhibition indexes were observed during maturation phase of composting (data not shown).

Once the antagonism assays were achieved, more representative strains were selected to be identified. Table 2 indicates all actinomycetes and fungi showing antagonistic effect at least toward 4 of the phytopathogen microorganisms tested. Several actinomycetes selected in Table 4 were identified as *Streptomyces* spp. while fungal strains were identified as *Aspergillus ochraceus* and *Phialemonium obovatum* (5-9-3-11-1-2 and 4-3-3-11-1-2 respectively).

Table 2. Strains showing antagonistic effect at least toward 4 phytopathogens

	Strain	Ecc 225	Pss 127	Xc 95	Fom 20474	Rs 2824	Pu 2365
Actinomycetes	3-2-1-13-1-3	-	+	+	+	+	+
	3-5-1-13-1-3	+	+	+	-	+	+
	4-6-2-13-1-3	+	-	+	+	+	+
	7-4-2-13-1-3	-	+	+	+	+	+
	10-3-2-13-1-3	-	+	+	+	+	+
Fungi	4-3-3-11-1-2	-	-	+	+	+	+
	5-9-3-11-1-2	+	-	-	+	+	+

4   Discussion

Due to environmental concerns there is a considerable interest in finding alternatives to chemical pesticides for suppression of soilborne plant pathogens (3). In relation to this subject, compost, the final product of the aerobic biodegradation of organic matter, exhibits marked suppressive activity (14). To date, *Bacillus* spp., *Enterobacter* spp., *Flavobacterium balustinum, Pseudomonas* spp., other bacterial kind and *Streptomyces* spp., as well as *Penicillium* spp., several *Thichoderma* spp., isolates of *Gliocladium virens* and other fungi have been identified as biocontrol agents in compost-amended substrates (5). This suppression capacity could be due to compost recolonization by effective biocontrol agents after peak heating occurred in the composting process (15). In this sense, biocontrol agents inhibit or kill pathogens in mature composts and thereby induce disease suppression (16). Results shown in this work support the importance of maturation phase in relation to antagonistic fungi isolation (data not shown).

In most cases, effective bacteria used as biocontrol agents of plant diseases belong to the genera *Bacillus, Pseudomonas* and *Streptomyces* (17). Several actinomycete strains showing a strong antagonistic capacity have been identified as *Streptomyces* spp. in this work. Similar results have been obtained by Zaitlin et al. (18) who reported active actinomycete strains against several pathogens *in vitro*.

A strain of *A. ochraceus* isolated from compost piles showed to be an effective biocontrol agent against several of the phytopathogens assayed. In this sense, Marois and Mitchell

374

(19) reported an increase in the ID50 value for *Fusarium oxysporum* f.sp. *radicis-lycopersici* when soil was fumigated and amended with *Trichoderma harzianum, Aspergillus ochraceus* and *Penicillium funiculosum.*
Bearing in mind the results obtained, several of the isolated strains showed an antagonistic broad spectrum and could be applied individually and/or in combination for the suppression of soilborne diseases. However, additional research is required in order to perform *in vivo* tests leading to ascertain the real role these strains can play as effective biocontrol agents.

Aknowledgements

This work has been supported by funds from a FEDER project (1FD97-1389) and a project from Spanish "Ministerio de Ciencia y Tecnología" (CICYT AGL2001-2815).

References

[1]   H.A.J. Hoitink, A.G. Stone, and D.Y. Han, HortScience, **32**, 184 (1997).
[2]   S.L.F. Meyer, and D.P. Roberts, J. Nematol. 34, 1 (2002).
[3]   R.P. Larkin, and D.R. Fravel, Plant Dis. **82**, 1022 (1998).
[4]   A. De Cal, S. Pascual, and P. Melgarejo, J. Phytopathology, **145**, 231 (1997).
[5]   K.A. El-Tarabily, M.H. Soliman, A.H. Nassar, H.A. Al-Hassani, K. Sivasithamparam, F. McKenna, and G.E. St.J. Hardy, Plant Pathol. 49, 573 (2000).
[6]   H.A.J. Hoitink, and M.J. Boehm, An. Rev. Phytopathol. 37, 427 (1999).
[7]   F. Suárez-Estrella, M.A. Elorrieta, M.C. Vargas-García, MJ. López, and J. Moreno, IOBC wprs Bulletin **24 (3)**, 109 (2001).
[8]   G.E.S.J. Hardy, and K. Sivasithamparam, Soil Biol. Biochem. **23**, 756 (1991).
[9]   M. Reche, Enfermedades de Hortalizas en Invernadero. *Ed.* Servicio de Extensión Agraria. Ministerio de Agricultura, Pesca y Alimentación, Madrid, España (1991).
[10]   M. de Boer, I. Van der Sluis, L.C. Van Loon, and P.A.H.M. Bakker, European J. Plant Pathol. **105**, 201 (1999).
[11]   W.M.M. Kirby, A.W. Bauer, J.C. Sherrits, and M. Turck, Amer J Clin Path 45, 493 (1966).
[12]   B. Landa, A, Hervás, W. Bettiol, and R.M. Jiménez-Díaz, Phytoparasitica **25 (4)**, 305 (1997).
[13]   C. Mónaco, A. Perello, and M.C. Rollan, 1994. Microbiologia **10(4)**, 423 (1994).
[14]   H.A.J. Hoitink, M.J. Boehm, and Y. Hadar, Mechanisms of suppression of soilborne plant pathogens in compost-amended substrates, in: Hoitink, H.A.J. and Keener H.M. (eds.), Science and engineering of composting: Design, environmental, microbiological and utilization aspects. Renaissance Publ., Worthington, Ohio, Estados Unidos (1993) pp.601-621.
[15]   C.G. Phae, M. Saski, M. Shoda, and H. Kubota, Soil Sci. Plant Nutr. **36**, 575 (1990).
[16]   H.A.J. Hoitink, Y. Inbar, and M.J. Boehm, Plant Disease 75, 869 (1991).
[17]   S.G. Edwards, T. McKay, and B. Seddon, Interaction of *Bacillus* species with phytopathogenic fungi. Methods of Analysis and Manipulation for Biocontrol Purposes. In: Blakeman JP, Williamson B, (Eds), Ecology of Plant Pathogens. Wallingford, Oxon, UK: CAB International (1994).
[18]   M.S. El-Abyad, M.A. El-Sayed, A.R. El-Shanshoury, and S.M. El-Sabbagh, Folia Microbiol. **41**, 321 (1996).
[19]   J.J. Marois, and D.J. Mitchell, Phytopaholology **71**, 167 (1981).

Application of arbuscular mycorrhizal fungal in vitro biofertilizers in agro-industries

Bert Bago* and **Custodia Cano**

Department of Soil Microbiology and Simbiotic Systems, Estación Experimental del Zaidín (CSIC). Calle Profesor Albareda, 1. 18008-Granada, SPAIN

*Corresponding author: e-mail: abago@eez.csic.es, Phone: +34 958 121600 ext. 33, Fax: +34 958 129600

The benefits of arbuscular mycorrhizal fungi (AMF) application as biofertilizers and bioprotective agents are enormous. However current methods for large-scale AMF production are based on open-air host plant growth on soil-based substrates. This has important disadvantages in terms of inoculum certification and quality. We present here first scientific evidence for application of AMF biofertilizers produced *in vitro*, which may render revolutionary advantages for early inoculation on either seed-raised or micropropagated plants.

Keywords Arbuscular mycorrhizal fungi; biofertilizers; *in vitro* mycorrhizas; micropropagated plants

1   Introduction

The relevance of arbuscular mycorrhizas in enhancing plant mineral nutrition, soil stability and agroecosystem sustainability is well recognized [1]. This mutualistic symbiosis is established between most of the land plants and a small group of fungi, the arbuscular mycorrhizal fungi (AMF, Glomeromycota), common inhabitants of soils (Figure 1). The prospective benefits that application of AMF as biofertilizers and bioprotective agents could report are enormous, and they perfectly fit within EU directives reinforcing and promoting sustainable agro-ecosystems management. Nevertheless the production of large-scale AM fungal inoculants of certified quality has been always hindered by the obligate biotrophic nature of these microorganisms, which are unable to complete their life cycle and multiply unless establishing a mutualistic symbiosis with a host root [2].

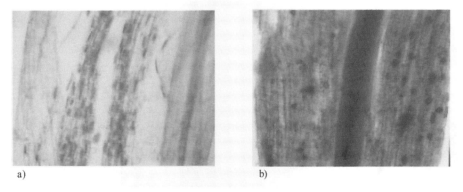

a)   b)

Fig. 1 Arbuscular- (left) and vesicular- (right) colonization of a root by arbuscular mycorrhizal fungi. Micropropagated olive-tree root colonized by isolates BEG 123 (a) or CIMA 09 (b).

Usual production protocols for currently-commercialized AMF-based biofertilizers involve culturing mycorrhizal plants in open-air containers, and using soil-type growing substrates. Once produced, such AMF-enriched substrates are collected, usually diluted and/or mixed with other substrates, and sold. Such soil-based AMF inoculants have clear disadvantages: i) it is very difficult to certify the absence of undesired and potentially harmful microorganisms (e.g. viruses, phytopathogenic fungi) in open-air produced inoculants; ii) the microscopic nature of AMF and the non-transparent nature of soil-based substrates makes it almost impossible for non-experts to confirm the product propagule-richness; iii) the up-to-date absence of strict EU directives for AMF biofertilizer production and commercialization allows fraud to occur; and iv) the lack of technical experts for evaluating AMF commercial inoculants makes it almost impossible to detect such potential frauds. We should add to all this the potential risk of introducing foreign, exotic AMF isolates, often contained in commercial biofertilizers, in distant agricultural/natural soils: these may compete with AMF natural populations, resulting detrimental for soil equilibrium [3].

2   Culturing arbuscular mycorrhizal fungi *in vitro*

The recently-developed technique of arbuscular mycorrhiza monoxenic cultures (AMMC, Fig. 2), consisting of the *in vitro* co-cultivation of AMF and root organ cultures [4], has greatly increased expectations in the possibility of large-scale production and application of AMF certified biofertilizers. *In vitro* production of AMF *via* AMMC overcomes problems usually presented by commercial, soil-cultured AM inocula. Very few of these new mycorrhizal products based on AMMC biotechnology are available by now in the market, and still they are mostly based on a few selected exotic isolates. We have carried out inoculation tests with seed-raised and micropropagated plants of commercial interest, and demonstrate that i) AMMC-based biofertilizers are equally or even more efficient than soil-produced ones; and ii) AMMC-based biofertilizers containing native fungal strains are far more efficient than those containing foreigner AMF isolates.

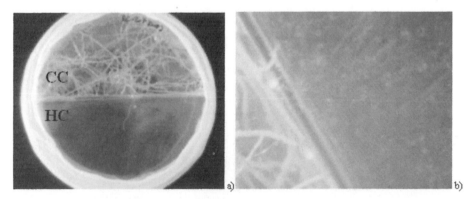

Fig. 2   Arbuscular mycorrhiza monoxenic cultures (AMMC) on bi-compartmented Petri plates (a). The culture compartment (CC) contains a root organ culture in symbiosis with an AMF. Extraradical AMF hyphae, but not roots, are allowed to cross the plastic barrier and develop in the hyphal compartment (HC), producing numerous spores (b).

3 Applying AMMC biofertilizers to nursery-produced plants

3.1 Biofertilization tests on seed-raised and micropropagated plants

Lettuce, leek and sunflower seeds were surface-sterilized following current protocols and pre-germinated *in vitro*. After one week germinated seed were transferred to 250 ml pots containing a steam-tindallized soil:quartz sand mixture (1:9). Micropropagated olive-tree, peach-tree and grapevine plants were *in vitro*-produced according to current industrial protocols. Plants were transferred from the *in vitro* flasks to 30 (olive-tree), 100 (fig-tree) or 250 (grapevine) ml pots containing an organic substrate mixture amended with a slow-release chemical fertilizer (Osmocote® 1g/L), and maintained in acclimation tunnels (1 wk). Four types of AMF inoculants were added to seed-raised or micropropagated plants upon transplant: 1) AMMC with foreign fungal strains (CIMA01, 02 and 08); 2) AMMC of an endemic south-eastern Spain fungal strain (CIMA09); 3) soil-produced inoculum of second endemic south-eastern Spain AMF strain (BEG 123); and 4) a mixture containing all the four AMMC (MIX). AMF inocula contained a minimum of 500 propagules per ml of each AMF isolate. Once acclimated, plants were moved to regular green-houses where they were watered thrice a week with tap water and maintained for two months.

a)

b)

Fig. 3 General view of seed-raised lettuce (a) and micropropagated grapevine (b) plants inoculated with AMMC biofertilizers containing either native or foreign AMF strains.

3.2 Results on seed-raised plants

The results obtained for lettuce and leek plants after two months culture with the different AMMC biofertilizers are shown in Table 1. Plants inoculated with AMMC were in general twice as big as non-mycorrhizal plants, this increase mainly corresponding to shoot fresh weight. Of special relevance were the results obtained for lettuce plants inoculated with CIMA 09, the AMMC-cultured endemic AMF, which rendered by large the best results. Percentages of mycorrhizal colonization were high (>50%) both for lettuces and leeks, the highest obtained in plants biofertilized with AMMC inoculants.

3.3 Results on micropropagated plants

The results obtained for olive-tree micropropagated plants after two months culture with the different AMMC biofertilizers are shown in Table 1 (results obtained for peach-tree and grapevines closely resemble these). No significant differences were observed for either shoot or root fresh weight between treatments, possibly due to the small volume of pots used. However, after two months of acclimation plant roots presented a very high percentage of mycorrhizal colonization. This has been shown to confer micropropagated plants of important advantages for survival/vitality compared to non-AM plants [5].

Table 1 Results of the inoculation of seed-raised and micropropagated plants with *in vitro*-produced AM biofertilizers. Different letters indicate statistically-different results.

Lettuce	Shoot fresh weight (g)	Root Fresh weight (g)	Shoot:Root ratio	% AM
WithoutAMF	14.4 a	9.3 a	1.5 a	0.0
CIMA 01	18.1 a	10.3 a	1.4 a	60.4 (5.7)
CIMA 02	19.0 b	14.0 b	0.8 c	63.6 (3.8)
CIMA 08	9.0 a	4.7 a	1.3 a	55.0 (12.8)
CIMA 09	22.1 b	11.5 a	1.5 a	51.5 (6.5)
BEG 123	19.3 b	11.3 a	1.3 a	49.0 (11.3)
MIX	16.9 ab	10.7 a	1.1 a	65.4 (5.6)
Leek	Shoot fresh weight (g)	Root Fresh weight (g)	Shoot:Root ratio	% AM
WithoutAMF	3.6 a	3.0 a	1.9 a	0.0
CIMA 01	7.7 b	5.9 b	1.6 a	64.7 (9.0)
CIMA 02	7.2 b	3.5 a	2.6 a	78.4 (2.4)
CIMA 08	7.1 b	5.0 ab	1.9 a	76.2 (4.6)
CIMA 09	8.1 b	5.0 ab	2.0 a	68.4 (4.9)
BEG 123	9.3 b	5.3 ab	2.4 a	52.7 (19.6)
MIX	5.6 ab	2.7 a	2.4 a	66.5 (1.3)
Olive tree	Shoot fresh weight (g)	Root Fresh weight (g)	Shoot:Root ratio	% AM
WithoutAMF	0.8 a	0.5 a	2.4 a	0.0
CIMA 01	0.9 a	0.6 a	2.3 a	70.2 (5.0)
CIMA 02	1.2 a	0.8 b	2.6 a	65.5 (5.8)
CIMA 08	1.0 a	0.6 a	2.4 a	70.7 (4.7)
CIMA 09	1.0 a	0.7 a	2.5 a	57.8 (3.5)
BEG 123	0.7 a	0.5 a	2.2 a	62.7 (15.8)

4   Conclusions

Huge prospect in the design and commercialization of *in vitro*-produced AMF biofertilizers are open by our confirmation of either seed-raised or micropropagated plants responsiveness to such a high quality inoculants. New mycorrhizal products based on AMMC biotechnology will be soon available in the market which will allow applying the immense potential of AMF-based *in vitro* biofertilizers.

Acknowledgements

This work was funded through Project AGL2001-1363 from the Spanish Ministry of Education and Science (MEC). C.C. was partially supported by Project REN2003-00968GLO (MEC; JM Barea, I.P.). We thank COTEVISA (L'Alcudia, Spain) for providing micro-propagated plants to carry out the inoculation tests.

References

[1]   S.E. Smith, D.J. Read (1997). Academic Press, London.
[2]   B. Bago, G. Bécard (2002) In: Gianinazzi, Schüepp, Barea, Haselwandter (eds). Birkhäuser-Verlag pp. 33-48
[3]   B. Bago, C. Cano (2005) In: S. Declerck, D.G. Strullu, A. Fortin (eds). Springer-Verlag (In press).
[4]   S. Declerck, D.G. Strullu, J.A. Fortin (2005) Soil Biology Series, Springer-Verlag (In press).
[5]   C. Azcón-Aguilar, A. Barceló, MT Vidal, G. de la Viña (1992) *Agronomie,* 12: 837-840

Biological control of flax scorch using *Glomus intraradices* and *Trichoderma atroviride*

E. Cariou-Pham[1*] **and S. Bonnan**[1]

[1] Institut Technique du Lin (ITL) au CETIOM, Centre de Grignon BP4, 78850 Thiverval-Grignon, FRANCE.

*Corresponding author: e-mail: emmanuelle.cariou@lin-itl.com, Phone: +33 130799589, Fax: +33 130799590

Flax scorch is a soilborne disease contaminating fields close to the coast in northern Europe. Our aim is to find new disease management controls. We evaluated the response of some fiber flax cultivars (sensitive and resistant) to seed coating (biotized seeds, provided by AGRAUXINE company) with a vesicular arbuscular mycorhizal fungus (*Glomus intraradices*) or an antagonist (*Trichoderma atroviride*). Two types of experiments were carried out. First, tests were undertaken in fields naturally contaminated by the pathogenic complex usually found in France which includes *C elegans* and several *Pythium* species. Second, plants cultivated in greenhouses, were inoculated with two pathogenic agents, *C elegans* and *P. sylvaticum*.
We showed that both in greenhouse and field trials, seed coating enhances growth and plant vigourness. In greenhouse seed coating also reduces root symptoms. Seed coating makes plants more resistant to flax scorch till plants were 8 weeks age but this effect was lost when plants were harvested.

Keywords *Linum usitatissimum*; flax scorch; *Chalara elegans*; *Pythium sylvaticum*; *Glomus intraradices*; *Trichoderma atroviride*; biological control.

1   Introduction

Flax scorch is a soilborne disease which contaminates more than 20 % of flax culture in northern Europe. Flax scorch is considered an important disease and is responsible for severe damage, up to total loss of harvest. Symptoms of flax scorch are the development of root necrosis and a reduction of plant growth. Two pathogenic fungi complexes are responsible for the disease. Either *Chalara elegans* (Deuteromycète) or *Pythium buismaniae* (Oomycète), is the main pathogenic agent. Both are associated with other species of *Pythium* like *P. sylvaticum* and *P. intermedium* [1]. Nowadays management control of the disease uses prophylactic means and only very few cultivars are resistant to this disease.

Our aim is to find a new disease management control based on the association of mycorhizal or antagonistic fungi and flax We evaluated the response of flax to seed coating with a vesicular arbuscular mycorhizal (MA) fungus (*Glomus intraradices*) or an Ascomycete antagonist (*Trichoderma atroviride*). MA fungus increases the resistance of plants to the disease by the induction of natural plant protection (*Fusarium oxysporum* on flax, [2]). *Trichoderma* are biological control agents of fungi pathogens (*Colletotrichum acutatum* and *Botrytis cinerea* on strawberry) thanks to their process of mycoparasitism [3]. Tests were carried out in naturally contaminated fields with the pathogenic complex mainly present on french fields which includes *C elegans* with several *Pythium* species. Plants cultivated in greenhouses, were artificially inoculated with two pathogenic agents, *C elegans* and *P. sylvaticum*.

2 Material and methods

2.1 Cultivar and coated seeds (biotized seeds)

The cultivars used are given in table 1. Coated seeds were made with *G. intraradices* (Schench & Smith) or the antagonist fungi *T. atroviride*. They were provided by AGRAUXINE company. Both were used in greenhouse and field experiments.

Table 1 Description of the cultivars used for our study.

Cultivar	Breeder	Inscription year to the french catalog	Resistance to flax scorch disease	greenhouse test (a)	field test
Aurore	SCA Terre de Lin	1997	resistant		+
Diane	SCA Terre de Lin	1995	resistant	+	+
Hermes	INRA & SCA Terre de Lin	1992	sensitive	+	+
Opaline	INRA & SCA Plessis-Belleville	1984	sensitive		+
Venus	SCA Terre de Lin	1997	sensitive		+
Viking	INRA & SCA Terre de Lin	1985	medium resistant	+	+

(a): '+' indicates cultivar used in test.

2.2 Greenhouse tests

Seeds were used naked or coated. In each case, plants were cultivated in presence or absence (control) of pathogenic fungi *C elegans* and *P. sylvaticum* (table 2). *C. elegans* and *P. sylvaticum* are mixed to the soil at a rate of 10^5 g spores/ml for *C.elegans* and $2,34.10^5$ cfu/ml for *P. sylvaticum*. 10 seeds were sown in each pot (capacity of 300 ml) repeated three times. Soil composition is of 3/5 from topsoil, 1/5 of peat and 1/5 of sand. Plants were cultivated in greenhouse at 14°C, 75 % of moisture and a light intensity of 34000 Lux.

Table 2 Pathogenic Fungi isolate description

Pathogenic fungi	Collection	Accessing number
Chalara elegans	ITL Collection	R 82.296
Pythium sylvaticum	ITL Collection	98.10.28.19

After six weeks cultivation, plant length were measured. The extent of root necrosis was determined using a picture of the total root system of the 10 plants from each pot. Pictures were analysed using the software Microsoft Photo Editor to make a black and white picture: black for root necrosis, and white for healthy roots. These pictures were analysed using: GIMP, version 2.0 for Microsoft (http://www.gimp.org) to quantify root necrosis.

2.3 Field trials

Flax plants were grown under field conditions at Quiberville (Normandy, France). Different seed coatings (biotized seeds) were tested on several cultivars. Seeds were sown on March 16, 2004, at a density of 2000 plants/m² using an experimental seeder (BAURAL). No fungicides were applied on field. The "Institut Technique du Lin" uses the sensitive

382

Opaline cultivar as a control for the presence of the disease. Two replications were done [4].
Three weeks after sowing, plant vigourness was determined with a code note ranging from 1 (the least vigorous plants) to 9 (the most vigorous plants). Every three weeks, average height of plants was measured. At the bloomer stage, [1], a resistance disease level was determined according to a scale of 1 (culture completely destroyed) to 9 (healthy culture).

3   Results

3.1   Greenhouse test

After harvesting, plant height and root necrosis length were measured. Results are given in Fig. 1.

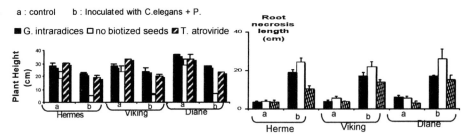

Fig. 1: Plant height (left) and root necrosis length (right) with or without pathogenic fungi in soil mixture(*C elegans* and *P. sylvaticum*) and different biotized seeds (*T atroviride* or *G intraradices*). Cultivars are Hermes, Viking and Diane.

In the presence of both pathogens *C. elegans* and *P. sylvaticum*, plant height (Fig1. left) without coated seeds is reduced at least 4 to 5 times compared to the control (without pathogens).   There is no significant differences between the cultivars. When comparing biotized seeds and non coated seeds, we obtained similar heights for inoculated plants as for control plants (without pathogens). The biotized seeds are efficient in reducing plant symptom after experimental inoculation of *C elegans* and *P. sylvaticum*. We observed a significant reduction of root necrosis in plants from coated seed (Fig.1 right). The effects of seed coating were similar, although for Hermes and Viking where root necrosis reduction length was more important with *T. atroviride*. Nevertheless this root necrosis reduction length is not sufficient enough to recover healthy plants from the control un-inoculated plants (a is compared to b).

3.2 Field test

Results of plant vigourness observation are given in Fig. 2. All cultivars with coated seeds are more vigorous than the others except for Diane, for which only seed coating with *T. atroviride* improves plant vigourness. Results of plant height are in Fig. 3.

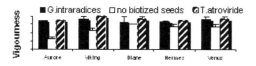

Fig. 2 Effects of biotized seeds on plant vigourness after germination. The cultivars Aurore, Viking, Diane, Hermes and Venus were sown on a naturally contaminated field by flax scorch disease.

Fig 3. Plant height at 8 (left) or 10 (right) weeks after sowing the cultivars Aurore, Viking, Hermes and Venus on a naturally contaminated field by flax scorch disease. Different seed coatings (control, G. *intraradices* or *T. atroviride*) were tested.

Seed coating improves the growth of plants 8 weeks after sowing for the resistant and medium resistant cultivars Aurore, Diane and Viking but not for the sensitive ones, Hermes and Venus, where no significant differences are noticed (Fig. 3 left). Ten weeks after sowing seed coating has no longer effects on plant height (Fig. 3 right).

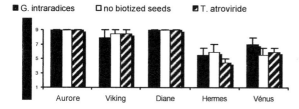

Fig. 4 Disease resistance (scale from 1 to 9) of cultivars with seeds coating (*G. intraradices* or *T. atroviride*) compared to control 10 weeks after sowing on a naturally contaminated field.

Disease resistance level appeared to be correlated to the use of resistant or sensitive cultivar without any significant effect of seed coating (Fig. 4). Resistant cultivars like Aurore and Diane get the highest mark 9 for disease resistance, so as Viking (medium resistant cultivar) whereas sensitive ones (Hermes and Venus) gets intermediate values (5 to 7) compared to Opaline (value of 3) a very sensitive cultivar used as a control for the presence of the disease in field trials.

3   Discussion and conclusion

Seed coatings with one or another of the fungi used, *T. atroviride* or *G. intraradices*, allow a reduction of root necrosis in greenhouse and improve plants height. These results are consistent with the field results at the beginning of cultivation with an improve of plant vigour and height. However this effect is lost 10 weeks later at the bloom plant stage where no differences are shown on disease resistance level. Dugassa *et al.*, (1996) showed that flax plants with MA gets their ethylene concentration increased in roots whereas auxines and gibberellins hormones concentration were increased in the vegetative part of the plant. This could explain the differences in growth between plants with MA and control. Also, MA are known to be inductors of defence mechanisms of the plants [5]. Indeed, mycorhization of plants by MA type fungi improve its respiratory activity [6], which indicates an increase of metabolic activity. This makes the plants able to react more quickly against pathogens. In some cases, roots MA show an increase in resistance gene expression [7]. Fungi of the genius *Trichoderma* are frequently used because they act like biological control agent by affecting pathogens directly by mycoparasitism [8] or indirectly, by the production of many antibiotics [9]. The use of *Trichoderma* species as antagonists of pathogenic *Pythium* species on flax, was already studied [10]. As in our experiment, poorly effects of seed coating were noticed by these authors in field test compare to greenhouse test. As MA or *Trichoderma* fungi seems to act very early after germination on plant vigour and health, seed coating could be used for flax control of damping off.

Acknowledgements

This work was founded by "Association de coordination Technique Agricole" (ACTA). We would like to thank BACHAR BLAL from AGRAUXINE company.

References

[1]   E. Cariou. La brûlure du lin : étiologie, épidémiologie et stratégie de lutte : Thèse : Institut National Agronomique Paris-Grignon , (2001), p.109.
[2]   Von Reichenbach G.H. et Schönbeck F. Influence of VA mycorrhiza on drought tolerance of flax (*Linum usitatissimum L.).* I. Influence of VAM on growth and morphology of flax and on physical parameters of the soil. Angew. Bot. **69** 49 (1995).
[3]   Sivan A., Elad Y. et Chet I. Biological control effects of a new isolate of *Trichoderma harzianum* on *Pythium aphanidermatum*. Phytopathology **74** 498 (1984).
[4]   Beaudoin X. Compte-rendu de l'Institut Technique du Lin (Paris, 1991), p.89 (1991).
[5]   Koide R.T et Schreiner R.P. Regulation of the vesicular-arbuscular mycorrhizal symbiosis. Anu. Rev. Plant Physiol. **43** 557(1992).
[6]   Dehne H.W. et Backhaus G.H. The use of vesicular-arbuscular mycorrhizal fungi in plant production. I. Inoculum production. Z. Pflkrankh. PflSchutz **93** 415 (1986).
[7]   Dugassa G.D., Von Alten H. et Schönbeck F. Effects of arbuscular mycorrhiza (AM) on health of *Linum usitatissimum* L. infected by fungal pathogens. Plant and Soil **185** 173 (1996).
[8]   Dennis C. et Webster J. Antagonistic properties of species-groups of *Trichoderma*. Trans. Br. Mycol. Soc. **57 (I)** 25 (1971).
[9]   Pieta D. Some aspects of using antagonistic microorganisms in plant disease control. Ann. UMCS Lublin, sectio E, Agricultura, 1 (1997).
[10]   Ligocka A., Paluszak Z., Sadowski S. et Dziedzic T. Enzymatic and antagonistic potential observed in flax-root-infecting fungi. Electronic Journal of Polish Agricultural Universities, Agronomy, **5** I (2002).

Characterization of *Brenneria* sp. from poplar cankers in Spain

Elena G. Biosca[*1,2], **Silvia Martín**[2], **Pedro Zuriaga**[3], **Carmina Montón**[4], **Laura López-Ocaña**[5], **María M. López**[2]

[*1] Departamento de Microbiología y Ecología, Universidad de Valencia, 46100 Burjassot (Valencia), SPAIN
[2] Departamento de Protección Vegetal y Biotecnología, Instituto Valenciano de Investigaciones Agrarias (IVIA), 46113 Moncada (Valencia), SPAIN
[3] Departamento de Sanidad Vegetal, Diputación General de Aragón, 44001, Teruel, SPAIN
[4] Laboratori de Sanitat Vegetal, DARP, 08040, Barcelona, SPAIN
[5] Colección Española de Cultivos Tipo (CECT), Universidad de Valencia, 46100 Burjassot (Valencia), SPAIN

[*] Corresponding autor: e-mail: elena.biosca@uv.es, Phone: +34 96 354 3194, Fax: +34 96 354 3202

This study reports the first description of *Brenneria* sp. causing bark cankers in poplar (*Populus x euroamericana*) in Spain. The disease is characterized by cankers with copious brownish exudations staining the trunk bark and reddish inner lesions progressing to necrosis in mature trees. A *Brenneria*-like bacterium was always associated with cankers from diseased poplar trees. Phenotypic characterization of poplar isolates by API galleries, Biolog system and fatty acid analysis revealed a high homogeneity among them and confirmed their identification as *Brenneria* sp. Pathogenicity of poplar isolates was demonstrated by inoculations on poplar leaves where they reproduced necrosis and exudates. Inoculation on poplar trunks confirmed the difficulty to reproduce exudative cankers on a short period of time.

Keywords *Populus* sp.; *Brenneria*; bark cankers; necrosis, characterization; pathogenicity.

1 Introduction

The poplar, *Populus* spp., is one of the leafy trees more cultivated in Spain, being its production of economical importance in several Spanish regions. These trees have different industrial applications as well as environmental wealth. In the last decade, poplar trees showing bark cankers with copious brownish exudations were observed in several plantations in Aragón, Spain. These exudates were more frequently observed in the lower part of the trunk in spring and autumn [1,2]. Other associated symptoms observed were chlorosis, defoliation, progressive loss of vigor and, in some cases, tree death [1, 2]. These symptoms were similar to those caused by some *Brenneria* species in woody plants [3–6]. These bacterial species were practically unknown in Spain until they were recently identified as responsible of bacterial cankers in *Quercus* spp. and *Juglans regia* [7–10]. In 2000, the causal agent of cankers in poplar was first identified as *Brenneria* sp. from samples taken from poplar cultivated in Teruel, Aragón, Spain [1, 2].
To confirm the etiology and the extent of this disease, surveys were carried out in 2001 in several poplar plantations in Aragón, showing trees with brownish oozing cankers on the trunk. We report here the first description of *Brenneria* sp. causing bark cankers in poplar (*Populus* spp.) in Spain and the phenotypic characterization of selected poplar isolates, comparing them with reference strains of other *Brenneria* species.

2 Materials and methods

2.1 Surveys and bacteriological analysis

Surveys were performed in 2001 in several poplar plantations in Aragón looking for trees with exudative cankers on the trunk. Samples were collected in different locations in Teruel and Zaragoza from 8 to 11 years old disease trees (*Populus x euroamericana*) showing symptoms (Table 1).

Table 1 Date, sampling site and reference of exudative bark canker samples of *P. x euroamericana* from Aragón, Spain.

Date	Sampling location			Clone
	Province	Village	Place	
8/2000	Teruel	Villel	Peña Rubia	I-214
9/2000	Teruel	Villel	Peña Rubia	I-214
7/2001	Teruel	Villel	Peña Rubia	I-214
7/2001	Teruel	Villel	Zamarra	I-214
7/2001	Teruel	Villel	Angosto	I-214
7/2001	Teruel	Alfambra	Prado	I-214
7/2001	Zaragoza	Alagón	Mejana Calvero	Eridano

Outer and inner bark tissue samples from cankers were analysed as previously described [10]. Briefly, isolations were made on King's B medium [11] plus cycloheximide (250 µg/ml) and after 48-72 h at 25ºC, *Brenneria*-like colonies were purified. Presumptive identification of bacterial isolates by conventional tests (Gram reaction, Kovacs' oxidase, O/F metabolism, aesculin hydrolysis, urease activity and levan production) was performed according to Biosca *et al.* [10].

2.2 Characterization of bacterial isolates

A collection of twenty seven poplar isolates from different origins and clones was selected for further characterization as described below. The reference strains of the six *Brenneria* species described (*B. alni, B. nigrifluens, B. paradisiaca, B. quercina, B. rubrifaciens and B. salicis*) [3] were also included for comparative purposes.

2.2.1 Physiological and biochemical characterization

It was performed by using the miniaturized API 20E, API 20NE, API 50CH and API ZYM systems (BioMérieux, France) as previously described [10]. Isolates were further characterized by using the BIOLOG-Microlog system as recomended by the manufacturer.

2.2.2 Fatty acid analyses

Qualitative and quantitative cellular fatty acid analyses of selected poplar isolates and reference *Brenneria* ssp. strains were done as described by Sasser [12].

2.2.3 Pathogenicity tests

The pathogenicity of selected poplar isolates from different origins was assayed on poplar leaves (27 isolates) and on trunk bark of young trees (7 isolates), using sterile saline buffered (SSB) as negative control. Poplar leaves were disinfected with 3% ethanol and washed twice with sterile distilled water before inoculation by transversal cuts with sterile scissors soaked in a bacterial suspension in SSB (10^9 CFU/ml). Trunk inoculations on one year old *P. x euroamericana* trees grown in pots were conducted as previously described for *B. quercina* [10], using five trees per each isolate. Reisolation of bacteria from symptomatic leaves or trunks was done on King's B agar.

3   Results

3.1 Surveys, symptoms and isolations

Along the surveys, external and internal bark canker samples from twenty six diseased trees from different poplar plantations in Aragón were obtained.

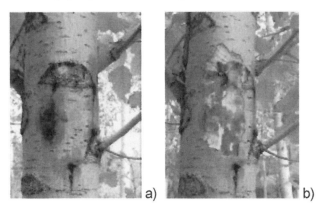

Fig 1 Bark cankers with exudates on trunk of a *P. x euroamericana* tree afected by *Brenneria* sp. a) external cankers with copious brownish exudations; b) inner bark tissue showing redish colour which progress to necrotic lesions.

The bacteriological analysis of infected tissue samples allowed the isolation of bacterial colonies with morphology similar to *Brenneria* sp. in King's B medium from all of them.

3.2 Characterization of bacterial isolates

3.2.1 Physiological and biochemical characterization

All selected poplar isolates exhibited the same biochemical and physiological characteristics and were identified as *Brenneria* sp according to Hauben *et al.* [3]. They also showed a high biochemical and physiological homogeneity.

3.2.2 Fatty acid analyses

The fatty acid qualitative composition of poplar isolates was similar to that of the reference strains assayed, which confirmed that they belong to the genus *Brenneria*. However, quantitative differences were observed between poplar isolates and the other *Brenneria* species.

3.2.3 Pathogenicity tests

Regarding pathogenicity tests, necrosis and exudates were observed within a week on poplar leaves inoculated with all poplar isolates, both on limbs and petioles, while no symptoms were observed on control leaves. Inoculation on poplar trunks produced a callus at the inoculation site in all assayed poplar isolates, which were reisolated, but cankers were not reproduced one month after inoculation. Trunk callus were not observed in control trees.

4 Discussion

The causal agent of cankers in poplar was first isolated in the summer of 2000 from *P. x euramericana* trees cultivated in Aragón (Spain) and identified as *Brenneria* sp. [1, 2]. The external symptoms of this disease were elongated cankers showing copious dark brownish and sticky exudations which stained the trunk bark in mature trees. Inner bark tissues showed a characteristic reddish colour which progressed to necrotic lesions. The external cankers were similar to those caused by some *Brenneria* species on the trunk of walnut, alder and oaks trees [4–6,10], but not reports were available about any *Brenneria* species affecting *Populus* spp.

During 2001, more surveys were done in different poplar plantations in the Aragón region, and poplar trees with brownish oozing cankers on the trunk were found. The bacteriological analysis of poplar samples with exudative bark cankers showed that *Brenneria*-like colonies were always associated with cankers from diseased trees, which confirmed the etiology of this symptoms in different poplar plantations in Spain. Biochemical and physiological characterization of selected poplar isolates by API and Biolog systems revealed a great homogeneity among them and confirmed their identification as a *Brenneria* species. These data were validated by fatty acid profile analysis. The comparative study of these poplar isolates with reference strains of the six *Brenneria* species described, showed phenotypic differences in some biochemical and physiological tests. Further, their fatty acid profiles were quantitatively different from those of reported *Brenneria* spp. The former may suggest that poplar isolates could be a new species or subspecies of genus *Brenneria*. Further characterization of selected poplar strains is still undergoing to check this hypothesis.

The results from the inoculation experiments on poplar trunks confirmed the difficulty to reproduce, within a short period of time, the exudative cankers observed in naturally infected poplar in inoculated trees, as previously reported for other woody plants [4, 7, 9, 10]. Because this pathogenicity assays can be influenced by several environmental factors, further studies are in progress to try to reproduce natural lesions. By contrast, the inoculations on poplar leaves showed the utility of this method for rapid pathogenicity tests, since the production of necrosis and exudates can be observed in a week.

This is the first description of *Brenneria* sp. causing cankers in poplar trees but the definition of its taxonomic position will require further investigations.

Acknowledgements

This study was supported by the grant 1FD97-0911-C03-02 of the FEDER program from the E.U.

References

[1]   E.G. Biosca, R. González, R., M.J. López-López, S. Martin, and M.M. López. Phytoma-España. **138**, 73 (2002).

[2]   E.G. Biosca, S. Martin, P. Zuriaga, and M.M. López. Phytoma-España. **139**, 9 (2002).

[3]   L. Hauben, E. Moore, L. Vauterin, M. Steenackers, J. Mergaert, L. Verdonck, and J. Swings. Syst. Appl. Microbiol. **21**, 384 (1998).

[4]   E.E. Wilson, M.P. Starr, and J.A. Berger. Phytopathology. **47**, 669 (1957).

[5]   E.E. Wilson, F.M. Zeitoun, and D.L. Fredrickson. 1966. Phytopathology. **57**,618 (1966).

[6]   G. Surico, L. Mugnai, R. Pastorelli, L. Giovannetti and D.E. Otead. Int. J. Syst. Bacteriol. **46**, 720 (1996).

[7]   M.M. López, R. Martí, C. Morente, N. Orellana, T. Ninot, and N. Aleta. Invest. Agrar. **2**, 307 (1994).

[8]   S. Soria, M.M. López, and M.J. López-López. Ecología **11**, 295 (1997)

[9]   R. González, M.J. López-López, E.G. Biosca, F. López, R. Santiago, and M.M. López. Plant Dis. **86**, 696 (2002).

[10]   E.G. Biosca, R. González, R., M.J. López-López, S. Soria, C. Montón, E. Pérez-Laorga, and M.M. López. Phytopathology. **93**, 485 (2003).

[11]   E.O. King, M.K. Ward and D.E. Raney. 1954. J. Lab. Clin. Med. **44**, 301 (1954)

[12]   M. Sasser, in: Methods in Phytobacteriology., Z. Klement, K. Rudolph and D.C. Sands, (eds.) (Akadémiai Kiado, Budapest, Hungary, 1990)

Environmental purification capacity and control of microbiological contamination of water resources in dairy mountain area

D. Trevisan[*], J.M. Dorioz, J. Poulenard, J.Y. Vansteelant

INRA, UMR CARRTEL 75, avenue de Corzent, 74203 Thonon les bains, FRANCE.

[*]Corresponding author : trevisan@thonon.inra.fr, Phone : +33 (0)4 50 26 78 30, Fax : +33 (0)4 50 26 07 60

The aim of this work is to achieve a better understanding of the behavior of fecal coliform populations on the vegetation and in the soil after grazing or manure spreading in environmental conditions, typical of vegetative growth period in mountain hay meadows. Changes in fecal coliform populations on the vegetation and in the soil were monitored *in situ*. Variations found in populations are related to the agricultural, soil, and climatic characteristics of plots and to the moisture regime of soils. These observations are compared with laboratory experiments on undisturbed soil microcosms and with rain simulation experiments. The data enable us to pinpoint the influence of various factors affecting the decline and transfer of fecal bacteria in the plant-soil system and consequently to discuss the risk management of water contamination by agriculture.

Keywords: Fecal coliform, Protozoa, Bacterial transfer, Grazing, Slurry spreading.

1   Introduction

In the agricultural mountain zone, problems associated with contamination of waters by fecal bacteria are often significant and frequently associated with grazing and manure spreading. Individually, the effect of the different factor which governs the survival or transfer of fecal microorganisms has been the subject of both laboratory and field experiments, mainly carried out using microcosms, often under conditions of disturbed soil, or isolated soil columns [1] These approaches provide some understanding of mechanisms. However the available results are difficult to extrapolate to field conditions [2], because the survival and the transfer of micro-organisms are inseparable phenomena. Our aims are to study the fate of the fecal bacteria combining laboratory experiments and field studies, within different situations representative of the agricultural, bioclimatic and soil conditions of mountain meadows of the northern Alps. The results should be useful for planning the management of water contamination risks.

2   Methods

2.1 Field monitoring of fecal bacteria.

Our experimental design is based on monitoring of the amounts of fecal coliforms (FC) on the vegetation and in the soil after grazing or manure spreading. Several meadows situated between 800 and 1700 m altitude and representative of the main agricultural and soil conditions of the northern French Alps were monitored. The studied soils can be classified into two main types: "shallow calcareous soils" (with a silty sand texture; a surface pH of 7-8; an alterite layer situated between 0.4 and 0.6m and a stone content of 10-20% at 0.1m and 20-40% at 0.6m); and "deep slightly acid soils" (with a silty clay texture, a pH of 6-6.7, a minimum alterite depth of 0.8m, and a uniform stone content of 5-10%). Samplings for moisture monitoring and microbiological analysis of soils were done on vegetation and different layers of soils. Soil moisture regimes were estimated with an adapted water balance model [3].

For the different plots, for the vegetation and soil layers, the variations M in the amount of bacteria were calculated as followed:

$$M = \frac{\log S_{d+n} - \log S_d}{t_{d+n} - t_d} \qquad (1)$$

where S_d and t_d represent respectively the content of bacteria measured (cfu.g^{-1}) and the number of days elapsed from slurry spreading to the sampling day d; S_{d+n} is the same measurement made at time d+n. Equation [1] gives a direct indication of the change in content of bacteria, a negative value of M representing the decay rate of FC during the time interval $\Delta t = t_{d+n} - t_d$. Analysis of variance was carried out to characterize the variability in M. Different linear models M=f(X) were tested after graphical analysis of the relationships existing between M (to be explained) and the independent variables X likely to influence the survival or transfer of the bacteria (X being the variables for the soil, weather or vegetation).

2.2 Rainfall simulation

To analyze soil bacteria transfer, simulated rainfalls have been made on different plots representative of the two soil types. E. Coli cells (cultivated from a bovine source) where previously suspended in the rain water to a final concentration of 10^3 cfu.ml^{-1}. The rain (20 mm.h^{-1}) was simulated with a droplet system and after a one hour rain event, soil sampling was performed at various depths to evaluate bacterial distribution.

2.3 Laboratory experiments

The survival of FC was studied using microcosms created from undisturbed soil samples contaminated by infiltration with slurry. We tested the effect of two initial concentrations of FC and two incubation temperatures. Soils were sampled in metal cylinders. After contamination, the cylinders had a initial FC content of 10^3 or 10^6 cfu. g^{-1} dry soil, corresponding to applications by spreading of 20T.ha^{-1} of slurry with a content of 10^5 and 10^8 cfu.g^{-1} respectively. The cylinders were maintained at constant temperature (5 or 20°C) and moisture content for three months. The mortality rates k of FC were calculated from:

$$\log(\frac{N_t}{N_0}) = kt \qquad (2),$$

where Nt is the FC content (log cfu.g^{-1} dry soil) at time t and No, the initial content.

Active protozoans were counted by a adapted method [4] whereby 20g of soil is placed in a Petri dish and covered with a lay of water. After four hours of contact (the time needed for liberation of unencysted protozoans from soil aggregates), replicates of the solution were removed and placed on slides for microscopic examination and direct counting of various active protozoans (ciliates, flagellates and amebas). Protozoan counts were expressed on a dry soil basis.

3 Results

3.1 Field study

3.1.1 Bacterial survival on soil surface

When the biomass B_u of the upper part of the canopy is small ($B_u < 100g.m^{-2}$) there is a rapid fall in bacterial populations (fig.1).

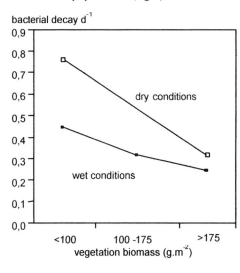

Fig. 1 Bacterial decay in relation to biomass canopy vegetation and weather conditions

These falls are also pronounced under extreme rainfall conditions (frequency of dry days - without rain - f_{dd} less than 0.4 or greater than 0.7). There is a significant relation between the M the evolution of FC at the soil surface and B_u the mean biomass of the upper part of the canopy and f_{dd} the frequency of dry (days without rain) during the time intervals Δt. Analysis of variance confirms the statistical validity of these trends. The mean biomass B_u has a significant influence on M ($P<0.0001$). The frequency of dry days is also a significant factor ($P=0.0004$). The FC count is maintained better when the canopy biomass is larger, suggesting a protective effect of the vegetation. The frequency of dry days has a particular effect on the value of M: during very dry ($f_{dd} > 0.7$) or very wet ($f_{dd} < 0.4$) spells, FC falls rapidly. In intermediate conditions on the other hand there is a smaller decrease in the bacterial count. The falls in numbers are thus most marked in extreme weather conditions, during periods of water stress, or, at the other extreme, of leaching through the crop.

3.1.2 Bacterial evolution in soil

Averaged over all soils and soil layers, there is a good correlation between the measured soil water contents and those calculated with CREAMS ($r^2 = 0.72$). The model estimates daily percolation (Pe) at the bottom of the upper and lower layers for different soil types. The factors which have a significant effect on M are S_d (log $cfu.g^{-1}$) the count at the start of the period Δt ($P<0.0001$), the soil type ($P=0.004$) and finally Pe, the mean depth of percolation water at the bottom of the upper soil layer (P=0.02). The model used, giving an r^2 of 0.78, is of the form:

$$Y = \alpha_0 + \beta_1 * S_d + \beta_2(\text{soil}) + \beta_3 * Pe$$

The bacterial numbers decrease more quickly when the initial count S_d is high (the coefficient β_1 is negative), suggesting competition effects. The dynamics of population change also vary with the nature of the soil. This result leads us to suppose that for the two soil types studied the bacteria experience a different physical and biological environment, probably more or less favorable to their survival or retention. β_2 is negative for silty-clay texture and positive for silty-sand soil, suggesting different competition modalities for the two soil conditions. The soil properties also affect the percolation rate whose effect is to reduce the bacterial content (β_3 is negative). Under the conditions of the study, in which the moisture stress is limited because of the constant high moisture content of the soil, this effect could be explained by increased export due to leaching.

3.1.3 Bacteria transfer after simulated rainfall

The results obtained after simulated rainfall suggest that bacteria transfer is controlled by textural and structural properties (fig. 2). One can observe that bacteria retention is greater in fine sized textural conditions and reduced in soils of high infiltration rate where high bacteria counts are registered at large depth. These trends corroborate the previous results obtained from field monitoring.

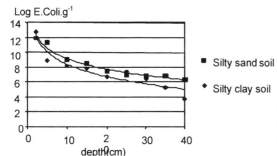

Fig.2. Bacteria counts after simulated rainfall in silty clay and silty sand conditions.

3.2 Microcosms experiments

The protozoans counts show a first phase of growth followed by a population decline. The behavior of protozoans has been described by the ratio Qmax/tQmax, where Qmax is the highest peak value observed and tQmax, the length of the growth phase. This ratio gives an indication of protozoan activity, i.e. the dynamics of cyst germination and cell multiplication involved in the growth of these populations. The coefficients k of FC mortality have been related to the protozoan growth descriptor Qmax/tQmax (fig. 3).

394

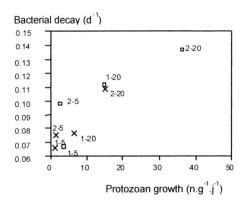

Bacterial decay (d^{-1})

Protozoan growth (n.g^{-1}.j^{-1})

Fig.3. Bacteria decay and protozoan growth. (□: silty soil, **X** : sandy soil ; 1 : initial concentration of 103 E.coli.g-1 ; 2 : 106.g-1 ; 5 : temperature 5°C, 20 :20°C).

FC mortality is lower in the silty sand soils: on average, k for the silty clay soils is 0.11 d^{-1}, compared with 0.075 d^{-1} for the silty sands. At a high initial concentration or at high temperature there is an increase in mortality, with k increasing on average by 0.02 d^{-1} and 0.04 d^{-1} respectively. The same factors affect the activity of protozoans. At 5°C they are relatively inactive whatever the initial bacterial concentration (the Qmax/tQmax ratios are very similar). However at 20°C there is more activity when the initial concentration is high. Finally there is a highly significant relationship between FC mortality and protozoan dynamics (r^2=0.79; P=0.0028) which suggests a predator-prey relation whose characteristics depend on soil conditions, temperature and bacterial concentration.

We find that the k values are similar to those of M calculated with eq. (1) for field conditions, assuming zero percolation (Pe=0) and therefore that variation in M must be solely due to bacterial death. For initial concentrations of 10^3 and 10^6 cfu.g^{-1}, M thus calculated varies between 0.05 and 0.20 d^{-1} for the silty sand soils and between 0.17 and 0.30 d^{-1} for the silty clays. In the absence of percolation, bacterial behavior in the field and in the microcosms is thus very similar.

4 Discussion

The dynamics of FC on the vegetation and in the soil are controlled by the interaction of weather and biological and water-related factors. Combining field and laboratory experiments we have observed:

- a protective effect of the plant canopy at the soil surface;
- a leaching of the bacteria through the soil governed by textural and structural soil properties ;
- competition phenomena between bacteria and protozoans controlled by soil properties and bioclimatic conditions.

The risk of microbiological contamination of aquifers by vertical transport of FCs depends on the numbers of bacteria present in the soil/plant system and on water transfer. These numbers are related to agricultural practices which determine the magnitude of the inputs into the system (a function of the type of organic matter applied and the rates of application) and which also determine certain factors relevant to survival, such as the state of the crop canopy, the trophic state of the soil and the physical state of the top few centimeters of soil. The transfers are linked to intrinsic soil properties such as retention and filtration capacity and to random phenomena connected with rainfall, such as moisture regime.

The risk is thus determined by the functioning of the entire soil/plant system, of which several parameters can be modified *a priori* with the aim of limiting transport into aquifers. The possibilities for intervention relate to the rate, positioning and timing of spreading and to the management of the crop canopy and the maintenance of the biological activity of the soil.

The inadvisability of applying large instantaneous doses of slurry has been mentioned [5] Apart from the fact that the dose influences the inputs into the system, it is likely that massive applications which result in thick deposits limit the penetration of bactericidal UV and slow down desiccation, which would reduce the buffering effects expected at the surface. In view of the fact that, because of the size of mountain farms, it is often difficult to increase the area over which to spread slurry, splitting the doses is another way of limiting the risks. For one thing, the reduction in the dose can favor the purifying effects which occur in the crop canopy, and also the repeated application of organic matter to the soil should favour the maintenance of predator populations adapted to the fecal bacteria [6] Timing of spreading should be planned so as to limit the frequency of critical periods when large bacterial numbers and rapid drainage occurs simultaneously [7]. One should try especially to restrict spreading to dry periods by paying attention to weather forecasts. This alternative is probably difficult to implement because in practice farmers prefer to spread in rainy periods to limit undesirable effects such as odors. The spreading sites with the least risk are those with short grass or a low growth rate (less intensive hay meadows, or frequent cuts) or with soil types with a high retention-filtration capacity for bacteria (deep slightly acid silty clays).

References

[1]   Smith M.S., Thomas G.W., White R.E., Ritonga D. (1985). Transport of *Escherichia coli* trough intact and disturbed soil columns. *J. Env. Qual.* **14:1**, 87-91.

[2]   Sjorgen R.E. (1994).Thirteen year survival study of an environmental *Escherichia coli* in field mini plots. *Wat, air and soil poll.* **81**, 315-335.

[3]   Trevisan D., Dorioz J.M., Pauthenet Y. (1994). Apport d'un modèle simplifié de simulation du bilan hydrique pour l'analyse de la satisfaction en eau de prairies irriguées des alpes internes. *Agronomie* **14**, 683-696.

[4]   Foissner, W. (1987). Soil protozoa: fundamental problems, ecological significance, adaptations in ciliates and testaceans, bioindicators and guide to the litterature. *Progr. Protistol.* **2**: 69-212.

[5]   Cote D. (1994). Problématique de la contamination de la nappe phréatique par lessivage d'ammonium et des bactéries fécales d'engrais de ferme. *Agrosols* **VII(1)**, 20-25.

[6]   McCambridge J., McMeking T.A. (1974). Protozoan predation of *Escherichia Coli* in estuarine waters. *Wat. Res*. **13**, 659-663.

[7]   Pasquarell G.C., Boyer D.G. (1995). Agricultural impacts on bacterial water quality in karst ground water. *J. Environ. Qual.* **24**, 959-969.

Group specific primers for studying *Methylobacterium* biodiversity on crop plants

Z.S. Omer[*1], R. Tombolini[2], and B. Gerhardson[1]

[1] The Mase Laboratories, P.O. Box 148, SE-751 04 Uppsala, Sweden
[2] Dip. Scienze e Tecnologie Biomediche Sez. Microbiologia Applicata
Via Porcell 4 09124 Cagliari (CA) Italy [1]

[*]Corresponding author: e-mail: Zahra.Saad.Omer@maselab.se, Phone: +46 (0) 18 67 48 55, Fax: +46 (0) 18 67 48 98

Pink-pigmented facultative methylotrophic bacteria belonging to the genus Methylobacterium (PPFM) are common inhabitants of plant phyllosphere. They also possess features that make them interesting to include in the screenings for agents exerting plant-growth promotion. By using specific, newly designed primers to amplify their 16S rRNA and T-RFLP technique, their biodiversity on crop plants was estimated in a pilot set of experiments. PPFMs population biodiversity was compared at different times on two plant species, that were grown in two different locations. The sampling time and the location were found to be factors linked to higher diversity in the population pattern.

Keywords: Pink-pigmented facultative methylotrophic bacteria; T-RFLP; biodiversity; *Triticum aeistivum*; *Trifolium pretense*

1 Introduction

The genus Methylobacterium contain strictly aerobic, facultative methylotrophic, gram-negative, rod-shaped bacteria that are able to grow on one-carbon compounds, as sole carbon and energy sources [1]. Bacteria belonging to this genus are also called pink-pigmented facultative methylotroph (PPFM) because of their typical pink pigmentation. They have been isolated from a wide range of environmental sources and of particular interest from the standpoint of plant microbiology is their consistent occurrence as epiphytes in the phyllosphere [2–4]. In a previous paper we reported their presence and variation on crop plants and found a pattern of population level that vary during the growing season [5]. The capability of selected isolates to colonize roots and phyllosphere of crop plant after inoculation in greenhouse also varied depending the plant species and the inoculation methods used [5].

From the 16S rRNA sequence information it has been shown that the genus Methylobacterium represents a line of descent in the alfa-2 subclass of the class Proteobacteria [6]. In the Ribosomal Database Project II (RDP II) [7] the Methylobacterium subgroup is part of the Rhizobium-Agrobacterium group. The Methylobacterium group, in turn, comprises four subgroups: M. extorquens subgroup, M. radiotolaerans subgroup, M. organophilum subgroup and Bosea thiooxidans subgroup. Bergey's taxonomy places the family Methylobacteriaceae, which includes only the genus Methylobacterium, within the order Rhizobiales and in a separate family (Family Bradyrhizobiaceae) the genera Bosea and Afipia, which on the basis of the 16S rRNA are grouped in the Bosea thiooxidans subgroup.

Two sets of primers for detection of Methylobacterium spp. have been previously published [8], which amplify fragments of about 350 bp and 260 bp. Such short PCR product fragments do not allow realistic analysis in order to evaluate the biodiversity within the genus Methylobacterium.

The objectives of the present paper are twofold: 1) Reporting about newly designed specific primers for the Methylobacterium group, improved in specificity and producing PCR product fragments of useful sizes for further comparative analysis like sequencing, DGGE, RFLP, T-RFLP etc., 2) Presenting the results of a preliminary analysis of phyllosphere samples of two crop plants for PPFMs biodiversity.

2   Materials and methods

2.1 Bacterial strains and culture conditions

The bacterial strains and media used in this work are listed in Table 1. Bacteria were grown either at room temperature or at 25 °C. Genomic DNA from the archea species Sulfolobus acidicaldarius, DSM 639, and Sulfolobus solfataricus, DSM 1617, was kindly provided by prof. Ralf Bernander, EBC, Uppsala University.

Table 1. Bacterial strains and media

Strains	Culture collections codes	Media
Methylobacterium extorquens (T)	ATTCC 43645, NCCB 78015, NCIMB9399, NCCB78015, JCM2802	AMS
Methylobacterium radiotolerans (T)	ATTCC 27329, LMG 2269	AMS
Methylobacterium rhodium (T)	ATTCC 14821, LMG 2275, NCIMB9421	AMS
Methylobacterium mesophilicum (T)	ATTCC 29983, LMG 5275	AMS
Methylobacterium organophilum (T)	ATTCC 27886, LMG 6083, NCCB78041	AMS
Methylobacterium zatmanii (T)	LMG 6087, NCIB12242	AMS
Bosea thiooxidans str. BI-42 (T)	DSM 9653	DSMZ 763
Bradyrhizobium japonicum (T)	LMG 6138, DSM 30131, ATCC 1032	TSA
Pseudomonas fluorescens (T)	CCUG 1253	TSA
Pseudomonas syringae pv. syringae	CCUG 14279	TSA
Pseudomonas savastanoi	CCUG 2171	TSA
Erwinia amylovora	CCUG 4906	TSA
Erwinia carotovora ssp carotovora	CCUG4907	TSA
Xanthomonas populi pv salicis	NCPPB 3426	TSA
Vibrio harvey		MM

2.2 Environmental samples

Environmental samples (leaves) were collected from red clover (in September and October) and from wheat plants (in September) grown in nearby fields in the area of Hammerskog (Uppsala, Sweden). Red clover samples were also collected from another location in September. Leaf samples were collected and transferred to the laboratory in plastic bags. DNA was sampled by suspending two grams (fresh weight) of the leaves in 25 ml of sterilized distilled water and then sonicated for ten minutes. The resulting suspensions were centrifuged and the pellets were used to isolate the DNA as described below.

2.3 Primers design and PCR conditions

Primers for 16S rRNA PCR have been designed by using the freeware software PRIMROSE version 1.1.7 (download: http://www.cf.ac.uk/biosi/researcWbiosoft/Primrose/) [9]. The target sequences used in Primrose software were represented by all the 16S rRNA aligned sequences longer than 1200 bp belonging to the Methylobacterium group in the RDP II apart from the ones in the Bosea thiooxidans subgroup. The criteria used for selecting among the possible primers were the maximum length of the PCR product, maximum number of Methylobacterium species targeted and minimum number of non-Methylobacterium species targeted. Selected oligonucleotides were further modified in size for making them compatible primers with the aid of the online program Primer3 (http://www-genome.wi.mit.edu/genome_software/other/primer3.html) [10]. Primer specificity has been checked by using the online software Probe Match available at the RDP II (http://rdp.cme.msu.edu/html/index.html) [3].

For total DNA extraction a small loopful from a freshly grown agar culture was thoroughly suspended in 400 µL of sterile purified water in eppendorf tubes. The suspensions were then treated with 200 µL of n-butanol by vortexing for one minute before adding 200 µL of 95 % ethanol. After 5 minutes of full speed centrifugation the organic and aqueous phases were removed and the bacterial pellets were washed with 500 µL of sterile purified water. Total DNA was extracted by using Bactozol^{TM} kit (Molecular Research Center, Inc., Cincinnati, USA) according to instructions of the manufacturer. PCR reactions were carried out in 0.2 mL tubes PuRe Taq ready-to-go PCR beads (Amersham Biosciences, Little Chalfont, England) which were reconstituted with 1 µL of 50 µM of the forward and reverse primers, 1 µL of template DNA and 22 µL of sterile purified water. Final concentration of each of the dNTP was 200 µM in 10 mM Tris-HCl, pH 9.0, 50 mM KCl and 1.5 mM MgCl$_2$. The mix contained as well non specified stabilizers, BSA and 2.5 units of puRe Taq DNA polymerase (Amersham Biosciences, Little Chalfont, England). PCR reactions were conducted for 30 cycles after a preheating at 95 °C for 5 minutes, including 30 sec at 95 °C, 30 sec at 60 °C (or 64 or 55 or 50, see results) and 30 sec at 72 °C. After the last cycle an elongation step at 72 °C for 10 min. was included.

2.4 T-RFLP analysis

When PCR products were to be used for T-RFLP analysis the forward primer (Meth404F) was labeled with fluorochrome 5',6-carboxy fluorescein (FAM) on the 5'-end. The amplified fragments were purified by using Wizard SV gel and PCR clean-up kit (Promega, USA) according to instructions from the manufacturer. And about 300 ng in 10 µL reaction volume were digested with one of the following restriction enzymes: Bsr I, Hinf I, ScrF I, Dde I, Rsa I, Hha I, Sau3A I, Alu I, Msp I and Hae III. After 2 hours of incubation at the optimal temperature one more microliter of the specific restriction enzyme was added and the tubes were further incubated for one hour. The fluorescently labeled terminal restriction fragments were analyzed by capillary electrophoresis on an automated sequence analyzer (ABI PRISM® 3700 DNA Analyzer, Applied Biosystems) in Genescan mode. The standard size marker GS-500 (Applied Biosystems) was included in each run.

3   Results and discussion

3.1 Primers specificity and T-RFLP setup

Three primers were selected: Meth404F, Meth565F and Meth1123R (first three entries in Table 2). The two forward primers when checked against the whole database of the RDP II proved to be the most specific while the reverse primer was selected as the best compromise between selectivity and fragment size of the amplified fragment (Table 2). In Table 2 is reported a comparison of the selectivity of the primers newly designed and the one previously published [8]. By in silico specificity tests (Probe Mach software in the RDB II) Meth404F hit the 16S rRNA related Bosea subgroup by allowing one single mismatch at position C-13 and Meth565F by allowing three mismatches.

The new primers produced the expected fragments for the Methylobacterium type strains and for all the Methylobacterium isolates collected in our laboratory (results not shown). In Table 3 the results of the PCR with the above primers are reported for bacteria ranging from archea to the closely related genus Bosea. At the annealing temperature of 60 °C (1.5 mM MgCl$_2$) Bosea sp. DNA is also hit by the primers pairs Meth404F-Meth1123R, Meth565F-Meth1123R and Meth404F- 1492R while the primers pair Meth565F-1492R amplifies also Vibrio harvey. By increasing the annealing temperature to 64 °C the primer set Meth404F-Meth1123R proved to be specific for the genus Methylobacterium (Table 3).

3.2 T-RFLP of the environmental samples

For T-RFLP, the restriction enzyme was chosen after having tested a number of 4-base-cutting enzymes with the Methylobacterium type strains, and HaeIII was selected as the one producing the most diversified T-RFLP signals (results not shown). We used the primer set Meth404F-Meth1123R and annealing temperature for the PCR of 60 °C in order to include in the target bacterial group the whole Methylobacterium group as defined by the RDP II and maximize the DNA product amount from the environmental samples.

T-RFLP profiles of Methylobacterium community of clover and wheat crops sampled at the same time and in the same locations (bordering fields) have been compared in order to address the question whether it exist a crop specificity for such bacteria. The results show that clover and wheat phyllosphere from the same location harbors the same principal components of the Methylobacterium populations (Fig. 1A and B). Only one peak in the wheat sample (T-RF=92, Fig. 1 A) is not shared with the clover sample (Fig. 1B). A sample from clover originated from a different location exhibited a slightly more different population pattern (Fig. 1C). A clover sample collected at the same location but at a different season produced a more different profile (Fig. 1D).

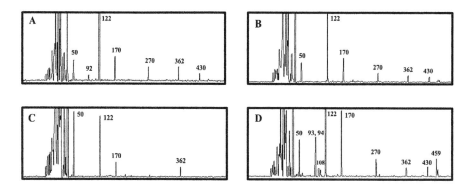

Figure 1. T-RFLP profiles of *Methylobacterium* group targeted PCR products of wheat and clover phyllosphere samples. A: Wheat sampled at location 1 at time 1; B: clover sampled at location 1 time 1; C: clover sampled at location 2 time 1; D: clover sampled at location 1 time 2.

Table 2. New and previously published primers specificity in terms of number of sequences hit in the RDP II defined bacterial groups

Name	Sequence	Reference	M. ext subgr.	M. radio. Subgr.	M. organo.S ubgr.	Bosea subgr.	Other Rhiz-Agr groups	Other α-proteob.
Meth404F	TTT TRT CCG GGA CGA TAA TG	This paper	14	15	3	0	0	0
Meth 565F	CCA CAG AAT KGC CTT CGA TAC T	This paper	17	11	3	0	0	0
Meth 1123R	CAC CTT CCT CGC GGC TTA	This paper	16	14	3	16	139	0 + 7*
1492R	TAC GGY TAC CTT GTT ACG ACT T	[11]	Universal primer					
2F 226-246	GAT CGG CCC GCG TCT GAT TAG	[8]	10	16	2	16	217	7
2R 439-459	CCG TCA TTA TCG TCC CGG ACA	[8]	15	6	3	0	0	0
1F 88-108	CTT CGG GTG TCA GTG GCA GAC	[8]	7	0	0	0	0	0
1R 432-452	TAT CGT CCC GGA CAA AAG AGC	[8]	13	6	2	0	0	0
3R 1153-1173	GGC TTA TCA CCG GCA GTC TCC	[8]	15	13	2	0	106	3 + 5*
3F 856-876	CGC CGT AAC GCA TTA AGC ATT	[8]	4	0	0	0		

* Other Proteobacteria
M=C,A; Y=C,T; K=G,T; R=A, G; S=G,C; W=A,T.

Table 3. PCR specificity of primers tested with different bacterial strains at different annealing temperatures.

Strains	404F-1123R				565F-1123R				404F-1492R				565F-1492R			
	50	55	60	64	50	55	60	64	50	55	60	64	50	55	60	64
Sulfolobus acidicaldarius DSM 639	-	-	-	-	-	-	-	-	-	-	-	-	-	-	-	-
Sulfolobus solfataricus DSM 1617	-	-	-	-	-	-	-	-	-	-	-	-	-	-	-	-
Vibrio Harvey MM30 (67)	-	-	-	-	-	-/+	-	-	-	+	-	-/+	-	-/+	+++	+
Lactobacillus fermentum ATCC 14931	-	-	-	-	-	-	-	-	-	-	-	-	-	-	-	-
Bosea thiooxidans str. BI-42 DSM 9653	-	+/+	-	-	+++	+++	+++	+	+	+++	+++	+	+	+++	+++	+
Erwinia amylovora CCUG 4906 (71)	-	-	-	-	++	-	-	-	-	-	-	-	-	+	-	-
Erwinia carotovara subsp. carotovora CCUG 4907	-	-	-	-	-	-	-	-	-	-	-	-	++	-/+	-	-
Pseudomonas fluorescens CCUG 1253	-	-	-	-	-	-	-	-	-	-	-	-	-	-	-	-
Pseudomonas savastonoi CCUG 2171	-	-	-	-	-	-	-	-	-	-	-	-	-	-	-	-
Ps. syringae pv. syringae CCUG 14279	-	-	-	-	-	-	-	-	-	-	-	-	-	-	-	-
Xantomonas populi pv. salicis NCPPB 3462	-	-	-	-	-	-	-	-	+	-	-	-	-	+	-	-
Bradyrhizobium japonicum LMG 6138	-	-	-	-	-	+	-	-	-	-	-	-	-/+	-	-	-

Acknowledgements

We thank Rolf Bernander for supplying bacterial isolates and The Swedish Farmers Foundation for Agricultural Research (SLF) and MISTRA for financial support.

References

[1]   P.N. Green, Methylobacterium, in: The Prokaryotes: An Evolving Electronic Resource for the   Microbiological Community (http://link.springer-ny.com/link/service/books/10125/) 3rd edition release 3.7 (Springer-Verlag, 2001).

[2]   W.A. Corpe, J. Microb. Meth. 3, 215 (1985).

[3]   W.A. Corpe, and S. Rheem, FEMS Microbiol. Ecol. 62, 243 (1989).

[4]   M.A. Holland, and J.C. Polacco, Annu. Rev. Plant Physiol. Plant. Mol. Biol. 45, 197(1994).

[5]   Z.S. Omer, R. Tombolini, and B. Gerhardson, FEMS Microbiol. Ecol. 47, 319 (2004).

[6]   A. Hiraishi, K. Furuhata, A. Matsumoto, K.A. Koike, M. Fukuyama, and K. Tabuchi, Appl. Environ. Microbiol. 61, 2099 (1995).

[7]   J.R. Cole, B. Chai, T.L. Marsh, R.J. Farris, Q. Wang, S.A. Kulam, S. Chandra, D.M. McGarrell,   T.M. Schmidt, G.M. Garrity, and J.M. Tiedje, Nucl. Acids. Res. 31, 442 (2003).

[8]   T. Nishio, T. Yoshikura, and H. Itoh, Appl. Environ. Microbiol. 63, 1594 (1997.

[9]   K.E. Ashelford, A.J. Weightman, and J.C. Fry, Nucl. Acids. Res. 30, 3481 (2002.

[10]   S. Rozen, and H.J. Skaletsky, in: S. Krawetz, and S. Misener (Eds.),   (Humana Press, Totowa, 2000), p. 365.

[11]   D.J. Lane, in: E. Stackebrandt, and M. Goodfellow (Eds.), (John Wiley & Sons, Chichester, 1991), p. 115.

Microbial biodiversity in soils under a conservation agriculture regime

A. Muñoz [*1], **M. Ramírez**[2] and **A. López-Piñeiro**[1]

[1] Department of Biology and Plant Production, [2] Department of Microbiology, University of Extremadura, Avda. Elvas s/n, Badajoz, Spain.

[*]Corresponding author: e-mail: anamg@unex.es, Phone: +34 924289426, Fax: +34 924289427

The preservation of soil microbial biodiversity is important for the correct functioning of biogeochemical cycles, which are essential for plant growth and the maintenance of terrestrial ecosystems. We analyzed the changes in soil microbial biodiversity under conservation agriculture management (direct seeding) with respect to traditional agricultural management. Direct seeding avoids tillage, leaving the stubble on the soil surface to produce a vegetation cover that protects the soil from rain and wind. It also stabilizes the humidity and temperature of the soil surface. This improvement in environmental conditions facilitates the development of organism communities (insects, fungi, and bacteria). We collected samples during three consecutive years from the surface (five centimetres) of three contiguous plots of land under different management: CT (conventional tillage), DS (direct seeding), and DSC (direct seeding with cover). The populations of bacteria and fungi were analyzed by culturing them in different culture media. In parallel, we analyzed the evolution of the soil parameters in the three plots. The results indicate that the conservation agriculture regime was more beneficial than the traditional land management. Microbial biodiversity was higher in soil under direct seeding, and there was a major improvement of the soil's physicochemical parameters in contrast with the deterioration observed under traditional management.

Keywords: Soil microbiology, conservation agriculture, direct seeding, biodiversity.

1 Introduction

The preservation of soil microbial biodiversity is needed for the correct functioning of biogeochemical cycles, which are essential for plant growth and the maintenance of terrestrial ecosystems.

A soil's microbial ecology depends on its physicochemical properties, climate characteristics, and vegetation, and each trophic level of the biological community has a clear impact on the overall dynamics of the ecosystem. In particular, microbial diversity and its stability in the soil ecosystem determine the soil's capacity to react to an external change, impact, or degradation such as agriculture induces.

The use of soils for agricultural production may considerably reduce their quality. Tillage, for instance, initiates processes that may damage the natural soil ecosystem. Hence, one modern alternative to conventional tillage is direct seeding. This management practice, which forms part of conservation agriculture, never tills the soil and leaves the last year's stubble on the field. It is widely used in semi-arid regions all over the world to control wind erosion and maintain soil humidity.

The aim of this work was to determine the effects of different soil management regimes on the microbe populations associated with the rhizosphere in a maize field, and on the soil properties which are important to soil conservation.

2   Materials and methods

The soil studied is a stony distric luvisol under three different management regimes of a maize crop. The experiment was conducted in two similar plots. One has been under direct seeding with cover (**DSC***) for nine years, and the other has been divided into three parts for three years: (a) maize under conventional tillage (**CT**); (b) maize under direct seeding (**DS**); and (c) maize under direct seeding with cover (**DSC**).

Soil samples were taken from 0 to 5 centimetres down in the A horizon. Samples were taken in triplicate approximately every two months for three consecutive years.

The bacterial and fungi populations were analyzed by culturing them in different culture media: **YEPD, Rose Bengal Agar, Peptone, TSA, Green Malachite**, special medium for **Azotobacter and Azomones,** and **Starch-Casein**.

More than 74 bacterial colonies with differing morphology were detected by direct observation on the plate cultures and under microscopy. Two diversity index were calculated: **Shannon-Weaver's and Simpson's**. The use of the number and counts of bacterial species as indicators of soil diversity is based on their primary role in the biological decomposition of organic matter. The evolution of the soil parameters humidity, compaction, and organic matter in the three plots over three years was studied in parallel. Organic matter: assayed by oxidation with potassium dichromate. Humidity: measured by a PR1 profile probe with electromagnetic sensors. Compaction: determined by a hand penetrometer.

3   Results and discussion

3.1 Soil parameters

The soil humidity and organic matter strongly correlated with the counts of soil microorganisms (Table 1).

Table 1. Pearson's correlation coefficient.

	Total microorganisms (cfu/g dry soil)
Humidity	0.620**
Organic matter	0.781**

**Significant correlation at 0.01 level (two-tailed)

Figures 1 and 2 show the mean values of the soil humidity and organic matter for the three consecutive years. Both parameters (percentages) were higher in the conservation agriculture management regimes with both short and long period application. With the nine-year direct seeding, soil humidity was 50 % higher than with conventional tillage and 40 % higher after only three years. The soil organic matter showed the same trend as the soil humidity, although the differences were not so high for the three- year plots, because the organic matter has a slower response than the humidity. The DSC*, however, contained almost 3 times more organic matter than the soil under conventional tillage. It is important to emphasize that the organic matter determines many physical and chemistry properties of the soil, facilitates the formation of the appropriate habitats for microorganism growth, and in general is important for crop growth and productivity.

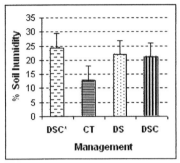
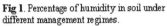

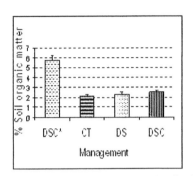

Fig 1. Percentage of humidity in soil under different management regimes.

Fig 2. Percentage of organic matter in soil under different management regimes.

3.2 Microbe population

Microorganisms are an essential part of a living soil, and of the utmost importance for soil health [5]. We therefore analyzed the bacteria and fungi populations associated with a maize crop rhizosphere.

The direct counts of microorganism colonies were tested statistically against homogeneity of the means. Table 2 gives the results for some groups of microorganisms according to the management of the soil. Most groups of microorganisms were significantly more abundant in the soils under a conservation agriculture regime than under conventional tillage. Similar results have been found in other studies dealing with other crop types [1].

Table 2. ANOVAs and Duncan's post hoc tests for the influence of soil management on the microorganism populations.

Management regime	Total fungi in Rose Bengal agar (cfu/g dry soil)	Orange-pigmented Arthrobacter in TSA (cfu/g dry soil)	Total bacteria in YEPD (cfu/g dry soil)	Total Streptomyces in starch-casein (cfu/g dry soil)
CT	$3.31 \cdot 10^6$a	$0.12 \cdot 10^6$a	$9.54 \cdot 10^6$a	$1.29 \cdot 10^6$a
DS	$4.22 \cdot 10^6$b	$0.26 \cdot 10^6$b	$15.97 \cdot 10^6$a	$2.11 \cdot 10^6$ab
DSC	$4.73 \cdot 10^6$b	$0.26 \cdot 10^6$b	$23.80 \cdot 10^6$b	$2.99 \cdot 10^6$b
DSC*	$6.88 \cdot 10^6$c	$0.48 \cdot 10^6$c	$29.05 \cdot 10^6$b	$4.62 \cdot 10^6$c
Levene sig.	0.127	0.159	0.192	0.080
F	28.274	10.321	10.634	10.511
Sig.	< 0.0001	< 0.0001	< 0.0001	< 0.0001

Data in a given column followed by the same letter were not significantly different at the $p < 0.05$ level.

Microbial biomass changes rapidly in response to conditions that will eventually result in changes in the soil organic matter content, and it has been used as a sensitive and early indicator of soil quality [2 y 3]. In the studied soils, the management conditions affected the organic matter content and therefore the microbial populations. Figures 3 and 4 show the differences in microbial biodiversity between the four management regimes. The soils under direct seeding had higher microbial biodiversities than the soil under conventional tillage, and these differences were already clear after only three years of application.

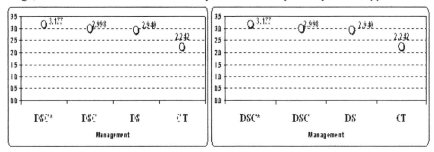

Fig 3. Simpson's diversity index of soils under different management regimes.

Fig 4. Shannon-Weaver's diversity index of soils under different management regimes.

4   Conclusions

Conservation agriculture systems improved the chemical, physical, and biological properties of previously tilled soil.

Conservation agriculture management (direct seeding) increased the soil microbe population, and reduced the impact of intensive agriculture on the maize rhizosphere ecosystem.

Microbial biodiversity in the soil with maize crops was higher under direct seeding or direct seeding with cover than under conventional tillage.

The improvement of the soil's physicochemical parameters under conservation agriculture management correlated with an increase in the counts of microorganisms in the soil.

Acknowledgements

This work has been financed by the Ministerio de Ciencia y Tecnología of the Spanish Government under the research Project AGL2000-0463-P4-05

References

[1]   Bezdicek, D.F.; Beaver, T. & Granatstein, D. Subsoil ridge tillage and lime effects on soil microbial activity, soil pH, erosion and wheat and pea yield in the Pacific Northwest, USA. Soil and Tillage Research, 74: 55-63 (2003).

[2]   Feng, Y.; Motta, A.C.; Reeves, D.W.; Burmester, C.H.; Van Santen, E. & Osborne, J.A. Soil microbial communities under conventional-till and no-till continuous cotton systems. Soil Biology and Biochemistry. 35: 1693-1703 (2003).

[3]   Franzluebbers, A.J., Langdale, G.W. & Schomberg, H.H. Soil carbon, nitrogen, and aggregation in response to type and frequency of tillage. Soil Science Society of America Journal 63, 349–355 (1999).

[4]   Huang, P.M., J.-M. Bollag & N. Senesi. (Editors). Interactions between soil particles and microorganisms: Impact on the terrestrial ecosystem. IUPAC. John Wiley and Sons, Ltd. Chichester, U.K. 582 pages (2002).

[5]   Nielsen, M.N. & Winding, A. Microorganisms as Indicators of Soil Health. National Environmental Research Institute, Denmark.Technical Report No. 388 (2002).

Molecular analysis of microbial communities in northern terrestrial systems

Vasara, R.E. [1,2*], **Martikainen, P.J.** [2], **Lipponen, M.T.T.** [1,2], **Kontro, M.H.** [3,4], **Nykänen, H.K** [2] **and Servomaa, K.** [1,2]

[1] North North Savo Regional Environment Centre, P.O. Box 1199, FIN-70211 Kuopio, FINLAND
[2] Department of Environmental Sciences, University of Kuopio, P.O. Box 1627, FIN-70211 Kuopio, FINLAND
[3] National Public Health Institute, P.O. Box 95, FIN-70701 Kuopio, FINLAND
[4] Department of Ecological and Environmental Sciences, University of Helsinki, Niemenkatu 73, FIN-15140 Lahti, FINLAND

*Corresponding author: e-mail: Ritva.Vasara@ymparisto.fi

The community structure of ammonia-oxidizing bacteria and methane-oxidizing bacteria was studied in terrestrial environments in Finland by using PCR based methods. The study included both mineral and organic soils, which were either natural or manipulated for agriculture or forestry. Compost samples were included as an example of an extremely manipulated terrestrial system. Despite the drastic changes in land use, we did not detect major differences in the diversity of ammonia-oxidizing bacteria between the studied environments. However, there were differences in the diversity of methane-oxidizing bacteria. Compost samples were analyzed also with 16S rDNA primers, and the total microbial community analysis revealed temporal differences in the population patterns during the composting process.

Keywords ammonia-oxidizing bacteria; methane-oxidizing bacteria; 16S rDNA

1 Introduction

The nutrient and gas dynamics of soils are controlled by microbial processes. Microbes involved in the biogeochemical cycling of nitrogen and carbon have influence also on atmospheric dynamics. Ammonia-oxidizing bacteria (AOB) and methane-oxidizing bacteria (MOB) impact atmospheric levels of nitrous oxide and methane, respectively. These greenhouse gases contribute to the global warming and their levels in the atmosphere are rising due to human actions. MOB oxidize methane in soils and thus reduce the atmospheric methane load. AOB are involved in biogeochemical reactions that release nitrous oxide to the atmosphere. AOB also contribute to the production of nitrate in soils, which enhances leaching of nitrogen to surface and ground waters.

AOB and MOB in Finnish soils are poorly known. It is not known how land use effects the activity and diversity of AOB and MOB and how it is reflected in the biogeochemical processes these microbes are involved in. Traditional culture-based microbiological methods are not suitable for the study of slow-growing AOB and MOB in environmental samples since culture-based methods favour certain species.

This study used molecular biological methods to evaluate the community structure of AOB and MOB in terrestrial environments with different land usages. PCR based methods were used in the analysis of AOB and MOB. Genes encoding key enzymes in ammonia oxidation and methane oxidation, i.e. ammonia monooxygenase (*amoA*) and particulate methane monooxygenase (*pmoA*), were analyzed. The study included both mineral and organic soils, which were either natural or manipulated for agriculture or forestry. Compost samples were included as an example of an extremely manipulated terrestrial system. Composting is based on microbial processes that are not fully controlled and can lead to emissions of nitric and nitrous oxides and methane. Compost samples were analyzed also with 16S rDNA primers to reveal total microbial community structure during composting process.

2   Materials and methods

Soil samples were collected from three mineral agricultural soils and a mineral forest soil. Organic soil samples included natural peatland, forested peatland and organic agricultural soil. Compost samples were collected from an experimental compost consisting of mechanical pulp amended with 5 % chicken manure. The composting process was followed for 120 days and samples for AOB and MOB diversity analyses were taken at different time points (9, 13, 28 and 90 days).

DNA was extracted from organic soil samples using a method modified from Ogram [1], Torsvik [2] and Zhou [3]. DNA extractions from mineral soil samples and compost samples were performed using UltraClean Soil DNA Isolation Kit (Mo Bio Laboratories, Solana Beach, CA, USA). Betaproteobacteria-specific *amoA* primers amoA-1F and amoA-2R [4] or primers A189 and A682 [5] specific for both *amoA* and *pmoA* were used for amplifications. Amplified DNA fragments were cloned and sequenced. Total microbial community structure during the composting process was analyzed using 16S rDNA primers 518R and GM5F [6, 7] and temperature gradient gel electrophoresis (TGGE). Major bands were excised, re-amplified and sequenced. Sequences were compared with reference sequences from the EMBL database and phylogenetic analyses were performed using EMBOSS [8] and PHYLIP [9] software packages.

3   Results and discussion

Despite the extremely drastic changes in land use, we did not detect major differences in the diversity of AOB between the studied soils. Nitrosospira-like amoA sequences were obtained from both natural and manipulated organic and mineral soils, and also at the early stages of the composting process. Nitrosomonas-like amoA sequences were detected only from two of the nutrient-rich mineral agricultural soils and from compost samples. However, there were differences in the diversity of MOB between the studied soils. pmoA-like sequences resembling those of type I and type II MOB were detected, including sequences from putative acidophilic as well as high-affinity MOB.

Aeration status, hydrological conditions and nutrient availability of the soils differed. Some of the soils had low methane concentrations while others had high methane concentrations. Also agricultural disturbances such as tillage, liming and fertilization change soil characteristics. These factors may explain the observed differences in the diversity of MOB. However, there were also pmoA sequence types that were found from several of the studied soils, and similar amoA sequences were detected from different soils. This indicates that many AOB and MOB are capable of surviving in very different environmental conditions.

The analysis of total microbial community structure during the composting process revealed temporal differences in population patterns. Some populations were detectable

410

during the whole composting process whereas others were observed only at certain time points.

References

[1]   A. Ogram, G.S. Sayler and T. Barkay, The extraction and purification of microbial DNA from sediments, J. Microbiol. Meth., **7**, 57 (1987).

[2]   V. Torsvik, J. Goksøyr and F.L. Daae, High diversity in DNA of soil bacteria, Appl. Environ. Microbiol., **56**, 782 (1990).

[3]   J. Zhou, M.A. Bruns and J.M. Tiedje, DNA recovery from soils of diverse composition. Appl. Environ. Microbiol., **62**, 316 (1996).

[4]   J.-H. Rotthauwe, K.-P. Witzel and W. Liesack, The ammonia monooxygenase structural gene *amoA* as a functional marker: Molecular fine-scale analysis of natural ammonia-oxidizing populations, Appl. Environ. Microbiol., **63**, 4704 (1997).

[5]   A.J. Holmes, A. Costello, M.E. Lidstrom and J.C. Murrell, Evidence that particulate methane monooxygenase and ammonia monooxygenase may be evolutionary related, FEMS M icrobiol. Lett., **132**, 203(1995).

[6]   G. Muyzer, E.C. de Vaal and A.G. Uitterlinden, Profiling of complex microbial populations by denaturing gradient gel electrophoresis analysis of polymerase chain reaction-amplified genes coding for 16S rRNA, Appl. Environ. Microbiol., **59**, 695 (1993).

[7]   C.M. Santegoeds, T.G. Ferdelman, G. Muyzer and D. de Beer, Structural and functional analysis of sulfate-reducing populations in bacterial biofilms, Appl. Environ. Microbiol. **64**, 3731 (1998).

[8]   P. Rice, I. Longden and A. Bleasby, EMBOSS: The European Molecular Biology Open Software Suite, Trends Genetics, **16**, 276 (2000).

[9]   J. Felsenstein, PHYLIP (Phylogeny Inference Package) 3.5c edn., Department of Genetics, University of Washington, Seattle, US, (1993).

MycorID: a molecular database based on ITS-RFLP analysis for identification of ectomycorrhizae at Iberian Peninsula

G. Marques [*] and **A. Nazaré-Pereira**

Centro de Estudos em Gestão de Ecossistemas, Universidade de Trás-os-Montes e Alto Douro, Apartado 1013, 5000-911,Vila Real, PORTUGAL

[*]Corresponding author: gmarques@utad.pt, Phone: +351 259 350635, Fax: +351 250 350480

Ectomycorrhizae are difficult to distinguish on the basis of morphological and anatomical features, which depends of the several factors, including the host plant. Molecular techniques based on PCR protocols, allowing differentiation between species, independent of host or environmental conditions, became indispensable for ecological surveys of the diversity of ectomycorrhizal fungi in natural ecosystems. The aim of this work was to develop a database of internal transcribed spacer (ITS)-RFLP patterns of a collection of some common fungal symbionts on forest ecosystems at Iberian Peninsula. This expansible reference database has actually the restriction fragments patterns of 125 ectomycorrhizal species, including fungi commonly found on young or mature stands, by the use of *Cfo*I, *Hinf*I and *Mbo*I endonucleases. This database is a precious support to ectomycorrhizae morphotyping in ecological studies or certification of inoculated plants at nurseries.

Keywords PCR, ITS-RFLP, molecular database, ectomycorrhizae, identification

1   Introduction

Ectomycorrhizal fungi (ECM) are major components of soil fungal community in most boreal and temperate forests. The importance of these symbiotic fungi for their host plants is well recognized, by the positive effects on nutrition and resistance to stress [1]. Worldwide, more than 5000 species of fungi form ectomycorrhizal associations, primarily of basidiomycetes but also some ascomycetes and zygomycetes, with most shrubs and trees [2]. In natural forest ecosystems, a wide range of ECM is typically found, forming mycelial networks between plants of the same or different species, which may have significant ecological role in plant community dynamics and ecosystem functioning [3]. Recently, an increasing attention has been given to the spatial and temporal variability of ectomycorrhizal communities [4, 5]. This knowledge is important, not only for our understanding of their ecological roles and interactions in ecosystems, but also as a prerequisite for better management and biotechnological exploitation of these fungi on forestry.

Most diversity studies of ECM communities are currently based on analysis of ectomycorrhizae in soil cores and morphotyping by detailed macro and microscopic observations [6]. With rare exceptions, this approach only allows grouping into mycorrhizal morphotypes, because ectomycorrhizas are difficult to distinguish on the basis of morphological and anatomical features, which depends of the several factors, including the host plant [7]. Morphotypes were not easily comparable among studies, and morphological characterisation alone often leads to false discrimination of the taxa. The need to increase the knowledge in ecophysiology of ectomycorrhizal species in field studies, and the development of commercial inoculation at nurseries, requires the use of more objective methods for ectomycorrhizae identification or characterisation.

A significant methodological advance has been the use of PCR-based methods. These techniques, allowing differentiation between fungal species, independent of host or environmental conditions, became indispensable for ecological surveys of the diversity of ECM in natural ecosystems [8]. In general, molecular analysis after stratification by morphotyping is the most used approach.

Molecular identification of ectomycorrhizae is based on selective amplification of fungal DNA by PCR, with subsequent analysis by restriction fragment length polymorphisms (RFLP) or sequencing [8]. The internal transcribed spacer (ITS) of the nuclear ribosomal DNA (rDNA) gene cluster has been widely used as the target sequence, because this noncoding region shows high resolving power for species discrimination [9, 10].

ITS-RFLP analysis has been a method of choice for identification of ECM samples because separates many species quickly and is relatively cheaply in relation to other molecular methods [8]. Typically, two or three enzyme restriction digests are used to distinguish most species [9, 10]. RFLP patterns from unknown ectomycorrhiza must be matched with patterns from known ectomycorrhizal fungal tissue, frequently from sporocarps. This approach depends the availability of an extensive ITS-RFLP reference database. Because intraspecific variation exists at ITS region in some species, mainly across large geographical scales, is necessary to construct this database at more reduced scale in order to minimize this problem [8].

The aim of this work was to develop an expansible database, based on ITS-RFLP patterns of a collection of some common fungal symbionts in forest ecosystems at Iberian Peninsula, to be used for ecological studies of ECM communities.

2 Material and methods

2.1 Fungal material

Sporocarps of several ECM species, collected in field surveys in Portugal and Spain, were used for molecular analysis after identification through a macro- and microscopic observations. Dried specimens deposited in the mycological herbaria of University of Trás-os-Montes e Alto Douro, Vila Real, Portugal, and of the Center of Forest Investigation of Lourizán, Pontevedra, Spain, were also used. With the aim to test intraspecific variability, several sporocarps (2 to 7) were analysed in some species. In total, 125 species from 32 genera were studied.

2.2 DNA extraction and PCR-RFLP analysis

DNA extractions were performed using a miniprep method of Gardes et al. [11] with some modifications. Dried sporocarp tissues (c. 30 mg) were crushed in CTAB buffer (1.4 M NaCl, 100 mM Tris-HCl, 20 mM EDTA, 2% CTAB), using a sterile glass micropestle. After incubation at 65 °C for 1 h, the samples were extracted with chloroform, centrifuged for 5 min at 13000g, precipitated with an equal volume of isopropanol by gentle mixing and incubation at –20 °C. The precipitated DNA was then collected by centrifugation at 13000g for 30 min at 4 °C, washed with ethanol and dried at room temperature. DNA pellet was dissolved in 100 ≥l TE buffer (1 mM Tris-HCl, 0.1 M EDTA) or ultra-pure water.

Amplifications were carried out in 25 ≥1 reaction volumes using the Pure-Taq[A] Ready-To-Go PCR beads (Amersham Biosciences) according to the manufacturer's instructions.
Aliquots of diluted DNA extracts corresponding to 0.1-10 ng DNA, were used as a template. Primers used were ITS1 and ITS4, targeting the ITS1-5.8S-ITS2 region of the ribosomal DNA operon [12]. PCR was performed in a Biometra Uno II[A] thermocycler, programmed for an initial denaturation at 95 °C for 2 min, followed by 30 cycles of denaturation at 94 °C for 1 min, annealing at 50 °C for 1 min and extension at 72°C for 2 min. The final extension step was at 72 °C for 10 min. All amplifications were made, at least, in duplicate. Amplified products were analyzed by electrophoresis in 1.5% agarose gels stained with ethidium bromide. Marker used was the 100 pb Ladder (Invitrogren). Fragment migration and lengths were estimated in scanned Polaroid[A] photos, using the software BandLead 3.0 [13] and Gel [14], respectively.
RFLP analysis of the PCR products was performed with restriction endonucleases Cfo I, Hinf I and Mbo I, for 3-4 h or overnight, following manufacturer's instructions (Invitrogen). The restriction fragments were separated electrophoretically on 2% agarose gels stained with ethidium bromide. Fragment lengths were estimated as previously described.

2.3 Construction of a database for ITS-RFLP matching

The restriction patterns obtained were entered into a database created in Microsoft Excel. Fragment patterns that differed at least in one enzyme were considered to be different genotypes. Fragments less than 100 base pairs were not counted in order to reduce the possibility of including PCR artefacts. For easy manipulation of the data and for matching an unknown sample, the program GERM 1.0 [15], is used. This program is available in www.tc.umn.edu/~dicki017.

3   Results and discussion

For most of the analyzed species, the size of ITS region amplified with the primers ITS1/ITS4 range from 600 to 900 pb. For exemplification, the mean values of the length of ITS region and the restriction profiles of some species analysed in this study are shown in Table 1.
Some genera and some species of the same genus, exhibit a uniform size of ITS region, but the combination of restriction patterns with all endonucleases, allowed the discrimination of most of the analyzed sporocarps. The results show that the intraspecific variation of the ITS was low (8%) in accordance with other similar studies [10, 16]. Thus, in general, each species is characterized by a specific ITS-RFLP pattern using, at least, three endonucleases, confirming the potentialities of this technique for fungal species discrimination. However, this technique does not allowed us to discriminate forms or varieties of the same species, eg. *Russula cyanoxantha* and *R. cyanoxantha* f. *peltereaui* (Table 1) and *Amanita muscaria* and *A. muscaria* var. *formosa* (data not show). In the present study, only the use of the enzyme Mbo I allowed to distinguish *Laccaria bicolor* from *L. amethystina* and *L. laccata* (Table 1). Other species taxonomically related could not be discriminated by PCR-RFLP analysis. In the other way, many ectomycorrhizae remains unmatched since these databases are usually biased for fungi that form visible sporocarps. In spite of these limitations, this technique is a very useful tool for ecological studies in ECM communities. Unmatched ectomycorrhizae could then be sequenced and their phylogenetic placement determined, at least a family level, by searching public databases [8].

414

Table 1 – RFLPs of ITS region with *Cfo* I , *Hinf* I e *Mbo*I endonucleases of some ectomycorrhizal fungi analysed in this study (in base pairs).

Species	ITS	CfoI	HinfI	MboI
Amanita caesarea	695	250 234 170	386 183 117	365 196 (155)
Amanita citrina	725	400 321	379 345	297 278 148
Amanita ponderosa	780	765	382 362	448 309
Amanita rubescens	750	400 305	378 326	326 301 (140)
Boletus edulis	800	359 276	270 245 220	298 194
Boletus erythropus	765	384 347	324 281	304 224 179
Boletus pinophilus	750	351 299 (100)	300 247 180	300 250 220
Laccaria amethystina	720	390 325	375 342	430 290
Laccaria bicolor	720	403 324	365 328	430 (170) (160)
Laccaria laccata	720	399 325	375 339	430 290
Leccinum lepidum	530	291 273	283 255	373 (104)
Lepista nuda	720	319 221 171	382 349	436 277
Paxillus involutus 1	790	448 161 106	428 296	272 161 124
Paxillus involutus 2	790	445 360	418 368	292 149 110
Russula cyanoxantha	715	340 240 148	347 315	280 240 185
R. cyanoxantha f. *peltereaui*	715	347 246 152	354 332	285 249 193
Russula virescens	690	332 277 (136)	334 334	275 240 177
Scleroderma citrinum	690	336 284	245 203 127	363 245
Scleroderma polyrhizon	685	276 201 151	373 308	306 167 148 116
Xerocomus badius	620	326 293	302 302	297 (186)
Xerocomus subtomentosus	805	415 385	435 333	316 157 (105)

An error of 25 pb is assumed in algorithms used for determination of similarities between samples running in different gels. Data between ( ) are faint bands.

By comparison of our results with other published databases [10], different restriction patterns were obtained in some species with one or more endonucleases, reinforcing the necessity to construct these databases at a regional scale.

In ecosystems with high ECM diversity, and when an extensive database is available, identification of ectomycorrhizae requires the use of computerized RFLP pattern matching programs. Our database (MycorID) uses the program GERM 1.01 developed for ectomycorrhizae identification [15]

To date, restriction fragment patterns of 125 fungal species from 32 genera are available. This molecular database is a precious support to ectomycorrhizae morphotyping and can be used in many situations, like the confirmation of the identity of mycelial cultures, certification of inoculated seedlings at nurseries and identification of the some common ectomycorrhizal symbionts in early development of seedlings or in mature forest stands.

References

[1]   S. E. Smith and D. J. Read (eds.), Mycorrhizal Symbiosis, 2nd edition, London, Academic Press (1997).

[2]   R. Molina, H. Massicote and J. M. Trappe, in: M. F. Allen (ed), Mycorrhizal Functioning: an Integrative Plant-Fungal Process. (Chapman and Hall, New York, 1992), chap. 11.

[3]   S. W. Simard et al., Can. J. For. Res. **27**, 331 (1997).

[4]   L. Jonsson, D. Anders, and B. Tor-Erik. For. Ecol. Manag. **132**, 143 (2000).

[5]   A. Dahlberg, New Phytol. **150**, 555 (2001).

[6]   R. Agerer (ed.). Colour Atlas of Ectomycorrhizae, Einhorn-Verlag, Schwäbisch Gmünd (1987-2000).

[7]   K. N. Egger, Can. J. Bot. **73,** 1415 (1995).

[8]   T. R Horton and T. D. Bruns, Molecular Ecology. **10**, 1855 (2001)

[9]   M. Gardes and T.D. Bruns, Can. J. Bot. **74**, 1572 (1996).

[10]   O. Kårén et al., New Phytol. **136**, 313 (1997).

[11]   M. Gardes et al., Can. J. Bot., **69**, 180 (1991).

[12]   T. J. White et al., in PCR Protocols. A Guide to Methods and Applications. M.A. Innis et al., (eds), Academic Press, Inc., California, pp. 315 (1990).

[13]   M. Aharoni, Band Leader Software (version 3.00), Magnitec Ltd. (1994-1997).

[14]   M. Lacroix, Gel (IBM version 1.2) (1989).

[15]   I. A. Dickie et al., Mycorrhiza **13**, 171 (2003).

[16]   K. Pritsch et al., New Phytol. **137**, 357 (1997)

Nitrification and denitrification associated with N_2O production in a temperate N- fertilized irrigated Uruguayan rice field

S. Tarlera[*1], **S. Gonnet**[2], **P. Irisarri**[2], **J. Menes**[1], **A. Fernández**[1], **G. Paolino**[1], **D. Travers**[1] and **E. Deambrosi**[3]

[1] Department of Microbiology, Facultad de Química and Facultad de Ciencias, Avda. Gral Flores 2124, CC 1157, Montevideo, URUGUAY
[2] Department of Plant Biology, Facultad de Agronomía, Avda. Garzón 780, Montevideo, URUGUAY
[3] INIA, Estación Experimental, Treinta y Tres, URUGUAY

[*]Corresponding author: e-mail: starlera@fq.edu.uy, Phone: +5982 924 4209, Fax: +5982 924 1906

Uruguay is the seventh largest rice exporting country in the world. Nitrogen is the single most limiting factor for rice production. Microbial soil processes, e.g. nitrification, denitrification and mineralization, influence the fate of the "mobile" N atom. There are gaps in the understanding of key processes that govern N cycling, availability and plant acquisition in irrigated rice systems. Nitrous oxide (N_2O), a greenhouse gas, is produced as a by-product during nitrification and occurs as an intermediate during denitrification. The use of fertilizers with inhibitors of nitrification has been proposed as a mitigation strategy. This study showed that less than 7 % of the total N_2O emission from ENTEC®-fertilized soils, containing an inhibitor of nitrification (DMPP: 3,4-dimethylpyrazole-phosphate) was due to nitrification. T-RFLP *amoA* analysis detected the presence of *Nitrosomonas* and *Nitrosospira* genera in urea-fertilized plots.

Keywords: nitrification, rice, T-RFLP

1   Introduction

Nitrous oxide (N_2O), a greenhouse gas, is produced as a by-product during nitrification and occurs as an intermediate during denitrification. Microbial processes in soils contribute about 70% of the atmospheric budget of N_2O. N_2O emissions from soils have greatly increased with increasing N inputs by fertilization of agricultural soils [1]. After the application of top dressings of ammonium fertilizers, aerobic nitrification is favoured producing nitrate which diffuses into the surrounding anoxic bulk soil where it is prone to denitrification. The use of nitrification inhibitors has been proposed as means of reducing N losses by denitrification and leaching.

Uruguay is the seventh largest rice exporting country in the world. Only one crop per year is grown during the southern hemisphere spring-summer season (October-March). The crop is broadcast seeded into dry soil and permanent flood is established 35–55 days after sowing. Nitrogen is the single most limiting factor for rice production. Several authors have found that N_2O losses occur mainly after fertilization application during a period no longer than five weeks after application [2], coincident for rice in Uruguay with the period before flooding. After flooding, waterlogged conditions will probably induce N_2O reduction to N_2 by denitrification. Highest nitrification activity in flooded rice soil fields has been reported to occur at the tillering rice stage and in the surface soil fraction (top 2–5 mm soil) [3].

The objective of this work was to evaluate the effect of a commercial fertilizer ENTEC[®], containing an inhibitor of nitrification (DMPP: 3,4-dimethylpyrazole-phosphate) on nitrification-denitrification dynamics in a rice soil during a period of high risk for N_2O emission and to examine the nature of the AOB communities inhabiting these soils.

2 Materials and methods

This work was conducted at the east of Uruguay (Treinta y Tres, Estación Experimental del INIA,Paso de la Laguna) during the 2004 growth season before flooding (October-December). This season was selected because soils are mainly aerobic and nitrification losses are maximal. A field experiment with *Oriza sativa* variety INIA Tacuarí broadcast seeded was laid out in a randomized complete block factorial design with three replicate plots (4 x 5 m each) of three treatments: control (no fertilizer), urea (18 KgN/ha at seeding, 23 KgN/ha at tillering and 23 kgN/ha at flowering), ENTEC[®], BASF (60 KgN/ha at seeding): $N-NH_4^+$ 18,5%, $N-NO_3^-$ 7,5% and DMPP (1 % content relative to $N-NH_4^+$ content). Three sampling times were established: 10 and 42 days (tillering) after sowing and one week after flooding (day 52 after sowing).

Ten soil cores were collected from each plot and homogenized. All assays were performed in triplicate. Routine soil textural and chemical analysis were determined by standard procedures. Potential nitrification rates (PNR) were determined by the shaken soil-slurry method adding 1.5 mM NH_4^+ as described by Mintie et al. [4]. Denitrification enzyme assay (DEA) was determined following standard procedures [5] by measuring N_2O formation in anaerobic slurries that received glucose (10 mM), NO_3^- (1 mM) and C_2H_2 (10%, v/v). Nitrous oxide production was measured in sealed jars exposed to 0, 0.1 and 5% C_2H_2 atmospheres in order to determine N_2O total production, N_2O coming from denitrification and total ($N_2O + N_2$) coming from denitrification as described by Merino et al. [6].

T-RFLP analysis (Terminal restriction fragment length polymorphism). DNA extraction from the samples was performed with Kit Mobio according to manufacture protocol. The DNA was amplified by PCR with specific primers for *amo* A genes and the primer amoA-1F were labeled with the dye 6-FAM (5-[6-carboxy-fluorescein). The amplification product was purified and digested with *Taq*I. The restriction fragments were separated on an ABI 373 automated sequencer. The results were analyzed using GeneScan and Genotyper software (Applied Biosystems) [7]. Statistical differences were assessed using two-factor analysis of variance (ANOVA) for sampling time and fertilization, followed by Tukey's test for mean multiple comparisons.

3 Results and discussion

3.1 Physical and chemical characteristics of soils

Soil characteristics are: silty loam in texture, 3.0–3.5 % organic C and 5.0–5.5 % organic matter. Soil moisture content decreased from an average of 27 %, 10 days after seeding to an average of 2.3 % at tillering. No significant differences in initial $N-NH_4^+$ values (mean: 150 ≥g/gdw) between control, urea and ENTEC[®] fertilized soils were detected. A ten-fold decline in $N-NH_4^+$ in all treatments was observed over the six-weeks sampling period resulting in a final mean value of 25 ≥g /gdw. Nitrate values also decreased with time from initial mean values of 33, 17 and 10 ≥gN-NO_3^-/gdw in ENTEC[®], urea and control soils respectively, to below detection limit (™1 ≥gN-NO_3^-/gdw) after flooding in all treatments.

Soil pH remained constant at 5.2. Average temperatures increased from 18.2 °C to 21.2 °C during the sampling period. 60.5 mm of rainfall accumulated from seeding up to day 10 and a further 23.9 mm up to day 42.

In the present investigation, mineral-N concentration decreased during the sampling period. This rapid decline in applied N- fertilizer in the inorganic fraction during the early growing season can be ascribed to rapid plant uptake, immobilization into the microbial biomass and loss of nitrogen through nitrification-denitrification reactions.

3.2 Nitrification and denitrification potentials

PNRs in fertilized and non-fertilized surface soils (Fig. 1), were calculated from the first 24 hours of incubation of the soil slurry experiments (linear rates of $NO_2^{\int}$ and $NO_3^{\int}$ accumulation in soil slurry experiments). Tukey´s comparison of the rate data indicated significant differences only for ENTEC® fertilization at both sampling times ($= 0.05$). Highest activity was observed for the ENTEC® fertilization treatment at both sampling times. Initial nitrification rates of ENTEC® fertilized plots were 3.4 and 5.7 greater than urea and control treatments, respectively. Six weeks later, an order of magnitude increase over controls and a 5.3 times increase over urea was detected. ENTEC® fertilization clearly stimulated nitrifying activity after six weeks. PNR measured in this study, which ranged from ca 26 to 400 nmol N /gdw.h, were in the same range or even higher than rates found for other non-flooded agricultural soils [4] consistent with the drier soil conditions of this sampling period. In a recent study, Nicolaisen et al. [3] obtained PNR for urea-fertilized flooded rice surface soil averaging 54 nmol N/gdw.h. Nitrification studies in rice fields have mostly being conducted with permanently flooded rice soil systems, where anaerobic conditions prevail.

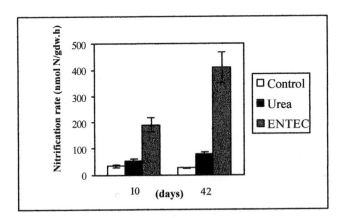

Fig. 1 Potential nitrification rates in fertilized and control rice surface soil 10 and 42 days after soewing. Error bars represent ± S. E.

DEA values obtained from control (113,6 o 4,3 nmol N/gdw.h) urea (142,9 o 20,7 nmol N/gdw.h) and ENTEC® (102,9 o 22,1 nmol N/gdw.h) plots 6 weeks after sowing did not differ significantly. Furthermore, DEA measurements 10 days later (one week after flooding), indicated similar mean values in the plots with the different treatments (data not shown). However, nitrate values sharply declined in all treatments after flooding.

Denitrifying communities were similarly active under denitrifying conditions compared to other rice submerged soils [2]. DEA and NPR values were similar in magnitude (Fig. 1), suggesting that denitrification can serve as a sink for nitrification.

3.3 Nitrous oxide production

To explore the effects of the nitrification inhibitor contained in ENTEC®, N_2O production was measured in fertilized (urea and ENTEC®) and control soils. ENTEC® has been reported to remain active for a period of 4-10 weeks [8]. DMPP is a nitrification inhibitor which blocks the first step of ammonia oxidation to nitrite in autotrophic ammonia oxidizers.

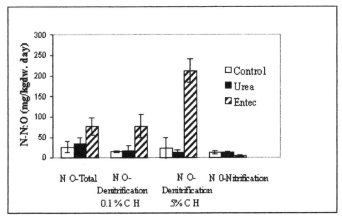

Fig. 2 Nitrous oxide production from different processes 42 days after fertilization. Error bars represent ± S. E

Results indicate that in control and urea-fertilized treatments, nitrification and denitrification processes were similarly active. A relatively minor contribution of nitrification to N_2O production was observed in ENTEC® -fertilized soils. However, a higher N_2O release from denitrification was observed in ENTEC® treatments not attributable to higher nitrate content at this sampling time (23 ENTEC® vs 21 urea ≥gN-NO_3^-/gdw). In addition, an even contribution of N_2O and N_2 to total denitrification was detected. Heterotrophic nitrification cannot be dismissed since reports suggest that a higher concentration of acetylene (®0.1%) is needed for complete inhibition ([9], Fig. 2). Further studies are needed to explore the importance of this process in this ecosystem.

3.4 Diversity of ammonia-oxidizing bacteria

The diversity of the ammonia-oxidizing community in rice surface fertilized soil on day 42 was assessed by the cloning-independent procedure TRFLP. Horz et al. [7] provided experimental evidence for the strict correlation between *amoA* based T-RFLP *Taq*I-fragments and phylogenetically defined subgroups of ammonia oxidizers. Amplification of *amoA* genes in samples from ENTEC® fertilized plots was not successful. T-RFLP analysis with *Taq*I profiles from triplicate urea treated soils consisted of either two fragments of 220 and 284-bp size or of the addition of a third fragment of 41 bp as shown in Fig. 2.

According to available nucleotide sequence information, the 220 and the 284-bp fragments suggest the presence of *Nitrosomonas* species and of *Nitrosospira* species, respectively [7, 4]. *In silico* analysis of the *amoA* database could not indicate a possible assignment for the 41 bp fragment. To our knowledge, only isolates and sequences affiliated with *Nitrosomonas* species have been recovered from surface, bulk and rhizospheric soil [3], [10]. Both genus differ, among other traits, in affinity for NH_4^+ and growth rates.

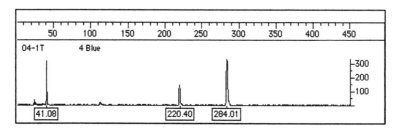

Fig. 3 T-RFLP profile from the digested (*Taq* I) *amoA* PCR product of an urea soil sample after 42 days. Fragment size of T-RFs are shown in base pairs (bp).

4   Conclusions

Less than 7 % of the total N_2O emission from ENTEC®-fertilized soils was due to nitrification. On the other hand, higher potential nitrification rates compared to urea treatments could be indicative of the activity of heterotrophic ammonia oxidizers. T-RFLP *amoA* analysis detected the presence of *Nitrosomonas and Nitrosospira* genera in urea-fertilized plots.

Acknowledgements

The support of CSIC (Comisión Sectorial de Investigación Científica) is gratefully acknowledged.

References

[1]   U. Skiba and K. A. Smith, Glob. Change Sci., **2**, 379 (2000)
[2]   L. Mei, L. Yang, D. Wang, B. Yin, J. Hu, and S. Yin, Soil Biol. Biochem., **36**,1057 (2004).
[3]   M. H. Nicolaisen, N. Risgaard-Petersen, N.P. Revsbech, and W. Reichardt, Microbiol. Ecol., **49**,359 (2004).
[4]   A. T. Mintie, R. S. Heichen, K. Cromack, Jr., D. Myrold and P. Bottomley, Appl. Environ. Microbiol., **69**, 3129 (2003).
[5]   J. M. Tiedje, Methods of Soil Analysis, Part2- Microbiological and Biochemical Properties, Madison, p.245 (1994).
[6]   P. Merino, J.M. Estavillo, G. Besca, M. Pinto and C. Gonzáles-Murua, Nutrient Cycling in Agrosystems, **60**, 9 (2001).
[7]   H-P.Hortz, J-H. Rotthauwe, T. Lukow, and W. Liesack, J. Microbiol. Methods, **39**,197 (2000).
[8]   F. Azam, G. Benckiser, C Müller, and J. Ottow, Biol. Fertil. Soils, **34**, 118 (2001).
[9]   M. Daum, W. Zimmer, H. Papen, K. Kloos, K. Nawrath and H. Bothe, Curr. Microbiol,. **37**, 281 (1998).
[10] A. Briones, S. Okabe, Y. Umemiya, N-B. Ramsing, W. Reichardt and H. Okuyama. Plant and Soil, **250**, 335 (2003).

Nitrifier bacterial activity linked to mineralization of soil organic matter: individual based simulations

A. Gras[*,1], **J. Valls**[2], and **M. Ginovart**[3]

[1] College of Agriculture of Barcelona, Technical University of Catalonia, Campus Baix Llobregat, Avda. Canal Olímpic s/n, 08860 Castelldefels (Barcelona), SPAIN
[2] Department of Physics and Nuclear Engineering, Technical University of Catalonia, Campus Baix Llobregat, Avda. Canal Olímpic s/n, 08860 Castelldefels (Barcelona), SPAIN
[3] Department of Applied Mathematics III, Technical University of Catalonia, Campus Baix Llobregat, Avda. Canal Olímpic s/n, 08860 Castelldefels (Barcelona), SPAIN

[*]Corresponding author: e-mail: anna.gras@upc.edu, Phone: +34 935521224, Fax: +34 935521001

The nitrifier bacteria in soil have been modelled using a discrete simulation methodology that stands on individual-based models. INDISIM-SOM is a simulator to deal with soil microbial activity in terms of the microscopic components. Exchanges of N between soil N pools are driven by micro-organisms. INDISIM-SOM has been specifically used to study the behaviour of the nitrifiers and their contribution on the N oxidation that takes place in short-term dynamics of organic and inorganic carbon and nitrogen in soil organic matter. The nitrifiers are controlled at each time step, using a set of time dependent variables for each biotic element and specific rules for their motion, uptake, metabolism, reproduction, viability and lysis. The simulation model has been calibrated using experimental data from laboratory incubations of three different Mediterranean soils.

Keywords: Nitrifier bacteria; Individual-based Model

1   Introduction

The micro-organisms are the first colonizers of soil, and they are mainly responsible for some of the steps of Carbon and Nitrogen cycling. Mineralization of soil organic matter (OM) is related to the net release of mineral carbon and nitrogen due the decay of SOM, and immobilization is related to the assimilation of inorganic compounds to the organic state. Micro-organisms mediate in both processes and a balance can be established between them. Nitrification plays an important role on the control of mineral N transformation from the less mobile NH_{4+} to NO_3^-, laying available N for plants. The nitrification process interfaces with other processes of the N cycle, a feature that contributes to make difficult its study. Nitrification is mainly associated with two bacteria groups and it could be described by growth functions. The ammonia oxidiser bacteria (oxidizing NH_4^+ to NO_2^-) and the nitrite oxidiser bacteria (oxidizing NO_2^- to NO_3^-) are the key agents of autotrophic nitrification The metabolism of each one of these groups is clearly related to the corresponding stages of nitrification [1].

There are different types of simulation models to study the transformations of soil C and N. These models have been mainly applied to the mineralization and immobilization of C and N. Conceptually modellers visualize the existence of different pools of organic matter (living or dead) which follow first or second order kinetic laws [2]. The last decade has witnessed the emergence of Individual based Models (IbM), referred by Grimm as "bottom up" approaches [3]. IbM stipulates that populations of organisms are modelled in terms of discrete individuals that are unique only in terms of characteristics. A bacterial IbM has been developed and called INDISIM, it studies bacterial populations describing the individual properties and the actions of bacteria [4]. The computer code INDISIM has been adapted to several biological systems with success. INDISIM -SOM is a new simulator that proceeds from INDISIM. INDISIM -SOM enables to attempt the mineralization and immobilisation of C and N, and the integration of the nitrification process in this context [5–7]. It is focused on the activity that microorganisms carry out on the different organic and/or mineral substrates, and is assume two different prototypes of microorganisms: decomposers and nitrifiers. It also takes into account the role of C and N during their microbial lives, thus linking the C and N cycles.

In this review will be illustrated the part of the simulator INDISIM -SOM that is referred to the nitrification process. The temporal evolutions of mineral N in ammonium and nitrate are studied. These state variables are allied to mineralization of SOM and the heterotrophic metabolism of decomposers because the nitrification is the second step of N evolution. The simulation results will be put side by side with experimental data from the evolution of three different soils.

2 Materials and methods

2.1 Data acquisition

Data from incubation experiments conducted under laboratory conditions concerned with three soils of Catalonia were collected by Vidal [8]. These Mediterranean soils are from Calaf, Sta. Maria de Miralles and Caldes de Montbui. The samples were from Ap horizons (0-20 cm) and were incubated for 90 days at a constant temperature of 30°C and a moisture content to 80% water-holding capacity. The soil samples were wetted daily to keep the field capacity and aerated to ensure their aerobic conditions. The specific properties of soils studied, are shown in Table 1. The analyses were performed on three replicates.

Table 1. Chemical and physical properties of the three soil samples: Calaf, Miralles and Caldes soil.

	Calaf	Miralles	Caldes
pH	7.20	7.30	7.20
Organic C (%)	2.17	1.16	0.91
C/N	9.90	10.60	9.10
Total N ($\geq$g g^{-1})	2200	1160	1000
Organic N ($\geq$g g^{-1})	2180	1150	982
Easily hydrolysable N ($\geq$g g^{-1})	151	86	54
N-NH$_4^+$($\geq$g g^{-1})	20	10	16
N-NO$_3^-$($\geq$g g^{-1})	33	23	17
C.E.C (meq/100 g)	19.62	22.75	9.59
Clay (%)	13	16	9

2.2 General outline of the simulation model INDISIM-SOM

INDISIM-SOM is discrete in space and time. The physical lattice is subdivided into squared spatial cells. The time evolution of the system is divided into equal intervals that we identify with time steps. The behaviour of microbial population is specified taking into account their motion, uptake, metabolism, reproduction, viability and lyses of each individual that makes up the system. The microbial biomass of the population is identified as live pool, and its biochemical composition is defined by its C/N ratio. At each time step and for each microorganism the simulator controls its own time dependent properties: biomass, position in the spatial domain, reproduction biomass and the state of its cellular cycle. These microorganisms, a group of decomposers and another group of nitrifiers, and particles of different substrates, are evolving in a two-dimensional spatial grid. Nine different types of substrates, five of which are organic compounds namely: polymerised carbon (C_P), polymerised nitrogen (CN_P), labile carbon (C_L), labile nitrogen (CN_L) and humic compounds (CN_H), and the others are mineral compounds, N_{NH4}, N_{NO3}, C_{CO2} and O_2. The simulation model considers the output flow of CO_2 from the system, the input flow of O_2, the diffusion of the organic labile forms and mineral compounds in the medium, and the hydrolysis processes of the polymerised substrates and the hummus to labile and mineral compounds. Diverse considerations make it possible to set specific rules for the simulated microbial population. The ammonifiers have been modelled as heterotrophic microorganisms as a representative of a group of diverse microbe. They use C_L and CN_L as energy sources, C_L as C source, and CN_L, N_{NH4} and N_{NO3} as N source. At the next section will be given more details about the nitrification 'submodel'. After modelling each part of the system and implementing the overall model in a computer code. At each time step a complete temporal and spatial description of the simulated system is obtained . Other general aspects and more specific details of this simulation model INDISIM-SOM are introduced and discussed in Ginovart et al. [8].

2.3 Nitrification model

The oxidation of NH_4^+ to NO_3^- is modelled as a single process mediated by a single nitrifier community so that the formation and oxidation of NO_2^- is not explicitly represented, in the same way as considered by Grant in his nitrification model [8]. In consequence has been modelled a unique group of bacteria responsible of nitrification, nitrifiers. For each nitrifier a set of time dependent variables are controlled: biomass, position in the spatial domain, reproduction biomass, and its cellular cycle state. At each time step, a individual may perform the following actions directed by the computer code: (I) *Motion*. The nitrifier has a position in the space and it can move at random from one spatial cell to another. (II) *Uptake*. The nitrifier uptakes some of the substrate particles surrounding it of N_{NH4}, CO_2 and O_2, from the spatial cell that it is occupying. The uptake depends of its maximum capability that is function of the microbial surface which makes contact with the external medium and also of the substrate availability [4, 7]. (III) *Metabolism*. The substrate particles uptaken are metabolised. The nitrifier cell is modelled entirely as autotrophic bacterium. The cell uses N_{NH4} for its cellular maintenance, but if it is not enough and is still more energy required it can be accomplished by using up its own biomass (endogenous respiration). It uses CO_2 and N_{NH4} to synthesize new biomass. The CO_2 is assimilated through the Calvin cycle. The decomposers will be able to use NO_3^- that is release into the medium to synthesize their biomass when others N source are expired. (IV) *Reproduction*. The bipartition model that it has been used in base bacterial simulator INDISIM, which is compatible with the I+C+D model [4]. (V) *Death and lyses*. When the individual has no chance to acquire its maintenance requirements successfully, neither from the external source nor from its own biomass, cellular lyses occurs, also is implemented a death probability to symbolize other death causes. Then its biomass is released to the medium as polymeric compounds and humus and ammonium, in order to balance C and N pools.

Most soils have the capacity to bind ammonium from aqueous solution to the soil surface lowering the free NH_4^+ concentration governing the rate of ammonia oxidation [9]. Therefore has been considered two different compartments of NH_4^+. One of them is the ammonium in solution which is diffused as other mineral and soluble organic compounds the other one is the ammonium which is adsorbed on the colloids surface (clay and organic matter) among other soil sites. Both forms of ammonium are in equilibrium and are subject to a process of adsorption and desorption that is modelled by two constants. Only the ammonium in soil solution is available for microorganisms.

3　　Results and discussion

The good fitting between the experimental results of soils and the simulated pools has been performed in order to calibrate the simulator INDISIM-SOM. With the same set of values for all the parameters considered in the nitrification model have been possible obtain the evolutions point out in Figure 1 [5–7]. Figure 1 shows the cumulative production of ammonium and nitrate. The simulated temporal evolutions keep up a correspondence with three sets of experimental data related to soils of Calaf, Miralles and Caldes. INDISIM-SOM is a simulation model that share de philosophy of the individual based models, and has been developed and is ready to study diverse variables related to soil microbial activity to deal with soil C and N turnover.

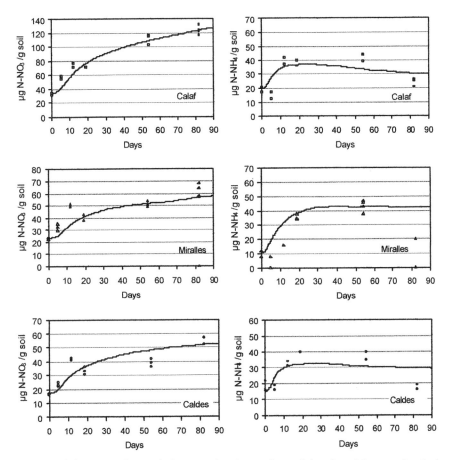

Fig. 1 Cumulative ammonium and nitrate produced: experimental data from laboratory incubations of the three sampled soils (points) and simulations results representing the N-NH₄ and N-NO₃ compounds (lines).

Acknowledgements

The financial supports of the Ministry of Education and Science of Spain Grant REN2000-0049-P4-04 and the DURSI Generalitat de Catalunya 2003ACES00064 are gratefully acknowledged.

References

[1]   W. De Boer and G.A. Kowalchuk, Soil biology and biochemistry **33**, 853 (2001).
[2]   S. Hansen, H.E. Jensen and H. Svendsen, Daisy Soil Plant Atmosphere System Model. NPO Report, Nº A10. The national Agency for Environmental Protection, Copenhagen (1990).
[3]   V. Grimm, Ecological Modelling **115**, 129 (1999).
[4]   M. Ginovart, D. López and J. Valls, Journal of Theoretical Biology **214**, 305 (2002).

426

[5]   A. Gras and M. Ginovart, in: Proceedings of the IASTED International Conference on Applied Simulation and Modelling, Rhodes, Greece, 2004, (ACTA Press), pp. 125-129.

[6]   A. Gras, INDISIM -SOM, un model discret per a l'estudi de la dinàmica de la material orgànica i la nitrificació en sòls. Ph.D. Thesis, University of Lleida, Spain, (2004).

[7]   M. Ginovart, D. López and A. Gras. Nonlinear Analysis: Real World Applications (in press).

[8]   R.F. Grant, (1994), Soil Biology and Biochemistry 3, 305 (1994).

[9]   N. Hommes, S. Rusell, P.Bottomley & Arp (1998). Applied and Environmental Microbiology, 64,4: 1372-1378.

Nitrogen biological fixation ability by *Rhizobium legominosarum* biovar *phaseoli* on cultivars of *Phaseolus vulgaris L.*

Abdollah Ghasemi Pirbalouti [a],[*], Iraj Allahdadi [b], Gholam Abbas Akbari [b], Ahamd Reza Golparvar [c] and Safar Ali Rostampoor [a]

[a] Department of Agriculture, Islamic Azad University of Shahrekord, Shahrekord, Iran. POBox: 166.
[b] Department of Agronomy and Plant Breeding, University of Tehran, Aboorayhan Compus, Tehran, Iran.
[c] Department of Agronomy and Plant Breeding, Islamic Azad University of Khorasgan, Khorasgan (Isfahan), Iran.

[*] Corresponding author. Tel: +98 (381) 33 31 001; Fax: +98 (381) 33 36 999- E-mail address: ghasemi955@yahoo.com

In order to evaluate of nitrogen biological fixation ability by different strains of *Rhizobium legominosarum* biovar *phaseoli* in common bean cultivars, a spilt plot experiment in randomized complete block design was conducted a field in ShahreKord, Iran. The factors were four bacterial strains: *L-78, L-47, L-125, l-109*, +N and –N controls and three cultivars: local cranberry bean, Talash cranberry bean and local red Mexican type. The results showed that seed inoculation with strain increased nodule dry weight, N total (shoot) and percentage of fixed N_2 in relation to bean rhizobia population naturalized and +N control. However, strains *L-109* gave highest nodule dry weight, N total and percentage of fixed N_2.

Keywords: Cultivars; Nitrogen fixation; Phaseolus vulgaris; Rhizobium legominosarum biovar phaseoli

1   Introduction

The common bean (*Phaseolus vulgaris* L.) is the most important legume for human nutrition, major protein and calorie source in the world [7]. It is now growing extensively in all major continental areas. In Iran, the crop occupies 0.125 million ha of area, the yield is moderate, on average 1470 kg ha^{-1}. *Phaseolus vulgaris* L. is a legume capable of symbiotically fixing atmospheric nitrogen gas (N_2). Nodules formed on the bean roots contain bacteria that convert N_2 from the air to plant – available form [5]. These bacteria can be used as an inoculum that is applied to the seed [6]. Board ranges of *Rhizobium* species are able nodule and fix N_2 with beans including *R. legominosarum* biovar *phaseoli* [10] *R. tropic* [11] and *R. etli* [17].

In Iran, the usual practices of bean cultivation dose not involve inoculation of seeds with a specific rhizobial inoculation and farmers depend on the application of inorganic nitrogen fertilizer to sustain growth to improve yield. In Iran soils, for example high rate of levels of inorganic nitrogen fertilizer used at the rates of 100–300 kg N ha^{-1}. Since chemical N fertilizers are expensive to most farmers. In addition, N fertilizer application had no significant effect on the dry matter yield and N content of *Phaseolus vulgaris* cv. B9 [18]. However, the most soils in Iran nitrogen have deficiency and N_2 fixation by *Rhizobium* bacteria can increase the yield at a low cost and preserve water resources from nitrate pollution.

Yield response of bean to inoculation with a specific *Rhizobium* is often variable and depends on environmental and agronomic [8]. The lack of response to inoculation can be attributed to intrinsic characteristics of both the host plant and bacteria, as well as the great sensitivity of the symbiosis to environmental stress, soil dryness and low soil fertility [9]. Other important factors to be considered are general limitation to N_2 fixation per se, like the high rate of levels used in intensive agriculture and the residual N remaining in soil [15]. Genotype variation in common bean for traits affecting nodulation and N_2 fixation has been found [4]. However, when highly effective *Rhizobium / phaseoli* combination have been selected, they provided bean yield of 60–70 % of those obtained with N fertilizer control under field conditions [16].

Determination of N_2 fixation effectiveness in the process of strain selection is normally several step procedures involving an initial selection under greenhouse conditions and a final testing in field trails [2, 5]. In the process of strain selection, variation in the efficiency of the strain * cultivars association was detected for parameters like total N, yield, plant growth, nodule number, weight and N fixed [16].

The aims of this work were to study (i) evaluate the effect of different bacterial strains on common bean yield and N_2 fixation rate (ii) estimate the best bacteria and cultivar's combination response of common bean cultivars to inoculation with different *Rhizobium legominosarum* biovar *phaseoli* strains.

2   Materials and methods

Field experiment was conducted at Shahrekord (latitude 32 ^{0} 44 $^/$ N, 2100 m asl), located at about 100 km of Isfahan and 500 km of the capital town of Iran on spring and summer 2002. The medial annual rainfall is about 337.2-mm year^{-1}. Average annual temperature is 11.9 °C. Soil samples were collected to a depth of 0–30 cm before planting. The experiment was performed in soil that had not been cultivated with beans for at least 5 years. C, N, P and K content, electrical conductivity (EC), pH and percentage of sand, silt and clay were determined (Table. 1).

Table 1. Physical and chemical characteristics of the soil of experimental site

P PPM	N total %	OC %	K PPM	EC ds/m²	pH	Clay %	Silt %	Sand %
24.4	0.163	1.47	371.4	0.66	7.68	13	40	47

Total N was measured calorimetrically following Kjeldahl digestion. Also, soil pH and EC were estimated using a glass electrode pH meter and an EC meter in 1:1 soil water suspension. Organic carbon was determined by modified Walkley and Black method [12]. The rhizobia population evaluated was approximately 10^3–10^4 cells g^{-1} of soil.

The experiment was arranged in a randomized complete block design with a split plot layout and four replications. Four bacterial strains of rhizobia (table 2.) and non-inoculated controls including application of nitrogen fertilizer treatment (40 kg N ha^{-1} as urea at sowing and 60 kg N ha^{-1} at 35 days after sowing) and without application of nitrogen fertilizer were assigned to main plots. Inoculation was added to the seeds with 15% (w/v) source solution to increase adherence. Three cultivars included local cranberry bean; Talash cranberry and local red Mexican were assigned as the subplots.

Table 2. Bacterial strains used in this work

Strain	Year and space	Source
L-78	Shahrekord (Chahar Mahal & Bakhtyiari) -2002	Institute of soil and water of soil biology part, Karaj, Iran
L-109	Touysercan (Hamadan)-2002	"
L-47	Feraydonshahr (Isfahan)-2002	"
L-125	Aleshtar (Lorastan)-2002	"

Each block consisted of six main plots spaced 2 m apart. The main plots had three subplots that consisted of five rows spaced 50 cm apart. Phosphate and starter N fertilizer were applied at rates 100 kg P ha^{-1} and 40 kg N ha^{-1} as urea before of planting, respectively. Four seeds of each genotype were planted per hole on 10 June 2002, weeding and thinned 20 days after emergence 135,000 plants ha^{-1} for red bean and 100,000 plants ha^{-1} for cranberry bean [7]. Topsoil of the experimental plot area was kept moist throughout the growing season when necessary.

The characteristics under investigation were grain yield, number of pods per plants, number of seeds per pods, number and weight nodule at 40–50% flowering (50 days after emergence) and percentage of nitrogen fixation. Percentage nitrogen fixed was estimated according [14] with following formula :

$$\text{Percent nitrogen fixed} = \frac{(\text{N content of inoculated plants} - \text{total N content of control plants})}{\text{Total N content of inoculated plants}} \quad 100$$

Plants were sampled at 25 days after planting, 45–50 days after planting and at physiological maturity (70–80 days after planting). Five plants were randomly samples from each treatment replicated and various parameters assessed. Nodule number and dry weight verified in 5 plants at early (45–50 days after emergence). Plant samples and nodules were dried to constant weight at 75 °C at 72 hr in oven. Shoots were dried for dry weight determination and for total nitrogen determination by the Kjeldahl method. Yield and its components were evaluated at final harvest and values were corrected for 13% moisture.

All data were subjected to analysis of variance using the statistical computer package SAS (Release 6.12, Statistical Analysis System Institute., 1996) and treatment means separated using LSD range test at $P \leq 0.05$ level.

3 Results

3.1 Seed yield

In the present study, high significant difference ($P \leq 0.01$) in seed yield was observed among seeds inoculated with different strains and non-inoculated controls (N fertilizer treatment and without N fertilizer treatment) (Table. 3). Seeds inoculated with *L-125* showed higher seed yield than other treatments (Table. 4), but similar result was observed in seed inoculated with *L-109, L-78* strains and non-inoculated control (N fertilizer treatment).

The results of analysis of variance indicated that there were high significant differences *(P≤0.01)* between cultivars in seed yield (Table. 3). The interaction between strain * cultivar effect was no significant difference, but the results table 5 showed the highest and lowest seed yield were associated to Talash * *L-125* strain and local red bean * control. However, local red bean * *L-125*, local red bean * *L-78* and local red bean* *L-109* similar combinations were observed in seed yield with Talash * *L-125* strain.

Table 3. Analysis of variance for different traits under seeds inoculated with different strains of *Rhizobium legominosarum* bv. *phaseoli*

S.O.V	d.f	Mean square					
		Shoot N content %	Nitrogen fixed rate %	Nodule number Plant^{-1}	Nodule dry weight g plant^{-1}	Shoot dry matter Kg m^{-2}	Seed yield Kg m^{-2}
Block	4	13.28	524.68	66.62	0.008	0.005	0.005
Strain	5	1.88*	217.71**	14.58 ns	0.013**	0.22**	0.016**
Error a	15	0.35	43.58	23.23	0.005	0.051	0.006
Cultivar	2	6.84*	2.54 ns	84.12 *	0.044***	0.38**	0.053**
Strain*Cultivar	10	0.27	24.89 ns	6.20 ns	0.002 ns	0.042**	0.002 ns
Error	36	0.656	43.76	19.00	0.003	0.019	0.003
cv %		8.86	35.02	28.1	28.01	22.16	25.78

ns, * and ** : Not significant, significant at 5% and 1% level of probability, respectively.

Table 4. Mean comparison of characteristic in different strains of bean (*Phasseolus vulgaris* L.) using LSD range test at *P ≤* 0.05 level

Strains	Nitrogen fixed rate %	Shoot N content % N total	Pod dry weight g plant^{-1}	Nodule dry weight g plant^{-1}	Shoot dry weight Kg m^{-2}	Seed yield Kg m^{-2}
Control	0.00	5.53	11.67	0.055	3740	1441
N 100	5.29	6.21	19.44	0.050	6530	2121
L-47	7.62	5.91	13.51	0.086	3960	1771
L-109	11.90	6.32	14.92	0.1260	4280	2103
L-125	10.71	5.98	18.92	0.1120	5120	2506
L-78	9.82	6.18	16.53	0.1210	5110	2202
L.S.D (0.05)	5.741	0.621	5.171	0.062	0.237	0.067

3.2 Shoot and pods dry matter

Shoot and pods dry matter were influenced *(P≤0.01)* by seeds inoculated with strains and non-inoculated controls (Table. 3). Seeds inoculated with strains showed lower shoot dry matter than non-inoculated N-fertilizer, but similar pods dry matter was observed in seeds inoculated with strains *L-109, L-125* and *L-78* and the non-inoculated N-fertilizer (Table. 4). There were high significant differences between cultivars and strains in shoot and pods dry matter as well as an interaction strain * cultivar for both parameters (Table. 3). Non–inoculated N-fertilizer treatment produced high shoot and pods dry matter in Talash cultivar (Table 5).

Table 5. Mean comparison of characteristic in strain *cultivar interaction effects

Cultivar	Strains	Seed yield Kg m⁻¹	Nodule number Plant⁻¹	Nodule dry weight g plant⁻¹	Pod dry weight g plant⁻¹	Nitrogen fixed rate %
	Control	1416.7	5.45	0.05	3.63	0.00
	N 100	2651.6	1.98	0.001	23.40	4.63
Cranberry bean	L-47	1946.7	6.73	0.010	17.40	7.32
Talash cv	L-109	2370.1	4.45	0.121	13.1	11.60
	L-123	2701.7	4.12	0.132	21.23	9.10
	L-78	1713.4	5.20	0.071	11.63	9.41
mean		*2134.2*	*4.6*	*0.069*	*17.67*	*6.99*
	Control	1995.1	6.11	0.081	17.13	0.00
	N 100	2172.2	1.63	0.091	13.45	7.11
Cranberry bean	L-47	2065.3	5.20	0.085	15.24	11.53
Local	L-109	1271.7	5.96	0.052	7.53	12.3
	L-123	2123.3	4.49	0.041	13.68	10.46
	L-78	2133.9	9.73	0.142	13.19	7.49
mean		*1969.42*	*5.52*	*0.082*	*15.04*	*8.16*
	Control	912.3	4.51	0.001	9.23	0.00
	N 100	1533.3	0.31	0.021	11.43	10.12
Mexican red bean	L-47	1301.7	5.73	0.063	7.30	4.04
Local	L-109	2663.2	7.37	0.0191	19.02	11.97
	L-123	2604.4	7.33	0.152	21.23	12.30
	L-78	2693.3	5.13	0.144	19.77	12.74
mean		*1968.95*	*5.32*	*0.072*	*14.77*	*8.54*
ANOVA	Cv.	$p < 0.01$	ns	$p < 0.01$	$p < 0.01$	$p < 0.01$
	St.	$p < 0.01$	$p < 0.05$	$p < 0.01$	$p < 0.01$	ns
Significance of F	Cv. *St.	ns	ns	ns	$p < 0.01$	ns
CV%		*35.02*	*28.1*	*28.01*	*23.59*	*25.78*

3.3 Nitrogen fixation rate

Table 3 shows that nodule dry weight (at 40–50% flowering), N total (shoot) and percentage of fixed N_2 were influenced by seeds inoculated with strains and non-inoculated controls (N-fertilizer and without N-fertilizer treatments). While, there were no significant differences between cultivars and strains in nodule number at 40–50% flowering. The results revealed that all treatments were capable of nodulation, however, strains *L-109* gave highest nodule dry weight, N total (shoot) and percentage of fixed N_2 (Table 2).

Furthermore, seed inoculation with strain increased nodule dry weight, N total (shoot) and percentage of fixed N_2 in relation to bean rhizobia population naturalized and N fertilizer treatment. The results of analysis of variance indicated that there were high significant differences between cultivars in nodule dry weight and N total (Table. 3). While, there were no significant differences between cultivars in nodule number and percentage of N_2 fixed.

In the present study, no significant interaction *Rhizobium* strain * common bean cultivar for nodule number and weight at 40–50 % flowering, N total in shoot and percentage of N_2

fixed were found (Table. 3). Table 5 shows the highest nodule number, nodule weight and percentage of N_2 fixed were associated to local cranberry cultivar * *L-78* strain, local red bean * *L-125* and local red bean * *L-78*, respectively.

4   Discussion

The results revealed that significant difference in seed yield, % nitrogen fixation, % nitrogen per plants, % seed protein, number and weight nod at 50% flowering (50 days after emergence) were observed among seed inoculated with different strains and non-inoculated controls (N fertilizer treatment and without N fertilizer treatment). Seed inoculated with *L-125* showed higher seed yield than other treatments, but similar result was observed in seed inoculated with *L-109*, *L-78* strains and non-inoculated control (N fertilizer treatment).

Seeds inoculated with *L-125* probably aided the establishment of the most efficient inoculated rhizobia, improving symbiotic performance and could increased nitrogen available for plant. In the Central Regional of Brazil, the Cerrados, where over $0.8 * 10^6$ ha are being cultivated with beans and soil N content is low, bean yield is generally increased by inoculation [8]. Hungria *et al.*, 2000 reported that the common bean inoculated with *PRF31, PRF55 and PRF81* showing high rates on N_2 fixation and inoculated with *PRF81* allowed yield increased of up to 906 kg ha^{-1}, compared with the non-inoculated (control) with a population of native rhizobia in Brazilian soils. Also, the study revealed that tepary bean inoculated with *R3254* and non-inoculated (control without inoculated and N fertilizer) the highest and lowest seed yield, respectively [18].

Also, the results revealed that all treatments were capable of nodulation, however, strains *L-109* gave highest nodule dry weight, N total (shoot) and percentage of fixed N_2 (table. 4). Furthermore, seed inoculation with strain increased nodule dry weight, N total and percentage of fixed N_2 in relation to bean rhizobia population naturalized and N fertilizer treatment. An efficient symbiotic was achieved with strain *L-109*, since total N content of plants inoculated with these strains was similar to those plants supplied with N – fertilizer treatment.

Hungria *et al.*, 2000 reported that under controlled greenhouse condition, inoculation with strains *PRF81* and *CIAT899* enhanced nodulation, compared to other treatments, resulting in accumulation of more N in common bean tissues. An efficient symbiotic was achieved with strains *PRF81* and *CIAT899*. Rates of N_2 fixation achieved with strain *R. tropici* type IIB *CIAT 899* were 112% greater than with *R. legominosarum* biovar *phaseoli* strain *USDA 2671* and could reflect the adapted symbiosis of Brazilian cultivar with *R. tropici* species [9].

These results confirm a report also with Iranian cultivar [1] and disagrees with obtained in other countries [13, 19]. However, the strains different substantially in symbiotic properties, such as, nodulation and N_2 fixation capacity, as well as in the synthesis of Nod factors after inoculation.

The difference in vegetative and reproductive (pods) growth between treatments reflected changes in assimilate partitioning which led to significant differences in harvest index [15]. The results showed that + N control gave highest shoot and pods dry matter (table. 4). The results of study by Tamimi *et al.*, 2002 showed that shoot dry matter was the highest for common bean inoculated with isolate *JOV1*. Rodriguez-Navarro *et al.*, 1999 working in greenhouse also found that non-inoculated + N control higher shoot growth than the plants inoculated with strains, but similar pods growth was observed in plants inoculated with strains *ISP-1* and *CIAT 899* and the non-inoculated + N plants. In tepary bean, inoculated with strain *R3254* and non-inoculated control (without N-fertilizer) demonstrated the highest and lowest shoot and pods dry matter, respectively [18].

A significant interaction *Rhizobium* strain * common bean cultivar for seed yield, shoot and pods dry matter were found [15]. While, other authors [3 , 18] found no interaction strain * cultivar effects on plant growth and seed yield. Therefor, plant breeding of bean can be done with and effective strain. The other investigation by Rodriguez-Navarro *et al.*, 1999 showed that the highest shoot dry matter was associated to *c.v Cnellini* * strain *CIAT* and *c.v Cnellini* * strain *ISP-1* and *c.v Bina* * non-inoculated + N treatment for pods dry matter.

Acknowledgments

This work received support from the vice chancellor for research at the University of Tehran, Aboorayhan campus, and the Soil Biology Part of Institute of Soil and Water of Iran. We thank Dr. Asadi Rahmani, Masc. Mehnat Kesh, Masc. Yadegari and Mc. Farzan for present study help.

References

[1]   Asadi Rahmani, H. (2000) Evaluation of N₂ fixation efficient with different strains of Iranian rhizobia on common bean (*Phaseolus vulgaris* L). Journal of Iranian Soil and Water Sciences. 9: 93-97.

[2]   Bergersen, F. J. (1980) Measurement of nitrogen fixation by direct means. In: Bergersen, J. Methods for evaluating biological nitrogen fixation. Wiley Chichester, Pages 65-110.

[3]   Buttery Brian, R., Soont, P and Van Berkum Peter, B. (1998) Effects of cultivar and strain of *rhizobium* on growth bean yield and nitrogen content of Common bean (*Phaseolus vulgaris* L.). Tektran. Agriculture Research Service.

[4]   Chavera, M. H., Graham, P. H. (1992) Cultivar variation in traits affecting early nodulation of common bean. Crop. Sci. 32: 1432-1436.

[5]   Giller, K. E. (2001) Nitrogen fixation in tropical cropping systems. 2nd edn. CABI publishing.

[6]   Giller, K. E. and Cadisch, G. (1995) Future benefits from biological nitrogen fixation – an ecological approach to agriculture, Plant Sci, 174: 225-277.

[7]   Graham. P. H., and Ranallin, P. (1997) Common bean (*Phaseolus vulgaris* L). Field Crops Research. 53:131-146.

[8]   Handerson, G. (1993) Methods for enhancing symbiotic nitrogen fixation. Plant and Soil. 152, 1-17.

[9]   Hungria, M. Andrade, S. Chueire, L. M. Probanza, A. Guttierrez-Manero, F. J. and Megias, M. (2000) Isolation and characterization of new efficient and competitive bean (*Phaseolus vulgaris* L.) rhizobia from Brazil. Soil Biology and Biochemistry. 32, 1515-1528.

[10]   Jordan, D. C. (1984) *Rhizobium* Conn. In: Krieg, N. G., Holt J.G. (Eds). Bergeys Manual of Systematic Bacteriology. Williams & Wikins, Baltimore, USA. Pages: 235-244.

[11]   Martinez-Romero, E., Segavio, E. Mercante, F. M. Feranco, A. A. Graham, P. H. and Pardo, M. A. (1991) *Rhizobium tropici* a novel species nodulating *Phaseolus vulgaris* L. beans and *Leucaena sp*. trees. International Journal of Systematic Bacteriology. 41: 417-426.

[12]   McKeague, J. A. (1978) Manual of soil sampling and methods of analysis, Soil Survey Committee, Canadian Society of Soil Science, Ottawa, Ont., Canada.

[13]   Olivera, L. A., Graham. P. H. (1990) Speed of nodulation and competitive ability among *Rhizobium legominosarum biovar phaseoli*. Archives of Microbiology. 153: 311-315.

[14]   Rennie, R. J. (1984) Comparison of N balance and 15 N dilution to quantity N fixation in field grown legumes. Agronomy Journal. 76: 785-790.

434

[15]   Rodriguez-Navarro. DN, Santamaria C, Temprano F, and Leidi EO (1999) Interaction effects between *Rhizobium* strain and bean cultivar on nodulation, plant growth, biomass partitioning and xylem sap composition. European Journal of Agronomy. 11, 131-143.

[16]   Santamaria, C., Rodriguez, D. N., Camacho, M., Daza, A., Temprano, F. and Leidi, E. O. (1997) Selection de cepas de *Rhizobium legominosarum bv. Phaseoli* effectivas con cultivars comercial de judia verda. Consejeria Agriculture y Pesca, Junta de Andalucia, Seville. Pages: 48-53.

[17]   Segavio, E., Young, J. P. W. and Martinez-Romero, E. (1993) Reclassification of American *Rhizobium leguminosarum* biovar *phaseoli* type I strain as *Rhizobium etli* sp. nov. I. International Journal of Systematic Bacteriology. 43: 347-377.

[18]   Shisanya, C. A. (2002) Improvement of drought adapted tepary bean (*Phaseolus acutifolius* A. Gray var latifolius) yield through biological nitrogen fixation in semi-arid SE-Kenya. European Journal of Agronomy. 16: 13-24.

[19]   Streit, W., Kosch, K. and Werner, D. (1992) Nodulation competitiveness of *Rhizobium legominosarum bv. Phaseoli* and *R. tropici* strains measured by glucoronidase gene fusions. Biology and Fertility of Soils. 14: 140-144.

[20]   Tamimi S. M (2002) Genetics diversity and symbiotic effectiveness of rhizobia isolated from root nodules of common bean (*Phaseolus vulgaris* L.) grown in the soils of the Jordan valley. Applied Soil Ecology. 19, 183-190.

Phenotypic and genetic diversity of fluorescent *Pseudomonas* recovered from different host plants

M. A. Argudín[1], C. Pérez[1], M. C. Mendoza[1], M. R. Rodicio[1] and A. J. González[*,2]

[1] Departamento de Biología Funcional, Universidad de Oviedo, C/ Julián Clavería 6, 33006 Oviedo, SPAIN
[2] Servicio Regional de Investigación y Desarrollo Agroalimentario (SERIDA), Carretera de Oviedo sn, 33300 Villaviciosa, Asturias, SPAIN

[*]Corresponding author: e-mail: ajgonzalez@serida.org, Phone: +34985890066, Fax: +34985891854

Seventy three isolates of fluorescent *Pseudomonas* recovered in Asturias (Spain) from plants with disease symptoms were characterized. Diversity was revealed by results of nutritional and biochemical tests, and by PCR amplifications of genes involved in the biosynthesis of levan (*lsc*) and phytotoxin production (*phtE*, *syrB1-syrB2*, and *syrD*). The isolates were grouped in two main clusters by comparison of 26 features. Cluster A (S = 0.90) included 35 isolates and a control strain of pathovar (pv) *phaseolicola*. Cluster B displayed a higher heterogeneity, with most isolates (29) grouping in one subcluster (B3) together with a control strain of pv *syringae* (S = 0.83). Three other subclusters included isolates that differed from pv *syringae* by being negative for lipodepsipeptide production (B1), positive for pectinolytic activity (B2), or negative for levan production and/or *lsc* genes (B4). However, they were still related to pv *syringae* at S = 0.78-0.72, and analysis of their 16S rDNA sequences supported their assignation to *P. syringae*.

Keywords: *Pseudomonas syringae*; pathovar; LOPAT; levansucrase; phytotoxin; cluster analysis

1 Introduction

Some *Pseudomonas* species show the characteristic property to synthesize yellow-green, fluorescent pigments under iron stress [1]. Among these, *P. syringae* is a well-known phytopathogen that causes important and common diseases on a wide range of unrelated plants [2]. On the basis of pathogenicity and host specificity the species has been subdivided into more than 40 pathovars (pv) [2, 3]. Identification of *P. syringae* at the pathovar level requires inoculation of a range of known hosts. In addition, pathovars often display other differentiating characters (nutritional, biochemical and genetic) that can assist identification. In fact, for general diagnosis in phytopathology laboratories, determinative keys of biochemical and nutritional tests remain the basis for identification [4]. Although strains within most pathovars exhibit rather narrow host ranges [2], pv *syringae*, with more than 80 plant species listed as hosts, constitutes an exception [2]. However, evidence supports that the pathovar is heterogeneous, and may be a repository of strains with limited host range.

In our laboratory, previous studies on fluorescent *Pseudomonas* causing damage to different plants with economic interest in Asturias (a northern region of Spain), have revealed *P. viridiflava* isolates (belonging to a biotype with unusual biochemical properties [5], and *P. syringae* [6; A. J. González, unpublished], as important pathogens. In the present work, 73 new isolates recovered from herbaceous and woody plants with disease symptoms were characterized by phenotypic and genotypic approaches. The obtained results, apart from a general interest, will be of value for the epidemiological surveillance of the phytopathogens.

2   Phenotypic characterization of the isolates

Seventy three *Pseudomonas* strains collected from different plants (*Phaseolus vulgaris, Vicia faba, Capsicum annuum, Actinidia deliciosa, Prunus avium* and *Malus domestica*) and three control strains (*P. s.* pv *phaseolicola* CECT 321, *P. s.* pv *syringae* CECT 4429 and *P. viridiflava* CECT 458) were used in this work. All of them generated fluorescent colonies on King B medium and displayed a strictly oxidative metabolism. The isolates were tentatively identified by means of the LOPAT determinative scheme (levan production, oxidase activity, potato soft rot, arginine dihydrolase activity, and tobacco hypersensitivity response) [7], with the aid of additional biochemical tests [4] (utilization of mannitol, *m*-inositol, erythritol, homoserine, sorbitol, D-tartrate, sucrose, L-lactate, trigonelline, quinate, betaine and adonitol as a sole carbon source, and gelatine and aesculine hydrolysis). Sixty seven isolates showed LOPAT profile 1a [+---+], characteristic of *P. syringae*, while the remaining six displayed two LOPAT profiles: [----+] (with three isolates, one from *P. vulgaris* and two from *V. faba*) and [+-+-+] (three isolates from *A. deliciosa*), which do not correspond to any of the established species of fluorescent *Pseudomonas*. According to results of additional tests, 35 isolates with LOPAT 1a (collected from *Phaseolus vulgaris*) could be assigned to pv *phaseolicola*, and 24 (recovered from *P. vulgaris, V. faba, C. annuum, P. avium* and *M. domestica*), to pv *syringae*. The remaining eight (from *P. vulgaris* and *A. deliciosa*) displayed unusual combinations of nutritional and biochemical features, a fact that hindered their identification at the pathovar level.

3   PCR detection of lsc genes

As indicated above, levan formation is an important taxonomic characteristic of *P. syringae*. Levan, like other extracellular exopolysaccharides is though to function as a virulence determinant in diseases caused by phytopathogenic bacteria [8]. It is a high molecular weight ß-(2,6) polyfructan with extensive branching through ß-(2,1) linkages, synthesized by the extracellular enzyme levansucrase (Lsc). Several isoenzymes of Lsc have been detected in *P. syringae*. For example, in pv *glycinea* three different isoenzymes, encoded by the *lscA, lscB* and *lscC* genes, have been identified [9]. By PCR amplifications with primers specific for each of these genes, the presence of multiple Lsc isoenzymes was also demonstrated in other pathovars [9]. In this work, the primers and conditions described in [9] were applied to analyse the distribution of *lsc* genes in the isolates under study.

Consistent with levan production, positive amplification of *lsc* genes was obtained for the 35 isolates assigned to pv *phaseolicola*, which were discriminated into two profiles: *lscA-lscB-lscC* (31 isolates) and *lscA-lscC* (four isolates). Two other *lsc* profiles were identified in *P. s.* pv *syringae*: *lscB-lscC* (18 isolates) and *lscC* (six isolates). The *lsc* profiles of the remaining isolates (of undetermined pathovar or with an atypical LOPAT profile) showed a higher variation, including *lscB-lscC, lscC* (both characteristic of pv *syringae*), and *lscA* (with four, three and two isolates, respectively), while five isolates gave a negative result for the three screened genes. Since two of the latter were levan producers, the existence of additional isoenzymes of levansucrase can not be ruled out. For the remaining three, the negative amplification of *lsc* genes was consistent with our failure to detect levan production.

4   Biological and/or molecular detection of phytotoxins

P. syringae pathovars are known to synthesize a wide spectrum of secondary metabolites that exhibit phytotoxic capabilities [10]. These include chlorosis-inducing phytoxins such as phaseolotoxin, a potent inhibitor of the enzyme L-ornithine-carbamyltransferase, common to the arginine biosynthesis pathway and the urea cycle, which is very specific to pv *phaseolicola*. In this work, primers targeting *phtE*, a gene of the biosynthetic cluster of phaseolotoxin [11], were used in PCR reactions to investigate the distribution of the cluster in the isolates under study. Only isolates previously assigned to pv *phaseolicola*, proved to be positive. Thus, like most pv *phaseolicola* isolates tested so far, those recovered in Asturias appear to be toxigenic. However, nontoxigenic strains have been reported [12].

P. syringae also produces the cyclic lipodepsipeptides known as syringomycins and syringopeptins, which induce necrosis in plant tissues based on their ability to form pores into the plasma membrane [10]. Production of lipodepsipeptide toxins is a distinctive feature of *P. s.* pv *syringae* and two closely related pathovars: pv *atrofaciens* and pv *aptata*. Besides being phytotoxic, lipodepsipeptides exhibit fungicidal activity toward a broad spectrum of filamentous fungi and yeasts, a property that has been conveniently exploited in bioassays of the toxins. In the present work, bioassays for lipodepsipeptides used *Rhodotorula mucilaginosa* (strain CECT 11016) as the indicator organism [13]. In addition, PCR amplifications with primers specific for the *syrB1* and *syrB2* genes of the syringomycin biosynthetic pathway [14], and for *syrD*, a gene required for secretion of both syringomycin and syringopeptin [15], were also performed.

Nineteen isolates of pv *syringae*, two isolates of undetermined pathovar, and four isolates initially classified as *Pseudomonas* sp (the three with pectinolytic activity and the levan negative isolate recovered from *P. vulgaris*), inhibited the growth of *R. mucilaginosa* and gave positive amplification for the screened genes. Positive amplification of *syrB1-B2* and *syrD* was also obtained with the remaining five isolates of pv *syringae*, and one isolate of undetermined pathovar, although they behaved as negative in the bioassay. In these isolates, the syringomycin biosynthetic pathway appears to be present, although mutations or deletions may block toxin production. Interestingly, the two levan negative isolates recovered from *V. faba* caused inhibition of *R. mucilaginosa* and gave positive amplification for *syrD* while being negative for *syrB1-B2*. Accordingly, they might produce syringopeptin but not syringomycin. The syringomycin biosynthetic pathway seems to be either absent or defective in other two isolates (from *P. vulgaris*) that were only positive for *syrD*. However, since they were negative in the bioassay, the syringopeptin biosynthetic pathway may be also inactive. Finally, only three *P. syringae* isolates of undetermined pathovar were unable to inhibit *R. mucilaginosa* and were also negative for the screened *syr* genes.

5   Cluster analysis

Results of all tests performed were included in a cluster analysis, using the Jaccard coefficient of similarity (S) and the unweighted pair group method with arithmetic averages (UPGMA). Consistent with previous results, the analysis revealed two major clusters (S = 0.36; Fig. 1). Cluster A included all pv *phaseolicola* isolates, which proved to be highly homogeneous (S = 0.90). In fact, they could be only distinguished on the basis of *lsc* profile (subclusters A1 and A2). Cluster B, showed a higher heterogeneity (with an overall

438

similarity of 72%), and was further differentiated into four subclusters. Most is olates fall within subcluster B3 (S = 0.83), together with *P. s.* pv *syringae* CECT 4429. Interestingly, the *syr* negative isolates, the isolates with pectinolytic activity, and the levan and/or *lscA-lscB-lscC* negative isolates grouped separately into subclusters B1 (S = 0.95), B2 (S = 0.90), and B4 (S = 0.82). However, they were still related to the pv *syringae* subcluster with S values of 0.78-0.72. In addition, analysis of their 16S rRNA genes supported their assignation to *P. syringae*. Finally, it should be indicated that the low level of similarity found between clusters A and B is consistent with differences revealed by DNA-DNA hybridization experiments between pv *phaseolicola* and pv *syringae*, which has led to reclassify the former as a pathovar of the genomespecies *P. savastanoi* [16].

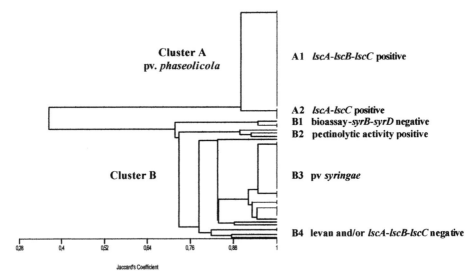

Fig. 1 Relatedness among phytopathogenic isolates recovered from plants with commercial interest in Asturias (Spain). The dendogram was obtained by comparison of 26 characteristics using the unweighted pair group method with arithmetic averages, and the Jaccard coefficient of similarity. Distinctive features of the different subclusters are indicated.

Acknowledgements

The financial support of the University of Oviedo (project MB-04-516-3) is gratefully acknowledged. M. A. A. is the recipient of grant UNIOVI-04-BECDOC-01 from the University of Oviedo.

References

[1]   R. Y. Stanier, N. J. Palleroni, and M. Doudoroff, J. Gen. Microbiol. **43**, 159-271 (1966).
[2]   J. F. Bradbury, Guide to plant pathogenic bacteria: CAB International Mycological Institute, Kew, UK (1986).

[3]   J. M. Young, Y. Takikawa, L. Gardan, and D. E. Stead, Ann. Rev. Phytopathol. **30**, 67-105 (1992).
[4]   J. M. Young, and C. M. Triggs, J. App. Bacteriol. **77**, 195-207 (1994).
[5]   A. J. González, M. R. Rodicio, and M. C. Mendoza, Appl. Environ. Microbiol. **69**, 2936-2941 (2003).
[6]   A. J. González, and M. Ávila, Plant Dis. **85**, 1287 (2001).
[7]   R. A. Lelliott, E. Billing, and A. C. Hayward, J. Appl. Bacteriol. **29**, 470-478 (1966).
[8]   T. P. Denny, Ann. Rev. Phytopathol. **33**, 173-197 (1995).
[9]   H. Lee, and M. S. Ullrich, J. Bacteriol. **183**, 3282-3292 (2001).
[10]   C. L. Bender, F. Alarcón-Chaidez, and D. C. Gross, Microbiol. Mol. Biol. Rev. **63**, 266-292 (1999).
[11]   N. W. Schaad, S. S. Cheong, S. Tamaki, E. Hatziloukas, and N. J. Panopoulos, Phytopathology **85**, 243-248 (1995).
[12]   A. Rico, R. López, C. Asensio, M. T. Aizpún, M. C. Asensio, S. Manzanera, and J. Murillo, Phytopathology **93**, 1553-1559 (2003).
[13]   F. -P. Hu, J. M. Young, and M. J. Fletcher, J. Appl. Microbiol. **85**, 365-371 (1988).
[14]   K. N. Sorensen, K. -H. Kim, and J. Y. Takemoto, Appl. Environ. Microbiol. **64**, 226-230 (1998).
[15]   A. Bultreys, and I. Gheysen, Appl. Environ. Microbiol. **65**, 1904-1909 (1999).
[16]   L. Gardan, H. Shafik, S. Belouin, R. Broch, F. Grimont, and P. A. D. Grimont, Int. J. Syst. Bacteriol. **49**, 469-478 (1999).

Population of *Aspergillus* flavus on pistachio buds and flowers

Sui Sheng T. Hua[1*], Cesaria E. McAlpin[2] and Siov Bouy Ly[1]

[1]USDA-ARS, Western Regional Research Center, Albany, California, USA
[2]USDA-ARS, National Center for Agricultural Utilization Resaerch, Peoria, Illinois, USA

[*]Corresponding author: e-mail: ssth@pw.usda.gov, Phone: 510-559-5905, Fax: 510-559-5777

California is the major state of pistachio production in US. Analysis of *A. flavus* population in pistachio orchard is essential for developing strategies to meet the mandatory levels of 2-4ppb aflatoin in the edible nuts Forty one isolates of *Aspergillus flavus* from pistachio orchard were analyzed for aflatoxin production, morphoty and vegetative competitive groups (VCG). All the s-type isolates produced aflatoxin. The percentage of sclerotium-producing isolates from California orchard was much lower than those reported. The population consists 27 VCGs determined by conventional complementation of *nit* mutants. The result gives further evidence that DNA fingerprinting using the repetitive DNA probe pAF28 could predict with 90% accuracy.

Keywords: aflatoxin; *Aspergillus flavus*; VCG; fingerprinting; population diversity; food safety.

1   Introduction

Both *Aspergillus flavus* Link and *A. parasiticus* Speare cause aflatoxin contamination in crops such as corn, cottonseed, peanuts and tree-nuts. These two fungal species are saprophytes which infect plants through woundings. *A. flavus* is typically the dominant aflatoxigenic species. *A. flavus* produces aflatoxin B_1 and B_2 and the species has been divided into S and L morphotypes based on sclerotial sizes Many isolates of *A. flvus* do not produce aflatoxin and are named atoxigenic strains. *A. parasiticus* strains produce aflatoxins B_1, B_2, G_1 and G_2 and atoxigenic isolate has rarely been described. Of the four aflatoxins, B_1 is the most potent carcinogen known[1, 2]. VCGs have also been used to examine the genetic diversity of *A. flavus* populations in an Arizona cotton field [3], in a Georgia peanut field [4], in samples of corn grown in Georgia [5] and in samples of peanuts grown in Argentina [6].

Aspergillus flavus has no known sexual stage; consequently, most studies on its genetic variability have been made mainly by characterizing isolates based on vegetative compatibility. Complementary nitrate-nonutilizing (*nit*) mutants are commonly used to identify compatible isolates. *Nit* mutants are selected for on a medium containing chlorate in which chlorate possibly acts as an analog of nitrate and is converted by nitrate reductase into toxic chlorite [7–10]. Three major types of *nit* mutants are designated as *niaD*, *nirA* and *cnx*. Complementation of paired *nit* mutants is indicated by an area of dense wild-type growth and sporulation where the two colonies come in contact resulting from hyphal anastomosis and nutritional complementation in the heterokaryon. When incompatible isolates are then paired on nitrate medium only sparse growth with little sporulation at the inoculated sites occur on agar plates.

VCG analysis becomes cumbersome for genetic analysis involving large number of isolates because it requires pairing each new mutant strain with a representative of each VCG determined for that population while eliminating isolates that are self-incompatible. In recent years the probe pAF28 has been used for strain identification and population analysis of *A. flavu.s* [11–13]. Strains belong to the same VCG shows near identical fingerprinting. California is the major state of pistachio production in US. The volume of production is the second largest in the world. Analysis of *A. flavus* population in pistachio orchard is essential for developing strategies to reduce aflatoxin in the edible nuts. This initial study was to collect *A. flavus* strains from buds and flowers. Forty one isolates were analyzed for aflatoxin production, morphoty , and vegetative competitive groups (VCG). Of special interest is to test that if the pattern of pAF28 hybridization can be used to correlate with the vegetative compatibility groups of *A. flavus* strains determined by conventional complementation of *nit* mutants.

2   Materials and methods

2.1 Isolation of fungal strains

Aspergillus flavus isolates were identified based on standard taxonomic systems. Pistachio buds and flowers were collected at from Wolfskill Grant Experimental Farm by random sampling. Buds were placed on salt agar (6% NaCl, 1.5% agar) supplemented with 100 mg/ml of chloramphenicol and incubated at 28 °C in the dark for 14 d.   *A. flavus* growing on the buds were transferred to dichloran rose bengal chloramphenicol (DRBC) agar. Purified isolates were maintained on potato dextrose agar (PDA) (Difco, Inc., Detroit, Michigan, USA). Fungal cultures were stored in 30% glycerol at –20 °C or in silica gel at 4 °C. Fungal spores were suspended in 0.05% Tween 80 and the number of spores was enumerated using a Beckman-Coulter Multisizer (Miami, Florida, USA).

2.2 Sclerotium formation

Malt extract agar (MEA), Czapek yeast agar (CYA) and PDA were inoculated with *A. flavus* and incubated at 28 °C in the dark for one month. The presence of sclerotia on agar plates were observed under a stereoscope and sclerotium-producing strains were classified according to sclerotium size. Strain L isolates produced very few sclerotia with diameter greater than 400 $\geq$ m and strain S isolates produced numerous sclerotia with diameter under 200 $\geq$ m.

2.3 Analysis of aflatoxin B_1

Aflatoxin was extracted from the fungal mat and agar by methanol and analyzed by high performance liquid chromatography (HPLC) on a Hewlett Packard model 1050 Chem Station (Hewlett Packard, Palo Alto, California). Aflatoxin B_1 was detected by a fluorescent detector with excitation at 365 nm and emission at 455 nm [14].

2.4 Isolation of *nit* mutants and Complementation test

Nitrate non-utilizating mutants were isolated on PDA supplemented with 40 g/l potassium chlorate (Fisher Scientific, Houston, Texas). Chlorate resistant mutants were purified on chlorate PDA as single colonies. The mutants were further tested on minimal salt media (MM) with either nitrate, nitrite, ammonia or hypoxanthine as nitrogen source. Three classes of mutants [10] were identified: *nia* D (nitrate non-utilizing), *nir* A (nitrate and nitrite non-utilizing), and *cnx* (hypoxanthine and nitrate non-utilizing). Spores of nitrate non-utilizing mutants were inoculated on PDA and incubated at 28 °C for two day. Agar plugs containing mycelia of each pair of *nit* mutants were placed 1 cm apart on MM with nitrate as the nitrogen source and were incubated at 8 ° C for 7–14 d. Complementation was visualized for dense hyphal growth and sporulation at the junction where the mycelia of the paired mutants came in contact and was recognized as vegetatively compatible.

2.5 DNA Fingerprints and VCG prediction

Restriction fragments of Pst I-digested *A. flavus* genomic DNA were separated by agarose gel electrophoresis and the fragments were transferred to Nytran N membranes (Schleicher and Schuell, Keene, New Hampshire) for hybridization. The probe pAF28 was labeled by the Digoxigenin Nonradioactive Nucleic Acid Labeling and Detection System (Roche Diagnostic Corp. Basel, Switzerland). Details of the probe preparation and hybridization procedure are described in McAlpin and Mannnarelli [11]. The membranes were exposed to Kodak BioMax MR film (Eastman Kodak, Rochester, New York) for 1–2 h. Bands from x-ray films were recorded. A phenogram based on the presence or absence of a band representing a known molecular weight at a specific position was generated using the NTSYS-pc ver. 2.01 [15] based on the Dice similarity coefficient. The phenogram was used for grouping identical or similar fingerprints (Similarity index, C > 80%) to predict the number of VCGs in the sample population.

3   Results

3.1 Relationship of Aflatoxin and sclerotiumproduction

Five (12%) of the *A. flavus* isolates were S type, producing plentiful small sclerotia and were all toxigenic. The most toxigenic strain was CA28 which produced 245 μg of AFB_1 per 5 mL of PDA culture. Eleven (27%) L type strains produced few large sclerotia on PDA comprised both toxigenic and atoxigenic isolates. An L type strain, CA19, produced moderate amount of aflatoxin (24 μg AFB_1) per 5 mL of PDA cultures. Other L type strains, CA18, CA 26 and CA47 produced small amount of aflatoxin (7.1, 4.2, 1.0 μg, respectively per 5 ml of PDA) and the remainder were non-toxigenic. Twenty five of the isolates that did not produce any sclerotium in MEA, CYA or PDA were all atoxigenic with the exception of CA14 which produced 23.1 μg/5 mL PDA (Table 1).

3.2 Vegetative compatibility groups.

All isolates were self-compatible and 27 compatibilty groups were identified in 41 isolates (Table 1). These included 6 VCGs containing two or more isolates and 19 VCGs represented by a single isolate (Table 1 and 2). VCGs J and K consisted of two *A. flavus* isolates each: CA 14 with CA 19 and CA13 with CA25, respectively. VCG A with 3 isolates included CA1, CA2, CA9. Dominant VCGs were VCG M with 5 isolates (CA16, CA17, CA24, CA30, CA37), VCG B and VCG E with 4 isolates each: (CA3, CA6, CA35, CA 47) and (CA7, CA38, CA41, CA 48), respectively.

Table 1. Isolates from a pistachio orchard in California

Isolates	VCG	Aflatoxin B$_1$ ($\geq$g/5 ml PDA)	Sclerotia		
			MEA	CYA	PDA
CA 1	A	nd*	--	--	--
CA 2	A	nd	--	--	--
CA 3	B	nd	--	--	--
CA 4	C	nd	--	--	--
CA 5	D	nd	--	L	L
CA 6	B	nd	--	--	--
CA 7	E	nd	L	--	L
CA 8	F	nd	--	--	--
CA 9	A	nd	--	--	--
CA 10	G	nd	--	--	--
CA 11	H	nd	L	--	L
CA 12	I	nd	--	--	--
CA 13	J	nd	--	--	--
CA 14	K	23.1	--	--	--
CA 15	L	nd	--	--	L
CA 16	M	nd	--	--	L
CA 17	M	nd	--	--	--
CA 18	N	7.1	--	--	L
CA 19	K	24.0	L	L	--
CA 20	O	nd	--	--	--
CA 21	P	nd	--	--	--
CA 23	Q	nd	--	--	--
CA 24	M	nd	--	--	--
CA 25	J	nd	--	--	--
CA 26	R	4.2	L	--	--
CA 28	S	244.99	S	S	S
CA 30	M	nd	--	--	--
CA 32	T	nd	--	--	--
CA 35	B	nd	L	--	--
CA 36	U	nd	--	--	--
CA 37	M	nd	--	--	--
CA 38	E	nd	--	--	L
CA 39	V	nd	L	--	L
CA 40	W	nd	L	--	L
CA 41	E	nd	--	--	L
CA 42	X	21.3	S	S	S
CA 43	Y	31.5	S	S	S
CA 44	Z	22.2	S	S	S
CA 45	AA	35.6	S	--	S
CA 47	B	1.0	--	--	L
CA 48	E	nd	--	--	--

* Aflatoxin B$_1$ is not detected by HPLC analysis, -- sclerotium was not observed.

3.3 VCG prediction

Fingerprints of *A. flavus* isolates by Southern hybridization to the DNA probe pAF28 were highly polymorphic. A phenogram was established based on cluster analysis showing strains with identical or similar (Similarity index, C > 80%) fingerprint profiles in which twenty five fingerprint groups, counting from the top to the bottom branch of the phenogram, were predicted for the 41 isolates. Table 2 compares the VCGs of *A. flavus* isolates by complementation to the predicted VCGs (Pre-VCG) by DNA fingerprinting analysis. The predicted VCGs 12 and 18 did not occur by complementation test of *nit* mutants.

Table 2. Comparison of dominant vegetative compability groups (VCG) in *Aspergillus flavus* isolates to the predicted- VCG (Pre-VCG) by fingerprinting.

Pre- VCG	Isolates	VCG	Isolates
1	CA1, CA2, CA9	A	CA1, CA2, CA9
4	CA3, CA6, CA35, CA47	B	CA3, CA6, CA35, CA47
11	CA7, CA38, CA41, CA48	E	CA7, CA38, CA41, CA48
12	CA10, CA20		
18	CA13, CA25	J	CA13, CA25
6	CA14, CA19	K	CA14, CA19
10	CA16, CA17, CA24, CA30, CA37	M	CA16, CA17, CA24, CA30, CA37
18	CA26, CA28		

4   Discussion

Sclerotium production in *Aspergillus* species was found to be a function of culture conditions and isolates identified here as being non-sclerotial may produce sclerotia on other media. Eighteen (44%) *A. flavus* isolates from California produced either L or S type sclerotia. The remaining isolates produced no sclerotia and abundant conidia. The percentage of sclerotium-producing isolates from California orchard was much lower than those reported by others: Bayman and Cotty [3] found that all *A. flavus* isolates examined from Arizona fields were sclerotial; Shearer et al [16] also reported that up to 92% of the *A. flavus/A. parasiticus* isolates from Iowa corn field produced sclerotia; Horn and Greene [4] found most isolates of *A. flavus*, from peanut field in Georgia produced sclerotia; and Wicklow et al [13] reported that 98% of the *A. flavus* from a cornfield in Illinois produced sclerotia.

This research gives further evidence that DNA fingerprinting using the repetitive DNA probe pAF28 could predict the VCG diversity of *A. flavus* populations with 90% accuracy.

Acknowledgements

The technical support of C. E. Platis, S. Hong , L. T. Fang and S. Kwong is gratefully acknowledged.

References

[1]   W. O. Ellis, J. P. Smith, Simpson, and J. H. Oldham. Crit. Rev. Food Sci. Nutr. **30,** pp. 403-439 (1991).

[2]   P. J. Cotty, P. Bayman, D. S. Egel, and K. S. Elias. The genus *Aspergillus* (Plenm Press, New York, 1994), pp. 1-27.

[3]   P. Bayman , and P. J. Cotty. Can. J. Bot. **69,** 1707-1711 (1991).

[4]   B. W. Horn, and R. L. Greene. Mycologia **87,** 324-332 (1995).

[5]   K. E. Papa. Mycologia **78,** 98-101 (1986).

[6]   V. M. Noval, and D. Cabral. Plant Dis. **86,** 215-219 (2002).

[7]   D. J. Cove. Heredity **36,** 191-203 (1976).

[8]   S. S. T. Hua. J. Gen. Microbiol. **123,** 355-357 (1981).

[9]   J. E. Puhalla. Can. J. Bot. **63,**179-183 1985).

[10]   C. Correll, C. J. R. Klittich, and J. F. Leslie. 1987. Phytopathology 77, 1640-1646 (1987).

[11]   C. E. McAlpin, and B. Mannarell. Appl. Environ. Microbiol. **61,** 1068-1072 (1995).

[12]   C. E. McAlpin, D. T. Wicklow, and C. Platis. Plant Dis. **82,** 1132-1135 (1998).

[13]   D. J. Wicklow, C. E. McAlpin, and C. E. Plastid. Mycol. Res. 102, 263-268 (1998).

[14]   S. S. T. Hua, J. L. Baker, and M. Flores-Espiritu. *Appl. Environ. Microbiol.* **65,** pp. 2738 - 2740 (1999).

[15]   F. J. Rohlf.   NTSYS-pc numerical taxonomy and multivariate analysis system. Version 2. 01. Exeter Software, Setauket,   NY. 1997.

[16]   J. F. Shearer, L. E. Sweets, N. E. Baker, and L. H. Tiffany. Plant Dis. **76,** 19-22 (1992)

Removal of microorganisms present in lettuces and soil irrigated with treated wastewaters

M. N. Rojas-Valencia[*], M. T. Orta-de-Velásquez, N. García-Ramíres, M. Martínez-Zamudio and V. Franco

Instituto de Ingeniería-UNAM. Edificio 5, Coordinación de Ingeniería Ambiental. Ap. 70-472, Coyoacán. CP 04510, México, D.F. MEXICO.

[*]Corresponding author: e-mail: nrov@pumas.iingen.unam.mx, Phone: (52) (55) 56-23-36-00 Ext. 8663. Fax: (52) (55) 56-16-21-64

The high contamination levels in Mexican wastewater due to the presence of pathogenic microorganisms pose serious health risks to those consuming fruits and vegetables irrigated with wastewater. For this reason, the aim of this work was, to determine the capacity of O_3 to destroy bacterias, protozoa and helminth eggs present in wastewater. The O_3 was applied at a concentration of 36.8 mgO_3/min at pH 7, for different lengths of time. Results showed that, destruction times required for 100% removal of the initial bacteria population was at 15 minutes, while helminths eggs and protozoa required 1 hour. Finally, lettuces and soil were irrigated with wastewaters treated with O_3. The analysis of microorganisms gave a reduction of the 100 % of bacteria faecal coliforms, *S. typhi* and *V. cholerae*, and significant reductions of protozoa and helminth eggs and showed an increment in leaves and root of lettuces of 6% and 16% respectively, with regard to the lettuces that were watered with wastewaters without treatment.

Keywords: Ozone, Lettuce, Pathogenic microorganisms

1 Introduction

Bacterial agents are one of the main aetiological agents causing infectious diarrhoea in practically all countries of the world, and Mexico is no exception. Crop irrigation with insufficiently treated wastewater may result in health risks. Given that the transmission paths of these agents are contaminated water and produce, the possibility of ingesting these agents when consuming raw fruits and vegetables, irrigated with inefficiently treated wastewaters, cannot be disregarded.

In the wastewaters of Mexico City, the recorded contents of pathogenic bacteria are extremely high: 4.1E-7 to 3.2E-9 MPN/100 mL of faecal coliforms, 6E-5 to 3E-9 MPN/100 mL of *S. typhi*, 2.1E-4 to 1.9E-7 MPN/100 mL of *Pseudomonas sp.* [1], and 4.2E-6 MPN/100 mL of *V. cholerae*[2].

Other microorganisms have also been found in such wastewaters: *Ascaris lumbricoides* 86.7 %; *Hymenolepis nana* and *H. diminuta* 5.9 %; *Trichuris trichiura* 4.8%; *Toxocara sp.* 2.2%; *Necator americanus* 0.4%; *Taenia sp.* 0.05%; and *Enterobius vermicularis* 0.04% [3]. In addition, amoeba of the following genera: *Acanthamoeba* (71.3%), *Vahlkampfia* (10%), *Pelomyxa* (6.4%), and *Mayorella* (5.0%), *Giardia* and *Entamoeba histolytica* [4] have also been isolated.

Given the diversity of microorganisms present in wastewaters and the risk this poses to public health, the aim of this project was to determine the disinfection capability of ozone, as well as the technical feasibility of applying an primary treatment method using ozone, in the destruction of micro-organisms that have shown marked resistance to the more commonly applied disinfectant agents. Also, the soil and crop the lettuces irrigation with wastewaters treated with ozone were analyzed.

2   Materials and methods

The primary treatment with ozone was applied to municipal wastewater samples, to determine its disinfection capability on helminth eggs, bacteria, and on the *Acanthamoeba* protozoa, as well as on biological pollution indicators such as total coliform (TC), and faecal coliform (FC) bacteria.

For isolating and quantifying helminth eggs present in the raw wastewater of the municipal treatment plant, a 5 L sample of water was taken and processed according to Mexican Standard NMX-AA-113-SCFI-1999. For isolating and quantifying *Acanthamoeba*, a specific nutritive agar for free-life amoeba was used, and the growth in the number of amoeba was quantified using a Nikon invertoscope (at 10x and 20x magnification), incubating the amoeba for 48 hr at 22°C [4].

For isolating and quantifying *V. cholerae*, two methods were employed: one was the Most Probable Number (MPN) method (with alkaline peptonated water as the culture medium) and the other was the Membrane Filter (MF) method, using Tiosulphate Citrate Bile-salts Sacarose (TCBS) selective agar. Both the MPN and MF methods were also used in the case of *S. typhi*, the medium for the MF method this time being sulphite bismuth agar. The MF method was used for the quantification of TC, the medium being M-ENDO agar. The three kinds of bacteria were incubated for 24 hr at 35° ± 2°C. FC were quantified in MFC medium, and were incubated in a water bath at 44.5° ± 2°C for 24 hr.

The ozone to be applied was produced by an Emery Trailgaz Labo 76 generator, using oxygen-enriched air as the feeding gas, a concentration of 36.8 mg O_3/min at pH7 was applied to all the samples. Simultaneously, determinations were made of the effect of ozone upon some physical-chemical parameters related to the disinfection process: alkalinity (pH), Biochemical Oxygen Demand (BOD_5), Chemical Oxygen Demand (COD), and organic nitrogen.

Finally, romaine lettuce was planted in a greenhouse, in 9 furrows, each 30cm wide by 250cm long. Three furrows were irrigated with untreated wastewater; another three were irrigated with raw wastewater that had been treated with ozone; and the remaining three furrows were irrigated with drinking-water (these last constituted the control group). The growth of the lettuce plants was evaluated by taking measurements of the length of the root and leaf, as the parameter.

3   Results and discussion

The results obtained from the raw wastewater samples taken from the municipal treatment plant (Cerro de la Estrella) are given in Figure 1. After (15 minutes) applying 36.8 mg O_3/min a pH 7, 100% of TC, *S. typhi*, and *V. cholerae* bacteria were found to have been destroyed.

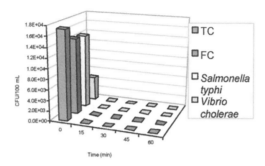

Fig. 1 Average microbiological results for the application of pH 7 and 36.8 mg O₃/min.

The results of the determination of the effect of ozone (36.8 mgO₃/min and pH 7) upon some ozone-demanding physical-chemical parameters are shown in figures 2. It was observed that, after 15 min, the TDS increased to almost double the initial value. At 15 min, the turbidity was reduced to a third of its initial value. By 30 min, there was a further small reduction in turbidity, and after 30 min there was hardly any change. The COD was reduced by 5% at 15 min, and by 10% at 60 min. The BOD₅ showed a final reduction of 24%. The total nitrogen concentration in the treatment plant's effluent water was very low, 0.78mg/L, and after 15 min of treatment a zero value of organic nitrogen was recorded.

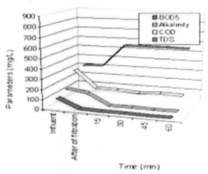

Fig. 2 Variation of the physical-chemical parameters in wastewater samples treated with ozone at pH 7 and 36.8 mg O₃/min, for different contact times.

As regards microbiological quality, as can be seen in figures 3 and 4 respectively, there was evidence of raised concentrations of micro-organisms both in the leaf and in the root of the lettuce plants irrigated with untreated wastewater (from 6E-4 to 4E-6, respectively). The lettuce plants irrigated with raw wastewater treated with ozone, presented low concentrations of the coliform (1E-1) and *V. cholerae* (2E-2) bacteria, but no Salmonella was found. The lettuce plants irrigated with untreated wastewater also presented greater concentrations of helminth eggs, on both the leaves 11 eggs/g and roots 23 eggs/g, when compared to the leaves and roots of the lettuce plants irrigated with ozonated wastewater, showed one reduction of 95 % on both the leaves and roots.

The soil irrigated with raw wastewater evidenced raised concentrations of total and faecal coliforms (from 4E-6 to 5E-7, respectively), the soil irrigated with wastewater treated with ozone clearly shows a reduction of these micro-organisms. The bacteria *S. typhi* and *V. cholerae* were not detected.

The soil irrigated with wastewaters showed 32 helminth eggs/g, while for the case of the soil irrigated with wastewater treated with ozone showed 5 helminth eggs/g (1 viable and 4 not viable). When making the analysis of *Giardia sp* and *Acanthamoeba sp*. their presence were observed in the soil irrigated with raw wastewater, while the soil irrigated with wastewater treated with ozone did not present these parasites.

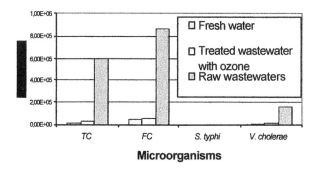

Fig. 3 Concentrations of microorganisms in the leaf of the lettuce plants irrigated with raw wastewaters, treated wastewater with ozone and fresh water.

The lettuce plants irrigated with ozonated wastewater, showed greater growth in both leaves (12cm) and roots (15). The lettuce irrigated with raw wastewater showed smaller growth (leaves 8cm and roots 9cm). In the control group, no bacterial growth was observed in any of the lettuce plants.

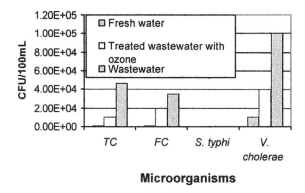

Fig. 4 Concentrations of microorganisms in the root of the lettuce plants irrigated with raw wastewaters, treated wastewater with ozone and fresh water.

4 Conclusions

The analysis of micro-organisms gave a reduction of the 100 % of bacteria faecal coliforms, *S. typhi and V. cholerae*, and significant reductions of protozoa and intestinal nematodes and showed an increment in leaves and root of lettuces of 6% and 16% respectively, with regard to the lettuces that were watered with wastewaters without treatment. According to these results, the use of ozone is one of the best alternatives for eliminating intestinal parasites that can be transmitted by hydraulic paths and help to improve the agricultural products exempt of problems of infections.

As with any other oxidizing agent, ozone has limitations in oxidizing both organic and inorganic matter, depending on the nature and concentration of the constituents of the wastewater under treatment. These constituents determine the effectiveness of ozone, because a certain proportion of the organic or inorganic matter might be resistant to oxidation and it can be used as crop nutrients.

The behaviour of the disinfection process under the conditions described in this paper, allow us to state categorically that the effluents produced following treatment using this method, do indeed comply with the Mexican quality standard (NOM-001-ECOL-1996) established for water destined for re-use in irrigation. The ozone treatment reduced bacterial count on the wastewater, soil and lettuce, thus better tasting lettuce, better appearance, water savings, longer shelf life of the fresh-cut lettuce.

This demonstrates that the application of ozone in the treatment of wastewater destined for re-use in agriculture, helps to improve both the microbiological quality and growth of lettuce plants.

Acknowledgements

The authors are grateful to the Programme of Support to Projects of Research and Technological Innovation (PAPIIT) for its financial support.

References

[1] Jiménez, B., Chávez, A., Maya C. and Jardines, L. (2001). Removal of microorganisms in different stages of wastewater treatement for Mexico City. Wat. Sci. Tech. **43**(10) 155-162.

[2] Orta de Velásquez, T., Rojas, V. N. and Vaca M. (2002). Destruction of helminth eggs (Ascaris suum) by ozone: second stage. Wat. Sci. Tech: Wat. Supply **2**(3) 227-233.

[3] Jiménez, B. and Chávez, A. (1998). Removal of helminth eggs in an avanced primary treatment with sludge blanket. *Environ. Technol.* **19,** 1061-1071.

[4] Matuz, M. D. (2001). Amebas de vida libre aisladas de aguas subterráneas del Valle del Mezquital, Hidalgo, México. Tesis Biólogo. FES-IZTACLA, UNAM. 1-53 pp. (In Spanish).

Soil amendment with sludge generated from metal finishing industries and its impact on metabolic quotient

Preeti Saxena[*]

School of Environmental Sciences, Jawaharlal Nehru University, New Delhi-110067, India.

[*]Corresponding author: e-mail: kpre_jnu@yahoo.com, Ph.No. 0091 11 26717676, Fax No. 0091 11 26717601

This experiment evaluated the effects of utilization of industrial waste, generated from metal finishing industries in agriculture field. Wastes have low pH (2.5-4.0) and high concentration of heavy metals (Cr, Ni). Composite waste samples were treated with lime to raise pH then lime stabilized wastes were mixed in two slightly alkaline soils at mixing rate, 0%, 10%, 20% and 30%. viz. cultivated and uncultivated land and then incubated at 28^0C with 50% water holding capacity for 120days and incubated soils were examined to monitor the changes in CO_2 evolution and metabolic quotient (qCO_2).

Metabolic quotient (qCO_2) is respiratory CO_2 release per unit biomass. The results of the studies on qCO2 –response upon heavy metal contamination, however are contradictory, since some authors reported an increase (Brookes and McGrath, 1984; Flie–bach et al., 1994) and some a decrease of the CO_2. A correlation analysis between bioavailable heavy metals and CO_2 evolution and metabolic quotient (qCO_2) evaluate the potential risk of these heavy metals on soil microflora.

1 Introduction

Agronomic research on the use of industrial sludge in agriculture field is gaining popularity as a mean of waste disposal and it must include an evaluation of their impact on soil chemical and microbiological properties. Agricultural uses of sewage sludge is common in practice since last few decades but utilization of industrial sludge is an upcoming alternative method of waste disposal.

A number of industrial activities, including metal finishing, electroplating, leather tanning, pulp production and mineral ore and petroleum refining, generate solid and aqueous waste products that are enriched with heavy metals including hexavalent chromium [Cr (VI)] (Forstner and Wittman, 1981) and Nickel [Ni (II)] etc. The disposal of these waste in agriculture field can potentially result in the release of [Cr (VI)], Nickel [Ni (II)] and other heavy metals. Cr (III) is strongly retained onto soil particles, Cr (VI) is very weakly adsorbed and is readily available for plant uptake and leaching to ground water (James and Bartlett, 1983).

Chaney and Ryan (1993) have stated that "all evidence available indicates that the specific metal adsorption capacity added with sludge will persist as long as the heavy metals of concern persist in the soil". These scientist rejects the arguments that the slow mineralization of organic matter in sludge could release metals into more soluble forms, often termed the sludge time bomb hypothesis given by Stigliani, 1988. In stead, they argue that the residuum of sludge decomposition can perpetually maintain heavy metal solubilities at very low levels. This could be termed the sludge protection hypothesis.

The sludge of metal finishing industries adds a significant quantity of heavy metals which affects the soil microbiological properties. Soil microbial biomass represents the main driving force performing the decomposition process (Kjøller et al., 2000) and can be

452

reduced by stressors such as heavy metals (Fritze et al., 1996). The microflora of heavy metal contaminated soils generally have altered microbial activities, manifested by higher metabolic quotients (qCO2) (Brookes, 1995) and less microbial synthesis per unit of organic substrate added (Chander and Brookes, 1991,1992), compared with values in the respective unpolluted soils. Among heavy metals Cr and Ni considered to particular toxic and responsible for serious biological degradation of soils.

Lime treatment of these waste is required to raise soil pH, immobilization of heavy metals and improvement of soil conditions to facilitate revegetation of contaminated soils. Liming carried out at different doses in waste and this lime stabilized waste is mixed with soils (Williamson and Johnson, 1981).

For our study we selected Wazirpur industrial area in Delhi, India, which is generating large amount of hazardous waste per day. Wazirpur industrial area is large polluting unit having clusters of more than 424 small scale industries. However, the relatively high nutrient content of the wastes, such as available nitrogen (>700 mg kg^{-1}), organic carbon (>4-5%), NH_4^+-N (>120 mg kg^{-1}) and NO_3^-N (250 mg kg^{-1}) make them potentially valuable for generated agriculture use. Toxic waste are also leached down the soil profile thus influencing the ground water quality. In each season (monsoon, winter summer) samples were collected and analyzed and a composite waste was mixed with two soils, cultivated (J.N.U.) and uncultivated (Chattarpur soil), in presence or absence of lime, and then incubated to monitor the changes of selected soil microbiological properties.

2   Material and methods

Study area: Wazirpur industrial area is located in North-West of Delhi, consist of three blocks A, B, and C, rich of small scale industries in which more than 80% are pickling and rolling industries and other "Metal Finishing industries" such as polishing and cutting etc and release a large amount of solid waste which are directly dumped on road side dump. This industrial sludge from the outside premises of each industry has been collected for this study.

Collection of industrial waste samples and preservation: Thirty samples were collected from July 2000 to May 2001 in each block A, B and C of Wazirpur industrial area in each monsoon, winter and summer season. For the physicochemical examination samples were collected in polythene bag and air dried and sieved (<2 mm) then stored at room temperature with proper labelling.

Soil Samples: Soil samples were collected in Delhi, cultivated land of vegetable garden in Chattarpur, New Delhi and uncultivated land of nursery in Jawaharlal Nehru University (J.N.U.), New Delhi. Samples were taken from the A horizon (0-20cm soil depth) in each monsoon, winter and summer season.

Physicochemical Examinations: pH, EC, CEC, Organic Carbon, Microbial biomass, Available Nitrogen were estimated as reported by Black C.A. (1982).

DTPA Method: The available fraction of heavy metals was estimated by chemical extractions using DTPA according to (Schalscha et al., 1982). Metal concentration in the extracts were determined by atomic absorption spectroscopy (Schimadzu AA 6800).

Respiration rate and Metabolic quotient (qCO_2) measurements

Respiration rate was estimated by trapping CO_2 in NaOH solution during the incubation period. The remaining alkali was titrated with 0.1M HCl after adding $BaCl_2$ solution correspondent to the NaOH equivalents for the precipitation of CO_2 (Terry.et.al 1979) Microbial biomass was determined by the fumigation-extraction (FE) technique (Anderson and Ingram, 1993). Metabolic quotient, qCO_2, was calculated as respiration rate per unit of microbial biomass by FE (Anvar and Dilly, 2002)

Incubation Study: Representative composite waste samples collected from three different blocks of Wazirpur industrial area having pH values lower than 3.5, were treated with lime at 0%, 0.5%, and 1.0% to neutralize the acidity of the material and then mixed with the soil samples collected from cultivated and uncultivated land. The waste of monsoon season was used for incubation study because it is properly homogenize due to heavy rain and true representative of this area. The mixing rate was 10%, 20% and 30%. Each combination of this mixture (200g) in triplicate were kept in ½ kg capacity cellophane bag at 28^0C for 120days. The moisture of the amended soils was maintained at 50% of water holding capacity. Sub samples were taken out at the regular interval of period viz. 0, 10, 20, 30, 45, 60 and 120 days and analysed immediately.

3   Results and discussion

The average values of results of 30 samples of each block in each season is shown in Table 1. In summer the pH was lowest due to lack of moisture in the environment .The salt concentration (EC) is maximum in monsoon due to mixing of waste with drain water and almost similar in all blocks and lowest in summer. Nutrient status is high due to mixing of sewage with industrial sludge and use if nitric acid in industries. According to USEPA (USEPA Clean Water Act 503 regulations: 1993) the use of sewage sludge will be permitted to accumulate chromium-3000 ppm and Ni-420 ppm and this waste contained chromium 3.0 and nickel 2.3 times higher concentrations.

Table 1. Physicochemical characterization of composite samples of three seasons of Wazirpur industrial area

S.N.	Name of Parameter	Types of Wastes		
		Mon	Win	Sum
1.	pH	3.47	3.21	3.00
2.	EC (mmho/cm)	1.92	1.54	1.82
3.	WHC (%)	51.00	54	56
4.	Organic Carbon (%)	2.50	2.8	3.4
5.	Available Nitrogen (mg kg^{-1})	494.2	550.0	520.0
6.	Cr (Total Conc.) (mg kg^{-1})	6100	5683.6	5575
	Cr (DTPA Ex.), (mg kg^{-1})	2.70	2.25	1.87
7.	Ni (Total Conc.) (mg kg^{-1})	630	565	585
	Ni (DTPA Ex.), (mg kg^{-1})	264	238	224

Table 2. Physicochemical and Microbiological Properties of two soils

Characteristics	J.N.U. (uncultivated soil)	Chattarpur (Cultivated soil)
pH	8.37	8.57
EC (mmho/cm)	0.116	0.119
WHC (%)	32	37
Organic Carbon (%)	0.36	0.52
Available Nitrogen (mg kg^{-1})	95.2	124.2
Cr (Total Conc.) (mg kg^{-1})	400	800
Cr (DTPA Ex.), (mg kg^{-1})	0.4	0.8
Ni (Total Conc.) (mg kg^{-1})	78.0	73.2
Ni (DTPA Ex.), (mg kg^{-1})	1.30	1.22
Particle Size Distribution		
Sand %	68.00	62.80
Silt%	21.90	23.20
Clay%	10.10	14.00
Microbiological characteristics		
Total Microbial Biomass ($\geq$g g^{-1})	150	65
Bacteria (Total plate count cfu/ml)	$10^{10}-10^{14}$	$10^{6}-10^{8}$
Fungi (Total plate count cfu/ml)	$10^{2}-10^{6}$	$10^{2}-10^{4}$

Table 3. Pearson correlation coefficient between pH, heavy metals and microbial parameters in waste amended soils

	0 -120 D % W	pH C	pH J	Chromium mg kg^{-1} C	Chromium mg kg^{-1} J	Nickel mg kg^{-1} C	Nickel mg kg^{-1} J
Microbial Biomass (μg g^{-1})	0	n.s.	n.s.	n.s.	n.s.	n.s.	n.s
	10	-0.47*	-0.56**	n.s.	0.46*	n.s.	n.s.
	20	n.s.	-0.66**	0.78**	n.s.	-0.79**	n.s.
	30	n.s.	n.s.	0.91**	n.s.	-0.44*	n.s.
Respiration rate (μg CO$_2$ g^{-1} h^{-1})	0-10 D 0–30 % W	-0.46*	-0.54**	n.s.	n.s.	n.s.	0.48*
qCO2	0-10 D 0-30 % W	n.s.	n.s.	n.s.	n.s.	-0.54**	n.s.

*, ** Significant at P<0.05 and 0.01, respectively. N.S. : not significant
C- Chattarpur soil (Cultivated land) , J- JNU nursery soil (Uncultivated land)

Note: ** indicates the relation between two parameters which are strongly related, ie with a confidence level at more than 99%. All above results were obtained at two tailed test, two tailed significance is the default, where for a strong relationship, the value must be between 0.010 to 0.000 means confidence in relation is between 99 (0.99) to 100 % (1.00)

Both soils (Table 2) are in alkaline range (8–9) and recommended for disposal of acidic waste. Chattarpur soil is having higher content of nitrogen, which is easily available to crop. Organic carbon mineralizes faster in sandy soils than clay soil hence organic carbon mineralization faster in JNU soil. Both soils come in category of sandy loam skeleton.

Simple correlation coefficients (Table 3.) between microbiological characteristics and pH were calculated. A significant correlations were found at 10% W.T. in both soils but at 20% W.T. significant in JNU soil. This different behavior of both soils is due to physicochemical and microbiological characteristics (Table 3). The incorporation of these acid waste in alkaline soils will results in permanent neutralization of the pH, even with low lime treatments. Simple correlation coefficients between microbial biomass and heavy metals (Table 3.) were calculated. A positive significant correlation were observed between microbial biomass and DTPA Ex. Cr in Chattarpur soil while significant at 10% W.T. and non significant at 20% W.T. in JNU soil. A negative significant correlation observed between microbial biomass and DTPA Ex. Ni in Chattarpur soil at 20% and 30% W.T. while completely non significant in JNU soil.

Soil respiration rate (Fig 1. a–d) were observed in waste incubated soil during 0–10 days. Respiration rate decreases in first 3 days in both soil without any amendment then increases at 4th days and in JNU soil respiration rate increases at higher rate. Respiration rate is increases at different rate of waste treatment in soil and it is 10%>20%>30% without any lime treatment. A correlation analysis between respiration rate and pH is significant in both soils (Table 3) and significant correlation coefficients (Table 3.) were calculated between respiration rate and Cr and in uncultivated soils.

Fig 1 Chronological changes in soil respiration rate (a–d) in lime treated sludge amended (0,10, 20 and 30%) cultivated (C-Chattarpur) and unclutivated (J-JNU) soils during incubation period (0–120 days).

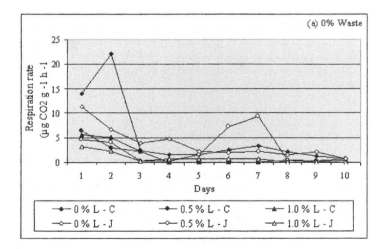

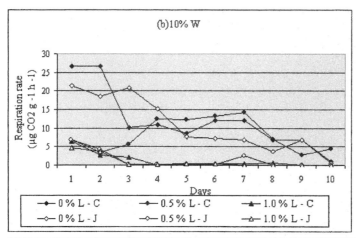

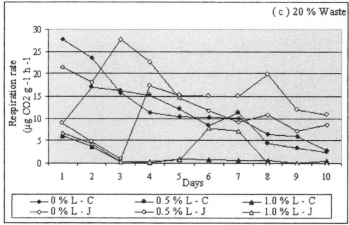

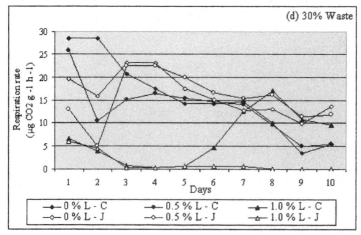

In our study qCO$_2$ (Fig 2a & 2b) decreases at 10% W.T. in Chattarpur soil, while increases in JNU soil, at 20% W.T. qCO$_2$ increases in both soils and decreases at 30% in both soil. Lime treatment in 30 % W amended soils did not change the qCO$_2$ values of the soils. Baath et al (1991) reported a slight decrease of qCO$_2$ with increasing heavy metal pollution, whereas Brookes and McGrath 1984 found a higher qCO$_2$ in metal polluted as compared to uncontaminated soils. According to Giancarlo Renella, 2005 organic matter input increased soil respiration, ATP contents and hydrolase activities in all soils, however the Cd-contaminated soils had significantly higher metabolic quotients as calculated by the CO$_2$-to-ATP ratio. A correlation coefficients (Table 3) between qCO$_2$ and pH, Cr, Ni were calculated and it is non significant between pH and Cr and negatively significant between qCO$_2$ and Ni in Chattarpur soil and non significant in JNU soil.

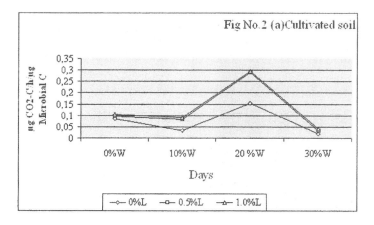

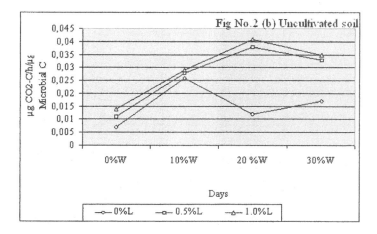

Fig 2. Microbial metabolic quotients (qCO2) in lime (L - 0,0.5,1.0) treated waste (W-0,10,20 and 30%) amended (a)cultivated (C-cultivated) and (b) uncultivated (J - J.N.U.) soils during 0–10 days.

458

Discussion

Decreasing respiration rates in soil under constant conditions is attributed most likely to nutrient depletion in available in straw fractions and the increasing dominance of recalcitrant compounds such as cellulose, hemicellulose and lignin (Neely et al., 1991). Concurrently, decreasing microbial activity with proceeding incubation was earlier reported by e.g. Ocio et al. (1991), Dilly and Munch (1996). Our findings are contrast to these results because mixing of nutrient rich waste in soils so nutrient depletion were observed in waste amended soil hence respiration rate is increasing in waste incubated soil without lime treatment. It may be high alkalinity of soils because at higher pH soil microbial activity in terms of fungal growth decreases. Thus this waste is harmless if mixed at lower concentration with out lime treatment.

Previous studies have shown that when soil microflora are exposed for long time to high metal concentration, it's qCO_2 increases, indicating a greater energy requirement for maintenance (Brookes, 1995) Our results confirm these findings, because qCO_2 is increases at higher dose of waste treatment 20%, this higher value shows more CO_2 –C was respired per unit of biomass due to supply of high concentration of heavy metals. Several microbial quotients were proposed to characterize the soil microbiota. The quotients have not been widely used, as doubts exits as to their ecological significance. Papers from Anderson and Domsch (1990), Wardle and Ghani (1995) and Dilly and Munch (1998) have tried to fill this gap. Metabolic quotient such as respiration rate per unit microbial C was suggested to evaluate the effects of environmental conditions (Flie—bach et al., 1994) or land-use on soil microbial ecophysiology and to quatify substrate utilization efficiency (Dilly et al., 1997; Dilly and Munch, 1998). Odum (1985) and Anderson and Domsch (1993) mentioned that high qCO_2 values are attributed to stress. Therefore, we conclude that microbial communities of waste incubated soils were under stress and modified metabolism by accelerating nutrient turnover. (Flie—bach et al., 1994) detected increased qCO_2 in soils polluted with heavy metals. Hence our data support the view of Anderson and Domsch (1994) to use the qCO_2 for assessing heavy metal effects only with similar soils. Our data further suggest that qCO_2 increase upto a certain increase of heavy metals in waste incubated soil but beyond 20% W.T. qCO_2 decreases in both soil.

4   Conclusion

The main conclusions are that the sludge studied here could be used in agriculture as no definitive indicators of soil stress were detected. These nutrients rich wastes have low pH and high concentration of heavy metals which can be control by lime treatment, to raise soil pH, immobilization of heavy metals and improvement of soil conditions to facilitate revegetation of contaminated soils.

Acknowledgement

We gratefully acknowledge the University Grant Commission, India for scholarship in Ph.D and financial grant for this research work.

References

[1]   Anderson, T.-H., Domsch, K.H., 1990. Application of eco-physiological quotients ($q\mathrm{CO_2}$ and qD) on microbial biomass from soils of different crop histories. Soil Biol. & Biochem 22, 251-255.

[2]   Anderson J.M., and Ingram J.S.I. (1993) Tropical soil biology fertility: A handbook of methods CAB International, Walling ford, U.K., 68-71.

[3]   Anderson, T.H.; Domsch, K.H., (1993) The metabolic quotient for CO2 (q CO2) as a specific activity parameter to assess the effects of environmental conditions, such as pH, on the microbial biomass of forest soils. *Soil Biol.Biochem.*, 25: 393-395.

[4]   Anderson, T.H.; Domsch, K.H., (1994) Physiological analysis of microbial communities in soil: Applications and limitations. In Beyond the biomass (K. Ritz, J. Dighton and K.E. Giller, Eds), pp.67-76 Wiley, Chichester.

[5]   Anvar Sh. Mamilov, Oliver M. Dilly. 2002 Soil microbial eco-physiology as affected by short–term variations in environmental conditions. Soil Biology & Biochemistry, 34, 1283-1290.

[6]   Baath E., Arnebrandt K. and Nordgren A.(1991) Microbial biomass and ATP in smelter-polluted forest humus. *Bulletin of Environmental Contaminated and toxicology* 47, 278-282.

[7]   Brookes P. C. and McGrath S. P. (1984) Effects of metal toxicity on the size of the soil microbial community. *Journal of Soil Science* 35, 341-346.

[8]   Brookes, P.C., 1995. Use of microbial parameters in monitoring soil pollution by heavy metals. *Biology and Fertility of Soils 19, 269-279.*

[9]   Black C.A. Methods of Soil Analysis. Part 2. 2nd ed. Agron. Monogr. 9. ASA and SSSA, Madison WI

[10]   Chander K. and Brookes P.C. (1991) Microbial biomass dynamics during decomposition of glucose and maize in metal-contaminated soils. *Soil Biology and Biochemistry23, 917-925.*

[11]   Chander, K., Brookes, P.C., 1992. Synthesis of microbial biomass from added glucose in metal-contaminated and non-contaminated soils following repeated fumigation. *Soil Biology & Biochemistry 24, 613-614.*

[12]   Chaney, R.L., and J.A. Ryan. 1993. Heavy metals and toxic organic pollutants in MSW – compost: Research results on phytoavailability, bioavailability, fate, etc. p.451-506. In H.A.J. Hoitink and H.M. Keener (ed.) Science and engineering of composting; design, environmental, microbiological and utilization aspects. *Renaissance Publ., Washington,Ohio.*

[13]   Dilly,O., Munch, J.C., 1996. Microbialbiomass content, basal respiration and enzyme activities during the course of decomposition of leaf litter in a black alder (Alnus glutinosa(L.)Gaertn.) forest . Soil Biol.and Biochem28, 1073-1081.

[14]   Dilly,O., Munch, J.C., 1998. Ratio between estimates of microbial biomass content and microbial activity in soils. Biology and Fertility of soils 27, 374-379.

[15]   Dilly, O.,Bernhard, M., Kutsch, W.L., Kappen, L., Munch, J.C., 1997. Aspects of carbon and nitrogen cycling in soils of the Bornhoved Lake district I.Microbiol characteristics and emission of carbon dioxide and nitrous oxide of arable and grassland soils. Biogeochemistry 39, 189-205.

[16]   Forstner,U., and G.T.W. Wittman. 1981 Metal pollution in the aqueous environment. *2 nd. Springer Verlag, New York.*

[17]   Fritz,e H., Vanhala, P., Pietikainen, J., Malkconen, E., 1996. Vitality fertilization of scots pine stands growing along a gradient of heavy matal pollution; short-term effects on microbial biomass and respiration rate of humus layer. Fresenius *Journal of Analytical Chemistry 354, 750-755.*

[18]   Flie–bach, A., Martens, R., Reber, H.H., 1994. Soil microbial biomass and microbial activity in soils treated with heavy metal contaminated sewage sludge. Soil Biology and Biochemistry 26, 1201-1205.

460

[19]   Giancarlo Renella, Michel Mench, Loretta Landi, Paolo Nannipieri. 2005. Microbial activity and hydrolase synthesis in long –term Cd-contaminated soils. *Soil Biology & Biochemistry* 37,133-139.

[20]   James, B.R., and R.J. Bartlett. 1983. Behaviour of chromium contaminated soil. *Environ. Sci. Technol. 30:248-251.*

[21]   Kjøller, A., Miller, M., Struwe, S., Wolters, V., Pflug A., 2000. Diversity and role of microorganisms. In Schulze, E.-D., (Ed.) Carbon and Nitrogen cycling in European Forest Ecosystem, *Ecological studies 142, Springer, Berlin pp- 382-402.*

[22]   Neely, C.L., beare, M.H., Hargrove, W.L., Coleman, D.C., 1991. Relationship between fungal and bacterial substrate-induced respiration, biomass and plant residue decomposition. *Soil Biol and Biochem23, 947-954.*

[23]   Ocio, J.A. Brookes, P.S., Jenkinson, D.S., 1991. Field incorporationof straw and its effects on soil microbial biomass and soil inorganicN. *Soil Biol.and Biochem23,171-176.*

[24]   Odum, E.P., 1985. Trends expected in stressed ecosystem. *Bio Sciences35, 419-422.*
Schalscha, E.G., M.Morales, I.Vergara and A.C. Chang (1982) Chemical fractionation of heavy metals in waste-water affected soils. *J. Water Pollu. Control Fe*d. 54: 175-180.

[25]   Stigliani, W.M. 1988. Changes in valued "capacities" of soils and sediments as indicators of nonlinear and time-delayed environmental effects. *Environmental Monitoring and Assessment* 10: 245-307.

[26]   Schalscha, E.G., M.Morales, I.Vergara and A.C. Chang (1982) Chemical fractionation of eavy metals in waste-water affected soils. *J. Water Pollu. Control Fe*d. 54: 175-180.

[27]   TerryR.E. (1979) Decomposition of anaerobically digested sewage sludge as affected by soil environment condition, *J.Environ.Qual. 8: 342-347.*

[28]   U.S.Environmental Protection Agency. 1993. Clean water act. Section 503. *Vol.58. No.32. USEPA, Washington, DC.*

[29]   Wardle, D.A., Ghani, A., 1995. A critique of the microbial metabolic quotient (qCO$_2$) as a bioindicator of disturbances and ecosystem development . *Soil Biol. And Biochem. 27, 1601-1610.*

[30]   Williamson,A. and M.S. Johnson. 1981 Reclamation of metalliferous mine wastes, pp.185-212. In N.W. Lepp (ed.). Effect of heavy metal pollution on plants Vol.2. Metals in the environment. *Applied Science PublishersLtd., Barking.*

Spatial properties in Individual-based Modelling of microbial systems. Study of the composting process

C. Prats, J. Ferrer, A. Giró, D. López, and J. Valls

Escola Superior d'Agricultura de Barcelona, Department of Physics and Nuclear Engineering, Universitat Politècnica de Catalunya, av. Canal Olímpic s/n, 08860 Castelldefels, Barcelona, Spain

Most of the real systems have complex spatial characteristics. The aim of this work is to develop some methods to be applied in space modelling of those systems. We use an INDividual DIScrete SIMulation, which has been adapted to study the composting process. This process has a special complexity related to the microbiological and chemical processes and to the physical structure and dynamics. In this paper we develop the model to study the spatial complexity, and we present a first application for a system without microbiological or chemical activity. The obtained results for the different parameters fit well with the expected behaviour.

Keywords: composting process, simulation, complex systems, diffusion, convection, heat conduction

1 Introduction

Most of the existing mathematical models in microbiology consider an uniform, homogeneous and isotropic culture medium. Nevertheless, in many real systems the spatial characteristics are complex and essential to understand the observed behaviours. An interesting example is the bacterial growth on agar plates with a low nutrient concentration and some certain spatial properties [1] (non-homogeneous medium), where the nutrient concentration gradient causes different growth behaviours. Spatial characteristics are also important in many systems of industrial interest; in several food microbiology cultures the growth takes place in mediums with different interacting phases. Other systems with great spatial complexity are the soil or the composting systems, where the environment plays an important role.

The main concern of this work is to develop some methods to be applied in the space modelling of complex systems like those aforementioned below. In concrete, we develop a specific space model to be used on composting processes modelling, due to the particular complexity of the compost systems. In this work we have used an Individual based Model to begin this study. It is done with the INDISIM (INDividual DIScrete SIMulation) methodology developed by [2].

The composting process is a basic tool in the treatment of organic waste, both urban and industrial. Its great complexity is related not only to the microbiological and chemical processes, but also to the physical structure and dynamics. This complexity causes an enormous difficulty in optimising the process for different waste types, because of the differences on the components and spatial structure. The current protocols are based both on the theoretical knowledge about the process and the experience obtained through many years. The development of mathematical models to study this process should help the theoretical comprehension and allow the virtual tests of the composting process.

The mathematical composting process modelling is divided in two parts. One of them takes into account the microbial behaviour and involves chemical processes [3]. The other, which is presented in this work, studies the spatial structure of the process. This model includes the transport phenomena (energy and matter) and the mass and energy exchanges between the three phases (liquid, solid and gas). In this way convection and conduction are modelled, as well as water evaporation and condensation. The custom forced convection of industrial composting is another important process to be modelled, as it is of great importance to control parameters like the temperature or the aerobic conditions. These processes are modelled in a two dimensional vertical grid with cubic spatial cells. Each cell contains the three phases and is controlled at each time step. Microorganisms are considered in dissolution in the liquid phase.

First of all, the model has been developed considering a material without microorganisms. Its suitability has been proved according to theoretical arguments. Later, and together with the IbM developed for the study of the microbial behaviour, the model will be readjusted to reproduce the behaviour of a composting tunnel correctly.

2 Bases for a spatial model of a composting process

The spatial structure of the composting processes is of great complexity: it is heterogeneous (with coexistence of solid, liquid and gaseous phases), anisotropic, and variable through time. Any mathematical model for describing such a process must be an important simplification of the real system.

In this sense, the spatial characteristics and physical processes of a composting tunnel have been studied. The most important physical processes have been selected and modelled. These processes are the matter and energy transport, and the mass and energy transfer between the three phases (liquid, solid and gas).

Although this model has been developed thinking on a composting system, it should provide new tools to be used in different systems.

2.1 Matter transport

The main matter transport phenomena in a composting system are diffusion and forced aeration. Equation 1 shows the one-dimensional Fick's law, which is considered to model the diffusion.

$$J_i = -D_i \cdot \frac{\Delta c_i}{\Delta x} \qquad (1)$$

The gases, water and soluble substrates will be diffused through the medium by means of Fick's law in two dimensions.

A composting system needs to be aerated several times during the process. This must be done in order to dry, cool and oxygenate the material. The forced aeration is essential to do the process in the correct way and ensure aerobic conditions to obtain a quality product (compost). It is usually carried out by injecting dry air from the bottom.

2.2. Heat transport

Two heat transport phenomena are considered. The first one is related with the heat conduction, which is described by the one-dimensional Fourier's law (Eq. 2).

$$\dot{q} = -k \cdot S \cdot \frac{\Delta T}{\Delta x} \qquad (2)$$

The second one is related to the matter transport, that implies a heat transport. Above all, the gases circulation (due to the diffusion and the forced aeration) produces an important heat transport usually related to convection. As the system is macroscopic, we can consider Eq. 3, which corresponds to a heat interchange in a gases mixture.

$$\left| q_{ij} \right| = m_i \cdot c_i \cdot \left| \Delta T_i \right| = m_j \cdot c_j \cdot \left| \Delta T_j \right| \tag{3}$$

2.3. Phase transitions

Water evaporation is crucial in a composting system. It is important for two reasons: it prevents the system from great temperature increases, and it keeps the system in aerobic conditions (a water excess leads to an anaerobic environment). The *exhaust air* in a composting system is near to saturation [4]. Then, it can be considered that liquid water will evaporate until the saturation pressure (Eq.4) is achieved. If vapour pressure exceeds the saturation value, condensation will take place.

$$P_{SAT} = \exp\left[60.433 - \frac{6834.271}{T} - 5.16923 \cdot \log(T) \right] \tag{4}$$

This evaporation (or condensation), produces a certain temperature increasing (or decreasing), which depends on the water latent heat (Eq.5).

$$L_{H_2O} = \frac{8.314 \cdot \log(\frac{P_{SAT}}{611.3})}{\frac{1}{273.15} - \frac{1}{T}} \tag{5}$$

In equations 4 and 5 P is given in Pa and T in K.

3 First spatial modelling and simulation

This spatial model has been tested with INDISIM, which is a discrete modelling and simulation methodology developed by Ginovart *et al.* [2], especially designed to simulate the growth and behaviour of microbial cultures. It has been developed with Compaq Visual Fortran Professional Edition 6.1.0, and has been successfully applied to other studies [1,5,6]. INDISIM enables the study of the evolution of a microbial culture based on the individual behaviour of the microorganisms, through a fixed period of time in a specific environment, in which space and time are discrete.

3.1 Modelling the space

A two-dimensional space is divided into spatial cubic cells. It represents a vertical slide of the composting system. Each cell contains a certain amount of mixed liquid, gases and solid (Fig. 1). The compounds of the three phases are:
- Gases: CO_2, O_2, ammonia, water vapour and inert gases.
- Liquid: water with some substrate particles and microorganisms in dissolution.
- Solid: basically structuring material.

464

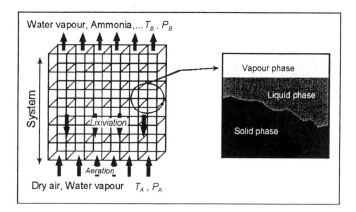

Fig. 1 Spatial grid used in the simulation. It is a two dimensional vertical grid divided into identical cubic cells. Aeration and lixiviation are also shown.

Each spatial cell has a defined state (pressure, temperature and composition) that may change through time because of different above mentioned processes. The cell state is updated each time step.

3.2. Modelling the spatial processes

The simulation's time-step is set to 30 minutes, as the global composting process requires 15 days at least. In this scale, we consider the modelled spatial phenomena as balances between the different components, each one being in thermodynamic equilibrium.

The diffusion is considered to take place between one cell and its first neighbour cells (Fig.2). A factor $1/\sqrt{2}$ is used to correct the diagonal flows. We impose forced aeration every 15 time steps (7h 30min). Aeration lasts one time step (30 min), and it produces a vertical linear pressure gradient. Therefore, we consider a vertical gas displacement until the system reaches this pressure profile.

Heat conduction is treated in the same way as diffusion. Heat transfer between neighbour cells is caused by the temperature gradients. The factor $1/\sqrt{2}$ is used again. This phenomena is a consequence of the physical contact between cells. At the same time, when a certain amount of substance i, Δm_i, goes from a cell to another, it transports an amount of heat, as it is shown in Eq. 6. This heat transport is considered both in matter diffusion and in aeration.

Water evaporation and condensation is taken into account cell by cell. Instantaneous values of temperature and composition will limit the maximum water vapour content allowed, n_{max}, by means of reaching the saturation pressure P_{SAT}. The ideal gases equation is used to relate P_{SAT} with n_{max}.

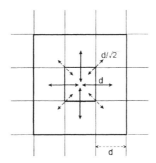

Fig. 2 Matter transport: diffusion between neighbour's cells. A factor $1/\sqrt{2}$ is considered in the diagonals (dotted arrows).

4   Results

A system without microorganisms has been simulated. Nevertheless, some water and heat generation have been considered, as if they were consequence of the microbial activity.

We have checked the correct behaviour of some parameters (temperature, pressure, liquid and gaseous water concentration) and their qualitative evolution along space and through time, for example:

  – the temperature and humidity decrease after an aeration, and increase between aerations
  – the liquid water mass can remain constant if there are the appropriate conditions (aeration frequency, air temperature, cells temperature, ...)
  – the pressure gradient between cells is reduced by means of diffusion

On further simulations, the microorganisms actions shall be taken into account [7].

Acknowledgements

The financial support of the Ministry of Education and Science of Spain Grant REN2000-0049-P4-04, the DURSI Generalitat de Catalunya 2003ACES00064, and the Ministry of Science and Education Plan Nacional I+D+i CGL 2004-01144.

References

[1]   M. Ginovart, D. López, J. Valls, M. Silbert. *Physics A.* **305**, 604 (2002)
[2]   M. Ginovart, D. López, J. Valls. *J. Theor. Biol.* **214**, 305 (2002)
[3]   A. Gras. INDISIM-SOM, un model discret per a l'estudi de la dinàmica de la matèria orgànica i la nitrificació en sòls. Ph.D. Thesis, University of Lleida, Spain (2004)
[4]   J. Kaiser. *Ecol. Model.* **91**, 25 (1996)
[5]   M. Ginovart, D. López, J. Valls, M. Silbert. *Int. J. Food Microbiol* **73**, 415 (2002)
[6]   J. Bermúdez, D. López, J. Valls, J. Wagensberg. *Comput. Appl. Biosci.* **5**, 305 (1989)
[7]   A. Gras, C. Prats, M. Ginovart. Poster in BioMicroWorld 2005, Badajoz, Spain, 15-18 March 2005.

Ultrastructural and cytochemical aspects of spores germination of *Mucor javanicus* wehmer

Vivian Karla Silva Barreira[1,2,4] Marcos Antônio Barbosa Lima[1,2,4]; Aline Elesbão do Nascimento[3,4,5];Kazutaka Fukushima[3,4], Galba Maria de Campos-Takaki[1,2,4,6a]

[1]Pós-Graduação em Biologia de Fungos - UFPE, Recife, PE – Brasil; [2]Laboratório de Imunopatologia Keizo Asami – UFPE - Recife, PE – Brasil; [3]Research Center for Pathogenic Fungi and Microbial Toxicoses-Chiba, Japan; [4]Núcleo de Pesquisas em Ciências Ambientais – UNICAP; Recife, PE – Brasil; [5]Departamento de Biologia - Recife, PE – Brasil; [6]Departamento de Química - UNICAP, Recife, PE – Brasil
[a]Corresponding author: email: takaki@unicap.br

The germination process in *Mucor javanicus* Wehmer IFO 4570, was evaluated through electron microscopy and morphometrical analysis. The ultrastructural analysis was carried out utilizing scanning electron microscopy. Alterations in the spore diameter and volume, during germination, were evaluated by using the morphometrical system of scanning electron microscopy micrographs. The microorganism was grown in synthetic *Mucor* medium during 9 hours at 28°C. Samples were collected each 1 hour of culture. The results obtained observe several modifications in the spore structure during the germination process. The morphometrical analysis revealed a spore volume increase of eight times during the first four hours of germination.

Key words: *Mucor javanicus*, germination, morphometric analysis

1   Introduction

Zygomycetes are of great interest to the industrial and economic areas due to its habilities involved in process of biodegradation and biodeterioration, and in secretion of secondary metabolites such as enzymes, organic acids, pigments, alcohols, steroids and vitamins (Alexopoulos et al. 1996).

The spore is an important vehicle for distribution or long-term survival of fungi. Spore germination is a key developmental step in the fungal life cycle and is critical for the establishment of the fungus in a new environmental niche. One of the impressive morphological developments in the life cycle of a mycelial fungus is the conversion of a dormant spore into an actively growing mycelium Many researchers have been devoted to characterize the phenomena (Alexopoulos et al. 1996; Bartinicki-Garcia, 1995a,b; Beckett et al. 1974; Edelman & Klomparens, 1995a,b; Klomparens 1990)

Mucor species are Zygomycetous fungi, which usually reproduce asexually, forming spores within a sporangium. Several species commonly spoil stored processed foods and fresh fruits and vegetables. Members are commonly found in soil, dung, decaying vegetation, and stored grains.

Some species are used commercially in the production of tofu, Chinese cheese and bean cake. Others cause zygomycosis in humans (Alexopoulos et al. 1996; Bartinicki-Garcia, 1995a,b; Beckett et al. 1974; Edelman & Klomparens, 1995a,b; Klomparens 1990).

Investigations related to fungal spore germination had been carried out using scaning electron microscopy and morphometrical analysis of germination in spores of *Mucor javanicus*.

2   Materials and methods

Microorganism: *Mucor javanicus* Wehmer IFO 4570 was obtained from Research Center for Pathogenic Fungi and Microbial Toxicoses, University of Chiba, Japan. The cultures were maintained on potato dextrose agar tubes.

Cultural Conditions: The strain 4570 was growing in potato dextrose agar medium (PDA) in Petri dishes during 5 days at 28^0C. After this time, the spores were collected and transfered for 100mL Erlenmeyrs flask, containing Synthetic *Mucor* Medium (SMM – Hesseltine and Anderson, 1957), pH 5,2. Pre-inoculum consisted of 10^7 spores/mL inoculated in Erelenmeyers flasks containing 30 mL of (SMM) in a reciprocating shaker for 9 hour at 28^0C.

Germination Study: Pre-inoculum consisted of 10^7 spores/mL inoculated in Erelenmeyers flasks containing 30 mL of (SMM) in a reciprocating shaker for 9 hour at 28^0C. Germination was determined by examination of at least 100 conidia by using an Ltda BX40 microscope (x1,000 magnification; Olympus Optical Co., Tokyo, Japan). The criterion used to measure germination was the emergence of germ tubes. The experiments were performed in triplicate. The spore samples were collected each one hour of incubation, centrifuged and spores were washed in saline phosphate buffer (PBS) and processed for electron microscopy.

Scanning electron microscopy (SEM): Spores were fixed in 2,5% gluteraldehyde in 0,1 M sodium cacodilate buffer at pH 7,4, washed, posfixed with 0,1% osmium tetroxide, washed with 0,1 sodium cacodilate buffer, dehydrated in an ethanol series (50%, 70%, 90%, 100%, 15 min each) were critical point dryer (HITACHI- HCP-2), mounted on aluminum stubs and coated with gold (JEOL-JFC-1100). Specimens were observed and photographed using a JEOL-JSM T-200 scanning electron microscope operating at 25 kV.

3   Results and discussion

Mucor javanicus biochemical, morphological and physiological features established a longstanding interest of both applied and theoretical research (Alexopoulos et al. 1996; Song et al. 2001; Wynn et al.2001; Lubbenhuse et al., 2003; Lubbenhuse et al., 2004).

Indeed, *Mucor javanicus* is being investigated as a possible host for the production of heterologous proteins. Thus, the environmental conditions defining the physiology and morphology of this dimorphic fungus have been investigated (Alexopoulos et al. 1996; Botha et al. 1997; Jackson et al. 1998; Lubbenhuse et al., 2003).

Considering that the development of any industrial process will be dependent on the information obtained for effective process optimization, the life cycle of the dimorphic fungus *Mucor javanicus* has been studied to understanding the growth and differentiation processes occurring in and between the different morphological forms of the organism (Song et al. 2001; Lubbenhuse et al. 2003). The present paper describes the Figures 1A–D showed scanning electron micrographs of ***Mucor javanicus*** spores during germination (T_0-T_3). The outer surface of spore appeared rugose. The spores swelled and became less rugose.

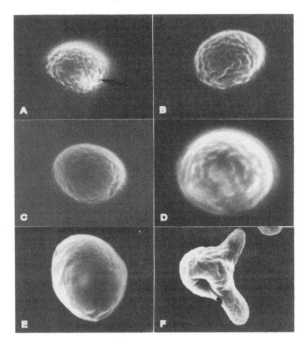

Figure 1. Scanning electron micrographs of *Mucor javanicus* during germination. A- T_0 spore; B- T_1 spore; C- T_2 spore; D- T_3 spore; E- T_4 spore; F- T_5 spore. X10.000.

The scanning electron micrograph in Figure 2A (T_4) depicted heightened surface irregularity and the beginning of outgrow (arrow). Further outgrow (T_5) is seen in Figure 2B as extensions in the spores reminiscent. The elongation continued up to T_9 (Figure 3A-D). The morphometrical analysis by using the electronmicrographs obtained by scaning eletron microscopy were submitted to the morphometrical analysis by usng the MOP-videoplan system for the morphometrical study osfarea, volume and diameter of spores. The spores modifications are presented in Table 1.

As seen from Table 1 the spore area during the beginning of germination a maximum increase of almost 4-fold was detected. For the spore diameter and volume increase of almost 3-fold and 8-fold were observed during the germination, respectively.

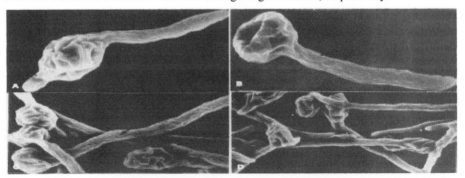

Figure 2. Electron micrographs of *Mucor javanicus* during germination. A- T_6 spore; B- T_7 spore; C- T_8 spore; D- T_9 spore. X10.000.

Table 1- Morphometrical analysis - MOP-Videoplan – of *Mucor javanicus* spores during germination

Time (h)	Area (≥m)	Diametre (≥m)	Volume (≥m)
To	3,618	1,926	4,615
T1	5,014	2,735	7,692
T2	7,103	2,974	13,009
T3	11,035	3,746	27,015
T4	13,101	4,392	32,915

* ≥m= micrômetro

In addition, spores represent a common reproductive mode in filamentous Fungi. Asexual sporulation strategies are nearly as varied as fungal species and are used as a model for understanding the mechanisms that control fungal sporulation. The vegetative growth begins with the germination of a spore. The spore germination leads to the formation of tubular hyphae that grow in a polar fashion by apical extension and branching to form a network of interconnected cells known as a mycelium. Substantial research aimed toward understanding basic cell biological problems associated with hyphal growth, including cell cycle, cytokinesis, and polarity determinants, has been performed. Indeed, the surface of spores of many fungi, including ascomycetes, basidiomycetes, and zygomycetes, is covered by a thin layer of regularly arranged rodlets (Alexopoulos etal. 1996; Aiken & Niederpreen, 1978; George et al. 1972; Grove 1972, 1979; Griffin, 1994; Hess, 1973; Jeffries & Young, 1975, 1976; Klomparens, 1990; Maia et al. 1993, 1994; Mckeown et al. 1996; Rivero & Cerda-Olmedo, 1994).

A modification in surface of *Mucor javanicus* spores, during germination, presents an interesting aspect of cell differentiation. The first stages involve modification in spore surface, increase in cell volume and altered shape. Germ tube emission and elongation may reflect internal chemical modification on the spore wall. The electron microscopy study permitted revealed the *Mucor javanicus* structural details.

The results obtained in this study are used as a basis to identify the specific biochemical and physiological aspects of spores germination in the specie studied, which are under investigation of the ultrastructural and the actin aspects.

Acknowledgements

The authors are grateful to CNPq, FINEP, FACEPE, and PADCT for the financial support.

References

[1] Aitken, W.B.; Niederpruem, D.J. (1970) Ultrastructural Changes and Biochemical Events in Basidiospore germination of *Schizophyllum commune. J. Bacteriol.*, **104**(2): 981-988.
[2] Alexopoulos, C. J., Mims, C. W., Blackwell, M. (1996) **Introductory Mycology**. John Wiley & Sons, Pub., 233p.
[3] Bartinicki- Garcia, S., Nelson, N., Cota Robles, E. (1968b). Electron Microscopy of Spore Germination and Cell Wall Formation in *Mucor rouxii. Arch. Microbiol.*, **63**,:. 242-255.

470

[4]   Beckett, A., Heath, I.B., McLaghlin, D.J. (1974) **An Atlas of Fungal Ultrastructure**. Longman editors, London, 221p.

[5]   Botha A, Kock JL, Coetzee DJ, Botes PJ. (1997) Physiological properties and fatty acid composition in Mucor circinelloides f. circinelloides. *Antonie Van Leeuwenhoek*.;71(3):201-6.

[6]   Edelman, R.E. & Klomparens, K.L. (1995a.) Zygosporogenesis in *Zygorhyncus heterogamus*, with a Proposal for Standardization of Structural Nomenclature. *Mycologia*,. 87:,304-318.

[7]   Edelmann, R.E.; Klomparens, K.L. (1995b)Low temperature scanning electron microscopy of the ultrastructual development of zygospores and sporangiospores in *Mycothypha africana*, and the effects of cultural conditions on sexual versus asexual ɛprodution. *Mycologia*, 90::212-218.

[8]   George, R.P.; Albrecht, R.M.; Raper, K.B.; Sachs, I.B.; Mackenzie, A.P. (1972) Scanning Electron Microscopy of Spores germination in *Dictyostelium discoideum*. *J. Bacteriol.*, 112(3): 1383-1386.

[9]   Griffin, D. H. 1994. Spore dormancy and germination,. In Fungal physiology, 2nd ed. John Wiley & Sons, New York, N.Y. p. 375-398.

[10]   Grove, D.N. - Form and Function of Zygomycetes Spore. **The Fungal Spore**. Ed. D.J. Weber & W.M. Hess, J.Wiley & Sons. New York, London..559-590 p, 1979.

[11]   Grove, S.M. (1972) Apical vesicles in Germination Conidia of *Aspergillus parasiticus*. *Mycologia*, 64:. 638-641.

[12]   Hess, W.M. & Weber, D.J. (1973) Ultrastructure of Dormant Spore and Germineted Sporangiospores of **Rhizopus arrhizus**. *Protoplasma*, 77:. 15-33.

[13]   Jackson, F.M.; Michaelson, L.; Fraser, T.C.M.; Stobart, A.K.; Griffiths, G. (1998) Biosyntesis of Triacylglycerol in the Filamentous fungus Mucor circinelloides. *J. Microbiol.*, 144, 2639-2645.

[14]   Jeffries, P. & Young, T.W.K. (1976) Physiology and Fine Structure of Sporangiospore Germination in *Piptocephalis unispora* Prior to Infection. *Arch. Microbiol.*, 107:. 99-107.

[15]   Klomparens, K. L. (1990) The development and application of ultrastructural research in mycology. *Mycopathologia*, 109:139-148.

[16]   Lubbehusen TL, Nielsen J, McIntyre M (2004) Morphology and physiology of the dimorphic fungus Mucor circinelloides (syn. M. racemosus) during anaerobic growth. *J. App. Microbiol. Biotechnol.* ; 63(5):543-8.

[17]   Lubbehusen TL, Nielsen J, McIntyre M. (2003) Aerobic and anaerobic ethanol production by Mucor circinelloides during submerged growth. *J. App. Microbiol Biotechnol..*;95(5):1152-60.

[18]   Maia, L.C., Kimbrough J.W., Benny G.L. (1994) Ultrastructure of Spore Germination in *Gigaspora albida* (Glomales). *Mycologia*, 3: 343-349.

[19]   Maia, L.C.; Kimbrough, J.W.; Benny, G. (1993) Ultrastructural studies of the spore wall of *Gigaspora Albida* (Glomales). *Mycologia*, 85:883-889.

[20]   Mckeown, T. A.; Moss, S. T.; Jones B. G. (1996) Ultrastructure of Ascospores of *Tunicaspora australiensis*. *Mycol. Res.*, 100:1247-1255..

[21]   Rivero, F.; Cerdá-Olmedo, E. (1994)Spore Germination in *Phycomyces blakesleeanus*. *Mycologia*, 86:781-786.

[22]   Schmidt, A. & Hall, N. (1998) Signaling to the actin cytoskeleton. *Ann. Rev. Cell Develop. Biol.* 14: 305-338.

[23]   Song Y, Wynn JP, Li Y, Grantham D, Ratledge C. (2001) A pre-genetic study of the isoforms of malic enzyme associated with lipid accumulation in *Mucor circinelloides*. *Microbiol..*, 147(Pt 6):1507-15.

[24]   Wynn JP, Hamid AA, Li Y, Ratledge C. (2001)Biochemical events leading to the diversion of carbon into storage lipids in the oleaginous fungi Mucor circinelloides and Mortierella alpina. *Microbiol..*, 147(Pt 10):2857-64.

Bioremediation

Activity of soil microbial communities to monitor the efficiency of a metal phytoremediation process with *Thlaspi caerulescens*

J. Hernández-Allica[1], O. Zárate[1], F. Blanco[1], J.M. Becerril[2], and C. Garbisu[*,1]

[1] NEIKER, Basque Institute of Agricultural Research and Development, c/ Berreaga 1, E-48160 Derio, SPAIN
[2] Department of Plant Biology and Ecology, University of the Basque Country, P. O. Box. 644, E-48080 Bilbao, SPAIN

[*] Corresponding author: e-mail: cgarbisu@neiker.net, Phone: +34 94 4034300, Fax: +34 94 4034310

The goal of any soil remediation process must always be not only to remove the contaminant from the polluted site but to restore soil health/quality as well. Although to date, much more emphasis has been placed on physicochemical indicators of soil quality, biological indicators are becoming increasingly used due to their being more sensitive to changes in the soil as well as to their capacity to provide information that integrates many environmental factors. The aim of this work was to evaluate the possibility of monitoring the efficiency of a metal phytoextraction process with the hyperaccumulator *T. caerulescens* by means of determining the changes observed in different biological parameters of soil health. Apart from confirming its great potential for metal phytoextraction, *T. caerulescens* growth has proved to have a beneficial effect on soil biological activity in the two polluted soils here studied. The revegetation of these soils with *T. caerulescens* could help activate the biochemical and microbial functionality of these soils.

Keywords: biological indicators; enzyme activities; hyperaccumulating plants; phytoextraction; soil health

1 Introduction

In the Basque Country (northern Spain), historically, metal mining activities have been an important component of its industrial development, leading to soil metal pollution. In the past few years, disquiet among ordinary citizens has grown and the public is now strongly demanding that cleanup measures be urgently introduced. In this context, phytoremediation, or the use of green plants to remove pollutants from the environment or to render them harmless, is currently being considered as a promising, cost-effective, aesthetically pleasing technology for the remediation of polluted sites.

Regarding metal pollution, there are at present two different strategies to phytoextract metals from soils: (i) continuous phytoextraction, through the utilization of plants that accumulate high concentrations of metals in their foliage (*i.e.,* metal hyperaccumulating plants), and (ii) induced phytoextraction, based on the application of mobilizing/chelating agents to the soil in an attempt to increase plant metal uptake, especially when the metal to be extracted is low in bioavailability [1]. Regarding continuous phytoextraction, the hyperaccumulator *Thlaspi caerulescens* has been extensively studied due to its remarkable capacity to phytoextract zinc (Zn) and cadmium (Cd) from polluted soils. However, it has been reported that metal phytoextraction using *T. caerulescens* appears feasible only when soils present moderate levels of Zn and Cd pollution [2].

Finally, it is most important to point out that the goal of any soil remediation process must always be not only to remove the contaminant from the polluted site but to restore soil health/quality as well. Although to date, much more emphasis has been placed on physicochemical indicators of soil quality, biological indicators are becoming increasingly used due to their being more sensitive to changes in the soil as well as to their capacity to provide information that integrates many environmental factors.

The aim of the current work was to evaluate the possibility of monitoring the efficiency of a metal phytoextraction process with the hyperaccumulator *T. caerulescens* by means of determining the changes observed in different biological parameters of soil health.

2   Materials and methods

Two soils, one heavily contaminated with metals (HPS) and the other showing relatively moderate levels of metal pollution (MPS), were collected (upper 0-20 cm) from an area formerly occupied by a nowadays abandoned Zn/Pb smelter in the province of Biscay (Basque Country, northern Spain). Immediately after sample collection, soils were sieved to <5 mm and stored at 4°C. Microcosm experiments were carried out in 250 ml plastic pots filled with 300 g (fresh weight) of soil in a growth chamber under the following controlled conditions: photoperiod 16/8 h light/darkness, temperature 22/18°C day/night. Half of the pots were fertilized with 120 mg kg^{-1} of N, P, and K, respectively. For each soil type (HPS, MPS), in half of the fertilized pots, 20 mg of *T. caerulescens* seeds were planted. The following three treatments were studied: (i) UP-NPK = unplanted (bare soil), non-fertilized pots, (ii) UP+NPK = unplanted (bare soil), fertilized pots, and (iii) P+NPK = planted (with *T. caerulescens*), fertilized pots. After germination, planted pots were homogenized to 4 plants per pot. Two months after germination, UP+NPK and P+NPK pots were again fertilized as before. Throughout the experimental period, soils were kept at 60-70% of their water holding capacity using deionized water. Four months after germination (approximately, the complete growing cycle of *T. caerulescens*), roots and shoots were harvested separately and their fresh weights recorded. Then, roots and shoots were washed thoroughly with deionized water, froze in liquid nitrogen, freeze-dried for 72 h, and their dry weights calculated.

For chemical analysis, soils were processed and analysed as previously described [3]. For analysis of biological parameters, soils were air-dried at 30°C for 48 h, sieved to <2 mm, and stored at 4°C. For dehydrogenase activity and fluorescein diacetate hydrolisis (FDA), soils were sieved to <2 mm in fresh and then stored at 4°C. All biological parameters (*i.e.*, dehydrogenase, —glucosidase, acid phosphatase, urease, arylsulphatase, FDA, potentially mineralizable nitrogen, basal and substrate induced respiration, and biomass carbon) were determined according to previously published methods [4, 5].

ANOVA analysis was performed on all data sets to establish significant differences among treatments using Microsoft StatView Software (Microsoft Corporation).

3   Results

Table 1 shows some physicochemical characteristics of HPS and MPS as well as of a control, non-polluted soil (NPS) collected in a grassland located in the vicinity of the Zn/Pb smelter. As seen in Table 1, HPS was severely contaminated, showing extremely high concentrations of total Zn, Pb and Cd. Although MPS was only moderately contaminated in terms of total Zn, Pb and Cd concentrations (Table 1), most importantly, 63, 9, and 71% of the total Zn, Pb, and Cd, respectively, was present in a relatively mobile form [as indicated by the metal concentration of the fractions obtained when soil samples were subjected to extraction with 1M Ca$_2$(NO$_3$)]. On the contrary, in HPS, these mobile fractions accounted for only 0.6, 5, and 14% of the total Zn, Pb, and Cd concentration, respectively.

Conversely, the amount of highly available metal (as indicated by the metal concentrations of the fractions obtained when soil samples were subjected to extraction with water) was very similar in both polluted soils.

Table 1 Physicochemical characteristics of HPS (heavily polluted soil), MPS (moderately polluted soil), and NPS (control, non-polluted soil).

	pH	OM (%)	C/N	Sand (%)		Total metal conc. (mg kg^{-1})		
				coarse	fine	Zn	Pb	Cd
HPS	6.7	4.75	1.4	22.9	50.2	18900	4930	15.1
MPS	5.6	3.54	13.4	12.6	59.0	1220	340	3.2
NPS	5.2	6.06	19.1	17.4	57.8	117	225	<0.8

Figure 1 shows that increasing levels of metal pollution led to lower values of all enzyme activities here studied (values of all enzyme activities adjust to the following trend: NPS>>MPS>HPS).

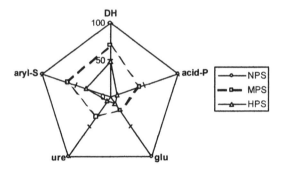

Figure 1: Sun-ray plot for enzyme activities in HPS (heavily polluted soil), MPS (moderately polluted soil) and NPS (non-polluted soil), scaled against the reference situation (*i.e.*, values found at the control, non-polluted soil = 100%). Acid-P: acid phosphatase; aryl-S: arylsulphatase; glu: — glucosidase; DH: dehydrogenase; ure: urease.

T. caerulescens plants grew well on both polluted soils (HPS, MPS). Furthermore, no significant differences in root or shoot biomass production (on a dry weight basis; DW) between both polluted soils were observed (data not shown). In both soils, *T. caerulescens* showed a remarkable capacity to both take up Zn from soils and then translocate this element to the shoots. Root and shoot Zn concentrations were significantly higher (p<0.01) in plants growing in HPS compared to MPS. In HPS, root and shoot Zn concentrations exceeded 2.1% of biomass dry weight. In MPS, Zn was efficiently translocated to the shoots where it reached a concentration of approximately 1.5% of shoot dry weight. In this same soil (MPS), root Zn concentration reached a value of 0.2% root dry weight. Cadmium concentration in roots and shoots of *T. caerulescens* were below 30 mg kg^{-1} DW in both polluted soils. In HPS, root Pb concentration was extremely high (but low in shoots). Interestingly, in planted pots, the amount of highly available metals (water extractable fraction) significantly increased (p<0.001) with fertilizer addition.

Both polluted soils showed no significant differences between treatments with respect to FDA, dehydrogenase, urease activity, and potentially mineralizable N. Arylsulphatase and acid phosphatase activities were significantly ($p<0.01$) inhibited by fertilizer addition (49 and 38% for arylsulphatase and acid phosphatase activity, respectively, as compared to values found in unplanted, non-fertilized pots) and plant growth (40 and 37% for arylsulphatase and acid phosphatase activity, respectively, as compared to values found in unplanted, non-fertilized pots). On the other hand, basal and substrate induced respiration, and biomass carbon were stimulated by plant growth.

4   Discussion

Despite the current great interest in improving metal extraction capacity of hyperaccumulating plants, their influence on soil microorganisms has been rarely investigated [6]. In fact, up to date, when evaluating the success of a phytoextraction process, emphasis has mostly been placed on contaminant removal. But, as abovementioned, the goal of any remediation process must always be not only to remove the contaminant from the polluted site but to restore soil health/quality as well.

In agreement with previous works [3, 7], Figure 1 shows that the presence of metals has a negative effect on all enzymes, with urease and —glucosidase being most sensitive. It has been reported that the nature and degree of this inhibition are highly dependent on soil type [3]. Then, the physicochemical characteristics of both polluted soils (lower OM contents and lighter textures than NPS) could, at least partly, explain this negative effect on enzyme activities. After all, soils with lighter textures have been reported to show increased susceptibility of their biological activity to metal inhibition, most likely due to their lower metal binding and buffering capacities [7]. Besides, it is a well-known fact that the free, hydrated metal ion is the most toxic form of metals present in the environment [8]. HPS and MPS are sandy soils which present extremely high concentrations of water soluble, bioavailable metals, thus resulting in microbial populations being severely exposed to these highly toxic, heavy metal forms.

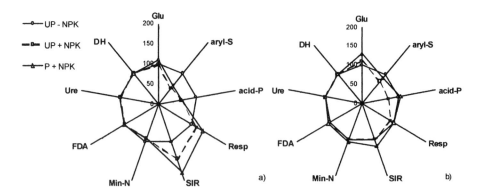

Figure 2: Sun-ray plot for enzyme activities in (a) HPS and (b) MPS, scaled against the reference situation (*i.e.*, values found at the unplanted, non-fertilized soil = 100%). UP, unplanted; P, planted; NPK, fertilized; acid-P: acid phosphatase; aryl-S: arylsulphatase; glu: —glucosidase; DH: dehydrogenase; ure: urease; FDA: fluorescein diacetate hydrolisis; SIR: substrate induced respiration; Resp: basal respiration; Min-N: potentially mineralizable N.

In this study, *T. caerulescens* has confirmed its great potential for metal phytoextraction, due not only to the remarkable Zn concentration values found in its harvestable parts but to the complete absence of stunted growth or phytotoxicity symptoms as well. In any case, for a phytoextraction process, the bioconcentration factor (BF = shoot metal concentration/soil metal concentration) is more important than shoot concentration *per se* [2]. In this study, *T. caerulescens* has shown a BF of 1.1 and 16.7 for the HPS and MPS, respectively. This latter BF value suggests that MPS could be phytoremediated with *T. caerulescens* in a reasonable number of years, of course, under the best estimates of *T. caerulescens* biomass production [2].

Growing *T. caerulescens* in MPS and HPS for phytoextraction purposes has also proved to have beneficial effects on soil biological activity and, therefore, soil health/quality. The main positive effect here observed relates to the improved C mineralization values reached in the presence of *T. caerulescens*, *i.e.* higher values of biomass C (data not shown), and basal and substrate induced respiration. This stimulating effect might be due to the presence of extra C from plant root exudates. Although it is true that MPS and HPS have lower OM contents than NPS, their OM values are very typical of grassland soils in the Basque Country. Notably, it has been suggested that the accumulation of OM in polluted soils might be due to heavy metals impeding mineralization cycles [9], especially in soils with lighter textures and high concentrations of bioavailable metal forms. The revegetation of MPS and HPS with *T. caerulescens* could indeed help activate the biochemical and microbial functionality of these soils.

References

[1]   D. E. Salt, R. D. Smith, and I Raskin, Annu. Rev. Plant Physiol. Plant Mol. Biol. **49**, 643 (1998).
[2]   F.J. Zhao, E. Lombi, and S.P. McGrath, Plant and Soil **249**, 37 (2003).
[3]   M.B. Hinojosa, J.A. Carreira, R. García-Ruíz, P. Dick, Soil Biol. Biochem. **36**, 1559 (2004).
[4] J.   W. Doran, and A. J. Jones (eds.), Methods for Assessing Soil Quality, SSSA Special Publication 49 (Soil   Science Society of America, Madison, 1996), pp. 410.
[5]   R. G. Burns, and R. P. Dick (eds.), Enzymes in the Environment, (Marcel Dekker, New York, 2002), pp. 614.
[6]   F. Gremion, A. Chatzinotas, K. Kaufmann, W. Von Sigler, H. Harms, FEMS Microbiol. Ecol. **48**, 273 (2004).
[7]   T.I. Stuczynsky, G.W. McCarty, and G. Siebelec, J. Environ. Qual. **32**, 1346 (2004).
[8]   J.S. Angle, R.L. Chaney, and D. Rhee, Soil Biol. Biochem. **25**, 1443 (1993).
[9]   K. Chander, and PC Brookes, Soil Biol. Biochem. **23**, 927 (1991).

Bioprocesses under practical aspects – Multimedia environmental impact assessment of in situ - soil remediation with genetically modified microorganisms

A. Rein[*,1], M. Bittens[2], and G. Teutsch[2]

[1] University of Tübingen, Center for Applied Geoscience, Sigwartstr. 10, 72076 Tübingen, GERMANY

[2] UFZ Centre for Environmental Research Leipzig-Halle, Permoserstr. 15, 04318 Leipzig, GERMANY

[*]Corresponding author: e-mail: arno.rein@uni.tuebingen.de, Phone: +49 7071 29 75496, Fax: +49 7071 29 5059

Bioremediation techniques are an option for the cleanup of contaminated sites in many cases. Requirements for an accepted implementation comprise a detailed and reliable risk analysis. For soils contaminated with polychlorinated biphenyls (PCBs), a plant-microorganism system for *in situ* - bioremediation has been developed based on genetically modified microorganisms (GMOs). Chlorobenzoic acids (CBAs) are the metabolites of concern. To evaluate potential effects to human health and the environment, a multimedia environmental impact assessment has been carried out based on laboratory and lysimeter experiments. Low- and moderately chlorinated PCB congeners are readily metabolised by the studied bacterial strains. Although the produced CBAs are mobile compounds showing an increased tendency to distribute into different environmental matrices, a clear risk reducing potential can be seen. Very low gene transfer rates have been observed. Horizontal spreading of the GMOs and potential impacts on the microbial communities in soil is expected to be very low

Keywords: remediation; biodegradation; environmental impact assessment; risk analysis; modelling; genetically modified microorganisms; PCB; CBA

1   Introduction

Contamination of soil and groundwater from anthropogenic sources is widely found in populated areas and identified as an important problem. In many cases, biological *in situ* - measures are an option for the remediation of contaminated sites. Requirements for an accepted implementation comprise a detailed and reliable risk analysis. For soils contaminated with polychlorinated biphenyls (PCBs), a plant-microorganism system for *in situ* - bioremediation has been developed. It consists of genetically modified microorganisms (GMOs) in conjunction with plant roots. The GMOs are *Pseudomonas fluorescens* strains which are genetically engineered to degrade PCB congeners *in situ*. Their metabolism requires root exudates and is therefore tightly coupled to plant rhizospheres. Laboratory and lysimeter experiments have been performed to investigate the efficiency of the remediation system and potential adverse effects. On the basis of these experiments, a detailed risk analysis has been carried out to draw conclusions for a potential field application.

2   Methods

A qualitative and quantitative analysis of potential effects to human health and the environment has been carried out, addressing a) PCB congeners of concern, b) GMOs used for bioremediation and c) chlorobenzoic acids (CBAs) as metabolites of concern (stable end products of the biphenyl pathway).

Pure culture assays have been performed to investigate degradation kinetics of the strains (cultures pregrown on biphenyl) [1]. The disappearance of individual PCB congeners has been measured versus time. Because of analytical difficulties, the production of metabolites has not been determined. Instead, full (equimolar) conversion of PCB into the corresponding CBA is assumed as a worst case scenario. The results have been analysed considering *Monod*-kinetics and exponential decay (first order kinetics). In lysimeter tests with contaminated soil, bacterial survival and growth has been studied over a period of 6 months. Estimations for field conditions (reduction of contaminant concentration C) have been made using degradation rate constants k and initial bacterial mass $B_{lab,0}$ from laboratory experiments, microbe mass B_{soil} observed in lysimeter tests and a factor for reduced degradation capacity under field conditions f_{RD} [2]:

$$C = C_0 * \exp\left(-k \times \frac{B_{soil}}{B_{lab,0}} \times f_{RD}^{-1} \times t\right) \tag{1}$$

Multimedia environmental computer modelling has been performed to evaluate fate and transport of PCBs and CBAs. Compound mass fluxes from soil into air, with leachate into the subsurface, and the uptake into plants have been calculated. In addition, a mass balance has been set up, including gains and losses for the considered compartments. Numerical modelling has been carried out according to the procedures used by the computer programs ARAMS v1.2.2, RISC v4.02 (modified according to the USEPA soil screening guidance [3]) and CemoS1 (for plant uptake, modified according to [4]). Generic site conditions have been considered. For the impact analysis, receptor point concentrations have been compared in soil, air, leachate and plants for two scenarios: a) conditions without biodegradation (no use of bacterial strains) and b) considering biodegradation with the studied microbes, i.e. depleted PCB concentrations and formation of CBA. The reduction of PCB concentration has been calculated and CBA formation and respective concentration determined accordingly for each studied environmental compartment.

Potential horizontal spreading of GMOs inoculated in the rhizosphere of willow plants has been investigated in lysimeter experiments and in a field release test (with non-GM derivates of the microbes). Furthermore, gene transfer rates have been determined in laboratory. Potential impacts on the function (e.g. enzymatic activities) and structure (genetic distribution) of indigenous microbial communities have been studied in lysimeter experiments.

3 Results and discussion

3.1 Degradation potential

15 of 29 PCB congeners that have been tested were depleted in the laboratory assays (low and moderately chlorinated PCBs). The potential to metabolise the commercial mixture Aroclor 1016 has been evaluated. PCBs have been used in technical mixtures for different applications (e.g., as flame retardants, as dielectric fluids in the electrical industry or as additives in oil) and were released into the environment. As low and moderately chlorinated PCBs have readily been degraded by studied the microbes, the degradation of the frequently used commercial mixture Aroclor 1016 has been chosen for the modelling

(low degree of chlorination). Around 39 % of the congeners included in Aroclor 1016 (mean value reported in literature) has been shown to be degradable by the studied microbes, 19 % were not capable of being depleted and for the remaining 42 %, no experiments have been performed.

3.2 Multimedia environmental modelling

Results of the multimedia environmental modelling are shown in Fig. 1. The modelling has been performed considering a generic environment (with soil properties reported for sandy loam soils, a soil pH of 6 and a contamination of 15 x 15 m, 1 m thick and located at a depth of 1 m). Contaminant input have been the degradable PCB congeners with concentrations according to the fractions reported for Aroclor 1016 (PCB concentration of 39 mg/kg soil in total). Degradation rate constants evaluated for the studied PCBs and observed bacterial numbers have been used ($B_{lab,0}$ = 1.52*10^{11} cells/l, B_{soil} = 2.6*10^{7} cfu/kg fresh weight) [2]. Optimum conditions have been assumed (f_{RD} = 1).

As it can be seen in Fig. 1a), reduction of PCB mass in soil is very slow for conditions without biodegradation. PCB-partitioning from soil with leachate, into ambient air and into plants can be seen. After 30 years, approximately 88 % of PCB mass remains in soil, 9 % and 3 % is distributed into leachate and air respectively and a very small percentage (lower than 0.1 %) is taken up by plants. Considering biodegradation with the studied microbes (see Fig. 1b), the PCB mass is significantly reduced in soil, leachate and air. Looking at the mass of CBA produced by the microbes (Fig. 1c), a small fraction is remaining in soil (peak in the first months) whereas high mass fluxes with leachate and into plants are obvious. About 87 % of the CBA mass is transported with leachtae, 13 % is taken up by plants.

Contaminant concentrations as a function of time are shown in Fig. 1d) to f). For the leachate, concentrations have been calculated directly below the unsaturated zone (1 m below the source zone). Without biodegradation, PCB concentrations in soil, leachate, air and plants slowly decrease. Considering biodegradation (Fig. 1e), concentrations in all compartments are rapidly reduced. High concentrations of CBA in leachate and plants are expected with a maximum in the first months (Fig 1f).

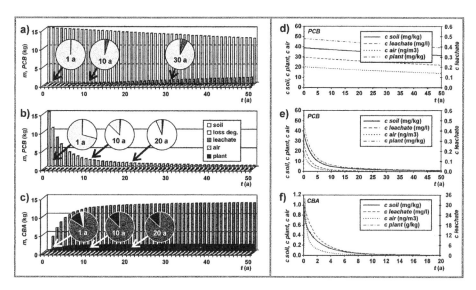

Fig. 1 Mass fluxes (a, b, c) and compound concentrations (d, e, f) in soil, leachate, air and plant. PCB mass balance and concentrations for conditions without biodegradation (a and d) and considering biodegradation (b and e). Results for CBA are shown in c) and f).

3.3 Potential for risk reduction

The risk potential for CBA is much lower than for PCB. Reference doses for dermal contact and ingestion are 160 to 290 times lower (based on data for Aroclor 1016 and on upper bound estimations for moderately chlorinated CBAs). For fish, no-observed-adverse-effect concentrations are 90 to 200 times lower (based on toxicological data for Bluegill, *Lepomis macrochirus* and Fathead Minnow, *Pimephales promelas*). As illustrated in Fig. 2a), receptor point concentrations of degradable congeners are reduced by 50 % after 2 to 2.5 years, by 75 % after 4 to 6 years and by 90 % after 7 to 13 years. The formation of CBA does not significantly influence the potential to reduce risks for human health and fish (for aquatic organisms, groundwater to river transport needs to be considered). Enhanced concentrations of CBAs plants indicate no impact on plant health as observed in lysimeter experiments. Figure 2b) shows the potential risk reduction for Aroclor 1016 in total (including PCB congeners that could not be degraded in the experiments and which were not analysed).

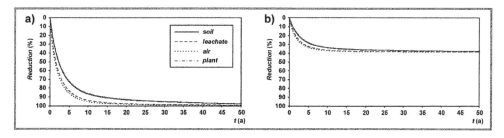

Fig. 2 Reduction of receptor point concentrations of PCBs and associated risks by the use of the investigated bacterial strains, a) degradable PCB congeners in Aroclor 1016, b) Aroclor 1016 in total.

482

3.4 Impact analysis of GMOs

Bacterial counts at the lysimeter tests found no significant hint of microbial dispersion into leaves, roots, root free soil and leachate. In the field-release test, the strains (non-GM derivatives of the GMOs) were identified in the rhizosphere of the inoculated willows only [5]. Therefore, horizontal spreading, transport in leachate and plant uptake are no pathways of concern for the considered GMOs. Concerning gene transfer, very low rates have been observed, as the introduced bph-trait is stably inserted into the chromosome of the microbes. Lateral transfer between homologue bacteria is assumed to occur with a frequency of 10^{-9} [6]. Looking at the function and structure of indigenous soil microflora, no impacts can be assigned to the presence of GMOs [5].

4   Summary and conclusions

Kinetical data from degradation experiments and data on the survival and growth of bacteria are required for the impact analysis of biological measures for soil remediation. Results of the experiments show that low- and moderately chlorinated PCB congeners are readily metabolised by the studied bacterial strains. Although the produced CBAs are mobile compounds showing an increased tendency to distribute into leachate and plants, a clear risk reducing potential can be seen. Impacts of GMOs on microbial soil communities and observed gene transfer rates are very low. There is no significant hint on horizontal spreading. Further experiments will be made to investigate effects of multiple congeners present in soil on degradation velocity.

Acknowledgements

This work was founded by the European Commission with grant QLK3-CT-2001-00101, "Testing Integrated GM -Rhizoremediation systems for soil bioremediation".

References

[1]   U. Karlson, M..M. Fernqvist, P. Mayer, S. Trapp, A. Rein, and M. Bittens, Degradation potential of PCB congeners by bacterial strains – Test method utilising SPME (In Preparation).

[2]   A. Rein, M. Bittens, G. Teutsch, S. Trapp, P. Mayer, and U. Karlson, Modelling of the degradation potential of PCB mixtures by soil bacteria (In Preparation).

[3]   US EPA, Soil Screening Guidance: Technical Background Document (EPA 540/R-95/126, Office of Solid Waste and Emergency Response, Washington, DC., 1996).

[4]   Trapp, S (2000): Modelling uptake into roots and subsequent translocation of neutral and ionisable organic compounds. Pest. Manag. Sci. 56, p. 767-778.

[5]   U. Karlson, D. Aguirre, R. Riviera, A. Rein, and M. Bittens, In situ – soil bioremediation based on genetically modified bacteria – Study of horizontal spreading, gene transfer and potential impacts on indigenous microbial communities (In Preparation).

[6]   G.M. Brazil , L. Kenefick, F. O'Gara, and D.N. Dowling, Investigating the possible lateral transfer of a chromosomally integrated bph cassette from a rhizosphere pseudomonad F113pcb in the sugar beet rhizosphere (In Preparation).

Characterization of *Cladosporium oxysporum* and *C. sphaerospermum* using Polyaromatic hydrocarbons (PAHs) as their sole carbon source in tropical coastal seawater

Á. M. Nieves-Rivera[*, 1], N. J. Rodríguez[1], F. M. Dugan[2], B. R. Zaidi[1], and E. H. Williams, Jr.[1]

[1] Department of Marine Sciences, University of Puerto Rico, Mayagüez, PR 00681-9013 USA
[2] Western Region P. I. Station, Washington State University, Pullman, WA 99164-6402 USA

[*]Corresponding author: e-mail: anieves@coqui.net, Phone: 787-265-3838, Fax: 787-265-5408

Two species of Cladosporium *(C. oxysporum and C. sphaerospermum)* were isolated from surface coastal seawater based on their ability to use the polyaromatic hydrocarbons (PAHs) naphthalene ($C_{10}H_8$) and phenanthrene ($C_{14}H_{10}$) as a sole carbon and energy source. Although both *Cladosporium* spp. are cosmopolitan species, both species are new records to mangrove forests of Puerto Rico. These two species may be of value in the bioremediation of natural oil spills or other contaminants in tropical environments.

Keywords: Cladosporium; naphthalene; phenanthrene; bioremediation; Puerto Rico; Caribbean

1   Introduction

Polyaromatic hydrocarbons or polycyclic aromatic hydrocarbons (PAHs) are formed primarily as products from the combustion of fossil fuels.   The United States Environmental Protection Agency lists 16 PAHs as priority pollutants [1].   In marine environments, due to their hydrophobic nature and low water solubility, PAHs are readily adsorbed by particulate matter and tend to accumulate in sediments [2, 3]. Because of their toxic, mutagenic, or carcinogenic properties, high concentrations of PAHs are harmful to the marine biota and human health [3].   In Puerto Rico, for example, PAHs pollution has induced nuclear mutations for chlorophyll deficiency in trees of Rhizophora mangle [4].

  The fate of most petroleum substances in the marine environment is ultimately defined by their transformation and degradation due to microbial activity [1, 2]. Microorganisms are the primary means of degrading PAHs naturally.   About a hundred known species of microorganisms (e.g., bacteria and fungi) are able to use oil components to sustain their growth and metabolism. In pristine areas, their proportions usually do not exceed 0.1 to 1.0 % of the total abundance of heterotrophic bacterial communities; in polluted areas, however, this proportion increases to 1.0 to 10.0 % [1, 2]. Most of the information on degradation of PAHs has being derived from pure cultures that were isolated from temperate environments [5–7].   Very little information is available on isolation of fungal strains capable of degrading PAHs in subtropical marine environments [8].

  Kohlmeyer & Kohlmeyer [9] and González et al. [10] have classified members of the genus *Cladosporium* as facultative marine fungi.   These fungi from freshwater or terrestrial areas are capable of growing in the marine environment. *Cladosporium oxysporum* and *C. sphaerospermum* are typically geophilic (soil loving) and cosmopolitan [11]. In this paper, we report the identification and characterization of *C. oxysporum* and *C. sphaerospermum* (from Cabo Rojo and Guayanilla, respectively, coastal waters of southwestern Puerto Rico)

484

two species that are capable of degrading PAHs naphthalene ($C_{10}H_8$) and phenanthrene ($C_{14}H_{10}$).

2 Material and methods

Guayanilla Bay (GB) is located on the south coast of Puerto Rico about 35 km east of the island's southwestern corner. This bay was the site of one of the largest petrochemical complexes in the world until these were shut down in 1982 [1]. Many studies have determined the fate of PAHs at this bay after closure of industrial complex [1, 2]. Bahía Sucia (BS) and Los Morrillos (LM) are part of the Boquerón Commonwealth Forest, Cabo Rojo, and were described by Nieves-Rivera et al. [12]. On March 18, 1973, the tanker Zoe Colocotronis ran aground on a reef 4.8 km off La Parguera (southern Puerto Rico) releasing 1.01 million gallons of Venezuelan (Tijuana) crude oil on BS shores [13].

Naphthalene (99 % purity), phenanthrene (> 96 % purity), and agarose type VII (low gelling temperature) were obtained from Sigma Chemical Co., St. Louis, MO, and Noble agar from Difco Laboratories, Detroit, MI. Seawater and sediment samples were collected from BS, GB, and LM bays, stored in a refrigerator and used within 2–3 weeks after collection of the isolates. Isolates were grown on modified seawater medium (SWM) [14] and on malt extract agar (MEA) [15]. Wet mounts were observed at 25–60 × and 100–1000 × with stereo and compound microscopes. Fungal isolates were deposited in the American Type Culture Collection, ATCC. Identification of *C. oxysporum* and *C. sphaerospermum* was based on deVries [11], Ellis [15], and Ho et al. [16].

3 Results and discussion

Characters of colonies and morphological structures of ATCC MYA-3068 and MYA-3069 on MEA (Figs. 1A-G) were generally consistent with descriptions of *C. sphaerospermum* and *C. oxysporum* respectively, in both Ellis [15] and Ho et al. [16]. Characters on SWM were indistinguishable from those on MEA with the exception of growth rate, which were faster in MEA (2 mm/week) versus the slower SWM (1 mm/week). MYA-3068 keys to *C. oxysporum* in Ho et al. [16] and Ellis [15] *C. oxysporum* resembles *C. tenuissimum*, but the latter lacks the intercalary nodes. The defining character of this species, rather long conidiophores routinely approaching 500 µm in length and possessing intercalary conidiogenous nodes, was observed in culture. Conidia of MYA-3069 (NJRR-1) were intermediate in size between those described for *C. sphaerospermum* in Ellis [15] and those in Ho et al. [16]. *Cladosporium sphaerospermum* has been isolated on extreme halophilic environments [17], in organic chemicals such as toluene [18], and after the fallout of Chernobyl [19].

Cladosporium oxysporum Berk. & M.A. Curtis, 1869.
Colonies on MEA [16] 43–44 mm diameter in 10 days at ca. 25 °C, deep olive green, lighter and concentrically banded toward margin, surface almost velutinuous to flocculose. Reverse dark greenish black, lighter at margin. Conidiophores macronematous, olive, smooth, mostly ca. 300-900 × 3.5–4.5 µm (up to 7.5 µm at nodes), straight, mostly unbranched, with (0–) 1–5 conspicuously swollen internodes and a swollen apical node (Figs. 1A–D). Conidia with 1-3 scars smooth, olive, mostly oval to elliptical, occasionally limoniform, mostly (3.5–) 4–6 × 2.5–3.5 µm, the larger conidia grading into ramoconidia (Fig. 1E). Ramoconidia smooth, olive, elliptical to cylindrical with 3–5 scars; up to ca.

20 × 4 μm (Fig. 1E). Scars protuberant, dark, on conidia, ramoconidia and conidiophores. Strain deposited as ATCC MYA-3068 (ÁMNR-7); seawater in *R. mangle* roots, Los Morrillos and Bahía Sucia, Cabo Rojo, Puerto Rico, 4 November 2002, Á. M. Nieves-Rivera.

Cladosporium sphaerospermum Penz., 1882.
Colonies on MEA [16] 25-26 mm and 36-37 mm on half strength V8 agar (½ V8 [20]) at 10 days at ca. 25 °C; dark olive green, velvety, powdery, and reverse blackish green. Conidiophores olivaceous, macronematous, straight to flexous, 0–1 (–3) branched, intercalary or terminal, smooth, septate, up to ca. 160 μm long, usually ca. 30 to 125 μm, up to 3 μm in diameter (slightly expanded at apices), not geniculate. Conidia in simple or branched chains, globose or subglobose to limoniform, mostly aseptate, moderately verrucose, olive, mostly 3.5–6.5 × 3.0–4.5 μm; abscission scars darkened, protuberant (Figs. 1F-G). Ramoconidia subglobose to cylindrical, olive, typically aseptate, 7.5–17.5 (– 25.0) × 3.5–4.5 μm wide (Fig. 1F–G). Hyphae septate, olivaceous, smooth, up to 3 μm wide. Strain deposited as ATCC MYA-3069 (NJRR-1); seawater in *R. mangle* roots, María Langa Cay, Guayanilla Bay, Guayanilla, Puerto Rico, 31 January 2002, N. J. Rodríguez.

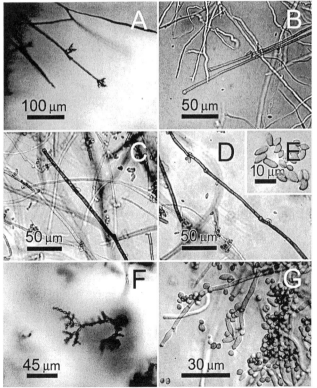

Figs. 1A–G Characteristic microscopic features of *Cladosporium oxysporum* (ATCC MYA-3068) and *C. sphaerospermum* (ATCC MYA-3069) from coastal seawater. A–D. Conidiophores of *C. oxysporum* on MEA. E. A group of oval to elliptical-shaped conidia and cylindrical ramoconidia of *C. oxysporum* at high magnification. F. Ramoconidia and conidia of *C. sphaerospermum* on ½ V8 agar. G. Conidiophores, ramoconidia, and conidia of *C. sphaerospermum* on MEA.

486

Figures 1A–G shows the arrangement of intact conidial chains on *C. oxysporum* and *C. sphaerospermum*. The top right and middle row show the variation in node placement and structure (an important character for identification of *C. oxysporum*). And the insert (Fig. 1E) and bottom right figures (Fig. 1G) show the conidia of *C. oxysporum* and *C. sphaerospermum*, respectively. Although both Cladosporium spp. are cosmopolitan species, they are new records for mangrove forests of Puerto Rico.

Coastal environments of Puerto Rico are prime repositories of PAHs because most industries as well as urban centers are located on the coast. For instance, for over 20 years GB was the location of one of the biggest concentrations of petrochemical industries in the world until it was shut down in 1982. Petrochemical industries and oil spills are the major source of organic pollutants which are of interest to this study. Although many studies have been conducted in areas along Puerto Rican coasts to determine the fate of pollutants by microbial degradation [1, 2], few have been conducted by using fungal isolates [8]. This study was baseline information on the conditions of GB and LM.

In conclusion, our particular interest was on the identifying potential bioremediation species that may be used in natural oil spills or other contaminant clean ups, which is a crucial part of environmental protection. The two species discussed above may play an important role in the future. However, we must discover, define, and experiment with additional naturally occurring species to ensure the protection of our tropical environments.

Acknowledgements

This work was founded through University of Puerto Rico Alliance for the Graduate Education and the Professorate fellowship (Grant No. NSF/AGEP–HRD # 0302696).

References

[1]   B.R. Zaidi, and S.H. Imam, Mar. Pollut. Bull. **38**, p. 737-742 (1999).
[2]   B.R. Zaidi, L.M. Hinkey, N.J. Rodríguez, N.S. Govind, and S.H. Imam, Mar. Pollut. Bull. **46**, p. 418-423 (2003).
[3]   J.C. Means, S.G. Wood, J.J. Hassett, and W.L. Banwart, Environ. Sci. Technol. **14**, p. 1524-1528 (1980).
[4]   J.E. Corredor, J.M. Morell, E.J. Klekowski, Jr., and R. Lowenfeld, Intl. J. Plant Sci. **156**, p. 55-60 (1995).
[5]   C.E. Cerniglia, and J.B. Sutherland, in: Fungi in Bioremediation (Cambridge Univ. Press, 2001), p. 136-187.
[6]   P.W. Kirk, and A.S. Gordon, Mycologia **80**, p. 776-782 (1988).
[7]   B. Qi, W.M. Moe, and K.A. Kinney, Appl. Microbiol. Biotechnol. **58**, p. 684-689 (2002).
[8]   C.T. Acevedo, Unpublished Ph. D. Thesis, Univ. Puerto Rico, Río Piedras (2001), 85 p.
[9]   J. Kohlmeyer, and E. Kohlmeyer, Marine Mycology: the Higher Fungi (Academic Press, New York, 1979), 690 p.
[10]   M.C. González, T. Herrera, M. Ulloa, and R.T. Hanlin, Mycoscience **39**, p. 115-121 (1998).
[11]   H. deVries, Contributions to the knowledge of the genus *Cladosporium* Link ex Fries. (Uitgeverrij & Drukkerij Hollandia, Baarn, The Netherlands, 1952), 121 p.
[12]   Á.M. Nieves-Rivera, T.A. Tattar, and E.H. Williams, Jr., Arboricult. J. **26**, p. 141-155 (2002).
[13]   J.E. Corredor, J.M. Morell, and C.E. del Castillo, Mar. Pollut. Bull. **21**, p. 385-388 (1990).
[14]   A.H. Bogardt, and B.B. Hemmingsen, Appl. Environm. Microbiol. **58**, p. 2579-2582 (1992).

[15]   M.B. Ellis, Dematiaceous Hyphomycetes (Commonwealth Mycological Institute, Kew, 1971), 608 p.

[16]   M.H.-M. Ho, R.F. Castañeda, F.M. Dugan, and S.C. Jong, Mycotaxon **72**, p. 115-157 (1999).

[17]   T. Kis-Papo, I. Grishkan, A. Oren, P. Wasser, E. Nevo, Mycol. Res. **105**, p. 749-756 (2001).

[18]   F.J. Weber, C.K.C. Hage, and A.M. De-Bont, Appl. Environ. Microbiol. **61**, p. 3562-3566 (1995).

[19]   N.N. Zhdanova, V.A. Zakharchenko, V.V. Vember, and L.T. Nakonechnaya, Mycol. Res. **104**, p. 1421-1426 (2000).

[20]   R.B. Stevens (ed.), Mycology Guidebook (Univ. Washington Press, Seattle, 1981), 712 p.

Comparison of microbial communities native to three differently polluted ecological niches in the industrial site of Bagnoli (Naples,Italy)

Anna Rosa Sprocati[*], Chiara Alisi, Flavia Tasso, Lia Segre and Carlo Cremisini

Unit for Environmental Protection and Development, Environmental Technology.
ENEA (International Agency for New Technology, Energy and the Environment), R.C. CASACCIA,
via Anguillarese 301- 00060 Rome, Italy

[*]Corresponding author: e-mail: sprocati@casaccia.enea.it *Phone* +39.06.30484495,
F*ax*+39.06.30484808

Industrial sites can be considered as a new source of biodiversity. Micro-organisms belonging to these biotopes represent potential solutions and cures for damaged areas. In the present work we explore different areas of a historical metallurgic site dismissed in the early 90s (Bagnoli-Naples, Italy), in order to compare the native microbial communities in relation with the contaminants and the ecotoxicity levels detected. All the three different areas studied show a biological respiratory activity and harbour diverse microbial communities, a fraction of which is culturable and harbour resistances to heavy metals. The biological respiratory activity, the metabolic profile at community-level and the total heterotrophic bacterial population result inversely correlated to the detected ecotoxicity. The most widespread resistances are to Cr^{6+} and Pb^{2+}, but also Cd^{2+} and Co^{2+}. As these metals are present in very low concentrations in nature, usually metabolic genes for these metals are unusual too. Thus these observed resistances are the result of a specific selective process exercised by the unusual excessive presence of these elements in this polluted site.

Keywords: bioremediation, polluted sites, heavy metals, microbial community

1   Introduction

Industrial sites can be considered as a new source of biodiversity, since peculiar biotopes have evolved in historical polluted areas. Micro-organisms belonging to these biotopes implemented adaptive strategies as an answer to chronic stress. The main adaptive responses include specific resistances or degradation competences towards heavy metals or pollutants; in some cases towards both of them. These pathways, often still unknown, contribute to amplify the metabolic networks of communities, helping to overcome metabolic bottlenecks in the biodegradation processes. These microorganisms represent potential solutions and cures for damaged areas and are, thus, strong candidates for potential biotechnological exploitation. Microbiological explorations of new niches contribute to find new metabolisms, but also to increase the still too scarce knowledge of the existing microorganisms. In the present work we explore three different areas of Bagnoli (Naples, Italy), a metallurgic site dismissed in the early 90s after a hundred years of activity, and now included in the "Priority National Plan for Industrial Sites Remediation". The metallurgic site was conceived as an integral cycle plant where steel was produced from mineral and fossil through 6 different process steps. Each step produced different types of contamination and all by-products were spread within the site, in different areas. The remediation plan foresees chemical and chemical-physical treatment in order to re-establish the civil use of the area (park and cultural facilities). With the present work, conducted with a research scope, we want to explore the microbiological state of the site, investigating, at the same time, the preliminary steps for a bioremediation feasibility study.

2   Aim

The AIM of the present work is to explore a few areas, within the metallurgic site, with different contamination typologies, in order to compare the native microbial communities considered by their composition and metabolic profiling, in relation to the contaminants and the detected ecotoxicity levels. Three different areas have been chosen: Agglomeration (AGL), polluted mainly by heavy metals; Oxygenation (OSS) polluted mainly by PHAs, PCBs and hydrocarbons; Lamination (LAM), polluted both by heavy metals and hydrocarbons.

3   Materials and methods

Chemical analyses have been carried out by spectrophotometry (Flame-AAS, Perkin Elmer 5100 Zla 213,9 nm for heavy metals; IR spectrophotometry for organic compounds). Microtox analysis have been carried out as in [1]. Respiratory activity was measured by incubating soil samples in closed bottles under stable conditions of pressure and temperature in the dark inside a Comput-ox WB500 computerized respirometer, N-Con Systems, Larchmont, NY. The community profiles were obtained by using Biolog ECO plates (Biolog Inc., Hayward, CA, USA) as described in [2]. T-RFLP profiles were obtained by following the method in [3]. The isolation of culturable populations was carried out on both Mineral Medium and Tryptic-Soy Broth at 28°C in orbitating shaker (180 rpm). Agar plates for total heterotrophic count were incubated at 28°C, 60% of humidity. Bacterial identification was carried out by using Biolog GN2 and GP2 plates (Biolog Inc., Hayward, CA, USA) according to the manufacturer' instructions. Solid media were amended with heavy metals ($Pb(NO_3)_2$, $ZnSO_4$, $CuSO_4$, $NiSO_4$, $CoSO_4$, $K_2Cr_2O_7$, $CdCl_2$) at increasing concentrations (ranging from 0,1 mM up to 8 mM) to determine the MIC (minimal inhibitory concentration).

4   Results and discussion

Chemical and ecotoxicological characterisation. Chemical analysis (tab.1) showed in all the three areas the presence of Cr, Zn, Pb and only for area AGL also As, at concentration higher than those allowed by law. Organic contamination by Hydrocarbon, PCBs and PHAs is almost absent for the AGL area, very high for the OSS area and medium for the LAM area. Furthermore the measured pH values for the three areas are very different: respectively 7, 9 and 11 for AGL, LAM and OSS.

Table 1 Chemical characterisation of the three areas

Sample	Cr	Hg	Zn	Co	Ni	Cu	As	Cd	Pb	PCBs	PHAs	Total Hydroc.	pH
AGL	210	0.6	353	5.9	29	129	118	2	345	0.067	0.382	30	7
LAM	215	0.3	177	4.1	80	117	41	1	227	1.241	5.167	358	9
OSS	240	0.6	214	5.4	54	65	34	1	240	1.703	1.270	1269	11

Heavy metals(mg/Kg) — Organics (mg/Kg)

As it has been proven that sole chemical analysis are not sufficient to describe the possible consequences on the ecosystem, ecotoxicological assays have been carried out in order to relate the pollutants concentrations with their bioavailability. Microtox assays have been carried out both on aqueous extract (for inorganic pollution) and on solid phase (for organic pollution) (tab. 2).

Table. 2. Ecotoxicological assay

MICROTOX	AGL	LAM	OSS
Aqueous extract (mg/L)	1014	Non tox	125.4
Solid phase (g/L)	146	55	<9

The results reveal that there is a correspondence, for the organic component, between the scale of measured concentrations and the level of toxicity: in decreasing order OSS, LAM and AGL. On the other hand, for the inorganic component the toxicity scale does not correspond to that of measured concentrations. The OSS area, although with the less amount of heavy metals, results the most toxic on aqueous extract, followed by AGL, while LAM presents a very low toxicity. This information suggests that in OSS the heavy metals present are more readily bioavailable, while in the LAM area –having comparable concentrations with respect to OSS- they are present in a less bioavailable state. This may be also linked to the different pH values.

Determination of the soil respiratory activity. Biological respiratory capacity has been detected in all the areas (Fig.1), but with different rate of oxygen consumption: AGL sample is the highest and fastest and reaches a first plateau after 23 hours, followed by LAM which reaches it after 90 hours, while OSS is the lowest.
Community-level physiological profiling through carbon-utilization analysis with Biolog ECO-plates. In Fig. 2 is shown the level of average colour development (AWCD) of the three communities. This measure gives a general indication on the reactive capacity of the community with respect to 31 different substrates.

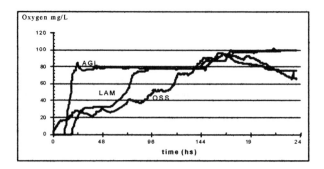

Fig.1 Biological respiratory capacity of soil samples

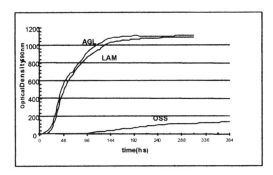

Fig.2. Average wells color development (AWCD)

AGL and LAM communit ies have a similar metabolic reactivity, high and rapid, while the OSS community shows a low and very slow reactivity which is characterised by a long lag phase. Both AGL and LAM communities are at their maximum reactivity after 144 incubation hours, while the OSS community has used few substrates and reaches after 384 hours a colour development which is only 1/3 of the other two communities. Also the variety of substrates used by the AGL and LAM communities clearly demonstrate a wider metabolic capacity –using 30 out of 31 substrates- compared to the 22 used by the OSS community. None of the communities is able to use D,L-பJ-Glycerol Phosphate. The use of the Biolog system for community-level physiological profiling has been questioned because the results are affected by a selective enrichment process of the microbial community originally present in nature. The substrates utilisation response is the result of a selectively- enriched subset of the natural community, due to the loss of unculturable strains. However, also in this case, these assays have an ecological relevance, since they can be considered very sensitive in detecting change in microbial communities [4].

Characterisation of the culturable fraction of the three microbial communities. The comparison between the results (fig. 3) of total heterotrophic bacterial population and pure colony morphotypes shows that the higher the abundance in the community, the lower is the biodiversity. The OSS community results the poorest in total number of bacteria and the richest in diversity (34 isolates on 10^2 UFC/mL). These data can be related with the respiratory activity which in OSS is very low.

At community-level, a molecular profiling has been performed by t-RFLP technique in order to estimate also the unculturable fraction of the population. Preliminary results show that AGL community is also the richest in fragments. These data stress the importance to carry out the microbiological studies through techniques which enable to overcome the limits of culturability, which is estimated to be limited to 1% of the putative 5% known microorganisms [5]

Identification of the isolates. All the 83 different colony morphotypes isolated have been tentatively identified by Biolog system, obtaining the following identification: 9 strains for AGL, 8 for LAM and 1 for OSS community. The AGL community includes the genus Rhizobium sp. radiobacter and *rizogenes, Pseudomonas sp. fulva maculicola* and *fluorescens,* and *Flavobacterium johnsoniae.* The LAM community includes *Achromobacter cholinophagus, xylosoxidans Acidovorax delafieldii, Pseudomonas sp., Pseudomonas fluorescens bt Fand bt G , Pseudomonas putida bt B, Micrococcus luteus.* The strain identified in OSS community is *Bacillus licheniformis.*

492

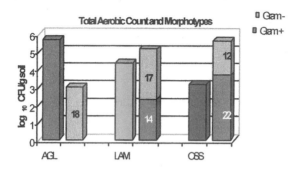

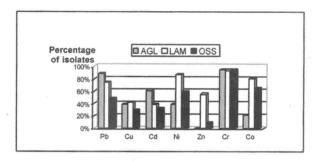

Fig.4. Distribution of metal resistances among the isolates

Distribution of the resistances to heavy metals All the isolates have been tested for resistance to Pb, Cu, Cd, Co, Ni, Zn, Cr. Minimum inhibitory concentration (MIC) defines the lowest concentration that causes the total growth inhibition. MIC has been utilized to describe the distribution of resistances between the isolates and among the communities (fig. 4), using E.coli resistances as a cut-off.

The most widespread resistance harboured among the isolates are Cr^{6+}, followed by Pb^{2+}. The LAM community, belonging to the area that does not show toxicity for inorganic contamination, shows the highest number of resistant strains with a wide spectrum of resistances: more than 50% are resistant to Cr^{6+}, Ni^{2+}, Co^{2+}, Pb^{2+} and Zn^{6+} and at least 40% are resistant to Cu^{2+} and Cd^{2+}. Considering. the ecology of heavy metals, Cr, Pb and Cd have in nature an average concentration smaller than 1 nM, and Co just a less higher, thus they are very unlikely ever to be useful or toxic and it would not pay to harbour metabolic genes for these metals [6]. Thus we can say that the observed resistances are due to a specific selective process exercised by the unusual excessive presence of these elements in this polluted site.

5   Conclusions and perspectives

All the three investigated areas - AGL, LAM and OSS- show a biological respiratory activity and harbour diverse microbial communities, a fraction of which is culturable. In each area the biological respiratory activity, the metabolic profile at community level and the total heterotrophic bacterial population result inversely correlated to the detected ecotoxicity. The OSS area shows the highest toxicity, the slowest soil respiratory activity, the lowest and restricted metabolic activity, the lowest microbial abundance.

The comparison of the three microbial communities shows that the higher the abundance, the lower is the biodiversity of the isolated strains. However, preliminary results by t-RFLP show that the area AGL, with the highest abundance, is also the richest in fragments, suggesting the existence of an important unculturable fraction of this community.

All the three communities show a wide range of heavy metals resistance. The most widespread resistances among the isolates are to Cr^{6+} and Pb^{2+}, in agreement with their concentrations in soil. Considering that Cr, Pb, Cd and Co are usually present in nature at very low concentrations and thus the metabolic genes for these metals are unusual and "useless" from an ecological point of view [6], we can say that the observed resistances are due to a specific selective process exercised by the unusual abundance of these elements and their bioavailability in the polluted areas.

The **future work** will be focused on the followings:
• resistances at community and isolates level to organic pollutants and their biodegradation potential estimated in laboratory scale, also in presence of heavy metals, for bioremediation purposes.
• identification of the unidentified strains by suitable molecular and biochemical techniques

Acknowledgements

Authors thanks: the Italian Ministery for the Environment and BagnoliFutura s.r.l. for the permission and the support during the sampling in the Bagnoli site; Roberto Ciccoli, Antonio Salluzzo and Sonia Manzo (laboratory ENEA - Portici) for the help in chemical and ecotoxicological analysis, Selene Grilli and Roberto Farina (laboratory ENEA - Bologna), for the help in respirometry assays;.

References

[1] Maffiotti A., Volterra L, Bona F., 1997. Analisi e recupero dei sedimenti marini. Pitagora editrice p. 152
[2] Garland J.L., 1996. Patterns of potential C source utilization by rhizosphere communities. Soil Biol Biochem. 28, 223-230.
[3] Osborn Mark A., Moore E. R. B., Timmis K.N., 2000. An evaluation of terminal-restriction fragment length polymorphism (t-RFLP) analysis for the study of microbial community structure and dynamics. Environmental Microbiology vol.2(1): 39-45.
[4] Garland J.L., 1999. Potential and limitations of BIOLOG for microbial community analysis. Microbial Biosystems: New Frontiers. In: VIII Int.Symp. on Microbial Ecology. Ed. Atlantica Canada Society for Microbial Ecology, Halifax, Canada.
[5] Convention on Biological Diversity, 2002. Montreal ISBN: 92-807-2172-0
[6] Nies D.H., 1999. Microbial heavy-metal resistance. Applied Microbiol Biotechnology 51, pp. 730-750.

Decolorization of Methyl Orange (as a model azo dye) by the newly discovered *Bacillus sp*

Pourbabaee, Ahmad Ali[1]; Malekzadeh[1] F; Sarbolouki, Mohamad N[2]; Mohajeri, Ali[3]

[1]Department of Biology, Science Faculty, Tehran Univrsity, Tehran, Iran
[2] Institue of Biochemistry and Biophysics, Tehran University, Tehran, Iran

1   Introduction

Azo dyes are the largest class of dyes commercially used in the textile industries [1]. Most of these compounds are highly resistant to microbial attack, and are therefore hardly removed from effluents by conventional biological wastewater treatments, such as activated sludge [2]. Not all dyes currently used may be degraded and /or removed with physical and chemical processes and sometimes the degradation products formed are even more toxic [3]. The new closed-loop technologies such as the reuse of microbia lly or enzymatically treatment of dyeing effluents could help reducing this enormous water consumption [3].

Bacterial decolorization under aerobic conditions usually results in adsorption of dyestuffs on bacteria rather than their oxidation [4]. White-rot fungi can degrade a wide variety of recalcitrant compounds by their extracellular enzyme systems [5]. However, it is difficult to keep them in functional form in the activated sluge systems, because of their special nutritional and environmental requirements. Moreover, bacterial degradation is much faster than fungal degradation of textile dyestuffs [6].

Thus, in this study our attention was focused on decolorization and biodegradation of methyl orange (MO) as a model compound for the azo dyes in order to understand the mechanism involved and the specific enzymes that are responsible for the azo dye metabolism by bacteria.

2   Materials and methods

Material
All chemicals and biochemicals (including culture media) of the highest available purity were obtained from the Merck, GmbH (Darmstadt, Germany) and Sigma, Inc, (St Louis, Mo, USA).

Methods
A modified method of Norteman et al, [8] was used for isolating and screening of the aerobic bacteria. Taxonomic identification including biochemical characterization and phylogenic analysis were performed by DSMZ (Deutsche Sammlung Von Mikroorganis-men und Zellkulturen GmbH, Braunschweig, Germany) under Id number: 03-1484.

Culture conditions
A loop full of 24h nutrient agar culture was transferred into a series of 250-ml Erlenmeyer flasks each containing 50ml of sterile nutrient broth medium. The flasks were incubated at 32°C in an orbital shaker (140 rpm, 24h)-(Orbital Incubator model SI50, Stuart Scientific, UK).

Decolorization experiment
Three ml methyl orange (MO) (stock solution $1000mgL^{-1}$) was added to 500-ml Erlenmeyer flasks each containing 100ml mineral salt (0.235 NaH_2PO_4; $0.07Mg$ SO_4, $7H_2O$; 0.014 $CaCl_2$; 0.001 $FeCl_3$, $6H_2O$ (g/l) (with or without carbon and nitrogen sources) and autoclaved. These flasks were inoculated either with the cell biomass or the overnight culture of the isolated strain and placed on the shaker incubator (3 days and 140 rpm, 32°C). After suitable intervals, samples were removed and processed for determining the extent of decolorization and chemical analysis. Decolorization of the methyl orange was measured following the method described by Nigam et al [9].

Chemical analyses
For chemical analysis of the biodegradation products, 10 Liter of dye (MO) treated or untreated were centrifuged at 4000g (ser. No. 331093. Code: 775, MSG, England), and the supernatants filtered through 0.45µm filters to remove the suspended cells and particles. For extraction, the pH of cell free solutions was adjusted oat 8.5 (using 10% NaOH) and dissolved in dichloromethane (1:1). The two phases were separated with a decantor after shaking. The organic phase dried with anhydrous sodium sulfate and the evaporated in a rotary evaporator after filtration.
Organic solid residues were dissolved in a small amount of dichloromethane and used first for thin-layer chromatography on silica gel (EtAc 30: 70hexane as mobil phase) and then for FT-IR, GC-Ms analyses. The same procedure was carried out with N, N-dimethyl 1, 4- phenylene diamine- dihydrochloride that results from the azo bond cleavage. The Infrared spectra were recorded on a Shimadzu-4300 FT-IR (Kyoto Japan). Mass spectra of the samples were determined by a Finnigan Mattsa 70 (USA) mass spectrometer.

Nucleotide sequence accession number
The nucleotide sequences was done by Instiue of microbiology, Viena university for the partial 16s-rRNA from *Bacillus Sp strain PS* and have been deposited in the EMBL Nucleotide Sequence Database under accession no: Aj515146.

3 Results and discussion

Isolation/ identification of the microbial strain
20 pure bacterial cultures were isolated from samples taken from the textile effluent. Decolorization tests showed that only one of the strains was able to decolorize methyle orang under aerobic conditions in presence of exogenous carbon source. Biochemical and phylogenic examination performed by DSMZ identification service in Germany showed that the PS strain is related to the Bacillus cereus group [10].

Decolorization of Methyl Orange
To determine the catabolic pathway of the azoic component, methyl orange (MO) was chosen as a model compound. The strain PS was able to grow in mineral medium containing MO in the presence of glucose. Decolorization of MO occurred after 2 days of incubation at 32°C under aerobic conditions (Fig 1). There have been some reports that suggest the decolorization of certain sulfonated azo dyes occur under aerobic conditions after 2 months [11]. Another report indicates that azo dyes are essentially nondegradable by bacteria under aerobic conditions [12].

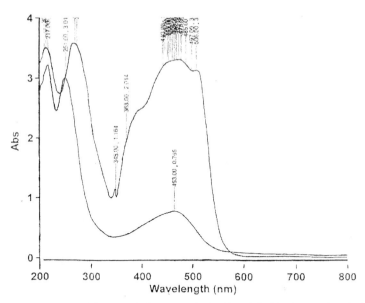

Figure 1. UV-VIS spectra of the a) methyle orange dye(MO) and the b) Biotreated methyle(BMO) after 30 min (by cells activated overnight in nutrient broth)

Kulla et al [14] describe a degradative pathway for sulfonated azo dyes by Pseudomonas strain previously adapted to grow on the corresponding carboxylated azo dyes [11].
Our experiments show that Bacillus sp strain PS is able to decolorize MO under aerobic conditions.
The TLC results and GC-Ms spectra show that the strain PS can convert about 98% of MO to N, N-dimethyl, 1, 4-phenylene diamine (Fig 2). Others have reported that anaerobic cleavage of the azo linkage by the reductase enzymes is the initial step in the biodegradation of the azo dyes [13]. The aerobic degradation of MO observed here suggests, in accordance with Kulla, that despite the presence of oxygen, the initial degradation step appears to be a reduction of the azo linkage by an oxygen-insensitive azo reductase [14].

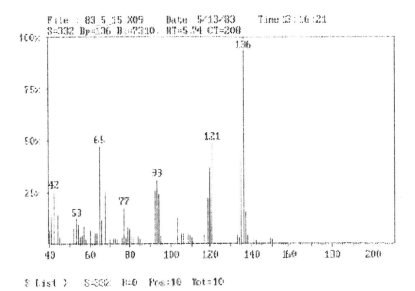

Figure 2. GC-Ms spectra showing the products (N, N-Dimethyl, 1, 4-phenylen diamine) formed after decolorization of methyl orange by strain PS.

In the presence of glucose, the N, N-dimethyl, 1, 4-phenylene diamine dihydrochloride disappeared by PS strain after 7 days of incubation on a shaker incubator (32°C and 140 rpm).

Therefore it is suggested that the expected reduction products of MO are completely metabolized as there are no traces observed in the TLC plate. Thus it seems that this is the first demonstration of a complete biological mineralization of a sulfonated azo dye by Bacillus cereus group. Haug [15] reported an anaerobic-aerobic degradation process of the sulfonated azo dye mordant yellow 3 by a bacterial culture. Therefore it seems likely that the mechanism proposed for decolorization of MO is similar to that of the other component of azoic dye. The proposed pathway for the biodegradation of methyl orange is schematically depicted in Fig [3].

We are currently attempting to identify the gene(s) and the expressed product(s) of the strain PS that are responsible for the formation of the mediators involved in biodegradation of MO.

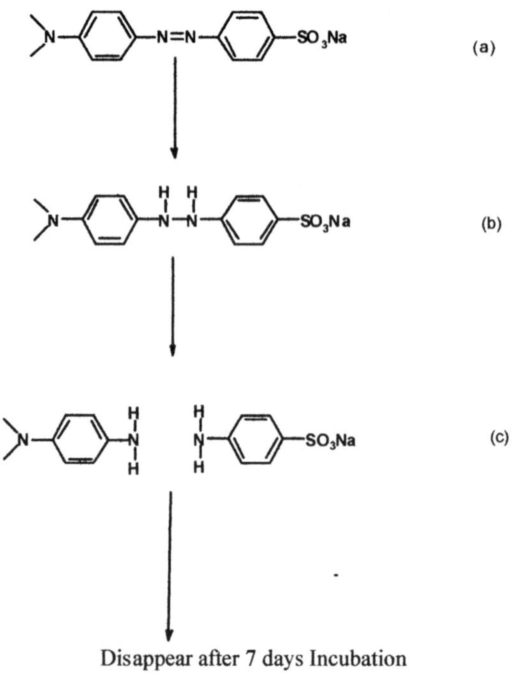

498

(a)

(b)

(c)

Disappear after 7 days Incubation

Figure 3. Proposed pathway for degradation of methyl orange by Bacillus Sp strain PS. a) Methyl orange; b) N, N-dimethyl, 1, 4-phenylene diamine Dihydrochloride; c) sulfanilic acid sodium salt

Acknowledgments

We wish to express our sincere gratitude to the research council of Azad and Tehran University for financing this research project.

References

[1]   Heinfling, A., Martinez, M. J., Martinez, A. T., Bergbauer, M. and Szewzyk, U. Applied and Environmental Microbiology 64(8), 2788-2793 (1998).
[2]   Seong, J. K and Makoto, S. Applied and Environmental Microbiology 65(3),1029-1035 (1999).
[3]   Abadulla, E., Tzanov, T., Costa, S., Robra, K ., Cavaco, A and Gubitz G. Applied and Environmental Microbiology, 66(8), 3357-3362 (2000).
[4]   Pagga, U., Brown, D. Chemosphere, 15,. 479-491 (1986).
[5]   Rogalski, J., Lundell, T., Leonowicz, A and Hatakka A., Acta Microbiol Polonica .40, 221-234 (1991).
[6]   Karapinar, K., Karagi, F., Mcmullan, G., and Marchant, R. Biotechnology Lett, 22, 1179-1181 (2000).
[7]   Arslan, I. Journal of Hazardous materials, B 85, 229-241(20001).
[8]   Norteman, B., Bauumgarten, H., Rast, G and Kanackmuss H. Applied and Environmental Microbiol, 52, 1195-1202(1986).
[9]   Nigam, P., Mcmullan, G., Banat,   I. and Marchant, R. Biotechnol Lett, 18, 117-120 (1996).
[10]   Pourbabaee, A. A., Sarbolouki, M. N, Malekzadeh,. Sent to: "Water Research", Sep. (2004).

[11]   Pasti, G., Paszczynski, S., Goszcynski, D and Crawfored, R.Applied and
         Environment Microbiology, 58(11), 3605-3613 (1992).
[12]   Silk, B., Matthias, C., Martina, L., Andreas, S and Hans, J. Applied and
         Environmental Microbiology, 64(6), 2315(1998).
[13]   Chung, K and Stevens, S and Cerniglia, C. Crit. Rev. Microbiol. 18, 175-
         190(1992).
[14]   Kulla, H., Klausener, F., Meyer, U., Ludeke, B and Leisinger, T. Arch .
         Microbiol, 57(135), 1-7(1991).
[15]   Huag, W., Schmidt, A., Nortemann, B., Hempel, D., Stolz, A and
         Knackmuss, H. Applied and Environmental Microbiology, 57(11), 3144-
         3149(1991).

Degradation of Polyethylene film strips by soil fungal isolates

Garima Sharma, Ben.M.J.Periera, Ramasare Prasad[1]

Molecular Biology and Proteomics Lab, Department of Biotechnology, Indian Institute of Technology Roorkee, Roorkee-247667, Uttaranchal, India.

[1] Corresponding Author: e-mail: rapdyfbs@iitr.ernet.in, Tel: +91-1332-285791 Fax: +91-1332-273560

Total twelve fungal isolates were isolated. Polyethylene degradation ability of these isolates was evaluated by weight loss, tensile strength measurement and Scanning electron microscopic study of plastic strips after 30 days incubation in pure shake culture conditions. Isolates number F4, F8 and F11 showed 63.7, 60, and 55 % reduction in tensile strength of strips, respectively compared to control, while no much change was observed with other isolates. All the isolates found to colonize the surface of plastic strips, but the extent of colonization varied a lot from isolate to isolate. However, the biodegradation signs were found only with isolates F4 and F8. SEM of surfaces of strips inoculated with these two isolates showed heavy colonization, striations, tearing and holes formation on the surface, confirming their biodegradation ability. These two isolates also showed high activities of esterase and lignolytic peroxidase in the culture broth. Since possible role of these enzymes has been suggest in polyethylene degradation though the exact mechanism yet not fully known. It is suggested the degradation abilities of the isolates F4 and F8 is attributed to their high extra cellular enzyme production which is discussed.

Keywords: Biodegradation, SEM, Esterase, Mn-peroxidase , polyethylene strips,   Fungi

1   Introduction

The drastic increase in the production and consumption of non-biodegradable plastic materials has not been accompanied by a corresponding development of procedures for its safe disposal or degradation. As a result plastic wastes, accumulating in the environment at a high rate and emerging as a major threat to terrestrial and water lives [1]. There are various methods of plastic waste disposal like mechanical, chemical recycling, incineration and landfills, but these methods have number of disadvantages and further add to environment pollution [2]. There could be two possible ways to overcome this problem, one to develop synthetic polymers susceptible to biodegradation and the other to develop the latent ability of microorganisms to degrade presently used plastics. Past two decades have shown a growing interest in the development of biodegradable plastics to enhance biodegradability of plastic product in landfills and compost [3, 4]. Due to its relatively resistant nature to microbial attack, the biodegradation of polyethylene found to be a very slow process under natural conditions [5]. However, it has been reported that pretreatment of plastic prior such as UV photo oxidation [6], thermal oxidation [7], or chemical oxidation [8] prior to exposure to microorganisms enhanced the biodegradation. Jen-hou and Schwartz [9] and Albertsson and Banhidi [10] reported the evidence of biodegradation of low density polyethylene (LDPE) and high density polyethylene (HDPE) films, respectively by microorganisms. During recent few years there has been several reports of polyethylene film degradation by fungi [11–15], bacteria [16–18] and streptomyces [19, 20] suggesting that degradation of polyethylene is plausible. It has been reported that polyethylene degrading ability by lignin-degrading fungus *Phanerochaete chrysosporium* and *Trametes versicokor* were due to there manganese peroxidase activities [12, 20, 21] and some cases due to extra-cellular esterase activity which help in colonization and biofilm formation by degrading plasticizers and other additive used [16].

Our emphasis has been on the second alternative and an attempt being made to isolate the potential fungi which has ability to degrade the commonly used polyethylene plastics. In the present paper, we describe the isolation of fungal isolates from the soil scrap of the plastic surface buried at dump sites. Degradation study plastic films were carried out under pure culture system. The biodegradable abilities of the fungal isolates were evaluated by changes in plastic strip weight and tensile strength measurement, SEM study and production of hydrolytic enzymes.

2   Materials and methods

2.1 Materials

All chemicals used were of analytical grade and purchased either from Sigma (USA), Merck (Germany), or Hi Media (India). Polyethylene film used in the study was type-2 (HDPE) carry bags, 20.3 × 25.4 cm size and 20 micron thick and was purchased from Carry n Packers, Surat, India.

2.2 Isolation, growth, and culturing of microorganisms

Polyethylene (HDPE) film degrading microorganisms were isolated from the soil adhering to the buried plastic waste collected from the dump sites spread plate method on Potato Dextrose Agar (PDA). Isolates were preserved at −80 °C in glycerol media as stock and were also maintained on PDA slants by sub culturing every 28 days for routine use. The polyethylene films were pretreated and disinfected as described [22]. Heat treated, disinfected and weighed plastic strips (size 4 × 1 inch) were put aseptically in broth containing 1.0% (w/v) malt extract in a nitrogen–free mineral salts solution (5.03g Na_2HPO_4, 1.98g KH_2PO_4, 0.2 g $MgSO_4.7H_2O$, 0.2g NaCl, 0.05g $CaCl_2.2H_2O$ and 1ml of trace element solution per liter of deionized water, pH 5.5). Plastic films containing culture media were then inoculated with 1 ml of spore suspension of the respective fungal isolates and were incubated with shaking at 120 rpm for 30 days. After 30 days incubation plastic strips were harvested , washed in 70% ethanol to remove cells mass from the residual films as much as possible and then dried at 45 °C overnight. For control sterile plastic strips were incubated in culture media under similar conditions without inoculation.

2.3 Evaluation of degradation of plastic films

Biodegradation was studied by weight loss, changes in tensile strength and SEM. After drying at 45°C overnight, the 30 days incubated strips were weighed after equilibration to room temperature and compared with those of the initial weight of the respective strips. The difference in the weight was determined to see the weight loss. Changes in the tensile strength were determined on an Instron Universal model 4301 at room temperature and 500 mm per min with a 5 cm gap. All samples were equilibrated to 50 % relative humidity for at least 40 h before analysis (ASTMI 882-83, Standard Test Method for Tensile Properties of thin Plastic Sheet). The strips were fixed in 7.5% glutaraldehyde for 2h, dehydrated by serial washing twice in 70% and 90% alcohol for 30 min each, and finally in absolute alcohol overnight. The strips were then dried under vacuum and shadowed with Gold and

502

examined under SEM (Leo 435 VP, England) and photographed. Since peroxidase and esterases has been suggested to involved to some extent in polyethylene degradation, the esterase and Mn–peroxidase activities were measured as per standard methods as described [12, 23].

3   Results and discussion

Biodegradation of plastic is found to be safe and environmental friendly approach for management of plastic waste. Therefore, there is always an emerging need to found more potent microorganisms. In the present work our emphasis was to isolate potential fungi which could degrade more commonly used polyethylene films (HDPE), due to its ability to grow and survive under most adverse conditions [24]. The soil adhering to plastic surface obtained from plastic waste dumpsites was chosen to isolate the fungi with the logic that this may most probably contain potent plastic degrading microorganisms. A total 12 fungal isolates were obtained, and named fungal isolates F1 to F12, respectively. The isolates were partially identified by their colony morphology, color, growth on selective media and some standard biochemical tests. The growth morphology of the isolates is summarized in **Table-1**. Since the two of the isolated F4 and F8 showed better colonization and biofilm formation on polyethylene films they were exclusive identified at National Facility for Microbial Culture Collection and Identification, IARI, New Delhi, India and identified to be *Aspergillus fumigatus* (ID.No 6125.05) and *Aspergillus flavus* (ID No 6127.05), respectively.

Degradation potentials of these isolates were carried in a pure culture biodegradation assay. There are several reports of biodegradation study in pure culture conditions [19, 16, 22]. The average weight data of the plastic strips before and after 30 days of incubation at 30°C showed loss in weight for fungal isolates F4, F6, F9, F10, and F12, while no significant weight notice with others **(Table 2)**. Similar observations have been reported in earlier studied and the variation in weight after incubation is mainly attributed to the microbial cell mass accumulation and biofilm formation on the plastic surfaces [13]. However, the tensile strength measurement data found to be more conclusive **(Table 3)**. There found to be 63.7, 60 and 55% reduction in the tensile strength of the plastic strips after incubation for 30 days with fungal isolates F4, F8 and F9, respectively. Reduction in tensile strength of plastic films after incubation with biodegradable microorganisms has been reported in numerous earlier studies [19, 22, 25].

Table 1. Colony morphology of fungal isolates

Organisms	Appearance		Color		
	Margin	Colony form	Mycelium	Spore	Media
Fungal Isolate F1	Irregular	Irregular	Dirty Green	Dirty Green	No pigment
Fungal Isolate F2	Regular	Regular	Brown	Brown	No pigment
Fungal Isolate F3	Regular	Regular	Light yellow	Brown	No pigment
Fungal Isolate F4	Thread-Like	Round	Yellow	Black	No pigment
Fungal Isolate F5	Thread-Like	Round	Cream	Cream	No pigment
Fungal Isolate F6	Thread- Like	Round	Cream	Cream	No pigment
Fungal Isolate F7	Round	Hairy	Light grey	Grey	No pigment
Fungal Isolate F8	Round	Irregular	Brownish Yellow	Brownish	No pigment
Fungal Isolate F9	Round(Hairy)	Hairy	White	White	No pigment
Fungal Isolate F10	Hairy(Thread Like)	Hairy	White	White	No pigment
Fungal Isolate F11	Thread-Like	Round	Green	Green	No pigment
Fungal Isolate F12	Thread-Like	Round	Yellow	Black	No pigment

Table2. The changes in the average weight of the plastic strips after 30 days incubation with respective fungal isolates

Organisms	Initial weight (in grams)	Weight after incubation (in grams)
Fungal Isolate F1	0.088	0.093
Fungal Isolate F2	0.087	0.091
Fungal Isolate F3	0.087	0.088
Fungal Isolate F4	**0.087**	**0.080**
Fungal Isolate F5	0.086	0.087
Fungal Isolate F6	0.086	0.084
Fungal Isolate F7	0.086	0.087
Fungal Isolate F8	**0.087**	**0.082**
Fungal Isolate F9	0.085	0.084
Fungal Isolate F10	0.088	0.086
Fungal Isolate F11	0.086	0.085
Fungal Isolate F12	0.087	0.085

Table3. Percentage reduction in tensile strength of plastic strips after 1 month of incubation with respective fungal isolates in comparison to control uninoculated strip.

Organisms	% reduction in Tensile strength
Fungal Isolate F1	7.2
Fungal Isolate F2	14.3
Fungal Isolate F3	5.15
Fungal Isolate F4	**63.7**
Fungal Isolate F5	11.43
Fungal Isolate F6	22.86
Fungal Isolate F7	5.72
Fungal Isolate F8	**60**
Fungal Isolate F9	17.15
Fungal Isolate F10	15.72
Fungal Isolate F11	55
Fungal Isolate F12	11.43

The colonization and roughness of plastic surface is indication of biodeterioration of polyethylene strips. SEM observations revealed that all the isolates found to colonize the surface of plastic strips, however, the clear degradation features such as shearing, tearing and holes formation on the surfaces could be seen only with isolates F8 **(Fig 1)**. This conclusion is based on the definition of Hueck, 1965 as cited by Eggins [26], of biodeterioration as " any undesirable change in the properties of a material of economical importance caused by the activities of organisms." The shearing, tearing and holes formation on the surfaces polyethylene films have been reported as evidence for degradation abilities of microbes in several studies in recent past [16, 27–29]. Polymer degradation abilities of microbes are suggested to be associated to their hydrolytic enzyme system [30}. The role of extra cellular esterase and lignotytic peroxidases mainly Mn Peroxidase in biodegradation of polythylene have been reported previously [12, 16, 20, 21] When extra cellular esterase activity was assayed in the cell free culture broths, significantly high esterase and Mn-perxidase activities were observed in the isolates F4 and F8 inoculated broths without use of any inducer in the medium (data not shown here). Since the isolates with high esterase and peroxidase activities also showed to have better colonization of surface degradation features as evidence from SEM and tensile strength. Since the major obstacle in polyethylene degradation is colonization of surface by microorganism due to hydrophobicity. In addition a number of plasticizers and additives are added to polymer for making smooth surface, increase durability, and making suitable for ink printing. It is also observed that change of surface from smooth to rough enhanced colonization and biodegradation. Though esterase do not play major role in polyethylene degradation unlike polyester based polymers, however, it is suggested that esterase help in initial stage by attacking small number of ester bond which are present due to plastisizers and making the strips vulnerable for more microbial biofilm formation and later degradation by Mn-peroxidase or other oxidative enzyme which play major role in polyethylene degradation. . FTR analysis of the polyethylene strip used in the present study found to have ester group (data not shown here) this support the role of esterase. Possible role of esterase and lignin peroxidases in polyethylene degradation has also been suggested in earlier studied [12, 16, 20, 21]. Thus it is concluded that the better colonization and degradation feature on polyethylene surface by fungal isolate F8 is due to their high

504

esterase and Mn-peroxides production ability. However, its need to be further investigated and study in progress.

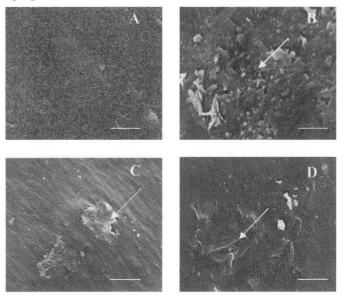

Fig.1. Scanning electron micrographs of plastic strips inoculated with fungal isolate (F8). A-SEM of control strip, B, C and D are SEM of inoculated strips. Colonization, shearing and tearing of strips are shown with arrows. (Magnification 1.5 KX) (Scale bar, 3µm)

Acknowledgements

Authors wish to thank the Ministry of Human Resource Development (MHRD, Govt. India); for its financial assistance in the form of Institutional fellowship to Ms Garima Sharma during the period of this work.

References

[1]   Shimao, M., Current opinion in Biotechnology **12**, 242 (2001).
[2]   Flechter, A. (4), In: Plastics from Bacteria and for Bacteria: PHA as Natural, Biodegradable Polyesters. (Springer Verlag, New York, 1993) ,pp. 77-93.
[3]   Gamal, R.S., Seliger, H., Polymer Degradation and Stability **83**, 101 (2004).
[4]   Masayuki, S., Current Opinion in Biotechnology **12**, 242 (2001).
[5]   Otake, Y., Kobayashi, T., Ashabe, H., Murakami, N., Ono, K. Journal of Applied Polymer Science **56**, 1789 (1995).
[6]   Cornell, J.H., Kaplan, A.M., Rogers, M.R., Journal of Applied Polymer Science **29**, 2581 (1984).
[7]   Volke-Sepulveda, T., Saucedo-Castaneda, G., Gutierrez-Rojas, M., Manzur, A., Favela-Torres, E., Journal of Applied Polymer Science **83**,305 (2002).
[8]   Brown, B.S., Mills, J., Hulse, J.M., Nature **250**, 161 (1974).
[9]   Jen-hou, L., Schwartz, A., Molekulargewichts Kunststoffe **51**, 317 (1961).
[10]   Albertsson, A,C, Banhidi, Z.G., Journal of Applied Polymer Science **25**, 1655 (1980).

[11]   Virginia, C. C., Rodolfo, L. M., The Philipine Journal of Science **125**, 117 (1997).

[12]   Liyoshi , Y., Tsutsumi, Y., Nishida, T., Journal of Wood Science **44**, 222(1998).

[13]   Jolanta, P., Bozena, N., Sylwia, T., Biotechnologia **4 (63)**, 214 (2003).

[14]   Kathiresan, K., Revista De Biologia Tropical **51**, 629 (2003).

[15]   Upreti, M.C., Srivastva, R.B., Current Science **84**, 1399 (2003).

[16]   Gilan(Orr), I., Hadar, Y., Sivan, A., Applied Microbiology and Biotechnology **65**, 97 (2004).

[17]   Orhan, Y., Jasna, H., Hanife, B., Acta Chim Slov. **51**, 579 (2004).

[18]   Hadad, D., Geresh, S., Sivan, A., Journal of Applied Microbiology **98**, 1093 (2005).

[19]   Byungate, L., Anthony, L.P., Alfred, F. & Theodore Jr, B.B., Applied and Environmental Microbiology **57**,   678 (1991).

[20]   Anthony, L., Pometto, H.I., Lee, B., Johnson, K.E., Applied and Environmental Microbiology **58**, 731 (1992).

[21]   Ehara, K., Yuka, L., Yuji, T., Journal of Wood Science **46**, 180 (2000).

[22]   Hanaa A. El-Shafei, Nadia H. Abd El-Nasser, Amany L. Kansoh & Amal M. Ali, Polymer Degradation and Stability **62**, 361(1998).

[23]   Kordel, M., Hofmann, B., Schomburg, D.,   Schmid, R.D., Journal of Bacteriology **173**, 4836 (1991).

[24]   Barratt, S.R., Ennos, A.R., Greenhalgh, M., Robson, G.D., & Handley, P.S., Journal of Applied Microbiology **95** , 78 (2003).

[25]   Imam, S.H., Gordon, S.H., Shorgren, R. L., Tosteson, T.R., Govind, N.S., Greene, R.V., Applied and Environmental Microbiology **65**, 431(1999).

[26]   Eggins, H.O.W., Mills, J., Holt, A., Scott, G., Biodeterioration and biodegrdation of synthetic polymers. In: Sykes and Skimmer (eds), Microbial aspects of pollution. The society of Applied Bacteriology Symposium, Series No 1 (Academic Press London New York, 1971) 267-279.

[27]   Yoko Tezuka, Nariaki Ishii, Ken-ichi Kasauya, Hiroshi Mitomo, Polymer Degradation and Stability

[28]   Kim, D. Y., Rhee, H.Y., Applied and Microbial Biotechnology **61**, 300 (2003).

[29]   Jeremy S. Webb, Marianne Nixon, Ian M. Eastwood, Malcolm Greenhalgh, Geoffrey D. Robson, Pauline S. Handley, Applied and Environmental Microbiology **66** , 3184 (2000).

[30]   Wales, D.S. and Sagar, B.R. (7th ed), Mechanistic aspects of polyurethane biodeterioration. In Biodeterioration   Houghton, D.R.,Smith, R.N. and Eggins, H.O.W.(Elsevier Applied Science, London,1988), pp. 351-358.

Enhanced biodegradation of polycyclic aromatic hydrocarbons using nonionic surfactants in soil slurries

M. Bueno-Montes and J. J. Ortega-Calvo.

Instituto de Recursos Naturales y Agrobiología, CSIC. Avda. Reina Mercedes, 10, 41012 Sevilla , SPAIN.

The biodegradation of polycyclic aromatic hydrocarbons (PAHs) in soil is often limited by the desorption rate of these compounds. The use of synthetic surfactants to improve bioavailability is an alternative. This research constitutes an integrated study on the effect of an environmentally friendly surfactant Brij 35 ($C_{12}E_{23}$), on the desorption and biodegradation of native PAHs in soil slurries. We employed an historically creosote-polluted clay soil (60 %), inoculation with a PAH-degrading bacterium (*Mycobacterium Gilvum* VM 552) representative of the microorganisms active during bioremediation, and optimal conditions for microbial activity, (shaking, slurrying and nutrients). Tenax extraction of sterilized soil slurries revealed the presence of a desorption-resistant PAH fraction of the same order of magnitude as the residual PAH after biodegradation in systems without surfactant. The results showed that the nonionic surfactant at levels above its critical micelle concentration (CMC) increased the rates of desorption and mineralization of the compounds and reduced the biodegradation-resistant PAH fraction. Polymerase Chain Reaction (PCR) of 16S rRNA gene segments showed that surfactant treatments stimulated the *Mycobacterium sp.* population. We suggest that the addition of biodegradable, non-toxic surfactant may be a useful tool improve the performance of bioremediation of PAH-contaminated areas

Keywords: Bioavailability, biodegradation, PAHs, surfactants.

1   Introduction

The presence of carcinogenic polycyclic aromatic hydrocarbons (PAHs) in soils poses a potential threat to human health. Microbial degradation of PAHs is thought to be the major process involved in effective bioremediation of contaminated soils and sediments [1]. Unfortunately, PAH removal during bioremediation is often incomplete and residual concentration after bioremediation are often too high to satisfy the standards for clean soil. In many cases, these high residual PAH concentrations are caused by the limited bioavailability of PAHs. Reduced bioavailability of pollutants is caused by the slow mass transfer to the degrading microorganisms [2].

In soils and sediments contaminant mass transfer is often described in terms of desorption kinetics. Generally, desorption of PAHs is biphasic, whereby a short period of rapid desorption is followed by a longer period of slow desorption [3–5]. This biphasic behavior also occurs during bioremediation. It is evident that the poorly bioavailable PAH fraction is constituted by slowly desorbing fraction.

The application of surfactants has become an interesting way to influence the mass transfer of PAH. Surfactants consist of a hydrophilic and a hydrophobic moiety and by this tend to concentrate at surfaces and interfaces thereby decreasing levels of surface tension and interfacial tension. Solubilization occurs above a specific threshold, critical micellar concentration (CMC), where surfactant molecules aggregate to micelles. Incorporation of hydrophobic compounds in the micelles is termed solubilization. Studies by Guha and Jaffé [6] indicated that a PAH such as phenanthrene partitioned into the micellar phase of some nonionic surfactants is, to some degree, directly bioavailable.

The aim of this study was to test the suitability of nonionic surfactant Brij 35 to reduce the biodegradation resistant PAH fraction in a clay-rich creosote contaminated soil.

2 Materials and methods

Soils. Soils used in this study were provided by EMGRISA (Madrid, Spain) from a wood-treating facility in Southern Spain, with a record of pollution by creosote of more than 100 years. The sample was prepared by air drying and sieving (2 mm mesh). The resulting soil sample had a moisture of 60 % clay. The soil was contaminated with total PAHs 4540.80 mg/Kg dry soil.

Biodegradation and desorption experiments. Bioavailability of target PAH was assessed by a series of biodegradation and desorption experiments. Biodegradation experiments with soil slurries (1 g/ 70 mL) were performed, as previously described [7], under conditions that optimized biological activity. The inoculated bacterial strain (*Mycobacterium Gilvum* VM552) was able to mineralize phenanthrene, fluoranthene and pyrene. The measurement of $^{14}CO_2$ production from each labelled PAH and the calculation of mineralization rates was performed as previously described [7, 8]. Aqueous stock solution (140 mg/mL) of Brij 35 was prepared in mineral salt sterilised medium. The kinetics of desorption was determined with Tenax®, according to Cornelissen et al. [4]. Analysis was carried out with HPLC.

3 Results and discussions

In biodegradation experiment, parallel measurements of $^{14}CO_2$ production from the five labelled PAH and of residual contents of indigenous PAH by HPLC analysis showed the existence of an indigenous population in the soil, able to slowly metabolise the PAH under laboratory conditions . In figure 1, mineralisation curves showed the sequential respiration of the PAH. After 42 days, a significant fraction of the PAH initially present still remained in the soil, showing a very slow rate of dissipation (9.04 ± 1.75 mg/kg dry soil phe). This biphasic ("hockey stick") kinetics, consisting in an initial period of fast degradation, followed by second, much slower phase, is typical of situations of a limited bioavailability during the second phase [9].

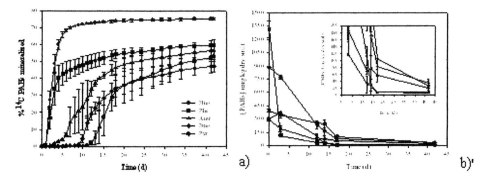

Fig.1 Mineralization of ^{14}C-labelled PAH (a) and residual PAH concentrations (b) in slurries of creosote-polluted soil by indigenous population.

508

The biodegradation of the PAH present in creosote-polluted soil was also determined in an experiment with well-aireated, soil slurries inoculated with *Mycobacterium* strains able to mineralize the target compounds. Figure 2 a) shows the same sequential pattern of compound mineralization was observed, as compared with biodegradation by indigenous population. However, the resulting levels, in mg/Kg dry soil in figure 2 b), were still significant: 7.32 o 0.15 phe, 87.56 o 2.23 ftne, and 36.72 o 2.06 pyr.

Desorption from soil could be described by first-order, two-comparments kinetic model. The desorption model suggests the existence, for each compound, of a rapidly and a slowly desorbing fraction. After 42 days (1008 hours) of continuous desorption, the concentration of PAH remaining in soil was similar to the residual PAH after active biodegradation in soil slurries (e.g. 3,10 ± 0,62 mg/kg dry soil for phen). **Fig.2** Mineralization of ^{14}C-labelled phen (a)

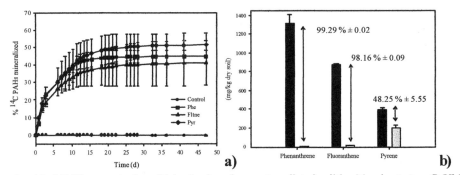

and residual PAH concentrations (b) in slurries of creosote-polluted soil by *Mycobacterium G.* VM 552..

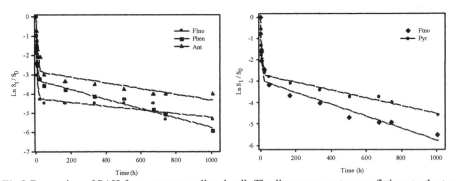

Fig.3 Desorption of PAH from creosote-polluted soil. The lines represent curve fittings to the two-comparments kinetic model.

Figure 4 a shows the mineralization of ^{14}C-phenanthrene in slurries creosote-polluted soil inoculated with *Mycobacterium* strain and with surfactant Brij 35 in different concentrations (>CMC). Brij 35 caused higher maximum mineralization rates than in the systems without surfactant. The presence of Brij 35 increased desorption of creosote PAHs (about 100 %, data no showed) and reduced the resistant fraction after bioremediation (Table 1). Figure 4 b shows PCR analysis of *Mycobacterium.* Brij 35 was found not to inhibit microbial growth at the concentrations studied. The intensity of the bands showed that *Mycobacterium* inoculated were in a concentration higher than the initial population $(D.O_{600nm} = 0.90; 10^7-10^8$ CFU/ml) after 4.2 days.

Table 1 Resistant fraction of PAH after bioremediation with diferent concentration of Brij 35. n.d. no detected. Detection limit: 2.53 mg/kg phe; 2.60 mg/kg ftne and 0.52 mg/kg pyr.

PAHs	Before bioremediation (mg/kg dry soil)	After bioremediation with Brij 35 (mg/kg dry soil)				
		Control	500 µg/ml	10000 µg/ml	20000 µg/ml	40000 µg/ml
Phe	1230.67 ± 13.69	19.28 ± 3.46	9.41 ± 2.41	n.d.	n.d.	n.d.
Ftne	866.57 ± 30.00	114.73 ± 30.69	74.13 ± 5.03	10.16 ± 1.70	7.81 ± 0.13	n.d.
Pyr	391.63 ± 35.49	88.64 ± 1.36	37.97 ± 0.59	n.d.	n.d.	n.d.
B(a)pyr	21.99 ± 1.31	20.87 ± 1.32	14.19 ± 2.54	18.60 ± 1.75	14.81 ± 0.45	14.58 ± 2.14

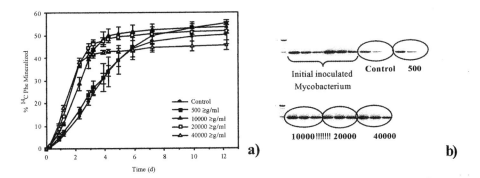

Fig.4 Mineralization of ^{14}C-phenanthrene in slurries creosote-polluted soil inoculated with *Mycobacterium* strain and with surfactant Brij 35 in different concentrations (>CMC) (a). PCR analysis of *Mycobacterium* in presence of Brij 35 (b).

Results show the potential of bioremediation to achieve, under optimized conditions, a significant reduction in PAHs levels. The presence of a desorption – resistant PAH fraction, as revealed by Tenax extraction, of the same order of magnitude as the residual PAH after bioremediation suggests that microorganisms alone can not overcome the bioavailability restrictions imposed by slow desorption. Our results suggest that the use of relatively high concentrations of a nonionic surfactant (Brij 35) is a reasonable alternative to improve bioremediation performance.

References

[1]   C. E. Cerniglia, Biodegradation. **3**, 351 (1992)
[2]   A. J. Beck, S. C. Wilson, R. E. Alcock, and K. C. Jones, Crit. Rev. Environ. Sci. Technol. **25**, 1 (1995)
[3]   G. Cornelissen, P. C. M. van Noort, and H. A. J. Gover, Environ. Toxicol. Chem. **16**, 1351 (1997)
[4]   G. Cornelissen, H. Rigterink, M. M. A. Ferdinandy, and P. C. M. van Noort, Environ. Sci. Technol. **32**, 966 (1998)
[5]   I. T. Yeom, M. M. Ghosh, and C. D. Cox, Environ. Sci. Technol. **30**, 1589 (1996)

510

[6]   S. Guha, P. R. Jaffé, and C. A. Peters, Environ. Sci. Technol. **32,** 2317 (1998)
[7]   J. J. Ortega-Calvo, I. Birman, and M. Alexander, Environ. Sci. Technol. **29,** 2222 (1995)
[8]   J. J. Ortega-Calvo, M. Lahlou, and C. Saiz-Jimenez, Int Biodeterior Biodegrad. **40,** 101 (1997)
[9]   M. Alexander, Environ. Sci. Technol. **34,** 4259 (2000)

Enumeration of naphthalene and phenanthrene degrading bacteria as an indicator of hydrocarbon pollution in surface waters of Guayanilla, Puerto Rico: Impact of seasonal variations

Nydia. J. Rodríguez[1], Ángel M. Nieves-Rivera[1], Baqar R. Zaidi[1]*, and Syed H. Imam[2]

[1] Department of Marine Sciences P. O. Box 9013, University of Puerto Rico, Mayaguez, PR 00681-9013
[2] Bio product Chemistry & Engineering Research, Western Regional Research Centre, ARS- USDA, Albany, CA 9710-1105 USA

*Corresponding author: e-mail: bzaidi@uprm.edu Phone: 1-787-265-5487, Fax: 1-787-265-5487

A study was conducted to determine the microbial potential of Guayanilla Bay surface water to use polyaromatic hydrocarbons (PAHs) e.g. naphthalene and phenanthrene as a sole carbon source. The capacity of surface water for utilization of these compounds was correlated with the population density of naphthalene and phenanthrene degrading bacteria found in coastal water samples as an indicator of hydrocarbon pollution. The number of naphthalene and phenanthrene degrading bacteria decreased with distance from the coast. No PAHs degrading bacteria were found in offshore samples used as control. The number of naphthalene degrading bacteria increased in all samples taken during rainy season. Heavy rainfall causes perturbation of sediments rich in organic matter as evidenced by increased turbidities, providing both the environment and the surfaces conducive for microbial growth and proliferation. These observations were further supported by the presence of low levels of dissolved oxygen in water.

Keywords: Polyaromatic hydrocarbons, biodegradation, naphthalene, phenanthrene, Guayanilla Bay

1   Introduction

Guayanilla Bay located on the south coast (18°N; 67.45°W) of Puerto Rico was the site of one of the biggest petrochemical complexes in the world until it was shut down in 1982. High levels of toxic compounds were found to be present in the Bay water around the petrochemical complex, but only few studies have been done to determine the fate of pollutants at this site or the recovery of the coastal environment of Guayanilla after the closure of the petrochemical complex [1, 2]. Thus, Guayanilla coastal environment is an ideal site to study the fate of toxic chemicals in tropical marine environments.

Most of the information on abiotic factors affecting PAHs biodegradation has come from temperate marine environments [3–7] and are not truly applicable to tropical marine environment. The present work focuses on the study of water quality variables, their effect on potential for PAHs biodegradation in the tropical coastal marine water of Guayanilla. This information will help in the evaluation of the fate of toxic compounds in seawater and predict the damage to coastal environments in Puerto Rico resulting from toxic compounds.

512

2 Materials and methods

Phenanthrene (> 96% purity), naphthalene (99% purity), agarose type VII (low gelling temperature) and cycloheximide were obtained from Sigma chemical Co., St. Louis, MO, and Noble agar from Difco Laboratories, Detroit, MI.

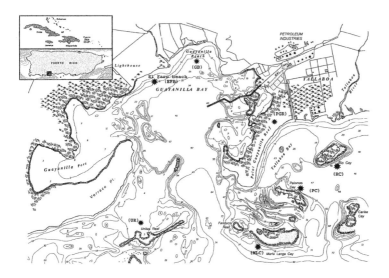

Fig. 1 Study sites (*)
Coral reef is a sensitive indicator of environmental stresses because of its response to the disturbances [8]. Four coral reefs Point Guayanilla (PGR), Río Cay (RC), María Langa Cay (MLC) and Unitas Reef (UR) were selected based on their locations. Additionally three coastal areas, Guayanilla beach (GB), El Faro beach (EFB) and a Paloma Cay (PC) seagrass bed were also sampled while an offshore station was used as control. Offshore station was used as a control because the area is not impacted by industrial pollution.

2.1 Sampling strategy

For water quality parameters the samples from seven stations were collected in April, June, August, and December. For toxic substrates utilization pattern and for enumeration of naphthalene and phenanthrene degrading bacteria the samples from seven stations were collected during month of December (dry period) and June (rainy season). The samples were taken during the moming's hours to minimize the effects of wind, tides, and temperature.

2.2 Water quality variables

Water and air temperature measurements were routinely obtained with a mercury thermometer. Turbidity was measured in Jackson Turbidity Units (JTUs) using a LaMotte Corporation water quality test kits. Briefly, a 50-ml sub-sample from the 1-liter water sample was placed in a 100-ml graduated cylinder. A second graduated cylinder was filled with 50-ml of distilled water. The LaMotte turbidity reagent was added by 0.5-ml increments until the distilled water was of same turbidity as the sample, as determined by a visual comparison of a black dot on a white surface in the bottom of the cylinders. The number of drops was converted to Johnson Turbidity Units (JTUs) using the conversion factor provided by LaMotte Company.

2.3 Enumeration of bacteria

The enumeration of bacteria capable of degrading naphthalene and phenanthrene was done by the agar overlay technique [9; Zaidi, 10].

3   Results

There were no significant differences between the sampling stations and control stations for temperature, salinity, and pH. However, the highest turbidity sample was either adjacent mangrove lagoon (Point Guayanilla reef) or near Guayanilla River (El Faro and Guayanilla beach). The lowest average DO value was found in Point Guayanilla reef, the sampling station with the highest turbidity. No significant difference in DO values was found in any of the other sampling stations including control.

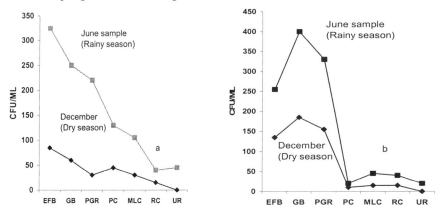

Fig. 2. Enumeration of naphthalene (a) and phenanthrene (b) degrading bacteria from sampling sites of El Faro beach (EFB), Guayanilla beach (GB), Paloma Cay (PC), María Langa Cay (MLC), Río Cay (RC), and Unitas reef (UR).

Degradation capacity of water samples collected from various locations was determined by estimating the presence of bacteria specific in their ability to degrade toxic compound. Bacterial colonies of at least two compounds, namely, naphthalene, and phenanthrene (10 µg/ml) were enumerated by agar overlayer technique during the month of June (rainy season) and December (dry season). In El Faro beach sample 2.4×10^2 cells/ml of naphthalene degrading bacteria were found in June samples compared to 8.5×10^1 in December (Fig. 2a). The bacteria in both and June and December samples declined as one moved farther from the coast. Thus, in June sample their number decline from 2×10^2 cells/ml found in El Faro beach water samples to 4.5×10^1 in Unita reef. Similarly, December sample the cell number declined from 8.5×10^1 in El Faro beach samples to non-detectable levels in Unitas reef. Phenanthrene degrading bacteria also declined from high 2.15×10^2 in Guayanilla beach to 1.5×10^1 in Unitas reef in June samples and from 1.85×10^2 cells/ml in Guayanilla beach samples to undetectable levels in Unita reef in

December (Fig. 2b). However, unlike naphthalene degraders more phenanthrene degrading bacteria were found in Guayanilla beach instead of El Faro beach samples both in rainy and dry season (Fi.g. 2b). The number of naphthalene degrading bacteria showed an overall increase in all sampling sites during rainy season compared to dry season. When each sampling site is compared , the average number of naphthalene degrading bacteria increased from 60 colonies to 380 in Pt. Guayanilla reef, from 120 colonies to 380 in Guayanilla beach anf from 170 to 480 colonies in El Faro beach samples (Fig. 2a). However, in contrast to naphthalene degrading bacteria only a slight overall increase was observed for phenanthrene degrading bacteria during the rainy season (Fig. 2b). No naphthalene or phenanthrene degrading bacteria were found in offshore (control) samples.

Through enrichment techniques, four bacterial strains capable of degrading naphthalene and phenanthrene were isolated, purified and identified by standard microbiological techniques. These bacteria were isolated from El Faro and Guayanilla beach. *Pseudomonas* strain N-1 was isolated from El Faro beach and is capable of degrading naphthalene, while *Pseudomonas* strain N-2 was isolated from Guayanilla beach and is capable of degrading phenanthrene. Another strain capable of degrading naphthalene was isolated from Guayanilla beach and belong to the genus *Aeromonas*. Another strain isolated from El Faro beach and capable of degrading phenanthrene remained unidentified.

4 Discussion

Over all no significant differences were found in temperature, salinity, and pH. As expected most turbid samples were from the ones either collected near the beach or mangrove channels. Generally turbidity is caused by suspended solids such as clay, silt, plankton, industrial waste or sewage in the water. Particularly, in coastal areas, the turbidity could also result from sediment input from land after heavy rainfall which was evidenced from samples collected in June. The samples were collected after unusually heavy rains of 9.8 inches during the month of May compared to 16.9 inches for the rest of 11 months combined. The dissolved oxygen (DO) values in these samples were also low, indicating higher microbial activity due to input and mixing of sediments after heavy rains. Thus presence of higher number of phenanthrene and naphthalene degrading in these samples in rainy season compared to dry season essentially confirm this observation.

Acknowledgements

The authors thank Mr. Peter Rocafort for his assistance in preparation of site map. The research was supported in part by Sea Grant # R-96-1-00, University of Puerto Rico, Mayaguez, PR.

References

[1]   B. R. Zaidi, and S. H. Imam, Mar. Pollut. Bull. 38, 737(1999).
[2]   B. R. Zaidi et al., Mar. Pollut. Bullet. 46, 418 (2003).
[3]   J. E. Baur, and D. G. Capone, Appl. Environ. Microbiol. 50, 81 (1985).
[4]   R. M. Atlas, Microbiol. Rev. 45, 180 (1981).
[5]   R. P. Kerr, and D. G. Capone, Mar. Env. Res. 26, 181 (1988).
[6]   G. Roubal, and R. M. Atlas, Appl. Environ. Microbiol. 35, 897 (1978).
[7]   M. P. Shiaris, Appl. Environ. Microbiol. 55, 1391 (1989).

[8]   J. Morelock, G. Galler, and K. Boulon, in: Proceedings of the symposium on energy industry and the marine environment in Guayanilla Bay, CEER University of Puerto Rico, US Dept. of Energy, Mayaguez, PR (1979).

[9]   A. H. Bogart, B. B. Hemmingsen, Appl. Environ. Mirobiol. **58**, 2579 (1992).

[10]   B. R. Zaidi, R. V. Greene, and S. H. Imam, (eds.), Biopolymers: Utilizing Natur's Advanced Materials (Oxford University, NY 1999), chap. 13.

Formation of Cr(V) and Cr(III) in *Arthrobacter oxydans* exposed to high concentrations of Cr(VI)

T. L. Kalabegishvili[1], N. Y. Tsibakhashvili[*,1], I. G. Murusidze[1]
D. T. Pataraya[2], M. A. Gurielidze[2], H.-Y. N. Holman[3]

[1] Andronikashvili Institute of Physics, Georgian Academy of Sciences, Tbilisi, Georgia
[2] Durmishidze Institute of Biochemistry and Biotechnology, Georgian Academy of Sciences, Tbilisi, Georgia
[3] Center for Environmental Biotechnology, Lawrence Berkeley National Laboratory, Berkeley, CA, USA

[*] Corresponding author: e-mail: ntsiba@gol.ge, Phone: +995 32 396716, Fax: +995 32 391494

Electron spin resonance (ESR) was applied to evaluate the potential of *Arthrobacter oxydans* to detoxify Cr(VI) from heavily contaminated environment under aerobic conditions. Results from our batch experiments show that *A. oxydans* reduces Cr(VI) through the formation of Cr(V) complexes at the surface of bacteria in the presence of oxygen. Further transformation of Cr(V) to Cr(III) takes place inside bacterial cells. Reduction of Cr(VI) by this mechanism is found to be efficient in the concentration range of 50-500 mg/L and becomes ineffective at higher doses (above 700 mg/L).

Keywords: *Arthrobacter oxydans*; ESR; Cr(V); Cr(III); bioremediation;

1 Introduction

Chromium is a widespread environmental pollutant. The two oxidation states of chromium commonly found in the environment are trivalent [Cr(III)] and hexavalent [Cr(VI)] chromium, which have widely contrasting toxicity and transport characteristic [1]. Cr(VI) compounds are highly water soluble and toxic species, while most Cr(III) compounds are less water soluble and its availability to organisms are limited. Since metal ions cannot be destroyed in environments, factors which influence the reduction of Cr(VI) to Cr(III) compounds can dictate the Cr toxicity to ecological receptors. Indigenous bacteria can detoxify chromium with a multiplicity of mechanisms effecting transformations between the soluble and insoluble forms [2]. Among chromium reducing bacteria *Arthrobacter* species is of interest because of its high potential for the reduction and immobilization of chromium in aerobic environments [3].

The reduction of Cr(VI) to Cr(III) may produce reactive intermediates Cr(V) and Cr(IV), which are known to play an important role in chromium-induced DNA damages [1]. The formation of these compounds and their persistence in environments need to be evaluated carefully should one wish to employ intrinsic microorganisms as a means of remediation chromium polluted subsurface sites. Today there are few studies that examine the formation of Cr(V) species in bacteria, and even fewer laboratory studies have been conducted to investigate how microorganisms respond to chromium stress at higher doses.

In the present work we focused on *Arthrobacter oxydans*, a widespread Gram-positive aerobic bacterium as our model. *A. oxydans* is isolated from Columbia basalt rocks (USA), that have been polluted with mixture of heavy metals, radionuclides and organic compounds [4]. Unlike our earlier work [5], we applied electron spin resonance (ESR) to study Cr(V) and Cr(III) formation by *A. oxydans* from Cr(VI) at higher doses. Specifically, this work is an extension of our studies [3–7], where the mechanisms associated with the Cr(VI) reduction by *A. oxydans* have been elucidated partly.

2 Experimental

2.1 Bacterial growth conditions and sample preparation

Cells of *A. oxydans* were grown aerobically as a batch culture in a nutrient medium as described in [5]. Cr(VI) [as K_2CrO_4] was introduced into the bacterial medium at the beginning of the stationary phas e of growth to provide the chromium concentration within a range of 50-1000 mg/L. Cells were harvested by centrifugation (10 000 rpm for 15 min at $4\pi C$) five days later. All chemicals were ACS-reagent grade and purchased from Sigma (St. Louis, MO, USA).

2.2 Electron Spin Resonance (ESR) Experiments

ESR was the key technique employed in this study. ESR measurements were carried out on the RE 1306 radio spectrometer (Russia) as described previously [5]. Registration of Cr(V) and Cr(III) were carried out at liquid nitrogen temperature (77 K) to avoid a decrease in sensitivity of ESR spectrometer caused by water content in bacterial samples.

3 Results and discussion

Our earlier work showed that during reduction of Cr(VI) to Cr(III) by *A. oxydans*, reactive intermediates (such as Cr(V), Cr(IV)) were generated [5]. In this work we investigated the mechanism of Cr(VI) reduction to Cr(III) by *A. oxydans* with a focus on the dose dependence of Cr(V) and Cr(III) formation. Results are presented in Figs. 1–2.
In Figure 1, Cr(V) and Cr(III) were detected concomitantly. The Cr(V) ESR line is characterized with a g-factor of 1.980 and a width of 12 G [5, 6]. The Cr(III) ESR signal in Fig. 1 with a g-factor of 2.02 and a line width of 650 G corresponds to the next oxidation state of chromium - Cr(III) [5]. The spectra in Fig. 1 illustrate that formation of Cr(V) and Cr(III) complexes increased with doses. The reduction of Cr(VI) began at the surface of the *A. oxydans* cells [5]. The surface macromolecules of *A. oxydans* could act as an electron donor to Cr(VI) to form a relatively stable Cr(V)-diol complexes. It is known that Cr(V) complexes can easily penetrate into the bacteria through their cell walls [1]. It is also known that the cell walls are impermeable to most of Cr(III) species. Exposure of cells to different model Cr(V) complexes resulted in similar chromium permeability under similar experimental conditions [1,8]. These circumstances suggest the following mechanism of the reduction of Cr(VI) to Cr(III) by *A. oxydans*. At first Cr(V) is formed at the surface of bacteria, then, penetrating through the cell-wall, Cr(V) is reduced to Cr(III) inside the bacteria. Consequently, one can presume that the formation of Cr(V) and its further reduction to Cr(III) must be two processes separated in space in our system. The dynamics of the formation and accumulation of Cr(III) and C(V) complexes provided by the ESR data, which are presented in Fig. 2, confirmed the above mentioned mechanism.

A rapid increase in the Cr(V) ESR signal intensity in the range from 500 to 700 mg/L of the Cr(VI) concentration (Fig. 2) implies an increase in the Cr(V) concentration. After such a significant "jump", the concentration growth saturates and remains almost unchangeable in the range from 700 to 1000 mg/L of the Cr(VI) concentration.

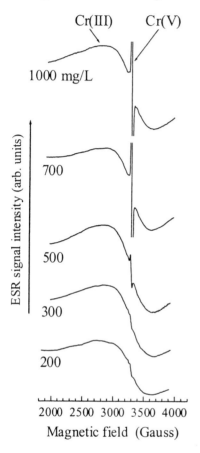

Fig.1. ESR spectra of the formation of Cr(V) and Cr(III) from *A. oxydans* cells recorded five days after growing in the nutrient medium containing different concentrations of Cr(VI). Chromate solution was added to the nutrient medium at the early stationary phase. Spectrometer settings: microwave frequency – 9.15 GHz, modulation frequency – 100 kHz, microwave power – 25 mW, modulation amplitude – 20 G, time constant – 0.3 s, field set – 3000, sweep width – 2000 G, scan time – 20 min. Samples were measured at 77 K. The spectra have common scales.

The Cr(III) ESR signal intensity in Fig. 2 shows that the Cr(III) concentration increases continuously within the range 50–500 mg/L, and the concentration of Cr(III) saturates earlier than that of Cr(V), before the Cr(VI) concentration reaches the value 700 mg/L. In the range of Cr(VI) concentrations from 700 to 1000 mg/L, both Cr(V) and Cr(III) remain almost unchangeable, as well as the concentration of Cr(VI) in the nutrient (concentration of Cr(VI) in the nutrient medium was measured by ion interaction reversed phase HPLC method. Data are not shown). This means that at higher concentrations of Cr(VI) both the formation of Cr(V) complexes and their reduction to Cr(III) are suppressed. We propose that the excessive amount of Cr(V) and Cr(III) formed in the interval 500–700 mg/L may have stressed the bacterial cells and made them loose their ability to accept the newly formed Cr(V) complexes and transform them into Cr(III). As a result Cr(V) accumulated, while the concentration of Cr(III) remained almost unchanged. Further increase in the Cr(VI) concentration (above 700 mg/L) apparently inhibited the mechanism of the reduction of Cr(VI)→Cr(V) too. Cr species, that penetrate bacterial cell began to react with intracellular reductants (both enzymatic and nonenzymatic), and generated unstable radical species. Any of these Cr(V/IV/III) or radical species might modify bacterial cell wall. On the other hand, they could also modify specific nucleotides, thus affecting gene expression. It is known that protein-Cr(III)-DNA, amino acid-Cr(III)-DNA, and DNA-Cr(III)-DNA cross-links form in cells exposed to Cr(VI) [1]. Destruction of the reducing powers inhibits metabolism of bacterial cells, and thus results in decrease and eventual termination of Cr(VI) reduction. Mutagenicity of Cr in prokaryotic cells is exhibited across a number of bacterial species (such as *Escherichia coli*, *Bacillus subtilis*, or *Salmonella typhimurium*) [9]. Certain bacteria have the ability to become tolerant to Cr(VI) uptake. Bacterial strains that were tolerant to Cr(VI) showed changes in the composition of cell walls and decreased $(CrO_4)^{2-}$ uptake [10]. An induced tolerance to Cr(VI) and other heavy metals by some bacteria is thought to be due to a co-expression of efflux proteins [11].

According to estimations for Cr(III), at 1000 mg/L of Cr(VI) in our experiments we have about 10^{20} spin per gram of dry sample that corresponds to 10^4 ppm, i.e., the concentration of Cr(III) inside the cells is about an order of magnitude higher than that of extracellular. This value of Cr(III) concentration should be considered as critical for the *A. oxydans*.

Our batch experiments reveal that *A. oxydans* is able to detoxify of Cr(VI) efficiently in the concentration range 50–500 mg/L. We observed that reduction of Cr(V) to Cr(III) proceeds with the formation of Cr(III) inside the cells. Specifically, we found that *A. oxydans* cells maintained the high survival at the concentrations of Cr(VI) (700–1000 mg/L) (the viability was detected by cell growth on agar plates with a cell suspension dilution. Data are not presented here). The obtained results indicate that exposure of *A. oxydans* to Cr(VI) above 500 mg/L for five days alter the primary function of the natural system. Cr(VI) transformation mechanism by *A. oxydans* at this concentration range (700–1000 mg/L) needs future investigations in detail. These experiments are underway.

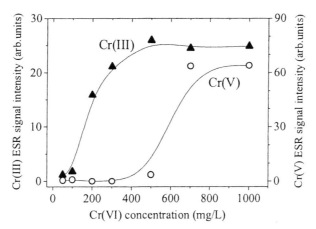

Fig. 2. Formation of Cr(V) and Cr (III) in *A. oxydans* are found to be concentration-dependent. Spectrometer settings: microwave frequency – 9.15 GHz, modulation frequency – 100 kHz, microwave power – 25 mW, time constant – 0.3 s.
For Cr(V): Field set – 3300 G, sweep width –100 G, modulation amplitude – 0.2 G, scan time – 5 min.
For Cr(III): Field set – 3000 G, sweep width –2000 G, modulation amplitude – 20 G, scan time – 20 min. Samples were measured at 77 K.

Acknowledgements

This work was funded through Project GE-B2-2597-TB-04 from the U.S. Civilian Research and Development Foundation (CRDF). We gratefully acknowledge Prof. M. Rukhadze for providing unpublished results on HPLC and Dr. M. Janjalia for her help in ESR measurements.

References

[1]   A. Levina et al., Chromium in Biology: Toxicology and Nutritional Aspects, in: Progress in Inorganic Chemistry, v.51 (Ed. By K.D. Karlin, John Willey & Sons, In, 2002), p.233.
[2]   M. Bruins, S. Kapil, and F. Oehme, Ecotox. & Environ. Safty. **45**, 198 (2000).
[3]   N. Asatiani et al., Curr. Microbiol. **49**, 321 (2004).
[4]   H.-Y. Holman et al., Geomicrobiol J. 16, 307 (1999).
[5]   T. Kalabegishvili, N. Tsibakhashvili, and H.-Y. Holman, Environ. Sci. & Technol.37, 4678 (2003).
[6]   N. Tsibakhashvili et al., J. Radioanal. Nucl. Chem. **259**, 527 (2004).
[7]   M. Abuladze et al., Fres. Environ. Bull., **11**, 562 (2002)
[8]   C. Dillon et al., Chem. Res. Toxicol. **13**, 742 (2000).
[9]   J. Chen, and O. Hao, Crit. Rev. Environ. Sci. & Technol. **28**, 219 (1998).
[10]   Y. Lu, and J. Yang, J. Cell. Biochem., **57,** 655 (1995).
[11]   G. Vincze et al. Bull. Environ. Contam. Toxicol. **65**, 772 (2000).

Metabolism of cyanate and cyanide in the alkalophilic bacterium *Pseudomonas pseudoalcaligenes* CECT5344

M. D. Roldán[1], V.M. Luque-Almagro[1], M.J. Huertas[1], C. Moreno-Vivián[1], M. Martínez-Luque[1], R. Blasco[2], and F. Castillo* ,[1]

[1] Departamento de Bioquímica y Biología Molecular, Edificio Severo Ochoa, Campus de Rabanales, Universidad de Córdoba, Córdoba, Spain.
[2] Departamento de Bioquímica y Biología Molecular, Facultad de Veterinaria, Universidad de Extremadura, Cáceres, Spain.

*Corresponding author: Francisco Castillo. Departamento de Bioquímica y Biología Molecular, Edificio Severo Ochoa, Campus de Rabanales, Universidad de Córdoba, Córdoba, Spain.
e-mail: bb1carof@uco.es Phone: 957-218318

Pseudomonas pseudoalcaligenes CECT5344 is an alkalophilic bacterium which degrades cyanate, — cyanoalanine, cyanide and cyano-metal complexes. This strain is the first bacterium described that degrades cyanide at alkaline pH, thus avoiding its volatilisation as HCN. Degradation of cyanide involved a mechanism to allow iron acquisition, scavenging of iron by siderophore-mediated competition, since cyanide has a strong affinity for iron. This strain also assimilates ammonium and nitrate as nitrogen sources, although cyanide addition inhibits these assimilatory processes. A cyanase activity which catalyzes cyanate conversion into CO_2 and NH_4^+ was induced in the presence of cyanide. At present studies related with simultaneous cyanate and cyanide metabolisms have not been reported.

1   Introduction

Cyanide is a toxic nitrogen compound for almost all organisms since it binds irreversibly to haem-proteins, such as the cytochromes involved in all known respiratory processes. Mining, metallurgic and jewellery industries produce residues containing high amounts of cyanide and its very stable metal complexes. At present, physicochemical treatments are available for these residues, but they are expensive and also present some collateral undesirable effects. Since cyanide is a natural biodegradable compound, biological treatments are more indicated to eliminate it from industrial effluents [1]. In the cyanide molecule, the oxidation state of C (+2, like that in CO) and N (-3, like that in NH_4^+) makes this compound a poor C source but a good N source for bacterial growth. Nevertheless, a microorganism can metabolize cyanide only when, in addition to a biodegradable pathway to convert cyanide into an assimilative product (NH_4^+), it also contains a cyanide-resistance mechanism (generally an alternative, cyanide-insensitive oxidase) and a system for taking up Fe^{3+} from the medium (siderophores), since Fe forms very stable complexes with cyanide and it is not available for organisms in the presence of cyanide. Treatments to eliminate cyanide from contaminated effluents are based on activated sludges or pure enzymes [2, 3]. Unfortunately, these systems are not useful for complex mixtures of cyanide and metals. Some phytopathogenic fungi, like *Fusarium solani* [4], are able to degrade cyanide, but bacterial biodegradation shows considerable advantages since bacteria are more easily manipulated both at the biochemical and genetic levels. In nature, there are organisms as many plants, bacteria, algae and fungi, which are able to produce cyanide. However contamination problems are mainly due to human activities. Thus, the jewellery industry of Córdoba (Spain) generates a cyanurated waste (residue) with about 20 g L^1 of free cyanide (2 mL of this residue are lethal for a human adult). In addition to free cyanide, this residue contains cyano-metal complexes, making it even more poisonous. In spite of cyanide toxicity, there are organisms able to survive in its presence and some of them are able to use it as a nitrogen source. Several enzymatic pathways have been described for cyanide degradation [1–5]. Most of cyanotrophic microorganisms are able to degrade cyanide at a neutral pH, but under this condition a high concentration of cyanide evaporates as hydrocyanic acid (HCN), a weak acid with a pK_a value of 9.2. Thus, it is very important to isolate cyanotrophic microorganisms that thrive at alkaline pH. In this sense, an autochthonous bacterium from the Guadalquivir River in Córdoba, which is able to degrade free cyanide and cyano-metalic complexes under alkaline conditions (up to pH 12) was isolated. This bacterium, identified as *Pseudomonas pseudoalcaligenes* CECT5344, tolerates and degrades high cyanide concentrations (up to 30 mM) and uses several nitrogen sources, such as cyanate, —cyanoalanine, cyanoacetamide, nitroferricyanide (nitroprusside) and cyano-metal complexes [6, 7]. As far as we know, there are no other previous reports concerning alkaline cyanide biodegradation in bacteria [6, 7], although several aspects of cyanide and cyanate metabolism have been studied at neutral pH in *Pseudomonas aeruginosa* and *Pseudomonas fluorescens* [8, 9]. Cyanate has been proposed as a degradation product of cyanide, although there are no evidences related to simultaneous metabolism of cyanate and cyanide.

2 Material and methods

Culture conditions.
P. pseudoalcaligenes was cultured in M9 minimal medium on a rotatory shaker at 230 rpm and 30°C as previously described [6, 7] in the presence of cyanate, ferricyanide, sodium cyanide or the jewellery residue. The jewellery residue contains 20 g. L^{-1} of free cyanide and cyano-metal complexes and was kindly supplied by GEMASUR (Spain). Nitrate and ammonium were used also as nitrogen sources when indicated.

Nitrate reductase assay.
Nitrate reductase activity was determined in cell-free extracts with reduced methyl viologen as artificial electron donor and the nitrite produced in the reaction was determined colorimetrically [6].

Chrome Azurol S agar medium.
CAS plates were prepared as previously described [10] with the green complex chrome azurol S.

Analytical determinations.
Bacterial growth was monitored by following the absorbance at 600 nm. Ammonium concentration was determined by the Nessler reagent as previously described [7].

3 Results and discussion

Growth of Pseudomonas pseudoalcaligenes *CECT5344 on cyanate, jewellery residue and —cyanoalanine.*

Pseudomonas pseudoalcaligenes CECT5344 is able to growth in the presence of several cyanurated compounds as the sole nitrogen source under alkaline conditions (Fig. 1). Thus, this bacterium showed an optimum pH for growth on M9 minimal medium of 9.0 (not shown). The growth of *P. pseudoalcaligenes* on cyanate, cyanide (present in the jewellery residue) or —cyanoalanine was achieved within 90 hours. The best growth rate was obtained in the presence of cyanate, although the highest absorbance at 600 nm at the end of the growth curve was achieved with the residue (Fig. 1). During growth on cyanate, a cyanase activity is induced in *P. pseudoalcaligenes*. This activity is responsible for the conversion of cyanate and bicarbonate into ammonium and CO_2 [6]. This enzyme was purified and immobilised in a biosensor able to detect ppm of cyanate [11]. Cyanase activity is induced in cyanide-grown cells of *P. pseudoalcaligenes*. In addition, a cyanide monooxygenase activity which converts cyanide into cyanate has been described in *Pseudomonas* sp. [3]. In this sense, cyanate could be a degradation product of cyanide in *P. pseudoalcaligenes*. However, the cyanide monooxygenase has not been detected in *P. pseudoalcaligenes*. On the other hand, *P. pseudoalcaligenes* CECT5344 grows with —cyanoalanine (Fig. 1) which is a nitrile-like compound and an enzymatic activity able to degrade —cyanoalanine to produce ammonium was detected in crude extracts of cyanide-grown cells of *P. pseudoalcaligenes* (not shown).

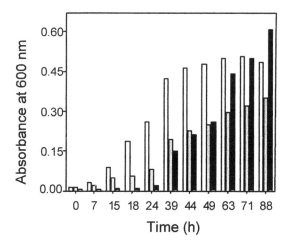

Fig. 1 *Growth of* P. pseudoalcaligenes *CECT5344 on cyanate, jewellery residue and —cyanoalanine.* Cells were cultured in M9 minimal media with sodium acetate as carbon source and with 2 mM of the nitrogen source: cyanate (white bars), —cyanoalanine (grey bars) or the residue (with 2 mm of free cyanide, black bars).

In addition, preliminary results indicate that *P. pseudoalcaligenes* accumulates 2-oxoglutarate in the culture media in response to cyanide (not shown). 2-oxoglutarate can chemically react with cyanide to produce its respective cyanhydrin, also a nitrile-like compound, which could be further metabolised. The presence of several cyanide degradation pathways in the same organism is not inusual. Thus, *Pseudomonas fluorescens* presents different degradation pathways for cyanide [12].

Regulation of cyanide assimilation in Pseudomonas pseudoalcaligenes *CECT5344.*

In addition to cyanurated compounds, *P. pseudoalcaligenes* uses nitrate, nitrite or ammonium as the sole nitrogen source. Cyanide and ammonium are assimilated simultaneously, whereas cyanide strongly inhibits nitrate and nitrite assimilation [6]. However, addition of cyanide to cells growing with ammonium or nitrate causes interruption of both assimilative processes (Fig. 2, a). In the case of nitrate assimilation, it has been observed that cyanide inhibits nitrate reductase, whereas azide, a cyanide analog, does not (Fig. 2, b).

524

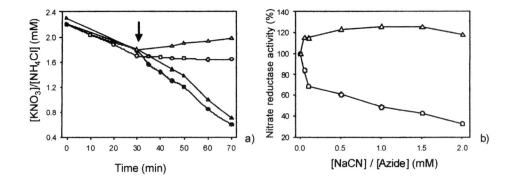

a) b)

Fig. 2 *Regulation of cyanide assimilation in* P. pseudoalcaligenes *CECT5344.* a) Cells were cultured with 10 mM ammonium chloride or 10 mM potassium nitrate and harvested when the cultures reached an absorbance at 600 nm of 1.0 (resting-cells). Addition of 2.5 mM nitrate or 2.5 mM ammonium chloride was performed and uptake of nitrate (circles) and ammonium (triangles) was analyzed. 1 mM NaCN was added at the time indicated by the arrow (open simbols). Controls without cyanide addition were performed (filled simbols). b) Cells were cultured with 50 mM sodium acetate and 10 mM potassium nitrate and at the end of the exponential phase of growth aliquots were taken. Nitrate reductase activity was assayed in the presence of NaCN (circles) or azide (triangles).

Production of siderophores by Pseudomonas pseudoalcaligenes *CECT5344.*
Cyanide produces very stable complexes with transition metals which are essential for protein function. For this reason, cyanide is toxic for living organisms. In this sense, a cyanotrophic organism must have a system for iron acquisition which is mainly based on the production of siderophores. *P. pseudoalcaligenes* is able to growth in the jewellery residue [7]. This bacterium is also able to grow in the presence of 1 mM ferricyanide as the sole nitrogen and iron sources (Fig. 3, a). The ability to grow in ferricyanide at pH 9.5 and pH 7.5 in the presence of 2,2'-bipyridyl or with other metal complexes such as the cyanocomplexes of Fe(II) and Cu(II) indicates that this bacterium grows by producing siderophores. In addition, cells of *P. pseudoalcaligenes* were cultured in plates with chrome azurol S (CAS) and decoloured halos were observed around the colonies as a consequence of the iron elimination of the green complex (Fig. 3, b).

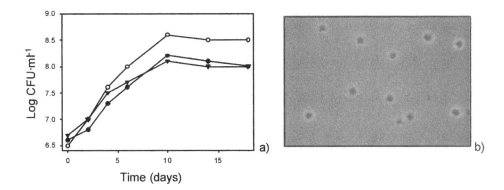

Fig. 3 *Production of siderophores by* P. pseudoalcaligenes *CECT5344.* a) Cells were cultured with sodium acetate as carbon source and with 1 mM ferricyanide as the sole nitrogen and iron sources at pH 9.5 (open circles) or at pH 7.5 (filled circles). A growth curve at pH 7.5 with 1 mM fericyanide and 100 μM 2,2'-bipyridyl, a powerful iron chelator, was also performed (filled triangles). b) CAS-plate medium with colonies of *P. pseudoalcaligenes* CECT5344.

Acknowledgements

This work was funded by the Ministerio de Ciencia y Tecnología (BMC2002-04126-C03-01 and -03) and the FEDER (1FD97-0653). VMLA was the recipient of a fellowship from the MEC (Spanish Ministry of Education) and MDR holds a Postdoctoral fellowship from the Junta de Andalucía (Spain). We also thank GEMASUR for providing the residue from the electroplating jewelry industry and for its fruitful collaboration.

References

[1]   J. L. Whitlock, Geomicrobiol. J. **8**, 241-249 (1991).
[2]   S. A. Raybuck, Biodegradation **3**, 3-18 (1992).
[3]   S. K. Dubey, and D. S. Holmes, World J. Microbiol. Biotech. **11**, 257-265 (1995).
[4]   A. Dumestre, T. Chone, J. M. Portal, M. Gerard, and L. Berthelin L., Appl. Environ. Microbiol. **63**, 2729-2734 (1997).
[5]   M. Barclay, A. Hart, C. J. Knowles, J. C. L. Mecussen, and V. A. Tett V. A., Enzyme Microb. Technol. **22**, 223-231 (1998).
[6]   V. M. Luque-Almagro, M. J. Huertas, M. Martínez-Luque, C. Moreno-Vivián, M. D. Roldán, L. J. García-Gil, F. Castillo F, and R. Blasco, Appl. Environ. Microbiol. **71**, 940-947 (2005a).
[7]   V. M. Luque-Almagro, R, Blasco, M. J. Huertas, M. Martínez-Luque, C. Moreno-Vivián, F. Castillo, and M. D. Roldán, Biochem. Soc. Trans. **33**, 172-173 (2005b).
[8]   K. Chapatwala, G. Babu, O. Vijaya, K. P. Kumar, and J. H. Wolfram, J. Ind. Microbiol. Biotechnol. **20**, 28-33 (1988).
[9]   J. K. Dhillon, and N. Shivaraman, Can. J. Microbiol. **45**, 201-208 (1999).
[10]   B. Schwyn, and J. B. Neilands, Anal. Biochem. **160**, 47-56 (1987).
[11]   V. M. Luque-Almagro, R. Blasco, J M. Fernández-Romero, and M. D. Luque de Castro, Anal. Bioanal. Chem. **377**, 1071-1078 (2003).
[12]   D. A. Kunz, R. F. Fernandez, and P. Parab, Biochem. Biophys. Res. Commun. **287**, 514-518 (2001).

Microbial ecology of authothermal aerobic digestion (ATAD): diversity, dynamics and activity of bacterial communities involved in treatment of a municipal wastewater.

Anna.V. Piterina* [1], **Ciaran McCusland** [2], **John Bartlett** [2] **and J.Tony Pembroke** [1].

[1] - Molecular Biochemistry Group, Department of Chemical and Environmental Sciences, University of Limerick, Limerick, Ireland   [2] - Institute of Technology Sligo, Sligo, Ireland.

*Corresponding author: Anna V. Piterina. Tel.35361202979. e-mail anna.piterina@ul.ie

Authothermal thermophilic aerobic digestion (ATAD), or liquid composting, is a versatile new thermophilic process for the treatment and stabilization of high strength wastes of slurry consistency. The process has initially been applied for the treatment of pharmaceutical and industrial wastewater but has more recently been used to stabilize municipal wastes. There have been a number of studies focusing on reactor performance and the pattern of inactivation of various pathogenic organisms, but little focus on investigating the microbial population associated with the thermophilic step where microbial activity increases the temperature to 55-65°C during the biodegradation of municipal sludge. We have investigated the thermotolerant and thermophilic bacterial community responsible for efficient performance of ATAD by analysis of a clone library constructed of ATAD bacterial community 16S rDNA genes.   A prevalence of thermophilic *Bacillus, Paenibacillus, and Peptostreptococcaceae species* was found.   8% of the clones were affiliated to *Symbiobacterium thermophiiums* often found associated with *Bacillus* species, while 40% of the clones have high similarity (>90%) to 16S rDNA sequences of compost-associated bacteria. Most of the phylogenetic affiliations discovered in this study have previously been characterized as physiologically active microorganisms associated with elevated temperature niches.   Monitoring of the operating parameters has demonstrated that there is on average a 35% removal of soluble COD, a 31% reduction in total solids (TS) and a 46% VS (volatile solids) reduction during the treatment of domestic waste sludge via ATAD.

1   Introduction

The stabilization and use of sludge biosolids generated from primary treatment of domestic or industrial wastewater is now a major issue in any wastewater treatment process. Autothermal thermophilic aerobic digestion (ATAD) is a novel approach to achieve a standard stabilized, disinfected biosolids with the potential for use as a fertilizer for land spread (Class A Biosolids) [1]. The ATAD process for treatment of domestic sludge generally operates in a semi-batch mode, with a feed cycle occurring once per day. In a municipal ATAD plant operating in Killarney in Ireland the processes utilizes multiple stage reactors with treatment initiated by thickening of sludge after secondary processing. Thickening is carried out by addition of polymers to achieve the consistency of 6% Total Solids (TS). This thickened sludge is pumped into a jacketed reactor (reactor 1A), where it is mixed and aerated by high efficiency air pumps. Here the process of sludge biodegradation begins and the temperature of the sludge mass begins to rise, aided by the enclosed jacketed reactor design. The resulting increase in microbial activity, heat release and organic matter decomposition makes the reactor self-sustaining from an energy perspective. After 24 hours, 1/9 of the reactor volume is pumped into a second jacketed reactor (reactor 2A) where the operating temperature rises, due to microbial metabolic activity, to between 55-65°C, this is termed the thermophilic stage (Figure1).

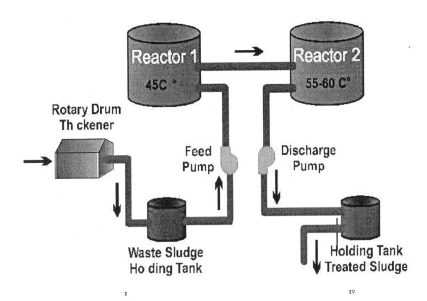

Fig. 1. Flow diagram of the ATAD process

Although there has been work carried out on understanding the design and operating parameters of the ATAD process [2, 3, 4, 5] there has been little published on the microbial populations present and in particular those contributing to the thermophilic stage, in the main due to the difficulty in cultivation of such organisms defined media. In general during the thermophilic stage, which occurs in reactor 2A, two distinct categories of microbes can be detected. One termed the 'process bacteria' mediate the biodegradation of the sludge and are responsible for the rise in temperature. These possible originate in the feed with thermotolerant or thermophilic subpopulations being selected as the temperature increases. A second population comprising of pathogens can also be detected which also originate from the feed but being thermo-labile are eliminated by the process conditions [1, 2, 3], a combination of temperature and holding time.

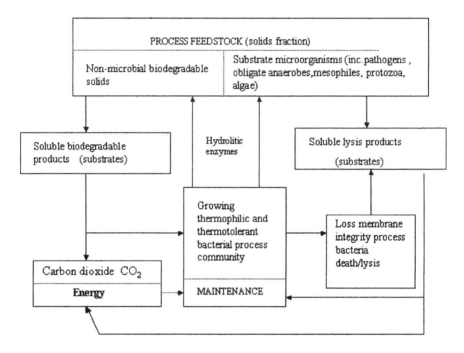

Fig 2. Overview of the process of solubilisation and biodegradation via bacterial communities within ATAD

Thus to achieve a safe Class A Biosolids as per US EPA regulations, free from pathogenic microorganisms, it is essential that the appropriate temperature/time exposure be achieved. In the present study we have analysed both the ATAD process performance and the microbial population dynamics (using 16S rDNA analysis of community DNA [6, 7, 8, 9,10] when the operational temperature is raised to the recommended limit of >60°C for thermophilic digestion.

2   Methods of investigation

2.1 Physico-chemical analysis.

All analyses were carried out on an ATAD reactor treating municipal domestic sludge in Killarney, Ireland and designed by Fucks GMBH, Germany. Total Suspended Solids (TSS), was determined as described by [9] on samples centrifuged at 13000 rpm for 10 min. The supernatant from the TSS assay was analyzed for soluble COD by the closed reflux titremetric method [9]. Volatile Solids (VS) was determined by evaporating a known volume (10 ml quantities) of the supernatant from the TSS assay and drying it in an oven at 105 °C to a constant weight. The TSS plus the (dissolved) soluble solid constitute the total solids (TS) in the slurry and this component was used to monitor the degradation of total waste load. Temperature was measured using a reactor specific probe designed for the bioreactor.

2.2 Community analysis

Non-clarified triplicate grab samples were taken from the middle of the bioreactor tank from the ATAD sludge treatment system, combined and transferred into 50-ml sterile tubes. For DNA isolation, 1.5 ml of each sample was centrifuged immediately at 14,000 x g for 5 min., decanted and the pellet transferred on ice for analysis or frozen and stored at − 80°C for later analysis. Crude DNA was isolated from thawed samples using a method successfully applied for analysis of thermophilic bacterial communities from environmental sites with high organic matter [12], with an important modification; the inclusion of a bead-beating step after the last thawing cycle. The 16S rDNA was selectively amplified from isolated 1/10 diluted (necessary for successful amplification reactions due to the presence of humic compounds which inhibit *Taq* polymerase) crude community DNA by PCR using universal oligonucleotide primers P4F, P5R [13] designed to anneal to the conserved V6-V8 region of bacterial 16S rDNA [13]. Products of three separate PCR reactions were combined and concentrated for cloning using a Wizard® SV Gel and PCR Clean-Up kit (Promega). DNA was ligated in the pGEM-T-Easy vector using the manufacturers protocol (Promega) [12]. Plasmid DNA from selected 16S rDNA clones, which contained inserts of the expected size, were purified and sequenced with universal primers via a LI-COR, Long Read 4200 DNA sequencer (MWG Biotech, Ebensburg, Germany). All sequences were checked for chimera formation with the CHECK_CHIMERA software of the Ribosomal Database Project (RDP) [13] and the phylogenetic affiliations of their 5' and 3' ends were compared. Phylogenetic trees were calculated by parsimony, neighbour-joining, and maximum-likelihood analysis with different sets of filters. The sequences were compared with known sequences using the ribosomal database project [13] and the EMBL nucleotide sequence database [14]. The SIMILARITY-RANK tool of the RDP and the FASTA search option for the EMBL database were used to search for close evolutionary relatives.

3 Results and discussion

3.1 System performance

Figure 3 (a) and (b) illustrates a temperature profile for the two ATAD reactors (1A and 2A) over a 6 month period. The temperature data presented are the daily temperatures taken just prior to reactor feeding. Figure 3 (a) illustrates the temperature profile one hour following the feed-cycle. These values represent the minimum and maximum temperatures found in each reactor during the process cycle. As illustrated in Figure 3, the multi stage reactors helps to minimize the temperature fluctuation in the final stage reactor (2A) (where the added sludge represents only 1/9 the volume and has been preheated in reactor 1A) which is essential if the appropriate temperature /time exposure is to be achieved. This is also essential for the removal /deactivation of both pathogenic organisms present and those indicative of fecal contamination [3].

530

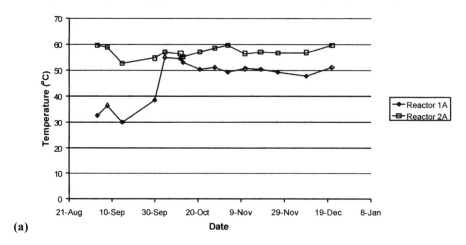

(a)

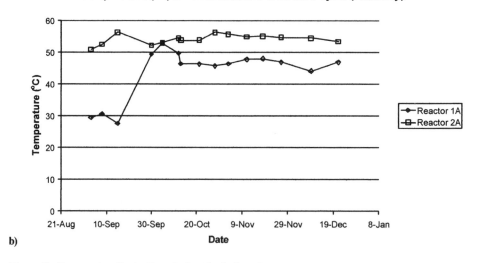

b)

Figure 3. Temperature fluctuations during the feed-cycle

Analysis of the waste loading to the ATAD reactor indicates an average COD loading value of $78g.l^{-1}$ over a 6-month period August -December 2004. The average solids and organics removal values for the ATAD process are illustrated in Table 1. The data presented are average values for the final 6 months of year 2004. The total solids content of the influent sludge varied throughout the year in this plant, with an annual average of 5.7% (w/w) for last 6 months of 2004. To initiate the ATAD process secondary sludge from the wastewater treatment plant is transported to a picket-fence thickener where it is thickened to 3% (w/w). Following thickening the sludge is fed to a belt filter press, and is further thickened to 5.7% (w/w). Zetag 7867, a high weight cationic polyacrylamide based polymer is then added to condition the sludge before pressure filtration. Because of the nature and level of thickening it is not feasible to evaluate the system performance in terms of suspended

Table 1. Performance of the multistage ATAD process treating domestic sludge

Sampling site	T$_m$(°C) operation	Aeration and mixing	TS% (w/w)	VS% (w/w)	VS (% of TS)	VLR (m^3 sludge) p.a.	tCOD (effluent) p.a.	tDS (effluent) p.a.	pH (range)
Feed Tank	5-10	-	5.7	4.5	75	8395	654	478	5.8-6.3
1A	40-50	+	4.7	3.2	68	****	*no data	394	6.7-7.9
2A	55-65	+	4.0	2.5	64	****	*no data	335	7.9-8.6
Effluent	25-30	-	3.3	1.9	59	****	327	252	7.3-8.2

solids (as filtration of suitably sized samples is exceptionally difficult) at this stage as would be done in a typical activated sludge system. With an average of 23 m^3 day^{-1} throughput of sludge, the annual volumetric loading rate for the Killarney ATAD process is 8395 m^3 annually
(wet weight) and 478 tonnes of dry solids, with an average volatile content of influent of 72%, this equates to 344 tonnes of volatile solids annually. The organic loading rate in terms of total COD assuming an average of 78g.l^{-1} is 1.79 tonnes day^{-1} or 654 tonnes of COD per annum.
The annual average total solids of the effluent which is used for land spread (following a storage period) was 3.23% (w/w) in 2004, and the average volatile solids was 1.92% (w/w); this represents a volatile content of 59.44%. The annual average total solids of the effluent from Reactor 2A was 3.96%, and the average volatile solids was 2.53%, a volatile content of 63.89%. The average COD value for the Killarney ATAD plant for July-December 2004 was 34 g. l^{-1}, 0.782 tonnes per day, or 285 tonnes per annum.
The average TS removal in the Killarney ATAD plant was 31% in the last 6 months of 2004 as shown at Fig.4, while the average VS removal was 42% (with substrate removal in the product storage considered) as shown at Fig.5. These figures rise to 44% and 55% for TS and VS respectively if substrate removal in the product storage tanks is taken into account as well. Analysis of the overall figures for solids removal during the ATAD process (including the storage tanks) illustrates that the largest fraction of substrate removal occurs in reactor 1A (40% of total TS and 47% of total VS removal). Extensive solubilisation of sludge components also occurred in this reactor as indicated in Figures. 2 and 5 and Table 2. In reactor 2A (the thermophilic reactor) some 30% of total TS and 30% of total VS removal was observed. A similar pattern of soluble COD removal was observed in reactor 1A. This level of biodegradation is slightly higher than in the thermophilic reactor (2A) and may be explained by a wide range of microbial species being active including a substantial mesophilic population.

532

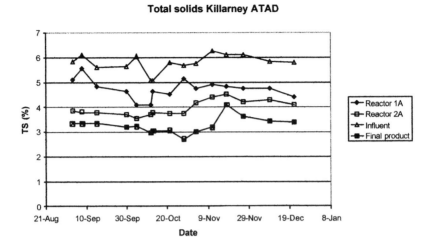

Figure 4. Total Solids removal during processing

Table 2. Average Total and soluble COD values

Sampling Source	Total COD (mg. l^{-1})	Soluble COD (mg. l^{-1})
Feed	92430	6057
1A	67697	14951
2A	46085	9742
Final Product	36731	5788

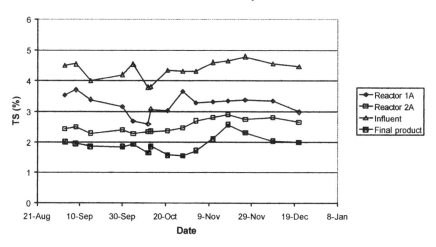

Figure 5. Volatile solids removal

3.2. Molecular analysis of microorganisms associated with ATAD:

A clone library of 16S rDNA was constructed from total community DNA isolated from ATAD reactor 2A. Nucleotide sequences were determined for 80 16S rDNA clones with no chimeric sequences detected amongst these clones. Using the BLASTn and/or SEQUENCE_MATCH algorithms [14], the majority of the 16S rDNA nucleotide sequences were shown to be >85% similarity to reference strains found in the GenBank and RDP databases. Phylogenetic analysis of clones from the bacterial16S rRNA gene library revealed a few representatives from a number of phylogenetic divisions, indicative of low genetic diversity as presented at Table 3. The phylotypes observed were in the main physiologically adapted to elevated temperatures and obligate or facultative aerobes. In general the genera detected are known to be associated with a variety of extracellular enzymatic activities similar to those detected previously in compost related genera [15]. In a small number of cases phylotypes were detected that clustered with obligate anaerobes in the low-G+C gram-positive *Clostridia* and *Peptostreptococcaceae* divisions. The presence of anaerobes within the ATAD reactor initially seemed unlikely because of the aggressive aeration employed and the lack of microbial floccule particles that could offer an anaerobic microenvironment. However given the elevated temperature, non perfect mixing, sludge viscosity and oxygen solubility at elevated temperatures it is possible that some anoxic areas may occur within the ATAD reactor. The contribution of such anaerobes is however unclear and indeed these cells need not be active to be detected. Recently however evidence has been presented for methanogenic activity during ATAD processing [16] suggesting that anaerobic biodegradation may play some role in the overall ATAD process.

534

Table 3. Closest phylogenetic affiliation of clones isolated from analysis of community DNA from ATAD reactor 2A via 16S rDNA analysis

Closest matches from databases	Accession Number	Similarity (%)
1. Bacillus lichenformis	AF2372616	96-100%
2. Paenibacillus larvae	AY 300801	92-98%
3.Paenibacillus sp. C-168	J16129	90%
4. Bacillus sp. MZLV	AF57343	90%
5. Bacillus sp. MSP06G	AB084065	88-95%
6. Uncultured bacterium clone KD 4-96	AY 218649	98%
7. Clostridium sp. CSBNIIRNA	X755909.1	90%
8. Uncultured bacterium tmbr-11-7	AF280819	90%
9. Peptostreptococcaceae bacterium 19gly3	AF350609	93-98%
10.Uncultured compost bacteria 4-28	AB034710	85%
11. Uncultured compost bacteria 4-11	AB034705	88%
12. Uncultured eubacteria WCHBI -54	AF050582	93%
13 Clostridiales bacterium NS5-4	AY466717	95%
14. Unclassified bacteria (multidrug resistant)	NC 004840	93%
15. Uncultured bacterium ARKCH2Br252	AF 46824	91%
16. Symbiobacterium thermophilium	AB0049131	90-99%

Acknowledgements

This work was founded through Higher Education Authority Ireland, PRTLI Programme on Biosolids

References

[1]   Bruce, A.M., F., Colin and P.J. Newman (Eds) (2003) Treatment of sewage sludge: Thermophilic aerobic digestion and processing requirements for land filling. Elsevier Applied Science Publishers.

[2]   LaPara, T.M., Konopka, A, Nakatsu, C.H., and Alleman, J.E. (2000)Thermophilic aerobic wastewater treatment in continuous-flow bioreactors. *Journal of Environmental Engineering ASCE* 126 (8): 739-744.

[3]   Ugwuanyi, J.O., Harvey, L,M., and B., McNeil (1999) Effect of process temperature, pH and suspended solids content upon pasteurization of a model agricultural waste during thermophilic aerobic digestion. *Journal of Applied Microbiology.* 87 (3): 387-398.

[4]   Lapara, T.M., Nakatsu, C.H., Pantea, L,M., and J.E., Alleman, (2001) Aerobic biological treatment of a pharmaceutical wastewater effect of temperature on COD removal and bacterial community development. *Water Resource* 35 (18):4417-25.

[5]   Lapara, T.M., Konopka, A., Nakatsu, C., and J.E., Alleman (2001) Thermophilic aerobic treatment of a synthetic wastewater in a membrane- coupled bioreactor. *Journal of Industrial Microbiology and Biotechnology* 26:203-209.

[6]   You, S.J., Ouyang, C.F., Lin, S.F., and W.T.,Liu (2000) Characterization of the microbial diversity in a biotreatments process using non-culture based methods *Water Science and Technology* 42:143-148.

[7]   Theron, J., and T., Cloete (2000) Molecular techniques for determining microbial diversity and community structure in natural environments. *Critical Review in Microbiology* 26(1):37-57.

[8]   Hugenholtz, P., and N., Pace (1996) Identifying microbial diversity in the natural environment: A molecular phylogenetic approach. *Trends in Biotechnology* 14(6):190-197.

[9]   Snaidr, J, Amann, R, Huber, I, Ludwig, W, and K.H., Schleifer (1997) Phylogenetic analysis and in situ identification of bacteria in activated sludge *Applied and Environmental Microbiology* 63:2884-2896.

[10]   Von Wintzingerode, F., Gobel, U.B., and E., Stackebrandt (1997) Determination of microbial diversity in environmental samples: pitfalls of PCR-based rRNA analysis. *FEMS Microbiology Reviews* 21:213- 229.

[11]   American Public Health Association (APHA) (Eds) (1995) Standard Methods for the Examination of Water and Wastewater. APHA Publishers, Washington DC.

[12]   Fortin, N., D., Beamier, K., Lee, and C., Greer, (2004) Soil washing improves the recovery of the total community DNA from polluted and high organic content sediments. *Journal of Microbiological Methods* 56:181-191.

[13]   Ribosomal Database Project (http://rdp.cme.msu.edu/)

[14]   EMBL database (http://www.ebi.ac.uk/embl/).

[15]   Dees, P., and W., Ghiorse (2001) Microbial diversity in hot synthetic compost as revealed by PCR-amplified rRNA sequences from cultivated isolates and extracted DNA. *FEMS Microbial Ecology* 35 (2):207-216.

[16]   Ugwuanyi, J.O, Harvey, L.M., and B., McNeil (2005) Effect of digestion temperature and pH on treatment efficiency and evolution of volatile fatty acids during thermophilic aerobic digestion of model high strength agricultural waste. *Bioresource Technology* 96:707-719.

Microbial function after assisted natural remediation of a trace element polluted soil

A. Pérez-de-Mora, P. Burgos, J.J. Ortega-Calvo, F. Cabrera, and E. Madejón

Instituto de Recursos Naturales y Agrobiología (CSIC), Av. De Reina Mercedes 10, 41012 Sevilla, SPAIN

We studied the effect of different amendments and/or a plant cover on soil microbial properties of a trace element polluted soil. The experiment lasted 30 months and was carried out in containers filled with ca. 150 kg of contaminated soil. The remediation measures consisted of the application of different amendments and/or development of a plant cover (*Agrostis stolonifera L.*). Seven treatments were established: four with organic amendments (leonardite LEO, litter LIT, municipal waste compost MWC and biosolid compost BC) and one inorganic amendment (sugarbeet lime SL), where agrostis was sown, and two controls without amendment addition (with *Agrostis* CTRP or without *Agrostis* CTR). Microbial function was analysed by means of microbial biomass C (MBC), microbial biomass C to total organic C ratio (MBC/TOC), enzyme activities (dehydrogenase, aryl-sulphatase, β-glucosidase, acid-phophatase and protease) and microbial heterotrophic potential (MHP) to MBC ratio (MHP/MBC). The MWC and BC treatments were the most effective in raising MBC, MBC/TOC, dehydrogenase and aryl-sulphatase activities and reducing the MPH/MBC ratio. Whereas β-glucosidase was higher in the amended treatments, acid-phosphatase and protease activities showed no significant differences between the control and the amended treatments. Assisted natural remediation can be a useful and reliable technique to improve soil microbial function in the mid-term. Further monitoring is necessary to evaluate the potential of this technique in long-term experiments.

Keywords: *Agrostis*; amendments; assisted natural remediation; enzyme activities; trace elements

1   Introduction

Trace elements are toxic to most living organisms at excessive concentrations [1]. Unlike organic contaminants, trace elements cannot be degraded, therefore remediation of trace element polluted soils is either based on the extraction or the stabilization of the contaminants. Assisted natural remediation (ANR) is an "in situ" stabilization technique based on the use of amendments to accelerate those processes (sorption, precipitation and complexation reactions) that take place naturally in soils to reduce mobility and bioavailability of toxic elements [2]. Due to their restricted nature, natural attenuation processes alone may not be sufficient in mitigating risks from trace elements. Moreover, ANR may enhance microbial activity, plant colonisation and development, and thus re-start the nutrient cycling in the affected soils.

Trace elements can exhibit negative effects towards soil biota affecting both microbial key processes and the number and activity of soil microbial populations [3, 4]. Microorganisms respond quickly to changes and can rapidly adapt to environmental conditions. Changes in soil microbial populations and activities can be sensitive and early indicators of both natural and anthropogenic disturbances [1].

This work aims to evaluate the effects of mid-term ANR on microbial function of a trace element polluted soil.

2   Materials and methods

Soil (pH= 3.41; As=120 mg kg^{-1}; Cd=2.43 mg kg^{-1}; Cu=78.3 mg kg^{-1}; Pb=201 mg kg^{-1}; Zn=226 mg kg^{-1}) was sampled in an area affected by the Aznalcóllar mine accident named "El Vicario", where the only remediation work carried out by the authorities was the removal of the sludge layer together with the first 15 cm of the top soil. The experiment was carried out for 30 months in containers, filled with polluted soil. Four organic amendments (MWC, BC, LEO and a LIT) and an inorganic amendment (SL) were mixed with the top soil twice at the rates of 100 Mg ha^{-1} (beginning of the experiment) and 50 Mg ha^{-1} (12 months later). Characteristics of the amendments are shown in Table 1. *Agrostis stolonifera* L., a trace element tolerant plant, was sown in the containers one month after the beginning of the experiment. Two control treatments were also established: CTRP soil without amendment addition but sown with *Agrostis* and CTR soil without amendment addition and without *Agrostis*. A soil sampling was carried out at the end of the experiment. Microbial biomass C (MBC) content was determined by the chloroform fumigation-extraction method modified by [5]. Enzyme activities were determined as described in [6]. Microbial heterotrophic potential (MHP) was estimated as detailed in [4].

Table 1. Characterization of the amendments.

	pH	EC		N		K	As		Cu	Mn	Pb	Zn
		dS m^{-1}		%		%			mg kg^{-1}			
MW	7.36	6.16	18.	1.04	0.44	0.43	8.37	1.49	362	252	385	396
BC	6.93	2.91	19.	1.31	1.24	0.93	5.63	0.73	121	257	137	258
LEO	6.08	17.4	28.	1.17	0.04	3.97	34.9	0.83	28.2	66.2	22.0	64.5
LIT	4.49	0.92	54.	0.90	0.04	0.19	1.90	nd	6.45	676	9.36	27.0
SL	9.04	-	6.7	0.98	0.51	0.53	1.63	0.43	51.0	297	39.2	138

EC electrical conductivity, TOC total organic carbon.

A normality test was carried out for all variables prior to analysis of the variance. The data was analysed by ANOVA, considering the treatment as the independent variable. Significant statistical differences of all variables between treatments were established by Tukey´s test when there was homogeneity of the variance and by Games-Howell's test in the opposite case.

3   Results and discussion

Mid-term and long-term studies are necessary to understand the effects of remediation techniques on soil restoration. Evaluating soil functionality is of utmost importance since soil microorganisms are essential for nutrient cycling and plant colonization and development. Soil quality assessment in remediated soils requires monitorization of both chemical and microbiological properties. Criteria for choosing the microbiological characteristics analysed were based on previous experience with their sensitivity to soil disturbance and attenuation processes.

In general, MBC was significantly higher in the amended treatments compared with the control treatments (Figure 1a). Highest values were found in the MWC and BC treatments, which can be attributed to both the addition of microorganisms with the amendments and

538

the incorporation of easily degradable organic matter and other nutrients, which stimulate the growth of the autochthonous microorganisms of the soil.

The MBC/TOC ratio is usually comprised between 10-40 mg g^{-1}. Values below 10 mg g^{-1} can be regarded as a sign of stress such as that produced by trace elements. After 30 months, the MWC, BC and SL treatments were the only ones that showed mean values above this level (Figure 1b). These treatments were therefore more efficient to raise the active soil organic matter.

Dehydrogenase activity has been used to assess heavy metal toxicity in soils and microbial activity in semiarid Mediterranean areas. This activity was significantly higher in the MWC, BC, LIT and SL treatments compared with the control treatments (Figure 1c). These results seem to be related with higher pH values and lower soluble trace element concentrations in these treatments [4]. Aryl-sulphatase has been suggested as the most sensitive in tracing trace element effects. At the end of the experiment, mean values of this activity were higher in the MWC, BC and SL treatments (Figure 1d).

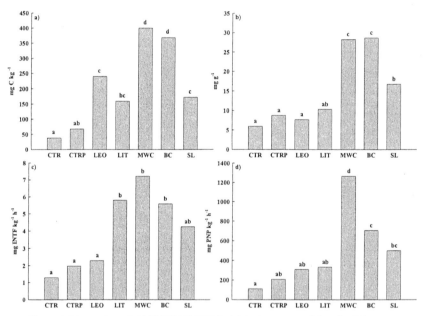

Figure 1. Mean values of a) MBC b) MBC/TOC c) dehydrogenase and d) aryl-sulphatase activities. Bars with the same letter do not differ significantly ($p<0.05$).

Data on β-glucosidase activity and trace elements is contradictory and both high and low sensitivity has been reported. In our study, β-glucosidase was higher in all amended treatments compared with the control treatments (figure 2a). Highest values were found in the LIT treatment probably due to its higher content in cellulose, the substrate of this enzyme. Acid-phosphatase was highest in the LEO treatment, and mean values in the control treatments were similar to those of the amended treatments (Figure 2b). These results seem to be related with feedback inhibition of the enzyme due to the available-P rather than with soil attenuation. The lowest activity was found in the SL, BC and MWC treatments, which were amended with P-rich materials (Table 1). Generally, protease activity did not show significant differences between amended and control treatments, although mean values were lowest in the CTR treatment (Figure 2c). These results could imply a recovery of the enzyme with time and thus a lower sensitivity to low pH and higher soluble trace element concentrations than other properties studied.

The MHP was estimated as the maximum rate of glucose mineralized by the soil microbial biomass after addition of glucose in excess to soil samples. The ratio MHP/MBC can be representative of stress conditions. Mean values of this ratio were, in fact, higher in the two control treatments compared with the amended treatments (Figure 2d). Significant differences with respect to the control treatments were observed for the MWC and BC treatments.

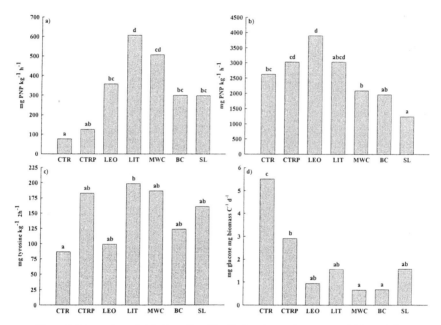

Figure 2. Mean values of a) β-glucosidase b) acid-phosphatase c) protease activities and d) MHP/MBC ratio. Bars followed by the same letter do not differ significantly ($p<0.05$).

4   Conclusions

Assisted natural remediation proved to be a reliable and effective technique to restore soil microbial function of a trace element polluted soil in the mid-term. In general, the two compost treatments gave the best results and were more efficient in enhancing soil microbial properties. Monitoring in the long-term should be encouraged, since mineralization of the organic matter could increase soluble trace element concentrations thereby reversing soil restoration and affecting negatively soil functionality.

540

References

[1]   K.E. Giller, E.Witter, S.P. McGrath, Soil Biol Biochem **30**, 1389-1414 (1998).

[2]   N.S Bolan, V.P Duraisamy, Aus J Soil Res **41**, 533-555 (2003).

[3]   E. Kandeler, C. Kampichler, O. Horak, Biol Fertil Soils **23**, 299-306 (1996).

[4]   A. Pérez de Mora, JJ. Ortega-Calvo, F. Cabrera, E. Madejón, Appl Soil Ecol **28**, 125-137 (2005).

[5]   E.G. Gregorich, G. Wen, RP. Voroney, RG. Kachanoski, Soil Biol Biochem **22**, 1009-1011 (1990).

[6]   K. Alef, P. Nannipieri (eds.), Methods in Applied Soil Microbiology and Biochemistry, Academic Press, London, 1995, chap. 7, pp. 311-373.

Microbiological characterization of ammonium oxidizing biofilms in rotating biological contactor (RBC) using different support materials

A. M. Lima [2], **V. Reginatto**[3,*], **W. Schmidell**[2], **H. M. Soares**[2] and **H. Hoffmann**[3]

[1] Federal University of Santa Catarina, Department of Food Science and Technology, Rod. Admar Gonzaga 1346 – Itacorubi , 88034-001 Florianópolis - SC, BRAZIL

[2] Federal University of Santa Catarina, Department of Chemical Engineering and Food Engineering, Campus Universitário - Trindade, P.O. box 476, 88040-970 Florianópolis - SC, BRAZIL

[3] Federal University of Santa Catarina, Department of Sanitary and Environmental Engineering, Campus Universitário - Trindade, 88040-970 Florianópolis – SC, BRAZIL

*Corresponding author: e-mail: valeria@cca.ufsc.br, Phone: +55 4833315368, Fax: +55 4833315390

Microbiological characteristics and efficiency of RBC reactors with disks of three different support materials: polyvinylchloride – PVC; polyestyrene – PE and polyurethane – PU were studied. Ammonium oxidation efficiency was monitored by analysis of ammonium, nitrite and nitrate. Biofilms microbiological characteristics were investigated by optical microscopy (OM). Ammonium-oxidizers bacteria (AOB) and nitrite-oxidizers bacteria population (NOB) were estimated by Most Probable Number (MPN) technique. PVC and PE reactors showed higher nitrification efficiency than PU in 330days. However in PU was observed an incomplete nitrification with nitrite formation. Biofilms microbiological structure in all support materials revealed a high abundance of filamentous bacteria. Attached (sessile) ciliates were observed mainly in PU reactor. MPN showed an increase of AOB mainly in PVC and PU reactors, while NOB lowest population was found in PU agreeing with the accumulation of nitrite in this reactor. These results indicated that PU could be the best material to develop an autotrophic nitrogen-removing biofilm.

Keywords: biofilms; ammonium oxidation; nitrifies

1 Introduction

Bioreactors with biofilms, particularly Rotating Biological Contactor (RBC), have been extensively used to remove organic carbon and nitrogen from wastewater with low C:N ratio. In most cases an aerobic oxidation of ammonium to nitrate (nitrification) is carried out by ammonium and nitrite oxidizers bacteria [1].

Moreover, this kind of reactor has been successfully used to establish an autotrophic nitrogen-removing biofilm, i.e., biofilms showing anaerobic ammonium oxidation (anammox) activity [2, 3]. In this process
ammonium removal is carried out by autotrophic bacteria belonging to the order of Planctomycetales using nitrite as the final electron acceptor [4]. However, Planctomyces seems to be dependent on the activity of aerobic ammonium oxidizers bacteria that will produce nitrite in oxygen limited conditions, e.g., at the oxic/anoxic interface, like biofilms [2]. In addition, Egli et al. (2003) revealed that the main groups of microorganisms constituting this kind of biofilms were ammonium-oxidizing bacteria from the Nitrosomonas europaea/eutropha group, nitrite-oxidizing bacteria mainly from the genus Nitrospira, anaerobic ammonium-oxidizing bacteria and filamentous bacteria from the phylum Bacteroidetes. However, there are no studies comparing the characteristics of the RBC formed biofilms in different support materials. The aim of this work is to study the

general microbiological characteristics and ammonium oxidation efficiencies of biofilms formed in RBC systems containing three different support materials.

2   Material and methods

2.1 RBC reactors

Three 5L RBC reactors with disks of three different support material, (Polyvinylchloride – PVC; Polyestyrene – PE and Polyurethane – PU) with superficial areas of $1.8m^2$, $1.8m^2$ and $0.775m^2$, respectively, were inoculated with sludge from an activated sludge system treating domestic wastewater containing 3.0gTSS/L. The reactors were fed daily with 5L of synthetic medium containing ammonium, bicarbonate and micronutrients [5] and pH was maintained between 7.5 and 8.0. Disks rotation speed were 2rpm. Increasing ammonium nitrogen loading rates (0.11; 0.28 and $0.55gN\text{-}NH_4^+/m^2.d$) were applied to the reactors with PVC and PE as support material, as well as to the reactor using PU (0.26; 0.64 and $1.3gN\text{-}NH_4^+/m^2.d$). Ammonium oxidation efficiency was monitored by periodic analysis of ammonium, nitrite and nitrate.

2.2 Analitycal determinations

Total suspended solids (TSS) and ammonium nitrogen ($N\text{-}NH_4^+$) were determined according to *Standard Methods* [6]. Nitrite was determined colorimetrically by the kit Nitriver (HACH® Company) and nitrate was determined by the method of salicylic acid [7].

2.3 Biofilms microbiological characterization

Biofilms were investigated by optical microscopy (Olympus BX 40). Most Probable Number (MPN) technique was used to estimate ammonium-oxidizers bacteria (AOB) and nitrite-oxidizers bacteria population (NOB) [8].

3   Results and discussions

3.1 RBC reactors performances

Figure 1 shows the oxidation of ammonium to nitrate (Fig.1a) and to nitrite (Fig.1b) during the increasing ammonium loading rate applied to the RBC with PVC, PE and PU as support materials. In the lowest ammonium loading rates (between 1 and 200days), high nitrification efficiency was observed, since ammonium influent concentration was almost totally converted to nitrate (Fig.1a). After 200days for the higher ammonium loading rate ($0.55gN\text{-}NH_4^+/m^2.d$ for PVC and PE and $1.3gN\text{-}NH_4^+/m^2.d$ for PU), a nitrite accumulation was verified at the beginning of the period in the three RBCs (Fig.1b). However, after an operation time of about 100days in this load PVC and PE reactors reestablished the nitrification, while in PU reactor was observed an increase of nitrite production until about 700mgN/L. The mean nitrogen removal during the period of the higher ammonium loading rate in PU reactor was still low (about 98mgN/L.d). However it indicates that an anaerobic ammonium oxidation process could be established in this kind of support material applying a relatively low nitrogen loading rate when compared with the literature. Pynaert et al.

(2003) working with RBC reactor containing PVC disks obtained 89% of nitrogen removal at surface loading rate of 8.3gN/m^2.d. Later the same author [3] observed a nitrogen removal of about 90% for ammonium nitrogen loading rate of 1.7gN-NH$_4$$^+$/m^2.d when reticulated polyurethane was attached to the disks to improve the available surface area.

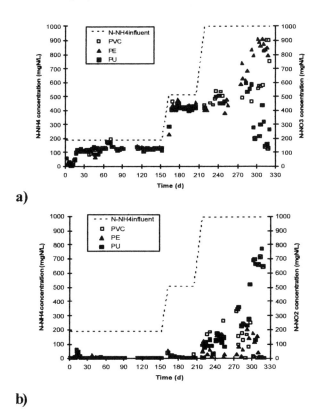

a)

b)

Figure 1. Oxidation of ammonium to nitrate (Fig. 1a) and to nitrite (Fig. 1b) during 330days of monitoring RBC reactors with PVC, PE and PU as support material.

3.2 Most Probable Number (NMP) Estimation

Results presented at Table 1 shows the estimated population of AOB and NOB in support materials in sludge used as inoculum and after 319days of reactors operation. It was observed an increase of AOB and a decrease of NOB population in the period, probably because NOB is apparently limited at higher ammonium concentrations. Higher AOB population was observed in PVC and PU. NOB lowest population was found in PU which is in agreement with the analytical determinations showing nitrite accumulation (Fig. 1b).

Table 1: Estimated population of AOB and NOB (MPN/gTSS) during the operation of RBC reactors

Sample	PVC		PE		PU	
	AOB	NOB	AOB	NOB	AOB	NOB
Inoculum	1.3×10^6	4.7×10^4	1.3×10^6	4.7×10^4	1.3×10^6	4.7×10^4
319 day	1.1×10^{10}	1.4×10^2	5.2×10^6	1.1×10^1	1.9×10^8	$<10^1$

3.3 Microscopic characteristics of the biofilms

The structure of the biofilm, the abundance of free filamentous bacteria and the appearance of protozoa/micrometazoa were analyzed by optic microscopy. The first observation was the surprising appearance of protozoa/metazoa during all operation time considering that its growth is limited by the use of synthetic wastewater and the high ammonium and nitrite concentration in the reactor. The role of Protozoa is often discussed especially for nitrifying reactors. van Dongen et al. (2001) attributed to free swimming ciliates a negative effect on the stability and efficiency of a continuous flow SHARON reactor, without sludge retention and a hydraulic retention time of one day. A decrease on nitrification efficiency was also observed in mixed heterotrofic-autotrofic systems promoted by excessive abundance of rotifers and flagellates [10]. On the other hand, Roessink and Eikelboom (1997) proved the importance of sessile ciliates for the effluent quality of mixed biofilms systems. Experiences with autotrophic biofilms are rarely but in this work no negative effect on nitrifying efficiency by protozoa was observed.

The abundance of organisms in PVC reactor was different from the others two, even if the loading rate was the same like for PE reactor. The main difference of PVC reactor is the higher abundance of free swimming ciliates, Metopus spec. and Euplotes spec. in the first phase, Aspidisca spec. in the second and third phase, rotifers in the lower loaded phase and nematodes in all phases. In PVC reactor, the nematodes were founded in the sludge and attached in the biofilm, probably contributing for the frequent lost of biofilm in this reactor. Attached (sessile) ciliates, normally typical for biofilm reactors like aerated filter, RBC and trickling filter [11], were only observed in the last phase of the PVC reactor and not observed in PE reactor but during all phases of PU reactor. Probably the fixation conditions and the protection for a biofilm formation were better on the porous PU material.

In addition, high density of filamentous bacteria was observed in the three RBC reactors, mainly during the lowest loading rates. This observation agrees with Egli et. al. (2003) that found filamentous bacteria in a nitrogen-removing biofilm formed in RBC reactors. Probably these bacteria formed the base for the development of the later biofilm. In the three reactors, filamentous bacteria could not be observed in higher loading by optical microscopy, probably due age and high density of biofilms.

4   Conclusions

RBC reactor of PVC and PE were efficiently used to oxidize ammonium to nitrate. Nitrite was produced in PU reactor for the higher ammonium loading rate, indicating the potential of this material to establish anaerobic ammonium oxidation process in biofilms. In PU reactor was also observed higher abundance of sessile ciliates and a denser biofilm probably due to the better condition of the porous PU material to attach microorganisms.

Acknowledgements

The support of CNPq is gratefully acknowledged.

References

[1]   D. N. Hiras, I. D. Mariotis, and S. G. Grigoropoulos, biores. technol. **93**, 91 (2004).
[2]   K. Egli, C. Bosshard, C. Werlen, P. Lais, H. Siegrist, A.J.B. Zehnder, and J. R. van der Meer, microb. ecol. **45**, 419 (2003).
[3]   K. Pynaert, B. F. Smets, D. Beheydt and W. Verstraete, environ. sci. technol. **38**(4) 1228 (2004).
[4]   L.G. J.M.Van Dongen, M.S.M. Jetten, M.C.M van Loosdrecht. The Combined Sharon/Anammox Process. Stowa: Foundation for Applied Water Research. IWA Publishing. London (2001).
[5]   J. L. Campos, A. Mosquera-Coral, M. Sánches, R. Méndez, J. M. Lema, wat. res. **36** 2555 (2002).
[6]   APHA – American Public Health Association, Standard Methods for the of Water and Wastewater. 19th ed, Washington, DC, (1995).
[7]   D.A. Cataldo, M. Haroon, L.E. Schrader, V.L Youngs. comun. soil sci. plant anal. **6** 71 (1975).
[8]   M. Alexander in: C. A. Black (ed.), Methods of soil analysis, part 2. Chemical and microbiological properties.American Society of Agronomy, Madison, Wis. 1467 (1982).
[9]   K. Pynaert, B.F. Smets, S. Wyffels, D. Beheydt, S. Siciliano, W. Verstraete, app. environ. microbiol. **69** 3626 (2003).
[10]   F.J.M Verhagen, H.J. Laanbroek, appl. environ. microbiol. **58** 1962 (1993).
[11]   R. Roessink, D.H. Eikelboom, wat. sci. technol. **36** 237 (1997).

Occurrence of two metabolic pathways in benzo[a]pyrene degradation by a Deuteromycete fungus *Fusarium solani*

C. Rafin[*,1], E. Veignie[1], P. Woisel[2], F. Cazier[3], G. Surpateanu[2]

[1]Université du Littoral Côte d'Opale (ULCO), 50 rue Ferdinand Buisson, BP 699, 62228 Calais Cédex, France
[2]ULCO, Laboratoire de Synthèse Organique et Environnement, MREI, 145 Avenue Maurice Schumann, 59140 Dunkerque, FRANCE
[3]ULCO, Centre Commun de Mesure, MREI, 59140 Dunkerque, FRANCE

Two distinct classes of metabolites in the oxidation of benzo[a]pyrene (BaP) by a Deuteromycete *fungus Fusarium solani* were isolated by high pressure liquid chromatography and characterized by UV-visible, gas chromatographic and mass spectrometric techniques. First, one metabolite, obtained in low amount, was identified as 6-hydroxybenzo[a]pyrene sulfate. This metabolite is known to be produced during BaP detoxification process mediated by cytochrome P450 monooxygenase and aryl sulfatases. The second metabolite, which was correlated with high and rapid BaP degradation, was partially identified as belonging to chemical classes of phthalates. To our knowledge, this class of metabolites has not been yet described as BaP metabolites by *F. solani*. The presence of phthalates as products of BaP ring fission indicates oxidation mechanism acting by free radical attack.

Keywords: Polycyclic Aromatic Hydrocarbons; Benzo[a]pyrene; Biodegradation; Fungi; Reactive Oxygen Species

1   Introduction

Benzo[a]pyrene (BaP), a polycyclic aromatic hydrocarbon (PAH) containing five fused benzene rings, is widely distributed in terrestrial and aquatic ecosystems due to a variety of anthropogenic activities. BaP has been classified by the US Environmental Protection Agency as a priority pollutant because of its carcinogenicity, teratogenicity and acute toxicity [1]. Among processes whereby PAHs are removed from the environment, microbial degradation plays a major role in the remediation of contaminated sites. However, bioremediation of PAH-polluted soil is severely hampered by the low rate degradation of high molecular weight PAHs due to site specific environmental factors, such as pollutant bioavailability, nutrients, redox potential, etc. and/or due to the scarcity of adequate microorganisms. Therefore, many investigations have focused attention on microorganisms able to degrade high molecular weight PAHs for cleaning up contaminated sites [2].
A diverse group of ligninolytic and non-ligninolytic fungi have been shown to be able to cometabolically oxidize BaP to both polar and non polar metabolites; however, only a few isolates can mineralise it [1]. But, despite promising results in laboratory experiments, which outlined the potential of various microorganisms to degrade BaP, these organisms may not be prevalent in soils where PAH/BaP remediation is necessary.
In our previous studies, the use of filamentous fungi isolated from PAH-contaminated soil has been investigated for bioremediation purposes. We recently isolated a Deuteromycete *fungus Fusarium solani* F33, that was able to mineralise [7,10-^{14}C] benzo[a]pyrene (BaP) rapidly at early stages of culture (15 hr) during the germination of fungal spores. BaP metabolization was also confirmed by isolating metabolic products identified as

benzo[a]pyrene quinones [3]. The present investigation extends our studies by isolating and identifying novel benzo[a]pyrene metabolites formed by strain F33 and demonstrated that *Fusarium solani* might have multiple metabolic pathways for BaP degradation.

2 Materials and methods

2.1 Chemicals and media

Benzo(a)pyrene (BaP) was purchased from Acros Organic (Noisy le Grand, France). Solvents and standard of dibutyl phthalate were obtained in the highest purity available from Merck (Darmstadt, Germany). Standard of 6-hydroxybenzo[a]pyrene sulfate was purchased from NCI Chemical Reference Standards Repository (Kansas City, USA). Malt Yeast Extract Agar (MYEA) medium contained 2.0% malt extract, 0.2% yeast extract and 1.5% agar in distilled water. The standard basal medium used for benzo[a]pyrene degradation studies was the Mineral salts Medium (MM) as described previously [3].

2.2 Microorganism and culture conditions

This study was carried out with *Fusarium solani* F33 previously isolated from petroleum-contaminated soil [4]. It was maintained on MYEA slants at 18°C and subcultured every 3 months.
Cultures for BaP metabolism were established as follows: BaP (0.252 g l^{-1} resulting in 10^{-3} mol l^{-1}) dissolved in 1 ml acetone was added to 250-ml empty Erlenmeyer flasks. After total evaporation of the organic solvent, the MM-medium (50 ml per flask) was added. The flasks were sterilized at 121°C for 20 min. Inoculation was performed by adding a spore suspension of *Fusarium solani* (aged 10 days), prepared as described previously [4], so as to obtain a final concentration of 10^4 spores per ml of culture medium. To detect abiotic BaP degradation, flasks without fungi were prepared and processed analogously. All treatments were incubated at 25°C with a 12 h photoperiod for 12 days on a reciprocating shaker (Laboshake, 90 min^{-1}). Triplicates were used to determine BaP metabolism at scheduled times.

2.3 Extraction and detection of BaP metabolites

At scheduled times, the cultures were lyophilized during three days. Total lyophilized cultures were scraped and extracted for 16 h in a Soxhlet apparatus with dichloromethane (DCM) in order to remove adsorbed and non-metabolized BaP on hyphae. Organic fractions were concentrated in 4 ml DCM/ethyl acetate (50:50, v/v). BaP metabolites detection was performed using HPLC Waters 2690 system fitted with a Waters Symmetry[R], C18, 5 μm, 100 Å column and a Waters 996 Photo Diode Array Detector. The separation was achieved with a 5 min linear gradient of acetonitrile/water (60:40 to 100:0, v/v) at a solvent flow rate of 0.3 ml.min^{-1} and ending with acetonitrile/water (100:0, v/v) during 50 min.
Ultraviolet and visible spectra of metabolites M1 and M2 were obtained using the above-mentioned diode array detector attached to the HPLC system.

2.4 Identification and quantification of metabolites M1 and M2

Tentative identification of metabolite M1 was performed using a GC/MS (4D Saturn, Varian Inc., CA, USA) with a capillary column (DB1MS, non polar column, 60 m x 0.25 mm x 0.25 μm, J &W Co., CA, USA) and a mass spectrometer detector being operated in electron impact mode. The oven temperature was set from 40°C (5 min) to 300°C (52 min) at 5°C/min ramp; this temperature being maintained for 30 min. Helium was the carrier gas and was regulated in pressure of 10 psi. The transfer line temperature was set to 310°C. Mass spectra were recorded at 1 scan s^{-1} under electron impact at 70 eV, mass range 40-450 amu. For qualitative analysis, the peak of M1 resolved in gas chromatography was identified by matching his fragmentation profile with those of the NIST92 library. The identity of M1 was confirmed by matching the retention time and fragmentation profile with commercial authentic standard of dibutyl phthalate.

Metabolite M2 was identified by matching the retention time and UV absorption spectrum of the commercial authentic standard with those of the resolved peak obtained by HPLC.

For quantitative analysis, at scheduled times, the peak area of metabolites was measured at 254 nm and compared with calibrated concentrations curve of standards.

3   Results and discussion

The purpose of this study was to determine the time course of BaP metabolites produced by *F. solani* and to identify novel metabolites formed by strain F33. At each scheduled time, the HPLC elution profiles of the cultures extracts (treatments and controls) were compared in order to quantify and identify the peaks derived from BaP metabolism. This comparison led to the detection of two principal metabolites, referred to as metabolites M1 and M2, respectively.

At the beginning of BaP degradation course, the production of M1 occurred rapidly during the first two days of culture (Fig. 1a). The quantity of M1 increased steeply, reaching a maximum level of 78 nanomoles after 2 days, but disappeared afterwards until 8th day. A second phase of M1 production was then observed between 8th and 12th day of BaP conversion. To elucidate the structure of the isolated metabolite M1, mass spectral parameters were determined. The mass spectrum of M1 contains a dominant fragment at m/z 149, characteristic of phthalic anhydride. Its UV spectrum shows also strong similarities with that of dibutyl phthalate, standard used in this study (Fig. 1b). Thus, we concluded that M1 belonged to chemical classes of phthalates. Nevertheless, in order to identify the nature of alkyl chains, ^{1}H-NMR analysis of purified metabolite needs to be performed. In our study, the weak amount produced did not allow to realize such precise identification

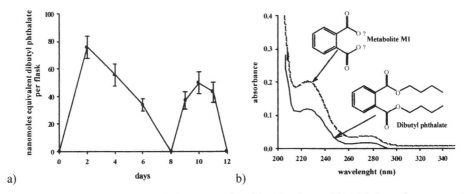

Fig. 1 a) Kinetic course of metabolite M1 produced by *F. solani* and b) UV absorption spectra of metabolite M1 and dibutyl phthalate

In addition to metabolite M1, small amounts of a second metabolite M2 were detected on day 6 (Fig. 2a). Then the quantity of M2 increased, reaching a maximum level of 170 picomoles per flask during the time of experiment. M2 was identified by its HPLC elution profile and its UV absorption spectrum, as 6-hydroxybenzo[a]pyrene sulfate, a well known transformation product of the microbial metabolism of BaP (Fig. 2b).

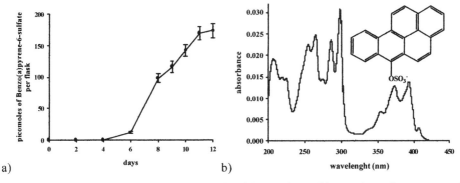

Fig. 2 a) Kinetic course of metabolite M2 produced by *F. solani* and b) UV absorption spectrum of metabolite M2

The identification of sulfate conjugates demonstrated that initial attack on BaP was apparently catalysed by a cytochrome P-450 monooxygenase. Many fungi have already been shown to oxidise BaP by cytochrome P-450, a mechanism similar to this observed in mammals, resulting in formation of polar metabolites (either trans-dihydrodiols via BaP epoxides, or phenols by a non enzymatic rearrangement) [1]. The conjugation reaction of the hydroxylated derivatives of BaP, leading to glucoside, glucuronide or sulfate conjugates is generally implicated in detoxification mechanism [5]. Nevertheless, metabolite M2 was detected in very low amount in this study. This suggests that either this proposed degradation pathway for BaP in *F. solani* is not the major phenomenon involved in BaP oxidation or the resulting oxidation products did not accumulate.

We also showed that *F. solani* could cleave the aromatic ring of BaP and formed metabolite M1, partially identified as a phthalate derivative. To our knowledge, such phthalate metabolites have not been yet described as BaP metabolites by fungi. However, production of phthalates has already been reported in degradation of other PAHs than BaP by lignolytic fungi [6, 7] or by spent mushroom compost [8]. Phthalic derivatives could then rapidly further be broken down into smaller molecules and/or carbon dioxide. This result correlated well with our previous studies in which BaP mineralization by *F. solani* was observed rapidly during the spores' germination process [3]. Phthalic derivatives were also previously reported as degradative products of PAHs by both ligninolysis, ozonation and photocatalytic oxidation. All these three mechanisms act by free radical attack on PAHs [9]. Therefore, another degradation pathway of BaP by *F. solani* may be proposed, independent of hemeperoxidases enzymes, and probably involving reactive oxygen species. This is in accordance with our previous results obtained by an indirect approach using inhibitors of such enzymes [10]. This unusual biodegradative strategy, leading to phthalic metabolites as BaP ring fission, will be further investigated as it might permit a rapid PAH degradation in contaminated soils.

References

[1]   A.L. Juhasz, R. Naidu, Int. Biodeterior. Biodegrad. **45**, 57 (2000)
[2]   C.E. Cerniglia, Biodegradation, **3**, 351 (1992)
[3]   E. Veignie, C. Rafin, P. Woisel, A. Hadj Sahraoui, F. Cazier, Polycyclic Aromat. Compd., **22**, 87 (2002)
[4]   C. Rafin, O. Potin, E. Veignie, A. Hadj Sahraoui, M. Sancholle, Polycyclic Aromat. Compd., **21**, 311 (2000)
[5]   G. Capotorti, P. Digianvincenzo, P. Cesti, A. Bernadi, G. Guglielmetti, Biodegradation, **15**, 79 (2004)
[6]   T. Cajthaml, M. Möder, P. Kacer, V. Sasek, P. Popp, J. Chromatogr. A, **974**, 213 (2002)
[7]   K.E. Hammel, B. Green, W. Zhi Gai, Proc. Natl. Acad. Sci., **88**, 10605 (1991)
[8]   K.L. Lau, Y.Y. Tsang, S.W. Chiu, Chemosphere, **52**, 1539 (2003)
[9]   Y. Zeng, P.K.A. Hong, D.A. Wavrek, Environ. Sci. Technol., **34**, 854 (2000)
[10]   E. Veignie, C. Rafin, P. Woisel, F. Cazier, Environ. Pollut., **129**, 1 (2004)

Oleophilic fertilizers and Bioremediation: A new perspective

J.R. Gallego[*,1], D. Menéndez-Vega[2], E. González-Rojas[2], J. Sánchez[2], M.J. García-Martínez[3], J.F. Llamas[3]

[1] Campus de Mieres. Univ. de Oviedo. C/ Gonzalo G. Quirós, S/N – 33600 Mieres (Asturias-SPAIN)
[2] Área de Microbiología. Univ. de Oviedo. C/Julián Clavería, S/N – 33006 Oviedo (Asturias-SPAIN)
[3] Dpt. Ing. Química y Combustibles. ETSMinas, U. Pol. Madrid. Ríos Rosas 21, 28003-Madrid (SPAIN)

[*]Corresponding author: e-mail: jgallego@uniovi.es, Phone: +34 985 458064, Fax: +34 985 458188

The role of oleophilic fertilizers in bioremediation is based on physic and chemical properties of their oleaginous components (usually oleic acid) that surround a core amended with sources of N and P. Since 1990 most uses of these products are related to the cleaning up of marine oil spills, but recently new applications have been suggested. The main scope of this work is the description of some pilot-scale and full-scale experiments in polluted groundwater and soils with S-200 series, a new generation of olephilic fertilizers. We also present a review of this technology and some unpublished results of the application of S-200 in Prestige oil spill.

Keywords: oleophilic fertilizers, bioremediation, soil, groundwater, Prestige, hydrocarbons.

1   Introduction

1.1 Relevance of oleophilic fertilizers

Oil biodegradation mainly occurs at the oil-water interface. Since oleophilic fertilizers are to adhere to oil and to provide nutrients to that interface, their use in marine oil spilled habitats is recommended, especially wherever rocky shores and/or wave action difficulties the effectiveness of other fertilizers (slow release compounds and water soluble nutrients). The best known application of oleophilic fertilizers (concretely Inipol EAP 22) took place during the remediation of Exxon Valdez oil spill in Alaska [1]. The original product (lauryl phosphate, urea, oleic acid, 2-butoxy-1-ethanol and water; C:N:P = 62:5:1) was designed by Elf Aquitaine after Amoco Cadiz oil spill in France in 1978. After ten years of research [2] it was applied in Alaska, firstly in some experiments in sandy and cobble beaches, where the product showed better results than other standard fertilizers, and then in extended areas of oiled sediments for three summer campaigns. Controversies about bioremediation effectiveness were solved by using GC-MS analytical techniques with appropriate chemical biomarkers (hopanes). However, some doubts about the toxicity of determinate components of Inipol, the solubility of Urea (used as N source) and the role of a supplementary carbon source such as oleic acid were pointed out.

After Valdez's experience Inipol and similar products as F1 or Inipol plus have been used in other experimental or real marine oil spills with acceptable results [3–5]. Recently in Europe, Erika and specially Prestige spillages have promoted new research in remediation of affected shorelines and, as a consequence, advances in the field of oleophilic fertilizers and bioremediation have been done [6].

1.2 Scope

S200 is a new oleophilic fertilizer listed in the NCP (National Oil and Hazardsous Substances Pollution Contingency Plan) in USA as a Bioremediation Agent. Its composition (urea, phosphoric ester, water, oleic acid and glycol ether) is quite similar to Inipol's but it has reduced toxicity, improved emulsification power and allows dilution and subsequent application in environments (mainly groundwater but also in soils) different than shorelines affected by oil spills. Thus, in this work we show some data and considerations related to the effectiveness of this product in Prestige polluted sites but also in hydrocarbon-affected groundwater and soils.

2 Field works and laboratory experiments

2.1 Prestige sites

Different works in bioremediation of the Prestige affected sites consisted on scale-pilot experiments, full-scale remediation and laboratory tests of different products and technologies, including oleophilic fertilizers. Whereas other researchers first reported S-200 advantages against commercial bioaugmentation products [7], in our bioremediation campaigns, S-200 was applied in different areas including cobble beaches, rocky areas in the upper tidal zone and sandy sediments. Depending on sites (different locations at Asturian and Galician coasts) and fuel quantities, two or three applications in periods of three to six months supposed a range of 0.6–1 litres of S-200 per 1 kg of hydrocarbons (Figure 1); control areas were established to quantify natural attenuation. Representative composite samples of fuel were taken periodically and results were referred to the original Prestige fuel (November 2002).

Fig.1. Application of S-200 in areas of Costa de la Muerte (Galicia-Spain) affected by the Prestige heavy fuel. The product was extended with a backpack powered sprayer.

2.2 Biotreatment of oily waters.

Medium-scale experiments with S-200 were designed in order to evaluate their potential to emulsify and degrade oily waters such as those produced in oil-water separators and in ship warehouses. To this scope, 100 ml of diluted S-200 was added in an aerobic bioreactor to 20 l of standard sea water polluted with a mixture of 1% of diesel fuel and 1% of Arabian light crude; a second bioreactor without S200 addition was used to control results. Bioreactors content was shaken at 60 rpm and hold on room temperature. Microbiological and chemical analyses (hydrocarbon degradation, pH, dissolved oxygen, etc.) monitored the evolution of the test.

2.3 Groundwater polluted by petrol station spillage

To study S-200 effectiveness in groundwater it was selected a petrol station where took place a spillage of 10.000 litres of diesel fuel. In this site, after more than one year carrying out a SVE (soil vapour extraction), a sampling campaign revealed that residual contaminants were still affecting soil and groundwater. A bioremediation approach was then suggested. S-200, in a dosage of 1 litre per 1 kg of hydrocarbons, was injected in piezometers in semi-continuous applications. After two months a new sampling campaign was made.

2.4 Analytical methods.

Liquid and solid samples were collected in appropriate bottles, sealed and preserved in darkness at 4°C. At any case after less than ten days, bottles were opened and the liquid ones circulated through a SPE cartridge (solid phase extraction – EPA Method 3535) to recover hydrocarbons with an ulterior washing with dichloromethane. Hydrocarbons from solid samples were extracted with Soxhlet and purified by liquid chromatography (EPA Method 3540C). Both solid and liquid extracts were injected into a gas chromatograph linked to a mass spectrometer (GC-MS). GC system HP 6890 Series was equipped with a capillary column (AT5 alltech); the column oven temperature was raised from 60°C to 300°C at 6°C/min. This equipment was linked to a mass detector (MSD HP 5973 Series) which allowed hopane and isoprenoid normalizated determinations (EPA Method 8270C). When microbiological determinations were performed, they consisted on measurement of microbial growth by plate counting in culture mediums with hydrocarbons as sole carbon source (0.13 % NH_4NO_3; 0.05% $MgSO_4 \cdot H_2O$; 0.02% $CaCl_2 \cdot 2H_2O$; 0.5% KH_2PO_4; 0.5% K_2HPO_4; 2% diesel fuel plus 3% agar). Additionally, isolation and study of selected strains were done (data not included).

3 Results

3.1 Prestige sites.

Table 1 shows a summary of the results obtained in one of the experiences with S-200 at Prestige sites. Basically, S200 improves ratios of alkane degradation compared with natural attenuation process and allows partial degradation of aromatics and sulfur compounds and thus detoxifies fuel residues. Visual results were not "flashy" because heavy –but no-toxic- oil fractions such as resines and asphaltenes are degraded very slowly in field. Other author's results fitted with these and thus, in September 2003, Spanish Department for Environment authorised S-200 application in areas still affected.

Table 1. Biodegradation percentages of hydrocarbon families obtained after four months and determined by GC-MS at Prestige-affected shorelines (Muxía, La Coruña, Spain). Experiments were performed during four months (June-September 2003) in 20 m^2 parcels of shore rocks

COMPOUNDS	CONTROL EXPERIMENT	S-200 (TWO APP., TOTAL: 0,6 l/m²)	S-200 (THREE APP., TOTAL: 0,9 l/m²)
Light n-alkanes (C_{11}-C_{20})	80	100	100
Heavy n-alkanes (C_{21}-C_{38})	20	59	67
Branched alkanes	30	35	48
Aromatics (Alkyl-PAHs)	5	25	45
Dibenzothiophenes (S-compounds)	0	26	45

554

3.2 Biotreatment of oily waters

Emulsifying power of S-200 was observed macroscopically and then microscopically as it is shown in Figure 2. This effect promoted a noticeable degradation of all hydrocarbon families (Table 2).

Table 2. Biodegradation percentages of hydrocarbon families determined by GC-MS in bioreactor experiments after one month.

COMPOUNDS	CONTROL	S-200
Light n-alkanes (C_{11}-C_{20})	20	54
Heavy n-alkanes (C_{21}-C_{38})	5	40
Branched alkanes	0	15
BTEX and derivatives	35	55
Aromatics (Alkyl-PAHs)	0	35
Sulfur compunds (dibenzothiophenes)	0	12

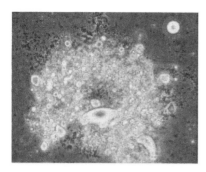

Fig. 2. Contrast-phase photograph (1000x – Nikon labophot) shows different bacteria degrading a mixture of emulsified hydrocarbons. Emulsification promotes bacterial adhesion to oil droplets and it is one of the benefits of using oleophilic fertilizers.

3.3 Experience 3: Petrol station.

Table 3 reports TPH removal in groundwater after S-200 application. Initial objetives were gained and final levels of pollution fulfilled legal requirements, thus remediation works finished. S200 injection enabled nutrient access to oily residues sorbed in soil, and increased surface emulsification in hydrocarbon droplets remaining in groundwater.

Table 3. TPH (mg/l) determinations by Infrared Spectrometry of groundwater in ten piezometers distributed along the contaminant plume.

Piezometer	P1	P2	P3	P4	P5	P6	P7	P8	P9
Before S-200 application	0,08	0,34	1,53	FP*	8,7	FP*	1,10	5,9	FP*
After two months	< 0,05	< 0,05	< 0,05	< 0,05	< 0,05	0,176	< 0,05	< 0,05	< 0,05

(*) Free phase

4   Conclusions

Oleophilic fertilizers, concretely S-200, are probably the best products to bioremediate shorelines affected by oil spills. Their effectivity, largely probed after Exxon Valdez oil spill until present day, is based on their resistance to wave action, their hydrophobicity and their emulsifying properties which facilitates bacterial activity. On the other hand, a new field of application of these products in polluted sites has been opened within the experiences explained in this work. Thus, groundwater and soil hydrocarbon-pollution can be treated with oleophilic fertilisers as an alternative to pump&treat methods, slow-release fertilizers and other commercial products. Other known experiences (data not included in this work) have shown S200 effectiveness in hydrocarbon-polluted "brownfields" by means of a simultaneous addition of a surfactant and an oxygen release compound to improve bioavailability of weathered hydrocarbons. However, further research on applicability of oleophilic fertilizers in landfarming techniques, especially to treat clay soils, or in 'in situ' soil remediation are required

Acknowledgements

Authors gratefully thank IEP Europe for technical support.

References

[1]   Bragg, J.R.; R.C. Prince; E.J. Harner and R.M. Atlas. Effectiveness of bioremediation for the Exxon Valdez oil spill.. Nature 368: 413-418 (1994).

[2]   Ladousse, A. and B. Tramier. Results of 12 years of research in spilled oil remediation Inipol EAP22. Proceedings of the 1991 Oil Spill Conf. pp. 577-581 (1991).

[3]   Santas, R.; A. Korda; A. Tenente; K. Buchholz and P. Santas. Mesocosm analysis of oil spill bioremediation with oleophilic fertilizers: Inipol, F1 or both?. Mar. Poll. Bull. 38(1): 44-48 (1999).

[4]   Coulon, F.; E. Pelletier; L. Gourhant and D. Delille. Effects of nutrient and temperature on degradation of petroleum hydrocarbons in contaminated sub-Antarctic soil. Chemosphere 58, 1439-1448 (2005).

[5]   Santas R. and P. Santas. Effects of wave action on the bioremediation of crude oil saturated hydrocarbons. Marine Pollution Bulletin 40 (5): 434-439 (2000).

[6]   Choi, S.C.; K. Kwon; J. Sohn and S. Kim. Evaluation of fertilizer additions to stimulate oil biodegradation in sand seashore mesocosms. Journal of Microbiology and Biotechnology 12 (3): 431-436 (2002).

[7]   Murado, M.A; J. Mirón; M. P. González; J.A. Vázquez; M.L. Cabo and J. Pintado. Prestige oil spill. Results of bioremediation assays on supra-tidal rocks of Sálvora Island (Galice, Spain). Interspill 2004, Trondheim (Norway) (2004).

On Site bioremediation and washing techniques in a cobble beach affected by Prestige oil spill

J.R. Gallego[*,1], L. Fernández[2], J.R. Fernández[2], F. Díez-Sanz[2], S. Ordoñez[2], E. González-Rojas[2], A.I. Peláez[3], J. Sánchez[3]

[1] Campus de Mieres. Univ. de Oviedo. C/ Gonzalo G. Quirós, S/N – 33600 Mieres (Asturias-SPAIN)

[2] Dpto Ingeniería Química. Univ. de Oviedo. C/Julián Clavería, S/N – 33006 Oviedo (Asturias-SPAIN)

[3] Área de Microbiología. Univ. de Oviedo. C/Julián Clavería, S/N – 33006 Oviedo (Asturias-SPAIN)

[*]Corresponding author: e-mail: jgallego@uniovi.es, Phone: +34 985 458064, Fax: +34 985 458182

Bahinas beach, located in the coast of Asturias (Northern Spain), was affected by Prestige oil spill in December 2002. The accumulation of fuel residues implied a severe contamination of cobble sediments situated in supratidal zone, and mechanical removal by water washing of sediments showed many difficulties on the immediate months following the spill. Therefore, the aim of this work was the selection of an effective method to remediate the polluted beach. To this scope, some experiments were carried out in order to select sources of nutrients, surfactants and bioremediation products. This approximation allowed us to evaluate the feasibility of using products such as commercial dispersants, oleophilic fertilizers, diesel fuel, biodiesel and other options.

Keywords: Bioremediation, chromatography, bacteria, Prestige, hydrocarbons.

1   Introduction

Shorelines affected by hydrocarbons are critical environments where a good selection of remediation techniques is necessary. In this sense, after Prestige fuel spill in Spanish coasts many different ideas have been applied to remediate sandy and rocky shorelines, taking into account that the special nature of Prestige fuel, very similar to Erika´s, made microbial degradation difficult because of the practical absence of easily-degradable compounds. Real-scale bioremediation in oil spills [1] has a short story that began in Alaska with Exxon Valdez spill where oleophilic fertilizers [2, 3] took an important role but, those special conditions pointed out in this case, and the different grain-size of sediments affected, makes necessary the study of new possibilities.

2   Experimental designs

2.1 Site description.

Bahinas beach is situated in the Cantabrian coast of Asturias (northern Spain). It is a medium grain-size sandy beach covered by a layer of pebbles and cobbles of an average diameter of 5 cm, at the west side of the beach the rock size grows to configure a small cliff; the total length of the beach is 400 m and in low tide width reaches 50 m. In December 2002 Prestige fuel left a considerable pollution of cobbles and rocky areas in the upper tidal limit. Remediation started applying hot pressurized water washing (100 bars and 110ºC). This process was very slow, not applicable for cobble areas, and generated important quantities of waste thus, a bioremediation approach was suggested. After initial laboratory experiments, combinations of surfactants, fertilicers and other products were selected to enforce bioavailability and emulsification of fuel residues and to optimise bioremediation effectiveness.

2.2 In situ treatment of shore rocks.

An area of 20 m^2 was divided in four parcels and different products and dosages suggested by manufacturers were applied (Table 1). Microbiological and chemical analysis were performed to follow biodegradation process. Experiments finished after 45 days.

Table 1. Description of treatments in rocky area.

Experience	Amendments	Dosification
Control parcel	-	-
Parcel A	Oleophilic fertilizer	Every fifteen days sprayed to get a final dossage of 1 litre per 1 kg of fuel.
Parcel B	Surfactant	Every fifteen days 1:1 mixture to get a total ratio of a litre per a m^2 of affected surface
Parcel C	Soy lecitin, diesel fuel, ammonium phosphate and nitrate	Weekly applications of 5 litres of water solution with 5 mg/l of lecithin, 5 ml/l of diesel, nutrients to get a C:N:P of 100:10:1

2.3 Cobble remediation.

Before implementing biotreatment of cobbles in water tanks, smaller containers were used to select amendments. Then, seven on site experiments with 20 kg of polluted cobbles inmersed in 13 liters of sea water were destributed as explained in table 2:

Table 2. On site experiments

Experiment	Amendments	Dosification
1	-	-
2	NH_4NO_3 and $(NH_4)_2PO_4$	C:N:P 100:10:1
3	NH_4NO_3, $(NH_4)_2PO_4$ and surfactant	C:N:P 100:10:1
4	Commercial bioaugmentation product (includes bacteria, nutrients and surfactant)	As proposed by manufacturer
5	NH_4NO_3, $(NH_4)_2PO_4$ and diesel fuel	C:N:P 100:10:1 1 ml/l diesel
6	NH_4NO_3, $(NH_4)_2PO_4$ and biodiesel	C:N:P 100:10:1 5 ml/l biodiesel
7	NH_4NO_3, $(NH_4)_2PO_4$,diesel fuel and soy lecithin	C:N:P 100:10:1 1 ml/l diesel 3 g/l lecithin

558

2.4 Chemical determinations.

Representative composite samples obtained by scraping were diluted in dichloromethane and then filtered. Extraction of organic phase with hexane to obtain asphaltenes was followed by liquid chromatography (EPA Method 3630) to obtain saturate, aromatic and resin fractions –SARA analysis-. Aliphatic and aromatic fractions were injected in a GC-FID equipment (HP 6890) and, when needed, in a GC-MS equipment with a mass detector MSD HP 5973. To evaluate biodegradation extents, different indexes were used: For alkane degradation C_{18}/Phytane index was calculated by GC/MS and beyond all special efforts were done to quantify toxic PAH and alkyl-PAH from the aromatic fraction with GC-FID [4]. Increasing proportion af asphaltene and resine fractions in SARA analysis in biodegraded samples was also studied. Oxygen, nitrogen and phosphorus consumption were also determined.

2.5 Microbiological methods.

Representative composite subsamples of fuel (1g) were added to 10 ml of a 0,1% $Na_2HPO_4 \cdot 12H_2O$ solution and vortexed vigorously for 10 minutes, then 1 ml of supernatant suspension was taken and 0.1 ml of successive dilutions of the culture were spread in triplicate TSB (tryptic soy broth) agar plates. Isolation of selected strains and further studies are not shown.

3   Results

3.1 In situ treatment of shore rocks.

Initial microbial counting reported averaged values of 10^3 CFU/g and control experiment did not go far beyond 10^5 CFU/g in weekly determinations. However, in the other parcels microbial growth was continuos and exponential for at least 20 days, specially in Parcel B where maximum levels reached 10^{10} CFU/g. Table 3 shows chemical data for aromatics; best results were obtained in Parcel A where a significative reduction of saturates (C_{18}/Phytane index showed more than a 50% of alkane degradation) and aromatics was observed.

Table 3. GC-FID determinations of selected aromatic compounds in rocky areas after 45 days. All results are expressed in μg/g of fuel. (1-MNA: 1- Methyl naphtalene; 1,2- DMNA: 1,2 Dimethylnaphtalene; 1-MFL: 1- Methyl fluorene; PHE: Phenanthrene; ANT: Anthracene; 1-MPHE: 1-Methylphenanthrene; PYR: Pyrene)

EXPERIMENT	1-MNA	1,2-DMNA	1-MFL	PHE	ANT	1-MPHE	PYR	Σ	% REDUCTION
Control (initial)	227	220	40	75	604	252	406	1824	0
Control (45 days)	204	201	39	72	590	121	243	1470	19.4
Parcel A	133	147	24	58	466	101	167	1096	39.9
Parcel B	181	183	32	64	514	115	194	1283	29.7
Parcel C	153	162	28	70	482	109	173	1177	35.4

3.2 Cobble remediation.

Microbial counting reported averaged values of 10^7 CFU/g at the beginning of the experiments and control experiment did not go far beyond 10^9 CFU/g in weekly determinations. Experiments 3, 5 and 7 showed maximum microbial growths reaching values closed to 10^{11} CFU/g fuel. Tables 4 and 5 show SARA and GC-FID results: The highest values of alkane degradation were reached in experiment 5 fo llowed by experiment 3. Nitrate and phosphate consumption was significative in most experiments with the exception of 1 and 4. Dissolved oxygen ranged from 1 to 4 ppm depending on the experience; hence, both aerobic and anaerobic populations degraded fuel utilizing ammonium and nitrate as source of nutrients and nitrate as alternative electron acceptor respectively, this was verified in experiences not shown in this work.

Table 4. GC-FID determinations of selected aromatic compounds in experiments with cobbles after 30 days. All results are expressed in µg/g of fuel.

EXPERIMENT	1-MNA	1,2-DMNA	1-MFl	PHE	ANT	1-MPHE	PYR	S	% REDUCTION
Control (initial)	66	138	204	196	42	590	90	1.326	0
1	51	125	194	185	41	575	86	1.257	5.3
2	33	69	98	132	24	352	69	777	42.5
3	48	82	105	115	27	361	84	822	40.6
4	50	98	126	170	38	559	56	1.097	18.3
5	28	72	97	120	27	348	61	773	42.7
6	38	102	111	138	38	493	52	972	27.7
7	35	96	99	136	38	459	50	913	32.1

Table 5. SARA analysis after 30 days of cobbles biotreatment.

EXPERIMENT	SATURATES (%)	AROMATICS (%)	RESINS (%)	ASPHALT ENES (%)
Control (initial)	26.0	37.9	18.3	17.8
1	25.9	36.5	18.9	18.7
2	24.1	24.1	26.3	25.5
3	24.5	26.3	24.4	24.8
4	25.6	29.2	22.8	22.4
5	24.2	23.8	26.4	25.6
6	25.1	27.0	24.1	23.8
7	25.0	27.3	23.8	23.9

4 Conclusions

Heavy fuel oils such as in the case of the Prestige can be naturally biodegraded in affected shorelines and thus, a bioremediation approach may improve natural processes despite these methods are limited by the recalcitrance of resin and asphaltene fractions. Experiments carried out in rocky areas demo nstrated the acceptable behaviour of oleophilic fertilizers for in situ bioremediation. This result corroborates the conclusions that other authors working in Prestige fuel bioremediation have reported [5]. As regards to on-site

560

cobble biotreatment, our research showed that microbial activity stimulated with nutrients and a surfactant (or even with a dissolvent such as diesel) removed an important amount of toxic compounds of fuel. Moreover, this approximation reduced fuel adhesion to cobbles and allowed subsequent water washing at low pressure and temperatures than used before, with an increasing effectiveness and a considerable reduction of the waste amount. More specific research on autochthonous microbial consortiums and optimization of nutrient sources is still being carried out.

Acknowledgements

This work was founded through Project CN-03-138, Government of Principado de Asturias (Spain)

References

[1]   Bragg, J.R.; R.C. Prince; E.J. Harner & R.M. Atlas. Effectiveness of bioremediation for the Exxon Valdez oil spill.. Nature 368: 413-418 (1994).

[2]   Zhu, X.; Venosa, A.; Suidam, M. & Lee, K.; "Guidelines for the bioremediation of marine shorelines and freshwater wetlands" Environmental Protection Agency. Office of Research and Development National Risk Management Research Laboratory. Land Remediation and Pollution Control Division, Cincinnati, USA (2001).

[3]   Head, I.M. & Swannell, R.P.J. Bioremediation of petroleum hydrocarbon contaminants in marine habitats. Curr. Op. Biotechnol. 10: 234-239 (1999).

[4]   Pollard, S.J.T; M. Whittaker & G.C. Risden. The fate of heavy oil wastes in soil microcosms I: a performance assessment of biotransformation indices. The Science of the Total Environment 226, 1-22 (1999).

[5]   Murado, M.A; J. Mirón; M. P. González; J.A. Vázquez; M.L. Cabo & J. Pintado. Prestige oil spill. Results of bioremediation assays on supra-tidal rocks of Sálvora Island (Galice, Spain). Interspill 2004, Trondheim (Norway) (2004).

Towards the applicability of rhodococci in monoaromatic compounds bioremediation

V. Jirku[*1], **A. Cejková**[1], **J. Masák**[1], **M. Pátek**[2], **and J. Nešvera**[2]

[1]Department of Fermentation Chemistry and Bioengineering, Institute of Chemical Technology,CZ-166 28   Prague 6, Czech Republic
[2]Institute of Microbiology, Academy of Sciences of the Czech Republic, CZ-14220 Prague 4, Czech Republic

[*] Corresponding author: e-mail: vladimir.jirku@vscht.cz

The capability of *Rhodococcus erythropolis* CCM 2595 (ATCC 11048) to utilize phenol, catechol, resorcinol, *p*-nitrophenol, *p*-chlorophenol, hydroquinone and hydroxybenzoate, respectively, or as respective binary mixtures with phenol, was described. This capability was found to depend on the substrate and its initial concentration. Some monoaromatic compounds have suppressive effect on the strain ability to utilize phenol in a binary mixture and easily utilizable monoaromatics are strong inducers of the phenol 2-monooxygenase (EC 1.14.13.7). The capacity of *R. erythropolis* to colonize a synthetic zeolite was demonstrated and the enhancement of phenol tolerance of biofilms utilizing phenol was observed. The effect of humic acids on phenol killing was described and discussed as well. To allow use of recombinant DNA technology for strain improvement, methods of genetic transfer (transformation and conjugation) in *R. erythropolis* were established.

Keywords: monoaromatic compounds, biofilm, humic acid, conjugation, 16S rRNA, *Rhodococcus erythropolis*

1   Introduction

Bioremediation of monoaromatic compounds has become an attractive alternative to the traditional physical and chemical decontamination methods that can be costly and can produce hazardous products. However, a limited capability of the degrading microorganism to utilize a wider range of monoaromatic homologues (including binary or complex mixtures), as well as a limited tolerance of degraders towards the cytotoxicity of these substrates are, among others, fundamental limitations on the biodegradation of these compounds. Therefore, efforts are focused on selecting strains with a disposition to tolerate the effect of both substrate interactions and substrate toxicity. In this context, the bacterial genus *Rhodococcus* has been shown to offer Gram-positive species manifesting not only a wide range of catabolic versatility], but also some cell properties required for a technological application of microbial degraders. In order to control properly these diverse capabilities, the development of both physiological and genetic tools enabling the construction of a tailor-made (rhodococcal) phenotype, is very important.

2   Materials and methods

Rhodococcus erythropolis CCM 2595 (ATCC 11048), was kindly provided by the Czech Collection of Microorganisms, Masaryk University Brno, Czech Republic. *Escherichia coli* XL1-Blue MRF' (Stratagene) and *E. coli* S17-1 (Simon 1983) were used for gene cloning and transfer, respectively. Precultivations of *R. erythropolis* were performed using a rotary shaker (90 rev min^{-1}, 20 °C) and basic salt medium [1]. The degradative function of suspended cells was tested using a bioreactor (Braun Biotech International, Germany) with an operating volume of 2.0 l, and with control of temperature (20 °C), pH (6.7) and rotation speed (120 rev min^{-1}). Cell growth/growth yield analysis was performed using Bioscreen C analyser (Labsystem Oy, Finland). Particles of synthetic zeolite MS 3A (Grace, USA) were colonized using a modified procedure according to Masák *et al.* [1]. The degradative function of rhodococcal biofilms was investigated using a jacketed, tubular reactor. Preparations of humic acids were kindly provided by the Institute of Inorganic Chemistry, Ústí nad Labem, Czech Republic. Phenol concentration in the cell-free medium was assayed using the 4-amino-antipyrine colorimetric method [2]. Phenol hydroxylase (phenol 2-monooxygenase, EC 1.14.13.7) was measured by assaying the oxidation of NADPH at 340 nm [U=0.1 ? A min^{-1}] [3]. The assessment of the colonization of zeolite surfaces was based on the image analysis of the carrier, using LUCIA 4.20 for Windows 2000 for light microscopy image processing. Ultrathin sections were examined in a transmission electron microscope Phillips EM 300 at 80 kV. DNA isolation, PCR, transformation of *E. coli*, DNA cloning and DNA analysis were done by standard methods [4]. The primers were designed according to the conserved regions of 16S rDNA sequences from five *Rhodococcus* type strains. The sequence of 881-bp 16S rDNA fragment was determined using the ABI Prism 2100 sequencer (Perkin Elmer) and deposited in GenBank under Accession No. AJ 620506.

3   Results and discussion

The potential of selected *Rhodococcus erythropolis* (from a screening of several other rhodococcal strains) to utilize a number of monoaromatic compounds, when added singly as sole carbon and energy source, is shown in Fig. 1. The decrease of specific growth rates of exponential growth had no uniform pattern on increasing the concentration of respective compounds. The inhibitory effect of the respective substrate concentrations starts in the range from 0.1 to 0.5 g l^{-1}. The individual effect of monoaromatic compounds on the lag period and growth yield was compared and no uniform capacity of *R. erythropolis* to utilize the range of substrates studied was found as well.

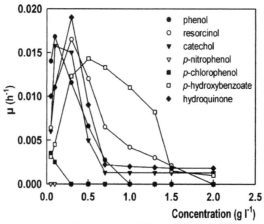

Fig. 1. Effect of substrate concentration on the specific growth rate of *R. erythropolis*.

The utilization of binary mixtures combining phenol with an additive (monoaromatic) substrate showed promoted growth yield only under the growth conditions combining phenol and resorcinol or phenol and hydroxybenzoate, respectively. More rapid utilization of these mixtures can be achieved by a pre-exposure to these substrates. The easily utilizable monoaromatics are strong inducers of phenol hydroxylase (EC 1.14.13.7). The induction capacity of others is much weaker and non-utilizable p-nitrophenol has no induction capacity at all. The enzyme induction was found to be a changeable marker as far as the effect of a long-term strain storage in the presence of phenol is concerned. The combination of phenol and resorcinol enhances the induction capacity of respective substrates.

The comparison between suspended and attached *R. erythropolis cells* provided results supporting the hypothesis that a multipoint, cell-support / cell-cell physical contact enhances cell capabilities to tolerate both the cytotoxic effects of xenobiotics and the stressing conditions of the environment, among others [5, 6]. The enhancement of phenol tolerance to those phenol concentrations, which almost or significantly inactivate the *R. erythropolis* growth in suspended cultures, was investigated comparing the specific uptake of phenol in suspended and biofilm cultures (Fig. 2). The phenol-utilizing biofilm (Fig. 3) can tolerate higher phenol concentrations. We were able to demonstrate a biofilm formation on the synthetic zeolite surface by most *Rhododococcus* strains tested, however, the colonization of zeolite was found to be fully reproducible in *R. erythropolis* only.

The precipitations associated with the outer part of cell wall of the *R. erythropolis* cells exposed to three dissolved humic acids, respectively (Fig. 4), encouraged the idea that such an attachment of a polydisperse, macromolecular compound could enhance the cell tolerance towards cytotoxic monoaromatics. Comparison of biomass yield in the cell populations exposed and not exposed, respectively, to three dissolved humic acids shows (Fig. 5), that the enhanced phenol tolerance in rhodococcal cells varies as a function of humic acid preparation and the initial concentration of carbon source (*R. erythropolis* has no capacity to utilize humic acids as sole as carbon source).

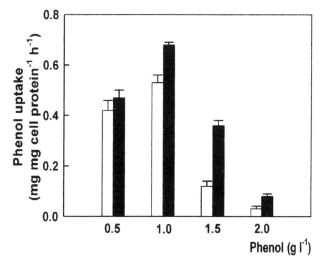

Fig. 2. Rate of phenol uptake in suspended (□) and attached (■) *R. erythropolis* cells. Error bars represent   total variations of two replicates.

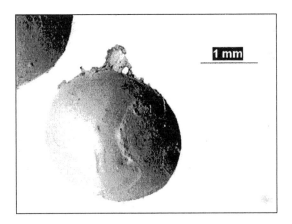

Fig. 3. Scanning electron micrograph of *R. erythropolis* biofilm formed on a zeolite MS 3A.

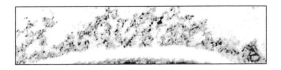

Fig. 4. Ultrathin section of *R. erythropolis* cell wall exposed to humic acid

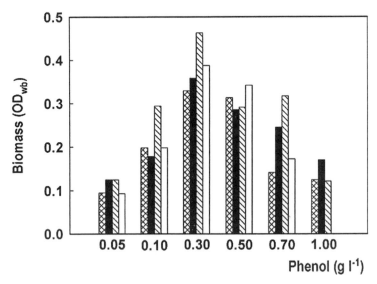

Fig. 5. Effect of phenol concentration on *R. erythropolis* growth yield in the presence (0.05 g l^{-1}) / absence (□) of humic acid S4, S5 (■) and PAB68.

The established methods for DNA manipulations in the characterized strain potentially allowing its purposeful modification have been as follows: established genetic transfer (electrotransformation, conjugation), cloning vector construction (pSRK21), plasmid integration into *R. erythropolis* chromosome by homologous recombination, the gene coding for 16S rRNA and identification, cloning and sequencing of the genes determining the phenol degradation pathway in *R. erythropolis*.

The plasmid pRE16SmobsacB was transferred by transformation into *E. coli* S17-1. The selected transformant was than used as a donor for conjugative transfer (mobilization) into *R. erythropolis*. The transconjugants were selected on the plates with kanamycin and nalidixic acid and integration of pRE16SmobsacB (which is unable to replicate in *R. erythropolis*) into the chromosome was confirmed by PCR. The system for positive selection of double recombination event based on the conditional lethal effect of the *sacB* gene [22] was also tested. The double recombination may result in complete removing the vector sequence from the chromosome and thus loosing the respective phenotypic markers (sucrose sensitivity and kanamycin resistance). The selection of clones growing on sucrose (appearing with the frequency of approximately 10^{-4}) and sensitive to kanamycin confirmed that this system may be used for the manipulations within the *R. erythropolis* chromosome.

Acknowledgement

The support of EUREKA: E! 3100 CAWAB, IAA4020213 and MSM 6046137305 projects is gratefully acknowledged.

References

[1]   J. Masák, A. Cejková, and V. Jirku, j. microbial. meth.**30,** 133 (1997).
[2]   R. V. Martin, nature, **21,**1419 (1949)
[3]   H. Y. Neujahr, and A. Goal, eur.j. biochem. **35,** 386 (1973)
[4]   J. Sambrook, and D. V. Russel, A Laboratory Manual: 3rd edition (Cold Spring Harbour Press, New York, 2001)
[5]   V. Jirku, J. Masák, and A. Cejková, j. microbial. Biotecjnol., **11,** 17 (2001)
[6]   G.A Junter, L. Coquet, S. Vilain,. and T. Jouenne, enz. microbial. Technol. **31,** 201 (2002)

Transport behaviour of chemotactic bacteria in model aquifers

P. Velasco-Casal., and J.J. Ortega-Calvo[*,1]

[1] Instituto de Recursos Naturales y Agrobiología , CSIC, Avda. Reina Mercedes 10, 41012 Sevilla, SPAIN

[*] Corresponding author: e-mail: jjortega@irnase.csic.es, Phone: +34 954 624711, Fax: +34 954 624002

This paper develops experimental protocols for examining the influence of the chemotactic response in the transport and biodegradation activity of the naphthalene degrading strain *Pseudomonas putida* G7 in saturated column experiments. The columns contained, as packing materials, isolated soil constituents and, as chemoattractant, naphthalene initially present as crystals or dissolved in model NAPLs. The results were compared with those obtained without chemoattractant and with a mutant (*P. putida* G7.C1) which was non-chemotactic but still able to degrade the compound. The experiments showed that the effect of chemotaxis on bacterial transport was strongly dependent on the formation of a gradient of naphthalene through the columns. We suggest therefore that bacterial transport through porous materials can be subtantially modified by chemotaxis.

Keywords bacterial chemotaxis; naphthalene

1   Introduction

The bacterial degradation of a number of toxic organic compounds,including toluene, naphthalene, and chlorinated biphenyls,has been extensively studied. Many of the enzymes involved in the degradation of these compounds have been purified, and the genes encoding these proteins have been cloned and sequenced [1-4]. Although the actual pathways of catabolism are well-known, an aspect of degradation that has been overlooked is chemotaxis. Chemotaxis enhances the ability of motile bacteria to locate and degrade low concentrations of organic compounds, and it is reasonable to expect that it also directs the movement of motile bacteria to toxic, but metabolizable, compounds present in contaminated environments. Naphthalene is a priority pollutant commonly found in industrial effluents and is a constituent of coal tar [5]. It is often used as a model compound for studies of in situ biodegradation of polyaromatic hydrocarbons (PAHs) because it is easily degraded by bacteria [6]. Naphthalene's relatively high solubility compared to those of other PAHs [7] and the fact that the naphthalene degradation genes are plasmid encoded [8-15] have contributed to a rapid pace of laboratory research on naphthalene degradation. In this study we examined the abilities of two *Pseudomonas* strains to respond chemotactically to naphthalene in two chemotaxis assays, and the effect of this chemotactic response in bacterial transport through saturated sand columns.

2   Materials and methods

The naphthalene-degrading strain used, *Pseudomonas putida* G7, as well as its naphthalene degradation plasmid-cured derivative *P. putida* G7.C1, were obtained from C. S. Harwood of the University of Iowa. All strains were motile by means of polar flagella. Cells were grown in a basal mineral salt medium at 30°C with shaking at 250 rpm. All compounds used as growth substrates or tested as possible attractants in chemotaxis assays were obtained from Sigma Chemical Co. (St. Louis, Mo.) and Panreac Química SA as well as the packing material used in column experiments. Salicylate was used as growth substrate at final concentrations of 5 mM. Naphthalene was provided as a carbon source by direct addition of naphthalene crystals to liquid basal medium.

Chemotaxis assay. Chemotaxis was tested with a modified capillary assay. This chemotaxis assay, developed as a modification of the classical capillary assay [16], allowed, respectively, qualitative and quantitative assessment of chemotaxis with a microscope and by plate counts. Cells grown to mid-logarithmic phase were harvested and resuspended in chemotaxis buffer to an A_{600} of approximately 0.020 (quantitative assay) and 0.4 (qualitative assay). The suspension of motile cells was placed in a small chamber formed by placing two 1-≥L capillary tubes between a microscope slide and coverslip. A heat-sealed 1-≥l capillary tube was immersed, open-side down, into the saturated naphthalene solution or control solution. And another capillay tube filled with phosphate buffer (pH 7.0), and then the open end was packed with finely ground crystals of naphthalene by pressing the capillary tube into a mound of crystals. A capillary was then inserted into a chemotaxis chamber that had been placed on a microscope stage. A positive chemotactic response was visualized by the accumulation of a cloud of motile cells around the mouth of the capillary tube. The chemotactic assay was quantified by counting the colony forming units grew in triptic soy agar plates, after pouring the contents of the capillary tubes on them.

Column experiments. Transport experiments were performed at 25 °C in percolated columns [17,18]. The test material (sand) was wet-packed in glass columns of 10 cm in length and 1 cm internal diameter. The amount of packing material present in each column was approximately 12 g. The columns were connected to a peristaltic pump, and suspensions of naphthalene-grown bacteria (OD_{600} = 0.3) were pumped through the columns at constant flow rates. This was 0.37 cm/min (PV, 2.677 mL). Column breakthrough of bacteria was followed photometrically at time intervals. The efficiency of bacterial removal was expressed as optical density (OD) at 600 nm in column effluents (C) divided by those in column influents (C_0).

3   Results and discussion

Chemotaxis assay. In the modified capillary assay, the quantitative test showed the positive response to naphthalene of wild type *P. putida* G7 but cells did not accumulate around the mouths of the tubes that contained only minimal medium. Strain G7.C1 did not respond to naphthalene (Fig. 1). Salicylate-grown cells of *P. putida* G7 accumulated around the mouths of 1-≥l capillary tubes that had been filled with a saturated naphthalene solution (Fig. 2a and 2b) and with naphthalene crystals (Fig. 2c-2e). The motility of cells near the tips of capillaries appear equal to or greater than those far away from the naphthalene crystals.

568

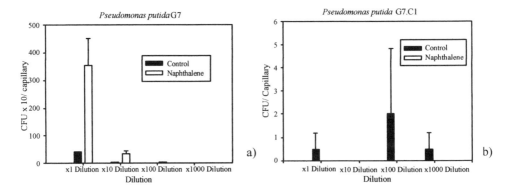

Fig. 1 Chemotactic responses of *Pseudomonas putida* G7 a) and *Pseudomonas putida* G7.C1 b) to naphthalene in modified capillary assay.

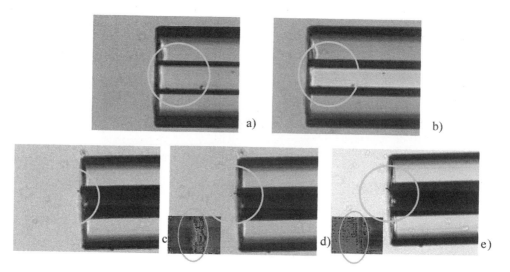

Fig. 2 Chemotactic responses of *Pseudomonas putida* G7 to naphthalene in modified capillary assays. Fig. a and c correspond to time zero and b, d and e to 75, 40 and 90 min, respectively. Crystals of naphthalene are visible inside the mouths of the capillary tubes (c to e).

Column Experiments. Naphthalene, initially present as crystals, promotes the transport of *P. putida* G7 through saturated sand columns (Fig. 3a). The final C/C_0 value is higher than the control. This effect can be attributed to the preference of bacterial cells to remain suspended in the fluid containing a high concentration of the dissolved chemoattractant. Transport is not significantly affected when naphthalene is initially dissolved in a NAPL (Fig. 3c), very likely because the aqueous concentration of the compound is markedly reduced due to partitioning. In contrast, naphthalene, iniatially present as crystals, does not promote the transport of the chemotactic-defficient mutant *P. putida* G7.C1 (Fig. 3b). These differences can be attributed to the chemotactic response of the wild strain.

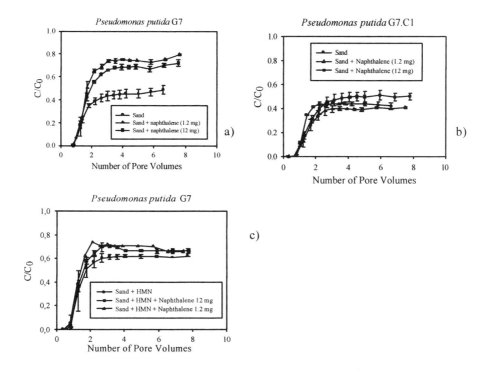

Fig. 3 Effect of naphthalene initially present as crystals (a and b) or dissolved in a NAPL (c) in transport of *Pseudomonas putida* G7 and *Pseudomonas putida* G7.C1 through saturated sand columns.

Our results indicate that low PAH concentrations may trigger chemotactic attraction which is relevant for situations of bioavailability restrictions. Indeed, the observed enhancement of bacterial transport through saturated sand columns caused by chemotaxis suggests that pollutant sensing by chemotactic bacteria can improve bioremediation performance in the subsurface. The selection of appropriate chemotactic strains seems therefore a reasonable step during the design of bio-augmentation strategies for decontamination of polluted aquifers.

Acknowledgements

We thank C. S. Harwood for the provision of bacterial strains. Support for this research was provided by a CSIC fellowship (I3P programme), the European Union (BIOSTIMUL project, QLRT-1999-00326) and the Spanish CICYT (BIO2000-1857-CE).

References

[1]   S. M. Resnick, K. Lee, and D. T. Gibson. J. In Microbiol. **17**, 438 (1996).
[2]   J. B. Sutherland, F. Rafii, A. A. Khan, and C. E. Cerniglia. Microbial transformation and degradation of toxic organic chemicals (Wiley-Liss, Inc., New York, 1995), p. 269-306.

570

[3]   J. R. Van der Meer. Antonie Leeuwenhoek. **159** (1997).

[4]   A. C. Grimm, and C. S. Harwood. Appl. Environ. Microbiol. **63**, 4111 (1997).

[5]   U.S. Environmental Protection Agency. Summary review of health effects associated with naphthalene. EPA/600/8-87/055F. Office of Health and Environmental Assessment, U.S. Environmental Protection Agency Washington, D.C. 1987.

[6]   K.-M. Yen, and C. M. Serdar. Crit. Rev. Microbiol. **15**, 247 (1988)

[7]   R. S. Pearlman, S. H. Yalkowsky, and S. Banerjee. J. Phys. Chem. **13**, 555 (1984).

[8]   N. W. Dunn, and I. C. Gunsalus. J. Bacteriol. **114**, 974 (1973).

[9]   R. W. Eaton. J. Bacteriol. **176**, 7757 (1994).

[10]   D. Ghosal, I.-S. You, and I. C. Gunsalus. Gene **55**, 19 (1987).

[11]   A. Platt, V. Shingler, S. C. Taylor, and P. A. Williams. Microbiology **141**, 2223 (1995).

[12]   C. M. Serdar. Ph.D. thesis. University of Texas at Austin (Austin, 1985).

[13]   M. J. Simon, T. D. Osslund, R. Saunders, B. D. Ensley, S. Suggs, A. Harcourt, W.-C. Suen, D. L. Cruden, D.T. Gibson, and G. J. Zylstra. Gene **127**, 31 (1993).

[14]   I.-S. You, D. Ghosal, and I. C. Gunsalus. Biochemistry **30**, 1635 (1991).

[15]   I.-S. You, D. Ghosal, and I. C. Gunsalus. J. Bacteriol. **170**, 5409 (1988).

[16]   J. Adler. J. Gen. Microbiol. **74**, 77 (1973).

[17]   J. J. Ortega-Calvo, C. Fesch, and H. Harms. Environ. Sci. Technol. **33**, 3737 (1999).

[18]   M. Lahlou, H. Harms, D. Springael, and J. J. Ortega-Calvo. Environ. Sci. Technol. **34**, 3649 (2000).

Microbial Biotechnology

Biomass selection and acclimatization for toluene degradation in a biofilter

I. Cano, A. Elías[*], A. Barona, J.C. Rodríguez, J.C., R. Proenza

Department of Chemical and Environmental Engineering. University of the Basque Country.
Alda Urquijo s/n 48013 Bilbao

[*] Corresponding author: e-mail: iapelsaa@bi.ehu.es, Phone: +34 94 6014086, Fax: +34 94 6014179

Keywords: biofiltration, toluene, sludge, *Pseudomona putida*.

Today, processes based on the biological degradation of contaminants are being used more extensively in industry. The ability of microorganisms to degrade chemical compounds and to adapt to different contaminant types is widely known. The present study focuses on the acclimation and development of active biomass for inoculation in a biofilter to be used for treating gas effluents contaminated with toluene.

In order to select an active biomass for toluene degradation in a biofilter, batch experiments have been carried out using biomass of two different origins. On the one hand, several pure microorganism strains have been purchased and, on the other hand, sludges from different industrial environments have been collected for acclimation experiments. After feeding toluene in the batch experiments, the biomass of industrial origin has shown lower degradation efficiency results than the pure strains, as far as the toluene biodegradation rate is concerned. The acclimation procedure established in this work led us to select active biomass for inoculation in a biofilter operating with low residence times for toluene.

1   Introduction

Biofiltration harnesses the efficiency of microorganisms to degrade contaminants to non-hazardous residues. It involves passing chemical-laden gases through a moist, porous medium containing active biomass. The gaseous contaminant is degraded by the microbes that form a film on the bed material and, in the case of toluene degradation, the final products are carbon dioxide and water.

When biomass acclimation is to be carried out, three related phenomena may take place; induction/depression of specific enzymes, biomass genetic changes which render new metabolic capabilities of the microorganisms and growth of specific biomass for degrading the feeding contaminant (toluene in this study).

The most important physical and chemical parameters for an effective and successful acclimation are the following:

❖   The natural physical state of the contaminant. The contaminant (hydrocarbons) in water emulsion is available for biomass and, consequently, microorganisms may develop more easily (1).

❖   The contaminant concentration. The influence of the contaminant concentration on the degradation rate is analysed by the Michaelis-Menten kintetics (1).

❖ Temperature range. This important parameter is related to the physical state of the hydrocarbon, the degradation rate and the microbial consortium composition. The optimum temperature ranges from 30 to 40 °C (2–4).

❖ Oxygen concentration. Hydrocarbon metabolic degradation involves the oxidation of the contaminant by oxygen consumption.

❖ Nutrient content. Although the C/H and C/P ratio required for biomass development is rather low, the periodical addition of inorganic salts containing nitrogen and phosphorous is essential.

❖ Acidity-alkalinity (pH). The support pH value near neutrality favours the development of bacteria and fungi (5, 6). Acid or alkaline values require the addition of NaOH or strong acids.

Based on the above parameters, batch experiments have been carried out in order to establish a methodology for the acclimation of biomass obtained from different sources. Subsequently, a biofilter support will be irrigated with an active solution containing the selected microorganisms. A toluene-bearing gas effluent will also be fed into the bioreactor in order to determine the elimination efficiency of the organic contaminant.

2   Experimental methodology

2.1 Toluene analysis by gas chromatography

In order to monitor the decrease in toluene concentration in the batch experiments and in the biofilter, a HP 6890 gas chromatograph was used. This device was equipped with two capillary columns connected in series (60 m, Hewlett-Packard HP- PLOTQ column and Molsieve column), a flame -ionization detector (FID) and a conductivity detector (TCD). Operating conditions were: injector temperature, 150 °C; oven temperature, an increasing rate of 30 °C min^{-1} to 250 °C; detector temperature, 250 °C; Helium carrier gas, 6 ml min^{-1}. A 250 ≥L gas-tight syringe (SGE, Melbourne, Australia) was used for sample injection.

2.2 Calibration curves

Calibration standards for toluene were prepared in the laboratory using standard 250 mL bottles capped with Mininert valves (Alltech, Carnforth, Lancashire) and provided with several small glass balls for gas homogeneity. A previously calculated liquid volume of toluene (PAI-ACS grade) was injected into the bottles with a syringe in order to prepare standard samples ranging from 1000 to 2000 ppmv.
Furthermore, a commercial standard sample supplied by Supelco (Bellefonte, PA) was used for additional calibration of toluene, CO_2, CO and O_2.

2.3 Preliminary batch experiments

Several batch experiments were carried out in the bottles in order to determine how many times a sample could be collected from each bottle with the gas -tight syringe before the septum is perforated. A toluene dose was injected into several shaken bottles containing fourteen glass balls, and a sample was collected with the syringe at different times. By measuring the concentration of toluene along time, the maximum number of samples that may be collected before the septum was replaced.

2.4 Determination of toluene solubility

In order to quantify the amount of toluene accessible to microorganisms (solubility of this contaminant in water and in nutrient solution), several glass bottles were filled with 50 ml of water or nutrient solution (mainly KH_2PO4, K_2HPO_2, $MgSO_4$ $7H_2O$, $CaSO_4$ $2H_2O$, $FeSO_4$ $7H_2O$ and $(NH_4)_2SO_4$) and then toluene volumes ranging from 1 to 3 $\geq l$ were injected into each bottle. Shaking was set at 150 rpm in a Selecta (Rotaterm) orbital shaker and room temperature was 23 ^oC. The amount of dissolved toluene was quantified by subtracting the amount measured in the headspace of the bottle once equilibrium was reached from the amount injected originally.

2.5 Contribution of endogenic respiration

In order to study the indigenous biomass in the samples, the endogenic respiration was determined by measuring the evolution of the concentration of oxygen and carbon dioxide along time in the absence of toluene.

3   Activation experiments

Biomass selection

The biomass selected for this study was of different origin. On the one hand, two pure microrganism strains were purchased from DSMZ (Deutsche Sammlung von Mikroorganismen und Zellkulturen Braunschweig, Germany). These pure strains were *Pseudonoma putida* PWWO and genetically modified *Pseudonoma putida* PWW53-4. On the other hand, indigenous biomass was obtained from one industrial sludge and two sludges collected from two wastewater treatment facilities.

Bearing in mind the different origins of the biomass, the activation of the microorganisms was carried out accordingly.

No special treatment was required for activation of the pure strains. Culture preparation was as recommended by the manufacturer.

The industrial sludge required cleaning for removing inert impurities from the medium. Hence, the sludge was first cleansed with a 5% NaCl solution and the supernatant containing the biomass was separated from the solid. The wastewater sludges simply required the collection of the supernatant. After the extraction of these supernatants, 0.03 g of glucose were added for 3-5 days to enable the microorganisms to develop.

After biomass activation, acclimation tests were carried out at 28 ^oC by injecting 1 $\geq l$ of toluene per day into 250 ml glass bottles capped with Mininert valves. The bottles containing the cultures were continuously shaken in an orbital shaker (at 150 rpm) and were opened to the atmosphere everyday for 20 minutes in order to ensure oxygen supply and to avoid carbon dioxide accumulation. After oxygenation, the bottles were capped and a new toluene dose of 1 $\geq l$ was injected. In these tests, the evolution of toluene and other compounds (as, for instance, oxygen and carbon dioxide) was monitored along time after the addition of the contaminant.

Preliminary results showed that biomass developed in the experimental conditions described, although differences in the acclimation period were observed.

As far as biomass from wastewater sludge is concerned, only two or three days were necessary to activate microorganisms when glucose was added as the only carbon source.

The acclimation procedure rendered an elimination efficiency of 100% when the third toluene dose (about 800 ppmv) was added.

The biomass of industrial origin required a brief activation period when glucose was added, but a very long acclimation period (several months) was necessary when the contaminant was fed into the bottles. Based on these results, the biomass was not selected for biofilter inoculation.

The pure strains were able to effectively degrade toluene after a 24-hour acclimation period. No addition of glucose was required in this case for microorganism activation.

Biofiltration experiments were carried out by inoculating the previously selected microorganisms or cultures into a compost-based support material, which rendered good results in other studies (7).

Acknowledgements

The authors acknowledge the Spanish Ministry of Science and Technology (MCYT PPQ2002-01088 with FEDER funding) and the Basque Government (ETORTEK program, BERRILUR project) for financial support for the project. We thank Dr. Jose Luis Sanz for skilful technical assistance.

Reference

[1] J. G. Leahy, R. R. Colwell "Microbial degradation of hydrocarbons in the environment", Microbiol Rev., **54 (3)**: 305–315, (1990).

[2] E. Acuña, R. Auria " Microbiological and kinetic aspects of a biofilter for the removal of toluene from waste gases " Biotechnology and Bioengineering, **63 (2)**: 175-184, (1999).

[3] E. I. García Peña, S. Hernández " Toluene biofiltration by the fungus *Scedosporium apiospermum* TB1 ", Biotechnol Bioeng.;**76 (1)**: 61-9, (2001).

[4] Sz-Chuwn John Hwang, Chi-Mei Lee " Biofiltration of waste gases containing both ethyl acetate and toluene using different combinations of bacterial cultures " J Biotechnol.; **105 (1-2)**: 83-94, (2003).

[5] L. Sene, A. Converti " Sugarcane bagasse as alternative packing material for biofiltration of benzene polluted gaseous streams: a preliminary study " Bioresour Technol.; **83 (2)**: 153-7, (2002).

[6] K. F. Reardon, D. C. Mosteller " Biodegradation kinetics of benzene, toluene and phenol as single and mixed substrates for Pseudomonas Putida F1 " Biotechnol. Bioeng.; **69 (4)**: 385-400, 2000.

[7] Elías, A., Barona, A., Arreguy, A., Rios, J., Aranguiz and I., Peñas, J., "Evaluation of a packing material for the biodegradation of H_2S and product analysis". Process Biochemistry, **37**: 813-820, (2002).

Cloning and overexpression of an alkaline and thermostable α-amylase from a native Iranian Bacillus species (*Bacillus* sp-GSH) isolated from the soil of Rasht

M. Shafiei[*,1], **A.-A. Ziaee**[1], **Z. Shaigan**[2], and **N. Ghaemi**[3]

[1]   Institute of Biochemistry and Biophysics, University of Tehran, Tehran, Iran
[2]   Faculty of Sciences, University of Guilan, Rasht, Iran
[3]   Faculty of Sciences, University of Tehran, Tehran, Iran

[*]Corresponding author: e-mail: mshafiei@ibb.ut.ac.ir, Phone: +982161113376, Fax: +98216404680

Thermostable α-amylases are important and interesting enzymes because of their usefulness in industrial applications. A novel alkaline and thermostable α-amylase was extracted from a native Iranian Bacillus species (*Bacillus* sp-GSH) collected from the soil of the city of Rasht in Gilan province in the North of Iran. Primers were designed by using genebank analysis of themostable α-amylases genes and, certain bases inserted as restriction sites. The PCR product was cloned in pBluscript ιι KS cloning vector following the use of genomic DNA of the above mentioned bacillus as a template. PBSHA-82 construct was made. Subsequently this gene was recloned in PET24a expression vector and named pESHA-83. For high and overexpression of the recombinant enzyme, this construct was used to transform BL21(DE3) E. coli strain. One of the well identified colonies harboring the recombinant enzyme was cultured in 2XYT medium and after gaining the value of 1 at OD_{600} , induction by IPTG and lactose at concentrations of 0.2 and 10 mM, respectively, was followed. After 24 h of induction by lactose, the supernatant was treated with 80 % of ammonium sulfate and the precipitate analyzed for enzymatic activity in 10 min at pH=7 for various temperatures using DNS method. Optimum pH activity was determined at 80^0C for 10 min and at various pHs. These data showed that this amylase is active up to 100^0C and pH=11.5.

Keywords α-amylase; bacillus; cloning; expression; thermostable; alkaline

1   Introduction

α-Amylases (EC3.2.1.1) are endo-acting enzymes that hydrolyze starch by cleaving α-1,4 glucosidic linkages at random [1]. Amylases can be derived from several sources, including plants, animals and microorganisms.

These enzymes are great significance nowadays in biotechnology with applications ranging from food, textile, paper and detergent industries [2]. Hyperthermophilic α-amylases could be advantageous not only for their extremethermostability but also because of their resistance to denaturing agents, solvents, and proteolytic enzymes [3]. Moreover alkalifilic type of α-amylases can be used as effective additives in dishwashing and laundry detergents under alkaline conditions [4]. Here we report the cloning and expressing the gene encoding alkaline and thermostable α-amylase from *Bacillus* sp-GSH in Escherichia coli. We also report some of biochemical characterization of the recombinant enzyme.

2   Materials and methods

2.1 Bacterial strains, plasmids and culture conditions

Bacillus sp-GSH was grown in a medium containing 0.5 % starch, 0.5 % soytone, 1.5 % tryptic soy broth, 0.5 % NaCl and 0.5 % Na_2Co_3 at 30°C for isolation of chromosomal DNA. E. coli TOP10 was used as a host of cloning and E. coli BL21(DE3) used for expression of the α-amylase gene. The plasmid pBluscript II KS was used for cloning, while PET24a for expression.
Recombinant E. coil's harboring α-amylase gene were cultured in LB medium containing the appropriate antibiotics, ampicilin (50 μg/ml) or kanamycin (50 μg/ml) and 2xYT medium containing kanamycin (50 μg/ml) for expressing the α-amylase.

2.2   Cloning of the α-amylase gene of *Bacillus* sp-GSH

The **amyl** gene from *Bacillus* sp-GSH was amplified by PCR using a forward primer (5'-AAT   TGGATCCGGAGGAAATATATGAAACAACAAAAACGGCTTTA-3'),   which encodes   a   BamHI   site   (underlined),   and   a   reverse   primer   (5'-TATAGAATTCCCGTCCTCTCTGCTCTTCTATCTTT-3'), containing an EcoRI site (underlined) [5]. The PCR protocol with the application of Taq DNApolymerase was done for 30 cycles as follows: 93°C for 1 min, 67°C for l min, and 72°C for 3 min.
The amplified fragment was ligated to the BamHI/EcoRI digested pBluscript II KS to have the pBSHA-82 construct.

2.3   Recombinant DNA techniques

Standard procedures for restriction endonuclease digestion, agarose gel electrophoresis, DNA Ligation, and other cloning related techniques were applied as described by Sambrook *etal* [6].
This construct was double digested with BamHI/EcoRI and the isolated fragment was recloned in PET24a expression vector, resulted in pESHA-83 costruct.

2.4   Plate assay of amylase expressing colonies

E. coli BL21(DE3) cells harboring pESHA-83 expression plasmid were grown on 2xYT-kanamycin plate containing 1 % soluble starch at 37°C for 24 h and α-amylase expressing transformants were detected by the addition of Lugol solution (I_2 0.5 %, KI 5 %).

2.5   Overexpression of α-amylase in E. coli

Plasmid pESHA-83 was used to transform E. coli BL21(DE3) containing lac UV5 promoter-driven T7 RNA polymerase. One percent of 0.5 Mac Farland of transformed cells were inoculated in 20 ml of 2xYT medium containing 50 μg/ml kanamycin for preculture. After incubation at 37°C for l h, l0 ml preculture was inoculated in 1 L 2xYT-kanamycin and after 15 h growth, the expression of α-amylase was induced for 24 h by the addition of l0 mM lactose and 0.2 mM IPTG.

2.6 Enzyme isolation

The cells were removed by centrifugation (15000 g; 10 min) and the supernatant was percipitated with 80 % (w/v) ammonium sulphate.
The percipitate was resuspended in 20 mM Tris-HCl buffer (pH 7.0) and dialyzed. This crude enzyme preparation was used for enzyme assay.

2.7 Enzyme assays

The α-amylase activity was determined by measuring the amount of reduced sugar that released during enzymatic hydrolysis of soluble starch. Standard assay mixture (total volume 1 ml) contained 450 µl of 1 % soluble starch in Tris-HCl buffer (20 mM pH 7.0) and 50 µl of appropriate amount of enzyme. After incubation at 80°C for 10 min, the reaction was stopped by adding 500 µl of dinitrosalicylic acid solution [7], then heated at 100°C for 5 min and the absorbance at 540 nm was determined.
One unit of amylase liberated 1 μmol^{-1} of reducing groups per minute.

2.8 pH and temperature studies

The optimal pH for α-amylase activity was determined at 80°C for 10 min in 50 mM sodium acetate (pH 3 to 6.0) and 50 mM Tris-HCl (pH 6 to 12).
All pH were adjusted at room temperature. The temperature for maximal activity was determined by performing standard enzyme assay at different temperatures.

2.9 Gel electrophoresis

Protein samples were analyzed by sodium dodecyl sulfate-polyacrylamid gel electrophoresis (SDS-PAGE, 12 % polyacrylamide) [8].
Reducing buffer (62.5 mM Tris-HCl; pH 7.0; 20 % glycerol; 2 % SDS; 5 % β-mercaptoethanol) at 80°C was used for protein samples denaturation.
Native PAGE was performed under the same conditions as described above except for the absence of SDS in the gel and buffer system. For activity staining the gel containing 1 % soluble starch was immersed in 100 mM Tris-HCl buffer (pH 7.0) containing 1% v/v Triton X-100 and shaked for 30 min to allow protein renaturation. Then the gel was washed with Tris-HCl buffer (pH 7.0) and incubated in the same buffer containing 1 % soluble starch at 80°C for 1 h then stained with iodine solution. Clear band on a dark background was confirmed as amylolytic activity of the enzyme.

3   Results and discussion

3.1 Construction of the expression plasmid pESHA-83

When the intermediate plasmid pBSHA-82 containing *amyl* was digested from upstream and downstreme of the gene by BamHI and EcoRI restriction enzymes, the fragment was released, then showed its insertion into the same endonuclease sites of the PET24a vector. The construction of pESHA-83 was also confirmed following the expression of the *amyl* gene under the control of T7 promoter.

580

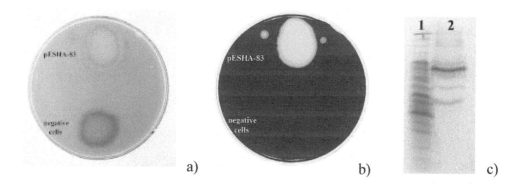

a)   b)   c)

Fig. 1 a) Escherichia coli BL21(DE3) transformed cells harbouring pESHA-83 and PET24a negative control grown on a 2xYT-kanamycin plate containing 1 % soluble starch. b) Same plate after addition of Lugol solution. c) SDS-PAGE analysis of the enzyme: Lane 1, crude enzyme; lane 2, after partial purification.

3.2 Expression and induction of amylase by lactose

The 2xYT-kanamycin plate containing bacteria harbouring pESHA-83 expression plasmid and the negative cells containing PET24a that were grown at 37°C for 24 h, exposed to I_2-KI reagent solution and the amylase activity of the cloned enzyme was very obvious in comparison to the negative cells (Fig 1a, b). After inducing with lactose (10 mM) and IPTG (0.2 mM) for 24 h, the crude enzyme of lactose induction was analyzed by SDS-PAGE and the major band clearly observed (Fig 1c). For demonstrating the amylolytic activity of the enzyme, the crude enzyme was resolved by SDS-PAGE containing 1 % soluble starch and after removing of SDS and renarturating by Triton X-100, the gel stained with iodine solution for a-amylase activity and the appeared band of the amylase coincided with the clear zone in the area exposed to the a-amylase enzyme.

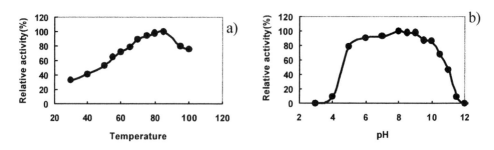

Fig. 2 a) Effect of pH on the relative activity of the amylase. b) Effect of temperature on the relative activity of the amylase.

3.3 Characterization of a-amylase

When the effect of pH on the activity of the a-amylase against starch was determined at 80°C in buffers with pH ranging from 3.0 to 12.0, the maximal activity observed at pH 8.0 and the enzyme was active up to pH 11.5 (Fig 2a).

Temperature dependence of a-amylase activity toward starch was determined by measuring the activity over a range of temperatures at pH 7.0. Maximal activity of the enzyme was at 80°C and it was active up to 100°C (Fig 2b).

Acknowledgements

We are grateful to Dr. Hamid Reza Karbalaei Heidari for his excellent guidance in enzymatic assays.

References

[1]   G. Dong, C. Vieille, A. Savchenko, and J. Gregory Zeikus, appl. env. mic. **63**, **9**, 3569 (1997).

[2]   R. Gupta, P. Gigras, H. Mohapatra, V. K. Goswami, and B. Chauhan, pro. bio. **38**, **11**, 1599 (2003).

[3]   W. J .Lim, S. R. Park, C. L. An, J. Y. Lee, S. Y. Hong, E. C. Shin, E. J. Kim, J. O. Kim, H. Kim, and H. D. Yun, res. mic. **154**, 681 (2003).

[4]   K. Igarashi, Y. Hatada, H. Hagiahara, K. Saeki, M. Takaiwa, T. Uemura, K. Ara, K. Oaki, S. Kawai, T. Kobayashi, and S. Ito, app. env. mic, **64**, **9**, 3282 (1998).

[5]   M. Shahhoseini, A.-A. Ziaee, A.-A. Pourbabai, N. Ghaemi, and N. Declerck, j. app. mic, **98**, 24 (2005).

[6]   J. Sambrook, and D. W. Russell: Molecular Cloning, A Laboratory Manual, (Cold Spring Harbor, New York 2001)

[7]   P. Bernfeld, met. enz. **1**, 149 (1955).

[8]   U. K. Laemmli, nature. **227**, 680 (1970).

Evaluation of performance and kinetics of mesophilic and thermophilic anaerobic digestion for treatment of palm oil mill effluent

W. Choorit[*1], P. Wisarnwan[1], and S. Echaroj[2]

[1] Department of Biotechnology, School of Agricultural Technology, Walailak University, Tasala, Nakhonsithammarat, THAILAND
[2] Energy for Environment Foundation, Bangkok, THAILAND

[*] Corresponding author: e-mail: cwanna@wu.ac.th, Phone: +66 075 672355, Fax: +66 075 672302

A comparative study was undertaken of anaerobic continuous stirred tank reactors (CSTRs) at mesophilic (37°C) and thermophilic (55°C) temperatures. The reactors were operated at different HRT. The results showed that the longer HRT achieved better process stability as total volatile fatty acid (TVFA) levels and TVFA/alkalinity ratio gradually increased response to the lower HRT. At HRT of less than 6 days, the mesophilic reactor could not be operated while the thermophilic reactor worked satisfactorily. TCOD removal in thermophilic reactor was slightly lower (63.3-75.4%) than that in the mesophilic reactor (64.6-80.0%) when the HRT was tested almost to the point of digester failure. The experimental data were evaluated to determine the kinetic parameter of the process. The results achieved methane yield coefficient of 0.378 and 0.333-l $CH_4/gTCOD_{removed}$ and biomass yield coefficient of 0.263 and 0.284 gVSS/gTCOD for mesophilic and thermophilic reactors, respectively.

Keywords palm oil mill; continuous stirred tank reactor; methane

1   Introduction

Anaerobic digestion involves a complex process in which several groups of bacteria sequentially convert organic pollutants into methane and carbon dioxide. This technology has been applied to treatment of many high-strength industrial wastewater. Despite the researches illustrated the performance of anaerobic treatment of palm oil mill effluent (POME) at the ranges of mesophilic and thermophilic temperatures [1-6], only a few studies have been conducted for comparison of the both conditions. Moreover, a scarce kinetic study has been reported for operating reactor system, especially in the completely mixed digester with suspended biomass. Therefore, the aims of this work were to compare the anaerobic digestion of mesophilic and thermophilic CSTRs for treatment of POME and evaluating the performance of the processes and the kinetic parameters by using experimental data.

2   Materials and methods

2.1 Wastewater

The characteristics of POME used in the experiment contained (mg/l) total chemical oxygen demand (TCOD) of 100,600 – 120,023; biological oxygen demand (BOD) of 62,500 - 69,215; total solid (TS) of 68,854 - 75,327; suspended solid (SS) of 44,680 - 47,140; mixed liquor volatile suspended solid (MLVSS) of 39,820 - 41,740; oil & grease of 8,845 - 10,052; total volatile fatty acid (TVFA) of 4,045 - 4,335; total Kjeldahl nitrogen of 1,345 - 1,493 and ammonia-nitrogen of 91 – 112 with pH of 4.24-4.66.

2.2 Experimental design

Experiments were studied using two fed-batch completely mixed reactors with 1.6 l working volume. Each reactor was controlled at temperatures of 37°C (mesophilic reactor) and 55°C (thermophilic reactor). The experiments were run continuously following feed flow rates of 106.7, 160.0, 200, 228.6, 266.7, 320.0 and 400 ml/d, which correspond to HRTs of 15, 10, 8, 7, 6, 5 and 4 days, respectively. As POME was diluted prior to feeding, influent TCOD concentrations were 66-68 g/l/day throughout the experiment. The steady-state conditions were achieved at each HRT as the levels of TVFA, TCOD removal, biogas production rate and composition vary less than 3% in consecutive days [7].

2.3 Chemical analysis

Gas volume was measured using a displacement of acidified water (pH 2-3) and analyzed the methane volume by KOH solution displacement in serum bottle [8]. Alkalinity was measured by direct titration method [9]. The parameters including BOD, TCOD, TVFA, pH, TS, SS, MLVSS, total Kjeldahl nitrogen and oil & grease were determined according to standard methods [10].

3   Results and discussion

3.1 Process stability

The steady-state performance of the mesophilic and thermophilc reactors at different HRTs are shown in Table 1. Fig. 1a shows that the levels of TVFA in the mesophilic reactor were fairly low at longer HRT operation and sharply increased at the HRT of 6 days. The TVFA concentration at HRT of 15 days was 63 mg/l, which increased to 641 mg/l at the lowest HRT. In contrast, the TVFA from 104 mg/l at HRT of 15 days was gradually increased to 765 mg/l at HRT of 4 days in the thermophilic reactor. During operation period, the pH values relatively constant, around 7.44–7.54 and 7.68–7.83 in the mesophilic and thermophilic reactors, respectively. In addition, the process is considered to be operated favorably as the TVFA/alkalinity were in the range of 0.02-0.20 (mesophilic) and 0.04-0.23 (thermophilic), always lower than the failure limit value of 0.3–0.4 [11].

584

Table 1 Reactor performance on steady-state conditions at different HRTs[a]

Reactors	HRT (days)	pH	Alkalinity (mg/l)	TVFA (mg/l)	Gas prod. rate (l/day)	CH₄ content (% vol.)	TCOD removal (%)	MLVSS (mg/l)
Mesophilic	15	7.44	2,720	63	1.58	72.0	80.0	9.99
	10	7.52	2,811	89	2.56	72.6	77.1	10.20
	8	7.49	2,869	123	3.75	71.1	73.3	10.38
	7	7.54	3,000	172	4.49	70.0	70.1	10.10
	6	7.52	3,154	641	4.62	68.7	64.6	10.02
Thermophilic	15	7.69	2,816	104	2.04	71.5	75.4	9.66
	10	7.74	2,962	162	2.73	72.2	71.4	10.29
	8	7.74	3,013	182	3.44	71.1	69.8	10.35
	7	7.80	3,126	270	4.32	70.8	67.8	10.23
	6	7.83	3,271	417	4.81	69.7	66.3	10.12
	5	7.79	3,408	654	5.96	70.3	65.2	10.60
	4	7.68	3,334	765	6.26	70.0	63.3	10.82

[a]Values are means of three replicates.

3.2 Process efficiency

Fig. 1b shows the variation of the process efficiency under the both temperatures. The percentage of TCOD removal decreased slightly from 75.4% to 63.3% was observed in the thermophilic reactor whilst the mesophilic reactor achieved the greater decreased range of 80.0-64.6%, especially a marked decrease took place at the lowest HRT (70.1% to 64.6%). The high MLVSS concentrations in the both reactors indicated that the processes are capable of holding high solids in the system and providing the high effluent TCOD concentration. The biogas production rate steadily increased in response to the HRTs above the minimum values. Although the failures were not occurred at the shortest HRTs in both cases, the TCOD removal efficiency were considerably decreased along with biogas production rate decelerated. This can be attributed to methane production is limited at the shortest HRTs, probably due to the toxicity of non-accumulated TVFA, especially the nondissociated VFAs [12]. The thermophilic reactor achieved the greater biogas content than that in the mesophilic reactor at the longer HRTs but slightly variations were observed at the shorter values. The methane contents in biogas varied by less than 4% in all cases, therefore, this may be attributed to the HRTs did not directly affect the gas composition.

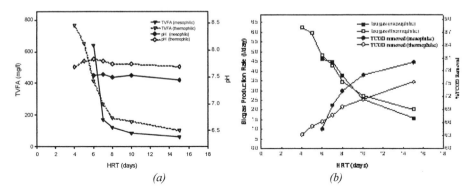

Fig. 1 TVFA levels and pH values of the POME operated under steady-state conditions on various HRTs in the mesophilic and thermophilic reactors (a). Biogas production rate and the percentage of TCOD removal under steady-state conditions on various HRTs (b).

3.3 Estimation of kinetic parameters

The models presented here describe the anaerobic digestion in a completely mixed reactor. For the system without cell recycle at steady-state, it is assumed that the concentration of biomass is neglected. Thus, a substrate balance model was proposed by considering a COD balance as follows [13]. $(TCOD)_0 = (SCOD)_e + (TCOD)_{biogas} + (TCOD_{VSS})_e + (TCOD)_m$, (1) Equation (1) can be transformed into the following: $qS_{t0} = qS_{Se} + q_{CH4}Y_{S/G} + q[S_{te}-S_{Se}] + k_mXV$,(2) Equation (2) can be rearranged to estimate the methane yield coefficient as follows $[S_{t0}-S_{te}]/(HRT)X = q_{CH4}Y_{S/G}/XV + k_m$ (3) By least square fitting the Eq (3), the values of $Y_{G/S}$ (l CH_4/gTCOD$_{removed}$) can be calculated as the inverse of $Y_{S/G}$, which was obtained from the slope of regression linear [14]. Fig. 2a, The regression coefficients of 0.976 and 0.984 achieved the methane yield coefficients of 0.378 and 0.333 l CH_4/gTCOD$_{removed}$ for the mesophilic and thermophilic reactors, respectively. The k_m obtained from the intercept of the straight line was 0.1818 and 0.0891 day^{-1}, respectively.

Monod equation which expresses the relationship between the specific growth rate and the substrate concentration can be applied for expression of yield and decay coefficients of biomass in the system. $\mu = (\mu_{max}) S_{te}/(K_s + S_{te})$ (4) At steady-state, the rate of change in biomass concentration could be expressed $q(X_0-X) = \mu XV - k_dXV$ (5) Equation (5) can be divided by the effluent biomass concentration and the flow rate, and substituted by Eq (4) as follows $S_{te} = Ks(k_d+1/HRT)/ \mu_{max} - k_d - (1/HRT)$ (6) The relationship between the rate of change in substrate and biomass concentrations at the steady-state can be expressed. $[S_{t0}-S_{te}]/HRT = \mu (X/Y)$ (7) By similar technique that used for the Eq (6), Equation (7) can be rearranged to estimate the biomass yield coefficient as follows $[S_{t0}-S_{te}]/(HRT)X = 1/(HRT)Y + Y/k_d$ (8) The regression coefficients of 0.986 (mesophilic) and 0.995 (thermophilic) achieved the yield coefficients of 0.263 and 0.284 gVSS/gTCOD, respectively. The k_d obtained from the intercept of the straight line were 0.030 and 0.034 day^{-1}, respectively (Fig. 2b).

586

4 Conclusion

The results showed that the longer HRT achieved better process stability as TVFA levels and TVFA/alkalinity ratio gradually increased response to the lower HRT. TCOD removal efficiency in the thermophilic reactor was slightly lower than that in the mesophilic reactor. The models were applied to evaluate methane yield coefficient and biomass yield coefficient for the mesophilic and thermophilic reactors, respectively.

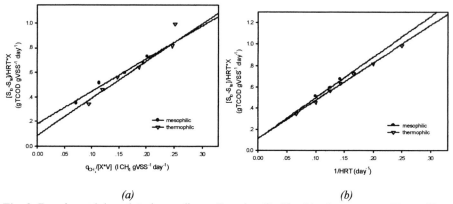

(a) *(b)*

Fig. 2 Experimental data plotted according to Equation (3). The kinetic parameters $Y_{G/S}$ and k_m can be determined for the mesophilic and thermophilic reactors (a).The kinetic parameters Y and k_d can be determined for the mesophilic and thermophilic reactors (b).

Acknowledgement

The authors wish to express their gratitude to the Southern Palm (1978) Co., Ltd, Thailand and Walailak University for their financial support.

References

[1]   A. Ibrahim, B.G. Yeoh, S.C. Cheah, A.N. Ma, and S. Ahmad, Water Sci.**17**, 155 (1984).
[2]   R.G. Cali, and J.P. Barford, Biomass 7, 287 (1985).
[3]   R. Borja, R. Padilla, and C.J. Banks, Biotechnol. Letters **15**, 761 (1993).
[4]   R. Borja, and C.J. Banks, Bioresource Technol. **48**, 209 (1994).
[5]   R. Borja, and C.J. Banks, Process Biochemistry **30**, 511 (1995).
[6]   L. Björnsson, B. Mattiasson, and T. Henrysson, Appl. Microbiol. Biotechnol. **47**, 640 (1997).
[7]   R. Borja, A. Martin, C.J. Banks, V. Alonso, and A. Chica, Environ. Pollution **88**, 13 (1995).
[8]   T.H. Ergüder, U. Tezed, E. Guven, and G.N. Demirer, Waste Managemaent **21**, 643 (2001).
[9]   S.R. Jenkins, J.M. Mongan, and C.L. Sawyen, J. Water Pollut. Control Fed. **55**, 448 (1983).
[10]   Standard Methods for the Examination of Water and Wastewater. (20th edn), Washington, D.C., USA. (1998).
[11]   B.E. Rittmann, and P.L. McCarty, Environmetal Biotechnology : Principles and . McGraw-Hill Companies, Inc. New York. pp 622-629 (2001).
[12]   B.K. Ahring, M. Sandberg, and I. Angelidaki, Appl. Microbiol. Biotechnol. **43**, 559 (1995).
[13]   R. Borja, A. Martin, B. Rincon, and F. Rapaso, J. Agric. Food Chem. **51**, 3390 (2003).
[14]   F. Rapaso, R. Borja, E. Sanchez, M.A. Martin, and A. Martin, Water Res. **38**, 2017 (2004).

Expression in *E. coli* of a recombinant form of pulmonary surfactant protein precursor proSP-B$_{\Delta C}$

A.G. Serrano, A. Palacios, J. M. Oviedo, E. J. Cabré, B. González, P. Estrada*, J. Pérez-Gil

Universidad Complutense, Facultad de Biología, Departamento de Bioquímica y Biología Molecular I, Ciudad Universitaria 28040 Madrid, SPAIN

*Corresponding author: e-mail: estrada@bbm1.ucm.es, Phone: +34 91 3944620. Fax: +34 91 3944672

The human wild-type proSP-B$_{\Delta C}$ sequence has been cloned in plasmid pProEX-1 isolated from *E. coli* GM48 cells (dam⁻). Putative recombinant plasmids isolated from DH5αF′ were checked for size and sequence and transformation of UT5600 cells was carried out. After sequence assessment, induction with IPTG allowed the production of 6xHis-tagged protein forming inclusion bodies which were solubilized in 8 M Urea, being the tagged protein refolded while decreasing the urea content by successive dialysis.

Keywords: pulmonary surfactant; SP-B; proSP-B$_{\Delta C}$

1   Introduction

Pulmonary surfactant is a macromolecular complex composed of phospholipids and proteins, which is synthesized and secreted by the alveolar type II cells. It plays a critical role in maintaining the structural integrity of the respiratory surface by reducing the surface tension at the alveolar air-liquid interface [1]. The hydrophobic surfactant protein B (SP-B) is essential to promote rapid formation and maintenance of the surface active film after birth. This 79-amino acid amphipathic protein is synthesized as a precursor preproSP-B which suffers ulterior proteolytic processing to remove N-terminal and C-terminal flanking propeptides [2]. The structural studies of the mature protein are hindered by its hydrophobic nature so that we chose to approach the study of the mature protein flanked by the N-terminal propeptide (proSP-B$_{\Delta C}$). To this goal, the proSP-B$_{\Delta C}$ sequence was cloned in a vector containing a 6xHis-tag to facilitate purification and the production of recombinant proSP-B$_{\Delta C}$ as inclusion bodies in *E. coli* was achieved.

2   Materials and methods

2.1. Bacterial strains, plasmids, medium and molecular biology procedures

Escherichia coli DH5αF′(Life Technologies) and GM48 (dam⁻) (New England Biolabs) were used as host strains for plasmid production. Cells were maintained in LB medium containing 1% (w/v) bactotryptone (Sharlau), 0.5% (w/v) yeast extract (Scharlau) and 1% (w/v) NaCl supplemented with 100 µg/mL ampicillin or 50 µg/mL kanamycin (Applichem). Solid LB medium contained 1.5% agar (Scharlau). *E. coli* UT5600 (New England Biolabs) were employed for protein expression. Cells were grown on Terrific Broth (TB; 1.33 % (w/v) bactotryptone, 2.66 % (w/v) yeast extract and 4 mL anhydrous glycerol (Fluka) in 900 mL water plus 0.17 M KH_2PO_4 and 0.72 M K_2HPO_4 in 100 mL water sterilized separately). Ampicillin was added to 100 μg/mL final concentration.

Plasmid pPROEX-1 containing a proSP-$B_{\Delta C}$ mutated sequence cloned in a MCS located after a 6xHis tag and a 60 bp spacer including a target for protease TEV and plasmid pENTR/D TOPO containing the proSPB$_{\Delta C}$ wild-type sequence, were a gift from Dr. T. Weaver (University of Cincinnati, USA).

Primers to generate the wt proSPB$_{\Delta C}$ sequence were designed based on reported sequence data (Genbank® Accession No. M24461). The forward primer was coincident with the 5′ propeptide sequence (5′-GGGAATTCGAGCCTGGACCACCTCAT) but extended to an overhanging sequence including an EcoRI site (underlined). In the reverse primer (5′-GCTCTAGATCACATGGAGCACCGGAGGAC) a XbaI site (underlined) and a stop codon were incorporated to facilitate subcloning in pPROEX-1. PCR was carried out in a Master Cycler personal (Eppendorf) with Pwo polymerase (Genaxis). Amplification conditions were 25 cycles of denaturation at 95 °C (1 min), annealing at 55 °C (2 min) and chain extension at 72 °C (1 min). Plasmid pENTR/D TOPO was employed as template. DNA fragments obtained were checked for size in agarose gel electrophoresis with Tris-Acetate-EDTA buffer, purified according to Sambrook et al. [3], digested with XbaI/EcoRI (Amersham Pharmacia Biotech) for 2 h at 37 °C and re-purified.

Electrocompetent GM48 cells prepared according to Sambrook et al. [3] were subjected to high-voltage electroporation. 1 μL pPROEX-1 and 50 μL cells were mixed in a sample holder (Biorad) at 4 °C where a mini-electrode applied high-voltage (2.5 Kv, 25 μF, 200 Ω, other conditions as prescribed by the supplier). After adding 1 mL cold LB to the sample holder, cells were grown at 37 °C for one hour, platted on LB containing 100 μg/mL ampicillin and cultured O/N at 37 °C. Plasmids pPROEX-1 with unmethylated -GATC-sequences were purified, digested with XbaI/EcoRI (conditions as described above) and checked for size in agarose gel electrophoresis. The DNA fragments lacking the insert were extracted and ligated with the PCR product (O/N, 16 °C in a water bath, T4 DNA ligase from Epicentre). Other DNA manipulations such as plasmids isolation, etc, were carried out by standard procedures.

200 μL of DH5αF′ competent cells prepared according to Hanahan [4] were transformed with 10 μL of the ligation product by the heat-shock procedure [3], including the addition of 1 mL SOB (2 % ,w/v bactotryptone, 0.5 %, w/v, yeast extract, 10 mM Na Cl, 2.5 mM KCl) and 10 μL of 2 M glucose. After 1 h incubation at 37 °C, 50 μL of the transformed cells were platted on LB containing Ap and incubated O/N at 37 °C. Plasmids from putative recombinants were purified and the presence and orientation of the inserted wt DNA were determined by a second PCR amplification with the same reverse primer described above and a new designed forward primer (5′ TCTCTCCGAGCAGCAATTC) complementary to the wt proSP-$B_{\Delta C}$ sequence but not to the mutated one. Once the correct structure of the construction was verified by sequencing, competent E. coli UT5600 cells were transformed as described above for DH5αF′ cells and the sequence was again assessed from putative recombinant plasmids.

2.2. His-tagged protein expression and purification

E. coli UT5600 cells transformed with recombinant pPROEX-1 were grown at 37 °C in 1 L TB at 200 rpm in an orbital incubator SI 50 (Stuart Sc.). When the mid-log phase was reached (A$_{600}$ = 0.5–0.7), isopropyl-β-D-thiogalactopyranoside (IPTG, from Genaxis) was added to a final concentration of 0.25 mM and incubation took place O/N at 37 °C. Cells were harvested by centrifugation at 5500 × g RCF for 20 min at 4 °C, suspended in 10 x (their wet weight in g) mL of 50 mM sodium phosphate buffer pH 7.0, 300 mM NaCl and frozen at least 3 h at –80 °C. After thawing in cold water, 1 mg/mL lysozyme (Merck), 0.1 mM phenylmethanesulfonyl fluoride (SIGMA) and 10 mM 2-mercaptoethanol (Merck) were added. Cytoplasmic content was released by sonication in conical-shaped tubes using as cell disruptor a Digital Sonifier 450 (Branson Ultrasonics Co.) equipped with a tip (50 % amplitude, 2 min bursts of 59.9 sec ON, 0.1 sec OFF with 1 min pause in between, 4 °C). After centrifugation (12000 × g, 20 min, 4 °C) took place, pellets containing inclusion bodies were suspended in 10 x (their wet weight in g) mL of 50 mM sodium phosphate buffer pH 7.0, 300 mM NaCl, 10 mM 2-mercaptoethanol, 8 M urea (buffer A), sonicated (50 %, 1 min burst of 1 sec ON, 1 sec OFF, 4 °C) and rocked smoothly O/N at 4 °C. A new sonication cycle was carried out and soluble fraction was collected after centrifugation (15000 x g, 20 min, 4 °C).

Purification of His-tagged protein was carried out by affinity chromatography in a cobalt-based Talon resin (BD biosciences) following a protein-resin incubation step (room temperature, 30 min). Other conditions as prescribed by the supplier. After washing the resin with buffer A containing 5 mM imidazole (SIGMA), the protein was eluted with the same buffer plus 200 mM imidazole. Fractions containing protein were pooled and subjected to successive dialysis (Spectum, MWCO: 12-14000, O/N, 4 °C) against buffer B (50 mM sodium phosphate buffer pH 7.0, 150 mM NaCl, 10 mM 2-mercaptoethanol) containing 5 M urea and 1 M urea respectively, followed by another dyalisis against buffer B without 2-mercaptoethanol and a final one against 5 mM Tris-ClH buffer pH 7.0, 150 mM NaCl. Protein aliquots were frozen at –80 °C until use.

Protein concentration was determined according with Lowry with bovine serum albumin (SIGMA) as standard [5]. SDS-PAGE was performed according to Laemli [6] and gels were either stained with Coomassie (Applichem) or subjected to electrotransfer (Transfer-blot SD, Biorad) and revealed with antipolyhistidine peroxidase conjugate (SIGMA) at a final dilution 1/4000.

3   Results and discussion

Pulmonary surfactant protein SP-B is very hydrophobic and can not be obtained as a soluble recombinant protein. Therefore, the recombinant form of the protein to be produced includes the N-terminal propeptide, which increases its solubility acting as an intramolecular chaperon [7]. We had available the plasmid pPROEX-1 (a gift from Prof. Tim Weaver, University of Cincinnati) containing the proSP-B$_{\Delta C}$ sequence with the following mutations: First, the pair $Q_{200}F_{201}$ had been substituted by $N_{200}G_{201}$ to create a hydroxylamine-cleavable target site at the edge between the N-terminal propeptide and the mature SP-B sequence and second, N_{134} had been replaced by Q_{134} to hinder a natural hydroxylamine target site in the propeptide sequence. We also obtained second plasmid containing the wt proSP-B$_{\Delta C}$ sequence (pENTR/D TOPO). In the present communication we report production of a recombinant protein with the sequence of wt proSP-B$_{\Delta C}$. To this purpose we have amplified the wt sequence from pENTR/D TOPO (Fig. 1, lane 1) obtaining a 700-800 bp DNA fragment of the expected size (789 bp) and cloned in pPROEX-1, which was previously devoted of the mutated insert. A crucial step was to

Fig. 1 Agarose gel electrophoresis of DNA fragments obtained from PCR amplification. *Lane 1)* DNA fragments with 700-800 bp, corresponding to proSP-B$_{\Delta C}$ sequence amplification; *lane 2)* DNA fragments with 200-300 bp, corresponding to wt mature SP-B sequence amplification. Lane M has been loaded with molecular weight markers.

1 2 M

recognize colonies bearing the putative recombinant plasmid from those bearing the non digested contaminating plasmid. Since both constructs have the bla gene and show the same restriction analysis, exhaustive DNA sequencing of all clones or a PCR differential amplification involving the 200 and 201 codons appeared the only ways to identify the desired clones. We chose the PCR method and proceed to design an upper differential primer showing the 6 bases in its 3′OH coincident with $Q_{200}F_{201}$ sequence (last codon of the N-terminal propeptide and first codon of mature SP-B), which should only anneal with wt sequence. Plasmids generating a 200-300 DNA fragment according to the expected size (264 bp) through differential PCR amplification were considered recombinants (Fig. 1, lane 2) and sequenced.

Moreover, the *Xba*I digestion of pPROEX-1 could not proceed initially since *Xba*I site - TCTAGA- overlapped with the methylase Dam target site -GATC-, being thus the -A- methylated. So, strain GM48 (dam⁻) was transformed with pPROEX-1, which could be digested with *Xba*I thereafter.

Recombinant protein purification was achieved from inclusion bodies solubilized in urea, by affinity chromatography through its His-tag. The elution profile from the cobalt-based resin in 200 mM imidazole is depicted in Fig. 2A. SDS-PAGE and Coomassie staining in Fig. 2B shows a major band of purified His-tagged-proSP-B$_{\Delta C}$, with an apparent molecular mass of about 31 kDa. The Western blot analysis in Fig. 2C, using anti-His-tag antibodies to detect protein, shows that the protein fraction obtained from inclusion bodies solubilized and incubated with the resin (S-I) possess lower electrophoretic mobility than the upper band of purified and dialyzed protein (P). This result seems to indicate that the His-tagged-protein adopts a particular conformation after urea removing, which has a substantially different electrophoretical mobility than the protein in 8M urea as is bound to the resin through the His-tag (this behaviour is shown in spite of the presence of SDS in the loading buffer). To this regard, preliminary results obtained in our lab indicate that the protein in 8 M or 5M urea shows the same mobility as in the S-I fraction and that soluble proSP-B$_{\Delta C}$ obtained in *E. coli* (under conditions preventing inclusion body formation) but denatured in 8 M urea, purified in the Cobalt-based resin and dialyzed, showed also the same mobility than protein in S-I fraction (results not shown). This could indicate that the post-dialysis conformation adopted by the protein might be different when the protein is produced as a soluble form or obtained from inclusion bodies. Differences in the establishment of certain disulphide bonds, as has been described for other proteins [8], can not be discarded.

591

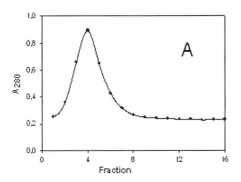
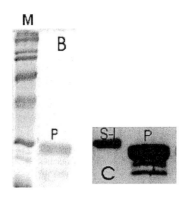

Fig. 2 Elution profile of affinity chromatography (A) and SDS-PAGE and Coomassie staining of purified His-tagged proSP-B$_{AC}$ (B), where M means molecular mass markers and P is the peak protein in A after several dialysis to eliminate the urea. (C) *Western blot* analysis of purified and dialyzed protein P and of the protein extracted from inclusion bodies by sonication and incubated with the cobalt-based resin just before loading into the column, S-I. Detection was carried out with anti-His-tag antibodies.

The appearance in P of several minor additional bands (Fig.2C), which do not appear in S-I, are due in our opinion to the presence of fragments originated by cut short translation events. This may happen since *E. coli* cells do not contain enough charged tRNAs recognizing codons used preferentially in eukaryotes, which in turn may lead to polypeptide elongation pausing [9] being noteworthy this effect when there is overexpresion of the hybrid protein, as is our case. These peptides may not be quantitatively relevant in the crude fractions but their transit through the resin and their concomitant concentration make them presumably to be revealed by anti-His-tag antibodies. However, we could not discard other possibilities such as degradation taking place along the purification process. In this sense, Cu^{2+} complexes of surfactant imidazole ligands were found to be highly active catalysts for the hydrolysis of 4-carboxyphenyl picolinate in revese micellar medium [10].

References

[1]   J. Goerke. Biochim. Biophys. Acta **1408**, 78 (1998)
[2]   F. Brasch, M. Ochs, T. Kähne, S. Guttentag, V. Schauer-Vukasinovic, M. Derrick, G. Johnen, N. Kapp, K-M. Müller, J. Richter, T. Giller, S. Hagwood, F. Bühling, J. Biol. Chem. **278**,49006 (2003)
[3]   J. Sambrook, E.F. Fritsch, T. Maniatis. In C. Nolan (ed) 3.Cold Spring Harbor Lab. Press. NY; 1989.
[4]   D. J. Hanahan. J. Mol. Biol. **166**,557 (1983)
[5]   O.H. Lowry, N.J. Rosebrough, A.L. Farr, R.T. Randall, J. Biol. Chem. **193**,256 (1981)
[6]   U. K. Laemmli. Nature. **227**,680 (1970)
[7]   A. Holzinger, K.S. Phillips, T.E. Weaver. BioTechniques **20**, 161 (1996)
[8]   J. S. Weissman, Kim, P. S. Science **256**, 111-113 (1992)
[9]   J. Parker, T. c. Johnston, P.T. Borgia, G. Holz, E. Remaut, W. Fiers. J. Mol. Biol. **258**, 10007 (1982)
[10]   K. Ogino, K. Nishi, H. Yamamoto, T. Yoshida, W. Tagaki. Tetrahedron Lett. **31**, 7023 (1990)

Influence of grape sanitary quality on yeast killer effect at the winery

F. Pérez-Nevado[*,1], J.A. Regodón Mateos[2], M. Ramírez Fernández[3]

[1] Department of Zootecnia, University of Extremadura, Crtra. de Cáceres s/n, 06071 Badajoz, SPAIN
[2] Department of Analytic Chemistry, University of Extremadura, Avda. Elvas s/n, 06071 Badajoz, SPAIN
[2] Department of Microbiology, University of Extremadura, Avda. Elvas s/n, 06071 Badajoz, SPAIN

[*]Corresponding author: e-mail: fpen@unex.es, Phone: +34 924286200, Fax: +34 924286286201

The effect of killer strains of *Saccharomyces cerevisiae* on the growth of sensitive and neutral wild strains during semi-continuous winery must fermentation was studied by monitoring yeast population dynamics. The capability of killer yeast strains to decrease sensitive strain populations depends on grape sanitary quality. In cold-settled fresh must from good sanitary quality grapes, we achieved inoculated killer yeast dominance (higher than 88% of the total yeast population) by using an initial proportion of 97%, although at the end of fermentation, wild *S. cerevisiae* strains had still not been totally eliminated, 33% of the remaining wild yeasts being killer-sensitive. The dominance of the inoculated killer yeast during the fermentation of cold-settled fresh must from poor quality grapes was greatly reduced. The large amount of suspended particles in the must would have neutralized K2 killer toxin, allowing a faster increase in the wild killer-sensitive *S. cerevisiae* population during fermentation.

Keywords fermentation; killer; *Saccharomyces cerevisiae*; yeast; wine; winery

1 Introduction

Wild wine yeasts with a killer phenotype are widespread in many wine regions of the world [1] [2]. Killer yeasts and their influence on wine fermentation have been reviewed by Gutiérrez et al. [1]. Inoculated killer wine yeasts may be effective in suppressing undesirable sensitive wild yeast strains to improve wine quality. The magnitude of the killer effect in wine fermentation depends on various factors [3] [4]: the initial ratio of killer to sensitive strains, the presence of protein-adsorbing substances, the environmental conditions and the growth phase of the sensitive cells, the presence of protective neutral yeasts, the susceptibility of sensitive strains to the killer toxins of different yeast strains, and the inoculum size and nitrogen availability. Several studies have been performed to quantify the effect of killer activity on winery fermentations with contradictory results. In addition, most data reported so far have come from experiments carried out at a laboratory level, in conditions very different from those of the highly variable winery environment. The effect of killer strains of *Saccharomyces cerevisiae* on the growth of sensitive and neutral wild strains during semi-continuous winery must fermentation was studied by monitoring yeast population dynamics.

2 Material and methods

Yeast strains: JP85R is a cycloheximide-resistant (CYH^R) mutant isolated in YEPD-CYH from JP85 [3]. E7AR1 is a K^+ CYH^R spore clone from the hybrid 7AR (cross JP88xJP85R) [5].

Wine fermentation trials: Winery fermentations were performed in a semi-continuous manner during two vintages (Figs. 1A and 2A). After 105 days following the end of fermentation wines were analyzed and sensorially evaluated (Table 1).

Determination of the percentage of CYH^R cells in fermentation trials: Samples from fermenting musts were diluted and plated on YEPD-agar to obtain single colonies. The detection of the CYH^R yeasts was accomplished by replica-plating the colonies to YEPD-CYH (2 mg mL^{-1} of cycloheximide) and YEPD-agar plates [5].

Plate assay for killer activity: Low-pH (pH 4) blue plates were seeded with 48-hour cultures of sensitive strains. Strains to be tested for killer activity were loaded onto the seeded agar to produce patches of about 5 mm diameter.

3 Results and discussion

During the first vintage period (Vintage 1), with grapes of good sanitary quality, we conducted must fermentations with the strains JP85R (K^- CYH^R) and E7AR1 (K^+ CYH^R). Fermentation rates were very similar for these two inoculated fermentations. However, the non-inoculated control fermentation started two days later (Fig. 1A). No killer yeast was detected among the isolated CYH^S wild yeasts in any of the three fermentations. As expected, inoculated CYH^R strains (killer or sensitive) dominated the two inoculated fermentations, their proportion always being higher than 88% (Fig. 1B). However, wild *S. cerevisiae* strains were not totally eliminated, with 4% and 10% remaining at the end of the E7AR1 and JP85R fermentations respectively, 33% of them being killer-sensitive. Therefore, we think the advantage of the inoculated killer strain was partially neutralized by the must's suspended particles in this commercial setting, in agreement with the data that we have reported elsewhere [4]. In the present study, the analytical results were similar for the three wines obtained in this vintage (Table 1). All of them had less than 1.5 g L^{-1} of reducing sugars and they were considered of good sensorial quality, although the control was less appreciated.

Table 1 Analyses of wines elaborated at industrial scale during two different vintages.

	Strain	Density 20/20	pH	Total acidity (g TH_2 L^{-1})	Volatile acidity (g AcH L^{-1})	Sugars (g L^{-1})	Total SO_2 (mg L^{-1})	Alcohol (% Vol.)	Acceptance (%)
Vintage 1	JP85R	992	3.48	5.31	0.27	1	12	12.3	62.3
	E7AR1	991	3.45	5.48	0.24	0.99	17	12.3	60.3
	Control 1	991	3.44	5.44	0.31	0.81	16	12.3	52.5
Vintage 2	E7AR1	994.2	3.57	5.47	0.45	1.2	110	12.4	45
	Control 2	997.2	3.98	5.17	0.57	1.9	117	12.3	18

594

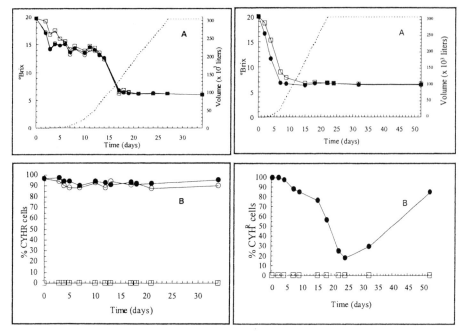

Figure 1. Fermentation kinetics (A) and temporal evolution of the percentage of inoculated CYH^R yeast cells (B) of winery must fermentation (Vintage 1) inoculated with JP85R (○), E7AR1 (●), and the non-inoculated control (□). (-----), must-wine volume.

Figure 2. Fermentation kinetics (A) and temporal evolution of the percentage of inoculated CYH^R yeast cells (B) of winery must fermentation (Vintage 2) inoculated with E7AR1 (●), and the non-inoculated control (□). (-----), must-wine volume.

In the second vintage period (Vintage 2), the sanitary quality of the grapes was not as good as in Vintage 1 (some mould and rot were found and the initial quantities of wild yeasts were greater), and hence the amount of suspended particles in the cold-settled must was higher. Again, the onset of inoculated fermentation was faster than the control (Fig. 2A). Nevertheless, this control fermentation started more rapidly than the control of the first vintage. This result indicates that the wild yeasts were more competitive during this vintage. In contrast to the case of Vintage 1, the proportion of E7AR1 decreased progressively to 27.8% of the total yeast population on day 24 (Fig. 2B). Probably the inoculated yeasts were somehow deteriorating biologically in the fermenting must-wine (approximately 5–8% of alcohol), becoming less competitive with respect to the new wild yeasts added with the contaminated fresh must. The large amount of suspended particles would neutralize K2 killer toxin, allowing a faster increase in the wild *S. cerevisiae* population. Surprisingly, just after we stopped adding new fresh must (day 24), there was an increase in the E7AR1 proportion to 82% on day 52 (Fig. 2B). During this increasing E7AR1 proportion phase, most particles of the must had already settled to the bottom of the vat because of the already low fermentation activity, and therefore it would be expected that the suspended sensitive yeasts would now be more susceptible to the killer yeasts. In fact, at the end of the control fermentation, 40% of the wild yeasts were sensitive and the rest neutral, while in the inoculated fermentation only 7% of wild yeasts were sensitive.

Another possibility is that other killer yeasts interact with the inoculated yeast's growth, allowing non-sensitive wild *S. cerevisiae* strains (neutral or sensitive to K2) to dominate. However, on analysing wild yeasts (*Saccharomyces* and non-*Saccharomyces*), we never found killer activity against E7AR1 under our assay conditions (low-pH blue plates). These results are in agreement with those of Gutiérrez et al. [1], who analysed the ecology of spontaneous fermentations, and found that *S. cerevisiae* with different killer phenotypes co-existed at the different fermentation stages. When wines of Vintage 2 were tested, as expected the E7AR1 wine was the most appreciated (Table 1), although the quality was lower than in the wines from the Vintage 1. This is consistent with the poor grape quality and the irregular dominance of the inoculated E7AR1 selected yeast strain.

4   Conclusions

In conclusion, in a semi-continuous winery fermentation, the addition of cold-settled fresh must from a poor quality harvest reduced inoculated killer yeast dominance over sensitive and neutral *S. cerevisiae* wild yeasts. This is an important issue and can explain different results obtained by various authors at the industrial level using killer yeasts as starter. In some of them, sensitive yeasts better adapted to these grape must conditions could even have dominated the fermentation process.

Acknowledgements

This work was partially financed by projects IPR98B024 (Government of Extremadura) and DIGESIC IFD97-1074 (Spanish Government). We thank the company Bodegas Coloma SA for supplying the grape juice and allowing us to use its equipment.

Referentes

[1]   A.R. Gutiérrez, S. Epifanio, P. Garijo, R. López, and P. Santamaría, Am. Journal. of Enol. Viticulture, **52**, 352 (2001).

[2]   J.A. Regodón, F. Pérez, M.E. Valdés, C. De Miguel, and M. Ramírez, Food Microbiol., **14**, 247 (1997).

[3]   G.A. Da Silva, Appl. Microbiol. Biotechnol., **46**, 112 (1996).

[4]   F. Pérez, M. Ramírez and J.A. Regodón, Antonie van Leeuwenhoek Journal of Microbiology, **79**, 393 (2001).

[5]   F. Pérez, J.A. Regodón, M.E. Valdés, C. De Miguel, and M. Ramírez, Food Microbiol., **17**, 119 (2000).

[6]   M. Ramírez, F. Pérez, and J.A. Regodón, Appl. Envirom. Microbiol., **64**, 5039 (1998).

Inhibitory effects of UV-absorption compounds in hemicellulose hydrolysate on xylitol production by yeast

ZHANG Hou-Rui* ZHENG Jian-Zhi HE Cheng-Xin FANG Hong CAI Ai-Hua

(Guangxi Institute of Botany, Guilin city, 541006, China)

)Corresponding author. Tel: +86-773-3550075, Fax: +86-773-3550067.
e-mail: zhhr@gl.gx.cninfo.net.cn

UV-absorptive compounds in hemicellulose hydrolysate are formed during hydrolysis. They are mainly the phenolics derived from lignin by lignin degradation, and the important inhibitor of microbial metabolism. For an effective xylitol production by yeast, it is necessary to reduce its concentration in hemicellulose hydrolysate. The present study showed that the Macroporous resin adsorption chromatography is an effective method to remove these types of inhibitors from hemicelluaose hydrolysate. The UV- spectrum peak at 286 nm is indicative of the presence of these compounds. Collecting the UV-absorptive compounds by Macroporous adsorption resin from sugar cane bagasse hemicellulose hydrolysate, its effects on xylitol production were studied in commercial xylose media by *Candida tropicalis* CGMC 2.1776. Adding the extractive UV-absorptive compounds to the media, the increase of the extractive UV-absorptive compounds concentrations in xylose media was companied by the decrease of xylitol yield, xylose consumption rate, but the cell densities were almost the same with the comparison. In a higher concentration of the extractive UV-absorptive compounds 6.0 g/l, the xylitol yield (g/g) decreased from 0.64 (without extractive UV-absorptive compounds) to 0.51 and the fermentation time was prolonged from 36h to 64h in the medium containing xylose of 150g/l, respectively. Our results first showed that the mainly toxic effect of UV-absorptive compounds in hydrolysate on xylitol fermentation was not the inhibition for yeast cell growth, but the inhibition of xylose utilization by from yeast cell, and the reduction of xylitol yield from xylose.

Keywords: Yeast; xylitol; Hydrolysate; Fermentation; UV-absorption compounds.

1   Introduction

Xylitol is a five-carbon polyalcohol that is found naturally in fruits and vegetables. It has received much attention for its use in some foods because of a number of advantageous natural properties. It is a natural sweetener of higher sweetening power than common polyols, indeed, it is as sweet as sucrose [1]. A further useful property is that it does not need insulin to regulate its metabolism and therefore can be used as a sucrose substitute in clinical diabetic foods. Currently, xylitol is manufactured by the chemical reduction of high-purified xylose derived from hemicelluloses hydrolysate of xylan-rich material. However, this is expensive due to difficulties in the separation and purification to remove by-products. Alternatively, the biotechnological production of xylitol by yeasts has become very attractive since it does not require pure xylose syrup as the chemical method does [2–4].

However, some by-products in hemicelluloses hydrolysis processes, such as furfural, acetic acid and lignin-degradation products, or lignin-derived phenolic compounds are formed. These are the inhibitor of microbial metabolism. In order to provide a substrate suitable for xylitol fermentation, it is necessary to reduce these kinds of toxic substance in hydrolysate[5].

Several strategies have been employed to limit the effect of inhibitors, including adaptation of microorganisms to the unfavorable environment[6], pretreatment of the raw material with ammonia[7], extraction with organic solvents[8,9], the use of ion exchange [10, 11], overliming [12], activated charcoal[10, 13, 14].

In previous studies, the attention has been focused on the inhibiting effect of acetic acid and furfural on yeast cell growth and fermentation. However, based on facts that xylitol yield from hydrolysate medium with different treatment is lower than those in synthetic xylose medium, Dominguez et al. [10] believed a possibility that the extraneous chemicals present in the hydrolysate could inhibit the release of the xylitol into the medium. We recently demonstrated that the yeast cell could utilize the high concentration of xylose in sugar cane bagasse hydrolysate treated with macro-porous adsorption resin and produce a higher concentration of xylitol with higher productivity [15]. In this report, we investigate the effects of UV-absorption compounds extracted by Macroporous resin adsorption chromatography from sugar cane bagasse hemicellusose hydrolysate on cell growth and xylitol production using *Candida tropicalis* AS2.1776 in the commercial xylose medium.

2   Materials and methods

2.1 Hydrolysis conditions

Dried sugar cane bagasse was impregnated for 12 h with sulfuric acid of 2.4%(w/v) at a solid: liquid of 1:6, and then hydrolyzed in 105°C for 2.5 hours. Following filtration, the liquid fraction was concentrated by evaporation under vacuum to 1/3 of initial volume.

2.2 Preparation of the Macroporous resin adsorption substances

Adsorption chromatographic separation was carried out in a column (2.5cm*50cm) filled with Macroporous resin HPD-500 (A polar macroporous resin, produced by CangZhou Bon Chemical Co., Ltd. China. http://www.bonchem.com/hpd500.asp). The column was washed first with alcohol 95 % (v/v) and second with 0.1M HCL, followed with water, and then with 0.1M NaOH, finally with water until the effluent pH was 7. Both the HCL and NaOH solution were each ten times the volume of the column.

The concentrated hydrolysate was applied to the column with a flow rate of 10ml/min. After washing with water at pH2 to remove the impurities the resin hadn't adsorbed, the column was eluted with 50% alcohol. The elution liquids collected were removed the alcohol and then dried under vacuum. This dried substance was named Macroporous resin adsorption substance (MRAS). It was used as a model compound to compare the UV-spectra with hydrolysate, to quantify the content of same compounds in hydrolysate, and to study its effects on xylitol production.

2.3 Microorganism and inoculums preparation

Yeast strain *Candida tropicalis* CGMC 2.1776 was chosen in this study and maintained at 4°C on malt extract agar slant. The inoculums were cultivated in a medium containing the following nutrients (g/l): xylose 15, glucose 15, yeast extract 5, peptone 5, and malt extract 3. The cultivation was carried out in 125 ml flask (containing 25ml of medium) on a gyratory shaker at 200 rpm and 30°C for 12 h..

2.4 Inhibitory effect of MRAS

The fermentation medium contained the same nutrients as inoculums medium, but the xylose concentration was increased to 150g/l and glucose was lacked. Each sample of MRAS was dissolved in 1ml water of pH 10 adjusted by KOH, and then added to the fermentation medium followed by the addition of 0.1M phosphonic acid to adjust the medium pH to 7. The 5% (v/v) of inoculums was transferred into a 250-ml flask containing 25ml fermentation medium with the supplement of MRAS. The cultivation was at 200 rpm and 30°C.

2.5 Analytical methods

For the spectral analysis of sugar cane bagasse hemicelluloses hydrolysate and MRAS, the pH of the samples was adjusted the pH to 10, and then diluted in water. The quantitative analysis and spectral scans of these solutions were done using a RUI-LI 1100 UV spectrophotometer (Beijing, China). Cell density was determined by means of a calibration curve relating absorbance at 625 nm (721-spectrophotometer, Shanghai, China) and represented as cell dry weight (g/l). Xylose, xylitol were quantified by high-performance liquid chromatography (HPLC) using Waters-510 apparatus, Waters-410 index refraction detector and Benson Carbohydrate Columns (BC-100, Ca^{+2}), and employing the following conditions: water eluant, 0.5ml/min flux; column temperature 90°C.

3   Results and discussion

3.1 The spectra character of MRAS and its concentration in hydrolysate

The spectra analyses (Fig) demonstrated that MRAS have the same uv-spectra with the sugar cane bagasse hemicellulose hydrolysate. They have higher absorption peak at 286nm in the strong absorption band of 268–312 nm. These results indicated that the most chemical components of MRAS extracted from sugar cane bagasse hydrolysate were the same with UV-absorptive compounds in hydrolysate.
After treating with HPD-500, there were few the compounds of absorption bands of 268-312 nm in hydrolysate. It is clear that a series of complex phenolic compounds degraded from lignin have been removed effectively by Macroporous resin adsorption chromatography from sugar cane bagasse hemicellulose hydrolysate since the absorption peak at 286 is a property of the phenolic compounds in alkaline condition.
Based on those facts described above, the MRAS could be used as a model compound for representing the concentration of the same type compounds in sugar cane bagasse hemicellulose hydrolysate by the absorption at 286 nm. After concentrating xylose concentration to 15% (w/v) under vacuum in this study, the content of *MRAS* in sugar cane bagasse hydrolysates is about 6.8g/l before passing HPD-500, and below 0.1 g/l after treatment.

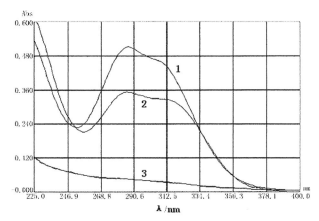

Fig 1 Spectra of uv-absorption compounds of sugar cane bagasse hydrolysate. **1**, MRAS; **2**, sugar cane bagasse hydrolysate hydrolysate; **3**, the hydrolysate after passing through Macroporous resin HPD-500.

3.2 Inhibitory effects of MRAS on xylitol production

Adding MRAS to xylose medium the fermentation times increased linearly with the increase of MRAS concentration, but the maximum cell densities was decreased appreciably (table). In the higher concentration of MRAS 6.0 g/L, the fermentation time was prolonged from 36h (lack of MRAS) to 64h, but the maximum cell density only decreased from 7.8g/L to 7.0g/L. These results suggested that the toxic effect of MRAS on xylitol fermentation by yeast is not mainly the inhibition of cell growth, but the xylos consumption rate by yeast cells.

Table. 1 The effect of MRAS in the xylose media on Xylitol production by *Candida tropicalis* CGMC 2.1776. P, xylitol concentration; Qp, xylitol volumetric production rate; Yp/s, xylitol yield from xylose; r_s, xylose consumption rate; t_f, fermentation time. Initial xylose concentration, 150g/L.

Treatment code	MARS (g/l)	P (g/l)	Qp (g/l/h)	Yp/s (g/g)	r_s (g/l/h)	Cell (g/l)	t_f (h)
1	0.0	97.7	2.69	0.64	4.16	7.8	36
2	0.5	97.0	2.49	0.64	3.85	8.0	39
3	1.0	88.0	1.91	0.58	3.26	7.8	46
4	2.0	85.0	1.57	0.57	2.78	7.0	54
5	4.0	80.0	1.33	0.53	2.50	6.9	60
6	6.0	77.0	1.20	0.51	2.34	7.0	64

Although, using hydrolysate as substrate, the inhibitory effects of toxic substance have been paid attention in xylitol production; the lower xylitol productivity and lower xylitol yield were generally explained as the inhibitor effect of acetic acid or furfural. The inhibitory effects of lignin degradation products were found, however, it was mainly considered as cell growth inhibitors. In fact, some detoxification methods removing acetic acid, furfural from hydrolysate were useful for improving the cell growth, but not xylitol production. For example, Pjrajo et al [8] found that xylose consumption rate can be significantly improved by overliming, but the carbon source was mainly converted to biomass. Treating with activate charcoal, passing through a cation-exchange resin, the sugar cane bagasse hydrolysate concentrated in vacuum can be utilized readily by yeast cell, but followed by the extraction with organic solvents (such as chloroform, diethyl ether, trichloroethylene, hexane), the xylitol production could be improved significantly. These results demonstrated directly that the toxic compounds in hydrolysate for xylitol fermentation not only include the cell growth inhibitors, but also xylitol-secreting inhibitors like those described in this study. The poor fermentability of hydrolysate for xylitol production is the result of the synergistic effects of the both toxic compounds, cell growth inhibitor and xylitol-secreting inhibitor.

In this study, we demonstrated that MRAS extracted with a polar macroporous resin (HPD-500) from sugar cane hydrolysate are mainly the UV-absorption compounds. They could greatly prolong the fermentation time and decrease xylitol yield in the concentration without obvious toxic effect on yeast cell growth. In order to reach a higher xylitol yield and productivity from hemicellulose hydrolysate by yeast, it is necessary to reduce UV-absorption compounds concentration in media.

Acknowledgements

This work was supported by a grant of the key project of Guangxi scientific and technological program (N0.0235005-4, Gangxi Autonomous Region), and by the National Nature Science Foundation of China (No.30450007)

References

[1]   Emodi, A., Xylitol, its properties and food applications. Food Technol. **32,** 28–32 (1978).
[2]   Parajó, J. C., Dominguez, H. Dominguez, J. M., Biotechnological production of xylitol. Part 1: Interest of xylitol and fundamentals of its biosynthesis, Bioresour. Technol. **65,** 191–201 (1998).
[3]   Parajó, J. C., Dominguez, H., Dominguez, J. M., Biotechnological production of xylitol. Part 2: Operation in culture media made with commercial sugars. Bioresour Technol. **65,** 203–212 (1998).
[4]   Nigam, P., Singh, D., Processes for fermentative production of xylitol – a sugar substitute. Process Biochemistry. **30,** 117–124(1995).
[5]   Parajó, J. C., Dominguez, H., Dominguez, J. M., Biotechnological production of xylitol. Part 3: Operation in culture media made from lignocellulose hydrolysates. Bioresour Technol.**66,** 25–40(1998).
[6]   SUN, K. S., XIA, L. M., Studies on xylitol fermentation of corn cobs hydrolysate by Candida Sp. Chemistry and industry of forest products (in Chinese). **22**(2), 26–30 (2002)..
[7]   Dominguez, J. M., Cao, N. J., Gong, C. S. et al., Dilute acid hemicellulose hydrolysates from corn cobs for xylitol production by yeast, Biore Technol. **61**, 85–90 (1997).
[8]   Parajó, J. C., Dominguez, H., Dominguez, J. M., Xylitol production from Eucalyptus wood hydrolysates extracted with organic solvents. Process Biochemistry.**32**(7), 599–604(1997).
[9]   Cruz, J. M.; Dominguez, H.; Parajo, J. C, Solvent extraction of hemicellulosic wood hydrolysates: a procedure useful for obtaining both detoxified fermentation media and polyphenols with antioxidant activity, Food Chemistry. **67,**147–153(1999).
[10]   Dominguez, J. M., Gong, C. S., Tsao, T. G,. Pretreatment of sugar cane bagasse hemicellulose hydrolysate for xylitol production by yeast. Appl. Biochem. Biotechnol., **57/58,** 49–56(1996).
[11]   Kim, S. Y., Oh, D. K., Kim, J. H,. Evaluation of xylitol production from corn cob hemicellulose hydrolysate by Candida parapsilosis. Biotech Letter, **21**, 891–895 (1999).
[12]   Roberto, I. C., Felipe, M. G. A., Lacis, L. S., et al.. Utilization of sugar cane bagasse himicellulosic hydrolysate by Candida guilliermondii for xylitol production. Biore Technol. 36, 271–275(1991).

[13]   Morita, T. A., Silva, S. S., Felipe, M. G. Effects of initial pH on biological syntheses of xylitol using xylose-rich hydrolysate, Appl. Biochem. Biotechnol. 84-86, 751–759(2000).

[14]   Mussatto S I , Roberto I C. Optimal Experimental Condition for Hemicellulosic Hydrolyzate Treatment with Activated Charcoal for Xylitol Production. Biotechol.Prog. 20,134–139(2004)

[15]   ZHANG, H. R., ZHENG, J. Z., HE, C. X., Utilization of sugar cane bagasse hydrolysate for xylitol production by yeast. Chinese Journal of Biotechnology. 18, 724–728(2002).

Production and Application of 5-Aminolevulinic Acid from *Rhodobacter capsulatus* SS3 Cultivating in Monosodium Glutamate Effluent

A. Chaikritsadakarn[1], P. Boonsawang[2], and P. Prasertsan[*1,2]

[1]The Joint Graduate School of Energy and Environment, King Mongkut's University of Technology Thonburi, Bangkok, 10140, Thailand.
[2]Department of Industrial Biotechnology, Faculty of Agro-industry, Prince of Songkla University, Songkhla, 90112, Thailand.

[*]Corresponding author : e-mail: poonsuk.p@psu.ac.th , Phone: +66 74 286369, Fax: +66 74 446727

Effluent from monosodium glutamate production was used as substrate for growth and 5-aminolevulinic acid (ALA) production from halotolerant photosynthetic bacterium *Rhodobacter capsulatus* SS3 under aerobic-dark condition at 37°C for 4 days. The monosodium glutamate effluent (MSGE) has an acidic pH (pH 3.60) and is high in organic matter and nutrients as it contains 220,100 mg/l COD, 814 mg/l reducing sugar, 3.72% v/v total nitrogen, 6.18 g/l Na and 11.56 % w/w glutamic acid. The bacterium was able to grow in MSGE, giving a dry cell weight (DCW) of 4.55 g/l after 36 h cultivation, but produced very small amount of ALA (0.20 μM). Both product yields were much lower than those obtained from cultivation in synthetic medium (GSY medium) with the values of 4.29 g/l DCW and 40 μM ALA. To improve ALA production in MSGE, the effect of COD concentrations (3,426, 7,644, 15,288 mg/l and 220,100 mg/l) was studied. The optimum COD concentration was found to be 15,288 mg/l, resulting in a 93 folds increase of ALA concentration to 18.6 μM. Further improvement of ALA production was achieved by supplementing MSGE with glycine, succinic acid, propionic acid, MgCl2 with the addition of levulinic acid, giving a 120 folds increase of ALA concentration to 23.74 μM.
The application of ALA as a herbicide was tested on giant sensitive plant Mimosa pigra Linn.. Six treatments were performed in which the effect of ALA in the culture broth of *Rhodobacter capsulatus* SS3 was compared with the commercial ALA at the same concentration (48 μM ALA) with and without 15 mM 2,2 dipyridyl (DPy). The effects of 15 mM DPy alone and water were included for comparison. Bioherbicide activity of ALA, expressed in terms of total tetrapyrrole, was found to be highest in the presence of both bacterial ALA and 15 mM DPy. These combined substances gave a 2.09 and 1.17 folds higher in herbicide activity than the culture broth and DPy alone, respectively. In term of photodynamic damage, this value increased from 30% to 100% when the concentration of ALA increased from 48 μM to 5 mM in the presence of Dpy.

Keywords: 5-aminolevulinic acid, production, application, monosodium glutamate effluent, bioherbicide

1   Introduction

Agro-industry plays a significant role in the economic growth of Thailand while being the source of high polluting wastewater. Monosodium glutamate (MSG) production was about 156×10^3 t/y and one plant generated 300-400 m^3/d effluent (MSGE) with the average values of 14,000 mg/l COD, 0.485 g/l glutamic acid, 3.72 total nitrogen and pH 3.6. In China, high content of organic matter could be reduced by cleaner production and comprehensive utilization [1]. Distillery stillage (DS) contained (mg/l); BOD 27,850, COD

184,000, reducing sugar 35,200, solid 2,345, nitrogen 2,700 and temperature of 92 °C with pH of 4.4 [2]. In addition, DS contains many yeast cells that can be the source of vitamins. An alternative approach for the treatment of these wastewaters is their use as a growth substrate for cultivation of photosynthetic bacteria to produce 5-aminolevulinic acid (ALA) which can be used as herbicide, insecticide, growth stimulator, and applied to medical field for cancer treatment. ALA from *Rhodobacter sphaeroides* can be produced in the effluent of swine waste from an anaerobic digester and enhanced by adding glycine (20 mM) plus levulinic acid (5-60 mM) [3].

This research work aims to utilize and treat the selected wastewater using the halotolerant photosynthetic bacterium *Rhodobacter capsulatus* SS3 to produce the valuable product 5-aminolevulenic acid (ALA).

2 Material and methods

2.1 Chemical composition of wastewater and synthetic medium

Glucose-salt-yeast extract medium (GSY) was the optimized synthetic medium for growth of *Rhodobacter capsulatus* SS3 and contained 50 mM glucose and 1.5 g/l yeast extract with 30 g/l NaCl [4]. MSGE and DS was kindly provided by Thai Fermentation Industry Co. Ltd., Rajaburi Province and Nateechai Co., Ltd., respectively. MSGE, DS and GSY media were analyzed for COD [5], nitrogen [6], reducing sugar [7], glutamic acid (analyzed by High Performance Liquid Chromatography method) and NaCl (analyzed by Inductively Coupled Plasma Optical Emisssion Spectrometry method).

2.2 Factors affecting ALA production from synthetic medium and wastes using *Rhodobacter capsulatus* SS3

2.2.1 Inoculum preparation
One loop of *Rhodobacter capsulatus* SS3 from GSY agar slant was inoculated into 100 ml of GSY medium and cultivated under aerobic-dark condition (150 rpm) at 37 °C for 24 h. The turbidity measured by spectrophotometer at 660 nm was adjusted to 0.5 using GSY medium before use as the inoculum.

2.2.2 Effect of cultivation media
A 10% starter culture of *R. capsulatus* SS3 was added to a 250 ml flask containing 100 ml each of GSY, MSGE and DS. Cultivation was conducted for 4 days under aerobic-dark condition on a shaker (150 rpm) at 37°C. Samples (10 ml) were taken every 6 h to measure for pH, growth (OD_{660}), dry cell weight, and extracellular ALA concentration [8]. The wastewater giving the highest ALA concentration was selected for further studies.

2.2.3 Effect of COD concentration of the selected wastewater
The selected wastewater was diluted to 3 levels with higher (medium A), equal (medium B) and lower (medium C) COD concentrations compared to that of GSY medium. Experiment was conducted as described above.

2.2.4. Effect of nutrient supplementation and levulinic acid (LA) addition in the selected diluted medium

To further improve the production of ALA from *R. capsulatus* SS3, effect of nutrients supplementation as well as levulinic acid addition were studied. The selected diluted wastewater without nutrients supplementation was used as a control.

2.3 Application of ALA as bioherbicide

Giant sensitive trees (*Mimosa pigra* L), a diicotyledonous (dicot) weed, were germinated in black plastic containers (10 cm dia) in a greenhouse. The bioherbicide substances were dissolved in a solvent mixture made up of acetone/ethyl alcohol/tween 80/water (0.45:0.45:0.1:9, v/v/v/v). Each container (10 plants) was sprayed with 10 ml of tested solution, which is equivalent to a spray rate of about 0.45 l/m^2. The solutions used were: 5 mM commercial ALA, culture filtrate (5 mM ALA content) of *R. capsulatus* SS3, or 15 mM 2, 2'-dypyridyl (DPy) or in combination (see Table 1). Each treatment was performed in triplicate. The seedling took place for 6 days, they were sprayed with the tested substances. After spraying, the plants were wrapped in aluminium foil and were placed inside a cardboard box which were wrapped in two layers of black plastic. The dark-boxes were incubated overnight for 17 h in a dark-growth chamber at 37 °C. After 17 h, a sample was taken from the plant to determine for tetrapyrroles accumulation by fluorescence intensity [9]. After sampling for tetrapyrrole content, the remaining treated plants were placed in the greenhouse and their growth were evaluated over another 6 days. Photodynamic damage was assessed as percentage of death, bleaching of the green leafy tissue, and dehydration of the tissues, in response to exposure to sunlight. The extent of photodynamic damage was related to the amount of accumulated tetrapyrrole [9].

Table 1 Fluorescence intensity of protochlorophyllide (Pchlide), Mg-protoporphyrin (monoester) [MP(E)], and protoporphyrin IX (Proto) of 6 days old weed *Mimosa pigra* L seedling treated with ALA (as bioherbicide) with and without , and placed in darkness at 37 °C for 17 h.

Sample	Fluorescence intensity				Protein	% Photo
	Pchlide	MP(E)	Proto	Total Tetrapyroles	(mg/ml)	dynamic damage*
Control : Sprayed water only	0.980	0.566	0.623	2.169	11.53	0
Treatment 1 : Sprayed 47 µM ALA and 15 mM 2,2'-Dipyridyl	1.040	1.532	0.516	3.088	11.85	20
Treatment 2 : Sprayed 47 µM ALA only	0.653	0.565	0.510	1.728	11.64	0
Treatment 3 : Sprayed 15 mM 2,2'-Dipyridyl	0.803	1.938	0.756	3.497	10.27	30
Treatment 4 : Sprayed culture broth (48 µM ALA)	0.788	0.564	0.598	1.95	10.83	0
Treatment 5 : Sprayed culture broth (48 µM ALA) and 15 mM 2,2'-Dipyridyl	1.018	2.419	0.656	4.093	11.87	30
Treatment 6: Sprayed 5 mM ALA and 15 mM 2,2'-Dipyridyl	-	-	-	-	-	100

- Not determined

* photodynamic damage was determined on the 8th day after ALA spraying.

3   Results and discussion

3.1 Chemical composition of wastewater and synthetic medium

The characteristics of GSY medium, monosodium glutamate effluent (MSGE) and distillery stillage (DS) were determined. MSGE had the highest concentrations of COD (220,100 mg/l), reducing sugar (814 mg/l) and total nitrogen (3.72 %v/v) compared to DS (115,600 mg/l COD, 371 mg/l reducing sugar, and 0.99% nitrogen) and GSY (7,740 mg/l, 121 mg/l, and 0.08%, respectively). Sodium content of these media were 6.18, 0.57 and 11.87 g/l, respectively while their glutamic acid content were 11.56, 0.73 and 11.0 %, respectively. Both wastewaters were more acidic (pH 3.60 and 4.24, respectively) than the GSY medium (pH 6.5). High COD (184,000 mg/l) and reducing sugar (35,000 mg/l) in DS were due to the presence of yeast cells as well as other nutrient residues. High content of sodium in MSG was directly related to the amount of salt (NaCl) added during the process while 11.87 g/l of Na was resulted from the addition of NaCl in GSY medium for ALA production from *R. capsulatus* SS3.

3.2 Factors affecting ALA production from *Rhodobacter capsulatus* SS3

3.2.1 Effect of cultivation medium

R. capsulatus SS3 grew better in GSY medium than in MSGE and DS giving the dry cell weights of 5.26, 4.55, and 4.00 g/l, respectively. This corresponded to the decrease of pH from 6.86, 7.20 and 6.07 to pH 5.03, 6.31 and 5.68, respectively. Extracellular ALA production was highest (40.05 μM) in GSY medium while only a small amount (0.20 μM) was detected in MSGE and none in DS (data not shown). MSGE was a better source of nutrients than DS as it contained higher organic matter (COD) especially glutamic acid. The consumption of the organic substances resulted in the COD removal of 29.58 % in MSGE which was higher than that in DS (20.16 %) but lower than in GSY (54.76 %). Between these two sources of wastewater, MSGE was selected for further studies.

3.2.2 Effect of COD concentration of the selected wastewater

MSGE was diluted 12, 23.5 and 46.5 folds to achieve COD concentrations of 15,288, 7,644 and 3,863 mg/l and assigned as medium A, B, and C, respectively. Medium A gave a better growth of *R. capsulatus* SS3 (1.96 g/l DCW) and extracellular ALA (18.61 μM) than medium B (0.98 g/l, 8.33 μM) and medium C (0.45 g/l, 0 μM), respectively (Figure 1). Therefore, the optimum COD concentration of MSGE was about 15,000 mg/l which was similar to the previous results on cultivating the photosynthetic bacterium *Rhodocyclus gelatinosus* in diluted tuna condensate with COD of 20,000 mg/l [10]. Cultivation of *R. Capsulatus* SS3 in this optimum COD concentration of the diluted MSGE gave lower extracellular ALA concentration (18.61 μM) than that in the synthetic GSY medium (40.05 μM ALA), although growth (1.96 g/l DCW) was more than 2 times lower (4.29 g/l DCW). Nevertheless, the ALA concentration obtained was 93 folds higher than that of the control (MSGE).

606

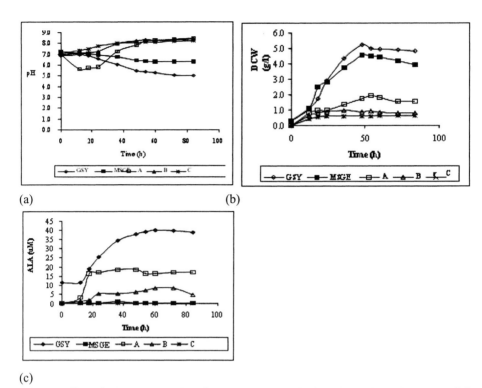

(a)   (b)

(c)

Figure 1 Effect of COD concentration of MSGE on (a) pH, (b) dry cell weight, and (c) extracellular ALA production from *Rhodobacter capsulatus* SS3 cultivated under aerobic-dark condition on a shaker (150 rpm) at 37 °C for 4 days.

        GSY contained 7,740 mg/l COD

        MSGE contained 220,100 mg/l COD

        Medium A contained 15,288 mg/l COD (12 folds diluted MSGE)

        Medium B contained 7,644 mg/l COD (23.5 folds diluted MSGE)

        Medium C contained 3,863 mg/l COD (46.5 folds diluted MSGE)

3.2.3 Effect of nutrients supplementation in the selected diluted MSGE

Further improvement of ALA production from *R. capsulatus* SS3 cultivated in the 12 folds diluted MSGE (medium A) was conducted with the supplementation of the optimized nutrients (so called medium E) and addition of 15 mM levulinic acid (LA) at 18 h cultivation [11]. The nutrients included 10 mM glycine, 40 mM succinic acid, 0.5 g/l propionic acid, 15 mM $MgCl_2$. Glycine and succinic acid are the substrates for ALA production via the Shemin (C_4) pathway [12]; propionic acid acts as an inhibitor of ALA dehydratase while $MgCl_2$ is a cofactor of ALA synthetase; and LA is an inhibitor of ALA dehydratase [8]. *R. capsulatus* SS3 cultivated in medium E gave a good growth (1.87 g/l) and produced higher extracellular ALA (23.74 µM) than that in the diluted MSGE (medium A) (1.96 g/l, 19 µM, respectively) (Figure 2). This resulted in 1.25 folds increase of ALA concentration or 120 folds increase compared to that obtained from cultivation in the MSGE.

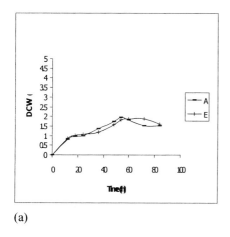
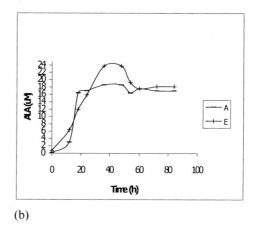

(a)   (b)

Figure 2 Effect of nutrients supplementation in the diluted monosodium glutamate effluent (MSGE) on (a) dry cell weight and (b) extracellular ALA production of *Rhodobacte capsulatus* SS3 cultivated under aerobic-dark condition on a shaker (150 rpm) at 37πC for 4 days

        Medium A : 12 folds diluted MSGE without nutrients supplementation

        Medium E : 12 folds diluted MSGE with nutrients supplementation

3.3 Application of ALA as bioherbicide

In the herbicidal mechanism, plants were treated with high concentration of ALA which converted to tetrapyrrole in plant cells and protoporphyrin IX (PPIX) was accumulated. When plants are exposed to light, the excess PPIX produces active oxygen which oxidizes unsaturated fatty acids on the cell surface and thereby damages the plants [9]. In this study, culture filtrate of *R. capsulatus* SS3 was tested as bioherbicide on weed, using commercial grade ALA as a positive control while using water instead of ALA solution as a negative control. Bioherbicide activity of ALA, expressed in terms of total tetrapyrrole, was found to be highest (4.093) in the presence of both bacterial ALA and 15 mM Dpy (treatment 5) (Table 1). These combined substances gave a 2.09 and 1.17 folds higher in herbicidal activity than the culture broth and DPy alone, respectively. In the presence of Dpy, the bacterial ALA demonstrated 1.32 folds higher herbicidal activity than the commercial ALA (treatment 5 and 1, respectively). For photodynamic damage, this value increased from 20-30% to 100% when the concentration of ALA increased from 48 ≥M to 5 mM in the presence of 15 mM Dpy. This was due to the effective dose for herbicide of ALA (5 mM) as well as the mixture of ALA and Dpy (the modulator). This result was higher than the 93% photodynamic damage of the same mixture (5 mM ALA and 15 mM Dpy) on cucumber seedlins [9]. Treatment with ALA alone (treatment 2 and 4) showed no effect.

4   Conclusion

Monosodium glutamate effluent (MSGE) was found to be a better substrate for growth and ALA production from *Rhodobacter capsulatus* SS3 than distillery stillage (DS), giving the extracellular ALA of 0.20 ≥M. The optimum concentration of organic matter (in term of

608

COD) of MSGE was about 15,000 mg/L. Supplementation of this diluted MSGE with 40 mM succinic acid, 10 mM glycine, 0.5 g/l propionic acid, 15 mM $MgCl_2$ and the addition of 15 mM levulinic acid at 18 h cultivation could further increase the ALA production to 23.74 ≥M. Hence, the optimization studies resulted in a nearly 120 folds increase of ALA production from *R. capsulatus* SS3. The ALA in the culture broth demonstrated herbicidal activity in the presence of 15 mM 2,2 dipyridyl and increase of ALA concentration to the effective dose (5 mM) could enhance the herbicidal activity to 100% photodynamic damage.

Acknowledgement

The authors would like to express sincere thanks to the Joint Graduate School for Energy and Environment (JGSEE) for the financial support of this project.

References

[1]   C.Yingxiang. Hunan Research Institute of Environmental Science, Changha, China. p. 1 (2003).

[2]   P. Prasertsan and P. Suksawat. Songklanakarin J. Sci. Tech. **4**, 138 (1982).

[3]   K. Sasaki, et al. J. Appl. Microbiol. Biotechnol. **32**, 727-731 (1990).

[4]   Madmarn, W. M. Sc. Thesis in Biotechnology, Prince of Songkla University, Thailand (2002)

[5]   APHA, AWWA and WPCF. Standard Method for the Examination of Water and Wastewater. 16ed. (American Public Health Association, Washington, D.C. 1985)

[6]   A.O.A.C. Official Method of Analysis of Association of Official Chemists. 15th ed. (Association of Official Analytical Chemists, Inc., Verginia. 1990).

[7]   N. Nelson. J. Biol. Chem. **153**, 375-380 (1944).

[8]   K. Sasaki, et al. J. Ferment. Technol. **65**, 511-515 (1987).

[9]   Rebeiz, C.A., Montazer-Zouhoor, A., Hopen, H..J., and Wu, S. M. J. Enzyme Microb. Tech. **6**, 390-401. (1984).

[10]   P. Prasertsan, M. Jaturapornpipat, and C. Siripatana. Pure & Appl. Chem. **69**, 2439 (1997).

[11]   S. Sattayasmithstid, and P. Prasertsan. International Conference on Engineering and Environment, ICEE Novi Sad 2005. Novi Sad, Serbia and Montenegro. 18-20 May, 2005.

[12]   Sasikala, Ch., Ramana, Ch.V., and Rao, P. R. Biotechnol. Prog. **10**, 451-459 (1994).

Protein profile of *Streptomyces clavuligerus* strains related to regenerating cells

M. G. Carneiro-da-Cunha[1,2] , **A. L. F. Porto**[2,3] , **J. L. Lima Filho**[1,2] **and G. M. Campos-Takaki**[4]*

[1]Biochemistry Department, UFPE; [2]Laboratory of Inmunopathology Keizo Asami-LIKA-UFPE; [3] Morphology and Physiology Animal Department, UFRPE; [4]Nucleous of Environmental Research, Chemistry Department-UNICAP.

*Corresponding author: e-mail: takaki@unicap.br

This paper describe the proteins isolated from *Streptomyces clavuligerus* cell walls and it's were correlated with regenerating colonies obtained by protoplasts formation. The proteins obtained were precipitated by ammonium sulfate in fractions of 0-40 and 40-60%, pointed greater total protein content in the regenerated colony denominated T_{4-8}. The electrophoresis analysis in SDS-PAGE showed the proteins profile with bands from 89 to 68 Kd whereas in the T_{4-8} regenerated colony. In relation to the antibiotic production was approximately 2.5 times greater for the colony T_{4-8} , considering as regenerated strain than for the wild strain.

Keywords: *Clavulanic acid;* Streptomyces clavuligerus*; Antibiotic production; Protein profile; Cell wall.*

1   Introduction

Streptomyces clavuligerus is a filamentous bacterium, industrially important, as it produces numerous antibiotics, including a —lactamase inhibitor of the clavulanic acid. The extracellular production of clavulanic acid is strongly dependent on inoculums activity, and is normally achieved after approximately 90 h of incubation, typically in the growth idiophase [6]. The procedures for *Streptomycetes* protoplasts isolation and regeneration, was early described by Sagara *et al.* [13], Okanshi *et al.*[10], as being one of the important process in genetic manipulation generating samples with specific characteristics, that can be directed towards the increase of antibiotic production. Sagara *et al.* [13] described the growth inhibition and a higher sensibility to lysozyme when the growth medium was supplemented with glycine. This amino acid also induces fragilization and morphological alterations in the cell wall, similar to the penicillin treatment [4].
Bacterial cells are encapsulated within a complex cell wall of which the major component is peptidoglycan, a polymer of carbohydrate chains connected by peptide crosslinks. Recent progress has been made in the synthesis of peptidoglycan intermediates that can be used to study enzymes which make peptidoglycan [8]. This paper describes the accumulation process of proteins into the cell wall of regenerated cells related to clavulanic acid production compared with the wild strain of *Streptomyces clavuligerus*

2   Materials and methods

Microorganisms: *Streptomyces clavuligerus* NRRL 3585 (ATCC 27064) was maintained at 28^0C in solid medium ISP-2 [14], modified by absence of glucose during 10 days, and the

610

Klebsiella pneumoniae (ATCC 29665) for antibiotic activity tests at 28 ^{0}C in nutritive agar (Difco).

Clavulanic acid production: The incubation were carried out using Erlenmeyer flasks (250 ml) containing 50 ml of cultivation medium (CM) described by Aharonowitz and Demaim [1], modified by the absence of 3-(N-morpholino) propanesulfonic acid (MOPS) and addition of K_2HPO_4 (0.435%). The cultivations were carried out in Fernbach (2800 ml) containing 450 ml of cultivation medium - CM or fragilization medium - CGM (CM added by 0.5% of glycine), with 10% of inoculum (OD of 0.1 at 600 nm), incubated in orbital shaker (200 rpm), during 96 h at 28 ^{0}C. The samples were collected each 24 h for the following determinations: antibiotic activity, protein content and protoplasts formation. The dry cell weight was determined at 105 ^{0}C until constant weight.

Cell wall remotion and Antibiotic Activity: Samples of 10 ml from fragilization medium (CGM) were centrifuged at 2000 x g for 10 min each 24 h, the cell mass washed with sorbitol solution at 20% (w/v), the mycelium resuspended in 5 ml of P medium [5], modified by replacement of the sucrose and 3-(N-Morpholino)propanesulfonic acid (MOPS) buffer by sorbitol and N-Tris[Hydroxymethyl]-2-aminoethanesulfonic acid (TES) respectively, containing lyzosyme (1 mg / ml), and filtered with Milipore membrane (0.45 μm). Colonies with higher antibiotic activity by diffusion disc method [5] were selected during 24, 48, 72 and 96 h, and were codified as T_{n-m} [n = culture growth time as: 1 = 24; 2 = 48; 3 = 72 and 4 = 96 h] and m = number of colony. The clavulanic acid production was expressed as g of antibiotic per g of biomass.

Protein Precipitation and SDS-PAGE: The proteins obtained during the cell wall remotion (supernatant), were precipitated to 40 and 60% of ammonium sulfate, dialyzed against Tris-HCl buffer 10 mM, pH 8.0 for 24 h. Proteins contents were measured by assay of Bradford [2]. The electrophoresis profile of the obtained proteins from cell walls were carried out by a discontinuous system according to Laemmli [7], applying phosphorylase B, bovine carbonic anhydrase, trypsin inhibitors and lysozyme as markers of molecular weight (BIO–RAD).

3   Results and discussion

In the forty regenerated colonies were selected by clavulanic acid production and the results showed that the regenerated colonies selected with 24 h (T_1) did not show antibiotic activity whereas, with 48 h (T_2) and 96 h (T_4), 90% of antibiotic activity and, with 72 h (T_3) only 40%, respectively. The best producer was the strain designated T_{4-8}, and the yield of clavulanic acid (0.140 g / biomass g, equivalent to 629.3 mg l^{-1}) showed 2.5 times higher than the wild strain (0.052 g / biomass g, equivalent to 237.65 mg l^{-1}), as illustrates the **Figure 1**. Although with a cheap culture medium, the soy medium, similar result   (220 mg l^{-1}) for wild strain, with glycerol feeding, was reported by Kuo-Cheng Chen *et al.* [6]. The authors using an expensive feeding (glycerol and ornithine), a further increase in clavulanic acid production to a maximum of 311 mg l^{-1} was also found, however those results are quite low when compared with 629.3 mg l^{-1} of the current work.

In relation to the precipitation with ammonium sulphate, the samples contents were distinguished by greater percentage of protein on fraction 0-40%, for wild strain and, on fraction 40–60% for regenerated samples(**Table 1**). The electrophoresis profiles analysis disclosed for the presence of bands with molecular weight at range from 89.0 to 68.0 for the wild strain while for regenerated samples, the band with higher molecular weight was around 60.0 Kd. In addition, it was yet observed that all regenerated samples showed bands with molecular weight below 21.5 Kd, which results did not happened with the wild strain.

However, for the majority of colonies was detected the presence of protein with molecular weight in the band from 59.0 to 21.5 Kd.

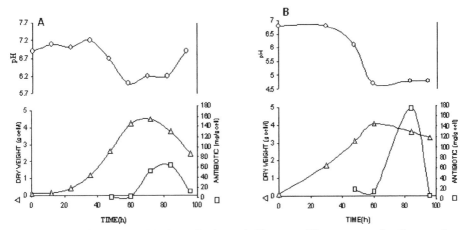

Figure 1 – Growth curves, antibiotic production and pH curves of *Streptomyces clavuligerus* cultures in cultivation medium (CM). **A** – Wild strain; **B** – T$_{4-8}$ Regenerated Colony.

Table 1 – Protein content on cell wall from wild strain and from regenerated colony grown during 96 h., and showed higher clavulanic acid activity

Strain/ Colony	Time	Total protein n g/ml	0–40% n g/ml	%	40–60% n g/ml	%
3585	48	241.4	79.0	32.7	30.0	12.4
	72	189.0	60.0	31.7	33.1	17.5
	96	404.2	252.2	62.4	35.7	8.8
T$_{4-8}$	48	372.6	68.6	18.4	162.6	43.7
	72	667.7	134.1	20.1	200.5	30.0
	96	781.6	52.5	6.7	180.4	23.1

(Column header spanning: **Protein by precipitation with (NH$_4$)$_2$SO$_4$**)

According to Rogers and Perkins[12], the cell wall synthesis does not occur simultaneously with the protein and nucleic acids However, it is connected with the formation of wall enzymes biosynthesis and with the supply of a limited number of amino acids and nucleo-tides. Although membrane-associated enzymes are intrinsically more difficult to study than cytoplasm enzymes because of their solubility properties, the principal reason for the pau-city of knowledge was that substrates to study the membrane-associated enzymes were not readily available [8]. With the regeneration process, the colony T$_{4-8}$ was the best producer of clavulanic acid, showing the higher protein content in cell wall extracts and bands from 45.0 to 14.4 Kd . It is possible to suggest that the formation and protoplasts regeneration processes with *Streptomyces clavuligerus*, allow the clavulanic acid production improve-ment. The results obtained with regenerated colonies from wild strain of *Streptomyces clavuligerus* suggest that the cell wall remotion and regeneration can be successfully used as an important method for increasing the yield of the antibiotic production.

612

In conclusion, the protoplast formation and regenerating cell wall induced to an increase of the clavulanic acid production, demonstrating to be an economical method and a supplied approach for the transfer to a scale production.

Acknowledgments

We are grateful to Dra. Carmen Donaduzzi for kindly supplied the *Streptomyces clavuligerus* strain, CNPq and CAPES Agencies for financial support

References

[1]   Aharonowitz, Y., Demain, A. L. (1978) Carbon catabolite regulation of cephalosporin production in *Streptomyces clavuligerus*. Antimicrob. Agents Ch. 14, 159-164.

[2]   Bradford, M. M (1976) A rapid and sensitive method for the quantitation of microgram quantities of protein utilizing the principle of protein-dye binding. Anal. Biochem. 72, 248-254.

[3]   Grove, D. C., Randall, W. (1955) Assay Methods of Antibiotics, a Laboratory Manual. Medical Enciclopedia 76-81.

[4]   Hames, W., Shleifer, K. H., Kander, O. (1973) Mode of action of glicine on the biosynthesis of peptidoglycan. J. Bacteriol. 116, 1029-1053.

[5]   Illing, G. T., Normansell, I. D., Peberdy, J. F. (1989) Protoplast isolation and regeneration in Streptomyces clavuligerus. J. Gen. Microbiol. 135, 2289-2297.

[6]   Kuo-Cheng Chen, Yun-Huin Lin, Jane-Yii Wu, Sz-Chwun John Hwang (2003) Enhancement of clavulanic acid production in Streptomyces clavuligerus with ornithine feeding. Enzyme Microb. Tech. 32, 152-156.

[7]   Laemmli, U. K. (1970) Cleavage of structural proteins during the assembly of the head of bacteriophage T4. Nature 227, 680-685.

[8]   Lazar, K. and Walker, S. (2002) Substrate analogues to study cell-wall biosynthesis and its inhibition. Curr. Opin. Chem. Biol. 6, 786-793.

[9]   Neves, A. A., Vieira, L M., Menezes, J. C. (2001) Effects of preculture variablity on clavulanic acid fermentation. Biotechnol. Bioeng. 72, 628-633.

[10]   Okanishi, M., Suzuki, K., Umezawa, H. (1974) Formation and reversion of streptomycete protoplaste cultural condition and morphological study. J. Gen. Microbiol. 80, 389-400.

[11]   Pridham, T. G., Anderson, P., Foley, C., Lindenfelser, L. A., Hesseltine, C. W. and Benedict, R. G. (1957) A selection of media for maintenance and taxonomic study of *Spretomycetes*. Antibio. Annu. 947-953.

[12]   Rogers, H. J., Perkins, H. R. (1980) Microbial Cell Walls and Membranes, Chapman and Hall (Eds.), New York, 564p.

[13]   Sagara, Y., Fukui, K., Ota, F. *et al.* (1971) Rapid formation of protoplasts of *Streptomyces griseoflavus* and their fine structure. Japanese Journal of Microbiology 15, 73-84.

Microfactories - Microbial Production of
Chemicals and Pharmaceuticals. Biopolymers

Isolation and screening of oleagenous microorganisms for the production of a nutraceutical- single cell oil

NIVEDITA SHARMA[*] and M. SANDHU

Department of Basic Sciences, Dr. Y.S. Parmar University of Horticulture and Forestry, Nauni, Solan, HP, India

[*]Corresponding author: e-mail: nivea_64@yahoo.co.in

In the present investigation, an attempt has been made to isolate single cell oil producing microorganisms from various sources. Seventy-one isolates were isolated and screened for single cell oil (SCO) production. Among them the isolated yeasts showed better production of SCO and therefore were selected for the further study. In total, 27 yeasts were isolated from different food sources capable of producing microbial fats. Yeast isolate strain no Y9 isolated from *Shorea robusta* sawdust was hyper producing yeast for microbial oil production (14.81%) and it was identified as *Rhodotorula* sp.Y9 The high yield of SCO from the *Rhodotorula*Y9 strain isolated from *Shorea robusta* is the main highlight of the present study.

Keywords: Single cell oil, Microbial fats, Yeast isolates, *Rhodotorula* sp.Y9

1   Introduction

Microorganisms are receiving increasing attention for their potential applications to the oils and fat industry either as means of producing high quality lipids and also to carry out selected transformation reactions that lead to high value lipid products (Ratledge, 1991). The microbial fats are relatively rich source of special oils like polyunsaturated fats (PUFAs),gamma linolenic acid, arachidonic acid etc. that are gaining popularity for their nutritional and therapeutic role (Certik and Shimizu ,1999 ; Aki et.al,2001)). Microbial fats rich in PUFAs have unique opportunity in the market place and the demand is likely to escalate further due to its high value applications such as nutritional supplement, fortification of food, health food and therapeutics. PUFA fortified foods are becoming very popular throughout the world among the health conscious consumers and thus are continuously in increasing demand and hence more production of PUFA would be required. A consistent search to isolate the new potent oleaginous microorganisms is the need of the hour. Therefore, in the present study an attempt has been made to isolate and screen the microorganisms from various sources for single cell oil production.

2   Materials and methods

Collection of samples : Different food samples, pressed oilcakes and saw dust of forest trees were collected.

Isolation: The oleaginous yeast were isolated from different samples on Yeast Extract Mannitol agar by serial dilution method and pure line cultures were established and stored at low temperature in the refrigerator for further use.

Screening : All the isolates of 1 O.D. @ 1% inoculum were grown in Yeast Extract Mannitol (YEM) broth for $37^{o}C$ for 5 days. The cells were harvested by centrifugation at 5000 rpm for 15 min and the pellets were collected for further use.

YEM broth : Yeast Extract (1.0g), Mannitol (10.0g), NaCl (0.1g), K$_2$HPO$_4$ (0.5g), MgSO$_4$ (0.2g)/1000ml , pH 6.0

Pretreatment: The method of Sattur and Karanth, (1987) was followed for refluxing the collected pellet of cells in reflaxometer with 50 ml of 0.1 NHCl. The cells of the isolates were boiled for 15 min and were washed twice with distilled water.The pellet was then filtered with preweighed Whatman No.1 filter paper and was dried in oven at 37^0 C.The weight of dried biomass was recorded.

Extraction of Single Cell Oil: Single cell oil extraction was done by modified Folch method, (1957) using petroleum ether (60-80 ^{0}C) solvent. The extraction period was 16 h as selected to be best for production of oil among 10,12,14 and 18 h of analysis. Extraction was done in Soxhlet apparatus following the procedure of Ranganna (1986) as given below:

Procedure: Apiece of filter paper was folded in_such a way so that to hold the sample containing_pretreated pellet and was tied tightly with thread. The sample was placed in the butt tubes of the apparatus. The extraction was done with petroleum ether for 16 h by gentle heating. Then the apparatus was allowed to cool down and the exraction flask was dismantled. Ether and oil extracted was collected in preweighed 50 ml beaker. The evaporation of ether was done on a water bath until there was no odour of ether remained. It was then allowed to cool at room temperature. The weight of the beaker containing oil was measured.and the heating was repeated until constant weight of the beaker was recorded.

Estimation of Single Cell oil: Estimation of oil obtained was done as (g oil/g dry biomass) x 100 given by Sattur and Karanth (1987).

Identification : The hyper oil producing yeast among all the isolates was identified as *Rhodotorula* sp.Y9 .

3   Results and discussion

In total, 27 yeasts were isolated form different sources viz. food, food/fruit processing waste, agricultural waste, forestry waste and oil cakes etc. Rainbow of different colors, ranging from cream, brown, white, pink, yellow and off white, was observed of different yeast colonies and

their cell shape varied from round to ovoid (Table 1). Out of these 27 yeast, many were found potential oleaginous yeasts producing single cell oil in the range of 11.12% to 14.81%. Maximum oil (14.81%) was produced by *Rhodotorula* sp.Y9 rendering it the best among all isolates. *Rhodotorula* sp.Y9 was isolated form sawdust of *Shorea robusta* and showed appreciable amount of oil produced, which could further be enhanced by optimizing the cultural conditions.

The oil of *Rhodotorula* sp.Y9 is of special interenst as it is rich in oleic palmitic, linolenic and stearic acid. It also contains polyunsaturated fatty acids (PUFA), which are the highly sought after in the field of medicines and pharmaceuticals (Jacob and Krishnamurthy, 1990). Moreover, ?-linolenic acid present in its lipids is currently used as a dietary supplement. Beside being the precursor of prostaglandin, it also helps in reducing the level of cholesterol in the blood (Satyanarayanan and Johri, 1992) thus playing a vital therauptic role.

Table 1: Isolation and Screening of single cell oil (SCO) producing yeasts from different sources and their morphological and cultural characteristics

Strain No.	Source	Shape	Colour	Texture	*Oil produced (%)
Y_1	(*Linum usitatissimum*) + (*Brassica campestris*)oil cake	Ovoid	Cream	Slimy	9.23
Y_2	Prunus armeniaca Squash	Round	Cream	Elevated	3.45
Y_3	*Linum usitatissimum* + *Brassica campestris*) oil cake+Oil	Ovoid	Brown	Slimy	8.05
Y_4	*Linum suitatissium* + *Brassica compestris*)oil cake+ H_2O	Irregular	Cream	Slimy	7.91
Y_5	Apple pomace	Ovoid	Cream	Slimy	6.24
Y_6	Bread	Ovoid	White	Elevated	4.82
Y_7	Paneer	Round	White	Elevated	7.62
Y_8	Curd + Oil	Round	Off white	Elevated	3.01
Y_9	Shorea robusta Saw Dust	Round	Pink	Smooth	14.81
Y_{10}	Curd	Ovoid	Cream	Filamentous	8.01
Y_{11}	Cheese	Round	Cream White	Elevated	5.57
Y_{12}	*Prunus armeniaca*	Round	Cream	Slimy	6.00
Y_{13}	Butter	Ovoid	White	Slimy	5.09
Y_{14}	(*Prunus armeniaca*)oil cake + Oil	Round	Brown	Slimy	4.81
Y_{15}	Apple Pomace	Ovoid	Cream	Smooth	5.01
Y_{16}	Molasses		Pale Yellow	Smooth	7.81
Y_{17}	(*Ephedra gerardiana*) + H_2O Oil Cake	Round	Cream	Lobate	7.41
Y_{18}	Molasses	Ovoid	Pale Yellow	Smooth	0.85
Y_{19}	(*Prunus persica*)Oil Cake + Oil	Ovoid	Cream Brown	Smooth	8.45
Y_{20}	Olive Pomace	Ovoid	Light Brown	Smooth	9.07
Y_{21}	(*Linum usitatissium*) + (*Brassica campestris*)oil cake + H_2O	Ovoid	Light Brown	Lobate	8.42
Y_{22}	(*Ephedra gerardiana*) Oil Cake + H_2O	Round	White	Smooth	5.05
Y_{23}	(*Ephedra gerardiana*)Oil Cake + H_2O	Round	Brown	Smooth	3.28
Y_{24}	*Prunus armeniaca* Squash	Ovoid	Off white	Smooth	2.77
Y_{25}	Butter	Round	Green	Smooth	5.32
Y_{26}	Bread	Round	Off White	Slimy	4.60
Y_{27}	Bread	Round	White	Slimy	3.00

*% oil = (g oil/g dry biomass) x100

Each experiment was conducted in triplicatesqW

The yeasts are always the organisms of choice for microbial oil production in industry over molds and bacteria because of their easy handling and less contaminating properties during widely adapted continuous culture technique of fermentation.

Due to high oil production and its special medicinal properties the isolated strain of *Rhodotorula* sp.Y9 is of considerable interest and can be recommended to biotech industries for high value lipid production after optimizing its environmental conditions for further enhancement in the yield.

References

[1]   Aki T., Y. Nagahata, K. Ishihara, Y. Tanaka,T.Morinaga, K. Higashiyama, K. Akimoto,

[2]   S.Fujikawa, S.Kawamoto, S. Shigeta, K. Ono and O.Suzuki... *Am.Oil Chem. Soc.* **78** (2001), p 599.

[3]   Certik M. and S. Shimizu 1999. *J. Biol. Sci. Bioeng.* **87**(1999), p 1

[4]   Folch J., Lees, M. and Sloane-Stanley, G. *J.Biol Chem* **226**(1957), p 497

[5]   Jacob, Z and M.N. Krishnamurthy. *J.Am. Oil Chem. Soc.* **67**(10) (1990), p 642

[6]   Ranganna S. *Handbook of analysis and quality control for fruit and vegetable products.Tata McGraw Hill Publisher Company, N.Delhi.* pp. 231-232.

[7]   Ratledge C. *Acta Biotechnol.* **11**(1991), p 429

[8]   Sattur, A.P. and N.G. Karanth . *J. Microbiol. Biotechnol.* **2**(2)(1987), p 116.

[9]   Satyanarayana T and B.N. Johri . *J. Microbiol.* **32**(1)(1992), p 1

Manufacturing and characterization of bacterial cellulose tubes using two different fermentation techniques

A. Bodin[1,2], **H. Bäckdahl**[1,3], **L. Gustafsson**[2], **Bo Risberg**[3] and **P. Gatenholm**[*,1]

[1] Chalmers University of Technology, Department of Chemical and Biological Engineering, Kemivägen 4, SE-412 96 Gothenburg, Sweden
[2] Chalmers University of Technology, Department of Molecular Biotechnology, P.O. Box 462, SE-412 96 Gothenburg, Sweden
[3] Vascular Engineering Centre, Institution of Surgery, Department of Surgery, Sahlgrenska University Hospital, Gothenburg, Sweden

[*] Corresponding author: e-mail: Paul.Gatenholm@chem.chalmers.se, Phone: +46 31 7723407, Fax: +46 31 772 3418

A modified defined media with 0.6% lactate resulted in high cellulose yield and was therefore used in the tube formation. Two fermentation techniques were compared, A and B, using a modified disc reactor. Scanning Electron Microscopy (SEM) showed that tubes from method A were more homogeneous with denser parts in the network. In contrast, the tubes grown according to method B have denser inner surface and a porous outer surface. Mechanical measurements could not be done on the tubes from method A due to their fragility. However the tubes from method B were measured and showed to be stiffer, stronger, and less extensible when compared with a BC pellicle. This can probably be explained by a slightly higher solid content of the tubes and presumably a more interconnected fibril network as a consequence. The BC tubes manufactured according to fermentation method B results in desired asymmetric structure and mechanical properties and are expected to be used as a scaffold for tissue engineered blood vessels.

Keywords bacterial cellulose; rotating disc reactor; lactate; tissue engineered blood vessel

1   Introduction

There is a growing need for substitute for small blood vessels 5mm or less in diameter especially for bypass surgery. At the moment the standard is to use the saphenous vein or internal mammary artery. The availability of these grafts as well as their fragility and durability after transplantation makes the method questionable. For larger vessels synthetic material has been used for replacement among which ePTFE and PET are the most acceptable ones. Still there is problem with the reactivity between components in the blood and the surface of the materials. The aim of this study was to compare two fermentation techniques for the production of bacterial cellulose tubes. The tubes produced by the two different fermentation techniques were evaluated with respect to their structure and properties in order to obtain the most suitable scaffold for tissue engineered blood vessels [1].

2   Experimental

The strain used for the synthesis was *Acetobacter xylinum* subsp. *sucrofermentas* BPR2001 and was purchased from the American Type Culture Collection with the code ATCCXXXX. Two different fermentation techniques were used for the manufacturing of bacterial cellulose tubes. A modified rotating disc reactor, where a gortex carrier is either rotating in the surface of the culture media, see Figure 1A, or submerged in the culture

620

media [2]. In the latter case air was also blown into the oxygen-permeable gortex tube, see Figure 1B. The gortex support had a diameter of 5mm and a length of 180mm. The tubes were grown semi-static with a rotation speed of 0,2 rpm. When air was introduced into the gortex tube it was done with a rate of 480ml/min. The cellulose tubes produced during fermentation were examined using a scanning electron microscopy (SEM) model LEO 982 Gemini filed emission SEM. Mechanical measurements of the tubes were done in a Krebssolution at a velocity of 0,25 mm/s using a tensile tester developed and build by the Universität Essen [3].

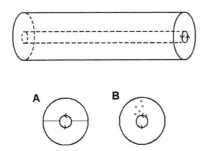

Fig. 1 Schematic figure of the cylindrical fermentor and the two fermentation techniques, method A and B respectively.

3 Results and discussion

3.1 Media development

In order to later follow the carbon consumption as well as cell proliferation a defined media was required. The defined media described by Son *et al* was further optimized with regards to cellulose yield by using three different concentration of either lactate or ethanol [4]. Exclusion of ethanol and addition of 0.6% lactate gave almost the same yield of cellulose, 4g/l, as in the complex media (CSL) described by Matsuoka *et al*, why this was used for fermentation [5].

3.2 Fermentation

Initially two different fermentation techniques were used for the manufacturing of bacterial cellulose tubes. We have found that the pellicle of bacterial cellulose fibrils is inhomogeneous, a finding also reported by Klemm *et* al [6]. The side facing the culture media is highly porous while the side at the air-liquid interface is much denser. This has probably to do with the slight drying seen at the surface with a subsequent contraction as a consequence. A hypothesis was therefore that the morphology of the tubes grown at the air-liquid interface would be more homogeneous and dense while the ones produced submerged would have a denser inner surface and a porous outer side. In this case a denser network results in a smoother surface. Smoother surfaces have been shown to improve adhesion and proliferation of endothelial cells (ECs) [7]. We are planning to seed ECs onto the luminal surface of the BC tubes to allow the formation of a monolayer of ECs prior to implantation. Hence, evaluation of the two fermentation methods was required before deciding which one to use and optimize for the production of tubular bacterial cellulose scaffolds.

3.3 Morphology

The tubes grown according to method A are more homogeneous, see Figure 2a and b. Denser parts in the network can be seen, something which probably has to do with a slight drying and a subsequent contraction as mentioned above. The tube grown according to method B has a denser inner surface but a porous outer surface, see Figure 2c and d respectively. As the tubes are later considered to be implanted and part of the body, it is advantageous if the body's own cells can enter the material. A porous network facilitates ingrowths of cells and is therefore believed to be a favourable property of the outer surface of the tube.

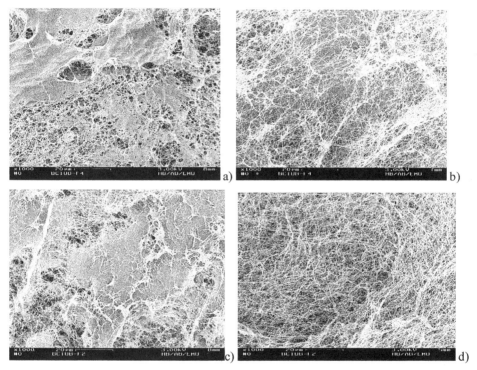

Fig. 2 SEM images of the inner (a) and outer (b) surfaces of the bacterial cellulose tubes grown according to method A. SEM images of the inner (c) and outer (d) surfaces of the bacterial cellulose tubes grown according to method B. All magnifications are 1000.

3.4 Mechanical properties

The mechanical properties of the bacterial cellulose tubes were compared with a pellicle of bacterial cellulose, which was grown in static condition. The tubes manufactured from method A was to thin and fragile and therefore mechanical properties could not be measured on those. The tubes grown according to method B are stiffer, higher Young's modulus, as well as stronger, higher stress at break (s_{max}), compared with the BC pellicle, see Figure 3a and b respectively. The pellicle on the other hand is more extensible, higher elongation at break (e), compared with the tubes, see Figure 4.

622

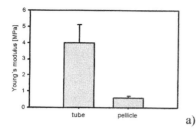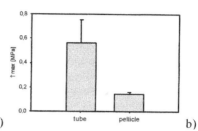

a)   b)

Fig. 3 Young's modulus of bacterial cellulose tubes and a pellicle of bacterial cellulose (a) and stress at break of bacterial cellulose tubes and a pellicle of bacterial cellulose (b).

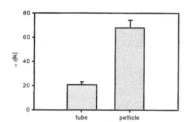

Fig. 4 Elongation at break of bacterial cellulose tubes compared with a pellicle of bacterial cellulose.

Higher Young's modulus and stress at break as well as lower elongation at break of the tubes might be due to a slightly higher solid content of the tubes (96% water content) compared with the BC pellicle (99% water content). Higher solid content probably results in a more interconnected fibril network and restrain in elongation of the micro fibrils in the BC tubes due to physical entanglements.

4   Conclusions

A lactate and ethanol deficient defined media resulted in nearly the same yield as in a complex media and can preferable be used for fermentation of bacterial cellulose tubes. The BC tubes are stronger, stiffer but less extensible than a pellicle of bacterial cellulose. BC tubes manufactured according to fermentation method B results in desired asymmetric structure and mechanical properties and is expected to be used as a scaffold for tissue engineered blood vessels.

Acknowledgements

This work was funded by Vinnova, the Swedish agency for innovation systems and our industrial partners, SCA Hygiene Products and EASTMAN Chemicals.

References

[1] Ma. Z. *et al.* Biomaterials. **26**, 2527 (2005).
[2] Krystynowicz. A. et al. Journal of Industrial Microbiology & Biotechnology **29** (4), 189 (2002).
[3] Borjesson. L. et al. Journal of Pharmacology and Experimental Therapeutics **291** (2) 717 (1999).
[4] Son, H.-J. et al. Biotechnology and Applied Biochemistry **33** (1), 1 (2001).
[5] Matsuoka, M et al. Biosci Biotech Biochem. **60**, 575 (1996).
[6] Klemm, D. et al. Prog Polym Sci. **26**, 1561 (2001).
[7] Xu. C. et al. Journal of Biomedical Materials Research, Part A. **71**A(1), 154 (2001).

Microbial biotechnology and bioengineering aspects for large-scale xylitol bioproduction from lignocellulosic residues

S. S. Silva[*,1], J. C. Santos[1], M. A. A. Cunha[1], W. Carvalho[1], B. F. Sarrouh[1], D. T. Santos[1], S. T. S. Ferreira[1], J. Polizel[1]; and R. Branco[1]

[1]Faculty of Chemical Engineering of Lorena- FAENQUIL/Department of Biotechnology, Rodovia Itajubá/Lorena km 74.5- 12600-000, Lorena, São Paulo, Brazil

[*] Corresponding author: e-mail: silvio@debiq.faenquil.br, Phone: +55 12 31595146

Keywords: xylitol, fermentation, sugar cane bagasse, fermentation parameters

Lignocellulosic residues, like sugar cane bagasse generated by the sugar-alcohol industries, is the most abundant biomass in Brazil. It is necessary to find new technologies to use this renewable biomass in different processes to produce economically valuable products. Xylitol is a anticariogenic poliol with negative dissolution heat and presenting several biomedical properties. It can be used successfully in food formulations, pharmaceutical industries and in medical applications. The xylitol microbial production using lignocellulosic residues as substrate, is a challenge for the modern biotechnology and has several economic advantages in comparison with the conventional chemical process. In this work, the most significant upstream parameters on xylitol production by biotechnological process are described and using of free and immobilized cells.

1   Introduction

The utilization of hemicellulosic biomass for production of chemicals and feedstocks is an important challenge for all countries. Brazil is well known for its abundance of renewable resources, such as agricultural crops and forests residues. The utilization of agroindustrial residues would help minimize environmental and energetic problems. Such residues could be used, for instance, to generate xylitol, a product with applications in the food and pharmaceutical industries and with biomedical properties. Xylitol is a natural sugar with sweetening power similar to. It is absolutely safe for the teeth, owing to its anticariogenic and cariostatic properties, and can be used by diabetics as a sugar substitute. At present, xylitol is chemically synthesized from xylose. This process is very costly, because it involves complex conditions and, at the end, xylitol gets mixed with sugars and polyols that hamper its purification. An alternative method is the microbial production of xylitol which is more interesting and attractive. The microbial production of xylitol using lignocellulosic residues as a substrate is a challenge for the modern biotechnology. The xylitol production by biotechnological means has several economic advantages in comparison with the conventional process. The efficiency and the productivity of this fermentation chiefly depends upon the microorganism and the process conditions employed. In the last ten years, the Bioprocess and Applied Microbiology Group of the Department of Biotechnology of the Faculty of Chemical Engineering of Lorena, São Paulo, has been studied the factors that affect the biosynthesis of xylitol from lignocellulosic residues. The major aim of this research project is to develop a low-cost technology for large scale xylitol production. The objective of this work is to present some results of advances achieved by our group, related to the microbial parameters on xylitol production by biotechnological processes using *Candida guilliermondii* FTI 20037 yeast cells in different experimental conditions.

Due to the limitation of scientific and technical information regarding the use of immobilized cells systems on the xylose-to-xylitol bioconversion, the authors have dedicated efforts to allow a better evaluation of this systems. Some studies have been developed such as use of Ca-alginate entrapped cells system and porous glass adsorbed cells system [12, 13] in xylitol bioproduction. Immobilized biocatalyst systems offers some advantages over free-cell systems, such as protection of the entrapped biocatalysts against adverse environmental conditions; cells reuse possibility, utilization of high cell densities that usually possibility higher processing velocities and high dilution rate in continuous operation. Moreover, cells or enzymes immobilization offers a promising potential for the improvement of the bioprocess efficiency. The main biotechnological aspects of this fermentation process will be presented.

2   Materials and methods

Microrganism

The yeast strain *Candida guilliermondii* FTI 20037 obtained from the Biotechnology Department of the Faculty of Chemical Engineering of Lorena, FAENQUIL, Lorena, S.P. – Brazil, was employed in all experiments [1]. The culture was maintained on malt extract agar slants at 4 $°C$. This culture has a particularity to growth on xylose as a carbon source.

Inoculum cultivation

A loopful of the stock culture was transfered into a 500 mL Erlenmeyer flask containig 200 mL of the following medium (SM): 5.0 $g.L^{-1}$ $(NH_4)_2SO_4$, 1.0 $g.L^{-1}$ yeast extract, 0.5 $g.L^{-1}$ $MgS0_4.7H_20$, 0.1 $g.L^{-1}$ $CaCl_2.2H_20$, 1.0 $g.L^{-1}$ KH_2PO_4 and 30 $g.L^{-1}$ xylose. The flask was incubated into a rotary shaker at 200 min^{-1} and 30° C for desired time (24–40 h). For the experiments using the sugar cane bagasse hydrolysate the cells were previously grown in a medium composed of hydrolysate supplemented with nutrients ($g.L^{-1}$): rice bran 20, $CaCl_2.2H_2O$ 0.1 and $(NH_4)_2SO_4$ 5. The cultivation was carried out in 125 mL Erlenmeyer flasks (containing 50 mL of medium) on a rotatory shaker set at 200 min^{-1}, at 30 °C for 48 h.

Obtention and preparation of sugar cane bagasse hydrolysate

The hemicellulosic hydrolysate was obtained by acid hydrolysis of sugar cane bagasse as previously described [12] .

Fermentation conditions

The fermentation were performed in a 125 mL Erlenmeyer flasks by shaking or in bench scale fermentors (1 and 2.5 L) using xylose or hemicelullosic hydrolysate treated and added of nutrient as described above. The fermentation system was equipped with pH, pO_2, temperature and aeration rate controllers. The fermentation conditions were changed according to the experiment.

Analytical methods

Samples of appropriate dilutions were prepared by filtration through a 0.22 micron filter (Waters Set-pak Cartridge -Millipore Corp., USA)

Xylose, glucose, arabinose, acetic acid and xylitol were analyzed in a high performance liquid chromatograph (HPLC), using a Bio-Rad Aminex HPX-87 H column at 45 $°C$ and 0.02 N H_2SO_4 as the eluent at a flow rate of 0.6 $mLmin^{-1}$.

Growth was monitored by measuring the culture turbidity at 600 nm. The cell mass was estimated using a relationship between optical density and dry cell weight.

The volumetric oxygen transfer coefficient (k_La) was determined under standard fermentation conditions by the gassing-out method as described by Pirt [2].

3   Results and discussion

The biological synthesis of xylitol is regulated by several physiological parameters. This biosynthesis occurs since the xylose-fermenting yeasts produce the xylose reductase enzyme that catalyzes the xylose reduction into xylitol as the first step in xylose metabolism [3]. Under all the conditions studied this strain of yeast (*Candida guilliermondii* FTI 20037) was able to excrete xylitol at differents rates [4–8]. This strain has shown a particularity to growth on a xylose-rich medium, which is not commum for most yeast strain. The use of sugar cane bagasse as the substrate (which is rich in xylose) in fermentative processes for xylitol production consists initially in releasing sugars from the hemicellulose portion through a mild acid hydrolysis process. This process is accompanied by the formation of considerable amounts of hemicellulose decomposition products, such as furfural, hydroxymethylfurfural, acetic acid and other products derived from lignin degradation. These chemical compounds interfere negatively with the yeast cell growth [3] and the additional xylitol fermentation [9–11]. Thus, the use of this biomass hydrolysate as a fermentation medium for microorganism growth is critical and several treatments are necessary for removing these products. The cell growth in this hydrolysate and the xylitol formation depend on the treatment and the fermentation conditions employed. In our work the treatment employed was able to reduce some toxic components in the hydrolysate, since it was observed a good performance of the *C. guilliermondii* in excrete xylitol in the broth [4–8]. The effect of some upstream parameters in this bioprocess was determined. The effect of the temperature was not so pronounced, best results on xylitol production were achieved on 30°C. For synthetic medium it is possible to conduct this fermentation process at low pH (3–4), but in hemicelullosic substrates the better fermentation runs occurred at a higher pH (5.3–6.0). This fact is due to the presence of inhibitors like acetic acid on hydrlysates and its effect on cellular membrane [4]. Xylitol production was stimulated by high xylose concentration, and the best Yp/s-value (0.83 gg^{-1}) were acheived using 70 gL^{-1} of initial xylose. The presence of glucose, acetic acid and furfural showed a negative effect on biosyntesis of xylitol [6]. The aeration rate proved to be the major physiological factor that affects the xylitol production, and the best results are achieved using a lower oxygen input or oxygen limitation. In our experiments, the maximum xylitol production (41.76 gL^{-1}) from sugar cane bagasse and the maximum xylitol volumetric productivity (0.87 gL^{-1}.h^{-1}) were attained under agitation set at 400 min^{-1} and aeration rate of 0.45 v.v.m [8].Under these conditions, by controlling the aeration rate the K_La for maximum xylitol production is near 27 h^{-1}. This result can be explain by the effect of the oxygen on the xylose metabolism and the regeneration of cofactors for the xylose reductase enzyme. A simple mathematical model adapted for this bioprocess showed that 10 to 30 % of xylitol production was not associated with cell growth. Experiments with cell recycle and using immobilized cells (Figure 1) has shown a increase in volumetric productivity. This is possible due a relationship between the oxygen uptake rate and the cell growth. Experiments using cell imobilization under differents oxygen input are under way in our laboratories.

According to the results, the biotechnological approach for xylitol production appears to be efficient and high xylitol production rates can be obtained under controlled fermentation conditions. A suitable control of the oxygen input permitting the xylitol formation from sugar cane bagasse hydrolysate is required for the development of an efficient fermentation process for large-scale applications. The Candida guilliermondii yeast used in this

626

bioprocess is potentially useful for xylitol production from hemicellulosic substrates, due to the high xylitol yields acheived that are comparable to that one obtained in synthetic medium. Actually, some downstream parameters are under study in our laboratories.

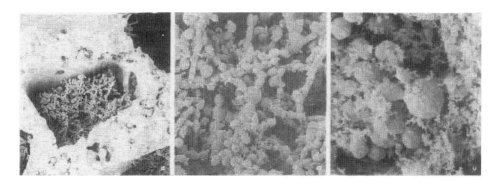

Fig. 1: Scanning electron microscope photograph showing *Candida guilliermondii* cells immolized in sintered glass after 200 h of fermentation.

Acknowledgments

CNPq, FAPESP, CAPES

References

[1]   Barbosa, M.F.S.; Medeiros, M. B.; Mancilha, I. M.; Scheneider, H.; Lee, H. Screening of yeasts for production of xylitol from D-xylose and some factors which affect xylitol yield in *Candida guilliermondii*. *J. Ind. Microbiol*. 3, 241-251, 1988.

[2]   Pirt, S.J. Principles of Microbe and cell cultivation.Blackwell Scientific Publications, 1975.

[3]   Silva, S. S.; Felipe, M. G. A. ; Mancilha, I. M. Factors that affect the biosynthesis of xylitol by xylose-fermenting yeasts- a review, *Appl. Biochem. Biotechnol*. 70-2, 331-339, 1998.

[4]   Morita, T.A, Silva, S.S; Effect of initial pH on biological synthesis of xylitol using xylose – rich hydrolysate *Applied Biochemistry and Biotechnology* vol.84-86, 751-759, 2000.

[5]   Rodrigues, R. C. L. B; Felipe, M. G.A ; Almeida, J. B; Vitolo, M.; Gómez, P. V. The influence of pH, temperature and hydrolisate concentration on the removal of volatile and nonvolatile compounds from sugarcane bagasse hemicellulosic hydrolysate treated with activated charcoal before or after vacuum evapoiration. *Brazilian Journal of Chemical Engineeringl*. 18, 299-311, 2001.

[6]   Silva, S.S.; Quesada-Chanto, A.; Vitolo, M. Up stream parametrs affecting the cell growth and xylitol production by *Candida guilliermondii* FTI 20037. *Zeitschrift für Natur*, 52 C, 359-363, 1997.

[7]   Mussato, S. I; Roberto, I. C. Hydrolysate detoxification with activated charcoal for xylitol production by *Candida guilliermondii*. *Biotechnology Letters.23*, 1681-1684, 2001.

[8]   Carvalho, W., Silva, S. S., Santos, J.C., Converti, A. Xylitol production by Ca-alginate entrapped cells: comparison of different fermentation systems. *Enzyme and Microbial Technology*, v. 32, p. 553-559, 2003.

[9]   Marton, J. M., Felipe, M G. A., Pessoa JR, A. Avaliação de carvões ativos e das condições de adsorção no tratamento do hidrolisado hemicelulósico de bagaço de cana empregando planejamento de experimentos. *Revista Analytica*, n. 3, Fev, 2003.

[10]   Converti, A; Perego, P.; Torre, P.; Silva, S.S. Mixed inhibitions by methanol, furfural and acetic acid on xylitol production by *Candida guilliermondii*. *Biotechnology Letters* vol .**22** : 1861-1865, 2000.

[11]   Converti, A; Domingues, J.M.; Perego, P.; Silva, S, S.; Zilli, M. Wood hydrolysis and hydrolysate detoxification for subsequent xylitol production. *Chemical Enginnering and Technology* vol. **23:** (11) –1013-1020, 2000.

[12]   Carvalho, W; Santos, J. C; Canilha, L; Almeida, J. B; Felipe, M. G. A; Mancilha, I. M; Silva, S. S. A study on xylitol production from sugarcane bagasse bagasse hemicellulosic hydrolysate by ca-alginate entrapped cells in a stirred tank reactor. Process Biochemistry. vol. **3:** 2135-2141, 2004.

[13]   Santos, J. C; Converti, A; Carvalho, W; Mussato, S. I; Silva, S. S. Influence of aeration rate and carrier concentration on xylitol production from sugarcane bagasse hydrolysate in immobiliced-cell fluidizad bed reactor. Process Biochemistry. vol. **40:** 113-118, 2004.

Neural network based software sensors: application to biosurfactant production by *Candida lipolytica*

C.D.C. Albuquerque[1,3] , G. M. Campos-Takaki[2*] , A. M. F. Fileti[3]

[1]Núcleo de Pesquisas em Ciências Ambientais-NPCIAMB,Departamento de Estatística e Informática-UNICAP; [2]Núcleo de Pesquisas em Ciências Ambientais-NPCIAMB, Departamento de Química - UNICAP;
[3]Departamento de Engenharia de Processos Químicos, Faculdade de , UNICAMP.

[)] Corresponding author: e-mail: takaki@unicap.br

This paper showed that on-line estimation and multi-step ahead prediction of emulsification activity and biomass concentration could be satisfactorily carried out employing well-trained feedforward backpropagation neural networks with one hidden layer. The results showed that neural 'software sensors' supplied for biomass concentration and emulsification activity on-line estimation and prediction within an acceptable variation of 5% of the experimental values. Coefficients of determination higher than 0.90 indicated excellent agreement of the neural network models with experimental test values, obtained for biomass concentration and emulsification activity.

Keywords: neural network, software sensor, biosurfactant, emulsification activity, biomass, *Candida lipolytica*

1 Introduction

The lack of reliable sensors for on-line measurements of primary process variables constitutes a great problem for bioprocess control and automation purposes [1]. To overcome this obstacle, numerous 'software sensors' have been proposed in the literature. 'Software sensors' are mathematical algorithms, which provide reliable real time estimation of unmeasured variables by using their correlation with available data. Well-designed software sensors are cheap and precious tools, able to work in parallel with, and eventually replace, a real sensor when it is affect by a fault or take off for maintenance [2]. Among conventional techniques used for development of 'software sensors', artificial neural networks have showed to be a powerful tool for modelling and control of complex bioprocess. The present work deals with the development of neural network based software sensors for real time estimation and prediction of biomass concentration and emulsification activity in a biosurfactant production process by *Candida lipolytica.*. Bioemulsifiers have received increasing attention in recent years because of their role in the growth of microorganisms on water- insoluble hydrophobic materials such as hydrocarbons and because of their commercial potential in the cosmetics, food and agriculture industries [3]. However, the most promising applications are cleaning of oil-contaminated recovery, recovery of crude oil from sludge, and bioremediation of sites contamined with hydrocarbons, heavy metals, an others pollutants [4]. The choice of inexpensive raw materials is important to the overall economics of the process because they account for fifty percent of the final product cost [5]. The yeast *Candida lipolytica* has been used with success, inclusive in this work, in the biopolymers production with emulsification activity, using culture medium of low cost, containing corn oil and sea water [6].

2   Materials and methods

2.1 Microorganism

Candida lipolytica UCP 988 were maintained at 4°C, on yeast mold agar (YMA) slants containing (w/v): yeast extract (0.3%), malt extract(0.3%), D-glucose(1%), tryptone(0.5%), and agar(1.5%). The pH was adjusted to 5.0 with HCl.

2.2 Biosurfactant production

Seed medium - SWDW-PASUG-2 – was composed of Sea Water (0.5% v/v), Distillated Water (0.5% v/v), Potassium Phosphate (2.628% w/v), Ammonium Sulphate (2.130% w/v), Urea (0.544% w/v), D-Glucose (5% w/v). The initial pH of the production media was adjusted to 5.3 with NaOH 40%. The inoculum for bioemulsifier production was prepared in four Erlenmeyer flasks with capacity of 500 mL containing 100 mL of SWDW-PASUG-2 medium Then this suspension was incubated at 28° C for 48 h at 150 rpm, and the culture having approximately 10^8 cells/mL, was used to inoculate the bioreactor at 10% v/v. Cultivations were conducted in a 5L bioreactor (BioFlo2000, New Brunswick) containing 4 L of SWDW-PASUCO-2 medium: Sea Water (0.5% v/v), Distillated Water (0.5% v/v), Potassium Phosphate (2.628% w/v), Ammonium Sulphate (2.130% w/v), Urea (0.544% w/v), Corn Oil (5% v/v). The initial pH of the production media was adjusted to 5.3 with NaOH 40%. Dissolved oxigen and pH of the media were not controlled during the experiments.

2.3 Analysis

Samples were taken every 0h, 4h, 18h and 24 h in the first day and every 24h during seven days, and biomass concentration (given as dry weight), emulsification activity, optical density and cell-free filtrate salinity were assayed off-line. Real time information was obtained by monitoring the temperature, agitation rate, pH and dissolved oxygen. Emulsification activity was evaluated according to the method described by Cirigliano and Carman [7]. Cell-free filtrates were prepared for each culture, and the emulsification activity for water-in-hexadecane emulsions, water-in-corn oil emulsions and water-in-canola oil were determined.

2.4 Training Data and Training Procedure

The data set required to train and test the neural software sensors was obtained from biosurfactant production experiments carried out using corn oil and sea water based media in a 5L bioreactor, under different temperature and agitation conditions. A 2^2 full-factorial design with three replicates at the center point was employed. Five experimental data set were obtained at 28°C and 150 rpm; 28°C and 300 rpm; 31 °C and 150 rpm; 31 °C and 300 rpm and 29.5 °C and 225 rpm [8]. Data was recorded from the available on-line sensors and from off-line sample analysis performed every 0h, 4h, 18h and 24 h in the first day and every 24h during seven days. Several neural network topologies with one hidden layer were tested in the biosurfactant production process for estimation and prediction of emulsification activity and biomass concentration [9]. The input process variables included pH, temperature, agitation rate, dissolved oxygen, optic density, salinity of the cell-free filtrate and also biomass concentration and emulsification activity at times (t-1) and (t-2), among other variables. The measurements patterns used to train and test the network were normalized between 0.1 and 0.9, because these values were found to improve convergence speed. The training and test sets were smoothed and expanded by interpolation using a piecewise smoothing cubic spline. Initially, all data of the learning set were presented to the network and a directed mode learning procedure was used. The procedure was performed ten times with different initial random weights, for each architecture, and the more accurate model was chosen. After the connection weight matrix determined in the step 1 was used to perform semi-directed learning with a prediction horizon of 1h or 2h and a moving window of 1h [10]. The training algorithm used was the Levenberg-Marquardt based backpropagation algorithm, in conjunction with Bayesian regularization. The sigmoid and linear functions were used, respectively, as activation functions in hidden and output layers. The root mean square error (RMSE) and the determination coefficient (R^2) were used to compare model performances.

3 Results and discussion

The contribution that neural networks can make in biosurfactant production scientific research was investigated. This investigation was concerned with evaluation of various modeling strategies, based on neural networks, when only small experimental data set was disponible. The various neural models were trained and tested with experimental process data. The noisy off-line data was first smoothed using piece smoothing cubic spline. The smoothing of relatively sparse off-line data was necessary not only to eliminate the noise related to the measurement errors but also to expand, by interpolation, the data set used for neural model training. Neural network were constructed using Neural Network Toolbox 4.01/ Matlab 6.01 (Mathworks). The performance of neural networks were evaluated based on the root mean square error (RMSE) and the determination coefficient (R^2), using normalized variable between 0.1 and 0.9. After the training was completed, the networks were tested by presenting experimental data sets which were not used during training. The proposed networks performance was verified on a number of test cases. Also the correct operation was analysed in the presence of measurement noise and disturbance. A comparison of operations speeds of all networks was made. To evaluate the performance for several architectures, various neural networks prototypes were introduced and various parameters were tried for each neural network. By using biosurfactant production as the example case, the effect of the number of hidden neuron on the goodness of fit in state variable estimation was investigated for each output variable (emulsification activity and biomass concentration) by multivariate regression [9]. Neural networks of 10-M-2 topology, with pH, temperature, agitation rate, dissolved oxygen, optic density, salinity of the cell-free filtrate and also biomass concentration and emulsification activity at times (t-1) and (t-2) forming the input vector, and emulsification activity and biomass, at time t, forming the output vector were employed. Each network was trained separately according with the methods section. The results of testing with two example experiments, evaluated statistically on the basis of the coefficient of determination (R^2) are shown in Fig 1. In all examined cases the optimum number of hidden neuron was 7, with an obvious increase in the calculation time and overfitting when too many hidden neurons were used. The smallest number of hidden neurons, which gave a satisfactory result, was chosen in order to minimize the calculation time [9]. In the present work, neural network of a 10-7-1 topology was successfully used for estimation of emulsification activity and biomass concentration, separately. A relatively good fit to the experimental off-line analyzed data was clearly evident, with $R^2 = 0.95$ for emulsification activity and $R^2 = 0.92$ for biomass concentration. The results of the testing procedure for the simultaneous prediction both of emulsification activity and biomass concentration two time–steps ahead were good. The testing procedure with a well trained neural network of 10-8-2 topology performed very nicely, especially in the prediction of the emulsification activity. Highly satisfactory coefficients of determination, $R^2 = 0.94$ for emulsification activity and R2= 0.90 for biomass, were obtained.. The results also showed that, in general, for neural models, the RMS values are slightly lower (1.9%–3.1%) and nearly independent of the horizon prediction.

This preliminary results showed that neural 'software sensors' supplied for biomass concentration and emulsification activity on-line estimation and prediction within an acceptable variation of 5% of the experimental values. Coefficients of determination higher than 0.90 indicated excellent agreement of the neural network models with experimental test values, obtained for biomass concentration and emulsification activity.

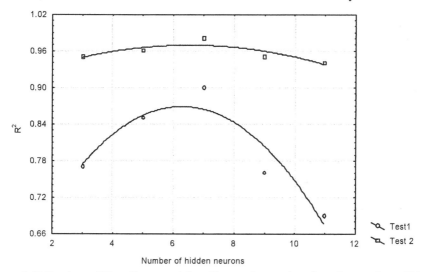

Figure 1. Estimation of biosurfactant activity with neural networks of varying number of hidden neurons, tested with two example cases.

Acknowledgments

This work was carried out with financial support by UNICAP, UNICAMP, FINEP/CT-PETRO, CNPq and CNPq /CT -PETRO.

References

[1]   J.Thibault, V.V.Breusegem and A.Chéury. *Biotechnology and Bioengineering*, **36**, 1041 (1990).
[2]   L.Fortuna., A. Rizzo, M.Sinatra and M.G. Xibilia. *Control Engineering Pratice*, **11**,1491 (2003).
[3]   E. Rosenberg. Critical *Reviews Biotecnology*. **3**, 109 (1986).
[4]   J.D. Desai and I.M. Banat. *Microbiol. Mol. Bio. Rev.*. **61(1)**,47, (1997).
[5]   Q-H Zhou and N. Kosaric. *JAOCS*, **72**, 67 (1995).
[6]   M.H. Vance-Harrop, L.A.Sarubbo, M.G. Carneiro da Cunha, N.Gusmão, G.M. Campos-Takaki. *Revista Symposium*, **2**,23,(1999).

632

[7]   M.C.Cirigliano and G.M.Carman. *Applied and Environmental Microbiology*. 747, (1984).
[8]   M.N. Karim and S.l. Rivera. Advances in Biochemical Engineering Biotecnology, **46**,1(1992).
[9]   S. Linko, J.Luopa and Y.-H.Zhu. *Journal of Biotecnology*, **52**, 257 (1997).
[10]   E.Latrille, G.Corrieu and J. Thibault. *Computers Chemical. Engineering*, **11/12**, 1171 (1994).

Microbial Physiology, Metabolism and Gene Expression

Conventional and real-time PCR assays for detection and quantification of the expression of fumonisin biosynthetic gene *fum5* in *Fusarium verticillioides*

E. López-Errasquín[1], B. Patiño[1], M. Jiménez[3], M. Jurado[2], A. González-Salgado[2], M.T. González-Jaén[2] and C. Vázquez[*,1]

[1] Universidad Complutense de Madrid, Department of Microbiology III, C/ José Antonio Novais 2, 28040 Madrid, SPAIN
[2] Universidad Complutense de Madrid, Department of Genetics, C/ José Antonio Novais 2, 28040 Madrid, SPAIN
[3] Universitat de València, Department of Microbiology and Ecology, C/ Dr. Moliner 50, 46100 Valencia, SPAIN

[*] Corresponding author: e-mail: covi@bio.ucm.es, Phone: +34 913944442, Fax: +34 913944964

Fumonisins are a group of mycotoxins that contaminate maize and related products and cause several animal and human diseases. Estimation of fumonisins production is expensive and labour consuming. In this work *fum5*, the key gene in the fumonisins pathway, has been used to develop a conventional PCR assay for specific detection of fumonisin-producing *F. verticillioides*. We have also developed a real-time PCR assay to quantify the level of expression of this gene and we have observed a correspondence between the level of *fum5* expression and the production of fumonisins in the culture measured by HPLC.

Keywords *fum5*, fumonisins, *Fusarium verticillioides*, real-time RT-PCR

1 Introduction

Fumonisins are mycotoxins produced by the maize pathogen *Fusarium verticillioides* (Sacc.) Nirenberg and several related species. Fumonisins are structurally similar to the sphingolipid intermediates sphinganine and sphingosine, and they disrupt sphingolipid metabolism by inhibiting the enzyme ceramide shyntase [1]. These mycotoxins have been associated with several mycotoxicoses, including equine leukoencephalomalacia, porcine pulmonary edema and experimental kidney and liver cancer in rats [2]. Fumonisin B_1 is the most abundant fumonisin in maize, whereas the less-hidroxylated fumonisins B_2, B_3 and B_4 typically occur at lower levels [3].

The fumonisin biosynthetic genes are clustered and a number of them have been identified and characterized [4, 5]. *Fum5* gene encodes a polyketide synthase required for fumonisin production that catalyses the synthesis of the fumonisin backbone from C-3 to C-20, which precursor feeding studies indicate is derived from acetate [6]. Conventional methods to detect, quantify and identify *F. verticillioides* include cultivation and taxonomic identification at the morphological level. This approach, however, is very time consuming and requires taxonomic skills. Alternative rapid molecular DNA-based tools to diagnose primarily toxigenic strains has been based mainly on *fum5* gene, the internal transcribed sequence (ITS) and the intergenic spacer region (IGS) of the rDNA [7–9].

The objectives of this work were to design *fum5*-specific primers to develop a real-time RT-PCR assay to quantify *fum5* gene expression, and determine if there was a good correlation between *fum5* expression and fumonisin (FB_1 and FB_2) production by *F.*

verticillioides strains, in order to be used to characterize the fumonisin-producing ability of *F. verticillioides* strains. The specificity of the assay was tested in a several closely related species by conventional PCR assay.

2   Material and methods

2.1 Fungal strains and growth conditions

For fumonisin production assays, the producing strains of *F. verticillioides* listed in Table 1 were cultured in fumonisin-inducing liquid medium [10]. Cultures were incubated at 20 °C under static conditions for 7 and 14 days. Mycelia were harvested by filtration through Whatman No 1 paper, immediately frozen in liquid nitrogen and kept at −80 °C for RNA isolation. Fumonisin production (FB$_1$ and FB$_2$) was analysed by liquid chromatography (HPLC). The species listed in Table 1 were also cultured in Sabouraud (Scharlau Chemie, Spain) liquid medium at 28 °C in an orbital shaker at 150 rpm for 3 days. Mycelia were harvested, frozen and kept as it's described above for DNA isolation

Table 1   *Fusarium* species and strains analyzed in this study indicating origin, host and ability to produce FB$_1$ and FB$_2$. (*) *F. verticillioides* strains supplied by A. Logrieco (CNR, Bari, Italy); *F. oxysporum* strain supplied by J. Tello (INIA, Madrid, Spain); (**) strains from *Gibberella fujikuroi* mating populations A-H. CECT = Colección Española de Cultivos Tipo; P = producing strain; NP = non-producing; NA = not analysed.

Specie	Strain	Host	Origin	FB$_1$/FB$_2$
F. verticillioides	GF 2	*Zea mays*	Spain	P
F. verticillioides	GF 16	*Musa sapientum*	Ecuador	NP
F. verticillioides	GF 15	*Musa sapientum*	Ecuador	NP
F. verticillioides	ITEM 1259 *	*Triticum* sp.	Italy	P
F. verticillioides	ITEM 2242 *	*Zea mays*	Russia	P
F. verticillioides	A-0149 **	*Zea mays*	USA	P
F. sacchari	B-3853 **	Lab. cross		NP
F. fujikuroi	C-1995 **	*Oryza sativa*	Taiwan	NA
F. proliferatum	D-4854 **	Lab. cross		P
F. subglutinans	E-2192 **	*Zea mays*	USA	NP
F. thapsinum	F-4093 **	Lab. cross		NA
F. nygamai	G-2150/5 **	Lab. cross		P
F. circinatum	H-69722 **	*Pinus patula*	South Africa	NP
F. oxysporum	r6	*Lycopersicon sculentum*	Spain	NP
F. graminearum	Fg 5		Italy	NP
F. sporotrichioides	CECT 20150	CECT		NP

2.2 RNA isolation, reverse transcription and real-time PCR

Fungal total RNA was isolated using the "Total Quick RNA Cells and Tissues" kit (Talent, Italy), according to the manufacturer's instructions, and stored at -80°C. DNAse I treatment to remove the chromosomal DNA contamination from the samples was performed using the "Deoxyribonuclease I, Amplification Grade" (Invitrogen, UK). First strand cDNA was synthesized using the "GeneAmp Gold RNA PCR Reagent Kit" (Applied Biosystems, USA). Real-time PCR reactions were performed in an ABI PRISM 7700 Sequence Detection System using the SYBER® Green PCR Master Mix (Applied Biosystems, USA) and the primer pairs PQF5-F (5'GAGCCGAGTCAGCAAGGATT3') and PQF5-R (5'AGGGTTCGTGAGCCAAGGA3') designed on the basis of *fum5* gene sequence (Accession no. AF155773) and PQTUB-F (5'CCCCGAGGACTTACGATGTC3') and PQTUB-R (5'CGCTTGAAGAGCTCCTGGAT3') on the basis of *tub2* sequence (Glass and Donaldson 1995).The results were normalized using the *tub2* (ß-tubulin) cDNA amplifications run on the same plate. Relative quantitation is the analytic method of choice for this study [11, 12].

2.3 DNA isolation and PCR amplifications

Genomic DNA of the strains was obtained using the "Genomix DNA Extraction" kit (Talent, Italy) following the manufacturer's instructions. *Fum5* gene was amplified by PCR in the species and strains listed in Table 1 using the primer pair PQF5-F and PQF5-R. The PCR reactions were performed in a Eppendorf Mastercycler Gradient (Eppendorf, Germany) using 50 ng of genomic DNA and EcoTaq DNA polymerase, following the manufacturer's instructions.

3   Results

In Figure 1 we can observe that a single fragment of about 70 bp was amplified in all the fumonisin producing strains of *F. verticillioides* and *F. nygamai*, but not in *F. proliferatum*, the other most important producer of fumonisins.

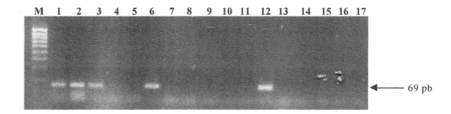

Fig. 1 PCR amplification using primers PQF5-F/PQF5-R, and DNA from *Fusarium verticillioides* strains used in this study: GF 2 (1), ITEM 1259 (2), ITEM 2242 (3), GF 16 (4), GF 15 (5) and A-0149 (6), *F. sacchari* (7), *F. fujikuroi* (8), *F. proliferatum* (9), *F. subglutinans* (10), *F. thapsinum* (11), *F. nygamai* (12), *F. circinatum* (13), *Fusarium oxysporum* f. sp. *radicis lycopersici* (14), *F. graminearum* (15) and *F. sporitrichioides* (16). M: DNA marker; Lane 17 = no template control. Control amplifications of the genomic DNA with the ß-tubulin gene were positive for all the strains analysed (data not shown).

The results of fumonisin (FB$_1$ and FB$_2$) production by *F. verticillioides* strains GF 2, ITEM 1259, ITEM 2242 and A-0149 are shown in Table 2. The highest production for both fumonisins was achieved by strain GF 2, followed by ITEM 1259, ITEM 2242 and A-0149. The results of the real time PCR analysis of *fum5* expression by the *F. verticillioides* strains in 7-day-old cultures are shown in Table 3. The non-producing strains GF 16 y GF 15 were used as negative control. The mRNA levels of *fum5* gene showed a good correspondence with FB$_1$ production, particularly those of 7-day-old cultures with FB$_1$ data of 14-day-old cultures (Pearson correlation coefficient of 0.53). The highest relative expression levels were achieved by GF 2 and ITEM 1259, which also showed the highest fumonisin values. No *fum5* amplification was obtained in both non-producing strains, as expected.

Table 2 Fumonisin production (µg/ml) of *F. verticillioides* strains GF 2, ITEM 1259, ITEM 2242 and A-0149 cultured for 7 and 14 days. ND = non detected.

CULTURE DAYS	GF 2		ITEM 1259		ITEM 2242		A-0149	
	FB$_1$	FB$_2$	FB$_1$	FB$_2$	FB$_1$	FB$_2$	FB$_1$	FB$_2$
7	2208,21	295,98	206,47	29,04	3581,58	51,40	64,3	ND
14	6009,55	504,56	7214,35	1227,97	1326,66	308,07	101,8	ND

Table 3 Quantitation of the relative transcript levels of *fum5* gene from the *F. verticillioides* fumonisin producing and non-producing strains cultured for 7 days. The results are averages of two repetitions. The values represent the number of times *fum5* are expressed in each sample compared to *F. verticillioides* A-0149 (set at 1.00). The range given for *fum5* is determined by evaluating the expression with (??C$_T$ + s) and (??C$_T$ – s), where s = the standard desviation of the ??CT value.

GF 2	ITEM 1259	ITEM 2242	A-0149	GF 16	GF 15
7927 (7168-8766)	762.8 (611-952)	65.13 (43.4-65.1)	1 (0.6-1.6)	0	0

4   Discussion

A pair of primers based on the polyketide synthase gene (*fum5*) of *F. verticillioides* has been developed to specifically detect both fumonisin-producing *F. verticillioides* and *F. nygamai* strains by conventional PCR (Figure 1). The specificity of these primers have been tested in several *F. verticillioides* strains from diverse origins and hosts and in a number of other *Fusarium* species (Table 1). The pair of primers designed did not allow detection of the producing strain of *F. proliferatum* included in the study, however, *fum5* and other several genes of the cluster of the biosynthetic pathway of fumonisins has been found in *F. proliferatum* and the non-producing *F. fujikuroi* [13], using an specific PCR assay and Southern-blot analysis. The ability to produce FB$_1$ by several fumonisin-producing *F. verticilliodes* strains has been characterized using a real time RT-PCR assay based on the expression of *fum5*, the key gene involved in fumonisin biosynthesis. Enzimes that catalyzes the first and committing step in mycotoxin-biosynthesis pathways have been used as a target to detect and quantify potential mycotoxins-producing *Fusarium* species [14, 15]. The expression levels of *fum5* were compared with production of fumonisins FB$_1$ and FB$_2$ in liquid cultures at different times and a good correspondence between *fum5* mRNA of 7-day-old cultures (Table 2) and FB$_1$ of 14-day-old cultures (Table 3) was found. The specificity of the *fum5* specific primers described here provides the basis for a simple, rapid, accurate and sensitive detection and identification method of fumonisin-producing *F. verticillioides* and the closely *F. nygamai*, and it represents an alternative to detection of fumonisin producing fungal species/strains by conventional methods, laborious and time-consuming. On the other hand, our real-time PCR assay offers a sensitive, rapid and easy alternative to the quantification of fumonisins in liquid cultures or, eventually, in agro-food products.

Acknowledgements

Supported by the CAM07G/0007/20031 and the AGLZ001-2927-C05-05 projects.

References

[1]   E. Wang, W.P. Norred, C.W. Bacon, R.T. Riley and A.H. Merrill, j. biol. chem. **266**, 14486(1991).
[2]   P.C. Howard, R.M. Eppley, M.E. Stack, A.Warbritton, K.A. Voss, R.J. Lorentzen, R. Kovach and T.J. Bucci, mycotoxins suppl. 1999, 45 (1999).
[3]   P.E. Nelson, A.E. Desjardins and R.D. Plattner, ann. rev. phytopathol. 31, 233 (1993).
[4]   J.A. Seo, R.H. Proctor and R.D. Plattner, fungal genet. biol. **34**, 155 (2001).
[5]   R.H. Proctor, D.W. Brown, R.D. Plattner and A.E. Desjardins, fungal genet. biol. **38**, 237 (2003).
[6]   Blackwell B.A., Miller J.D. and Savard M.E., j. AOAC int. **77**: 506 (1994).
[7]   B.H. Bluhm, J.E. Flaherty, M.A. Cousin and C.P. Woloshuk, j. food protect. **65**, 1955 (2002).
[8]   B. Patiño, S. Mirete, M.T. González-Jaén, G. Mulé, M.T. Rodríguez and C. Vázquez, j. food protect. **67**, 1278 (2004).
[9]   M.T. González-Jaén, S. Mirete, B. Patiño, E. López-Errasquín and C. Vázquez 2004, EJPP **110**, 525 (2004).
[10]   M.J. Hinojo. Ph thesis. Universitat de València (2003).
[11]   User Bulletin 2 ABI PRISM 7700 Sequence Detection System (1997).
[12]   D.G. Ginzinger, exp. hematol. **30**, 503 (2002).
[13]   S. Mirete. Ph thesis. Universidad Complutense de Madrid (2003).
[14]   F.M. Doohan, G. Weston, H.N. Rezanoor, D.W. Parry and P. Nicholson, appl. environ. microbiol. **65**, 3850 (1999).
[15]   S.G. Edwards, S.R. Pirgozliev, M.C. Hare and P. Jenkinson, appl. environ. microbiol. 67, 1575 (2001).

Induction of chromosome replication as a heat stress response in *Escherichia coli*

R. González-Soltero, A. Jiménez-Sánchez, and E. Botello [*]

University of Extremadura. Department of Biochemistry and Molecular Biology and Genetics. Avda Elvas s/n, 06080 Badajoz, SPAIN.

[*]Corresponding author: e-mail: ebotello@unex.es. Phone: +34 924289300 ext. 9050, Fax:+34924271304.

An upshift in the growth temperature of an *E. coli* culture causes induction of extra rounds of chromosome replication. This heat-induced replication (HIR) initiates at *oriC*, is transitory, and requires RNase H1 and RecA proteins but neither RNA polymerase activity nor *de novo* protein synthesis. HIR is not induced by the heat shock response and cannot be considered as SDR but as a recombination-dependent replication. It is SOS induction-independent, and needs RecA homologous recombinase and structural stabilization activities in a RecD exonuclease-dependent way, and RecBCD helicase function. DSBs are not generated during HIR, and possibly no D-loop structures. We suggest a structural stabilization of a heat opened structure in *oriC* via RecA and RecBCD, although other implications in the maintenance of replications forks during HIR elongation cannot be excluded.

Keywords *Escherichia coli*; replication; heat stress; recombination; HIR; RDR

1   Introduction

Bacteria have several stress responses that provide ways in which their viability can be increased. Temperature is a major environmental factor which, when altered, requires adaptive responses from bacterial cells. Among the metabolic changes cells suffer when shifted to a higher temperature, it is important to consider the effect on chromosome replication [1].

Chromosome replication in *Escherichia coli* initiates within a specific nucleotide sequence, called *oriC*, is bidirectional, and finishes in the *terC* region opposite *oriC* [2]. However, evolution allows for redundancy in vital functions. For this reason, *E. coli*, in addition to the normal mode of replication, has alternative ways to initiate replication [3, 4]. These modes of replication differ from each other, but all of them need to achieve local duplex opening where replication fork assembly takes place (Fig. 1).

A temperature upshift of 10°C or more in the growth temperature of an *E. coli* culture causes induction of extra rounds of chromosome replication. This heat-induced replication (HIR) initiates at *oriC,* is transitory, and requires RNase H1 and RecA proteins but neither RNA polymerase activity nor *de novo* protein synthesis [4]. The seminal work by Kogoma *et al.* focused attention on the interplay between replication and recombination [3]. They established that replication could occur in the absence of origin function. One process, called induced-stable DNA replication (iSDR), requires RecA, RecBCD, and PriA proteins, and initiates at D-loop structures. Another, called constitutive-stable DNA replication (cSDR), requires RecA and PriA proteins, and initiates at R-loops. Both processes require the function of enzymes that were essential to early steps of recombination for these *oriC*-independent modes of DNA replication. These observations were summarized under the name of 'recombination-dependent replication' (RDR) [3].

There are similarities and differences between HIR and iSDR. Neither require either RNA polymerase activity or protein synthesis, and both require RecA function. HIR requires RNase H1 functionality, but iSDR is independent of this activity; HIR initiates only at *oriC*, but iSDR initiates at both *oriC* and *terC* locations; and, finally, a temperature upshift induces a transient initiation of HIR, but SOS induction causes chromosome replication to proceed at a low kinetic rate for a period longer than 6 h in the absence of protein synthesis. Therefore, although HIR could be regarded as a recombination-dependent replication, it cannot be considered to be SDR [3, 4]. In the present work, we further characterized HIR dependence on other recombination proteins in order to elucidate the HIR mechanism.

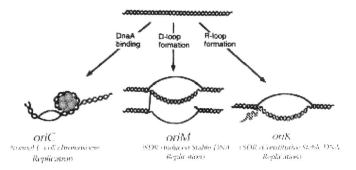

Fig. 1 Three forms of chromosomal replication and DNA duplex opening [5].

2   Results and discussion

2.1 HIR is not related to SOS induction although it needs RecA homologous recombinase and structural stabilization activities

RecA is a multifunctional and ubiquitous protein. After binding to single-stranded DNA regions, RecA changes to a conformational active state that promotes autocatalytic cleavage of the LexA repressor. In *E. coli*, SOS response is induced as a result of this RecA coprotease activity. Furthermore, RecA is required for homologous recombination and recombinational repair. RecA has also been described as being able to "sense" that replication is blocked, and to maintain the structural integrity of the replication fork until replication can be resumed [6].

Firstly, we tested whether or not HIR is SOS induction-dependent. A *lexA3* (Ind⁻) mutant, that is defective in SOS induction, showed wild type levels of HIR (Table 1). Therefore, a relationship between SOS response and HIR induction can be rejected. However, Botello and Jiménez-Sánchez proved that a *recA430* mutant with 40% of recombinase activity is unable to induce HIR [4]. In the present work, we further characterized HIR dependence on RecA recombination activity, analyzing the existence of HIR in different *E. coli* strains with partial RecA recombinase or coprotease activities (Table 1).

Table 1 HIR is not related to SOS induction although it requires RecA recombinase activity. A structural role can be assigned to RecA in HIR. The strains in the table are AB1157 derivatives.

Strain	Relevant Genotype	RecA Activity	HIR induction (i%)
AB1157	Wild type	Wild type	21.9
GY773	lexA3 (Ind⁻)	Wild type	20.6
JC10289	ΔrecA306	Null mutant	0
GY8322	⇐recA306 /miniF (recA+)	Wild type	22.8
IC400	recA430	SOS deficient / 40% recombinase activity	0
GY8202	⇐recA306 /miniF (recA428)	SOS constitutive / recombinase deficient	0
RG623	⇐recA306 /pGB2 (recA694)	SOS proficient / 75% recombinase activity	26
GY8201	⇐recA306 sfiA /miniF (recA423)	Homologous recombination proficient but deficient in recombinational repair	19
RG840	⇐recA306 /p7.7 (recAS25P)	Deficient in RecA-associated structural activities	0

DNA synthesis was determined by acid-insoluble incorporation of [³H]-thymidine. The number of origins per chromosome was determined by runout replication experiments after adding rifampicin to a mid-log growing culture (2^n). The relative number of origins activated by the heat, i, was obtained from the algorithm $\Leftarrow G=[2^n(i+1)$ n $\ln 2/2^n-1)]-1$, where $\Leftarrow G$ is the relative proportion of DNA synthesized after shifting the temperature to 41°C together with rifampicin addition, and n the number of overlapping replication cycles obtained from the runout replication at 30°C [7].

The results showed that HIR requires 40–75% from RecA recombinase activity but no coprotease activity. A recA423 mutant that is deficient in recombinational repair yields wild type levels of HIR. Therefore, RecA recombinase activity needed for HIR is homologous recombination but not recombinational repair. However, the recently described recAS25P mutant that is deficient in the structural stabilization of stalled replication forks but proficient in homologous recombination fails in HIR induction. We propose a structural stabilization of a heat-opened structure in oriC via RecA, although other implications of RecA activity in elongation cannot be excluded.

2.2 HIR partially requires RecBCD helicase activity

RecA recombinase and RecBCD helicase activities are required for homologous recombination when double strand breaks (DSBs) are present [8]. RecA and RecBCD play essential roles in the formation of D-loop structures that can be used for the initiation or restart of replication [5, 8]. We studied the implication of the RecBCD complex in HIR (Table 2).

Table 2 HIR partially requires RecBCD helicase activity. Bacterial strains are AB1157 derivatives.

Strain	Relevant Genotype	Rec - Pathway	HIR induction (i%)
JC5519	*recB21 recC22*	RecBCD helicase-	12
RG805	*recD1903::Tn10*	RecBCD exonuclease -	22.5
JC7623	*recB21 recC22 sbcB15 sbcC201*	RecFOR / QJ	22.8
JC8679	*recB21 recC22 sbcA23 rpsL*	RecET	18.1

HIR partially required RecBCD helicase activity but not exonuclease. When *recBC* mutation is suppressed by *sbcBC* or *sbcA* mutations, other recombination pathways are induced, such as RecFOR/QJ in the case of *sbcBC*, or RecET in the case of *sbcA*, and normal levels of HIR induction were restored.

In order to determine whether a temperature upshift induces the formation of DSBs, we measured the amount of linear DNA by PFGE at 30°C, and after a temperature upshift to 41°C in a *recB recC* and *recA recD* background (Table 3). The existence of greater amounts of linear DNA after the upshift would correlate with the increment of DSBs, and this could indicate the formation of Dloop structures. Therefore, a similar role for RecA and RecBCD in iSDR could be assigned during HIR initiation.

Table 3 Similar levels of linear DNA were found at 30°C and after the temperature upshift. The preparation of plugs and PFGE migration was performed as described previously [8].

Relevant Genotype	% linear DNA	
	30°C	41°C
Wild type	2.7 [8]	not determined
recB recC	20.36	17.05
recA::Cm recD::Tn10	17.33	16.58

No differences in the levels of linear DNA at 30° and after the shift to 41° were found. Therefore, we can exclude the formation of D-loop structures as an initiation mechanism in HIR.

2.3 RecA is needed for HIR only when RecD exonuclease is present

RecD is a potent double-stranded DNA exonuclease that degrades linear DNA. RecA deficient strains present a phenotype of asynchrony. It has been proposed that this asynchrony may be due to a selective degradation of chromosomes by the action of RecD. Although a *recA* mutant strain is HIR deficient, a *recA recD* double mutant showed wild type levels of HIR (Table 4).

Table 4 HIR requires RecA activity when RecD is present. Bacterial strains are AB1157 derivatives.

Strain	Relevant Genotype	HIR induction (i%)
JJC357	*ΔrecA938::Cm*	0
RG805	*recD1903::Tn10*	22.5
JJC432	*recA::Cm recD::Tn10*	21.5

We can exclude the function of RecA as being needed for HIR unless RecD exonuclease is present. RecA seems to play a role in stabilizing structures degraded by RecD during recombination. We can conclude that in order to observe HIR induction a faultless recombination machinery seems to be needed which guarantees that replication will proceed and that chromosomal integrity will be maintained.

3   Conclusions

Inter-dependencies between DNA replication, recombination, and repair suggest that in *E. coli* these systems are interconnected in order to maintain genomic integrity [3]. The heat-induced replication that we characterize, HIR, needs RecA recombinase and RecBCD helicase activities, although no DSBs are generated during the process and possibly no D-loop structures. Since HIR is induced by a temperature upshift that could open the *oriC* region, it would need the structures that allow replication to proceed in order to maintain the integrity of replication forks and chromosomal stability. RecA and RecBCD functions during HIR are possibly related to the stabilization of structures where the HIR replisome is assembled in *oriC* after the heat treatment, or to the maintenance of replication forks during HIR elongation.

Acknowledgements

This work was funded by grant BCM2002-00830 from MCYT, Spain. R. Gonzalez-Soltero is the recipient of a fellowship from the Spanish Government.

References

[1]   J. L. Ingraham, and A. G. Marr, In: Escherichia oli and Salmonella: Cellular and Molecular Biology, (ASM Washington, D.C., 1996), pp. 1570-1578.
[2]   W. Messer, FEMS Microbiol. Rev., **26**, 355 (2002).
[3]   T. Kogoma, Microbiol. Mol. Biol. Rev., **61**, 212 (1997).
[4]   E. Botello, and A. Jiménez-Sánchez, Mol. Microbiol., **26**, 133 (1997).
[5]   T. Asai, and T. Kogoma, J. Bacteriol., **176**, 1807 (1994).
[6]   J. Courcelle, and P. C. Hanawalt, Annu. Rev. Genet., **37**, 611 (2003).
[7]   A. Jiménez-Sánchez, and E. C. Guzmán, CABIOS, **4**, 431 (1988).
[8]   M. Seigneur, V. Bidnenko, S. D. Ehrlich, and B. Michel, Cell, **95**, 419 (1998).

Medium effect on heterologous protein leaky expression in *Escherichia coli*

Hasan Mirzahoseini[*], Delavar Shahbazzadeh, Somayeh Enayati, Mehdi Razzaghi Abyaneh[†], Freidoon Mahboudi.

Biotechnology Dept. & Mycology Dept.[†], Pasteur Institute of Iran, No: 69, Pasteur Ave., Tehran 13164, I.R.Iran.

*Corresponding author: e-mail: mirzahoseini@yahoo.com, phone: +98216480780, Fax: +98216465132

Re-culture of the E. coli/BL21(DE3)plysS clone bearing pET-1005 vector (cDNA-hbFGF gene inserted in pET3a plasmid), after 3 years, was cultured in the rich medium, Luria-Bertani(LB) and showed, surprisingly, an almost identical band for hbFGF (human basic fibroblast growth factor) for negative control (without induction by IPTG) on SDS-PAGE test. We first re-transformed the BL21(DE3)plysS cell with pET-1005 plasmid that gave rise to the same results. Presupposing that the enriched LB medium containing lactose, we went to minimum M9 medium supplemented with glucose, and by which leaky expression was eliminated resulted in no SDS-PAGE band in negative control sample. To reconfirm this result, we took in to account a number of other experiments.

Keywords: leaky expression, medium, *E. coli, lac* operon.

1 Introduction

Unlike systems based on *E. coli* promoters (*e.g., lac, tac, p_L*), the pET system uses the bacteriophage T7 promoter to direct the expression of target genes [1].

In this system, the T7 RNA polymerase gene is under the control of the *lac*UV5 promoter. In terms of potential industrial applications, the *lac*UV5 promoter, a derivative of the *lac* promoter, allows an efficient control of protein expression through direct induction with non-hydrolysable lactose analogue, isopropyl ß-D- thiogalactoside IPTG, if an adequate strain is used [2]. *E. coli* prefers glucose over other carbon sources. When *E. coli* is starved for glucose it synthesizes an unusual nucleotide: *cyclic 3'5'adenosine monophosphate* (cyclicAMP, or CAMP). Bound cAMP/CRP complex activates transcription of the gene that is controlled by *lac*UV5 promoter (T7 RNA polymerase). Even if the stimulatory effect on the *lac*UV5 promoter be small, the increase can be highly amplify in the T7 recombinant expression system, in which a gene on a multicopy plasmid is transcribed by the highly processive T7 RNA polymerase [3].

We had earlier [4] overproduced the bFGF hormone in a phage-T7 expression system. In this study, after 3 years, we observed leaky expression in the uninduced *E. coli*/BL21(DE3)plysS cell expressing bFGF (basic fibroblast growth factor) is controlled by the *T7* promoter. Such residual expression of the target gene was considerable. To overcome this problem, the influences of various media and medium composition on the target gene expression and leaky expression were evaluated. Moreover, this paper will put some emphasis on the analysis of possible mechanisms accounting for leaky expression, which can provide the theoretical basis for further industrialization of the fermentation process.

2   Materials & methods

2.1. Cultivation conditions

The *E. coli* strains BL21 (DE3) pLysS (F⁻, *ompT*, *hsdSB*, *gal*, dcm) were obtained from Novagene. All chemicals were from standard commercial sources, unless noted. The bFGF-cDNA has already amplified and ligated into the *Nde*I and *BamH*I sites of plasmid pET–3a (pET-1005) [4]. Cultures were routinely inoculated with seed stocks derived from early log phase cells (A_{600} =0.1–0.5) frozen in 15% glycerol at -80°C. Frozen seed stocks were used to inoculate shake flask (0.2- 0.4% inoculum) containing 5ml of LB medium and the appropriate antibiotic (Ampicillin, 50μg/ml). Cells were grown in 37°C and either induced with IPTG or left uninduced. Upon induction with IPTG, T7 RNA polymerase, which is under the control of the *lac*UV5 promoter, is synthesized, and directs the transcription of the target gene. To determine relative production of heterologous protein, three hours after induction, whole induced and uninduced cell lysates were analyzed by sodium dodecyl sulfate polyacrylamide gel electrophoresis (SDS-PAGE) using 18% polyacrylamide gels according to the method of Laemmli, [5].

2.2. Influence of the medium composition on the bFGF expression

In order to evaluate the influence of the glucose on the target gene expression, we used the LB and M9 (12.8 g/l Na_2HPO_4. $7H_2O$, 3 g/l KH_2PO_4, 0.5 g/l NaCl, 1 g/l NH_4Cl, 0.24 g/l $MgSO_4$ and 0.01 g/l $CaCl_2$) media, which supplemented by 0.4% glucose, named LB-Glu and M9-Glu respectively. In addition, instead of M9-glucose, the M9–glycerol (6μl/ml glycerol as carbon source) medium, named M9-Gly, was used to determine if glucose inhibits target gene expression or lactose induces leaky expression.

3   Results

The effects of some culture conditions on expression stability, which included the catabolic repression caused by glucose, were studied and the following results obtained.

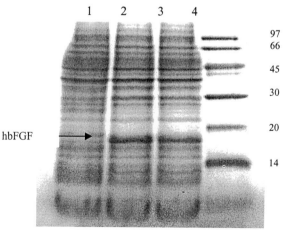

Fig. 1: SDS-PAGE (18%) of the *E. coli* BL21 DE3)plysS clone bearing pET-1005 vector (cDNA-bFGF gene inserted in pET3a plasmid).
Lane 1: *E. coli* BL21 DE3)plysS harboring pET3a plasmid (as negative control); lane 2: the cells were induced by IPTG; lane 3: the cells without induction, and lane 4: protein size marker.

To address the possibility that LB medium may contain sufficient lactose to allow *lac* operon induction, the strain BL21 (DE3) plysS (pET-1005) was grown and induced by IPTG. The induced and uninduced cells were analyzed for production of bFGF protein by SDS-PAGE.

For cells grown in LB medium, uninduced bFGF protein became detectable, as well as induced cells, and was present at high levels in cultures (Fig. 1). In contrast, no leaky production of bFGF protein could be detected in the culture grown in M9-Glu and M9-Gly media. When a leaky LB culture was supplemented with glucose (LB-Glu) and grown, the supplemented culture did not produce bFGF (Table 1).

Table 1. The effect of four media composition on hbFGF leaky expression.
Induce = induced by IPTG, and uninduce = uninduced by IPTG.

	LB		M9 + Glu		LB + Glu		M9 + Gly	
	Induce	uninduce	induce	uninduce	induce	uninduce	induce	uninduce
hbFGF production	+	+	+	—	+	—	+	—
Leakage	Yes		No		No		No	

4 Discussion

Although pET vectors have been used extensively to produce high level of recombinant protein in *E. coli*, problem of expression instability do sometimes arise. The pET expression host strains are λ DE3 lysogens in which the gene for T7 RNA polymerase is controlled by the *lac*UV5 promoter. In contrast to the wildtype *lac* promoter, the *lac*UV5 promoter is relatively insensitive to cAMP/CRP mediated stimulation. However, more recently it has been demonstrated that when λ DE3 hosts are grown to stationary phase in media lacking glucose, cAMP mediated derepression of both the wildtype and *lac*UV5 promoters occurs [3].

Because levels of leaky expression were strongly influenced by medium composition, we surmised that leaky expression was at least partially regulated by nutrient availability.

We have previously overexpressed the bFGF hormone in the T7 expression system [4]. In this work, we have studied the effects of various culture conditions on the expression stability and analyze possible mechanisms relating to target gene leaky expression in detail.

At present, many factors influencing expression stability has been studied which included plasmid load, replication patterns, and copy numbers of plasmid, substrate type, medium composition, culture conditions, growth rate, and expression level. In general, all of these factors can be classified as two aspects, namely, cell internal factors and culture conditions [6].

Here, we describe a few modifications of the mediums that effectively minimized leaky (uninduced) expression. We found that addition 0.4% glucose to cells virtually eliminated leaky expression. This result led us to evaluate the possible role of lactose and glucose in leaky expression. Recently, it has been shown that derepression of the *lac* operon in the absence of inducer may be part of a general cellular response to nutrient limitation [3]. Obviously, all of these indicated that culture conditions are very important factors influencing the target gene leaky expression. In light of the results displayed in Table 1, it is obvious that the presence of glucose in the media has a positive effect on the substantial reduction in basal expression of target gene. All these could be attributed to the catabolite repression caused by glucose on the *lac*UV5 promoter, which drives that T7 RNA polymerase gene in the BL21 (DE3). Catabolite repression inhibits the production of the T7RNA polymerase in the cells, which, in turn, reduces the basal expression [6]. Furthermore, the requirement of cAMP in promoting high levels of leaky expression from both the wt *lac* promoter and *lac*UV5 promoter (theoretically not subject to cAMP regulation) is puzzling. Earlier in vitro studies by others showed that stimulation of the *lac* and *lac*UV5 promoters occurred in the presence of cAMP and CRP. Low levels of T7 RNA polymerase were detectable in cases where the gene for the T7 enzyme was cloned under the *lac*UV5. Basal levels of RNA polymerase present in the uninduced cell can direct sufficient expression of target genes [7].

5   References

[1]   Robert Mierendorf, Keith Yeager and Robert Novy, Innovations. **1(1)**, 1-3 (1994).

[2]   Filipe J. M. Mergulhao, Gabriel A. Monterio, Gen Larsson, Maria Bostrom, Anne Farewell, Thomas Nystrom, Joaquim M.S. Carbral and M. Angela Taipa, Biotechnol. Appl. Biochem. **38**, 87-93 (2003).

[3]   Trudy H. Grossman, Ernest S. Kawasaki, Sandhya R. Punreddy, Marcia S. Osburne, Gene, **209**, 95-103 (1998).

[4]   Hasan Mirzahoseini, Farideh Mehraein, Eskandar Omidinia and Mohamad R. Razavi, World J. Microbiology & Biotechnology. **20**, 161-165 (2004).

[5]   Laemmli U. K., Nature, **227**, 680-685 (1970).

[6]   Yankai Zhang, Li Taiming and Jingjing Liu, Protein expression and purification **29(1)**, 132-139 (2003).

[7]   Bindu D. Paul, Vaidyanathan Ramesh, Valakunja Nagaraja, Gene, **190**, 11-15 (1997).

Modulation of gene expression by 3'-UTR regions in yeast: a promoter-independent alternative

S. Seoane-Rosende, A.M. Rodríguez-Torres, and M.A. Freire-Picos*

Area de Bioquímica, Department of Cellular and Molecular Biology. University of A Coruña. Facultad de Ciencias, Campus da Zapateira S/N, 15071 A Coruña. SPAIN.

*Corresponding author: e-mail: mafreire@udc.es, Phone: +34981167000, Fax: +34 9811670 65.

Changing the 3' untranslated region (3'-UTR) of the *KlCYC1* gene allows modulation of gene expression at the transcriptional level in two yeast species, *Kluyveromyces lactis* and *Saccharomyces cerevisiae*. These relative mRNA changes are achieved without altering the promoter or any growth condition. The comparative analysis performed revealed higher mRNA expression with the *KlCYC1* distal 3'-UTR region than with the proximal region in *S. cerevisiae*. As an application to modulate gene expression by changing the 3'-UTR region, an expression plasmid was constructed to clone coding regions under *GAL1* promoter and using *KlCYC1* distal 3'-UTR as terminator sequence.

Keywords: gene expression; 3'-UTR; mRNA

1 Introduction

In many organisms the use of promoter regions as a way of modulating gene expression is a commonly assumed mechanism, especially from the perspective of recombinant gene expression. However, there is an increasing number of works showing that the sequences downstream of the coding region (3'-UTR or 3'-UnTranslated Region) are especially important for mRNA export, translation modulation, mRNA stability, or local translation [1–3]. These processes are crucial in determining the final levels of a gene product.

Considering the mRNA processing mechanisms that allow mRNA 3'-end formation, most eukaryotes, from humans to yeast, express some genes producing several transcripts due to alternative (or multiple) processing of their 3'-UTR region. In yeast, the transcripts that present this specific processing may differ in their 3'-end either by a few nucleotides or by several hundreds [4]. These 3'-UTRs present several sequences that match with consensuses for processing elements [4, 5].

As a model to study alternative RNA processing, the gene *KlCYC1* encoding for cytochrome *c* in the yeast *Kluyveromyces lactis* was used. In this species, the gene is transcribed in two mRNAs of 1.5 and 1.1 kb, the shorter one being predominant only at early growth stages. When the gene is expressed in *Saccharomyces cerevisiae* the two transcripts are correctly processed, indicating that the signals governing this processing are conserved among yeast species [6].

As part of the functional characterisation of the *KlCYC1* 3'-UTR regions, unidirectional deletions of the complete 3'-UTR region and the distal 3'-UTR region were performed, expressing them in *S. cerevisiae* [7].

In this work a summary of the effects of expressing *KlCYC1* with the separate 3'-UTR regions in *K. lactis* and *S. cerevisiae* is presented. As a practical approach of this analysis, a new expression vector was constructed with *KlCYC1* distal terminator.

650

2   Materials and methods

2.1 Plasmids

Plasmid pCT5: contains distal *KlCYC1* 3'-UTR. This region was initially amplified with oligos that create *Sph*I at both sites [7]. Plasmid PCT5 was used as source of *KlCYC1* distal terminator for plasmid pSS6.

E. coli strain DH10B (GiBCo BRL) was used as host for cloning reactions. Enzyme reactions were carried out by standard protocols or manufacturer recommendations. *E. coli* transformation and DNA ligations were performed as in Sambroock and coworkers [8].

Construction of plasmid pSS6: cloning strategy shown in Figure 2. Plasmid p416GAL1 [9] was digested with *Xho*I and *Kpn*I to remove *CYC1* terminator. The *KlCYC1* distal terminator was cloned, after treatment of vector and fragment with Mung-bean nuclease to create compatible blunt ends for ligation. The plasmid was sequenced to check correct sequence at the ligation points.

The yeast strains, growth conditions and Northern experiments shown in Table 1 were used to conduct a comparative analysis of two previously published works [6, 7].

3   Results and discussion

3.1 Expressing *KlCYC1* with different 3´-UTR regions

The positions and components of the *KlCYC1* 3'-UTR region are shown in Figure 1. The two *KlCYC1* transcripts expressed in *S. cerevisiae* with either the complete or the distal 3'-UTR region are also shown [6-7].

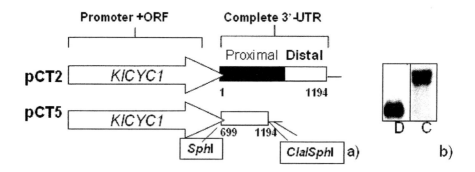

Fig. 1. a) Map of *KlCYC1* with details of the complete 3'-UTR and plasmid pCT5 with only the distal 3'-UTR region. b) Relative expression and size difference of *KlCYC1* mRNAs obtained either with the complete (C) or the distal (D) 3'-UTR expressed in *S. cerevisiae*.

Table 1 contains a summary of the results of *KlCYC1* gene expression with the complete or deleted (proximal or distal) 3'-UTR regions in two yeast species. Only the expression results at logarithmic phase are shown, since expression of *KlCYC1* in *S. cerevisiae* is not growth-phase regulated [6]. A value of 1 was used to indicate the total signal of the two *KlCYC1* wild type transcripts, to give the relative expression of cytochrome *c* mRNA independently of the number of transcripts.

Table 1 Comparison of relative *KlCYC1* mRNA levels when the gene was expressed with plasmids containing different 3′-UTR regions. *The two transcript signals were added to give a value of 1 (the total wild type transcript). *K. lactis* strain MW190-9b, *S. cerevisiae* strain ZW13.

	Plasmid	*KlCYC1* 3'-UTR	Relative mRNA signal	References
K. lactis	pKCYC1	Complete*	1	[6-7]
	pTU1	Proximal	0.79	[7]
	pTU2	Distal	0.23	[7]
S. cerevisiae	pCT2	Complete*	1	[6]
	pCT4	Proximal	0.1	[6]
	pCT5	Distal	1	[7]

Expression of *KlCYC1* with the distal 3′-UTR in *S. cerevisiae* produced a strong mRNA signal, higher than each of the two individual transcripts obtained when the gene was expressed with the wild type 3′-UTR. This result indicates that this is an optimal processing sequence when expressed in *S. cerevisiae*. However, the relative mRNA signal obtained with the proximal 3′-UTR was clearly poor (ten times lower).

Comparison of proximal and distal 3′-UTR expression levels reveals that it is possible to modulate mRNA levels, by changing the 3′-UTR region, without altering any other important element, such as the promoter (that is invariant in all constructs) or the growth media conditions.

When the experiment was performed in *K. lactis*, additional regulation of gene expression depending on the growth phase and independent of promoter-dependent regulation was found. Another consequence, derived from the fact that the final product was a cytochrome *c* protein, was that the changes in cytochrome *c* expression in *K. lactis* were correlated to changes in the respiro-fermentative metabolism [7].

These results show that 3′-UTR regions play an important role in the modulation of gene expression, which may be applied to modulate expression of recombinant genes in yeasts.

3.2. A plasmid with *KlCYC1* distal 3′-UTR.

To construct a new yeast vector to express genes in *S. cerevisiae* with *KlCYC1* distal 3′-UTR (positions 699 to 1194), the Mumberg and coworkers plasmid p416GAL1 [9] was chosen as starting material. This plasmid was designed for heterologous expression in yeast. It contains 461 bp *GAL1* promoter followed by a Multiple Cloning Site (MCS) and *S. cerevisiae CYC1* terminator.

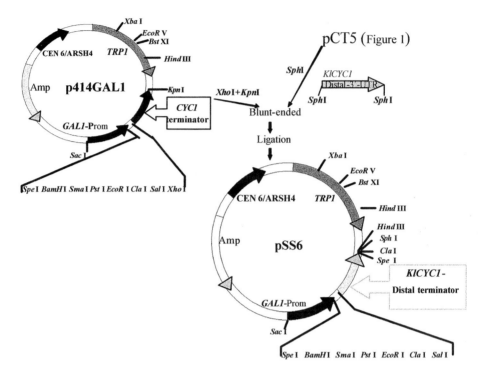

Fig. 2 Cloning strategy used to construct the pSS6 vector for expressing genes in *S. cerevisiae* with *KlCYC1* distal 3'-UTR. Unique restriction sites (suitable for cloning) are underlined.

The *GAL1* region present in this plasmid allows modulation at the expression level of any coding region cloned in the MCS. Using a *lacZ* fused version, changes were found from 0.1 to 181 —galactosidase units, depending on the presence of glucose or galactose in the growth media [9].

The *CYC1* gene terminator has been extensively characterised by the Sherman group [10-11] and the elements governing mRNA 3'-end processing are well known. The gene is transcribed in a single 3'-end mRNA. The 260 bp. 3'-UTR is commonly used as terminator region in many plasmids [9].

Heterologous expression experiments have shown that *KlCYC1* mRNA levels are higher (both transcripts) than *S. cerevisiae CYC1* mRNA [12]. On the other hand, the distal 3'-UTR seems to undergo more efficient processing than the proximal region (Table 1). These data suggest that it would be interesting to have a *S. cerevisiae* vector to express genes with *KlCYC1* 3'-UTR distal region. The strategy followed for replacing *S. cerevisiae CYC1* for *KlCYC1* distal terminator is shown in Figure 2. The resulting plasmid was named pSS6.

4   Conclusions

The data analysed in this work show the importance of 3'-UTR regions in modulating gene expression. The results suggest the use of 3'-UTR, combined with specific promoters, as tools to modulate gene expression. Hence, this would allow modulation of gene expression in a more precise way. The new vector, pSS6, may be used to clone any given coding sequence in *S. cerevisiae* and modulate its expression, with the advantage of a 3'-UTR that will help to optimise expression under inducing conditions. A similar approach could be taken with the proximal UTR, although in this case lower (sometimes desirable) expression would be obtained.

Acknowledgements

This work was funded through projects MCYT BMC 2000-0133, XUGA PGIDT03BTF10301PR and grant UDC03.IP: A.R.T. (A. Rodríguez Torres) from University of A Coruña, Spain. S. Seoane was awarded a fellowship from the Excma. Diputación Provincial de A Coruña.

References

[1]   N. Proudfoot, Curr. Opin. Cell Biol. **16**, 272-278 (2004).
[2]   R.J. Sims 3rd, S.S., Mandal, and D. Reinberg, Curr. Opin. Cell Biol. **16**, 263 (2004).
[3]   Y-H. Huang, and J.D. Richter, Curr. Opin. Cell Biol.. **16**, 308 (2004).
[4]   J.H. Graber, G.D. McAllister and T.F. Smith, *Nucleic Acids Res*. **30**, 1851 (2002).
[5]   K.A. Sparks, and C.L. Dieckerman, Nucleic Acids Res. 26, 4676 (1998).
[6]   M.A. Freire-Picos, L.J. Lombardía Ferreira, E. Ramil, M. González Domínguez and M.E. Cerdán. *Yeast* **18**, 1347 (2001).
[7]   S. Seoane, B. Guiard, A.M. Rodríguez Torres, and M.A. Freire-Picos, J. Biotechnol. **118**, 149-156. (2005).
[8]   J. Sambroock, T. Maniatis and E.F. Frisch. Molecular Cloning: A Laboratory Manual. CSH laboratory press. (Cold Spring Harbor, NY 1989).
[9]   D. Mumberg, R. Muller, and M. Funk, Nucleic Acids Res. **22**, 5767 (1994).
[10]   Z. Guo, and F. Sherman, Mol. Cell. Biol. **16**, 2772 (1996a).
[11]   Z. Guo, and F. Sherman, TIBS **21**, 477 (1996b).
[12]   M.A. Freire-Picos, M.I. González-Siso, E. Rodríguez-Belmonte, A.M. Rodríguez Torres, E. Ramil, and M.E. Cerdán, Gene **139**, 43 (1994).

Molecular properties of two nitroreductases from *Rhodobacter capsulatus* B10 involved in 2,4-dinitrophenol reduction

E. Pérez-Reinado[1], R. Blasco[2], F. Castillo[1], C. Moreno-Vivián[1], M. D. Roldán[1,*]

[1]Departamento de Bioquímica y Biología Molecular, Edificio Severo Ochoa, Campus de Rabanales, Universidad de Córdoba, 14071 Córdoba, Spain
[2]Departamento de Bioquímica y Biología Molecular, Facultad de Veterinaria, Universidad de Extremadura, Cáceres, Spain
Corresponding autor:

[*]Corresponding author: e-mail: bb2rorum@uco.es. Phone and fax: +34 957 218588

Rhodobacter capsulatus cometabolizes polynitrophenols by reducing them to 2-amino derivatives by a 27 kDa NAD(P)H, FMN-linked nitroreductase. The genomic sequence of *R. capsulatus* SB1003 allows to isolate two genes encoding NAD(P)H-nitroreductases, *nprA* (homologous to the nitroreductase gene *nfnB* from *E. coli*) and *nprB*. Phenotypic analysis of mutant strains lacking either *nprA*, *nprB* or both genes revealed that NprA and NprB are essential for reduction of 2,4-dinitrophenol. The *nprA* gene is inducible by nitro-compounds, salicylate and paraquat whereas *nprB* gene expression is constitutive. Purification and biochemical characterization of NprA indicates that the enzyme has a broad substrate specificity. Regulatory *mar/sox/rob* boxes present in the promoter of *nprA* confirm that its expression is induced by nitroaromatics and suggest an additional control depending on oxidative stress.

1.1 Metabolism of polynitrophenols by Rhodobacter capsulatus

R. capsulatus is extremely resistant to the uncoupler 2,4-dinitrophenol. In the presence of this compound, the bacterium induces a 2-nitroreductase activity and reduces 2,4-dinitrophenol to 2-amino-4-nitrophenol, which is stoichiometrically released into the medium under light-anaerobic conditions. The process is a type of cometabolism [1] since it enables the bacterium to grow by fixing N_2 dissolved in the culture medium. The nitroreductase is a 27 kDa NAD(P)H, FMN-linked enzyme whose activity increases in cells growing with 2,4-dinitrophenol and decreases in the presence of ammonium, which reversibly inhibits 2,4-dinitrophenol uptake [2-5].

Diazotrophic growth of *R. capsulatus* is strongly inhibited by 2,4-dinitrophenol, 2,6-dinitrophenol and 2,4,6-trinitrophenol (picric acid) since ammonium photoproduced from N_2, which can be easily detected by culturing the cells in the presence of the glutamine synthetase inhibitor L-methionine-D,L-sulfoximine (MSX), could be never detected in cultures containing nitroaromatics and MSX. Uptake of nitro-compounds was followed by releasing of the corresponding 2-amino derivative, which could be re-assimilated only under a partially oxic atmosphere.

As previously described with 2,4-dinitrophenol [2], uptake of 2,6-dinitrophenol and picric acid depends on the nitrogen source present in the medium, being strongly, but not completely, inhibited in the presence of ammonium or glutamine.

1.2   Isolation and characterization of nitroreductase genes from *R. capsulatus*

The availability of the complete sequence of *R. capsulatus* S1003 genome (www.integratedgenomics.com), a strain very similar to the B10 strain, allowed us to study the putative nitroreductase gene(s) involved in nitroaromatics cometabolism. The analysis of this sequence revealed the presence of three genes encoding putative nitroreductases: RRC01791 and RRC03929, which encode two NAD(P)H-linked, oxygen-insensitive enzymes (NprA and NprB), and a third one related with cobalamine synthesis. From the sequence data, NprA is homologous to the oxygen-insensitive nitroreductase NfnB from *E. coli*, but shows scarce similarity to NprB. To test if these genes were involved in nitroaromatics cometabolism, the corresponding gene regions of the *R. capsulatus* B10 were amplified, isolated and sequenced. After this, deletion mutants were generated by insertion of antibiotic resistance cassettes: the strain *nprA*⁻ (which only presents NprB activity), the strain *nprB*⁻ (which only presents NprA activity) and the double mutant *nprA*⁻*B*⁻ (which does present neither NprA nor NprB activities).

To generate the mutant *nprA*⁻ a 1.35 kb fragment was amplified from the *R. capsulatus* genome which contains the whole *nprA* gene with its promoter and part of the adjacent genes. A 519 bp *Hind*III/*Sal*I fragment in the middle of the *nprA* gene was replaced by a *Hind*III/*Sal*I fragment containing a kanamycin resistance cassette as a marker. The final construct (pMO8-AKm) was transferred from the S17-1 strain of *E. coli* to *R. capsulatus* B10 by conjugation. The flanking regions with sequences of the adjacent genes allowed the double recombination event to get the mutation *nprA*⁻.

To clone the *R. capsulatus* B10 *nprB* gene, a 1.95 kb fragment obtained by PCR was sequenced, verifying that the sequence matched perfectly with those the RRC03929 gene from *R. capsulatus* S1003.

To obtain the *nprB* mutant, two fragments containing part of the *nprB* gene were amplified generating a new *Bam*HI restriction site. The fragments were cloned into the pGEM-T plasmid and the gentamicin and spectomycin resistance genes were inserted in the *Bam*HI site. A final construct including the *nprB* gene, with a 323 bp deletion and the antibiotics cassette insertion, was transferred to the wild type strain B10 of *R. capsulatus*, obtaining the strain *nprB*⁻, and to the strain *nprA*⁻, obtaining the double mutant *nprAB*⁻.

Tables 1 and 2 shows the comparison of the NprA and NprB proteins, respectively, with other bacterial nitroreductases. The comparisons of the deduced amino acid sequences were made by the ClustalW method used by the BioEdit Program.

Table 1. Comparison of NprA with other bacterial nitroreductases

PROTEIN	ORGANISM	AA	%IDENTITY	%SIMILARITY	ACCESSION N
Hypothetic nitroreductase	A. tumefaciens	209	58	75	AE009377
Hypothetic nitroreductase	C. crescentus	210	59	72	AE005819
RNR	E. cloacae	217	29	53	AY013713
NR	E. cloacae	217	29	52	M63808
Cnr	S. typhimurium	217	28	51	AE008722
FRase I	V. fischeri	218	27	47	D17743
NfnB (NfsB)	E. coli	217	27	50	AAN79139
DrgA	Synechocystis sp	201	13	43	BA000022

AA: amino acid residues

Table 2. Comparison of NprB with other bacterial nitroreductases

PROTEIN	ORGANISM	AA	% IDENTITY	% SIMILARITY	ACCESSION N
Hypothetical nitroreductase	R. palustris	169	40	56	BX572604
Hypothetical nitroreductase	C. crescentus	181	33	45	AE005933
YdjA	E. coli	183	24	41	AAC74835
YdjA	S. typhimurium	183	35	50	AE008756

AA: amino acid residues

The *nprA* gene codes for a polypeptide of 210 amino acids with a monomeric molecular mass of 23 kDa. Comparison of the sequence with other nitroreductases reveals several conserved motifs such as ^{13}K and the sequences ^{32}AAQMAPTS y ^{154}PMEGFD, which can be involved in linking the FMN moiety. The NprA protein shows homology with yet uncharacterised nitroreductases from *Agrobacterium tumefaciens* y *Caulobacter crescentus* and with the well characterized enzymes from *Enterobacter cloacae*, *Salmonella typhimurium*, *Vibrio fisheri*, *E. coli* (NfnB), *Pseudomonas putida* and *Synechocystis* sp. PCC6803. Besides, NprA showed homology with an NADH oxidase from *Thermus thermophillus* and with NAD(P)H-flavin reductases from *Bacillus* strains.

The *nprA* promoter shows putative −31 −10 boxes as well as putative regulatory elements belonging to the *mar/sox/rob* regulons [6]. On the other hand, the start codon AUG was preceded by a typical Shine-Dalgarno sequence for ribosome binding.

2 Regulation of cometabolism of nitrophenols by *R. capsulatus*

2.1 The nitrogen regulon of *R. capsulatus*

R. capsulatus is a phototrophic bacterium which uses oxidized and reduced nitrogen sources. Since assimilation of inorganic nitrogen is a process with a high energy cost, the enzymes involved are subjected to a fine control both at the level of protein synthesis and activity. The master piece of this regulatory network, the so-called nitrogen regulon, is the intracellular carbon/nitrogen balance, reflected by the 2-oxoglutarate/glutamine ratio [7].

Thus, nitrogen fixation is subjected to an ammonium switch-off/on mechanism [8], which is due to the reversible ADP-ribosylation of the Arg^{103} in the Fe-S protein of the nitrogenase complex [9]. Besides, the operons encoding the nitrogenase (*nif*) and the electron supply to nitrogenase reaction (*rnf*) were under control by two bicomponent cascade systems (NtrB/NtrC and NifA/NifL), which requires a $\uparrow^{54}$ specific promoter and that activates gene transcription in response to a high intracellular C/N balance and low pO_2, respectively [7, 10].

Like nitrogen fixation and nitrate assimilation, nitrophenol cometabolism is affected by ammonium, which strongly and reversibly inhibits nitrophenol uptake and decreases nitroreductase activity *in vivo* [2–5]. By analysing the phenotypes of deletion mutant strains affected in Nif, Ntr and Rnf proteins, we demonstrated that both the regulatory proteins and the electron supply membrane complex Rnf were involved in nitro- phenol cometabolism [11].

2.2 Phenotypic analysis of the nitroreductase mutants of *R. capsulatus* with and their relationships with nitrophenol cometabolism

Previous results have showed that addition of ammonium does not repress completely nitroreductase activity in *R. capsulatus*, which suggested that other nitroreductase activity could be involved. According to this, the cometabolism of nitrophenols takes place at the same rate in both *nprA⁻* and *nprB⁻* mutants, whereas the double mutant *nprAB⁻* cometabolizes 2,4-dinitrophenol, 2,6-dinitrophenol and picric acid to a lower rate. These results suggest that the bacterium is able to thrive by using only one of the two putative main nitroreductases.

The above results were corroborated by analysing the nitroreductase activity in crude extracts from the wild type and mutant strains. When the mutant strains were assayed for nitroreductase activity, the *nprB⁻* strain (which possess the NprA enzyme) showed a higher nitroreductase activity than the *nprA⁻* strain (which present the NprB protein). The double mutant exhibit only a negligible nitroreductase activity. By contrast, in the presence of ammonium the time course of nitroreductase levels showed a very low activity in the *nprB⁻* mutant whereas similar high values in the wild type and *nprA⁻* mutant were observed.

The expression patterns of *nprA* and *nprB* genes were carefully examined with respect to the nitrogen source and the redox state of the cultures by using *lacZ* fusions. —galactosidase activity was induced by aromatic compounds (salicylate, paraquat), specially if they contains nitro groups (polynitrophenols, nitrobenzene), whereas in its absence a negligible — galactosidase activity was always observed. The higher induction levels were obtained in cells cultured with inorganic nitrogen sources and the lower activities were observed in nitrogen-rich media (ammonium or glutamine). As the promoter of *nprA* gene does not belong clearly to the $\uparrow^{54}$ family, it may be concluded that the lower expression observed in

658

nitrogen-rich media cannot be due to the NtrC gene repression, but to the inhibition of nitrofenol uptake caused by ammonium or glutamine [2, 11], which probably decreases the concentration of the inducer inside the cells. The role of nitroaromatics as inducers agree with the presence of a *mar* regulatory element in the promoter of *nprA* gene, which like other *mar* promoters, responds positively to the presence of salicylate and 2,4-dinitrophenol [12, 13].

Induction of the *nprA* gene was higher in media containing malate as electron donor than in those containing butyrate or caproate. Repression by a reductive environment agrees with the presence of a *sox* regulatory element in the *nprA* promoter, which is also activated by paraquat [14–16].

As expected, analysis of *nprB-lacZ* fusions revealed that *nprB* gene is expressed constitutively, independently from the nitrogen source or the redox environment.

In order to further biochemical characterization of the NprA enzyme, we obtained a His$_6$-NprA fusion protein to identify possible substrates. A 657 bp fragment from the genomic DNA of *R. capsulatus* containing the *nprA* gene was amplified and cloned into the pGEM-T and pQE32 vectors to obtain a His$_6$-NprA fusion protein, which was subsequently overproduced in *E. coli* and immunodetected in SDS-PAGE gels, showing the expected monomeric molecular mass of 24 kDa. The recombinant protein was purified by affinity chromatography in Ni-NTA agarose and its corresponding kinetic, molecular and spectroscopic properties were analyzed. The enzyme uses NAD(P)H as electron donor, contains 1 mol FMN per mol of enzyme, exhibits a broad substrate specificity (specially concerning nitroaromatics) and is strongly inhibited by dicumarol and $CuSO_2$. As previously described, NprA present a diaphorase activity that can be detected by the tetrazolium salts assay [2].

In view of the above results, it can be concluded that *R. capsulatus* express at least two nitroreductase enzymes that catalyze nitroaromatic cometabolism. NprA is an NAD(P)H, FMN-linked nitroreductase belonging to the B family of nitroreductases. The expression of *nprA* gene is induced by several aromatic compounds, specially 2,4-dinitrophenol and nitrobenzene, salicylate and paraquat, indicating that *nprA* is under control of the *mar/sox/rob* regulon. Also, nitrophenol cometabolism is dependent on the nitrogen source, probably at the level of the xenobiotic transport and or electron supply through the Rnf complex. By contrast, the second nitroreductase, NprB, is a constitutive enzyme which can be ascribed to a group of bacterial uncharacterized nitroreductases.

Acknowledgements

This work was funded by the Ministerio de Ciencia y Tecnología (BMC2002-04126-C03-03) and Junta de Andalucía (CVI0117). EPR was the recipient of a fellowship from the MEC (Spanish Ministry of Education) and MDR holds a Postdoctoral fellowship from the Junta de Andalucía (Spain).

References

[1]   M. Alexander, Science, **211**, 132 [1981]
[2]   R. Blasco, F. Castillo, Appl. Environ. Microbiol., **58**, 690 [1992]
[3]   R. Blasco, F. Castillo, Appl. Environ. Microbiol., **59**, 1774 [1993]
[4   R. Blasco, F. Castillo, Pest. Biochem. Physiol., **58**, 1 [1997]
[5]   R. Blasco, P.J. Aparicio, F. Castillo, Arch. Microbiol., **163**, 248 [1995]

[6]   T.M. Barbosa, S.B. Levy. 2002, Mol. Microbiol. **45**,191, [2002].

[7]   R.G. Kranz, D. Foster.Harnett, Mol.Microbiol., **4**, 1793 [1990]

[8]   W.G. Zumft, F. Castillo, Arch. Microbiol., **117**, 53 [1978]

[9]   M.R. Pope, S.A. Murrell, P.W. Ludden, Proc. Natl., Acad., Sci. USA, **82**, 3173 [1985]

[10]   B. Masepohl, W. Klipp, Arch Microbiol., **165**, 80 [1996]

[11]   L.P. Sáez, P. García, M. Martínez-Luque, W. Klipp, R. Blasco, F. Castillo, J. Bacteriol., **183**, 1780 [2001].

[12]   M.N. Alekshum, S.B. Levy, J. Bacteriol. **181**:4669, [1999].

[13]   S.P. Cohen, S.B. Levy, J. Foulds, J.L. Rosner, J. Bacteriol. **175**, 7856 [1993].

[14]   K.W. Jair, X. Yu, K. Skarstad, B. Thony, N. Fujita, A. Ishihama, R.E. Wolf Jr., J. Bacteriol. **178**:2507, [1996].

[15]   R.G. Martin, W.K. Gillette, S. Rhee, J.L. Rosner, Mol. Microbiol. **34**:431 [1999].

[16]   R.G. Martin, W.K. Gillete, J.L. Rosner, Mol. Microbiol. **34**:431 [2000].

Nitrate assimilation in *Rhodobacter capsulatus* E1F1: purification and biochemical characterization of nitrite and hydroxylamine reductases

P.   Cabello[1], M. F. Olmo-Mira[2], M. Martínez-Luque[2], C. Pino[2], F. Castillo[2], M. D. Roldán[2], and C. Moreno-Vivián[2*]

[1] Departamento de Biología Vegetal, Área de Fisiología Vegetal, Edificio Celestino Mutis, Campus de Rabanales, Universidad de Córdoba, Córdoba, Spain.
[2] Departamento de Bioquímica y Biología Molecular. Edificio Severo Ochoa, Campus de Rabanales, Universidad de Córdoba, Córdoba, Spain.

[*] Corresponding author: e-mail: bb1movic@uco.es. Phone and fax: +34 957 218588

Rhodobacter capsulatus E1F1 is a phototrophic bacterium that assimilates nitrate under light-anaerobic growth conditions. We have recently cloned a 18-kb nas gene region with 14 genes involved in the nitrate assimilation process, including the hcp gene encoding a hybrid cluster protein, and the putative structural genes coding for the catalytic subunits of nitrate reductase (*nasA*) and nitrite reductase (*nasB*). Recombinant His$_6$-tagged forms of R. capsulatus Hcp and NasB proteins have been overexpressed in *E. coli*, purified by affinity chromatography and biochemically characterized. These results indicate that *NasB* is an assimilatory NADH-nitrite reductase with siroheme, FAD and [4Fe-4S], and that the Hcp protein is a hydroxylamine reductase probably involved in detoxification and assimilation of hydroxylamine.

Keywords: hybrid cluster protein, hydroxylamine, nitrate, nitrite, nitrite reductase.

1   Introduction

Bacteria can use nitrate reduction for different metabolic purposes. Thus, nitrate is used as a nitrogen source for growth, as an alternative electron acceptor to generate ATP during anaerobic respiration, or as an electron sink to dissipate excess reducing power under some metabolic conditions. In accordance with this, assimilatory (Nas), respiratory (Nar) or periplasmic dissimilatory (Nap) nitrate reductases can be present in some bacteria [1]. *Rhodobacter capsulatus* E1F1 is a phototrophic bacterium that assimilates nitrate and nitrite only under light-anaerobic growth conditions [2]. The assimilatory NADH-nitrate reductase of this strain has been purified and characterized. The enzyme is a heterodimer composed of a 90 kDa catalytic subunit with a molybdopterin guanine dinucleotide cofactor and one [4Fe-4S] centre, and a 45 kDa diaphorase subunit with FAD [3]. In this strain, we have recently cloned and sequenced a 18-kb nas gene region, which includes 14 genes probably involved in the nitrate assimilation process. In addition to putative regulatory genes, the structural genes coding for an ABC-type nitrate transport system (*nasFED*) and the catalytic subunits of both nitrate reductase (*nasA*) and nitrite reductase (*nasB*), are present in this gene region. The *hcp* gene encoding a hybrid cluster protein (formerly prismane protein) is also found in this nas gene cluster [4]. It has been suggested that the *E. coli* Hcp protein, which contains one [2Fe-2S] center and one unusual hybrid cluster [4Fe-2S-2O], participates in nitrate/nitrite respiration [5], and more recently, it has been described that this protein has hydroxylamine reductase activity [6]. To investigate the physiological role of the *R. capsulatus* E1F1 Hcp and NasB proteins, we have overexpressed the His$_6$-tagged recombinant forms of these proteins in *E. coli*. These proteins have been purified by nickel-NTA affinity chromatography and biochemically characterized. These studies indicate that the NasB protein of the phototrophic bacterium *R. capsulatus* E1F1 is an assimilatory NADH-nitrite reductase with FAD, [4Fe-4S] and siroheme, which is similar to the enzyme described in non-phototrophic organisms, and that the Hcp protein is a hydroxylamine reductase probably involved in detoxification and assimilation of hydroxylamine, a highly toxic product that could be formed during nitrite reduction.

2   Material and methods

Rhodobacter capsulatus E1F1 was grown at 30 °C on minimal RCV medium under phototrophic (light-anaerobiosis) or heterotrophic (dark-aerobiosis) conditions, as previously described [2]. D,L-malate (4 g l^{-1}) was routinely used as the carbon source, and L-glutamate (1 g l^{-1}), KNO_3 (1 g l^{-1}), NH_4Cl (1 g l^{-1}), NH_2OH (1 mM) or KNO_2 (1 mM) were used as the nitrogen sources. Cells were harvested by centrifugation at 20,000 x *g*, 15 min, washed and resuspended in 50 mM Tris-HCl buffer (pH 8.0), and broken by cavitation at 90 W for 15 s (3 pulses of 5 s) in a VibraCell sonifier.

Cell growth was followed turbidimetrically measuring the absorbance of the cultures at 680 nm for *R. capsulatus* or at 600 nm for *E. coli* strains. Protein was estimated according to Lowry *et al.* [7]. Labile sulfur and iron determinations were performed following described methods [8,9]. Nitrate and nitrite reductase activities were routinely assayed with reduced methyl viologen as artificial electron donor, as described previously [10]. Hydroxylamine reductase activity was assayed as indicated in [4].

Routine DNA manipulations (DNA isolation, restriction enzyme analysis, agarose gel electrophoresis, cloning procedures and PCR amplifications) were performed using standard methods [11]. To express and purify the recombinant His_6-tagged Hcp and NasB proteins, the corresponding *hcp* and *nasB* genes were fused in-frame to the pQE32 plasmid and the resulting constructs were used to transform *E. coli* JM109. When the *E. coli* cultures reached an absorbance at 600 nm of about 0.6, 1 mM IPTG was added to induce His_6-NasB or His_6-Hcp expression and the cultures were grown for a further 3 h. After disrupting the cells, the soluble extracts were loaded onto a nickel-NTA agarose (Qiagen) column, and proteins were eluted with an imidazole gradient. Proteins were resolved by SDS-PAGE and the gels were used in Western-blot analysis or were stained with Coomassie Brilliant Blue as previously described [4].

3   Results and discussion

Rhodobacter capsulatus E1F1 can grow under phototrophic conditions (anaerobiosis-light) with nitrate as the sole nitrogen source. Induction of the assimilatory nitrate and nitrite reductases is observed during bacterial growth (Fig. 1). Similarly, this strain was also able to grow phototrophically with hydroxylamine as a nitrogen source and the cells showed hydroxylamine reductase activity. However, the strain E1F1 did not grow with these nitrogen sources under aerobic heterotrophic conditions (not shown). Degenerated primers were used to amplify by PCR a 400 bp DNA fragment corresponding to the putative *nasA* gene coding for the nitrate reductase from *R. capsulatus* E1F1. This fragment was used for the screening of a genomic library of this strain and a clone carrying a 18 kb nas gene region, which includes 14 genes probably required for nitrate assimilation, was isolated [4]. The *hcp* gene coding for a putative hybrid cluster protein, and the *nasB* gene probably encoding the catalytic subunit of the nitrite reductase were included in this region (Fig. 2).

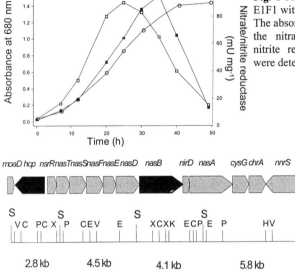

Fig. 1 Phototrophic growth of *R. capsulatus* E1F1 with nitrate as the sole nitrogen source. The absorbance at 680 nm (open circles), and the nitrate reductase (filled circles) and nitrite reductase (filled squares) activities were determined at the indicated times.

Fig. 2 Physical restriction map of the 18 kb *nas* region of *R. capsulatus* E1F1. Genes coding for the hybrid cluster protein (*hcp*) and the nitrite reductase (*nasB*) are shown in black. *Cla*I (C), *Eco*RI (E), *Eco*RV (V), *Hin*dIII (H), *Kpn*I (K), *Pst*I (P), *Sal*I (S), and *Xho*I (X).

EPR studies and crystal structures of the Hcp proteins indicate that they contain a typical [2Fe-2S] or [4Fe-4S] center and an unusual hybrid [4Fe-2S-2O] [5,12]. Despite of these studies, the physiological function of the Hcp proteins remains unknown, although it has been proposed that they can play a role in anaerobic nitrate respiration [5] or act as a hydroxylamine reductase [6]. To analyze if the *nasB* and hcp gene products of *R. capsulatus* E1F1 are involved in nitrite and hydroxylamine assimilation, we have fused in-frame these genes to the sequence encoding a polyhistidine tag in the pQE32 expression vector. The *E. coli* JM109 cells were transformed with these constructs, and the expression of the recombinants proteins were achieved by addition of IPTG. The corresponding cell extracts were loaded onto a niquel-NTA affinity chromatography and the His_6-tagged proteins were eluted by an imidazole gradient and detected by western blots analysis with anti-His antibodies and by Coomassie stain.

The biochemical characterization of the His_6-Hcp protein revealed that contained about 6 iron and 4 sulfide atoms per molecule and showed hydroxylamine reductase activity with methyl viologen (MV) and bromophenol blue as reductants, forming ammonium in vitro. The apparent K_m values for NH_2OH and MV were 1 mM and 7 µM, respectively, at the pH and temperature optima (9.3 and 40 °C). The activity was inhibited by oxygen and by iron and sulfide reagents (Table 1). On the other hand, the His_6-tagged NasB protein showed nitrite reductase activity in vitro with both NADH and MV. The activity presented values of pH and temperature optima of 9.0 and 30 °C, respectively, and the apparent K_m values for nitrite and NADH were 500 and 20 µM, respectively. Activity was also inhibited by Fe and S reagents, such as o-phenantroline, dipyridyl, cyanide, dithioerythritol and p-hydroxymercurybenzoate (Table 1). The optical spectrum of the purified protein revealed features suggesting the presence of flavin, iron-sulfur cluster and siroheme. Therefore, these results indicate that the phototrophic bacterium R. capsulatus E1F1 possesses an assimilatory NADH-nitrite reductase, which is similar to that described in non-phototrophic organisms, and that the Hcp protein is involved in assimilation of hydroxylamine, a highly toxic product that could be formed during nitrite reduction.

Table 1. Kinetic and molecular parameters of the nitrite reductase (NasB) and hydroxylamine reductase (Hcp)

	Nitrite reductase (NasB)	Hydroxylamine reductase (Hcp)
Molecular mass, kDa	90	55
Optimum pH	9.0	9.3
Optimum temperature, °C	30	40
Apparent K_m for substrate, ≥M	500 (NO_2^-)	1000 (NH_2OH)
Apparent K_m for reductant, ≥M	20 (NADH)	7 (Methyl viologen)
Electron donor (% activity)		
Methyl viologen	100	100
Benzyl viologen	77	21
Bromophenol blue	42	33
NADH	79	0
NADPH	24	0
Inhibitors (% activity)*		
None	100	100
Dithioerythritol	48	71
o-phenanthroline	28	7
2,2'-dipyridyl	32	46
Cyanide	31	70

*All of them at a final concentration of 2 mM

Acknowledgements

This work was supported by grants from Ministerio de Ciencia y Tecnología (project BMC2002-04126-CO3-03) and Junta de Andalucía (group CVI 0117).

References

[1]   C. Moreno-Vivián, P. Cabello, M. Martínez-Luque, R. Blasco, and F. Castillo. J. Bacteriol. **181**, 6573 (1999).

[2]   C. Moreno-Vivián, F. Castillo, and J. Cárdenas. Photosynth. Res. **3**, 313 (1982).

[3]   R. Blasco, F. Castillo, and M. Martínez-Luque. FEBS Lett. **414**, 45 (1997).

[4]   P. Cabello, C. Pino, M. F. Olmo-Mira, F. Castillo, M. D. Roldán, and C. Moreno-Vivián. J. Biol. Chem. **279**, 45485 (2004).

[5]   W. A. M. van der Berg, W. R. Hagen, and W. M. A. M. van Dongen. Eur. J. Biochem. **267**, 666 (2000).

[6]   M. T. Wolfe, J. Heo, J. Garavelli, and P. W. Ludden. J. Bacteriol. **184**, 5898 (2002).

[7]   O. H. Lowry, M. J. Rosebrough, A. L. Farr, and R. J. Randall. J. Biol. Chem. **193**, 265 (1951).

[8]   H. Beinert. J. Biol. Inorg. Chem. **5**, 2 (2000).

[9]   U. A. Nilsson, M. Bassen, K. Sävman, and I. Kjellmer. Free Radic. Res. **36**, 677 (2002).

[10] C. Moreno-Vivián, F. J. Cejudo, F. Castillo, and J. Cárdenas. Arch. Microbiol. **136**, 147 (1983).

[11] J. Sambrook, E. F. Fritsch, and T. Maniatis. Molecular Cloning: A Laboratory Manual, 2nd ed (Cold Spring Harbor Laboratory Press, Cold Spring Harbor, 1989).

[12] D. Aragão, S. Macedo, E. P. Mitchell, C. V. Romão, M. Y. Liu, C. Frazão, L. M. Saraiva, A. V. Xavier, J. LeGall, W. M. A. M. van Dongen, W. Hagen, M. Texeira, M. A. Carrondo, and P. Lindley. J. Biol. Inorg. Chem. **8**, 540 (2003).

Proteases of *Trichoderma* strains from Hungarian winter wheat rhizosphere

A. Szekeres[*,1], **B. Leitgeb**[2], **Z. Pénzes**[3], **L. Kredics**[4], **L. Hatvani**[1], **Z. Antal**[4], **L. Manczinger**[1] and **C. Vágvölgyi**[1]

[1] Department of Microbiology, University of Szeged, P.O. Box 533, H-6701 Szeged, HUNGARY
[2] Institute of Biophysics, Biological Research Center of the Hungarian Academy of Sciences, Temesvári krt 62, H-6726 Szeged, HUNGARY
[3] Department of Ecology, University of Szeged, P.O. Box 51, H-6701 Szeged, HUNGARY
[4] Microbiological Research Group, Hungarian Academy of Sciences and University of Szeged, P.O. Box 533, H-6701 Szeged, HUNGARY

[*]Corresponding author: e-mail: szandras@bio.u-szeged.hu

The species belonging to the filamentous fungal genus *Trichoderma* are well known to be potential biopesticide candidates for agricultural application. Several *Trichoderma* strains have been reported to be effective in promoting plant growth and in controlling plant diseases, and the action of fungal hydrolytic enzymes is considered as the main mechanism involved in the biocontrol process including antagonism, mycoparasitism and competition. In the present study, information was collected about the abilities of the extracellular protease secretion of Hungarian *Trichoderma* isolates from winter wheat rhizosphere.

Keywords *Trichoderma*; protease

1   Introduction

There is a number of potential biocontrol agents among the members of the filamentous fungal genus *Trichoderma*. The proposed mechanisms resulting in biocontrol are including the stimulation of the defensive mechanisms of the plants, competition for the substrate as well as antibiosis by the production of antifungal metabolites and mycoparasitism by the action of cell-wall degrading enzymes [1–3]. The ability of *Trichoderma* strains to produce extracellular proteases has been known for a long time [4]. The involvement of *T. harzianum* proteases in the biocontrol of *Botrytis cinerea* was suggested by Elad and Kapat [5]. Szekeres *et al.* reported that the improved production of extracellular proteolytic enzymes by *T. harzianum* caused enhanced *in vitro* antagonistic abilities against plant pathogenic fungi [6]. Geremia *et al.* isolated and cloned the gene *prb*1 of *T. harzianum* encoding a basic proteinase related to mycoparasitism [7], while Flores *et al.* demonstrated that the over-expression of this gene resulted in improved biocontrol activity, suggesting the importance of proteases in the degradation of the protein components of the host cell wall and in the lysis of whole host cells [8]. In the case of *T. harzianum*, several acidic, neutral and basic extracellular proteases have been detected by Delgado-Jarana *et al.* [9]. Specific trypsin- [10–12] and chymotrypsin-like [11, 13] activities were also detected in the case of *Trichoderma* strains and some of them were purified and characterized [10, 11, 13]. Studies are available about the effects of different environmental factors, e.g. low temperature [14], low water potential [15] and the presence of metal ions [16] on the *in vitro* activities of trypsin- and chymotrypsin-like proteases. On the other hand, the production of extracellular proteases may also represent potential virulence factors in the case of *Trichoderma longibrachiatum* strains emerging as opportunistic pathogens of humans [17].

The aim of this work was to examine the secretion of trypsin-, chymotrypsin- and chymoelastase-like proteases in the case of *Trichoderma* isolates derived from Hungarian soil samples and to study the distribution of their production within the population.

2 Methods

2.1 Isolation of *Trichoderma* strains from Hungarian soil samples

Soil samples including winter wheat seedlings were collected from five agricultural fields in random sampling order with a square sampler (5 cm x 5 cm). The chopped roots of wheat were placed to Petri dishes containing selective medium (5 g peptone, 1 g KH_2PO_4, 10 g glucose, 0.5 g $MgSO_4 \times 7H_2O$, 0.5 ml 0.2% dichloran-etanol solution, 0.25 ml 5% Rose Bengal and 20 g l agar per litre). Growing *Trichoderma* strains were transferred from the plates onto yeast extract medium (2 g yeast extract, 5 g KH_2PO_4 and 20 g agar per litre). All isolates were grown from single conidia for the further enzyme assays.

2.2 Culture conditions and measurement of extracellular proteases

Protease production of the isolated *Trichoderma* strains was examined in liquid yeast extract - glucose (10 g glucose, 10 g KH_2PO_4 and 5 g yeast extract per litre) and minimal (2 g glucose, 1 g KH_2PO_4, 1 g $MgSO_4 \times 7H_2O$ and 1 g $NaNO_3$ per litre) media. The experiments were carried out in three replicates and standard deviation (SD) values were determined. In 50 ml Erlenmeyer flasks, these solutions were inoculated with conidial suspensions of the *Trichoderma* strains to a final concentration of 10^5 conidia/ml and incubated on a shaker at 200 rpm and 25 °C. Samples were collected from the crude culture supernatants after 6 d of incubation.

The extracellular enzyme activities of trypsin-, chymotrypsin- and chymoelastase-like proteases were measured with the chromogenic substrates *N*-benzoyl-L-Phe-L-Val-L-Arg-*p*-nitroanilide, *N*-succinyl-L-Ala-L-Ala-L-Pro-L-Phe-*p*-nitroanilide and *N*-succinyl-L-Ala-L-Ala-L-Pro-L-Leu-*p*-nitroanilide (Sigma), respectively. The substrates were dissolved in small aliquots of dimethyl-sulfoxide and diluted to 2 mg/ml in Sorensen phosphate buffer (pH 6). One hundred μl substrate solution was added to 100 μl aliquots of the samples in the wells of microtiter plates, resulting in a substrate end concentration of 250 μg/ml. Control mixtures were prepared by adding 100 μl phosphate buffer to the samples. The optical densities of the reaction mixtures were determined at 405 nm with a Labsystems Uniskan II microtiter plate spectrophotometer after 2 h of incubation at 35 °C.

3 Results and discussion

Seventy-nine *Trichoderma* strains were isolated from roots of winter wheat from nineteen test holes of five agricultural fields in southern Hungary during January, 2002 (Table 1).

Table 1 Geographic origin of test holes and the number of isolated *Trichoderma* strains.

Agricultural field	Test hole	No. of isolated *Trichoderma* strains
Algyo	I., III.	4, 3
Deszk	I., III., IV., V.	5, 5, 3, 4
Kunszentmiklós	I., II.	13, 2
Rúzsa	I., II., III., IV., VI., VII.	10, 1, 2, 6, 11, 2
Tiszasziget	I., II., III., IV., V.	5, 6, 5, 5, 5

The Rose Bengal applied in the media is able to inhibit the growth of zygomycetous fungi, the fast growth of which could hamper the recognition of *Trichoderma* colonies on Petri-plates. The average number of isolated strains per test hole was the highest for the samples from Kunszentmiklós (7.5) and the lowest for those of Algyo (3.5).

The isolated *Trichoderma* strains were involved in the enzyme assays and the amounts of the examined enzymes showed high variability among the isolates (Fig. 1). Constitutive secretion at a moderate level (OD_{405} below 0.4) was found to be common, however, strains with high levels of enzyme production (OD_{405} above 1.8) could also be found with low frequencies.

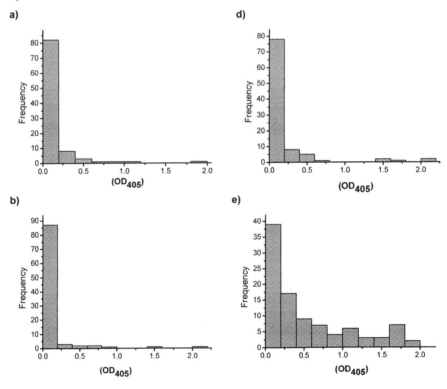

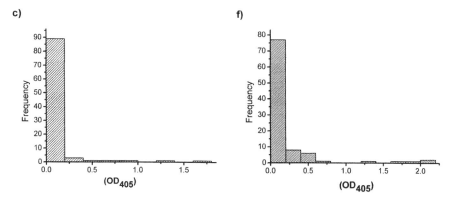

Fig. 1   The distributions of activities of trypsin- (a, d), chymotrypsin- (b, e) and chymoelastase-like (c, f) proteases in the Hungarian *Trichoderma* population on minimal (a, b, c) and yeast extract (d, e, f) media. Frequencies are the numbers of strains falling into certain bins.

The histograms of the activities of trypsin- and chymoelastase-like proteases were similar on minimal and yeast extract media, while differences in heterogeneity could be observed in the case of chymotrypsin-like proteases. The distributions of the enzyme activities in the population were exponential-like, and the same strains fell in the equivalent bins of different proteases in most of the cases (data not shown).

It would be important to adopt the practice of sustainable agriculture, using strategies that are environment-friendly, less dependent on agricultural chemicals and less damaging to soil and water resources. Before planning the practical application of *Trichoderma* strains as biocontrol agents, detailed ecophysiological examinations should be carried out including the monitoring of their extracellular proteolytic system, as the biocontrol process involves high-level secretion of proteases. Results of previous studies demonstrate that strains with increased biocontrol effectiveness can be produced by improving the secretion of proteases via UV-mutagenesis [6] or the elevation of the copy number of a protease-encoding gene [8]. An alternative for strain improvement is searching for natural isolates with superior biocontrol abilities. Examining the ecophysiological properties and the abilities for the production of extracellular enzymes within natural *Trichoderma* populations may provide important sets of data for statistical analysis aiming the study of the relationship between these parameters and biocontrol ability. Screening of field isolates and the selection of those tolerating local climatic conditions may prove to be useful in finding potential candidates for the purposes of practical application in a certain agricultural area.

Acknowledgements

This work was supported by grants F037663 of the Hungarian Scientific Research Fund and grant OMFB-00219/2002 of the Hungarian Ministry of Education.

References

[1]   Q. Wang, Y. Yang, and R. Wang, phys. stat. sol. (a) **155**, 289 (1996).
[2]   T. Benítez, J. Delgado-Jarana, A. Rincón, M. Rey, and C. Limón, recent. res. devel. microbial. **2**, 129 (1998).

668

[3]   L. Manczinger, Z. Antal, and L.Kredics, acta microbiol. immunol. hung. **49**, 1 (2000).
[4]   C. R. Howell, plant dis. **87**, 4 (2003).
[5]   R. Rodriguez-Kabana, W. D. Kelley, and E. A. Curl, can. j. microbiol. **24**, 487 (1978).
[6]   Y. Elad, and A. Kapat, eur. j. plant pathol. **105**, 177 (1999).
[7]   A. Szekeres, L. Kredics, Z. Antal, F. Kevei, and L. Manczinger, fems microbial. lett. **233**, 215 (2004).
[8]   R. Geremia, G. H. Goldman, D. Jacobs, W. Ardiles, S. B. Vila, M. Van-Montagu, and A. Herrera-Estrella, mol. microbiol. **8**, 603 (1993).
[9]   A. Flores, I. Chet and A. Herrera-Estrella, curr. genet. **31**, 30 (1997).
[10]   J. Delgado-Jarana, J. A. Pintor-Toro, and T. Benítez, biochim. biophys. acta **1481**, 289 (2000).
[11]   B. Suárez, M. Rey, E. Monte and A. Llobell, iobc wprs bull. **24**, 369 (2001).
[12]   T. Uchikoba, T. Mase, K. Arima, H. Yonezawa and M. Kaneda, biol. chem. **382**, 1509 (2001).
[13]   Z. Antal, L. Kredics, L. Manczinger and L. Ferenczy, iobc wprs bull. **24**, 337 (2001).
[14]   Y. E. Dunaevsky, T. N. Gruban, G. A. Beliakova, and M. A. Belozersky, biochemistry (Moscow) **65**, 723 (2000).
[15]   Z. Antal, L. Manczinger, G. Szakács, R. P. Tengerdy, and L. Ferenczy, mycol. res. **104**, 545 (2000).
[16]   L. Kredics, Z. Antal, and L. Manczinger, curr. microbiol. **40**, 310 (2000).
[17]   L. Kredics, I. Dóczi, Z. Antal, and L. Manczinger, bull. environ. contam. toxicol. 66, 249 (2001).
[18]   L. Kredics, Z. Antal, A. Szekeres, L. Manczinger, I. Dóczi, F. Kevei, and E. Nagy, acta microbiol. immunol. hung. **51**, 283 (2004).

The yeast external invertase as a reporter to study regulation of *Candida albicans* promoter sequences in *Saccharomyces cerevisiae*

P. Roig, V. Maneu, M.L. Gil, and D. Gozalbo[*]

Departament de Microbiologia i Ecologia, Facultat de Farmàcia, Universitat de Valencia, Avgda. Vicent Andrés Estellés s/n, 46100 Burjassot, SPAIN

[*]Corresponding autor: e-mail: daniel.gozalbo@uv.es, Phone:+34 963 543026, Fax: +34 963 544299

We have studied the regulation of three promoters of *Candida albicans* genes (*SSB1*, *UBI3* and *UBI4*) by measuring their ability to direct the synthesis of invertase in *Saccharomyces cerevisiae*. The invertase coding region of the yeast *SUC2* gene was ligated downstream from the *C. albicans* promoters in a yeast centromer-derivative plasmid. The resulting plasmids were transformed into *S. cerevisiae* SEY2101, a strain unable to produce invertase. The expression directed from the *C. albicans* promoters was measured by determining the invertase activity in whole yeast cells. Results showed a correlation between the invertase levels detected and the previously described regulation of these genes in *C. albicans*, indicating the validity of the yeast *SUC2* gene as a potential reporter to study expression from *C. albicans* promoters in *S. cerevisiae*.

Keywords *Candida albicans*; *Saccharomyces cerevisiae*; gene regulation; invertase; reporter protein; yeast transformation; promoter; centromer plasmid

1   Introduction

Most *Candida albicans* genes are functional in *Saccharomyces cerevisiae*, and complementation of *S. cerevisiae* mutant phenotypes has been used to show the function of *C. albicans* cloned genes [1, 2]. The study of the expression of these genes is more complex, and the use of suitable reporters may be very helpful. Although gene reporter systems for *C. albicans* have been described, expression of heterologous genes in this species is often difficult, due to the non-canonical genetic code and codon bias in *C. albicans* [1]. The expression in *S. cerevisiae* of homologous reporter genes under the control of *C. albicans* promoter sequences may represent an alternative to study their regulation. This approach relies on the proper interplay between *C. albicans* cis- and *S. cerevisaie* trans-acting regulatory transcriptional elements, and may avoid the problems derived from the codon usage. In this work we have evaluated the usefulness of the *S. cerevisiae* external invertase as a reporter protein to study regulation of *C. albicans* promoter sequences. This enzyme, encoded by members of the glucose-repressible *SUC* gene family, is secreted into the yeast cell wall, enabling the cells to use sucrose or raffinose as fermentable carbon source [3]. The promoter sequences of three *C. albicans* genes (*SSB1*, *UBI3*, and *UBI4*) have been used in our study. These genes are functional when expressed in *S. cerevisiae* and their expression is differentially regulated in *C. albicans* [2, 4–6].

2   Materials and methods

2.1 Plasmid contructions and yeast transformation

Plasmids pIN1, pIN3 and pIN4 are pLC7-derivatives containing the *SUC2* coding sequence for external invertase under the control of the promoter regions of the *C. albicans SSB1*, *UBI3*, and *UBI4* genes, respectively. Using engineered oligonucleotides as primers, promoter sequences of *SSB1* (from positions −10 to −451 relative to the translational star site), *UBI3* (from positions −10 to −451), and *UBI4* (from positions −6 to −553) were PCR-amplified from pSSB1, pPR3 and pPR4 (plasmids containing the *C. albicans SSB1*, *UBI3* and *UBI4* cloned genes, respectively [2, 4, 5]), and ligated into the *Eco*RI *Hin*dIII digested pLC7, a yeast centromer-derivative plasmid containing the *SUC2* gene [7]. The resulting plamids contain the *SUC2* coding region under the control of the *C. albicans* promoter sequences, which replace the *SUC2* promoter contained within an *Eco*RI *Hin*dIII fragment in pLC7. Plasmids were transformed into *Saccharomyces cerevisiae* SEY2101 strain (*MATα ura3-52 leu2-3 leu2-112 ade2 SUC2Δ9*), and transformants carrying pIN1, pIN3, pIN4 or pLC7 were isolated in selective medium lacking uracil (0.67% yeast nitrogen base w/o amino acids, 2% glucose, 1% agar for plates) supplemented with leucine (30 mg/L) and adenine (20 mg/L). All manipulations were performed according to standard procedures [2, 8, 9].

2.2 Determination of invertase activity and statistical analysis

Secreted (external) invertase activity was determined in intact whole cells from exponentially growing cultures [9,10]. For invertase assays, yeast transformants were cultured in modified YPD medium (1% yeast extract, 2% peptone, 4% glucose) or in YPS medium (1% yeast extract, 2% peptone, 2% sucrose). Starvation and temperature shift assays were performed as previously described [4,8,10]. One unit of invertase activity hydrolyzes 1 ≥mol of sucrose per minute of reaction. Relative activity is expressed as miliunits (mU) per mg (dry weight) of cells. Statistical differences were determined using the Student's two tailed t-test. Significance was accepted at $P < 0.05$.

3   Results and discussion

3.1 SSB1 promoter confers temperature-dependent production of invertase

C. albicans SSB1 gene encodes a chaperone involved in protein translation at low temperatures and its expression is regulated by the temperature of growth [2]. To check the effect of the temperature in the expression levels directed from the *SSB1* promoter, we determined the invertase activity in *S. cerevisiae* SEY2101 transformants carrying pIN1 cultured at different temperatures (23, 28 and 37 °C). Results showed an increased production of invertase as temperature decreases, particularly at 23 °C (relative activity per cell duplicates from 37 °C to 28 °C, and increased 10-fold at 23 °C). This temperature-dependent regulation of *SSB1* promoter was only observed in cells grown with sucrose as fermentable carbon source (Fig. 1), probably due to an effect of glucose either on transcription from *UBI3* promoter and/or on invertase mRNA stability at 23 °C (see below). Transformants carrying pLC7 showed a glucose-repressed pattern of invertase production (62 mU/mg in cells from YPS cultures, and 2.5 mU/mg in cells from YPD cultures), regardless of the temperature of growth (not shown). These results indicate that *SSB1* promoter is regulated by temperature, as in *C. albicans* cells [2], although the increased invertase production at 23 °C may be somehow regulated by glucose repression.

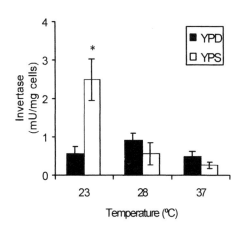

Fig. 1. Effect of the temperature of growth on the external invertase activity in *S. cerevisiae* SEY2101 transformants carrying pIN1. Cells were cultured at 23, 28 or 37 °C in YPD or YPS up to the exponential growth phase. The invertase activity was measured in cells from 10 ml aliquots. Data of relative activity (mean values of triplicate determinations ± standard deviations) are from one representative assay of two. (*) $P < 0,05$ with respect to the 28 °C and 37 °C samples

3.2 UBI3 promoter directs constitutive invertase production

UBI3 encodes an essential ubiquitin fusion protein involved in ribosome biogenesis and its expression is regulated coordinately with the cell growth; arrest of growth in response to stress conditions (temperature upshif or starvation), results in a rapid decrease in *UBI3* mRNA levels [4,6]. To check whether expression of *SUC2* under the control of the *UBI3* promoter is regulated in a similar way, we determined the effect of temperature shifts (see Fig. 3 for details) and starvation on the invertase activity in *S. cerevisiae* SEY2101 transformants carrying pIN3. Results obtained following temperature shifts showed that invertase activity per cell remains constant (90 and 120 mU per mg of cells from glucose and sucrose containing-medium, respectively), independently of the temperature. The higher activity detected in the presence of sucrose may be due to an increased stability of invertase mRNA and/or increased translational efficiency under derepresssing conditions [11, 12]. Control assays with pLC7 transformants showed the glucose-repressed invertase production pattern (not shown). Starvation assays showed a similar result (Fig. 2a): invertase activity remains constant per cell, even in absence of growth, due to the enzyme stability, indicating that under these conditions there is no expression from *UBI3* promoter. Control pLC7 transformants showed a derepression pattern of invertase production upon starvation (not shown). These results indicate that *UBI3* promoter confers constitutive expression in growing yeast cells, and that production of invertase is coordinately regulated with cell growth, which is consistent with the role of the encoded gene product, and agrees with the expression pattern described in *C. albicans* [4].

672

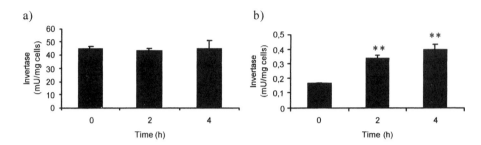

a)

b)

Fig. 2. Effect of starvation on external invertase activity in *S. cerevisiae* SEY2101 transformants carrying pIN3 (a) and pIN4 (b). Exponentially growing yeast cultured at 28 °C in YPD were collected by centrifugation, resuspended in one volume of prewarmed sterile water and incubated at 28 °C. At the indicated times of starvation, invertase activity was determined in cells from 0.5 ml aliquots (pIN3) or 10 ml aliquots (pIN4). Data of relative activity (mean values of triplicate determinations ± standard deviations) are from one representative assay of two. (**) $P < 0.01$ with respect to the 0 h sample.

3.3 UBI4 promoter confers stress-dependent production of invertase

Expression of the polyubiquitin (*UBI4*) gene occurs in stressed *C. albicans* yeast cells, either by starvation or temperature shifts [4,5]. To confirm this regulation, we determined the effect of temperature shifts and starvation on the invertase activity in *S. cerevisiae* SEY2101 transformants carrying pIN4. Temperature shifts in YPD medium caused an increase of the relative invertase activity per cell, particularly at 42 °C (3.5-fold after 4 h), and a moderated increase at 20 °C (50% increase) (Fig. 3); similar results were obtained in YPS medium (60% increase at 42 °C, and 30% increase at 28 °C; not shown). Starvation assays also showed that invertase per cell increases (2.5-fold after 4 h) (Fig. 2b). These results confirm that *UBI4* promoter is weakly expressed in growing cells (200-fold weaker than the *UBI3* promoter) and that is upregulated by stress.

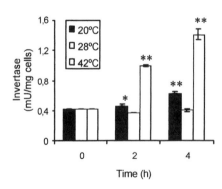

Fig. 3. Effect of temperature shifts on the external invertase activity in *S. cerevisiae* SEY2101 transformants carrying pIN4. Cells were cultured at 28 °C in YPD up to the exponential growth phase, and then shifted to 20, 42 °C or maintained at 28 °C. At the indicated times after the shifts, the invertase activity was measured in cells from 10 ml aliquots. Data of relative activity (mean values of triplicate determinations ± standard deviations), are from one representative assay of two. (*) P <0,05 and (**) P <0,01 with respect to the 28 °C sample.

Overall, our results confirm preliminary observations [8], and show the validity of the yeast invertase as a reporter activity to study expression from *C. albicans* promoters in *S. cerevisiae*, with the advantage of a high enzyme stability and an easy quantification by assaying invertase activity in intact cells. In addition, expression of invertase as a reporter protein in *S. cerevisiae* eliminates the effects of both the non-canonical genetic code of *C. albicans* and differences in codon usage among both species [1], although glucose effect on invertase mRNA stability and/or translational efficiency represents a drawback as it may influence invertase production. Besides, the sequence of the yeast *SUC2* gene coding for internal invertase has been used to determine the ability of the *C. albicans* glyceraldehyde-3-phosphate dehydrogenase, encoded by the *TDH1* gene, to direct the secretion of invertase fusion proteins into the yeast cell wall, as well as to study expression from *TDH1* promoter [9, 10]. This shows that invertase coding sequences may represent a potential useful tool to study gene expression as well as subcellular distribution of gene products.

Acknowledgements

The support of Generalitat Valenciana (grant GRUPOS03/172) is gratefully acknowledged.

References

[1]   M. D. De Backer, P. T. Magee, and J. Pla, Annu. Rev. Microbiol. **54**, 463 (2000).
[2]   V. Maneu, P. Roig, and D. Gozalbo, Res. Microbiol. **151**, 739 (2000).
[3]   D. Gozalbo, and L. del Castillo, FEMS Microbiol. Rev. **15**, 1 (1994).
[4]   P. Roig, et al., Yeast **16**, 1413 (2000).
[5]   P. Roig, and D. Gozalbo, Fungal Genet. Biol. **39**, 70 (2003)..
[6]   P. Roig, and D. Gozalbo, FEMS Yeast Res. **2**, 25 (2002).
[7]   L. del Castillo, A. Nieto, and R. Sentandreu, Gene **120**, 59 (1992).
[8]   P. Roig, and D. Gozalbo, Int. Microbiol. **5**, 33 (2002).
[9]   M .L. Delgado, M. L. Gil, and D. Gozalbo, Yeast **20**, 713 (2003).
[10]   M. L. Delgado, M. L. Gil, and D. Gozalbo, FEMS Yeast Res. **4**, 297 (2003).
[11]   S. Mormeneo, and R. Sentandreu, J. Bacteriol. **152**, 14 (1982).
[12]   A. Parets-Soler et al., J. Gen. Microbiol. **133**, 1471 (1987).

Towards understanding the acetic acid resistance in *Gluconacetobacter europaeus*

J. Trcek[*,1,2], H. Toyama[2], J. Czuba[3], A. Misiewicz[3], and K. Matsushita[2]

[1] Limnos, Podlimbarskega 31, Ljubljana, Slovenia
[2] Dep. of Biol. Chem., Fac. of Agric., Yamaguchi University, Yamaguchi 753-8515, Japan
[3] Institute of Agricultural and Food Biotechnology, Rakowiecka 36, 02-532 Warsaw, Poland

[*]Corresponding author: e-mail: janja.trcek@guest.arnes.si, Phone: +386 1 365 1507, Fax: +386 1 365 1507

High acetic acid resistance strains (up to 10 %), identified as *Gluconacetobacter europaeus*, *Gluconacetobacter intermedius* and *Acetobacter pasteurianus* have been analysed in this study by different molecular-biological approaches. The profiles of membrane proteins from *Ga. intermedius* and *Ga. europaeus* harvested at different points of ethanol oxidation phase were compared. Concomitant with the increased amount of acetic acid, the expression of a ca. 95 kDa protein substantially increased. The N-terminal sequence of the protein was determined. Low temperature difference spectra of membranes from all three species showed a presence of cytochrome a_1 which is a putative ba_3 ubiquinol oxidase. The morphology of cells was compared by scanning electron microscopy. Further, a transposon mutagenesis procedure was established with the aim to study the acetic acid resistance on the genetic level.

Keywords: acetic acid bacteria, *Gluconacetobacter europaeus*, *Gluconacetobacter intermedius*, *Acetobacter pasteurianus*, acetic acid resistance

1   Introduction

Acetic acid inhibits growth of the majority of microorganisms at the minimal concentration of 0.5 %. Although most of acetic acid bacteria exhibit natural resistance to this acid, tolerance differs among the species. Strains resistant to extremely high concentrations of acetic acid (> 10 %) have been isolated from European submerged vinegar bioreactors in the last years and described as novel species (*Gluconacetobacter europaeus*, *Gluconacetobacter intermedius*, *Gluconacetobacter oboediens* and *Gluconacetobacter entanii*). Since such a high tolerance to acetic acid has potential applications for industry (overexpression of resistance ability in strains used for vinegar production as well as in the other bacterial strains), it is important to elucidate the mechanism(s) of acetic acid resistance. For this purpose we are trying to characterize the strains resistant to high concentrations of acetic acid by physiological, biochemical and genetic approaches in more details.

2   Materials and methods

2.1 Strains and growth conditions

All strains used in this study were isolated from industrial vinegar bioreactors and grown as described previously [1, 2].

2.2 Scanning electron microscopy

Bacteria were grown in 20 mL of AE medium containing 3 % of ethanol and 3 % of acetic acid. Biomass was harvested from the late logaritmic phase (Klett Units around 100 and acetic acid concentration ca. 5 %), washed twice with 50 mM KPB (pH 6.5) and resuspended in 1 mL of 2.5 % glutaraldehyde (in 50 mM KPB, pH 6.5). After 1 h of incubation at room temperature, the cells were three times washed with 50 mM KPB (pH 6.5), resuspended in 1 mL of 50 % ethanol and incubated for 15 min at room temperature. The cells were further dehydrated with 60 %, 70 %, 80 %, 90 %, 95 %, 99 % and 100 % of ethanol. Samples were transferred into filter pockets, and incubated in t-butylalcohol at 37 ^{o}C twice for 1 hour. The samples were dried by freeze-drying, mounting on metallic stubs and coated with gold using a ion sputter apparatus. The samples were observed with scanning electron microscope (JEOLSM 6100) operating at 10 kV.

2.3 Low temperature spectra

Low-temperature difference spectra were performed as described previously [3].

2.4 Transposon mutagenesis

Transposon mutagenesis of Gluconacetobacter spp. was performed using Tn5-based delivery plasmid pCM639 [4]. Conjugation was performed on GY agar medium at 30 ^{o}C for 12 hours. Recombinants resistant to tetracycline (20 $\geq$g/ml) and chloramphenicol (50 $\geq$g/ml) were selected on the same medium. From randomly picked mutants, genomic DNA was isolated, restricted by PstI and transfer to the membrane by Southern transfer. Restricted plasmid pCM639 was used as a probe in hybridization experiments. The pre-hybridization, hybridization and detection of signals were performed as described by supplier of ECL Direct Nucleic Acid Labelling and Detection system (Amersham Biosciences).

3   Results and discussion

3.1 Cell morphology

The cell morphology of all strains studied in this work was examined by scanning electron microscopy (Fig. 1). Cells of *A. pasteurianus* form significantly shorter rods in comparison to *Ga. europaeus* and *Ga. intermedius*. The length of *A. pasteurianus* rods is 0.5 - 0.8 µm, whereas of *Ga. intermedius* and *Ga. europaeus* 1.2-1.7 µm. Among the cells of *A. pasteurianus* some amorphous materials were visible which might represent polysaccharides. Cells of *Ga. intermedius* seemed to be covered with a spongy soft material.

676

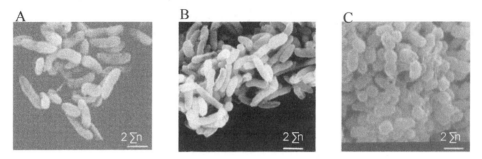

Fig. 1: Micrograph of *Ga. intermedius* JK3 (A), *Ga. europaeus* V3 (B), and *A. pasteurianus* KPP/584 (C) cells by SEM showing a difference in cell morphology between both species of genus *Gluconacetobacter* and species of the genus *Acetobacter*. Cells were harvested at the late exponential phase from medium AE containing 3 % of acetic acid and 3 % of ethanol .

3.2 Low temperature spectra

A potential difference in terminal oxidases of ethanol oxidase respiratory chain was checked among all three species. Low temperature difference spectra of membranes from all three species showed a large enhancement of the signal at 589 nm due to the reaction with cyanide, suggesting a presence of cytochrome a_1 which may be ba_3 ubiquinol oxidase.

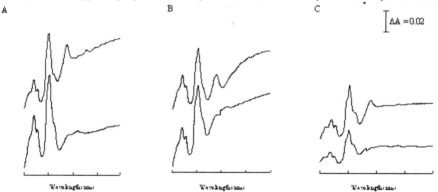

Fig. 2: Low temperature reduced-minus-oxidized difference spectra of the membranes from *Ga. europaeus* (A), *Ga. intermedius* (B) and *A. pasteurianus* (C). Membranes were prepared from the cells grown until the late exponential phase in AE broth containing 1 % of ethanol and 1 % of acetic acid. For analysis the membranes with protein concentration of about 8 mg/mL were used.

3.3 Profiles of membrane proteins

The profiles of membrane proteins from *Ga. intermedius* and *Ga. europaeus* harvested at different points of ethanol oxidation phase were compared. Concomitant with the increased amount of acetic acid, the expression of a ca. 95 kDa protein substantially increased. Two different aeration conditions were compared: buffled flasks (high aeration) and normal flasks (low aeration). The N-terminal sequence of the protein was determined but showed no significant homology to any sequences deposited into GenBank/Swissprot/ PIR/PRF/PDB database

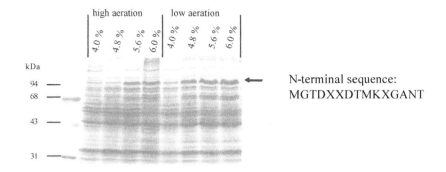

high aeration   low aeration

N-terminal sequence:
MGTDXXDTMKXGANT

Fig. 3: The profiles of membrane proteins from *Ga. intermedius*. Cells were grown in AE broth containing 3 % of ethanol and 3 % of acetic acid and harvested at different percentages of total acidity (indicated above lanes). Two different aeration conditions were compared. The membrane proteins (60 μg per lane) were separated by SDS-PAGE on 12.5 % gel.

3.4 Transposon mutagenesis

We established an efficient conjugation protocol for introducing the Tn5 based transposon tool pCM639 [4] into *Ga. europaeus* and *Ga. intermedius* (Table 1). The suicide plasmid conferred a tetracycline resistance to the recipient cells. The results of hybridization analysis (Fig. 4) showed a random insertion of transposon tool into the genome of *Gluconacetobacter europaeus*. Searching for acetic acid sensitive mutant(s) is presently under the procedure.

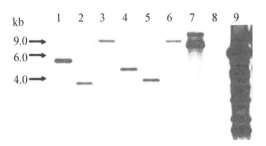

Fig. 4: Southern blot analysis of seven randomly picked, tetracycline resistant clones of *Ga.europaeus* after conjugation with *E. coli* containing plasmid pCM639. Chromosomal DNA was digested with *Pst*I. Lanes: 1-7, *Ga. europaeus* tetracycline resistant clones; 8, *Ga. europaeus* wild-type (control); 9, *Pst*I-restricted plasmid pCM639.

678

Table 1: Frequency of transposon insertions in *Ga. intermedius* and *Ga. europaeus*. The *E. coli* SM10 *pir* with plasmid pCM639 was used as a donor. Each conjugation experiment was repeated twice.

Strain	Frequency of transposon insertion
Ga. intermedius	8.3×10^{-5}
Ga. europaeus	1.2×10^{-5}

Acknowledgements

This work is supported by Grant in Aid for JSPS postdoctoral fellows (awarded to K.M. and J.T.) and MESS Republic of Slovenia (research project L4-5007 to J.T.). J.T. is a recipient of JSPS postdoctoral fellowship.

References

[1]   M. Sievers, S. Sellmer, and M. Teuber, *Acetobacter europaeus* sp. nov., a main component of industrial vinegar fermenters in central Europe. System. Appl. Microbiol. **15**, 386-392 (1992).

[2]   J. Trcek, P. Raspor, and M. Teuber, Molecular identification of *Acetobacter* isolates from submerged vinegar production, sequence analysis of plasmid pJK2-1 and application in development of a cloning vector. Appl. Microbiol. Biotechol. **53**, 289-295 (2002).

[3]   K. Matsushita et al., Change of the terminal oxidase from cytochrome a_1 in shaking culture to cytochrome *o* in static culture of *Acetobacter aceti*. J. Bacteriol. **174**, 122-129 (1992).

[4]   D. A. D'Argenio et al., *Drosophila* as a model host for *Pseudomonas aeruginosa* infection. J. Bacteriol. **183**, 1466-1471 (2001).

Transcriptome dynamics of ethanologenic yeast in response to 5-hydroxymethylfurfural stress related to biomass conversion to ethanol

Z. L. Liu[*] and P. J. Slininger

National Center for Agricultural Utilization Research USDA-ARS, Peoria, IL 61604 USA

[*]Corresponding author: Email: liuzl@ncaur.usda.gov; Telephone: 309 681 6294; Fax: 309 681 6693

Over 200 genes were identified as statistically significantly affected by 5-hydroxymethylfurfural to ethanologenic yeast among 6,388 genes of *Saccharomyces cerevisiae*. Transcriptome dynamics of the identified genes were described during the course of fermentation. Genes in a wide range of functional categories showed distinct patterns of expression compared with that of the control under normal conditions. Results of this study strongly suggested that cells of the ethanologenic yeast went through an adaptation process at the genome level and showed altered metabolic activities to cope with the HMF stress at the earlier cell growth phase.

Keywords: bioethanol, gene expression, 5-hydroxymethylfurfural, stress response

1   Introduction

Renewable energy is a major concern for the future considering the growing limitation of natural resources and continuing demand of society to meet energy needs. Biomass including lignocellulosic materials and agricultural residues is an attractive potential low cost feedstock for bioethanol production [1]. Biomass pretreatment using dilute acid hydrolysis generates inhibitory compounds, which interfere with the subsequent fermentation. One of the most potent inhibitory compounds is 5-hydroxymethylfurfural (HMF) which damages cell growth, reduces enzymatic and biological activities, breaks down DNA, and inhibits protein and RNA synthesis [2, 3, 4].
Genomic expression of the unicellular yeast *Saccharomyces cerevisiae*, as a widely used model system, has been widely studied including its environmental stress responses [5, 6]. However, relatively little information is available regarding the inhibitory stress of HMF involved in bioethanol fermentation, especially for ethanologenic yeast. To promote efficient and economic bioethanol production, we study stress tolerance mechanisms to improve performance of ethanologenic yeast. We previously demonstrated that ethanologenic yeast exhibited a dose-dependent inhibition of furfural and HMF [2]. In this study, we investigate genome-wide gene expression of ethanologenic yeast in response to the HMF stress for bioethanol fermentation.

2   Materials and methods

2.1 Yeast strain and culture conditions

Ethanologenic yeast *S. cerevisiae* NRRL Y-12632 obtained from the Agricultural Research Service Culture Collection, Peoria, IL, USA was used. Cultures were routinely maintained and cultured on a synthetic complete medium consisting of 6.7 g/L yeast nitrogen base without amino acids, 20 g/L dextrose, and the 16 amino acids as previously described [7]. The medium was amended with 30 mM of HMF. Cultures without inhibitors served as a control. Culture growth was monitored by sampling 150 µl of cell suspension and read in a 96-well microplate at OD$_{600}$ using a plate reader. Cultures were treated with 30 mM of HMF 6 hr after the initial incubation, and samples collected at 0, 5, 10, 30 min, and 1, 2, 4, 6, 8, 16, and 40 hr after the HMF treatment. Cells were sampled and harvested as previously described [8]. Cell samples were stored at -80 °C until RNA extraction. Replicated experiments were carried out for each treatment and sample collections.

2.2 Total RNA isolation.

Total RNA was extracted using a hot phenol method based on Schmitt et al. (9) with modifications. RNA was further purified using a nucleic acid purification column. Total RNA yield, integrity, and purity were measured by gel electrophoresis and spectrophotometry.

2.3 DNA Oligo gene microarray.

A recent version of *S. cerevisiae* 70-mer oligo set representing 6,388 genes was applied and Codelink activated slides were used for microarray fabrication. Oligo microarray was designed with two replications with spots printed in separated blocks on the array. The 70-mer DNA oligo microarray consisted of a total of 13,440 elements including spiking controls, positive controls, and negative controls.

2.4 Exogenous nucleic acid controls.

Control genes were selected and validated by sequence blast search and application tests. PCR and DNA sequencing were done to verify the selected gene sequences. Three DNA oligo sequences were designed and synthesized with 5' amino-linked for selected genes of CAB (photosystem I chlorophyll ab binding protein), MSG (major latex protein), and RBS1 (ribulose bisphosphate carboxylase small chain 1 precursor) from *Glycine max*. These oligos were printed in each sub-block of the array with a total of 32 spots each serving as exogenous nucleic acid controls. The mRNA of the three genes was transcribed *in vitro* and quantification verified separately for each by gel electrophoresis and spectrophotometry. A control mix consisting of 10, 100, and 1000 pg/µl each of the MSG, CAB, and RBS1 transcript was used as a spike-in control for each RNA probe labeling reaction.

2.5 RNA probe labeling and hybridization.

The RNA probe was labeled using a protocol originally developed by Hegde *et al*. [10] with modifications. Total RNA of 15 µg was used for each sample labeling. AA-dUTP was incorporated in the first strain cDNA in a reverse transcription reaction and subsequently labeled with Cy3 or Cy5 mono reactive dye. The labeled probe was quantified using a NanoDrop spectrophotometer, and labeled cDNA examined using sGel, an in-house developed slide gel electrophoresis system. RNA collected at 0 time point was used as a reference and applied to each microarray experiment. High quality RNA with purity greater than 1.8 was used. Efficiency of cDNA synthesis and fluorescent probe labeling were evaluated using sGel electrophoresis and by NanoDrop spectrophotometry.

Probes with matched labeling efficiency and proper length of cDNA populations were used to perform hybridization. Probes with matched quality and quantity of labeled cDNA populations were paired to perform hybridization using a hybridization chamber at 42 °C overnight.

2.6 Normalization, data acquisition, and analysis.

Data acquisition was completed using Axon Instruments GenePix 4000B scanner and GenePix Pro. Signal intensities of microarray raw data were normalized using three positive control genes. Data were filtered using stringent restrictions and analyzed using GeneSpring. Saccharomyces Genome Database and Gene Ontology were used for gene annotation and standard nomenclature.

2.7 HPLC analysis of fermentation kinetics

Samples were analyzed for glucose consumption, ethanol production, and HMF conversion using a Waters HPLC equipped with an Aminex HPX-87H column and a refractive index detector as previously described [2, 7].

3 Results and discussion

3.1 Yeast growth and fermentation kinetics

Under defined conditions using a synthetic medium, yeast cell growth reached stationary phase 16 h after treatment with 30 mM of HMF, which showed a 4 h delay from that of a control which reached stationary at 12 h (Fig. 1).

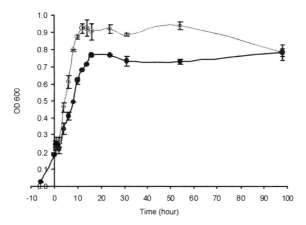

Figure 1. Cell growth of *Saccharomyces cerevisiae* NRRL Y-12632 in the presence of 30 mM of 5-hydroxymethylfurfural treated six hours after incubation (●) compared with a non treated control (○) as measured by OD$_{600}$ in a defined synthetic medium under laboratory conditions.

However, the OD value never reached as high as the control during the course of fermentation. Living cells and biomass from HMF treated cultures also appeared to be less than the control. There was a transient pause of cell growth observed following the HMF addition. It seemed that cell density never fully recovered and that the final active cell population could be reduced by the HMF treatment. In the absence of HMF, glucose consumption was exhausted at 12 h and ethanol yield reached the maximum 14 h after the treatment point. In the presence of HMF, ethanol yield reached the maximum at 16 h and glucose remained detectable till 16 h after treatment. HMF was almost depleted at 16 h after treatment and its conversion product 2,5-bis-hydroxymethylfuran (furan 2,5-dimethanol, FDM) reached the maximum at 14 h. The addition of HMF 6 hours after the initial incubation appeared to also slow down glucose consumption and ethanol production. However, there was no final ethanol yield reduced by addition of HMF in the culture.

3.2 Transcriptome dynamics

Among 6,388 genes, we identified 202 genes statistically significantly affected by HMF during the first two hours of the HMF addition. These genes showed consistently induced or repressed expression from 10 minute to 2 hours compared with the control (Fig. 2A). However, expression profile of these genes changed dramatically at a later phase of the fermentation at 16 and 40 hours in the presence of HMF (Fig. 2B). Under the normal conditions, many genes were repressed while a small number of genes were more enhanced expressed. By contrast, some of the repressed genes at the earlier hours were significantly enhanced in expression at 16 and 40 hours in the presence of HMF. Similarly, some of the earlier induced genes were significantly repressed at the later hours. These observations suggested a dramatic alteration of metabolic activity caused by HMF. Since the ethanol production reached the maximum at 16 h for cultures amended with HMF, the altered metabolic activities were important and impacted the final ethanol yield.

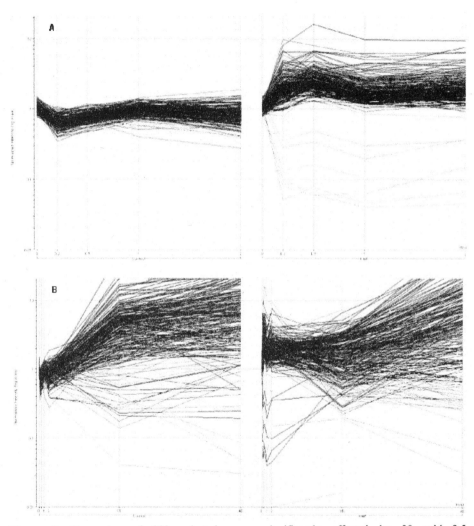

Figure 2. Expression of 202 selected genes significantly affected by 30 m*M* 2,5-hydroxy methylfurfural (HMF) treatment (right panels) compared with the control (left panels) showing dynamic changes (red, induced; green, repressed) in response to exposure of HMF during first two hours (A) and 40 hours (B).

Yeast cells were reported to have a similar number of genes for repressed and induced expression in response to defined environmental stress conditions [5, 6]. Unlike those observed using laboratory strains, we found that the number of significantly differentially expressed genes affected by HMF was not equal for repressed and induced expressions. Yeast cells responded immediately for at least 10 minute after the HMF addition with significantly induced or repressed gene expression. After a brief pause of cell growth, yeast cells were able to grow more vigorously to complete the fermentation process and produce ethanol. During the course of the fermentation, 100% of HMF was converted to

FDM. HMF conversion showed a similar trend to that of the glucose consumption after the cell recovery. These observations strongly suggested that yeast cells went through an adaptation process at the genome level and altered metabolic activities to cope with the HMF stress at the earlier phase of the cell growth.

Acknowledgements

Gene clones of CAB, MSG, and RBS1 were kindly supplied by Dr. M. R. Band, University of Illinois at Urbana-Champaign. Technical assistance by A. C. Cash and M. Shea-Andersh is greatly acknowledged.

References

[1]   R. J. Bothast, and B. C. Saha, Adv. Appl. Microbiol. **44**, 261 (1997)
[2]   Z. L. Liu, P. J. Slininger, B. S. Dien, M. A. Berhow, C. P. Kurtzman, and S. W. Gorsich, J. Ind. Microbiol. Biotechnol. **31**, 345 (2004).
[3]   B. Sanchez and J. Bautista, Enzyme Microb. Technol. **10**, 315 (1988)
[4]   M. J. Taherzadeh, R. Eklund, L. Gustafsson, D. Niklasson, and G. Liden, Ind. Eng. Chem. Res. **36**, 4659 (1997)
[5]   H. C. Causton, B. Ren, S. S. Koh, C. T, HArbison, E. Kanin, E. G. Jennings, T. I. Lee, H. L. True, E. S. Lander, and R. A. Young, Mol. Biol. Cell **12**, 323 (2001)
[6]   A. P. Gasch, P. T. Spellman, C. M. Kao, O. Carmel-Harel, M. B. Eisen, G. Storz, D. Botstein, and P. O. Brown, Mol. Biol. Cell **11**, 4241 (2000)
[7]   Z. L. Liu, P. J. Slininger, and S. W. Gorsich, Appl. Biochem. Biotechnol. **121-124**, (2005)
[8]   Z. L. Liu, and P. J. Slininger, Proceedings of Genome 2004, Hinxton Cambridgeshire, UK, 14-17 April 2004 61 (2004)
[9]   M. E. Schmitt, T. A. Brown, and B. L. Trumpower, Nucleic Acids Res. **18**, 3091 (1990)
[10]   P. Hegde, R. Qi, K. Abernathy, C. Gay, S. Dharap, R. Gaspard *et al.*, BioTechniques **29**, 548 (2000)

Medical Microbiology

Detection and characterization of integrons, transposons, plasmids and genomic types in multidrug resistant clinical isolates of *Salmonella enterica* serovar Virchow.

I. Rodríguez[1], A. Herrero[1], N. Martínez[1], M. C. Martín[2], M. R. Rodicio[1], and M. C. Mendoza[*,1]

[1] Department of Functional Biology, University of Oviedo, C/ Julián Clavería 6, 33006 Oviedo, SPAIN

[2] Institute of Lactic Products of Asturias (CSIC), Crta. Infiesto s/n, 33300 Villaviciosa, Asturias, SPAIN

[*]Corresponding author: e-mail: cmendoza@uniovi.es, Phone: +34 985103560, Fax: +34 985103148

Thirteen multidrug resistant (to 7 or more antimicrobial drugs) isolates of *Salmonella enterica* sv Virchow were analyzed by different molecular approaches. PCR amplification allowed the detection of several resistance (R) determinants and of integron and transposon sequences. The isolates were grouped into 11 resistance genotypes and 4 integron profiles. All of them were positive for class 1 integrons (with 3 different arrays of gene cassettes in the variable regions), and three isolates also contained the same class 2 integron. Specific sequences of transposons Tn*21*, Tn*7*, Tn*3* and Tn*9* could be detected in some isolates. In addition, R-genes, integron and transposon sequences were carried by conjugative plasmids in 12 of the 13 isolates. Genomic typing with RAPD and *Xba*I-PFGE macrorestriction discriminated the isolates, which proved to be highly heterogeneous, into 5 RAPD- and 11 *Xba*I-profiles.

Keywords *Salmonella;* antimicrobial resistance; integrons; transposons; plasmids; PFGE

1 Introduction

The extensive use of antimicrobial agents during the last decades has increased the number of resistant bacterial strains in different environments. The wide-spread of resistance (R) determinants is mainly the result of their mobility, mediated by genetic elements such as gene cassettes in integrons, transposons and plasmids. These elements represent an efficient route of acquisition and dissemination of such determinants within and/or between bacterial populations [1, 2]. For these reasons, current epidemiologic studies are not only focused on phenotypic aspects but also on the detection and characterization of the genetic background of the R-phenotypes. Integrons are elements that are able to capture gene cassettes downstream of a single promoter. These recombinant systems have been found associated with transposable elements [2, 3]. In addition, integrons and transposons can be present in conjugative plasmids, regarded as the main vehicles for the transfer of R-determinants among Gram-negative bacteria including *Salmonella* [4, 5].

Salmonella enterica sv Virchow (*S.* Virchow) is the third or fourth most prevalent serovar causing human illness in Europe [4, 6] and its incidence is even higher in countries like Australia [7] and Israel [8, 9]. In the latter country it has been also reported as an important cause of bacteraemia in children [8, 9]. However, while the genetic bases of antimicrobial resistance in the most prevalent non-typhoid *S. enterica* serovars (Enteritidis and Typhimurium) have been the subject of many reports, the information on *S.* Virchow is comparatively limited. For this reason, the present study aimed to identify the R-genes present in multidrug resistant isolates of this serotype, and to determine their possible association with mobile genetic elements. In addition, the genomic relationships among the R-isolates were established.

2 Multidrug-resistance phenotypes and genotypes

Thirteen multiresistant clinical isolates of *S.* Virchow were used in this work. The following antimicrobial agents (amount in μg): ampicillin (Ap 10), cefotaxime (Ctx 30), chloramphenicol (Cm 30), gentamicin (Gm 10), kanamycin (Km 30), nalidixic acid (Nal 30), piperacillin (Pip 100), spectinomycin (Sp 100), streptomycin (Sm 10), sulfadiazine (Sul 300), tetracycline (Tc 30) and trimethoprim (Tp 5), were tested. The selected isolates proved to be resistant to 7 or more agents (Table 1). In order to determine the R-genotypes, the presence of 11 R-determinants *tem1*, *catA1*, *aac(3)-II*, *aphA1*, *aadA1*, *strA*, *strB*, *sul1*, *sul2*, *tetA(A)* and *dfr1*, was investigated by PCR amplification, using previously described primers and conditions [10–12]. A total of 11 different R-patterns were found (Table 1).

Table 1 General features of *S.* Virchow isolates analyzed in this work

R-Pattern[a]		Conjugative plasmid[d]	Genomic type	
			RAPD	PFGE
I	P1: Ap Cm Gm Pip Sul Sm Sp Tp/ G1: *tem1 catA1 aac(3)-II sul1 sul2 aadA1 strA strB dfrA1*	+	S2 C1	X1
II	P2: Ap Cm Gm Pip Sul Sm Sp Tc Tp/ G2: *tem1 catA1 aac(3)-II sul1 sul2 aadA1 strA strB tetA(A) dfrA1*	+	S1 C2	X2
III	P3: Ap Cm Km Pip Sul Sm Sp/ G3: *tem1 catA1 aphA1 sul1 sul2 aadA1 strA strB*	+	S1 C2	X3
IV	P4: Ap Cm Pip Sul Sm Sp Tp/ G4: *tem1 catA1 sul1 sul2 aadA1 strA strB dfrA1*	+	S1 C2	X4
V[b]	P3: Ap Cm Km Pip Sul Sm Sp/ G5: *tem1 catA1 aphA1 sul1 sul2 aadA1 strB*	+	S1 C2	X3
VI	P5: Cm Km Sul Sm Sp Tc Tp/ G6: *catA1 aphA1 sul1 sul2 aadA1 strA strB tetA(A) dfrA1*	+	S1 C3	X5
VII[c]	P6: Ap Ctx[I] Km Nal Pip Sul Sm Sp Tc Tp/ G7: *tem1 aphA1 sul1 sul2 aadA1 strA strB tetA(A) dfrA1*	- / +	S1 C4	X6 / X9
VIII	P7: Ap Ctx[I] Nal Pip Sul Sm Sp Tc Tp/ G8: *tem1 sul1 sul2 aadA1 strA strB tetA(A) dfrA1*	+	S1 C4	X7
IX	P8: Ap Ctx[I] Nal Pip Sul Sm Sp Tp/ G9: *sul1 sul2 aadA1 strA strB dfrA1*	+	S1 C4	X8
X	P6: Ap Ctx[I] Km Nal Pip Sul Sm Sp Tc Tp/ G10: *tem1 aphA1 sul1 sul2 aadA1 strA tetA(A) dfrA1*	+	S1 C5	X10
XI	P7: Ap Ctx[I] Nal Pip Sul Sm Sp Tc Tp/ G11: *sul1 sul2 aadA1 strA strB tetA(A) dfrA1*	+	S1 C4	X11

[a]: R-patterns were defined on the basis of the R-phenotype (P) and R-genotype (G); [b]: Two isolates sharing the same features; [c]: Two isolates showing the same R-pattern but differing in other features; [d]: Plasmid sizes ranged between 200-300 kb; [I]: Intermediate resistance.

3 Detection of integrons, transposons and conjugative plasmids

Class 1 to 3 integrons were detected by PCR amplification and *Hinf*I-digestion, as described (White, 2001). All 13 isolates carried class 1 integrons (class 1), assigned to 3 different types according to their variable regions (size and gene-cassette content). Three isolates were also positive for the same class 2 integron (class 2). In total, 4 integron profiles (IP) were defined (Table 2): IP1 (class 1:1000 bp/*aadA1* and class 2:2300 bp/*dfrA1-sat1-aadA1*); IP2 (class 1:2300 bp/*sat-smr-aadA1* and class 2:2300 bp/*dfrA1-sat1-aadA1*), IP3 (class 1:1600 bp/*dfrA1-aadA1*), and IP4 (class 1:2300 bp/*sat-smr-*

aadA1).Class 1 and 2 integrons are frequently associated with transposons of the Tn*21* and Tn*7* families, respectively. For this reason, the presence of transposon-related sequences in the *S.* Virchow isolates was also tested by PCR ([12], Table 2). All isolates, except those assigned to IP3 contained the *tnpR*, *tnpA* and *merA* that respectively encode the transposase, resolvase and a mercurial-resistance determinant of Tn*21*. The IP3 isolates lacked either *merA* or *merA* and *tnpA*. By contrast, the *tnsD* and *tnsE* genes, whose products are involved in Tn*7* transposition, could be detected only in the isolates carrying the class 2 integron (IP1 and IP2).

Other R-determinants were also found in association with transposons. This was the case of *tem1* (that encodes the TEM-1 ß-lactamase), *catA1* (for a chloramfenicol acetyl transferase), *aphA1-strA/B* (conferring kanamycin and streptomycin resistance), and *tetA* (involved in tetracycline resistance), which are usually associated with Tn*3*, Tn*9*, Tn*5* and Tn*10*, respectively. PCR amplifications performed in the present work revealed that i) 7 out of the 10 *S.* Virchow isolates carrying *tem1*, were also positive for *tnp3*, the gene that encodes the Tn*3* transposase; ii) all *catA1* positive isolates were also positive for *IS1*, the insertion sequence flanking Tn*9*; iii) none of the *aphA1-strA/B* were positive for the gene encoding the transposase of Tn*5*; iv) none of the *tetA(A)* positive isolates contained the IS*10* and *tetR* sequences screened as markers for Tn*10* [14]).

Table 2 Relationships between R-patterns and mobile genetic elements.

R-pattern	IP	Integrons		Transposons[a]								
		Class 1	Class 2	Tn21			Tn7		Tn3		Tn9	
		VR/gene cassettes	VR/gene cassettes	tnpR	tnpA	merA	tnsD	tnsE	Tnp3	tem1	IS1	catA1
I	2	2300/sat-smr-aadA1	2300/dfr1-sat1-aadA1	+	+	+	+	+	+	+	+	+
II	1	1000/aadA1	2300/dfr1-sat1-aadA1	+	+	+	+	+	+	+	+	+
III	4	2300/sat-smr-aadA1	-	+	+	+	-	-	-	+	+	+
IV	3	1600/dfr1-aadA1	-	+	+	-	-	-	-	+	+	+
V	4	2300/sat-smr-aadA1	-	+	+	+	-	-	-	+	+	+
VI	1	1000/aadA1	2300/dfr1-sat1-aadA1	+	+	+	+	+	-	-	+	+
VII, VIII, X	3	1600/dfr1-aadA1	-	+	-	-	-	-	+	+	-	-
IX	3	1600/dfr1-aadA1	-	+	-	-	-	-	+	-	-	-
XI	3	1600/dfr1-aadA1	-	+	-	-	-	-	+	-	-	-

: For transposon screening (see text), the following primers (5'-3') were designed in the present work: *tnsD*-F/R ACAGGGATTGGCTAGTTCACTGG/CTTTGTGCTTCCTCAGTTATCCG; *tnsE*-F/R GGCGGATCGTGTGATTGAGTTTG/GCGTACTAC CATCACCTGCCTCC; *tnp3*-F/R GTGCTGACTGGCAGGCAAATCGG/GCCTGAAAATCAACCAGTCTGGC *IS1*-F/R CGAGATTTTCA GGAGCTAAGGAAGC/GTAGCACCAGGCGTTTAAGGGCACC; *tnp5*-F/R TGCTGGCCATTGAGGACACCACC/TGACATTCATCCGGG GTCAGCAC; *IS10*-F/R CTACCGAGTAACACCACACCGCTC/TGAAGCATCAGGGCGATTAGCAGC.

Transfer of R-determinants was studied by conjugation experiments, using the *S.* Virchow isolates as donors, and the riphampicin-resistant *Escherichia coli* K-12 strain J53 as recipient. Conjugations were performed in Luria Bertani broth for 4 h at 37°C, and transconjugants were selected on Eosin Methylene Blue agar containing riphampicin (50 ≥g/ml) plus either trimethoprim (5 ≥g/ml) or ampicillin (50 ≥g/ml). For 12 of the 13 isolates, transconjugants were obtained that showed the same R-phenotype, R-genotype and integron-profile as the donor isolates. Accordingly, the same R-plasmid (200-300 kb) could be detected in each donor strain and its transcojugants [12].

4   Genomic typing

Two molecular typing methods (RAPD and *Xba*I-PFGE) were used to classify the 13 *S.* Virchow isolates into genomic groups. Two-way RAPD fingerprinting assays were performed with primers and PCR conditions previously described [15]. With primer S only two profiles (S1 and S2, Fig. 1a) could be identified, being S1 (with 12 isolates) the predominant. Primer OPB-17 showed a higher discriminative power generating 5 profiles (C1 to C5, Fig. 1a), being C2 and C4 (each with 5 isolates) the most common. When results with the two primers were combined, the 13 isolates were discriminated into 5 RAPD profiles (Table 1). In addition, PFGE-macrorestriction with the *Xba*I endonuclease was performed using a standard protocol [16] and the CHEF-DRIII SYS220/240 apparatus (Bio-Rad laboratories). By these means, a total of 11 *Xba*I-PFGE profiles could be discriminated (Fig. 1b; Jaccard's Coefficient of similarity of 0.52-0.96). Thus, genomic typing confirmed the high diversity of the isolates, already revealed by the R-patterns, and the integron- and transposon-profiles. This stands out in clear contrast with what has been reported for other *Salmonella* serovars, including the pandemic Enteritidis and Typhimurium.

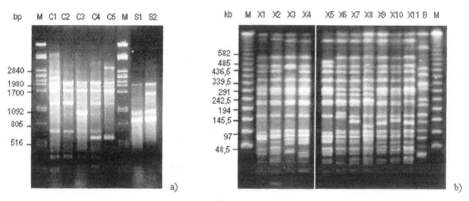

Fig. 1 Typing of the *S.* Virchow isolates by RAPD and *Xba*I-PFGE. Panel a) RAPD-profiles. Lane M, Size marker: λ DNA digested with *Pst*I; lanes C1-C5, profiles generated with the OPB-17 primer; lanes S1-S2, profiles generated by the S primer. Panel b) *Xba*I-PFGE Lane M, λ ladder PFGE Marker (New England BioLabs); X1-X11, different *Xba*I-PFGE profiles identified in this work.

In conclusion, here was shown that a set of different mobile genetic elements (integrons/gene cassettes, transposons, and plasmids) containing R-determinants are spread among clinical isolates of *S.* Virchow. Underlying this situation is the problem of both vertical and horizontal dispersion of R-determinants among bacteria and the transmissibility of these bacteria through the food chain.

Acknowledgements

This work has been supported by grant PI020172 from the "Fondo de Investigación Sanitaria". I. Rodríguez and A. Herrero are recipients of grants from the "Fundación para el Fomento en Asturias de la Investigación Científica Aplicada y la Tecnología" (FICYT).

References

[1]   T. F. O'Brien, Clin. Infect. Dis. **34** (Suppl 3), S78 (2002).
[2]   R. Cantón, T. M. Coque, and F. Baquero, Curr. Opin. Infect. Dis. **16**, 315 (2003).
[3]   D. A. Rowe-Magnus, and D. Mazel, Int. J. Med. Microbiol. **292**, 115 (2002).

[4]   E. J. Threlfall, I. S. T. Fisher, C. Berghold, P. Gerner-Smidt, H. Tschäpe, M. Cormican, I. Luzzi, F. Schnieder, W. Wannet, J. Machado, and G. Edwards, Eurosurveill. **8**, 41 (2003).

[5]   A. C. Fluit, and F. J. Schmitz, Clin. Microbiol. Infect. **10**, 272 (2004).

[6]   M. C. Martín, M. A. González-Hevia, J. A. Álvarez-Riesgo, M. C. Mendoza, Eur. J. Epidemiol. **17**, 31 (2001).

[7]   A. M. Sullivan, L. R. Ward, B. Rowe, J. B. Woolcock, and J. M. Cox, Lett. Appl. Microbiol. **27**, 216 (1998).

[8]   Z. Shimoni, S. Pitlik, L. Leibovici, Z. Samra, H. Konigsberger, M. Drucker, V. Agmon, S. Ashkenazi, and M. Weinberger, Clin. Infect. Dis. **28**, 822 (1999).

[9]   P. Yagupski, N. Maimon, and R. Dagan, Int. J. Infect. Dis. **6**, 94 (2002).

[10]   C. Levésque, L. Piché, C. Larose, and P. H. Roy, Antimicrob. Agents Chemother. **39**, 185 (1995).

[11]   B. Guerra, S. M. Soto, J. M. Argüelles, and M. C. Mendoza, Antimicrob. Agents Chemother. **45**,1305 (2001).

[12]   B. Guerra, E. Junker, A. Miko, R. Helmuth, and M. C. Mendoza, Microb. Drug Res. **10**, 77 (2004).

[13]   P. A. White, C. J. McIver, and W. D. Rawlinson, Antimicrob. Agents Chemother. **45**, 2658 (2001).

[14]   C. Morsczeck, D. Langendörfer, and J. M. Schierholz, J. Biochem. Biophys. Methods. **59**, 217 (2004).

[15]   S. M. Soto, B. Guerra, M. A. González-Hevia, M. C. Mendoza, App. Environ. Microbiol. **65**, 4830 (1999).

[16]   T. M. Peters, C. Maguirre, E. J. Threlfall, I. S. T. Fisher, N. Gill, and A. J. Gatto, Eurosurveill. **8**, 46 (2003).

Effect of the essential oils of Brazilian species of the genus *Cunila* on the growth and biofilm formation by *Aeromonas*

R. da Luz[1], S. Echeverrigaray[1], S.O.P. Costa[1,2], A.P.L. Delamare[1,2]
[1] Instituto de Biotecnologia, Universidade de Caxias do Sul, R. Francisco G. Vargas 1130, 95001-970 Caxias do Sul, RS, BRAZIL.
[2] Instituto de Biociências, Universidade de São Paulo, São Paulo, BRAZIL.

The present work aimed to study the potential of the essential oils of five Brazilian species of *Cunila* (*C. spicata*, *C. incisa*, *C. angustifolia*, *C. menthoides*, and two chemotypes of *C. galioides*) on the *Aeromonas* control and biofilm formation. Essential oils were obtained by hydrodistillation, and the volatile constituents determined by gas chromatography. Minimal inhibitory concentration, growth depletion, and inhibition of biofilm formation were evaluated on microtitle plates using serial dilutions of the essential oils. The results showed that the most effective essential oil in the control of *Aeromonas* growth was that from *C. galioides* chemotype citral, characterized by high concentration of neral and geranial. Conversely, the most effective oils for the inhibition of biofilm formation were those of *C. angustifolia*, *C. menthoides* and *C. spicata*. Differences were detected in the biofilm inhibition between *A. hydrophila* and *A. caviae* isolates.

Keywords: *A. caviae*, *A. hydrophila*, monoterpenes, antimicrobial activity.

1 Introduction

Aeromonas are an ubiquous group of Gram negative aquatic bacteria associated with several fish, reptile, and mammal diseases. In humans, *Aeromonas* are implicated in gastroenteritis and other infections including septicemia, meningitis, and wound infections. These bacteria are transmitted by food and water, and their pathological potential is associated with the production of toxins, invasins (proteases, lipases, DNAses) and biofilms [1]. A greater risk of *Aeromonas* infection is reported in young children, elderly people, and immunocompromised patients [2–4]. Although they are sometimes considered as opportunistic human pathogens and reported to be susceptible to food processing procedures, the Food and Drug Administration now considers *Aeromonas hydrophila* as an emerging foodborne pathogen of concern [5].

The antiseptic qualities of aromatic and medicinal plants and their extracts have been recognized since antiquity. Plant volatile oils are generally isolated from nonwoody plant material by distillation methods, usually steam or hydrodistillation, and are variable mixtures of principally terpenoids. Terpenes are among the chemicals responsible for the medicinal, culinary and fragrant uses of aromatic and medicinal plants. Most terpenes are derived from the condensation of branched five-carbon isoprene units and are categorized according to the number of these units present in the carbon skeleton [6].

The antimicrobial properties of plant volatile oils and their constituents from a wide variety of plants have been assessed and reviewed [7–9]. Investigations into the antimicrobial activities, mode of action and potential uses of plant volatile oils have regained momentum with the revival in the use of traditional approaches to protecting humans, livestock and food from disease, pests and spoilage. The organisms against which volatile oils have been tested include wide range of food spoiling and food poisoning organisms, and animal, human and plant pathogens [10].

Cunila species (Lamiaceae) are aromatic plants used in Brazilian popular medicine as stimulants, antiespasmidics, antipyretics, and in the treatment of chronic cough and infections of the respiratory tract [11]. This genus is formed by 22 species, with two centers of origin, one in Mexico and the other in the highlands of South Brazil. Depending on the species, the essential oils of *Cunila* are rich in menthofuran, citral, ocimene, menthene, 1,8-cineole, sabinene, and other monoterpenes, some of which are known as antimicrobial compounds [10–13].

The present work aimed to evaluate the potential of the essential oils of five Brazilian species of *Cunila* (*C. spicata*, *C. incisa*, *C. angustifolia*, *C. menthoides*, and two chemotypes of *C. galioides*) on the control of Aeromonas growth and biofilm formation.

2   Material and methods

Eight *Aeromonas* isolates, being four representants of each *A. hydrophila* and *A. caviae*, were used. Essential oils of five species of *Cunila* (*C. spicata*, *C. incisa*, *C. angustifolia*, *C. menthoides*, and two chemotypes of *C. galioides*) collected in their area of occurrence in Rio Grande do Sul State, Brazil, were evaluated. Essential oils were obtained by hydrodistillation of 200g (dry weight) of the aerial part of flowering plants using a Clevenger apparatus. The volatile constituents were determined by gas chromatography using a mass detector (GC-MS), and quantified by gas chromatography using a ionizing detector.

Minimal inhibitory concentration, growth depletion, and inhibition of biofilm formation were evaluated on microtitle plates using serial dilutions of the essential oils, 1% Triton-X-100, and LB medium. Bacterial strains were pre-culture on LB medium for 18h at 30°C, and 5≥l used to inoculate 100≥l LB medium, supplemented with different concentrations of the essential oils, previously loaded on the microtitle plates. The plates were incubated at 30°C for 24h.

Growth was spectrophotometrically evaluated at 560nm on a microtitle plate reader (Metrolab, Ind.). Biofilm assay followed the procedure described by Heilmann et al. [14]. All experiments were conducted twice with three replications per experiment. Percentage of bacterial growth and biofilm formation on increasing concentrations of the essential oils was calculated considering the absence of essential oils as 100%.

3   Results and discussion

The essential oils of the five species of *Cunila* evaluated exhibited complex profiles with more than 30 compounds. The main constituents of the essential oils used in the experiments were: *C. galioides* chemotype citral -citral 68.5%, terpinene 4.4%, and ocimene 3.4%; *C. galioides* chemotype menthene –menthatriene 20.2%, menthadiene-1-ol 11.5%, 1,8-cineole 10.7%, and terpinene 7.8%; *C. incisa* – 1,8-cineole 42.9%, terpineole 14%, sabinene 6.7%, terpinene 4.8%, and cimene 4.4%; *C. angustifolia* –sabinene 40.5%, terpinene 13%, and limonene 10.5%; *C. menthoides* –isomenthone 88.9%, menthone 4.7%, and pulegone 1.8%; and *C. spicata* –1,8-cineole 36.7%, linalool 29%, terpineole 11%, and sabinene 6.5%.

The results obtained (Table 1) showed that, considering the overall response of the eight *Aeromonas* isolates, the most effective essential oil was that from *C. galioides* chemotype citral, characterized by high concentration of neral and geranial, the two isomers of citral. In this case MIC values range from 0.50 to 1.25 mg l^1, concentrations comparable to the essential oils of other Lamiaceae species already used as antimicrobial agents. Aldehydes are known to possess powerful antimicrobial activity. It has been proposed that an aldehyde group conjugated to a carbon to carbon double bond is a highly electronegative arrangement. Such electronegativity may interfere in biological processes involving electron transfer and react with vital nitrogen components inhibiting bacterial growth [15]. The efficiency of citral on *Aeromonas* and other bacteria was reported by Dorman and Deans [10].

Table 1. Minimal inhibitory concentration (MIC, 100%) and partial inhibitory concentration (PIC, 50%) of *Cunila* essential oils on the growth of *Aeromonas hydrophila* and *A. caviae* isolates. Data expressed in mg l^{-1}.

A. hydrophila	J8		J9		J15		J18		Mean	
	50%	100%	50%	100%	50%	100%	50%	100%	50%	100%
C.galioides citral	0.90	1.25	0.60	1.20	0.40	1.00	0.40	1.25	0.57	1.17
C. incisa	1.25	2.20	1.50	2.30	1.25	4.20	1.05	2.80	1.26	2.87
C.galioides menthene	3.95	>5.00	3.70	>5.00	3.20	>5.00	3.05	>5.00	3.47	>5.00
C. angustifolia	3.75	5.00	4.40	>5.00	3.50	>5.00	4.45	>5.00	4.02	>5.00
C. menthoides	1.90	4.35	3.50	4.85	3.10	4.70	2.65	4.70	2.78	4.65
C. spicata	1.85	2.75	2.90	4.65	1.45	4.15	1.70	4.15	1.97	3.92

A. caviae	J16		ATCC15468		IOC		J11		Mean	
	50%	100%	50%	100%	50%	100%	50%	100%	50%	100%
C.galioides citral	0.30	0.62	1.15	1.60	1.50	1.85	0.40	0.50	0.84	1.14
C. incisa	0.45	1.50	1.50	1.80	1.50	1.80	1.45	1.75	1.22	1.72
C.galioides menthene	0.40	0.80	4.00	>5.00	>5.00	>5.00	4.25	>5.00	3.41	3.95
C. angustifolia	0.20	0.50	>5.00	>5.00	>5.00	>5.00	4.30	>5.00	3.62	3.87
C. menthoides	0.90	1.60	3.05	3.95	3.60	4.70	3.30	4.35	2.71	3.57
C. spicata	0.62	1.00	3.10	4.00	3.30	4.05	3.10	3.70	2.53	3.19

The other oils with important antimicrobial activity were those of *C. incisa*, *C. spicata*, and *C. menthoides*. The first two oils have high concentration of 1,8-cineole, a monoterpene with recognized antimicrobial activity [16], and *C. menthoides* oil is characterized by a very high content of isomenthone and menthone, compounds that inhibits the growth of *A. hydrophila* and other Gram negative bacteria [10].
Investigations into the effects of terpenoids upon isolated bacterial membranes suggest that their activity is a function of the lipophilic properties of the constituent terpenes, the potency of their functional groups and their aqueous solubility. Their site of action appeared to be at the phospholipid bilayer, affecting the electron transport, protein translocation, phosphorylation steps and other enzyme-dependent reactions [17]. However, considering the heterogeneous composition of *Cunila* essential oils and the antimicrobial activities of many of their components, it seems unlikely that there is only one mechanism of action or that only one component is responsible for the antimicrobial action.

The isolate J16 (*A. caviae*) was sensitive to all the essential oils tested, whereas J11 (*A. caviae*) exhibited a particularl sensibility to the oil of *C. galioides* citral. This differential response indicates that sensitibity/resistance data should be considered with caution, and that the evaluation of just one or few strains of a given species is not conclusive.

Sub-inhibitory concentrations of all the essential oils under analysis inhibited biofilm formation. This inhibition was not correlated with antimicrobial activity. The most effective oil for biofilm inhibition was that obtained from *C. angustifolia* (Figure 1-C). This oil rich in sabinene, terpinene and limonene, exhibited the lowest antimicrobial activity, indicating that biofilm inhibition is not associated with growth depletion. The major componentes of this oil were among the less effective antimicrobial terpenes [10] on *A. hydrophila* and other Gram negative bacteria.

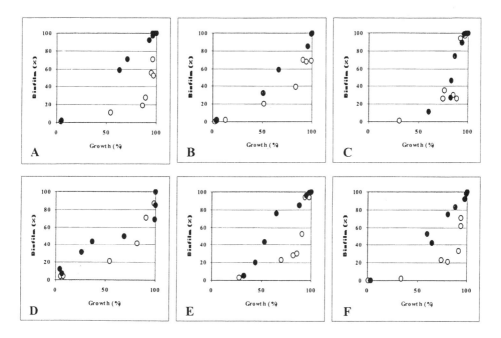

Fig. 1- Bacterial growth and biofilm depletion by increasing concentrations (0 to 5.0 mg l^{-1}) of the essential oils of A- *C. menthoides*, B- *C. incisa*, C- *C. angustifolia*, D- *C. galioides* chemotype citral, E- *C. galioides* chemotype menthene, F- *C. spicata*. (●) *A. caviae*, (○) *A. hydrophila*.

Remarkable differences were observed on the inhibition of biofilm formation between *A. caviae* and *A. hydrophila* isolates, specially in the presence of the essential oils of *C. menthoides*, *C. galioides* chemotype menthene, and *C. spicata* (Figure 1). Whereas biofilm inhibition is directy related to growth depletion on *A. caviae* isolates, *A. hydrophila* biofilms were drastically inhibited (>70%) on less than 1.2mg l^{1} of the essential oils, concentrations in which growth was slightly inhibited (<20%). These results indicate a differential response of these highly related *Aeromonas* species to terpenes, fact that may reflect a difference in the molecules and interactions involved in biofilm formation.

In general, the present results showed the potential of essential oils of *Cunila* species, especially that of *C. galioides* chemotype citral, for the control of *Aeromonas*. Moreover, they indicate that these oils can be used to reduce biofilm formation, a character that posse many problems both on food and medical control of microorganisms and is considered as one important virulence factor on several bacterial including *Aeromonas* [1, 18].

Aknowledgments

This work was supported by FAPERGS and UCS.

References

[1]   S.L. Abbott, L.S. Seli, J.M. Catino, M.A. Hartley, J.M. Janda. *J. Clin. Microbiol.* **36** (1998), 1103..

[2]   J.M. Janda, S.L. Abbott. *Clin. Infect. Disc.* **27** (1998), 332.

[3]   M.J. Figueras, L. Soler, M.R. Chacón, J. Guarro, A.J. Martinez-Murcia. *Int. J. . Evol. Microbiol.* **50** (2000), 2069.

[4]   S.M. Kirov. (2001) *Aeromonas* and *Plesiomonas* species. In Food Microbiology: fundamentals and frontiers eds. Doyle, M.P., Beuchat, L.R., Montville, T.J. pp. 301-327. Washington D.C.: ASM Press.

[5]   J.H. Isonhood, M. Drake. *J. Food Protect.* **65** (2002), 575.

[6]   M.M. Cowan. *Clin. Microbiol. Rev.* **12** (1999), 564.

[7]   M.A. Janssen, J.J.C. Scheffer, A.B. Svendsen. *Pharmaceutische Weekblad* **9** (1987), 193.

[8]   J.V. Larrondo, M. Agut, M.A. Calvo-Torras. *Microbios* **82** (1995), 171.

[9]   S. Pattnaik, V.R. Subramanyam, C.R. Kole. *Microbios* **86** (1996), 237.

[10]   H.J.D. Dorman, S.G. Deans. *J. Appl. Microbiol.* **88** (2000), 308.

[11]   C.M.O. Simões, L.A. Mentz, E.P. Schenkel, B.E. Irgang, J.R. Stehmann. Plantas da medicina popular no Rio Grande do Sul, Editora da Universidade /UFRGS, Brazil. (1994).

[12]   S.A.L. Bordignon, E.P.Schenkel, V. Spitzer. *J. Essent. Oil Res.* **10** (1998), 317.

[13]   S. Echeverrigaray, F. Fracaro, A.C.A. Santos, N. Paroul, R. Wasum, L.A. Serafini. *Biochem. Syst. Ecol.* **31** (2003), 467.

[14]   C.O. Heilmann, C. Schweitzer, N. Gerke, D. Vanittanakom, D. Mack, F. Gotz. *Mol. Microbiol.* **20** (1996), 1083.

[15]   V. Moleyar, P. Narasimham. *Food Microbiol.* **3** (1986), 331.

[16]   S. Inouye, T. Takizawa, H. Yamagushi. *J. Antimicrob. Chemoth.* **47** (2001), 565.

[17]   K. Knobloch, A. Pauli, N. Iberl, H.M. Weis, N. Weigand. (1988) Modes of action of essential oil components on whole cells of bacteria and fungi in plate tests. In *Bioflavour* ed. Schreier P. pp. 287–299. Berlin: Walter de Gruyther.

[18]   R.M. Donlan, J.W. Costerton. *Clin. Microbiol. Rev.* **15** (2002), 167.

Human and bovine constitute reservoirs of different sub-populations of *Staphylococcus aureus* in possession of the highly prevalent enterotoxin gene cluster egc_{like}.

J. M. Fueyo[1], **M. R. Rodicio**[1], **M. C. Mendoza**[1] and **M. C. Martín**[*,2]

[1] Department of Functional Biology, University of Oviedo, C/ Julián Clavería 6, 33006 Oviedo, SPAIN

[2] Institute of Lactic Products of Asturias (CSIC), Crta. Infiesto s/n, 33300 Villaviciosa, Asturias, SPAIN

*Corresponding author: e-mail: mcm@ipla.csic.es, Phone: +34985892131, Fax: +34985892233

A high proportion of *Staphylococcus aureus* contains enterotoxin gene clusters, *egc_{like}*, carrying five or six genes (*seg-sei-sem-sen-seo* ± *seu*) and forming part of a pathogenicity island, υSaβ1, which also contains the *lukE-lukD* leukotoxin genes. The present study was aimed at identifying the genomic relationships between *S. aureus* of human (58 isolates from nasal exudates of healthy carriers) and bovine (38 isolates from milk samples of cows with subclinical mastitis) origin in possession of *egc_{like}*. The relationships were traced using two procedures: determination of the toxin-genotype screening a total of 25 genes by PCR-amplification, and genomic macrorestriction analysis performed with *Sma*I-PFGE. Results showed a higher diversity of types in human than in bovine isolates; that none of the types was common to both sets; and that in all bovine but only in some human isolates *egc_{like}* could be part of υSaβ1. Results support that humans and bovines are reservoirs of different sub-populations of *S. aureus* carrying *egc_{like}*.

Keywords *Staphylococcus aureus,* enterotoxin gene cluster, human carriers, mastitis, superantigens, leukotoxins, PFGE, PCR, pathogenicity islands, plasmids

1   Introduction

Staphylococcus aureus is both a commensal and an extremely versatile pathogen in humans and warm-blooded animals. The primary habitat in humans is the mucosae of the nasopharynx where it exists as a persistent or transient member of the normal microbiota without causing any symptoms. However human carriers are the major infection source of *S. aureus*. The diseases caused by this microorganism can be grouped in three basic syndromes. (i) Superficial lesions such as skin abscesses and wound infections (ii) Deep-seated and systemic infections such as osteomyelitis, endocarditis, pneumonia, and bacteraemia (iii) Toxemic syndromes such as toxic shock, scarlet fever, scalded-skin, and food poisoning. For this complex set of diseases, *S. aureus* produces and secretes numerous and specific pathogenicity factors, including a wide variety of exoproteins that contribute to both colonization and disease process [1–3]. Only part of the isolates secrete leukotoxins (bi-component toxins called LukE/LukD, LukS-PV'/LukM, LukS-PV/LukF-PV) and/or pyrogenic toxin superantigen (PTSAg) including toxic shock syndrome toxins (TSSTs), staphylococcal enterotoxins (SEs), and exfoliative toxins (ETs). *S. aureus* is also worldwide recognized as a pathogen causing intramammary infections (mastitis) in cattle where both leukotoxins and PTSAgs are supposed to play an important role in the initiation and/or exacerbation of the disease [4].

Genes encoding PTSAgs have different genetic supports including prophages, plasmids, and pathogenicity islands [1–3]. An enterotoxin gene cluster (*egc*) encoding a putative nursery of superantigens, with five genes (*seg, sei, sem, sen, seo*) and two pseudogenes (ψ *ent1* and ψ *ent2*) has been identified [5] and located in the sequenced genome of methicillin-resistant strains as part of a pathogenicity island (*ν*Saβ), which also contains the genes encoding the bi-component leukotoxin LukE-LukD [3]. An *egc*-derivative encoding SEG, SEI and SEN variants and the new SEU (*seu* is the result of a 15 bp insertion into ψ *ent1*) was described two years later [6]. The aim of this study was to identify genomic relationships between *S. aureus* of human and bovine origin in possession of the highly prevalent enterotoxin gene clusters (*egc*$_{like}$) collected and partially analysed in previous studies [7, 8]. Here, we labeled *egc*1 and *egc*2 to the cluster carrying *seg-sei-sem-sen-seo* on its own or including *seu*, respectively.

2 Materials and methods

2.1 *S. aureus* isolates

All the 96 *S. aureus* isolates analyzed in this study were collected in the Principality of Asturias (PA), Spain, 58 from nasal cavities of healthy carriers, and 38 from milk samples of dairy cows diagnosed with subclinical mastitis. All of them were analyzed for the presence of five major serological toxins (SEA to SED and TSST-1) by reversed passive latex agglutination and also for their genes by polymerase chain reaction (PCR). Other toxin genes were tested only by PCR. Control strains with distinctive toxin gene profiles reported in [7-8] were also included.

2.2 Determination of toxin gene profiles

Detection of exotoxin encoding genes was done by conventional- and multiplex-PCR using 25 primer pairs for the *lukS-PV/lukF-PV* (*lukPV*), *lukM, lukE-lukD* (*lukED*), *tst, eta, etb, etd, sea-see, seg-ser* and *seu* genes, and employing genomic DNA as template [7-8]. In all PCR-assays each isolate was tested at least twice, and both positive and negative control strains were always included.

2.3 Macrorestriction-pulsed field gel electrophoresis (PFGE) analysis

Total DNA from each *S. aureus* isolate was analyzed by macrorestriction-PFGE performed with *Sma*I by means of the CHEF-DRIII SYS220/240 (Bio-Rad laboratories, S.A., Madrid, Spain), basically using a consensus protocol as in [7-8]. The genetic similarity between *Sma*I-profiles was determined by the unweighted pair group method with arithmetic averages (UPGMA) and the Jaccard´s similarity coefficient (*S*).

3 Results and discussion

The relationships between human (H) and bovine (B) isolates were traced using two genetic procedures and combining the results: toxin gene profile (TG) and macrorestriction genomic analysis performed with *Sma*I-PFGE. Results compiled in Table 1 show all different combinations of TGs, inferred genetic elements, *Sma*I genomic profiles, as well as the number of isolates from each group sharing the same features.
All isolates were negative for *eta, etb, see, sek, sep* and *seq*. All H-isolates were also negative for *etd, lukM*, and all except one also for *lukPV*. All B-isolates were negative for *sea, sed, sej, seh*, and *ser*, and all except one for both *seb* and *etd*. Other toxin-genes

appeared at different frequencies and TGs. The H-isolates generated 15 TGs, the B-isolates 5 TGs, while only one TG was common to both groups (TG1b: *seg-sei-sem-sen-seo, lukED*). On the other hand, the TGs were used as markers of the genetic elements in which the toxin genes could be inserted [2, 3, 7]. In addition to *egc$_{like}$* and *lukED* with υSaβ, the following associations could be traced: *tst* with SaPI2$_{like}$ (16 H), *sea* with a prophage (11 H), *sec-sel-sem* with SaPI4$_{like}$ (10 H), *sed-sej-ser* with plasmids (5 H); and *sec-sel-tst* with SaPIbov (10 B). It is remarkable that while all B-isolates were *lukED*-positive, 60.3% of the H-isolates was *lukED*-negative, suggesting in this case that the *egc$_{like}$* support is other than υSaβ.

Table 1. Toxin gene profiles and relationships between human and bovine *S. aureus* isolates

Toxin gene profiles				Origin		Genomic types
TG	PTSAgs	Leukotoxins	Genetic elements	H	B	*Sma*I-profiles
TG1	seg-sei-sem-sen-seo		egc1	7		S31, S37, S58, S60
TG1b	seg-sei-sem-sen-seo	lukED	υSaβ1	15	4	S31, S37, S51-S56, S58, S60, S87, S88, S90
TG1c	seg-sei-sem-sen-seo	lukED, lukM	υSaβ1 other		1	S83
TG2	seg-sei-sem-sen-seo-seu		egc1	2		S59
TG2b	seg-sei-sem-sen-seo-seu	lukED	υSaβ2	1		S57
TG2c	seg-sei-sem-sen-seo-seu	lukED, lukM	υSaβ2, other		22	S80, S82-S84, S89, S91, S92
TG3	seg-sei-sem-sen-seo-seu, tst.		egc2, SaPI2	1		S9
TG3b	seg-sei-sem-sen-seo-seu, tst.	lukED	υSaβ2, SaPI2	1		S9
TG4	seg-sei-sem-sen-seo-seu, tst, sea		egc2, SaPI2, prophage	10		S3-S5, S7
TG4d	seg-sei-sem-sen-seo-seu, tst	lukPV	υSaβ2 SaPI2, other	1		S9
TG5b	seg-sei-sem-sen-seo, sea	lukED	egc1, prophage	1		S18
TG6	seg-sei-sem-sen-seo-seu, tst, seh		υSaβ2, SaPI2	3		S11
TG7	seg-sei-sem-sen-seo, seb	lukED	υSaβ1, other	3		S23-S25
TG8	seg-sei-sem-sen-seo, seb, etd	lukED	υSaβ1, others		1	S85
TG9	seg-sei-sem-sen-seo, sec-sel		egc1, SaPI4	7		S31-S33
TG10	seg-sei-sem-sen-seo, sec-sel, sed-sej-ser		υSaβ1, SaPI4, pUO-SED1	3		S31
TG11b	seg-sei-sem-sen-seo, sed-sej-ser	lukED	υSaβ1, pUO-SED1	2		S42
TG12	seg-sei-sem-sen-seo, seh		egc1, other	1		S61
TG13c	seg-sei-sem-sen-seo-seu, tst-sec-sel	lukED, lukM	υSaβ2, SaPIbov, other		10	S80-S82

PFGE analysis (Fig.1) differentiated the isolates into 39 *Sma*I-profiles, 27 generated by H-isolates and 12 from B-isolates. None of the *Sma*I-profiles included isolates from both sets. The control strains generated different *Sma*I-profiles. A dendrogram was constructed on the basis of the coefficient of similarity between *Sma*I-profiles (Fig. 2). At a cut-off point of *S* = 0.7 eight clusters and several branches were revealed (5 H, 2 B, and 1 HB, see Fig. 2).

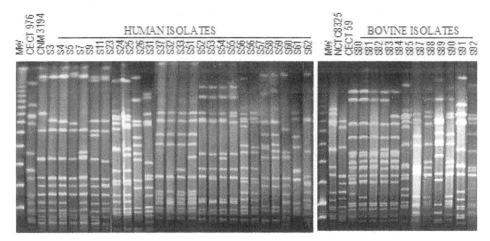

Fig. 1. Macrorestriction genomic profiles generated by _Sma_I in _S. aureus_ isolates from human and bovine origin. Lanes MW, Lambda ladder PFGE Markers (New England Biolabs). NCTC 8325, used as quality control in PFGE assays. Fragment size from top to bottom in kb are 674, 361, 324, 262, 257, 175, 135, 117, 80, 60 and < 60. Relationships between TGs and _Sma_I profiles are compiled in Table 1.

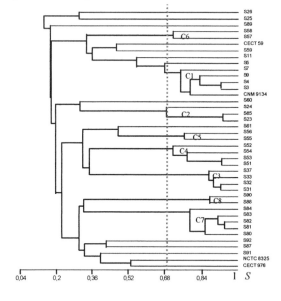

Fig. 2. Dendrogram showing relatedness among _Sma_I macrorestriction fragment profiles generated from _S. aureus_. Cluster analysis was performed by the Jaccard similarity coefficient (_S_) and the unweighted pair group method. At a cut-off point of $S = 0.7$, eight clusters (labeled C1 to C8) were revealed. Profiles S3 to S61 and clusters C1 to C6 were generated from human isolates, but C2 grouped also 1B isolate. Profiles S80-S92 and clusters C7 and C8 were generated from bovine isolates. The isolates falling into each cluster and branch, as well as their relationship with TGs, are compiled in Table 1.

Even considering the size of each set (58 vs 38 isolates) the diversity of TGs and _Sma_I-profiles was remarkably higher in H- than in B-isolates (15 and 27 vs 5 and 12). H-isolates were more frequently positive and carried a higher number of classical SE-genes than B-isolates (32 vs 11, and 5 vs 3, respectively). Conversely, the frequency of leukotoxin genes was strikingly higher in B-isolates (100% for _lukED_, 87% for _lukM_, and 0% _lukPV_) than in H-isolates (39.7% for _lukED_, 0% for _lukM_, and 1.7% for _lukPV_). This fact could be in relation with the important role of leukotoxins, especially LukS-PV'/LukM, in bovine mastitis [4]. On the other hand, H and B -isolates generated _Sma_I-profiles with low similarity (< 25%). As an exception, a single B-isolate showed = 70% similarity with three H-isolates (the four falling into C2, Fig 2). These 4 isolates have also in common that they were *egc1*, *seb*, *lukED* positive, but the B-isolate was also *etd*-positive. Presumably, the latter could be of human origin and could have reached the milk through manual manipulation. Also it is noteworthy that 10 bovine isolates, and only these, were *tst-sec-sel* positive, a gene grouping related with SaPIbov [2,3]. These results further support the hypothesis that humans and bovines constitute reservoirs of different sub-populations of *S. aureus* species carrying egc_{like} groupings.

Acknowledgements

We thank J. Sierra (Hospital Clínico, Barcelona), A. Vindel (Centro Nacional de Microbiología, Instituto Carlos III, Majadahonda, Madrid), and CECT for reference strains; and J. Muñiz (Laboratorio Interprofesional Lechero y Agroalimentario de Asturias) for bovine isolates. This work has been supported by a grant PI020172 from the "Fondo de Investigación Sanitaria".

References

[1]   M. M. Dinges, P. M. Orwin, and M. Schlievert, Clin. Microbiol. Rev. **13**, 16-34 (2000).
[2]   R. P. Novick, Plasmid **49**, 93-105 (2003).
[3]   H. Schmidt, and M. Hensel, Clin Microbiol. Rev. **17**, 14-56 (2004).
[4]   P. Rainard, J. C. Corrales, M. B. Barrio, T. Cochard, and B. Poutrel, Clin. Diagn. Lab. Immunol. **10**, 272-277 (2003).
[5]   S. Jarraud, M. A. Peyrat, A. Lim, A. Tristan, M. Bes, C. Mougel, J. Etienne, F. Vandenesch, M. Bonneville, and G. Lina, J. Immunol. **166**, 669-677 (2001).
[6]   C. Letertre, S. Perelle, F. Dilasser, and P. Fach, Appl. Microbiol. **95**, 38-43 (2003).
[7]   J. M.Fueyo, M. C, Mendoza, M. A. Alvarez, and M. C. Martín, FEMS Microbiol. Lett. **243**, 447-54 (2005).
[8]   J. M. Fueyo, M. C. Mendoza, M. R. Rodicio, J. Muñiz, M. A. Alvarez, and M. C. Martín, J. Clin. Microbiol. **43**,1278-1284 (2005).

Improved cytocompatibility of titanium alloy by coating with pure titanium films using sputter-deposition

T. Sonoda*, **T. Saito, A. Watazu, K. Katou, T. Yamada,** and **T. Asahina**

National Institute of Advanced Industrial Science and Technology(AIST), 2266-98 Anagahora, Shimoshidami, moriyama-ku, 463-8560 Nagoya, JAPAN

*Corresponding author: e-mail: tsutomu.sonoda@aist.go.jp, Phone: +81 52 736 7124, Fax: +81 52 736 7400

In order to improve the biocompatibility of Ti-6Al-4V alloy, coating of the alloy substrates with pure titanium films by magnetron DC sputtering in Ar gas was examined, because the pure titanium barrier layer prevents such harmful substances from leaching out to biological tissues. The effects of the pure titanium barrier layer on the cytocompatibility of the titanium alloy were investigated. Under visual observation, the obtained pure titanium film appeared to be uniform and adhesive. Under SEM, the surface of the deposited pure titanium film was observed to consist of fine particles accumulating and aggregating. According to AES in-depth profiles of Ti and Al for the deposited titanium film and Ti-6Al-4V alloy substrate, it was confirmed that no aluminum atoms diffused to the film surface from the alloy substrate. Thus it was expected that the obtained titanium film could serve the improvement of the biocompatibility of the alloy as a barrier layer which prevents aluminum atoms from leaching out to biological tissues. Based on the results of the in-vitro cytotoxicity test, it was found that the viability, i.e.,the ratio of living cells to the sum of living cells and dead cells was higher for the titanium film than for the alloy. Therefore the deposited titanium film exhibited better cytocompatibility than the Ti-6Al-4V alloy, concluding that the cytocompatibility of the alloy was improved by coating with the titanium film.

Keywords titanium alloy; pure titanium film; biocompatibility; cytocompatibility; sputter-deposition

1 Introduction

Ti-6Al-4V alloy attracting attention as a biomaterial features excellent mechanical properties and corrosion resistance, and super plasticity, so enables the forming of denture bases of complicated shapes. However, this alloy contains aluminum and vanadium liable to do serious harm to human bodies[1,2], so actual use will require prevention of direct contact with biological tissues.

On the other hand, pure titanium has been utilized as biomaterials in dental or orthopedic field such as denture clasps and artificial dental roots or artificial joints, fixing plates and screws for broken bone[3], and its prostheses or implants exhibit excellent biocompatibility[4].

In this study, therefore, in order to improve the biocompatibility of Ti-6Al-4V alloy, coating of the alloy substrates with pure titanium films by magnetron DC sputtering in Ar gas was examined, because the pure titanium barrier layer prevents such harmful substances from leaching out to biological tissues. Furthermore the effects of the pure titanium barrier layer on the cytocompatibility of the titanium alloy were investigated.

2 Experimental

A planar magnetron sputtering system with a 200mm-diameter, 130mm-high stainless steel chamber was used. The planar target used for this study was a 100mm-diameter pure titanium disk. Ti-6Al-4V alloy substrates($20 \times 20 mm^2$, thickness 0.55mm) were mounted on the water-cooled substrate holder. The magnetron sputtering to deposit pure titanium films was carried out in the atmosphere of argon (Ar).

The sputtering conditions examined in this study were as follow. Discharge voltage and discharge current for DC sputtering the pure titanium target were 350V and 1.0A respectively. The pressure of the argon sputter gas was around 1.0 Pa. Depositing time for sputtering the pure titanium target was 20min. Figure 1 shows a schematic outline of the DC sputtering system used in this study.

The thickness of deposited films was measured by tracing the substrate-film step using a surface roughness tester[5]. The surface morphology of the obtained films was studied on SEM images. In-depth profiles of the obtained films were analyzed by Auger electron spectroscopy(AES) with an ion sputter etching method using argon ion beam.

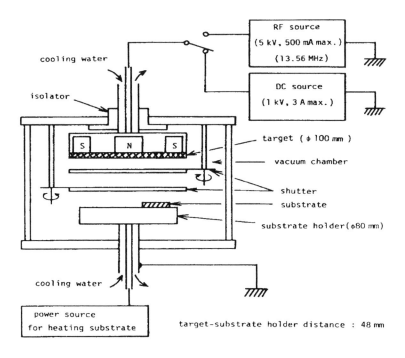

Fig. 1   Schematic outline of the DC sputtering system used in this study.

The cytotoxicity test was carried out for the pure titanium film deposited onto the Ti-6Al-4V alloy substrate by sputtering as well as the received alloy as reference material. These metal specimens were immersed into 20ml of Dulbecco's modified Eagle medium(DMEM) at 37°C for one month with no shaking. Human hepatoma HepG2 cells were used for detection of cytotoxicity of the specimens. Equal amounts of cells(2.5×10^5 cells/ml, 2ml/well) were seeded on the 6-well tissue culture dish and pre-cultured in DMEM containing 5% fatal bovine serum(FBS) at 37°C in 5% CO_2 atmosphere for 24 hours. Then the cells were washed in the DMEM for 30minutes and the culture medium was changed to the DMEM used to soak the metal specimens for one month. The DMEM was supplemented with 5% FBS. After 3 days of incubation at 37°C in 5% CO_2 atmosphere, the released lactate dehydrogenase(LDH) activity is measured using KYOKUTOMTX"LDH" kit on the both sides of supernatant and attached cells to determine the number of dead and living cells respectively. The controls of positive and negative were DMEM containing 5% FBS and 0% FBS without metal extract respectively. The test was performed by contrast with the reagent blank using a microplatereader(OD590).

3 Results and discussion

3.1 Appearance and surface morphology

Under visual observation, the obtained pure titanium films appeared to be uniform and adhesive, and their thickness was approximately 3μm. Under SEM, the surface of the deposited pure titanium film was observed to consist of fine particles accumulating and aggregating. Figure 2 shows a surface morphology of the obtained pure titanium films.

Fig. 2   Surface morphology of the pure titanium films obtained in this study. (SEM image)

3.2 In-depth profiles for pure Ti coatings

Figure 3 shows AES in-depth profiles for the obtained pure Ti coatings. According to these in-depth profiles of Ti(titanium), Al(aluminium) and O(oxygen) for the deposited titanium film and Ti-6Al-4V alloy substrate, it was confirmed that no aluminium atoms diffused to the film surface from the alloy substrate. Thus it was expected that the obtained titanium film could serve the improvement of the biocompatibility of the alloy as a barrier layer which prevents aluminium atoms from leaching out to biological tissues.

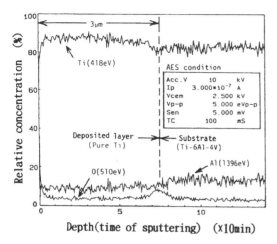

Fig. 3 AES in-depth profiles of Ti, Al and O for the titanium film deposited onto Ti-6Al-4V alloy substrate in Ar atmosphere by DC sputtering.

3.3 Cytotocompatibility

Figure 4, 5 amd 6 show LDH activity of living cell, LDH activity of dead cell, and viability, i.e.,the ratio of living cells to the sum of living cells and dead cells respectively, where samples No. 1 , 2, 3 and 4 are of 5%FS/DMEM, of 0%FBS/DMEM, of Ti-6Al-4V alloy substrate(5%FBS/DMEM) and of the obtained pure Ti film(5%FBS/DMEM) respectively.

Based on these results of the in-vitro cytotoxicity test, it was found that the number of dead cells obtained by measuring LDH activity of supernatant was smaller for the pure titanium film deposited onto the Ti-6Al-4V alloy than for the Ti-6Al-4V alloy, that the number of living cells including the proliferated cells was slightly larger for the deposited titanium film than for the alloy, and thereby that the viability, i.e.,the ratio of living cells to the sum of living cells and dead cells was higher for the titanium film than for the alloy.

Therefore the deposited titanium film exhibited better cytocompatibility than the Ti-6Al-4V alloy, concluding that the cytocompatibility of the alloy was improved by coating with the pure titanium film.

706

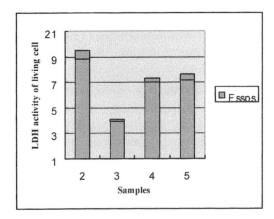

Fig. 4 LDH activity of living cell, where samples No. 1 , 2, 3 and 4 are of 5%FS/DMEM, of 0%FBS/DMEM, of Ti-6Al-4V alloy substrate(5%FBS/DMEM) and of the obtained pure Ti film(5%FBS/DMEM) respectively.

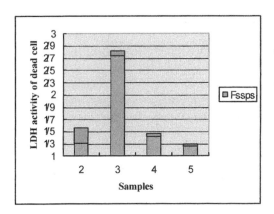

Fig. 5 LDH activity of dead cell, where samples No. 1 , 2, 3 and 4 are of 5%FS/DMEM, of 0%FBS/DMEM, of Ti-6Al-4V alloy substrate(5%FBS/DMEM) and of the obtained pure Ti film(5%FBS/DMEM) respectively.

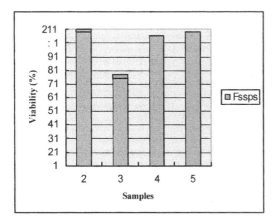

Fig. 6 Viability(%), i.e.,the ratio of living cells to the sum of living cells and dead cells, where samples No. 1 , 2, 3 and 4 are of 5%FS/DMEM, of 0%FBS/DMEM, of Ti-6Al-4V alloy substrate(5%FBS/DMEM) and of the obtained pure Ti film(5%FBS/DMEM) respectively.

4   Conclusions

Coating of the alloy substrates with pure titanium films by magnetron DC sputtering in Ar gas was examined in order to improve the biocompatibility of Ti-6Al-4V alloy, because the pure titanium barrier layer prevents such harmful substances from leaching out to biological tissues. The effects of the pure titanium barrier layer on the cytocompatibility of the titanium alloy were investigated. Under visual observation, the obtained pure titanium film appeared to be uniform and adhesive. Under SEM, the surface of the deposited pure titanium film was observed to consist of fine particles accumulating and aggregating. According to AES in-depth profiles of Ti and Al for the deposited titanium film and Ti-6Al-4V alloy substrate, it was confirmed that no aluminium atoms diffused to the film surface from the alloy substrate. Thus it was expected that the obtained titanium film could serve the improvement of the biocompatibility of the alloy as a barrier layer which prevents aluminium atoms from leaching out to biological tissues. Based on the results of the in-vitro cytotoxicity test, it was found that the viability, i.e.,the ratio of living cells to the sum of living cells and dead cells was higher for the titanium film than for the alloy. Therefore the deposited titanium film exhibited better cytocompatibility than the Ti-6Al-4V alloy, concluding that the cytocompatibility of the alloy was improved by coating with the titanium film.

References

[1]   S. G. Steinemann and S. M. Perren. *Titanium Sci. Technol.* **2**, 1327 (1985).
[2]   P. Galle. *Compt. rend.* **299**, 536 (1984).
[3]   C. B. Johansson, J. Lausmaa, T. Roestlund, P. Thomson. *J. Mater. Sci. Mater. Med.* **4**, 132 (1993).
[4]   A. Montague, K. Merritt, S. Brown, J. Prayer, *J. Biomed. Mater. Res.* **32**, 529 (1996).
[5]   N. Numoto. *J. Surface Finishing Society of Japan* **43**, 739 (1992).

Is intestinal microbiota bound up with changing lifestyle?

Epp Sepp[*1], Tiia Voor[2], Kaja Julge[2], Krista Lõivukene[1], Bengt Björkstén[3], Marika Mikelsaar[1]

[1]Department of Microbiology, University of Tartu, Ravila St 19, Tartu 50411, ESTONIA
[2]Children's Clinic of Tartu University Clinics, Lunini St 6, Tartu 50411, ESTONIA
[3]Institute of Environmental Medicine, Karolinska Institutet Box-285, S-171, 77, Stockholm, SWEDEN

[*]Correspondence author e-mail: epp.sepp@ut.ee, Phone +372 7 374 171, Fax+372 7 374 172

Previously clear differences between microbiota of Estonian and Swedish children were demonstrated. The aim of the present study was to compare the intestinal microbiota in Estonian and Swedish 5-years-old children born in different years. The faecal microbiota of 5-year-old Estonian children born at 1993-1994 or 1996 and Swedish children born at 1996-1997 was studied by quantitative culturing of faecal samples. In 5-years-old Estonian children born earlier the counts of gut microorganisms were higher than in their Estonian and Swedish counterparts born later. There were no differences in gut microbiota between countries in 1996-1997 born children at the age of 5 years. Our data indicate that the influence of industrialization could have changed gut microbiota in former socialistic countries after regain independence. This may be related to changes of lifestyle and level of hygiene.

Keywords: gut, microbiota, lifestyle change, former socialistic country

1 Introduction

The rich and complex intestinal microbiota are remarkably stable, but many factors such as diet and environment may temporarily change its composition. In various geographical regions the differences of diet emerge, usually play an important role in the composition of intestinal microflora [1, 2]. Swedish children and American adults consuming a Western diet is related to more anaerobic bacteria in gut microbiota than Japanese, Chinese and Estonian subjects, yet Japanese living for many years in Western countries have obtained a new type of gut microbiota [1, 3, 4]. On the other hand, in modern developed Western countries the environment factors as the higher hygienic lifestyle may cause shifts in intestinal microbiota [5, 6]. Children of developed countries with the antroposophic lifestyle showed a higher proportion of acetic acid in gut [5]. Similarly, children living in countries with lower level of industrialization had a significantly larger number of lactic acid bacteria as enterococci and lactobacilli [4, 7].

In former socialistic countries as Estonia the diet and environment have step-by-step changed after sovereignty. Unfortunately, some unwanted shifts eg the increase of allergic diseases is probably associated with the changes in lifestyle [8, 9]. However, there is little information available concerning the intestinal microbiota during the process of industrialization of the societies in East Europe. The aim of this study was to compare the composition of intestinal microbiota in five-year-old Swedish children with the same data from Estonian children born at different years after deliberation.

2 Material and methods

Altogether 34 children were divided into three study groups according to year and country of birth. Group 1 comprised 19 Estonian children born at the Maternity Hospital of Tartu University between October 1993 and February 1994 and group 2 comprised 7 Estonian children born in 1996. Group 3 consisted of 8 Swedish children born at the Linköping University Hospital between March 1996 and April 1997. Faecal samples of children were collected at the age of five-years and investigated by quantitative bacteriological methods.

Bacteriological analyses

Stool samples (1-2 g] were collected and kept in domestic refrigerator at 4° C for no more than 2 hours before transportation to the laboratory, where they were frozen at -70° C until analysed.

Weighed samples of faeces were serially diluted (10^{-2}–10^{-9}) in pre-reduced phosphate buffer (pH 7.2) in the anaerobic glove box (Sheldon Manufacturing Inc. with a gas mixture: 5% CO_2; 5% H_2, 90% N_2). A quantitative analysis of the faeces was performed using seeding of serial dilutions on nine freshly prepared media. Yeast extract agar was employed for total aerobes counts, yeast extract agar with 6,5 per cent of sodium chloride for staphylococci, Endo agar for enterobacteria, de Man-Rogosa-Sharpe agar (MRS; Oxoid) for microaerophiles as lactobacilli and streptococci, Wilkins-Chalgren agar (Oxoid) for total anaerobes, Wilkins-Chalgren agar with vancomycin and nalidixic acid supplement (Oxoid) for gram-negative anaerobes as bacteroides, Wilkins-Chalgren agar with colistin and nalidixic acid supplement (BBL) for gram-positive anaerobes, bifidobacteria and eubacteria, Cefoxitin-Cycloserine-Fructose agar (Oxoid) with egg yolk and sodium taurocholate for *Clostridium difficile* and Sabouraud dextrose agar with penicillin (50000 U/L) and streptomycin (40000 U/L) for yeasts and fungi were applied in our study. The total counts of spore-forming clostridia were estimated on Wilkins-Chalgren agar after ethanol treatment [10].

The colony counts of the different dilutions were recorded and from the highest dilutions with growth all the colonies with different morphology were analysed for identification by standard methods. The microorganisms were identified mostly on genus and species level. After the identification of microorganisms, which grew as single colony in the last emerging dilutions, the quantitative composition of faecal microbiota was determined. The number of the various species or genus were given as colony forming unit per gram faeces (CFU/g) and expressed in log_{10}. The detection level was µ3log CFU/g.

Statistical analysis

The statistical analyses were performed using "SigmaStat" (Jandel Scientific, address USA) and "Excel" (Microsoft Corp) software programs, employing Fisher test (the prevalence of microorganisms) and Mann-Whitney U rank sum test (the counts of microorganisms). P values less than 0.05 were considered to be statistically significant.

Ethical considerations

Informed consent was obtained from the parents of the babies. The study was approved both by the Institutional Review Boards of Tartu and Linköping Universities.

3 Results

The counts of microorganisms varied considerably in Estonian children (Group 1 *versus* Group 2) born in different years (Table 1). In Group 2 the counts of anaerobes were decreased (p=0.001) as compared to Group 1. The counts of gram-positive anaerobic cocci (p=0.02) and bacteroides (p=0.008) were decreased in Estonian children born later than in earlier born ones.

The similar trend was seen in Estonian children born in 1993-94 as compared to Swedish children born 4 years later: the counts of aerobes (p=0.02) and anaerobes (p=0.004) were significantly higher in the former. Particularly, coagulase-negative staphylococci (CONS; p=0.03) and bacteroides (p=0.02) were outnumbered of Swedish children (Table 1).

However, there were no statistically significant differences between the counts of gut bacteria of Estonian and Swedish children born in 1996-1997 (Table 1).

Table 1. Gut microbiota of Estonian and Swedish children (prevalence of colonization- number of children and counts of bacteria-log CFU/g)

Micro-organisms	Estonian children				Swedish children	
	Born in 1993-94 (Group 1)		Born 1996 (Group 2)		Born 1996-97 (Group 3)	
	No of children	Counts log; CFU/g (median/range)	No of children	Counts log; CFU/g (median/range)	No of children	Counts log; CFU/g (median/range)
Aerobes	19	9.3[a] 5.3-10.5	7	7.3 6.5-8.9	8	7.2[a] 3.3-9.6
CONS	10	7.6[a] 3.8-10.2	6	5.3 4.3-7.1	6	5.1[a] 3.0-6.8
S. aureus	3	4.4 3.6-.3	4	6 5.3-7.3	2	5.1 3.0-7.6
Enterococci	15	7.3 3.3-10.4	4	6.6 6.5-7.8	3	5.3 4.3-7.3
Enterobacteria	17	7.6 4.3-9.6	7	7.3 4.2-8.9	7	6.3 3.0-7.3
Streptococci	11	8.6 4.3-10.1	2	6.7 6.3-7.0	3	8.3 7.0-9.6
Candida	5	5.5 4.3-7.3	2	4.9 3.6-5.8	1	3.6
Anaerobes	19	10.6[c/b] 8.6-11.2	7	8.9[c] 7.5-10.3	8	9.3[b] 6.8-10.4
Lactobacilli	11	5.6 4.3-9.3	4	4.7 3.9-6.3	2	6.2 6.0-6.3
Anaerobic-cocci	19	9.6[a] 8.3-10.8	3	8.3[a] 7.3-9.3	6	9.2 8.3-10.1
Bifidobacteria	8	9.3 6.3-10.6	4	8.3 7.5-9.2	7	8.8 6.6-10.1
Eubacteria	8	8.5 7.3-9.6	1	10.3	2	9.5 8.8-10.2
Bacteroids	19	9.9[b/a] 7.3-10.9	6	8.9[b] 6.8-9.9	8	8.8[a] 6.8-10.3
Clostridia	17	6.6 3.8-10.3	6	7 3.3-8.8	2	6.9 5.3-8.5

[a] p<0.05; [b] p<0.01; [c] p=0.001

4   Discussion

In this study we compared the intestinal microflora of Estonian and Swedis h 5 years old children born at different years in the nineties. In our previous papers we have shown the differences in the gut microbiota composition between Estonian and Swedish children during first two years of life. [4, 11, 12]. Our present data indicate that certain environmental exposure in Estonia is associated with increasing industrialization after deliberation in 1991 and have caused the shifts in the composition of gut microbiota. The nature of these shifts contains the decrease in the load of both aerobic and anaerobic microorganisms. The reasons for described changes have not yet been assessed. However, we propose that during industrialization the clear changes of diet and environment have occurred in Estonia. Mainly, the hygienic and food standards have improved; our standards conform to Europe Unions directives and Estonian food law was accepted in 1999 [http://www.tervisekaitse.ee]. Food, bottle water, also betters conditions for personal hygiene can be associated with improved life in Estonia.

In 5-years-old Estonian children born at the early nineties the significantly higher counts of colonization resistance granting anaerobes (gram-positive anaerobic cocci, bacteroides) were present than in their Estonian and Swedish counterparts born later. In 1996–1997 born children of both countries in neonates age differences were seen [12] and after 3 months of life the differences had disappeared (unpublished data). This change of microbiota may be influence the immune system [13]. The human immune system has been designed to protect the integrity of the host by maintaining the healthy "*status quo*", but its variation brings upon allergic diseases. The prevalence of allergic diseases was lower in Estonia than in Sweden, but now it has increased in Es tonia [14, 15]. Allergic people have lower counts of microorganisms in gut than healthy ones [16, 17]. The decreased stimulation of the immune system by lower pressure of indigenous intestinal bacteria responding to the so-called "hygiene hypothesis" may explain the regional differences of allergic diseases in Western and Eastern countries [18–21]. The studies in Germany after re-unification have provided evidence that environmental and lifestyle factors are important in the development of allergic diseases [22]. Also the higher cultural adaptation has been correlated with higher rates of allergic diseases among Turkish preschool children living in Germany [8]. This correlation suggests that environmental rather than genetic differences are responsible for differences observed [8]. However, the ethnicity is an important determinant of atopic disorders and is independent on the external environment during childhood [23].

In conclusion, this study demonstrates the influence of industrialization on gut microbiota in former socialistic countries after regaining independence. In association with the lifestyle changes the Estonian children born in the early nineties had higher rate of anaerobic bacteria than later born 5-year-old children. The gut microbiota of later born Estonian children now resembles to Swedish children. The life style changes have disturbed the gut microbiota, which might be one reason for the increasing trend of allergies.

References

[1]   S.M. Finegold, H.R. Attebery and V.L. Sutter, Am. J. Clin. Nutr., **27**, 1456–1469 (1974).
[2]   S. Salminen, E. Isolauri and T. Onnela, **41**, 5–15 (1995).

[3]   B.S. Reddy and E.L. Wynder, J. Natl. Cancer Inst., **50**, 1437–1442 (1973).

712

[4]   E. Sepp, K. Julge, M. Vasar, P. Naaber, B. Björkstén and M. Mikelsaar, Acta Paediatrica, **86**, 956–961 (1997).

[5]   J.S. Alm, J. Swartz, B. Björksten, L. Engstrand, J. Engström, I. Kühn, G. Lilja, R. Möllby, E. Norin, G. Pershagen, C. Reinders, K. Wreiber, and A. Scheynius, Ped. All. Immunol., **13**, 402-411 (2002).

[6]   T. Dunder, L. Kuikka, J. Turtinen, L. Räsanen and M. Uhari, Allergy, **56**, 425-428 (2001).

[7]   V.C. Aries, J.S. Crowther, B.S. Drasar, M.J. Hill and R.E.O.Williams, Gut, **10**, 334–335 (1969).

[8]   C. Grüber, S. Illi, A. Plieth, C. Sommerfeld and U. Wahn,. Clin. Exp. All., **32**, 526-531 (2002).

[9]   E. von Mutius, C. Braun-Fahrländer, R. Schierl, J. Riedler, S. Ehlermann, S. Maisch, M. Waser and D. Nowak, Clin Exp Allergy, **30**, 1230-1234 (2000).

[10]   Marler, L.M., Siders, J.A., Wolters, L.C., Pettigrew, Y., Skitt, B.L. and Allen S.D, J. Clin. Microbiol. **30**, 514–516(1992).

[11]   E. Sepp, Dissertationes Medicinae Universitatis Tartuensis, **43** (1998).

[12]   E. Sepp, P. Naaber, T. Voor, M. Mikelsaar and B. Björkstén, Micr. Ecol. Health ., **12**, 22-26 (2000).

[13]   P. Brandtzaeg, Ann. New York Acad. Sci., **964**, 13-45 (2002).

[14]   B. Björkstén, D. Dumitrascu, T. Foucard, N. Khetsuriani, N. Khaitov, M. Leja, G. Lis, J. Pekkanen, A. Priftanji and M.A. Riikjärv,. Eur. Resp. J., **2**, 432-437 (1998).

[15]   A. Raukas-Kivioja, E. Raukas, H.M. Loit, J. Kiviloog, E. Rönmark K. Larsson, B. Lundbäck, Clin. Exp. Allergy. **33**, 1342-1348 (2003).

[16]   M. Matsumoto, H. Ohishi, K. Kakizoe and Y. Benno, Micr. Ecol. Health Dis., **16**, 13-17 (2004).

[17]   S. Watanabe, Y. Narisawa, S. Arase, H. Okamatsu, T. Ikenaga and M. Kumemura,. Allergy Clin. Immunol., **111**, 587-591 (2003).

[18]   P.G. Holt, P.D. Sly and B. Björksten, Ped. Allergy Immunol., **8**, 53-58 (1997).

[19]   A.E. Wold, Allergy, **53**, Suppl. 46, 20-25 (1998).

[20]   D.P. Strachan, Throax., **55**, Suppl. 1, 2-10 (2000).

[21]   Ö. Strannegård and I.-L. Strannegård, Allergy, **56**, 91-102 (2001).

[22]   S.K. Weiland, E. von Mutius, T. Hirsch, H. Duhme, C. Fritzsch, B. Werner, A. Hüsing, M. Stender, H. Renz, W. Leupold and U. Keil, Eur. Respir. J., **14**, 862-870 (1999).

[23]   A. Hjern, B. Haglund and G. Hedlin,. Clin. Exp. Allergy, **30**, 521-528 (2000).

PCR procedures for detection of emergent multidrug resistant lineages of *Salmonella* serovar Typhimurium. Evolution of a lineage carrying the plasmid pUO-StVR2

A. Herrero[1], I. Rodríguez[1], N. Martínez[1], M. A. González-Hevia[2], M. Bances[2], M. R. Rodicio[1], and M. C. Mendoza[1]

[1] Department of Functional Biology, University of Oviedo, C/ Julián Clavería 6, 33006 Oviedo, SPAIN
[2] Laboratory of Public Health (LSP). Oviedo, Asturias, SPAIN

A total of 184 out of 257 *S.* Typhimurium isolates recorded in Asturias over 2001-2002 exhibited resistance (R) to ampicillin. Of these, 94, 10 and 49 showed traits (such as ampicillin-R determinant and class 1 integron content) respectively associated to three major multidrug resistant lineages: L1 (corresponding to the pandemic DT 104 clone), L2 (with [1,4,(5),12:i:-] isolates) and L3. The L3-isolates displayed an R-genotype consistent with the presence of a virulence-resistance plasmid (pUO-StVR2), derived from pSLT and first detected in 1998 in our laboratory. These isolates generated five PFGE-profiles, with similarity of 92-72%, with only two found in previous years. The obtained data indicate an increase in the incidence of L3 (preceded by L1 and followed by L2), and a L3 evolution, suggested by the genomic *Xba*I-polymorphisms, the different plasmid profiles, and the acquisition of new resistance determinants (trimethoprim-R). However, a possible horizontal spread of pUO-StVR2 can not be ruled out.

Keywords *Salmonella;* serotype Typhimurium; antimicrobial resistance; integrons; plasmids; PCR; emergent types; PFGE

1   Introduction

In Spain, as well as in other European countries, *Salmonella enterica* serovar Typhimurium (*S.* Typhimurium) is the second most frequently isolated serovar as cause of food-borne gastroenteritis, only preceded by *S.* Enteritidis [1, 2]. The incidence of resistance (R) to antimicrobial drugs and multidrug resistance (MDR) in *S.* Typhimurium is high, with plasmids, transposons and integrons frequently acting as vehicles in the spread of R determinants [3, 4]. Over the 90′s at least three multidrug-resistance (MDR) lineages (L) have emerged in Asturias (a northern region of Spain). L1, which contains the R-genes [*bla*$_{PSE-1}$-*aadA2-floR-tetA(G)-sul1*] located in a chromosomal island (SGI-1), harbouring two class 1-*sul1* integrons: 1200 bp/*bla*$_{PSE-1}$ and 1000 pb/*aadA2* and associated to the DT-104 pandemic clone [4, 5]. L2, which includes strains of the monophasic variant [1,4,(5),12:i.-], carries a class 1 integron: 1900 pb/*dfr12-aadA2*, and displays the R-genotype [*bla*$_{TEM-1}$-*cmlA1-aadA1a-sul1/2-dfrA12-aac(3)-4-tetA(A)*] $\pm$ other R-genes, with the integron and all of the R-determinants located in plasmids [5-7]. L3 carries the R-genes [*bla*$_{OXA-30}$-*aadA1a-tetA(B)-sul1/2-catA1-strA/B*] inserted in a hybrid plasmid (pUO-StVR2) that contains the class 1 integron: 2000 pb/ *bla*$_{OXA-1}$ −*aadA1a* [5, 8, 9]. pUO-StVR2 apparently derived from pSLT, the 94 kb virulence plasmid of serovar Typhymurium that belongs to the incompatibility group IncFII and contains the *spvABCDR* locus, responsible for an increase in the bacterial growth rate in mice during the systemic phase of disease [10].

In this work we aimed to determine the incidence of the three MDR lineages in Asturias over the period 2001-2002, paying special attention to L3, that was first discovered in this region. For this, we took advantage of the fact that although strains of the three lineages are resistant to ampicillin, the responsible R-determinants are different (bla_{PSE-1}, bla_{TEM-1} and bla_{OXA-30}), and that each lineage is also characterized by distinct integrons. In addition, we carried out a survey of the genomic types containing pUO-StVR2, in order to trace a possible evolution of the L3 lineage.

2 Incidence of *S.* Typhimurium MDR lineages

In this work, 257 *S.* Typhimurium isolates (108 and 104 recovered in 2001 and 2002, respectively) were analysed using procedures as described [5, 7-9] Of these, 248 were of clinical origin while 9 were recovered from food. Attending to the antigenic formula 240 were identified as biphasic [4,(5),12:i:1,2] and 17 as monophasic [4,(5),12:i:-]. 184 out of the 257 isolates (71.6%) were ampicillin-R (tested according to NCCLS methods). They were screened for the presence of the three different *bla*-genes by multiple PCR, and for class 1 integron profile (IP) by single PCR amplification of the variable regions (VR). Ninety four, 37 and 49 isolates generated the amplicons expected for bla_{PSE-1}, bla_{TEM-1} and bla_{OXA-30}, respectively (Figure 1a) while in the remaining 4 isolates the ampicillin-R determinant could not be identified. In addition, 154 ampicillin-R isolates (83.7%) were positive for class 1 integrons and, according to the size of the VRs, these could be further differentiated into four IP (Fig. 1b): IP1: 1200 bp + 1000 bp; IP2: 1900 bp + 150 bp; IP3: 2000 bp; and IP4: 1600 bp (94, 10, 49, and 1 isolates, respectively). The presence of the bla_{PSE-1}, *aadA2*, *dfr12-aadA2* and bla_{OXA-30}-*aadA1a* gene cassettes, within the 1200, 1000, 1900 and 2000 bp VRs characteristic of IP1, IP2 and IP3, respectively, was confirmed by nested-PCR, using the amplicons of the VRs as templates and primers specific for the relevant R-genes. Both traits, *bla*-gene and IP, together with the antigenic formula, allowed the distribution of most isolates into three MDR lineages (L1, L2 and L3; with 94, 10 and 49 isolates, respectively) (Figure 1c). It is noticeable that the incidence of bla_{OXA-30}-containing isolates is higher that of bla_{TEM-1}-containing isolates, although the latter is the most common ß-lactamase encoding gene found in European countries [7], and that the L3 lineage is the second in frequency in Asturias, preceded by L1 (DT104, which is also the most frequent in Europe) [2, 3].

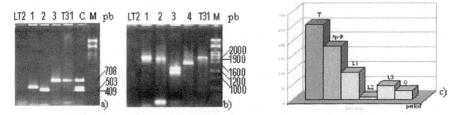

Fig. 1. Adscription of *S.* Typhimurium isolates to lineages. a) and b) Detection of *bla*-genes and class I integrons, respectively. M: ? digested with *PstI*. LT2: *bla*-InC1-negative. T31 (LSP31/93): bla_{OXA30}-InC1(2000)-positive. a) 1–3: Representative isolates leading to bla_{TEM-1}, bla_{PSE-1}-and bla_{OXA30} amplicons, respectively, C: positive control. b) 1-4: Representative isolates yielding amplicons corresponding to VR. c) Distribution of Typhimurium isolates (T=257). The 184 ampicilin resistant (Ap-R) isolates were classified in three lineages: L1, L2, L3 and O (other) attending to the *bla*-genes and class I integron content.

3   Resistance pattern and plasmid analysis of the bla_{OXA-30}- positive isolates

The 49 L3-isolates and two control strains: LT2 (the type strain of *S.* Typhimurium whose complete genome, including the virulence pSLT plasmid, has been sequenced) and LSP31/93 (a pUO-StVR2 positive isolate recovered in Asturias in 1993), were tested for plasmid-profile and R-pattern. Plasmid extraction was performed by the Kado and Liu method. All the L3 isolates and the control LSP31/93 carried a plasmid of the size expected for pUO-StVR2 (ca. 140 kb), while LT2 contained the 94 kb virulence plasmid, as expected. Apart from pUO-StVR2, one or more smaller plasmids of unknown function were detected in some of the L3 isolates (Fig. 2a). Antimicrobial susceptibility was then evaluated by a disk diffusion technique. The tested antimicrobials and concentrations in μg were ampicillin 10 (AMP), aztreonam 30 (ATM), cefotaxime 30 (CTX), ceftazidime 30 (CAZ), chloramphenicol 30 (CHL), co-amoxiclav 30 (AMC), gentamicin 10 (GEN), nalidixic acid 30 (NAL), spectimomycin 10 (SPT), streptomycin 10 (STR), sulphadiazine 300 (SU), tetracycline 30 (TET) and trimethoprim 5 (TMP). In order to determine the R-genotypes, the presence of 10 R-determinants *aadA1a*, *catA1*, *dfr1*, *dfr12*, *strA*, *strB*, *sul1*, *sul2* and *tetA(B)*, apart from those encoding for ampicillin-R, was investigated by PCR amplification. The 49 isolates showed the R-pattern expected for carriers of pUO-StVR2, with the exception of two isolates that were also trimethoprim-R (MIC>124 μg/ml). One of them was *dfr12*-positive but in the other (negative for *dfr1* and *dfr12*), the responsible gene has not yet been identified. In the following step, Southern hybridization assays confirmed that specific probes for *spvC* and bla_{OXA-30} (used as virulence and resistance markers of pUO-StVR2, respectively) mapped on the ca. 140 kb plasmid in all the L3 isolates and in the control strain LSP31/93. However, only the *spvC* probe hybridized with the pSLT virulence plasmid of LT2, as expected (Fig. 2b).

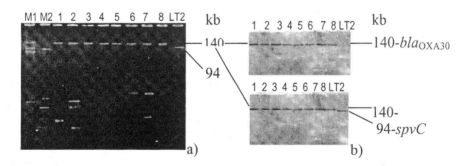

Fig. 2. Plasmid analysis of bla_{OXA-30} positive isolates. a) Different plasmids profiles obtained from representative isolates (1-8), b) Hybridizations of panel a) with bla_{OXA30} and *spvC* probes. M1 and M2: size standard plasmids from *Escherichia coli* V517 (NCTC 50193) and 39R861 (NCTC 50192).

4   Dispersion of pUO-StVR2 between different genetic types of *S.* Typhimurium

The 49 isolates assigned to L3, together with the control strains, were subjected to macrorestriction with *Xba*I endonuclease followed by PFGE. The isolates were discriminated into five *Xba*I-profiles or genomic types (X1 to X5) with LSP31/93 generating the X1 profile. This profile was the second in frequency (14 isolates), after the predominant X2 (28 isolates). In contrast, LT2 yielded a distinct profile, here termed X0. When the *Xba*I profiles were hybridized with selected plasmid probes it was found that *spvC* and *bla*$_{OXA-30}$ mapped on a fragment of ca. 35 kb, and *tetA(B)* on a different one of ca. 60 kb. Accordingly, more than two *Xba*I recognition sites seem to be present within pUO-StVR2. As expected, only *spvC* mapped on the ca. 94 kb fragment corresponding to the pSLT virulence plasmid, which is linearized by *Xba*I digestion. Similarity between the *Xba*I profiles was then evaluated by the unweighted pair group method of analysis with arithmetic averages (UPGMA), based on the Jaccard's coefficient of similarity (S). S values between the X1-X5-profiles ranged from 0.92 to 0.72. At a cut off of S=0.68 all the *Xba*I-profiles (including X0) were clustered. Thus, a high genetic similarity apparently exists among all the L3 isolates and LT2.

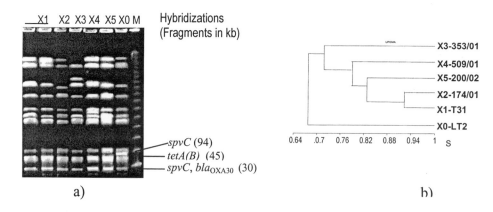

a)   b)

Fig. 3. *Xba*I-macrorrestriction-PFGE and dendogram of similarity of the pUO-StVR2 isolates. Panel a) *Xba*I-PFGE. Lane M: λ ladder PFGE Marker (New England BioLabs), X1-X5: different *Xba*I-PFGE profiles identified in this work. The fragments on which pUO-StVR2 probes mapped are indicated. b) Dendogram showing the similarity between *Xba*I profiles.

The presence of pUO-StVR2 in isolates assigned to different genomic types (S=0.72) suggests a possible horizontal transfer of the conjugative plasmid. Such transfer might have occurred in the livestock, a fact supported by the identification of the prevalent X2 profile in *S.* Typhimurium isolates recovered from minced meat in 1998 and from chicken in 2003, in our laboratory [11, unpublished]. The 5 different *Xba*I profiles detected, of which only one (X1) had been identified in 12 strains of previous years [5, 9], also suggest an evolution of the lineage. This is also supported by the different plasmid profiles and by the acquisition of at least two antimicrobial R-determinants (both conferring trimethoprim-R) in isolates carrying pUO-StVR2. The genetic background of trimethoprim-R is now under investigation, but preliminary results located *dfr12* on a second conjugative plasmid compatible with pUO-StVR2.

References

[1]   M.L. Güerri, A. Aladueña, A. Echeita, and R. Rotger, Int. J. Antimicrob. Agents. **24**, 327-333 (2004).

[2]   E. J. Threlfall, I. S. T. Fisher, C. Berghold, P. Gerner-Smidt, H. Tschäpe, M. Cormican, I. Luzzi, F. Schnieder, W. Wannet, J. Machado, and G. Edwards, Eurosurveill. **8**, 41 (2003).

[3]   E.J. Threlfall, FEMS Microb. Rev. **26**, 141-148 (2002).

[4]   D. Boyd, G.A. Peters, G, A., A. Cloekaert, K.S. Boumedine, E. Chaslus-Dancla, H. Imberechts and M.R.   Mulvey, J. Bacteriol. **183**,5725-5732 (2001).

[5]   B. Guerra, E. Junker, A. Miko, R. Helmuth, and M. C. Mendoza, Microb. Drug Res. **10**, 77 (2004).

[6]   M.A. Echeita, A. Aladueña, S. Cruchaga and M.A. Usera, J. Clin. Microbiol. **37**, 3425 (1999).

[7]   B. Guerra, S. M. Soto, J. M. Argüelles, and M. C. Mendoza, Antimicrob. Agents Chemother. **45**,1305 (2001).

[8]   B. Guerra, S. Soto, S. Cal and M.C. Mendoza, Antimicrob. Agents Chemother. **44**, 2166-2169 (2000).

[9]   B. Guerra, S. Soto, R. Helmuth and M.C. Mendoza, Antimicrob. Agents Chemother. **46**, 2977-2981 (2002).

[10]   R. Rotger and J. Casadesús, Antimicrob. Agents Chemother. **2**, 177-184 (1999).

[11]   A. Del Cerro, S.M. Soto and M.C. Mendoza , Food Microbiol. **20**, 431-438 (2003).

Analytical Techniques, Imaging Techniques, Microscopy

Biotechnology, microbiology and secondary school

J.M. Fernàndez-Novell[1], D. Cifuentes[1,2], J.C. Ferrer[1], J.J. Guinovart[1,2]

[1]Departament de Bioquímica i Biologia Molecular and [2]Institut de Recerca Biomèdica de Barcelona-Parc Científic de Barcelona,Universitat de Barcelona, Spain

Few concepts in biotechnology and microbiology are covered in Spanish secondary schools. Moreover, many secondary school teachers are not familiar with the present state of these disciplines; for instance, the first transgenic plant was created in 1983 when most of them had already finished their studies. However, it is of capital importance that these teachers can encourage their students to think about the rapid progress that biotechnology and microbiology have experienced and the significant impact that they have in our society. The avalanche of new information produced by biotechnology has many implications: from scientific, economical and health issues to those related to environmental, sociological and ethical questions. To bring biotechnology close to secondary school students and give them the opportunity to debate and gain an informed opinion about these issues, the Department of Biochemistry and Molecular Biology of the University of Barcelona designed a course specifically oriented to science teachers of secondary schools. The course, which comprised practical classes and lectures, was designed with two objectives in mind. It aimed to bring teachers up to date in relevant aspects of biotechnology and microbiology and to provide exciting issues for further discussion in secondary school courses.

1   Introduction.

The impressive progress that biotechnology and microbiology have experienced in the last two decades has economical, health, environmental, and ethical implications in our society. Just a few examples: in 1983 the first transgenic plant was created, in the mid-1990s DNA microarrays emerged as an indispensable tool to monitor the expression of genes thanks to advances in genome sequencing, and nowadays many microorganisms, particularly bacteria, are been used as detoxification agents, which promote the transformation and/or immobilization of pollutants. All this new information reaches the general public only through the daily news, which generally offer poor or biased scientific coverage.

Mass-media usually pay more attention to sensationalistic aspects. A clear example of this is the media treatment that received the sheep Dolly, the first cloned mammal. Another example, this one closer to Spanish teachers and students, is Aznalcollar ecological disaster. In that toxic spill, televisions and newspapers covered the initial flood of poisonous mining compounds to the river. However, they failed to explain the efforts taken to ameliorate the effects of the spill, like bioremediation of the river bed and adjacent contaminated soil, driven by bacteria and transgenic plants.

In Spanish secondary schools, the students aged 12 to 18 can only gain knowledge of the very few concepts in biotechnology and microbiology that are covered in biology courses. The secondary school curriculum on biotechnology is rather limited. Science teachers are confronted with the need to promote in their classes discussions on transgenic plants and their economical and environmental impact or on medical applications of biotechnology such as gene therapy, in many cases without the necessary knowledge to understand themselves these questions. Analogously, curriculum on microbiology covers only the simple description and classification of microorganisms, while crucial current questions such as the discovery and responsible utilization of antibiotics or bioremediation are left out. Teachers have to explain how microorganisms can contribute to clean a petrol spill on the sea or to incorporate heavy metals into biogeochemical cycles, again without an adequate knowledge. One of the possible causes is the inevitable delay between the scientific discovery and the incorporation of the newly acquired knowledge to secondary school text book.

In Spain, university science degrees are structured into three periods (cycles). Biotechnology is a second cycle degree, the third one corresponding to doctorate studies. Microbiology is only an optional course, not compulsory to obtain the degree, in biology, pharmacy, medicine or veterinary medicine studies. As a result, many science teachers in secondary schools are not able to acquire adequate knowledge in biotechnology and microbiology. Consequently, they are not prepared to answer the questions that their pupils pose on the, many times highly publicized, issues that arise from the current applications and the practical use of these disciplines.

In order to update microbiology and biotechnology knowledge of secondary school science teachers, the Department of Biochemistry and Molecular Biology of the University of Barcelona [1,2] designed the course "Avenços en bioquímica i biotecnologia" (Advances in biochemistry and biotechnology), which is included in the official training program of secondary school teachers of the Catalan Government (Col·legi Oficial de Doctors i Llicenciats en Filosofia i Lletres i en Ciències de Catalunya, Generalitat de Catalunya). The Catalan Board of Teachers sponsored partially the updating course.

The course, which was specifically addressed to science secondary school teachers [3], consisted of lectures and practical sessions. It had a duration of 5 days (33 hours), beginning every day at 8.30 h and finishing at 15 h, and there were 24 places available (Table 1).

The laboratory sessions (Table 1), which were explained and monitored by young researchers coursing their PhD. degrees, introduced the alumni to new techniques in biochemistry and promoted discussion on relevant related issues.

Table 1.- Schedule

	Monday	Tuesday	Wednesday	Thursday	Friday
8:30 – 10.00	Presentation Lab safety	Lecture	Lecture	Lecture	Lecture
10.00 – 11.30	Recombinant DNA	Protein expression	SDS-PAGE	Cell culture	Internet
11.30- 12.00	Coffee break	Coffee break	Coffee break	Coffee break	Coffee break
12.00 – 13.30	Recombinant DNA	Protein expression	Fermentor	Cell culture	Discussion
13.30 – 15.00	Lecture	Lecture	Lecture	Lecture	

Practical sessions. In order to optimize the available space and equipment, alumni were divided in groups. The practical exercises were performed in groups of six students [4] conducted by two PhD students, who shared the teaching responsibility. At the beginning of every session the students were handed out a protocol that included a full description of the exercise and its scientific fundaments.

Recombinant protein expression, purification and detection:

In these set of experiments the alumni developed skills in DNA manipulation, bacterial culture and transformation and recombinant expression of proteins.

First, participants began with the characterization of the pGFPCR plasmid (Fig. 1), which contains a gene coding for the green fluorescence protein (GFP). They where introduced to DNA digestion with restriction enzymes and obtained a restriction map of the plasmid (Fig. 2, left pannel).

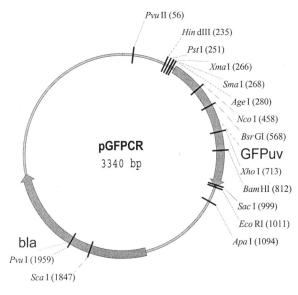

Fig.1. pGFPCR vector map, showing unique restriction sites.

The plasmid was transformed in *E.coli* DH5α by the heat shock method.

The transformed cells were plated in agar plates containing the appropriate antibiotic (ampicillin) and incubated overnight at 37 °C to allow bacterial growth and colony formation.

pGFPCR positive colonies were easily detected as those emitting green light, by visual inspection of agar plates in a UV transiluminator. This is a very simple procedure but represents a clear and illustrative evidence for the alumni that they succeeded in expressing a recombinant protein.

Positive colonies were selected to inoculate liquid media, which was incubated overnight at 37 °C in an orbital shaker. Cells from these liquid cultures were collected by centrifugation, disrupted by sonication and loaded in a SDS-PAGE. After Comassie blue staining, the E. coli proteome was visualized and pupils discussed the results. By comparison with a control sample of bacteria carrying an empty vector, they were able to identify GFP (Fig. 2).

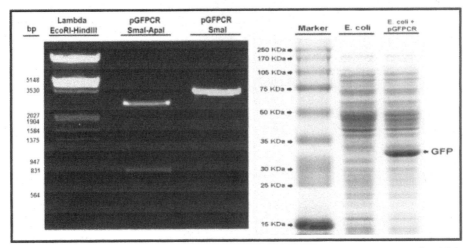

Fig 2. pGFPCR restriction map (left panel) and SDS-PAGE showing GFP overexpression (right panel)

Large scale bacterial growth:

Secondary school teachers' vision of bacterial growth usually is associated to disease, food waste and dirt. But they often forget that bacteria are an indispensable part of intestinal micro biota and protagonists of several processes, such as bioremediation or food and brewage production.

In an attempt to correct this situation, another practice consists of the use of a bioreactor to grow a bacterial culture under controlled conditions. With the use of a fermentor the alumni monitor important parameters of bacterial growth, such as pH, temperature, stirring, foaming, and aireation. The different phases of a growth curve are also studied by means of absorbance measures.

Cell culture:

The course includes a short practice with mammalian cell cultures just to show another option to produce proteins of pharmacological interest that due to post-traductional modifications cannot be expressed in bacteria. The alumni are introduced to basic manipulations of eukaryotic cell lines, which include trypsinization of cells and measurement of viability.

Bioinformatics:

The burst of scientific knowledge in recent years has lead to a flood of biochemical data (DNA and protein sequences, full sequenced genomes, myriads of articles in thousands of different journals). In a practical session in the computer room, students are introduced to the most used scientific databases and web resources (Medline, pubmed, genbank, expasy, protein databank).They are asked to find bacterial homologs of an eukaryotic protein using some basic bioinformatics tools like Blast.

The lectures. They consist of a 60-70 minute presentation by the lecturer followed by 10-20 minutes of questions and discussion at the end. The presentations are interactive and encourage the active participation of the alumni.

These sessions are held at the beginning and the end of the day and are delivered by professors of the University of Barcelona and by invited speakers from other universities. The topics covered are shown in Table 2.

Table 2. Lectures: topics and points of interest.

Topics	Points of interest
New technology	- Nucleic acid structure. The DNA book of life - PCR applications - Diagnosis and DNA chips
Biotechnology	- Molecular bases of anti-microbial action - The use of the large fermentor - Bread, wine, beer and biochemistry
Plants biochemistry	- Transgenic plants, present and future - Can plants produce antibodies?
Gene therapy	- Tools in genetic engineering - Transgenic animals

Course feedback. The feedback that we have received for the alumni has been very positive. Most of them expressed that they had gained useful knowledge, especially in biotechnology, genetic tools and microbiology. Participants explained that the lectures were of great help as many of the topics discussed could be addressed in their own classrooms, such as transgenic food, genes and clones and bioremediation. They also reported that the practical work was very useful to clarify the lecture issues. Finally, they also expressed that the coursed had provided them with adequate tools to address with greater confidence aspects of biotechnology and microbiology that frequently appear in the news.

Acknowledgments

We thank Anna Adrover for her skilful assistance in preparing laboratory materials.

References

[1]   M. Martínez, B. Gros, T. Romaña. The problem of training in Higher Education. Higher Education in Europe, vol XXIII, n. 4, 483-495. (1998)

[2]   Fernández-Novell, J.M., Cid, E., Gomis, R., Barberà, A. & Guinovart, J.J. A Biochemistry and Molecular Biology Course for Secondary School Teachers. BAMBED **32**, No 6, 378-380 (2004)

[3]   A. Corda, T. Ruzzon, S. Lercari, S. Ucelli. The role of scientific institutions in the promotion of Biotechnology to the public (school, the mass-media, entrepreneurs etc). Biochemical Education **26** 52-55. (1998)

[4]   R.G. Dennick, K. Exley. Teaching and learning in groups and teams. Biochemical Education **26** 111-115 (1998)

Ciliates as potential biosensors for heavy metal pollution

J.C. Gutiérrez*, S. Díaz, A. Martín-González, R. Ortega, A. Gallego, F. Amaro, V. Campos and M.P. Plana

Departamento de Microbiología-III. Facultad de Biología. C/. José Antonio Novais, 2. Universidad Complutense (UCM). 28040 Madrid. SPAIN.

*Corresponding author: e-mail: juancar@bio.ucm.es, Phone: + 34 91 3944968, Fax: +34 91 3944964

Ciliates can be valuable eukaryotic microorganisms for use as whole-cell biosensors or as a potential cellular source of molecular biomarkers/ biosensors to detect pollutants (such as heavy metals) in environmental samples. Here, we report the advantages of using ciliated protozoa in biomonitoring of heavy metals, in comparison with other micro-organisms. We report several examples of the suitability of ciliates as potential whole-cell or molecular biosensors to detect bioavailable heavy metals in environmental samples.

Keywords: ciliates; heavy metals; biosensors.

1 Introduction

The classical biosensor concept involves the existence of two components; a bioreceptor (biological material) and a physic-chemical transducer. The bioreceptor might be a biomolecule or a whole cell that recognizes the target (heavy metal), whereas the transducer converts the recognition event into a measurable signal.

Recently, the concept of the whole-cell biosensor has been introduced by several authors [1, 2], as a very useful alternative to classical biosensors. A whole-cell biosensor uses the whole prokaryotic or eukaryotic cell as a single reporter incorporating both the bioreceptor and transducer elements. In general, living systems to be use as whole-cell biosensors are experimentally modified to incorporate the transducer capacity.

In environmental biomonitoring, global parameters such as bioavailability, toxicity and genotoxicity can not be tested using molecular recognition or chemical analysis, but can only be assayed using whole cells. Obviously, in these bioassays the question to be resolved is not "What toxicants does the sample contain ? ", but rather "How toxic is the sample ?". Two types of bioassays using whole-cell biosensors may be considered: "turn off" and "turn on" assays. In "turn off" assays, the sample toxicity is estimated from the degree of inhibition of a cellular activity that is normally continuous (e.g. inhibition of growth, respiration or metabolism, motility or the biosynthesis of a specific molecule), and is based on the measurement of a decrease in growth rate, light (fluorescence / bioluminescence) emission, color-less cell population, motility, etc. as a function of sample toxic concentration. For instance, a good example of a "turn off" assay using ciliates, may be a reported rapid bioassay to detect mycotoxins using a melanin precursor overproducer mutant of the ciliate *Tetrahymena thermophila* [3]. On the other hand, in "turn on" assays a quantifiable molecular reporter is fused to a specific gene reporter (like a metallothionein promoter) (see Fig. 1), known to be activated by the target chemical or environmental pollutants (such as a heavy metal).

2 Advantages of using ciliates as whole-cell biosensors

Ciliates are eukaryotic microorganisms, therefore they have all the advantages showed by microorganisms (e.g. a rapid growth rate, easy manipulation, specific mutants with several gene reporters or the possibility to use genetically engineered microorganisms), but, furthermore, they present at least two additional advantages:

a)- In contrast to bacteria and yeasts, ciliates are unicellular organisms without a cell wall in the vegetative stage. A major limitation to the use of bacteria or yeasts as whole-cell biosensors is the uncertainty concerning diffusion of substrates and products or pollutants through the cell wall, which may result in a slower or less effective response. The absence of a cell wall in these eukaryotic microorganisms might offer a higher sensitivity to environmental pollutants, and, therefore, a faster cellular response.

b)- They are eukaryotic microorganisms with some metabolic traits that are more similar to those of human cells than that are bacteria or yeasts. Genome projects in two model ciliates often used in ecotoxicological studies, *Tetrahymena thermophila* and *Paramecium* [4, 5] have shown that they share a higher degree of functional conservation with human genes than do other microbial model eukaryotic microorganisms. This is shown by better matches of relevant ciliate coding sequences to those in humans, as compared with non-ciliate microbial models. Therefore, it seems to be more reasonable to use these eukaryotic cells in ecotoxicological studies as models for humans, and represents an alternative to animal tests.

3 Features that make ciliates suitable as whole-cell biosensors for heavy metal monitoring

3.1 Heavy metal bioaccumulation

Bioaccumulation is the most common heavy metal resistance mechanism among ciliates [6, 7]. Such metallic bioaccumulation can be revealed by fluorescence microscopy. In 2003, we reported for the first time in ciliates [8] the use of specific heavy metal fluorophores to distinguish ciliates exposed to sub-lethal metallic concentrations from controls. Two specific fluorophores have been applied to diverse ciliates after Zn or Cd treatment; TSQ [N-(6-metaoxy-8-quinolyl-p-toluenesulfonamide)] which is selective for Zn^{2+} in the presence of physiological concentrations of Ca^{2+} and Mg^{2+} ions, and bis-BTC (tetraammonium salt) which binds Cd^{2+}/Cd^0. Results from fluorescence microscopy suggest that this method is only useful to locate cytoplasmic metallic deposits when cells are exposed to high heavy metal concentrations. The fluorescence of cell populations treated with heavy metals might be measured by flow cytometry, so obtaining a quantification of heavy metal bioaccumulation at cell population level. This methodology might be a useful tool to detect heavy metals in urban wastewater treatment plants using ciliates as whole-cell biosensors.

3.2 Fluorescent detection of ROS (Reactive Oxygen Species) generated in ciliates by heavy metals.

Heavy metals may induce (directly or indirectly) oxidative stress in both prokaryotic and eukaryotic cells. Heavy metals with redox activity (Cu, among others) can directly give rise to ROS by Fenton/Haber-Weiss reactions or auto-oxidation. Besides, heavy metals without redox activity (like Cd or Zn) can also indirectly generate oxidative stress by blocking or decreasing cellular antioxidant defenses. These defenses may be enzymatic (antioxidant enzymes, such as; glutathione peroxidase, catalase or superoxide dismutase) or non-enzymatic (glutathione and metallothioneins, which have a protective effect against oxidative stress). The formation of hydrogen peroxide (H_2O_2) and superoxide anions as a result of the action of heavy metals has been analyzed in several ciliates by using three different fluorophores ϱ', 7′-dichlorofluorescein diacetate, dihydrorhodamine 123 and dihydroethidium [8]. Both fluorescence microscopy with quantitative image analysis on fixed cells and flow cytometry using living cell populations, have been shown to be useful tools to distinguish between controls and cells treated with heavy metals.

4 Ciliate metallothioneins: candidates for use as molecular biosensors and/or as specific gene reporters for whole-cell biosensors

Metallothioneins (MTs) is a common name for a superfamily of ubiquitous low molecular weight (7-10 Kda), cysteine-rich proteins that have the capacity for high affinity binding of heavy metal ions (Cd, Zn or Cu, mainly). Fifteen MT families have been characterized (http:// www.unizh.ch/~mtpage/MT.html). Ciliate MTs are included in family 7, and, at present, at least three different isoforms of CdMTs (Cd- inducible MT) and other three different isoforms of CuMTs (Cu- inducible MT) have been reported in different ciliate species. Recently, we have isolated the third CdMT isoform (MTT3) from distantly related ciliates (including *Tetrahymena thermophila*), which shows a high degree of sequence conservation with other previously reported ciliate CdMT isoform (MTT1). MTT3 and MTT1 CdMTs present the same length (162 amino acids), being the most large CdMTs until now reported. As a consequence of their unusual length these ciliate CdMTs have a considerably higher molecular weight ; about 11Kda for MT-1 (*T. pyriformis*) and 17.3 Kda for MTT3 and 16 Kda for MTT1(both from *T. thermophila*). Ciliate CuMTs have about > 10 – 11 Kda. Therefore, both ciliate Cd / CuMTs present molecular weighs considerably higher with regard standard MTs (7 – 10 Kda). As it is known, MTs are cysteine-rich proteins and all Cys residues are involved in the metal-binding capacity of these proteins. The highest number of Cys residues correspond to MTT1 isoform (48 residues), while MTT3 has 41 Cys residues. The MT-1 isoform (with 31 residues) has a similar number of Cys residues than CuMT isoforms (28 or 32 residues) (data from *Tetrahymena* species). On the other hand, in general, standard MTs only have about 18–23 Cys residues.

The CCC motif is characteristic and almost exclusive of ciliate CdMTs, excepting for one CdMT of the annelid *Eisenia fetida* (with only one CCC motif) (P81695) and two CuMTs, one from the arthropod *Callinectes sapidus* (AAF08966) (with two CCC motifs) and another one from the yeast *Yarrowia (Candida) lipolytica* (P41928) (with only one CCC motif).

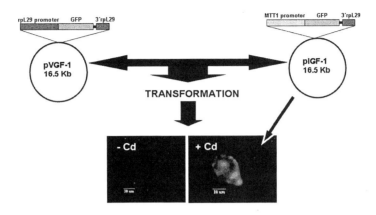

Fig. 1 Schematic representation of a possible "turn on" bioassay using a ciliate whole-cell biosensor. From plasmid pVGF-1 (16.5 Kb) (which presents the gene fusion between the ribosomal protein L29 promoter and the quantifiable reporter gene – green fluorescent protein- GFP), the plasmid pIGF-1 was obtained (in which the rpL29 promoter was substituted by the MTT1 promoter- a MT isoform from the ciliate *Tetrahymena thermophila*). By cell transformation (electroporation) this last plasmid can be introduced into the cell. So, the positive transformed cells might be used as whole-cell biosensors, which might over-express the GFP under Cd (or other heavy metal) exposure. Other gene constructions might be carried out (e.g. the promoter of MTT3 isoform also could be used).

In standard MTs all Cys residues are known to participate in the metal binding, so they incorporate up to 7 divalent metal (Cd^{2+} or Zn^{2+}) or 12 monovalent Cu atoms per molecule. The seven atoms of bound Cd are arranged in two separate polynuclear metal clusters, one containing three (β-domain) and the other four metal ions (α-domain), which satisfy the $Metal_7$ $(Cys)_{20}$ stoichiometry for MTs. If this stoichiometry is maintained for ciliate CdMTs, and we assume that all Cys residues are involved in metal binding, we might conclude that these more-rich Cys MTs present a considerable higher metal chelating capacity with regard standard MTs. Therefore, the ciliate MT-1 isoform, with 31 Cys residues, had a theoretical chelating capacity of about 11 Cd^{2+} ions. This theoretical data has been corroborated by an experimental result obtained in MT-1 of *T. pigmentosa* [9], which showed that this MT binds 12 Cd^{2+} / mol of protein. The theoretical binding capacity for MTT3 and MTT1 isoforms is about 14 and 17 Cd^{2+}, respectively. With regard ciliate CuMT isoforms, which are more similar to standard MTs, and according to the CuMT stoichiometry [$Metal_{12}$ $(Cys)_{20}$], although Cu^+ binds in multiple stoichiometries with a minimum of 7 Cu^+ / mol, the theoretical chelating capacity of them, is likewise higher than that for standard MTs. The smallest *Tetrahymena* CuMT isoform (containing 28 Cys residues) might bind up to 17 Cu^+ ions, and the largest one (with 32 Cys residues) about 19 Cu^+ ions. Therefore, we can conclude that both ciliate Cd / CuMTs isoforms present a higher theoretical metal binding capacity with regard standard MTs. This point is important to consider these proteins as a potential source of molecular biosensors for detecting bioavailable heavy metals in environmental samples. The promoter regions of Cd and CuMT isoforms of *Tetrahymena* have numerous motifs for binding transcriptional activator proteins, which corroborates the fast (CdMTs are over-expressed within a few minutes under heavy metal exposure) and intense expression of these genes under the presence of heavy metal.

All these properties make possible that these proteins can be very good candidates to be used as the biological element (bioreceptor) in the design of molecular biosensors. Probably, ciliate MTs are better than standard MTs to be used as molecular biosensors, because they have a higher molecular size, they have a considerably higher theoretical Cd-binding capacity and their biosynthesis is induced very fast with a high expression level.

Ciliate CdMTs can offer us many interesting possibilities for heavy metal detection in polluted environmental samples, for instance, the promoter of *T. thermophila* MTT1 isoform has proved to be a robust inducible-repressible promoter which facilities gene knockouts, conditional expression and over-expression of both homologous and heterologous genes [10]. This promoter might be used in "turn on" assays using whole-cell biosensors, after fusion to a quantifiable reporter gene (like green fluorescent protein) (Fig. 1). Ciliate CdMT fragments might be used as bioreceptors in the design of classical biosensors, after studying the complexation of the immobilized oligopeptide with heavy metal ions. Likewise, gene fragments of the ciliate CdMTs could be used in the construction of macro- or microarrays, so the fast expression of these genes under heavy metal exposure might, in the next future, provide a good molecular tool to detect the presence of heavy metals in polluted environmental samples.

Acknowledgements

Research related with several topics reported in this proceeding are supported by grant projects I+D (BOS2002-0167) and 07M/0029/2002 (CAM).

References

[1]   S.F. D´Souza, Biosensors Bioelectron. **16**, 337 (2003).
[2]   S. Belkin, Curr. Opin. Microbiol. **6**, 206 (2003).
[3]   A. Martín-González, L. Benítez, T. Soto, J. Rodríguez de Lecea, and J.C. Gutiérrez, Cell Biol. Intern. **21**, 213 (1997).
[4]   P. Dessen, M. Zaguiski, R. Gromadka, H. Plattner, R. Kissmehl, E. Meyer, M. Bétermier J.E. Schultz, J.U. Linder, R.E. Pearlman, C. Kung, J. Forney, B.H. Satir, J.L. Van Horten, A-M. Keller, M. Froissad, L. Sperling, and J. Cohen, Trends Genet. **17**, 306 (2001).
[5]   A. Turkewitz, E. Orias, and G. Kapler. Trends Genet. **18**, 35 (2002).
[6]   A. Martín-González, S. Díaz, C. Jareño, and J.C. Gutiérrez, in: Recent research developments in microbiology (Research Signpost, India), p. 93.
[7]   A. Martín-González, S. Borniquel, S. Díaz, R. Ortega, and J.C. Gutiérrez, Cell Biol. Intern. (in press) (2005).
[8]   J.C. Gutiérrez, A. Martín-González, S. Díaz, and R. Ortega, Europ. J. Protistol. **39**, 461 (2003).
[9]   F. Boldrin, G. Santolito, E. Negrisolo, and E. Piccinni, Protist. **154**, 431 (2003).
[10]   Y. Shang, X. Song, J. Bowen, R. Corstanje, Y. Gao, J. Gaertig, and M.A. Gorovsky, Proc. Natl. Acad. Sci. USA, **97**, 3734 (2002).

Concentration, detection and identification of Infectious Enteroviruses in Sewage

D. Papaventsis[*,1,4], **N. Siafakas**[1,2], **P. Markoulatos**[2], **V. Sopidou**[3], **C. Economou**[3] and **S. Leveidiotou**[4]

[1] Department of Virology, National Reference Enteroviruses Centre, Hellenic Pasteur Institute, 127 Vassilisis Sofias Ave, Athens 11521, GREECE
[2] Microbiology-Virology Laboratory, Department of Biochemistry & Biotechnology, University of Thessaly, Ploutonos 26 & Aeolou str., Larissa 41221, GREECE
[3] General Department of Essential Medical Modules, Technological Educational Institution, Ag. Spiridonos & Dimitsanas, Aigaleo 12243, GREECE
[4] Department of Microbiology, Medical School, University of Ioannina, Panepistimioupoli Ioanninon, Ioannina 45332, GREECE

[*] Corresponding author. Mailing address: 37, Apostolopoulou street, Pefki - Athens 15121, Greece. e-mail: dpapaventsis@yahoo.gr, Phone: +30 210 6126876, Fax: +30 210 8324135.

This work presents a new approach for the molecular 'serotyping' of enteroviruses concentrated and isolated from sewage. Viruses were adsorbed to cellulose nitrate membrane filters, isolated from the membrane filter using the VIRADEN method and identified by RT-PCR, followed by 5?-UTR RFLP analysis and partial sequencing of the VP1 protein coding region. All 42 isolates belonged to the same genetic sub-cluster of non-polio enteroviruses. Partial VP1 sequencing revealed that isolates belonged to serotypes CBV4 (33 isolates, 79%), CBV1 (8 isolates, 19%) and Echovirus7 (1 isolate, 2%). The methodology presented here is highly promising for environmental surveillance of enterovirus circulation and epidemiology. 5?-UTR RFLP analysis with HpaII is useful for initial sub-classification, while partial VP1 sequencing is efficient for molecular 'serotyping'.

Keywords Environmental Microbiology, Municipal Wastewater Treatment, Enterovirus, Coxsackievirus, Polymerase Chain Reaction (PCR)

1   Introduction

Many human viruses may infect the gastrointestinal tract and can be excreted in the feces into the environment. Once in the environment, viruses can reach water supplies, recreational waters, crops, and shellfish through sewage, land runoff, solid waste landfills, and septic tanks. Enteroviruses are the viruses most often detected in polluted water. This is partly due to the fact that they can easily grow in conventional cell lines [1]. Human enteroviruses (family *Picornaviridae*) consist of 65 immunologically distinct human serotypes dispersed in five *Human Enterovirus Species* (*A-D* and *Poliovirus*). Serotypes included are Polioviruses (PV, 3 serotypes), Coxsackie A viruses (CAV, 23 serotypes), Coxsackie B viruses (CBV, 6 serotypes), Echoviruses (ECV, 28 serotypes) and enteroviruses (ENV) 68-71 and 73 [2]. Infections with the members of this group of viruses, even the most pathogenic ones, are characterized by a significant percentage of sub-clinical manifestations.

The examination of stool samples from patients identified through acute flaccid paralysis (AFP) surveillance, links enterovirus and in particular poliovirus isolates to specific individuals. However, the examination of composite human faecal samples links enterovirus isolates from unknown individuals to populations served by the wastewater system. Environmental surveillance is essential for monitoring and assessing the extent or duration of epidemic enterovirus circulation in human populations. It can provide valuable supplementary information, particularly in urban populations where AFP surveillance is absent or questionable, persistent virus circulation is suspected, or frequent virus re-introduction is perceived [3]. This work presents a new approach for the molecular 'serotyping' of enteroviruses concentrated and isolated from sewage.

2   Materials and methods

2.1 Materials

Two raw sewage samples were studied. Samples were collected during the peak morning flow from the Nicosia sewage treatment plant in Cyprus. The Nicosia sewage treatment plant is a waste stabilization pond plant that receives about 13,000 m3 of sewage effluents per day.

2.2 Methods

Viruses present in the raw sewage samples were first adsorbed to specific cellulose nitrate membrane filters (0.22-μm-pore-size, syringe-driven hydrophilic Polyether Sulphonate (PES) filter units; SLGP R25 LS, Millipore Corp.), by using the VIRADEN method as previously described [4, 5]. Viruses adsorbed on the membrane filter, were detected and counted on a BGM cell monolayer, based on their cytopathogenic effect. Viral RNA extraction was carried out by the method described by Casas et al. [6] and it was applied directly to the VIRADEN final products. Isolated RNA was converted to cDNA by reverse transcription.

The cDNA produced was amplified by Polymerase Chain Reaction (PCR). The antisense primer UC53 (5'-TTGTCACCATAACCAGCCA -3'; positions 583 to 601 in the genome of CAV9 reference strain Griggs) and the sense primer UG52 (5'-CAAGCACTTCTGTTTCCCCGG-3'; positions 167 to 187 in the genome of CAV9 reference strain Griggs) were selected to be homologous to the corresponding parts in the highly conserved 5-UTR enterovirus genomic region [7]. These primers yielded amplicons that were 435 bp long. In addition, the sense primer 292 (5'-MIGCIGYIGARACNGG-3'; positions 2612 to 2627 in the genome of PV1 reference strain Mahoney) and the antisense primer 222 (5'-CICCIGGIGGIAYRWACAT-3'; positions 2969 to 2951 in the genome of PV1 reference strain Mahoney) were used to amplify a portion of the gene encoding VP1 capsid protein [8]. These primers yielded amplicons that were approximately 340 bp long.

An initial Restriction fragment length polymorphism analysis (RFLP) of the UC53/UG52-produced RT-PCR amplicons was carried out as described previously [9]. We used the restriction enzyme HpaII (New England Biolabs, Beverly, Mass.) The results were analyzed by using the GelPro Analyzer software (Media Cybernetics, Silver Spring, Md.). Moreover, since the Viral Protein 1 (VP1) gene contains important serotype-specific epitopes, the partial VP1 sequences of the environmental isolates were also obtained for serotypic identification. We utilized the model proposed by Oberste et al. [8]. The serotypic identity of each isolate was deduced by comparison of the partial VP1 sequences with the sequences of the corresponding genomic regions of all human enteroviruses which are available in the GenBank database (BLAST, version 2.2.8). A phylogenetic analysis was carried out by pairwise comparison of the partial VP1 sequences of our isolates and the corresponding sequences of reference and wild-type human enterovirus strains of the same serotype, which are available from GenBank (www.ebi.ac.uk/ClustalW). A dendrogram was plotted in the PHYLIP format output by using the TreeView software (version 3.0).

3   Results

In total, 42 enteroviruses were isolated (results are shown in Table 1). Initial sub-grouping based on HpaII restriction profile showed that all isolates belonged to the same genetic sub-cluster of non-polio enteroviruses. Partial VP1 sequencing revealed that isolates belonged to serotypes CBV4 (33 isolates, 79%), CBV1 (8 isolates, 19%) and Echovirus7 (1 isolate, 2%).

Table 1 Total number of isolates, RFLP sub-cluster classification, partial VP1 serotype and sequence identity results. For CBV4, CBV1 and ECV7 abbreviations please refer to the text.

Year	Isolates number	HpaII cluster	Partial VP1 Serotype	No	%	Sequence identity
2004	42	I	CBV4	33	79	93-96%
			CBV1	8	19	82-95%
			ECV7	1	2	82%

It is worth to mention that all isolates of the same serotype had highly similar VP1 sequences (range 82-96%). Moreover, the HpaII digests predicted the genetic sub-cluster in all cases. Finally, phylogenetic analysis revealed that our isolates correlate with strains of the same serotype isolated during the last four decades in Europe, Asia, Northern Africa and the Middle East, implying possible epidemiological relationship. In particular, CBV4 environmental isolates showed the greatest levels of VP1 alignment with other enterovirus strains isolated elsewhere in Europe, in Asia, in northern Africa, and in the Middle East (Fig. 1), implying a possible epidemiological relationship of all these isolates [10].

734

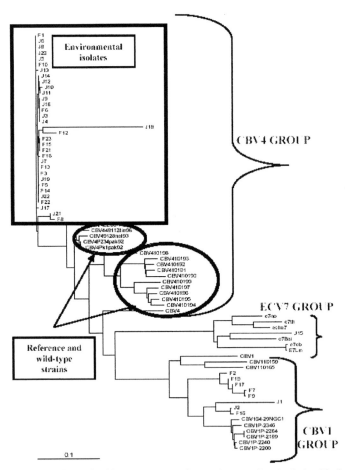

Fig. 1 Dendrogram showing the relationships among enterovirus environmental strains isolated in the present study and reference and other clinical strains.

CBV4: Coxsackie virus ? serotype 4
CBV1: Coxsackie virus ? serotype 1
ECV7: Echovirus serotype 7
J1-J22, F1 -F23: environmental isolates
Accession numbers of reference and wild-type human enterovirus strains used for the phylogenetic analysis:
JVB:X05690; P234pak92:AF160018; Pk1pak92:AF160019; BE00-117:AF521311
48112fin98:AF160025; 9128net93:AF160021
10199:AY373132; 10198:AY373131
10197:AY373130; 10196:AY373129
10195:AY373128; 10194:AY373127
10193:AY373126; 10192:AY373125
10191:AY373124; 10190:AY373123
e7no:AF295
462; e7th:AY342733
echo7:AY036578; e7Bai:E7241447
e7ob:AF081324; E7Lin:AF465516
JAPAN: M16560;
P-2264/CB1/Kanagawa/2003: AB162736;
P-2199/CB1/Kanagawa/2003: AB162733;
P-2240/CB1/Kanagawa/2003: AB162735;
04-29NGC1: AB167989;
P-2200/CB1/Kanagawa/2003: AB162734;
P-2346/CB1/Kanagawa/2003: AB162737;
10159: AY373092; 10165: AY373098

4   Discussion

An effective surveillance scheme is a prerequisite for environmental epidemiological investigation of enterovirus circulation. In the present study we attempted to contribute to this goal by developing a simple and effective method for detection and identification of infectious enteroviruses in sewage by utilizing the VIRADEN method (cellulose nitrate membrane adsorption and direct cell culture from the membrane) and RT-PCR. In our procedure, no special equipment or decontamination procedures were necessary, and all isolates from environmental specimens could be identified within a few days from sampling.

Partial sequencing of the VP1-encoding gene gave excellent results regarding serotyping of the isolates. This certified the effectiveness of the specific serotyping system and also showed that the VIRADEN obtained cultures were pure and each contained only one virus clone.

An interesting finding of the present study was that our environmental isolates, and in particular CBV4 isolates, showed the highest levels of VP1 alignment with other enterovirus strains of the same serotype isolated elsewhere in Europe, in North America, in Asia Manor [10], in northern Africa, and in the Middle East [11], implying the possible epidemiological relationships of all these isolates. Cyprus is located in the middle between Europe, Asia and Africa, and travelling from one region to the other could be the way of viral importation to the island.

In conclusion, virus concentration by cellulose nitrate membrane adsorption and cell culture (VIRADEN), followed by detection with RT-PCR and serotyping by partial VP1 sequencing, seems to be a very promising approach for environmental surveillance of enterovirus circulation and epidemiology. RFLP analysis of the 5-UTR genomic region can also be used for the initial sub-classification of enterovirus isolates. The application in a permanent basis of such an environmental surveillance system for enteroviruses would have great impact for the protection of Public Health.

Nucleotide Sequence Accession Numbers

The sequences reported in this study were deposited in the GenBank sequence database under accession numbers **AY634232 - AY634273**.

Acknowledgements

Dr **G.T. Papageorgiou**, Head of the Water Microbiology – Virology Laboratory, State General Laboratory, Nicosia, Cyprus, is gratefully acknowledged for providing final VIRADEN products.

This study was co-funded by 75% from the European Union and 25% from the Greek Government under the framework of the Education and Initial Vocational Training Program–'Archimedes'.

References

[1]   R. Maier, I. Pepper, and C. Gerba (eds.), Environmental Microbiology, (Academic Press, London, 2000), p.472-484.

736

[2]   Anonymous, Guidelines for environmental surveillance of poliovirus circulation, (World Health Organization Department of Vaccines and Biologicals, Geneva, Switzerland, 2003), p.1.

[3]   B.N. Fields, D.M. Knipe, P.M. Howley, R.M. Chanock, J.L. Melnick, T.P. Monath, B. Roizman, and S.E. Straus (eds.), *Virology*, (Lippincott-Raven, Philadelphia, 1996), p.655-712.

[4]   G.T. Papageorgiou, L. Moce-Llivina, C.G. Christodoulou, F. Lucena, D. Akkelidou, E. Ioannou, and J. Jofre, Appl. Environ. Microbiol. **66**, p.194-198 (2000).

[5]   L. Mocé-Llivina, J. Jofre, and M. Muniesa, J. Virol. Methods **109**, p.99-101 (2003).

[6]   I. Casas, L. Powell, P.E. Klapper, and G.M. Cleator, J. Virol. Methods **53**, p. 25-36 (1995).

[7]   N. Siafakas, A. Georgopoulou, P. Markoulatos, N. Spyrou, and G. Stanway, J. Clin. Lab. Anal. **15**, p. 87–93 (2001).

[8]   M.S. Oberste, W.A. Nix, K. Maher, and M.A. Pallansch, J. Clin. Virol. **26**, p.375-377 (2003).

[9]   N. Siafakas, P. Markoulatos, C. Vlachos, G. Stanway, G. Tzanakaki, and J. Kourea-Kremastinou, Mol. Cel. Probes. **17**, p. 113-123 (2003).

[10]   M.N. Mulders, M. Salminen, N. Kalkkinen, and T. Hovi, J. Gen. Virol. 81, p.803-812 (2001).

[11]   C.F. Yang, T. Naguib, S.J. Yang, E. Nasr, J. Jorba, N. Ahmed, R. Campagnoli, H. Van Der Avoort, H. Shimizu, T. Yoneyama, T. Miyamura, M. Pallansch, and O. Kew, J. Virol. **77**, p.8366-8377 (2003).

Effect of inoculum on the survival of pathogenic agents in different composts

D. Zarina[1], L. Dubova[1], A.Berzins[1], U.Viesturs[*1,2], and S.Strikauska[3]

[1] Institute of Microbiology and Biotechnology, University of Latvia, Kronvalda Boulv., LV-1586 Riga, LATVIA

[2] Institute of Wood Chemistry, 27 Dzerbenes Str., LV-1006 Riga, LATVIA

[3] Latvian University of Agriculture, 2 Liela Str., LV-3001 Jelgava, LATVIA

* Corresponding author: e-mail: lumbi@lanet.lv, koks@edi.lv, Phone: +371 7034884, +371 7553063, Mob. phone: 9284923, Fax: +371-7550635, +371 7034885

The subject of this study was to test the possibilities of dissemination of obligatory or facultative pathogens through composts, when the composting process is speeded up by the addition of an inoculum supplement. Composting experiments were conducted to determine suitability, when the composting process was undertaken in different seasons, namely, in early spring, late autumn and early winter in different equipment: in a Bin and under field conditions in a windrow [1-4]. The volume of the composting Bin (Biolan Juwel BIO) was 600 L, and the size of the windrow was 3 x 3 x 12 m. The composting materials were the sewage sludge from the wastewater treatment plant "Daugavgriva" (Riga) or the municipal biodegradable organic solid waste and sawdust. The compost quality was assessed by chemical and microbiological analyses, seeds germination tests and Toxkit microbiotests. In the compost, the number of bacteria, fungi, *Salmonella* sp. and *Staphylococcus aureus* was estimated.

Keywords: composting; inoculum; sewage sludge; municipal solid waste (MSW); survival of pathogens

1   Introduction

Composting is a very old biological process, converting organic matter in the presence of suitable amounts of air and moisture into a humus-like product, in which the microbial activity is essential for the decomposition and bio-oxidation of organic materials.

 Solid-state bioconversion (SSB) has been usually employed, with more rational use of the selected microbial strains together with appropriate parameter settings, to promote growth and bioconversion. The basic principles of composting and SSB are similar, however, in recent years, researchers have focused their attention on applying the SSB approach to optimize the composting process for the treatment of wastewater sludge [5]. The use of a composting bioreactor or windrows can be potentially connected with the utilization of organic waste [6]. The quality of compost is related to its agronomic and commercial value as an organic solid conditioner. The compost, used as a soil amendment, should not contain great amounts of such methabolic compounds or microorganisms that may be harmful for soil. It should be stable enough, without phytotoxic compounds. Low-quality compost can enrich soil and water with harmful compounds.

2   Aim

The aim of this study was to overlook the beginning of composting processes in different seasons by inoculum supplement and in different equipment, and the survival of pathogenic microorganisms in different composts. The role of inoculum was also investigated.

3   Material and methods

3.1 Composting

Composting processes were realized in a composting Bin (Biolan Juwel BIO) and a windrow [1-4]. The volume of the composting Bin was 600 L, and the sizes of the windrow were 3 x 3 x 12 m. Experiments were started in different seasons, namely, in early spring and summer, late autumn and early winter.

3.2 Materials for composting

1) Sewage sludge from the treatment plant "Daugavgriva" (Riga) mixed with sawdust was used. To realize the solid-state bioconversion of the organic waste mixture, i.e. sewage sludge and fine dispersed wastewood, microorganisms were added. Two strains of *Trichoderma*: *Trichoderma lignorum* and *Trichoderma viride,* and associations of bacteria regulating the circulation of nitrogen–ammonification and nitrification processes were applied.

2) Sewage sludge from the treatment plant "Daugavgriva" (Riga) mixed with sawdust was used. To realize the solid-state bioconversion of the organic waste mixture, i.e. sewage sludge and fine disperse wastewood, microorganisms were added. Two strains of *Trichoderma*: *Trichoderma lignorum* and *Trichoderma viride* were applied, but without adding the associations of bacteria regulating the circulation of nitrogen.

3) Municipal biodegradable organic solid waste mixed with sawdust was used. To promote the composting process, *Trichoderma viride* and *Trichoderma lignorum* were added [7].

The quality of the compost was assessed by chemical analyses, microbiological analyses, seeds germination tests and Toxkit microbiotests [8, 9]. The concentrations of ammonia, nitrite and nitrate were determined by a FIAstar 5020 Analyser. Total nitrogen was measured by a "Bushi" Kjeldahl Line; total carbon was determined by the modified Tjurin's method [10]. Then C:N ratios were calculated.

3.3 Microbiological analyses

The quantification of the total number of microorganisms, fungi, *Staphylococcus aureus* and *Salmonella* sp. in the compost was performed. The plate count method was applied for estimation of the total number of microorganisms and fungi per 1 g of dry compost. The medium for quantification of the total number of bacteria was Bacto nutrient agar (DIFCO LABORATORIES, USA); Chapec medium was used for fungi, and RidaCount® Tests media (R-Biopharm AG, Germany) for *Salmonella* and *Staphylococcus aureus*.

3.4 Biotesting

Germination tests were performed with cress salad. Microbiotests were performed with Rototoxkit F™, Thamnotoxkit F™ and Ostracodtoxkit F. The Rototoxkit F™ and Thamnotoxkit F™ measured the lethal effect of toxicants after a 24-h exposure to the rotifers *Branchionus calyciflorus* and *Thamnocephalus platyurus*, respectively, which were freshly hatched from cysts. Ostracodtoxkit F with the benthos organism *Heterocypris incongruens* was a "direct contact" chronic toxicity microbiotest. After 6 days, the crustacean morbidity and growth intensity were compared with the control.

4   Results

The compost used as a soil amendment should contain plant nutrient elements, but should not contain compounds, which are harmful for the environment. Low-quality compost can enrich soil and water with harmful compounds and pathogens. Toxic compounds could develop in the compost during the composting process.

Table 1 shows that the composting process of sewage sludge and sawdust in a windrow that began in late autumn showed a decrease in the number of both fungi and bacteria. However, at the same time, the number of *Staphylococcus aureus* decreased only from 600 to 390 cells per gram of dry compost. Inmature compost contained more bacteria, including *Staphylococcus aureus*.

Table 2 shows that the microbiological analysis of the municipal biodegradable organic waste compost that was started in early winter showed that pathogenic bacteria remained intact during windrow composting.

Table 1 Changes in the main parameters of compost during composting of sewage sludge and sawdust in a windrow.

Sample	Sampling time	C:N ratio	DW, %	Number of bacteria (cells per 1 g of dry compost)			
				Total number of bacteria	*Staphylococcus aureus*	*Salmonella sp.*	Fungi
Immature compost	June 2004	31.3	30	5.50×10^6	600	0	1.50×10^3
Mature compost	December 2004	27.1	44	3.85×10^5	360	0	5.35×10^2

Table 2. Changes in the main parameters of compost during composting of municipal biodegradable organic waste in a windrow.

Sampling time	pH	C:N ratio	DW, %	Number of microorganisms (cells per 1 g of dry compost)				Toxicity (EP, %)		
				Total number of bacteria	Staphylococcus aureus	Salmonella sp.	Fungi	Algaltoxkit	Protoxkit	Rotoxkit
10.03 .04	6.6 7	16.3	43.8	4.56 x 10^7	2.77 x 10^3	456	1.33 x 10^3	54.5	11. 4	0
22.03 .04	7.1 1	23.1	46.7	3.53 x 10^8	9.53 x 10^3	107	2.29 x 10^3	68.1*	26. 4	0
14.04 .04	7.2 2	15.4	44.6	1.04 x 10^7	8.28 x 10^3	156	2.35 x 10^3	13.5	6.5	0
12.07 .04	7.2 0	19.4	42.1	1.13 x 10^7	1.42 x 10^3	340	2.69 x 10^3	0.7*	2.5	0

＊ algae growth

The number of *Salmonella* and *Staphylococcus aureus* varied during the composting process, but finally decreased. The number of *Salmonella* and *Staphylococcus aureus* decreased from 456 to 340 and from 2.7 x 10^3 to 1.42 x 10^3 cells per 1 g of dry compost, respectively. As can be seen fom Table 2, during the estimated composting time, an increase in the number of fungi from 1.33 x 10^3 to 2.69 x 10^3 cells per 1 g of dry compost was revealed.

A comparison of these results with those obtained in 2002 and 2003 has shown that the season is of great importance for the beginning of the composting process. In 2002 and 2003, the composting process began in early summer; the number of *Esherichia coli*, *Salmonella* sp., and *Clostridium perfringens* at the thermophilic regime was 0. The total number of bacteria during the composting process increased, but at the same time, the number of fungi decreased. The C:N ratio was 20:30, pH 7.2–7.8, and humidity about 60 % [4].

Obviously, the activity of the microbiological processes decreased in winter months and increased slowly in the spring. Therefore, the quality of the compost was poor.

The determination of the toxicity of the municipal organic waste compost with a battery of microbiotests at the beginning and during the composting process showed that the toxicity level decreased from 11.4% to 2.5% in the Protoxkit and increased from 54.5% to 0.7% in the Algaltoxkit.

Table 2 shows that the cause of this toxicity growth in the Algaltoxkit could be possibly ammonia ions, which were determined in a compost water extract with the Nesler's reagent. Toxicity was not determined with Rotoxkit microbiotests.

The samples of the compost from the Bin, when composting process was began in early spring, were collected twice, each time in tree depths, as is shown in Table 3.

Table 3. Composting processes in a Bin.

Sample	Sampling time	C:N ratio	pH	DW, %	Number of bacteria (cells per 1 g of dry compost)				Toxicity (EP, %)		
					Total number of bacteria	Salmonella sp.	Staphylococcus aureus	Fungi	Protoxkit	Ostracodtoxkit	Rotoxkit
1-Bin-1	September 2004	49.0	8.3	55.6	3.48×10^7	0	8.9×10^3	5.80×10^3	10.4	5.2	30
2-Bin-1		50.0	8.4	51.1	2.60×10^7	0	3.7×10^3	3.65×10^3	15.3	24.2	36
3-Bin-1		49.7	8.5	49.8	2.30×10^7	0	2.7×10^3	1.14×10^4	27.9	25.2	40
1-Bin-2	December 2004	28.4	8.3	54.4	6.58×10^7	0	3.1×10^2	6.98×10^4	2.4	10.0	10
2-Bin-2		28.2	8.6	53.6	4.76×10^7	0	1.4×10^2	6.25×10^4	18.0	24.2	25
3-Bin-2		36.8	8.1	54.0	5.73×10^7	0	1.5×10^2	6.30×10^4	12.5	15.5	18

A four-month interval was between both sampling times. The analyses of the first three compost samples from the Bin (collected in September) showed a very high C:N ratio in the compost. This fact indicated the lack of nitrogen in the composting material; therefore, microbiological processes were delayed. To promote the microbiological processes, brewer's yeast was added, thereby reducing the C:N ratio. After correcting the C:N ratio, the total number of bacteria and fungi was increased. The number of *Staphylococcus aureus* decreased from 8.9×10^3 to 1.4×10^2 cells per 1 g of dry compost. The highest number of *S. aureus* was in the sample from the upper layer collected in September, but the lowest number – in the middle layer collected in December. In this compost, *Salmonella* sp. was not detected.

The results with inoculum added are shown in Table 4.

Table 4 Composting process in a bioreactor with and without the nitrificators association.

Variants	Dry weight, %	C:N	Total number of bacteria	Fungi	Activity of catalase, ml O$_2$ / 1 min
			cells per 1g of dry compost		
With the nitrificators association	34.5	19.5	4.04E+06	1.09E+04	3.6
Without the nitrificators association	38.0	25.9	1.48E+06	7.34E+03	4.93

Compost samples from the Bin were collected in three depths – the upper - 1, middle - 2 and bottom - 3 layers. Table 3 shows that, in all the three layers, the pH value, moisture and number of bacteria were similar, although the number of fungi and toxicity detected with microbiotests differed. The toxicity level was between 25.2% (Ostracodtoxkit) and 40.0% (Rotoxkit). Hence, the compost was slight acute hazard. The seeds germination intensity was in the range 80–100%, as is shown in Table 5.

Table 5 Composting processes in a Bin.

Sample	Sampling time	Heavy metals mg kg^{-1}			Seeds germination tests		
		Zn	Pb	Cd	Germination, %	Roots length (average) mm	% from the control
1-Bin-1	September 2004	72.5	3.2	0.11	100	20.6	32.8
2-Bin-1		81.4	4.0	0.18	100	17.5	27.8
3-Bin-1		63.1	4.5	0.15	80	12.8	19.5
1-Bin-2	December 2004	83.2	4.4	0.13	100	56.4	86.7
2-Bin-2		86.4	5.1	0.15	90	38.5	59.2
3-Bin-2		78.5	5.6	0.09	100	47.6	73.2

However, the length of the primary root was significantly shorter than in the control (F = 117.53 > F$_{crit}$ 2.5787). The results allow concluding that conventional seeds germination tests could not provide complete information about the compost quality. In the compost, where some toxic compounds can be formed, seeds germinated, but their growth and development were hindered.

5   Discussion and conclusions

The number of *Salmonella* and *Staphylococcus aureus* decreased during composting, but the composting conditions in the windrow could not provide the loss of these bacteria.

Hygienic requirements were met by specifying the areas, where sewage sludge should be used as a fertilizer. In this case, the anaerobic fermentation process was necessary to prevent the spread of pathogenic microorganisms in sewage sludge.

During composting, a more significant effect on the compost toxicity level was determined with Toxkit microbiotests. Seeds germination tests did not show any toxic effects. Therefore, for the assessment of the compost quality, Toxkit microbiotests could be considered as more suitable indicators with comparison to conventional seeds germination biotests. The immature compost hindered the growth and development of plant roots.

The addition of inoculum showed some intensification of the process. However, it seems that this method could be recommended mostly for degradation of specific target contaminants [4].

The most significant factor to produce a high-quality compost is the season, when the composting process is started.

Acknowledgements

The support of the study by Grants No. 01.0398 and 01.0370 of the Latvian Council of Science, the International Project "Biofertilizer Production from Renewable Organic Waste" and the International Project LIFE is gratefully acknowledged.

References

[1]   U. Viesturs, and M. Leite, in: Advances in Bioprocess Engineering (Kluwer Academic Publishers, Dordrecht – Boston – London, 1996), pp. 81-87.

[2]   U. Viesturs, and M. Leite, Appl. Biochem. Microbiol. 33 (3), 213 (1997).

[3]   A. Berzins, M. Toma, M. Rikmanis, and U. Viesturs, Acta Biotechnol. 21 (2), 155 2001).

[4]   U. Viesturs, Dz. Zarina, L. Dubova, and A. Berzinš, LLU Raksti 12 (307), 1 (2004).

[5]   R. P. Tengerdy, in: H. W. Doclle, D. A. Mitchel, and C. E. Rolz (eds.), Solid Substrate Cultivation (Elsevier Science Ltd., London, 1992), pp. 269-282.

[6]   A. H. Molla, A. Fakhru'l-Razi, M. M. Hanafi, and Md. Zahangir Alam. International Biodeterioration and Biodegradation 53, 49 (2004).

[7]   R. Bendere, Dz. Zarina, D. Arina, and L. Dubova, in: Proceedings of the 2nd World Conference and Technology Exhibition on Biomass for Energy, Industry and Climate Protection, Rome, Italy, 10-14 May 2004, VB1.63 (2004).

[8]   EPA 712-C-96-154. Seed Germination / Root Elongation Toxicity Test / Ecological Effects Test Guidelines.

[9]   G. Persoone, B. Marsalek, I. Blinova, Dz. Zarina, L. Manusadzianas, G. Nalecz-Jawecki, L. Tofan. N. Stepanova, L. Tothova, and B. Kolar. Environmental Toxicology 18 (6), 395 (2003).

[10]   G. J. Rinkis, H. K. Ramane, and T. A. Kunickaja, Methods for Analysis of Soils and Plants (Zinatne, Riga, 1987) (in Russian).

Microbial diversity and chimerae from PCR amplified products

J.M. Gonzalez[*,1], **J. Zimmermann**[1], and **C. Saiz-Jimenez**[1]

[1] Instituto de Recursos Naturales, CSIC, Apartado 1052, 41080 Sevilla, Spain

[*]Corresponding author: e-mail: jmgrau@irnase.csic.es, Phone: +34 954 624711, Fax: +34 954 624002

Advances in environmental microbiology have generated a completely new perspective on microbial diversity. PCR amplification of highly homologous genes (i.e., 16S ribosomal RNA) from complex DNA mixtures is known to generate a significant proportion of chimeric sequences. Chimerae could lead to false diversity estimates. In this study, we utilize a novel algorithm for the detection of chimerae and the risk involved in false discrimination of sequences from actual microorganisms. The proposed strategy considers the actual variability existing in different regions throughout the analyzed sequences. A significant fraction of the microbial diversity of our planet remains to be discovered and so there is a need for further molecular surveys of environmental microbial communities and the analysis of the obtained results.

Keywords: PCR, DNA amplification, environment, microbial community, biodiversity

1 Introduction

The discovery of novel microorganisms represents an important research line within the microbiology. However, the cultured microorganisms only represent a minor fraction of the total microbial diversity in our planet [1, 2]. The use of molecular techniques to retrieve and detect nucleic acid sequences from microorganisms in the environment has been essential to establish the fundations of modern environmental microbiology. Microorganisms are highly dverse in nature and there are many microbes still to be discovered.

The modern view of microbial diversity was pioneered by phylogenetic perspectives of microbial evolution [3, 4]. Three major domains of life: Bacteria, Archaea, and Eukarya were defined [4]. In the following years, an increasing number of novel, divergent microorganisms were discovered. For instance, since the study of Woese et al. [3] structuring the known microbial world in 12 phyla, the number of bacterial phyla has increased up to 52 [2]. In addition, about half of these detected bacterial phyla have been proposed only based on DNA sequences retrieved from the environment.

Molecular methods for the survey of microorganisms from the environment generally involve the use of DNA amplification techniques such as the polymerase chain reaction (PCR). Amplification by PCR often lead to a fraction of chimeric sequences [5, 6]. Chimeric sequences, or simply chimerae, are those generated as a result of amplification from two different DNA templates (Figure 1). Thus, chimerae are formed by sequences from two (or more) different microorganisms, and could represent the identification of novel, non-existing microorganisms. Environmental microbial studies should check for the possibility of obtaining chimerae during the surveys. So far, different software packages and algorithms have been proposed for this task [7–11].

In spite of the existence of different procedures for the detection of chimeric sequences, recent analyses of common DNA databases have unveiled the presence of numerous chimerae [11, 12]. The potential risk of getting increasing numbers of chimerae in DNA databases together with the risk involved in discarding real sequences as potential chimerae represent actual problems that need to be brought out. The objective of this study is to emphasize the importance of these risks and to analyze their importance within the current knowledge of microbial diversity in our planet.

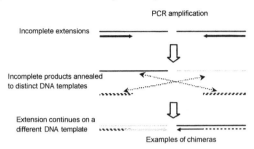

Fig. 1 Scheme explaining the formation of chimeric sequences during PCR amplification from a two different DNA templates with elevated homology.

2   Materials and methods

The retrieval of DNA sequences from microorganisms during molecular surveys of microbial communities from the environment is generally based on the amplification of specific DNA fragments or genes from total DNA extractions. The most frequently used gene for the detection and identification of microorganisms is the 16S ribosomal RNA gene (16S rRNA). The methodology used during these surveys implies the use of an amplification reaction (i.e., polymerase chain reaction or PCR), an electrophoretic method for the visualization of the amplified DNA fragments (i.e., Denaturing Gradient Gel Electrophoresis or DGGE, terminal Restriction Fragments Length Polymorphisms or t-RFLP, Single Strand Comformation Polymorphisms or SSCP), cloning into a DNA library, a screening procedure, DNA sequencing, and bioinformatic analysis of sequences. These methodologies have been recently described in detail [13–15].

Sequences as well as those obtained from microbial isolates are submitted to DNA databases (i.e., GenBank) and are publically available to the scientific community. Sequence retrieval and homology searches can be carried out online, for instance at the NCBI (National Center for Biotechnology Information; http://www.ncbi.nlm.nih.gov/).

Evaluation of sequences for the potential generation of chimerae during molecular surveys of complex microbial communities can be performed using several of the methods available for this objective [7–11]. These methods are based on the detection of closely related sequences to two halves (5' vs. 3' ends) of the sequence to be tested, resulting in ambiguous determinations and depending on sequence variability could result in misinterpretations. The only exception to this basic procedure is the method recently proposed by the authors [11] using the software Ccode. The procedure used by Ccode takes advantage of DNA sequences available in databases to obtain the closest relatives to the sequence to be analyzed (query sequence) and analyze the variability within the known sequences to obtain a decision whether the query sequence is or not a chimera. The basic structure of its procedure is described in the following (also see [11]).

A chimera is an artifact generated from the synthesis of a single DNA molecule from two, or more, original template sequences. The problem is to find a procedure to detect and discard those chimeric sequences. The best alternative to approach the existence of a chimera would be to evaluate it based on the existing variability within the phylogenetic group to be considered. Besides, rRNA sequences show different variability in different zones of the molecule which makes the case even more complicated. Variability within a phylogenetic group can be determined by inspecting the differences existing between pairs of sequences. The solution comes as a comparative analysis of the sequences in fragments of selected length. Each fragment is referred to as "word" and its length should be selected for each type of analysis. First, a set of related sequences, clearly defining the surroundings of the evaluated sequence, is selected and it is about to define the variability allowed within the phyletic group. These sequences are named reference sequences. Second, pairwise comparison between the query and the reference sequences will describe the variability added by the sequence to be evaluated (query sequence) to the group. Considering that the sequence is a chimera, significant variability will be added in the area with a different origin and thus a chimera can be detected. Not very good yet... Chimerae should require the existence of several continuous words showing significantly high variability above the level defined by the reference sequences. Thus, a quotient between the average variability between query and reference strains (avgQ) to the average variability between reference sequences should be around one for non chimeric sequences while it should be higher than one for chimeric sequences. A 95% confidence limit [16] should statistically determine the significance of the estimated comparisons. A second analysis that could be performed to statistically check comparisons of variability is a test of analysis of variance [16]. The software Ccode performs these analyses as part of the flow chart of Figure 2 and presents with the results of two independent statistical methods.

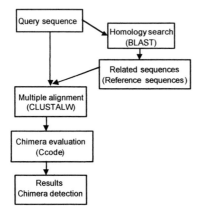

Fig. 2   Flow chart of the steps involved in chimera evaluation using Ccode.

Ccode is freely available on line at http://www.irnase.csic.es/download. Ccode does not find the closest relatives to a query sequence, it just analyzes a set of given reference sequences and compare them to a query sequence. Reference sequences can be obtained by homology search using the Blast algorithm [17] at the NCBI (see above). An accurate multiple alignment of reference and query sequences must be performed; this can be done by using the software ClustalW [18].

3   Results and discussion

The experimental protocol required to carry out a molecular survey of complex microbial communities is well stablished. Thus, there is no evidence that in the next decade this methodology might radically change. As a consequence, environmental microbiologists are left with the need of improving the methods of analysis in order to detect chimeric sequences generated in the course of past and present molecular surveys of microbial communities.

CCODE run: ./ccode025 -INitratiruptor1 -ONitratiruptor2 -p10

Query sequence (Number of bases): AB175499 (1231)
Reference sequences (Number of bases; Percent similarity with Query):
  AB105048 (1231; 88.5)
  AY691430 (1231; 85.6)
  AB167820 (1231; 84.1)
  AB091292 (1231; 84.0)
  AJ575809 (1231; 84.6)
Word length = 123.

Bases	AvgQ (sdQ)	AvgR (sdR)	Ratio	Anova[1, 13]
123	33.80 (3.27)	39.90 (14.62)	0.85	0.820
246	39.80 (4.32)	32.30 (16.14)	1.23	1.008
369	5.80 (3.63)	6.00 (3.27)	0.97	0.012
492	13.40 (2.07)	14.10 (5.80)	0.95	0.066
615	20.80 (2.77)	21.50 (6.15)	0.97	0.057
738	11.00 (3.39)	9.30 (4.55)	1.18	0.540
861	13.00 (3.46)	16.80 (9.04)	0.77	0.799
984	18.00 (4.95)	20.50 (4.90)	0.88	0.861
1107	9.40 (5.37)	12.30 (5.14)	0.76	1.032
1230	11.40 (4.28)	13.70 (3.97)	0.83	1.065

Q -> Comparisons between query sequence and reference sequences.
R -> Comparisons between reference sequences.
Significance at P<0.05 level is indicated by *.
Number of pairwise comparisons between reference sequences = 10, and between query and reference sequences = 5.
Results of ANOVA are given for the degrees of freedom between brackets.

a)

CCODE run: ./ccode04 -ltmp/align_18 -Otmp/res_10_18 -p10 -c

Query sequence (Number of bases): AF068790 (1409)
Reference sequences (Number of bases):
  AF068789 (1409)
  AF068795 (1409)
  AJ535664 (1409)
  AF299131 (1409)
  AJ309655 (1409)
  AJ309654 (1409)
  AF299126 (1409)
  AJ132722 (1409)
  AF299124 (1409)
Word length = 140.

Bases	AvgQ (sdQ)	AvgR (sdR)	Ratio	Anova[1, 43]
140	24.11 (2.93)	14.50 (5.41)	1.66*	26.192*
280	56.44 (3.13)	19.69 (9.74)	2.87*	122.986*
420	14.00 (2.40)	4.89 (2.29)	2.86*	111.958*
560	4.78 (3.90)	7.19 (4.17)	0.66	2.473
700	7.22 (6.34)	9.75 (5.89)	0.74	1.289
840	3.33 (3.28)	5.25 (3.32)	0.63	2.416
980	1.22 (1.56)	2.06 (2.11)	0.59	1.225
1120	11.56 (7.25)	16.89 (5.58)	0.68	5.833
1260	8.89 (4.76)	10.00 (2.86)	0.89	0.819
1400	5.00 (3.39)	7.19 (2.72)	0.69	4.240

Q -> Comparisons between query sequence and reference sequences.
R -> Comparisons between reference sequences.
Significance at P<0.05 level is indicated by *.
Number of pairwise comparisons between reference sequences = 36, and between query and reference sequences = 9.
Results of ANOVA are given for the degrees of freedom between brackets.

b)

Fig. 3   Example of the output of Ccode during the evaluation of chimerae. Two sequences were analyzed. In A, a real DNA sequence, and in B, a chimeric sequence successfully detected by Ccode. The observation of several continuous 'words' showing significant variability is a characteristic result for chimerae. Reference sequences define the variability for the phyletic group as well as the variability over the sequence length. Query and reference sequences are indicated.

The detection of chimeric sequences using previous software algorithms resulted in ambiguous results [7–10]. The use of the strategy presented in this study allows a more precise determination of chimerae among environmental sequences. Figure 3 shows two examples of how Ccode works: a chimera showing a portion of the sequence with significantly higher variability than defined by the reference sequences and a non-chimera, that is, a real sequence not showing significant deviations from the level determined by reference sequences. Requirements for this analysis are accurate, chimera-free DNA databases, the adequated selection of representative reference sequences together with an accurate, unambiguous multiple alignment among reference and query sequences.

As has been recently reported, DNA databases are at present not entirely free of chimerae [12]. This is a serious problem since we would not be able to detect chimerae if they are inadvertedly included among the reference sequences during a chimera evaluation. Submissions of environmental sequences to DNA databases should be thoroughly screened for their potential chimeric origin. DNA databases should be analyzed exhaustively for the presence of chimerae and detected chimerae should be eliminated from the databases. These two steps must be primary objectives in order to guarantee chimera-free DNA databases.

Variability among sequences is limited to the set of known sequences. In the last years, mainly because of the application of molecular techniques for the detection and identification of microorganisms, the number of microbial species, and even divisions, is increasing rapidly. This apparently ensures a sufficient number of sequences in databases to secure a reasonable level of variability to account for novel, different species to be accepted. Nevertheless, the possibility of rejecting sequences from actual microorganisms is inversely proportional to the known portion of microbial diversity. As more microorganisms are known, better representation of natural intersequences variability can be obtained. Consequently, assuming our planet holds a huge microbial diversity and most of it remains to be discovered, the risk of discriminating sequences from actual microorganisms as potential chimerae is significant. Learning how to avoid these false positives is important. Figure 4, for instance, shows an example of significant variability of words within an non-chimeric sequence. A chimera is usually well defined if a continuous set of aberrant words show significance. Otherwise, when single words show statistical significance, the result could suggest interspecies variability above the level defined by its closest relatives. Thus, it is important to highlight two different concepts. First, a chimera is a non-real, artificial sequence composed by two, or more, fragments with different origins. Second, a novel, divergent microorganism can result in bringing up additional variability in specific portions of the sequence (words) above the known variation level set by reference sequences. In spite of our increasing knowledge of microbial diversity on our planet, recent studies have reported the existence of radically divergent microbial groups (for example, [19, 20]) even at the Kingdom level, so the discovery of microorganisms and/or their sequences quite divergent from current known diversity and stablished phylogenies is likely.

```
-------------------------------------------------------------------------------
CCODE run:  ./ccode04 –IAY705461_1 –OAY705461_2 -p10 -c
-------------------------------------------------------------------------------

Query sequence (Number of bases): AY705461 (1473)
Reference sequences (Number of bases; Percent similarity with Query):
    Z95723 (1473; 85.5)
    Z95708 (1473; 85.5)
    Z95722 (1473; 84.3)
    AY587229 (1473; 81.9)
    AF498719 (1473; 82.1)
    AY587230 (1473; 82.1)
    AF498733 (1473; 81.6)
Word length = 147.
=================================================================
Bases   AvgQ (sdQ)      AvgR (sdR)      Ratio   Anova[1, 26]
=================================================================
147     22.57 (9.57)    22.95 (9.23)    0.98    0.009
294     38.71 (8.28)    35.52 (25.83)   1.09    0.101
441     17.86 (3.44)    17.29 (10.96)   1.03    0.018
588     27.71 (9.41)    28.33 (16.95)   0.98    0.008
735     39.14 (4.63)    25.00 (13.58)   1.57*   7.157*
882     23.86 (2.54)    19.10 (11.41)   1.25    1.170
1029    24.57 (4.58)    19.86 (9.22)    1.24    1.663
1176    25.14 (6.84)    18.62 (10.58)   1.35    2.306
1323    9.14 (1.95)     4.48 (2.60)     2.04*   18.803*
1470    12.29 (2.43)    12.71 (8.14)    0.97    0.018
=================================================================
Q -> Comparisons between query sequence and reference sequences.
R -> Comparisons between reference sequences.
Significance at P<0.05 level is indicated by *.
Number of pairwise comparisons between reference sequences = 21, and
between query and reference sequences = 7.
Results of ANOVA are given for the degrees of freedom between brackets.
```

Fig. 4 Example of non-chimeric, real sequence analyzed but showing variability above the defined by reference sequences. Typical example of novel environmental sequence, above defined variability, corresponding to a new, previously unknown microorganism.

The generation of chimerae in the course of environmental molecular surveys of microbial communities is a fact and care should be taken to limit their occurrence, for instance by avoiding excessively high cycle numbers or too short extension periods during amplification, and evaluating the generated sequences for the presence of chimerae. An increase in our knowledge of microbial diversity on Earth will reduce the possibilities of discriminating actual sequences as chimeric and increase the chances of detecting a real chimera as such. While, at present, the retrieval of environmental sequences (i.e., from uncultured microorganisms) is essential, the development of cultures and culturing procedures to isolate and analyze the physiology of novel or so far uncultured microorganisms is of great priority in order to confirm the existence of microorganisms and to associate microorganisms with their physiology and metabolism.

As the discovery of novel microbial species and novel environmental sequences retrieved from nature progresses over time, an adequate management of DNA databases becomes increasingly important as educational, research, and archiving tools for future initiatives must have a solid base of the actual information gathered through years of study on the microbial diversity on Earth.

Acknowledgements

J.M.G. acknowledges support from the "Ramon y Cajal" programme, Spanish Ministry of Education and Science (MEC), and J.Z. from the Marie Curie Programme. This work was funded through MEC projects REN2002-00041/GLO, REN2003-02854, and BTE2002-04492-C02-01.

References

[1]   D. M. Ward, R. Weller, M. M. Bateson, *Nature* **345**: 63 (1990).
[2]   M. S. Rappé, and S.J. Giovannoni, *Annu. Rev. Microbiol.* **57**: 369 (2003).
[3]   C. R. Woese, *Microbiol. Rev.* **51**: 221 (1987).
[4]   C. R. Woese, O. Kandler, M.L. Wheelis, *Proc. Natl. Acad. Sci. USA* **87**: 4576 (1990).
[5]   F. von Wintzingerode, U. B. Göbel, and E. Stackebrandt, *FEMS Microbiol. Rev.* **21**: 213 (1997).
[6]   G. C. T. Wang, and Y. Wang, *Microbiol.* **142**: 1107 (1996).
[7]   J. F. Robinson-Cox, M. M. Bateson, and D. M. Ward, *Appl. Environ. Microbiol.,* **61**: 1240 (1995).
[8]   G. A. Komatsoulis, and M. S. Waterman, *Appl. Environ. Microbiol.,* **63**: 2338 (1997).
[9]   J. R. Cole et al., *Nucleic Acids Res.* **31**: 442 (2003).
[10]   T. Huber, T., G. Faulkner, and P. Hugenholtz, *Bioinformatics* 20: (2004).
[11]   J. M. Gonzalez, J. Zimmermann, and C. Saiz-Jimenez, *Bioinformatics* 21: 333 (2005).
[12]   P. Hugenholtz, and T. Huber, *Intl. J. Syst. Evol Microbiol.,* **53**: 289 (2003).
[13]   C. Schabereiter-Gurtner et al., *J. Microbiol. Methods* **45**: 77 (2001).
[14]   J. M. Gonzalez et al., *J. Microbiol. Methods* **55**: 459 (2003).
[15]   J. M. Gonzalez and C. Saiz-Jimenez, *J. Sep. Science* 27: 174 (2004).
[16]   R. R. Sokal, and F. J. Rohlf, Biometry (W.H. Freeman and Co., New York, 1981).
[17]   S. F. Altschul et al., *J. Biol. Mol.* **215**: 403 (1990).
[18]   J. D. Thompson, D. G. Higgins, T. J. Gibson, *Nucleic Acids Res.* **22**: 4673 (1994).
[19]   S. Huber et al., *Nature* **417**: 63 (2002).
[20]   A.-L. Reysenbach, M. Ehringer, and K. Hershberger, *Extremophiles* 4: 61 (2000).

Screen-printed metal oxides-based enzyme biosensors for food analysis

P. Kotzian[1], **P. Brazdilova**[1], **K. Vytras**[1]*, and **K. Kalcher**[2]

[1] University of Pardubice, Department of Analytical Chemistry, CZ-53210 Pardubice, CZECH REP.
[2] Karl-Franzens' University, Institute for Chemistry – Analytical Chemistry, A-8010 Graz, AUSTRIA

An overview is presented which summarizes our recent accomplishment in the development of sensors and biosensors based on heterogeneous carbon ink screen-printed electrodes modified with both manganese dioxide or ruthenium dioxide. These electrodes were investigated as amperometric sensors for determination of hydrogen peroxide using flow injection analysis. With additions of enzymes (glucose oxidase, sarcosine oxidase, or xanthine oxidase, resp.) immobilized onto the electrode surface *via* Nafion films, corresponding biosensors for the quantification of glucose, sarcosine, and hypoxanthine were developed. Examples of applications of these biosensors in food analysis are given.

Keywords: Screen-printed electrodes; Modification; Biosensors; Oxidases; Amperometry; Flow-injection analysis

1 Introduction

An enzyme biosensor is an analytical device that combines an enzyme with a transducer to produce a signal proportional to target analyte concentration. In amperometric biosensors, an output is represented by produced current, which is measured. Since the pioneer concept by Clark and Lyons [1], the field of biosensors has greatly expanded; lot of examples can be found in corresponding literature, *e.g.* [2–4]. Among these devices, glucose biosensors incorporating glucose oxidase (GOx) have remained the most attractive, mainly due to the significance of strict diabetes mellitus control, which enhances their potential commercialization [5]. Heterogeneous carbon materials have often been used in electrochemical (bio)sensors because of their availability in a variety of forms, low cost, broad exploitable potential window, low background current, chemical inertness, ease of chemical derivatization and modification, and suitability for various applications. Among the various carbon-based electrodes available for the development of these sensors, carbon paste electrodes (CPEs) and screen-printed carbon electrodes (SCPEs) have got widespread popularity due to their ease of preparation and modification, ease of surface renewal and reproducibility in case of CPEs, and mass production of highly reproducible electrodes in case of SCPEs [6, 7]. One of the analytes detected at CPEs and SPCEs is hydrogen peroxide because of its environmental, biological and industrial importance. For its determination, modified electrodes are appreciated as they reduce the H_2O_2 overvoltage. These electrodes can be prepared using various techniques, including formation of heterogeneous layers (*e.g.*, thorough mixing of the modifier with carbon paste or carbon ink). A very recent example is the use of manganese dioxide, intensively studied by our group since the mid-1990s (see [8] and references therein).

This work is aimed to present such newly developed SPCEs as sensors used for measurements in flow systems like flow injection analysis (FIA), etc. In corresponding biosensors, enzymes (glucose oxidase, sarcosine oxidase, hypoxanthine oxidase) are properly immobilized onto the electrode fabricated by screen-printing technology using carbon ink modified by either manganese dioxide or ruthenium dioxide as a mediator; an advantage of the other is its higher stability at more negative potentials applied and when used in more acidic media. Applications in food analysis (fruit and wine samples, fish) were studied and the results are presented here as well.

2   Hydrogen peroxide sensors

Hydrogen peroxide represents an important intermediate in environmental and biological systems as a product of numerous enzymatic reactions (especially, when some oxidases are applied). However, its direct oxidation at carbon electrodes requires relatively high positive potentials, which in turn causes interferences to many other oxidizable components. Therefore, the main aim is to reduce the overpotential by means of a mediator introduced as a modifier to the electrode surface. Very recent examples are represented with the use of either manganese [8, 9] or ruthenium [10] dioxides .

Hydrogen peroxide reacts chemically with the mediator (MnO_2 or RuO_2) producing water, molecular oxygen and manganese oxides of lower oxidation states (Mn^{II} and Mn^{III}) in the first case, or metallic ruthenium in the second one. The later is electrochemically reoxidized, which produces the catalytic current. The oxidative current is then directly related to the hydrogen peroxide concentration and, in general, the compound can be determined in concentration levels up to 1000 mg L^{-1} although the linear dependence of the signal exhibits two diffe rent slopes with an intercept at 1 mg L^{-1} [10].

Such a basic amperometric H_2O_2 unit is also able to detect the substance produced in numerous biocatalytic reactions. Thus, when combined with a proper biocatalytic layer, e.g., enzymes (oxidases), it may serve as a "support" of different biosensors.

3   Glucose biosensors

It should be mentioned that the development of different glucose biosensors is, in general, very popular because of the two main reasons: First, the detection of ß-D-glucose is important in the area of industrial quality control and processing applications and, especially, in clinical diabetes diagnosis and treatment. Secondly, the corresponding enzyme, glucose oxidase (GOx), is relatively cheap and stable, which gives a great chance to be applied by all beginners in the field of biosensors.

Our group has been actively engaged in the development of sensors and biosensors for several analytes based on heterogeneous carbon electrodes modified with manganese oxide since the mid-1990s, [8]. Very recently, RuO_2-modified biosensors that have also been developed [10]. The method is based on the biocatalytic oxidation of glucose in the presence of the enzyme mentioned, with the simultaneous formation of hydrogen peroxide and D-glucono-1,5-lactone (gluconate) as products,

$$Glucose + O_2 \, \rho \quad Gluconate + H_2O_2.$$

The coenzyme and redox reactive center of GOx, flavin adenine dinucleotide (FAD), is reduced by ß-D-glucose to form gluconolactone and the reduced form of the flavin moiety ($FADH_2$), which then undergoes an oxidation in the presence of oxygen giving FAD and H_2O_2. Therefore, the monitoring of the H_2O_2 formation is usually exploited for the determination of glucose.

The glucose biosensor described above was applied to determine glucose content in white wine grapes and tangerines; measurements were performed using the spiked samples and the results are listed in Table 1.

Table 1   Determination of glucose in real fruit samples and corresponding recovery studies [10]

Sample	Spike, mg L^{-1}	Expected, mg L^{-1}	Found, mg L^{-1}	Recovery, %
Wine grapes	0.0		23.2	
	33.0	56.2	54.2	96.4
Tangerines	0.0		11.1	
	25.0	36.1	35.7	98.9

4   Sarcosine biosensors

Sarcosine belongs to the family of biogenic amines. According to the structure, these represents aliphatic, alicyclic and heterocyclic organic bases with low molecular mass that possess biological activity. Their occurrence is extremely widespread in a variety of food products ranging from fish and seafood to meat products, milk products, beer, wine, sauerkraut, grapes, olives, *etc.* Biogenic amines are generated by microbial spoilage of food high in protein content or through processing, ripening and storage of fermented foodstuffs. Determination of biogenic amines is usually carried out by methods of high performance liquid chromatography (HPLC). Electrochemical enzyme probes based on the oxygen consumption or the hydrogen peroxide production are also widely used. Amine oxidases are ubiquitous water-soluble enzymes, which catalyze the oxidative deamination of amines to the corresponding aldehydes, ammonia and H_2O_2 according to reaction

$$RCH_2NH_2 + O_2 + H_2O \rho \; RCHO + NH_3 + H_2O_2.$$

For sarcosine, it can simply be re-written as

$$Sarcosine + O_2 + H_2O \rho \; HCHO + Glycine + H_2O_2.$$

The optimized MnO_2-based biosensor was employed for analyses of real food samples (see Table 2). After pre-treatment, each of the samples was divided into two parts. The first one was measured directly, the second was spiked with sarcosine, and the dependence of the sarcosine concentration on the storage time was followed [9]. In not spiked samples, a time dependence resulting in increase of the sarcosine content was observed; particularly pronounced was this effect with fish. On the other hand, spiked samples from meat and cheese indicated some irregularities (which could be explained by possible side reactions with food additives [9]).

754

Table 2 Changes in the found sarcosine content in filtered homogenates of the food samples [9]

Sample	Not spiked			Spiked*		
	Starting value	After 3 days	After 5 days	Starting value	After 3 days	After 5 days
Fish[a]	BL[c]	22.8	33.8	4.4	31.8	67.6
Cheese[a]	5.0	BL[c]	23.9	17.6	5.8	29.9
Meat[b]	2.5	11.8	19.5	19.9	14.8	19.9

*) Spiked with sarcosine, [a]) 20 mg L^{-1}; [b]) 50 mg L^{-1}; [c]) Below detection limit.

5 Hypoxanthine biosensors

The determination of hypoxanthine has considerable importance for the quality control of fish products in food industry. Adenosine triphosphate (ATP) in fish tissue begins to degrade after death; in such case, ATP is decomposed quickly to adenosine diphosphate, adenosine monophosphate, and inosine monophosphate, which is further decomposed to inosine (HxR), hypoxanthine (Hx), and uric acid. In general, the enzymatic decomposition of HxR or Hx is the rate-determining step, depending on the kind of fish. Therefore, Hx is accumulated in the fish tissue and its amount can be used as an indicator of the fish freshness. Xanthine oxidase (XOx) catalyzes the oxidation of hypoxanthine gradually to xanthine, which is followed by the oxidation to uric acid. In each step, hydrogen peroxide is involved as a product. The overall reaction discussed above could be described by the scheme

$$\text{Hypoxanthine} + O_2 + H_2O \,\rho\ \text{Uric acid} + H_2O_2.$$

The optimized hypoxanthine biosensor based on the RuO_2 modification of the screen-printed electrode was applied to analyze the fish sample [10,11]. A growing amount of hypoxanthine was found when the fish sample was kept at room temperature (25 °C). A pronounced increase of the sensor response during 24 h could be related to continuous degradation of the meat tissue. To verify the influence of the matrix, hypoxanthine was determined also in freshly prepared homogenates of salmon meat samples spiked with known additions of the substance. The results (see Table 3) can be considered quite satisfactory.

Table 3 Determination of hypoxanthine in salmon meat and corresponding recovery studies [10]

Sample	Spike, mg L^{-1}	Expected, mg L^{-1}	Found, mg L^{-1}	Recovery, %
Salmon	0.0		8.1	
meat	2.5	10.6	10.6	100.0
	5.0	13.1	13.5	103.0

6   Conclusion

Both the modifiers (MnO_2 and RuO_2) have been found suitable for the construction of simple screen-printed biosensors for determination of different analytes. MnO_2-modified sensors are very simple to prepare; additionally, they are cheap and the modifier is non-toxic. These sensors can be applied satisfactorily in both neutral and slightly alkaline media assuring their compatibility with oxidase catalyzed reactions and hence suitability for biosensor applications. On the other hand, sensors based on RuO_2 modification are not affected by neither less positive potentials nor higher acidity of the samples measured, which can be appreciated as a big advantage.

Acknowledgements

The supports of the Czech Ministry of Education, Youths and Sports (project No. MSM0021627502) and of the CEEPUS program (project PL-110) are gratefully acknowledged.

References

[1]   L. C. Clark, Jr., C. Lyons. *Ann. NY Acad. Sci.* **102** (1962), p. 29.
[2]   A. Mulchandani, K. R. Rogers (eds.). *Enzyme and Microbial Biosensors.* Humana Press, Totowa (NJ), 1998.
[3]   A. O. Scott (ed.). *Biosensors for Food Analysis.* Royal Society of Chemistry, Cambridge, 2000.
[4]   D. Diamond (ed.). *Principles of Chemical and Biological Sensors.* Wiley, New York, 1998.
[5]   J. K. Kirk, C. C. Rheney. *J. Am. Pharm. Assoc.* **38** (1998), p. 210.
[6]   K. Kalcher, J.-M. Kauffmann, J. Wang, I. Svancara, K. Vytras, C. Neuhold, Z. P. Yang. *Electroanalysis* **7** (1995), p. 5.
[7]   I. Svancara, K. Vytras, J. Barek, J. Zima. *Crit. Rev. Anal. Chem.* **31** (2001), p. 311.
[8]   N. W. Beyene, P. Kotzian, K. Schachl, H. Alemu, E. Turkusic, A. Copra, H. Moderegger, I. Svan- cara, K. Vytras, K. Kalcher. *Talanta* **64** (2004), p. 1151.
[9]   P. Kotzian, N. W. Beyene, L. F. Llano, H. Moderegger, P. Tunon-Blanco, K. Kalcher, K. Vytras: *Sci. Pap. Univ. Pardubice, Ser. A* **8** (2002), p. 93.
[10]   P. Kotzian, P. Brazdilova, K. Kalcher, K. Vytras: *Anal. Lett.* **38** (2005), p. 1099.
[11]   P. Brazdilova, P. Kotzian, K. Vytras: *Bull. Potrav. Vysk.* **44** (2005), p. 75.

The study of the fermentative growth of *Saccharomyces cerevisiae* S288C using auxo-accelerostat technique

Kaja Kasemets[1], Ildar Nisamedtinov[2], Kristo Abner[2] and Toomas Paalme*[2]

[1] National Institute of Chemical Physics and Biophysics, Akadeemia tee 23, 12618 Tallinn, Estonia

[2] Tallinn University of Technology Department of Food Processing, Ehitajate tee 5, 19086 Tallinn, Estonia

A novel cultivation method – *auxo-accelerostat* was applied for determination of the culture characteristics of *Saccharomyces cerevisiae* S288C during fermentative growth. Fermentative growth in aerobic conditions was obtained keeping glucose concentration in range of 3-6% and biomass concentration 0.2-07 g dwt l^{1}. Absence of ATP production in respiratory process was demonstrated by analysing glycerol to ethanol production ratio. Lack of respiratory metabolism enabled to calculate Y_{ATP} in aerobic conditions. The effect of pH, T, pO_2, yeast extracts concentration, ethanol and weak organic acids on growth yield (Y_{ATP}), specific ATP production rate (Q_{ATP}) and specific growth rate (μ) were studied. The results demonstrated that *auxo-accelerostat* is very efficient method for strain characterisation and optimisation of fermentation processes as well as in physiological studies of yeast growth and metabolism.

Keywords auxostat; auxo-accelerostat; Saccharomyces cerevisiae; fermentative growth, Y_{ATP}

The strains of haploid yeast *Saccharomyces cerevisiae* are often used as a model in physiological studies because the sequence of its genome has been determined and the proteins coded by most of the ORF-s are known or at least identified. All this information is available in genomic databases in the internet. On the other hand, the information about quantitative physiological and technological properties of the strains is often poor and not well accessible. This is caused by the lack of effective cultivation methods for growth characterisation, especially at surplus substrate concentrations, common for initials steps of most cultivation processes as well as for several production processes in food industry, carried out by the fermentation.

We have developed the *auxo-accelerostat* technique [1] and used this method to study the effect of environmental factors (T, pH, pO_2, etc.) on μ, Y_{ATP}, Q_{ATP} of lactic acid bacteria [2] which all have only fermentative metabolism. In this work we used the method for growth characterization of yeast *Saccharomyces cerevisiae* that has the respiratory metabolism that is repressed with high concentrations of glucose. Therefore, the fermentative growth of *Saccharomyces sp.* can be obtained even at the presence of oxygen. Obtaining completely fermentative growth is important in case we want to determine the growth yield (Y_{ATP}) and specific ATP production rate (Q_{ATP}) as key parameters to characterize the effect of environmental conditions on the energetic efficiency of cell metabolism and compare it with different organisms. During respiratory growth the P/O ratio (the value showing the number of ATP molecules produced per oxygen atom reduced) is not exactly known and can change significantly with environmental conditions. In case of fully fermentative growth ATP is produced in the reactions of phosphorylation at the level of substrates which is stoichiometry well fixed. The most straightforward method to obtain fermentative growth in case of *Saccharomyces sp.* is to use anaerobic growth conditions. This however can bring the decrease of growth yield and growth rate, as oxygen is required for synthesis of some important cellular components such as unsaturated fatty acids and sterols [3]. To prevent using media containing sterols and unsaturated acids in high amounts, and anaerobic conditions having unpredictable effect on yeast growth the high glucose concentrations should be used. In batch culture of *S. cerevisiae* on glucose mineral media the growth yield (Y_{XS}) starts to decrease already at the cell density of 0.7 g L^{-1}. Thus, to prevent the effect of the biomass concentration, it should be kept below this value, which makes the determination of Y_{XS} very inexact, and other parameters for characterisation of growth efficiency should be used. For calculation of Y_{ATP} the determination of the specific rate of glucose consumption is not required as Y_{ATP} can be calculated on the basis of specific ethanol and glycerol production rate, as suggested by [4]. For more precise calculation of Y_{ATP} the biomass composition should also be taken into account:

$$Q_{ATP} = Q_{eth} - Q_{glr} + B/\mu \qquad Y_{ATP} = \frac{\mu}{Q_{ATP}} \qquad (1)$$

where Q_{ATP}, Q_{glr}, Q_{eth} – the specific ATP, glycerol and ethanol production rates (mmol g dwt^{-1} h^{-1}), μ – the specific growth rate (h^{-1}), B - the factor derived from stoichiometric matrix of the biomass composition [5] ($B = o_8 - o_1 - o_2 - o_3 - o_4 - o_5 - o_9$ (mmol key intermediate g dwt^{-1}), o_1, polysaccharides; o_2, peptidoglucans; o_3, His, dAMP, dGMP, dUMP and dCMP; o_4, Phe, Trp and Tyr; o_5, glycerolphosphate; o_9, Lys, Ser, Met, Leu, lauric acid, palmitoleic acid and oleic acid). Generally, as our analyses have shown, the effect of B is relatively small (about 5%) and the effect of variation of biomass macromolecular composition is insignificant.

To visualize the effect of environmental conditions in batch culture the several growth curves at different (initial) environmental conditions (*T, pH, pO₂, S_i, P_i*, etc.) should be obtained. The problem is not only that more than ten cultivations should be carried out to obtain the environment effect curve, but also in difficulties of ensuring that the other batch culture related effects would not affect the results. For example, the exponential phase can be too short for precise measurements in batch cultures.

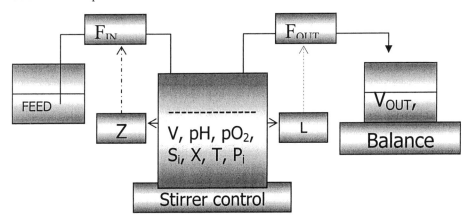

Fig. 1 Simplified scheme of *auxo-accelerostat* (one environmental parameter is changed with constant change rate while others are kept constant); Z – biomass controller, L – level controller, F_{IN} – feeding pump, F_{OUT} outflow pump, V_{OUT} – total outflow volume.

The *auxo-accelerostat* cultivation (Fig. 1) technique has many advantages compared to the regular batch. It allows keeping the cultivation conditions constant (or in neutral change range) during the experiments; to choose the optimal biomass density for experiment, start experiments with a steady-state culture and keep it further during change of environmental factor in the quasi-steady-state and obtain an environment response curve in single experiment.

The *auxostat* technique is based on the fact that the optical density (turbidity), CO_2 concentration in the exhaust gas, O_2 concentration in the exhaust gas, pH, pO_2 etc. is proportional to biomass concentration at well-defined environmental conditions.

Keeping one of parameters listed above constant by means of controlling dilution rate, steady-state culture can be obtained. After steady state (specific growth rate = dilution rate) is obtained, the desired environmental parameter (for example temperature) will be changed at fixed change rate while all the other parameters are kept constant or at least in the "neutral range".

The effect of temperature on μ, Y_{ATP} and Q_{eth} obtained in the single *auxo-accelerostat* experiment is described on Figure 2.

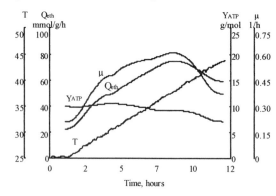

Fig. 2 The effect of change of temperature on culture characteristics during fermentative growth of *Saccharomyces cerevisiae* on mineral media with glucose (50 g l^{1}).

In *auxo-accelerostat* experiments a new technologically important environmental factor – the change rate of environmental conditions is introduced. Our experiments with growth inhibitors of yeast such as weak organic acids clearly demonstrated that change rate of their concentration affect the culture behaviour and growth characteristics (Fig. 3). This clearly indicates that change in cultivation conditions may have dramatic effect on culture performance, especially in the experiments of expression of recombinant proteins when induction with IPTG, methanol, temperature etc. is involved.

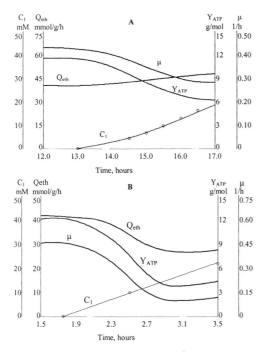

Fig. 3 Effect of slow (A) 5.4 mM h^{-1} and fast (B) 10.8 mM h^{-1} change rate of formic acid (C$_1$) concentration in the culture medium.

Going back to quantitative characterisation of fermentative growth of yeast it should be exclusively sure that respiration is repressed in aerobic conditions. Here we suggest using the ratio of glycerol to ethanol production rate as indicator. According to the stoichiometric model of yeast growth we can calculate the expected ratio (Y_{GE}) in case of fully fermentative growth:

$$Y_{GE} = \frac{o_{10} + 2o_{11} + o_6 + o_7 + o_8 + o_9 - o_{NADH}}{1000000 / Y_{ATP} + o_1 + o_2 + o_3 + o_4 + o_6 + o_7 + 2o_9 + o_{10} + 3o_{11} - o_{NADH}} \quad (2)$$

where o_{1-5} is requirement of key intermediates as given above in Eq. (1); o_6, Cys, Gly, AMP and GMP; o_7, Phe, Trp and Tyr; o_8, Ala, Ile, Leu and Val; o_9, Lys, Ser, Met, Leu, lauric acid, palmitoleic acid, oleic acid,; o_{10}, Asp, Ile, Met, Thr, Asn, UMP and CMP; o_{11}, Glu, Arg, Gln, Pro and Lys; o_{NADH}, requirement of NADH for biomass synthesis. Y_{GE} should remain in range of 0.11–0.14 in case of the regular yeast macromolecular composition and growth yields in fermentative conditions ($Y_{ATP} = 10$-13 g mmol ATP^{-1}). In our experiments this ratio was very close to the expected value indicating the reliability of the method for determination of the specific ATP production.

With increasing temperature, concentrations of growth inhibitors, or decreasing the pH and pO$_2$ the Y_{ATP} started to decrease earlier than the specific ethanol production rate Q_{eth}, which can be explained by the requirement of additional energy to keep the cell homeostasis etc. With increasing NaCl, biomass and propanol concentration (Fig. 4), however, Q_{eth} started to decrease before the decrease in Y_{ATP} was seen. This suggests that, in these conditions, the inhibitor acted either directly or by inhibiting the synthesis of biomass through feedback mechanism on ATP production rate.

760

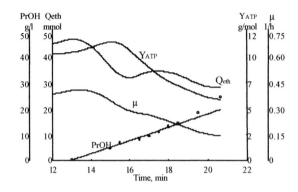

Fig. 4 Effect of propanol (PrOH) concentration on the fermentative growth of the *Saccharomyces cerevisiae*.

Our *auxo-accelerostat* experiments have demonstrated the efficiency of the method in studying the transition processes of the cell metabolism, especially if combined with studies of gene and protein expression. In addition, the culture characteristics obtained in this work using the method have great practical value in optimisation of ethanol fermentation processes using *Saccharomyces cerevisiae*.

Acknowledgements

This work was founded through Estonian Science Foundation grant 5160.

References

[1]   K. Kasemets, M. Drews, I Nisamedtinov, K. Adamberg and T. Paalme, Journal of Microbiological Methods **55**, 187 (2003).

[2]   K. Adamberg, S. Kask, T-M. Laht and T. Paalme, International Journal of Food Microbiology **85**, 171 (2003).

[3]   P.J. Rogers, P.R. Stewart, J Bacteriol. **115**, 88 (1973).

[4]   C.Verduyn, E. Postma, W.A. Scheffers and van J.P. Dijken, J. Gen. Microbiol. **136**, 405 (1990).

[5]   S. Cortassa, J.C. Aon and M.A. Aon, Biotechnol. Bioeng. **47**, 193 (1995).

Methods in Basic and Applied Microbiology.

Microbiology Education

Comparison of different analytical methods for determination of type B trichothecenes in wheat and ochratoxin A in beer

F.M. Valle-Algarra[1], A. Medina[2], J.V. Gimeno-Adelantado[1], A. Llorens[2], M. Jiménez [2]*, and R. Mateo[1]

[1] Departamento de Química Analítica, Facultad de Química, Universidad de Valencia, Dr. Moliner 50, E-46100 Burjassot, Valencia, Spain.
[2] Departamento de Microbiología y Ecología, Facultad de Biología, Universidad de Valencia, Dr. Moliner 50, E-46100 Burjassot, Valencia, Spain.

*Corresponding author: e-mail: misericordia.jimenez@uv.es, Phone: +34 963543144, Fax: +34 963543202

Type B trichothecenes were determined in wheat by gas chromatography with electron capture detector. Various SPE procedures for clean-up and two reagents for derivatization were tested. Two columns (HP-1701 and HP-5) were compared for separation of perfluoroacyl derivatives. Regarding SPE procedures, MycoSep 225 column, and cartridges made with alumina-charcoal-silica and alumina-charcoal-C_{18} silica provided the best results as recovery values for deoxynivalenol at 1.0 mg/kg spiking level were 89.6%, 87.3% and 86.1%, respectively. HP-1701 column can separate 3- and 15-acetyldeoxynivalenol derivatives while HP-5 cannot, although this last column provides lower bleed and better sensitivity
A new sample treatment was proposed for analysis of ochratoxin A (OTA) in beer by liquid chromatography using fluorescence detection. Degassed beer was mixed with lead hydroxyacetate. The precipitate was separated and the acidified liquid was extracted with chloroform. The separation was optimized with regard to composition and flow of the mobile phase. At 0.1 ng/ml spiking level, high recoveries were accomplished using immunoaffinity columns and lead hydroxyacetate (83.8% and 96.1%, respectively). Although not specific, clean-up with the last reagent is cheap.

Keywords: beer; chromatography; derivatization; ochratoxin A; solid-phase extraction; type B trichothecenes.

1   Introduction

Mycotoxins are secondary metabolites produced by certain filamentous fungi. Trichothecenes are sesquiterpenoid mycotoxins, mainly produced by various species of *Fusarium* [1]. OTA is a widely distributed mycotoxin produced by some species of the genus *Aspergillus, Penicillium verrucosum* or the genera *Petromyces* and *Neopetromyces* [2–4]. Different extraction and clean-up procedures are usually used for determination of these toxins.

Clean-up for analysis of type B trichothecenes includes solid-phase extraction (SPE) with various mixtures of solid phases like charcoal-alumina-Celite 545 [5], cation exchange resin [6], alumina-charcoal [7, 8], C_{18} [9], MycoSep 225 column [10] and Florisil [11, 12]. Current reported separation or detection methods for analysis of type B trichothecenes include enzyme-linked immunosorbent assay (EIA) [13], thin-layer chromatography (TLC) [14], liquid chromatography (LC) or gas chromatography (GC).

Various extraction and clean-up protocols for OTA in beer have been reported. Some procedures use addition of $NaHCO_3$ and NaCl to samples [15] or dilution with polyethyleneglycol 8000-NaHCO$_3$ [16], followed by clean-up on immunoaffinity columns (IAC). Jorgensen (1998) [17] added degassed samples directly to IAC columns. The most widely technique used for analysis of OTA is LC with fluorescence detection (LC-FLD) [16]. Alternative detectors, such as mass spectrometers (LC-MS-MS) [18] have also been used. Other analytical techniques are TLC, GC-MS of the trimethylsilyl derivative [19], capillary electrophoresis with laser induced fluorescence [20] and EIA [21].

GC has been the most widely used technique for trichothecene analysis, but it requires derivatization of free hydroxyl groups to prepare trimethylsilyl ethers or perfluoroacyl esters [22]. The fluorinated derivatives can be detected selectively at very low levels using GC-MS or GC-electron capture detection (GC-ECD) [23].

Our aim was to perform a comparative study of different clean-up procedures and GC or LC-FLD analytical conditions to determine these mycotoxins in a great number of wheat and beer samples. Owing to the large variety of existing methods, it is necessary their evaluation and discussion to choose the most suitable, easy-to-use, rapid, accurate, and specific method.

2   Materials and methods

2.1 Chemicals and reagents

Trichothecene and OTA standards, pentafluoropropionic anhydride (PFPA), polyethyleneglycol 8000 and 4-dimethylaminopyridine (DMAP) were from Sigma (Sigma-Aldrich, Alcobendas, Spain). Toluene, lead hydroxyacetate solution (25% PbO) and NaHCO$_3$ were from Panreac (Barcelona, Spain). Acetonitrile, methanol, chloroform, acetic acid, phosphoric acid (85%) and dichloromethane were from J.T. Baker (Deventer, Holland). Aluminium oxide 90 was from Merck (Darmstadt, Germany). C$_{18}$, Florisil and silica were from Waters (Milford, Ma, USA). Heptafluorobutyric anhydride (HFBA), Celite 545 and activated charcoal (Norit) were from Fluka (Sigma-Aldrich). Glass microfibre filters (GF/C) and filter papers (Whatman No 4) were from Whatman (Maidstone, UK). Pure water was obtained from a Milli-Q apparatus (Millipore, Billerica, Ma, USA). OchraTest immunoaffinity columns were from Vicam Science Technology (Watertown, Ma., USA) and MycoSep 225 tubes were from Romer Laboratories (Union, Mo., USA).

2.2 Preparation of standard solutions

Each trichothecene standard was dissolved in acetonitrile (1.0 mg/ml) and stored at -20°C in a sealed vial until use. Working standards (0.02 - 10.0 µg/ml) were prepared by appropriate dilution with acetonitrile and used to obtain calibration curves. A stock solution of OTA (about 500 mg/l) was prepared. Working standards were prepared by evaporation of stock solution aliquots followed by dilution with LC mobile phase and used to calibrate the detector response. The concentration of the stock solution was determined as described in the literature [24].

2.3 Equipment

The GC system was a HP-6890 Plus gas chromatograph with a [63]Ni ECD (Hewlett-Packard, Avondale, Pa, USA).

The LC system consisted on a Waters 600 pump and a Waters 474 scanning fluorescence detector (Waters). Separation was performed on stainless steel LiChrospher 100 C_{18} reversed-phase column (250 x 4 mm, 5 $\geq$m particle size) connected to a guard column (4 x 4 mm, 5 $\geq$m particle size) filled with the same phase. The column was kept at 30°C.

2.4 Extraction and clean up

2.4.1. Trichothecene extraction
Fifty g of wheat sample, previously found to be free from type B trichothecenes using the method of Mateo *et al.* [11], was ground to a powder and 5 g of the milled product was placed into a 100-ml Erlenmeyer flask. Fortification of samples at 0.1 or 1.0 mg mycotoxin/kg level was accomplished by adding appropriate aliquots of each standard solution. Solvent was evaporated and, after adding 40 ml of acetonitrile-water (84:16, v/v), the mixture was blended for 10 min. After filtration through Whatman No. 4 filter, the extraction mixture was stored in a screw-capped glass bottle at -20°C.

2.4.2 Trichothecene clean-up
The following SPE clean-up procedures were assayed.

2.4.2.1 MycoSep 225 column
Five ml of the filtrate sample extract in acetonitrile-water (84:16, v/v) was placed into a MycoSep 225 tube [10]. One ml volume of the purified extract was transferred to a vial.

2.4.2.2 Florisil cartridge
A previous procedure [12] was slightly modified and n-hexane was used instead of light petroleum.

2.4.2.3 Celite 545-alumina-charcoal mixture
'Made-in-laboratory' cartridges were prepared using 5-ml syringes. A glass microfibre filter was placed at the bottom. Then, a mixture of packing bed was placed on it. It was made of a first layer of 0.1 g of Celite 545 and second layer of 1.5 g of a homogenised mixture containing alumina–charcoal–Celite 545 (5:7:3, w/w/w). Another glass microfibre filter was placed on the bed top. The procedure was described by Mossora *et al.* [5].

2.4.2.4 Alumina-charcoal
A solid-phase extraction cartridge was prepared as indicated previously (see 2.4.2.3), but the packed material consisted of about 1 g of a homogeneous mixture containing alumina–charcoal (100:15, w/w). Three milliliters of sample extract was passed through the column. The filtrate was collected in a vial. The cartridge was rinsed with 2 ml of acetonitrile–water (84:16, v/v) and the rinsing liquid was combined with the filtrate. The rinsing volume was slightly different from the volume used in the literature [7, 8].

2.4.2.5 Alumina-charcoal-C_{18} silica
Cartridges were prepared as indicated previously (2.4.2.3), but the packed material consisted of 1.16 g of alumina-charcoal-C_{18} silica (75:1:40, w/w/w). Three ml of sample extract was passed through the prepared cartridge and collected in a vial. The cartridge was rinsed with 2 ml acetonitrile-water (84:16, v/v).

2.4.2.6 Alumina-charcoal-silica

Cartridges were prepared as indicated previously (2.4.2.3), but the packed material consisted of about 1.5 g of a mixture containing alumina-charcoal-silica (90:1.5:5, w/w/w). The extract was cleaned-up as described above (section 2.4.2.5).

2.4.3 OTA extraction and clean-up

2.4.3.1 Proposed method (lead hydroxyacetate)

About 250 ml of beer was degassed in ultrasonic bath for 40 minutes. Seventy ml of degassed beer was alkalinized with 0.8 ml of 1 M NaOH. After shaking, 2x1 ml of a 25% aqueous solution of lead hydroxyacetate was added. The mixture was centrifuged (4500 rpm) for 10 minutes. Fifty-two ml of supernatant was acidified to pH 2.8-3 with H_3PO_4. This solution was filtered through glass microfiber filter Whatman GF/C. The filtrate was extracted with 3x10 ml chloroform. The organic extracts were pooled and evaporated at 40°C. The residue was dissolved in 2 ml of methanol containing 5% (v/v) acetic acid and transferred to a vial. The solvent was evaporated at 50°C under N_2 and the residue was solved in 250 µl of mobile phase.

2.4.3.2 IAC procedure

The AOAC Official Method [17] was followed, but using double sample amount [25].

2.5 Derivatization and GC-ECD analysis for trichothecenes

Solvents from clean-up were evaporated at 50°C under N_2 stream. One hundred µl of a 2 mg/l solution of DMAP in toluene-acetonitrile (80:20, v/v) and 50 µl of HFBA (or PFPA) were added to dry extracts. After capping, the reaction mixture was heated at 60°C for 60 min. After the mixture had cooled, 1 ml of a 5% (w/v) aqueous solution of $NaHCO_3$ was added and the vial was vortexed. The two layers were allowed to separate. The top layer was analyzed by GC-ECD.

The GC-ECD determination was carried out using two different sets of chromatographic conditions. In both cases 1 µl of solution was injected in splitless mode. The injector and detector temperatures were 250°C and 300°C, respectively. The first procedure used a HP-5 fused-silica capillary column. The oven temperature program was: 90°C held for 1 min, 40°C/min to 160°C, 1.5°C/min to 182,5°C/min to 240°C, and then 40°C/min to 275°C/min, held for 2 min. Helium at a constant pressure of 42.1 kPa was used as carrier gas. The second procedure used a HP-1701 fused-silica capillary column. The oven temperature program was: 90°C held for 1 min, 40°C/min to 160°C, 5°C/min to 173, 2°C/min to 195°C, 5°C/min to 240°C/min, and then 40°C/min to 270°C/min held for 1 min. Helium at a constant pressure of 103 kPa was used.

2.6 OTA analysis

The mobile phase was acetonitrile -water 40:60 (v/v) acidified at pH 3.0, using phosphoric acid as a modifier, with a flow-rate of 1.4 ml/min in isocratic conditions. One hundred µl of concentrated extract was injected in the chromatograph. The excitation and emission wavelengths were 333 and 460 nm, respectively.

3   Results and discussion

3.1 Trichothecene analysis

The clean-up methods for type B trichothecenes in wheat provided very different recoveries (Table 1). The higher recoveries were obtained with MycoSep 225, alumina-charcoal-C18 silica and alumina-charcoal-silica cartridges. Following these protocols, the extract can be transferred directly to the vial for concentration and further derivatization. Consequently, these procedures are relatively easy and fast. Among these three procedures, MycoSep 225 column provided good results in terms of recovery for DON, 3- and 15-ADON. As for NIV, the best results were obtained with the alumina-charcoal-C_{18} silica cartridge. The main disadvantage of MycoSep 225 column with respect to 'made-in-laboratory' cartridges is its high cost. When 'made-in-laboratory' cartridges were used, recoveries were roughly 73-87% for DON, 102-110% for NIV and 58-70% for 3- and 15-ADON depending on the clean-up procedure, and reproducibility was similar to that obtained with MycoSep 225 column.

Table 1. Influence of the clean-up procedure on the recovery of type B trichothecenes in wheat (spiking level: 1.0 mg/kg), using HP-5 column. Derivatization was carried out with PFPA (n = 5).

| SPE method | Trichothecene | | | | | | | |
| | DON | | NIV | | 15-ADON | | 3-ADON | |
	Recovery (%)	RSD (%)	Recovery (%)	RSD (%)	Recovery (%)	RSD (%)	Recovery (%)	RSD (%)
MycoSep 225 column	89.6	8.3	75.6	7.8	75.1	8.9	72.6	9.4
Florisil cartridge	55.1	16.0	70.7	21.3	68.2	10.4	66.6	8.0
Celite 545-alumina-charcoal	42.2	34.4	47.3	16.2	29.0	14.3	21.8	16.7
Alumina-charcoal	131.9	58.0	121.0	25.5	103.9	28.5	108.5	43.3
Alumina-charcoal-C_{18} silica	86.1	11.1	101.8	9.8	69.4	11.0	65.3	9.7
Alumina-charcoal-silica	87.3	19.3	105.7	20.8	69.9	14.9	68.4	19.4

The SPE procedures that can be chosen in terms of best reliability, low cost and less analysis time are those involving alumina-charcoal-C_{18} silica and alumina-charcoal-silica cartridges. Both procedures have been compared. More irregular baseline and a greater number of neighbour peaks appeared working with alumina-charcoal-silica cartridge. This behaviour caused that the RSDs of recoveries were higher than those attained with alumina-charcoal-C_{18} silica cartridges. Limits of detection (LOD) (signal/noise ratio of 3:1) were practically the same for NIV, 3-ADON and 15-ADON, regardless of the derivatization reagent using clean-up on alumina-charcoal-C_{18} silica cartridge (Table 2). In the case of DON it was observed that, for the same amount of standard, signal obtained with HFBA was superior to that obtained with PFPA. Calibration lines were linear in the 0.006-0.625 mg/l range for both types of derivatives although r^2 was higher when PFPA was used. PFPA was selected as derivatization reagent for further analysis as HFBA is more expensive, has less stability against moisture while recoveries were not significantly better.

Table 2. Limits of detection (LOD), and coefficients of determination (r^2) of the linear calibration lines for type B trichothecenes as both pentafluoropropionyl and heptafluorobutyryl derivatives separated by GC-ECD with HP-5 and HP-1701 capillary columns. Trichothecene range: 0.006-0.625 mg/l. LOD were calculated using alumina-charcoal–C18 silica to clean-up a blank wheat sample.

| | | Derivatization reagent | | | |
| | | PFPA | | HFBA | |
Trichothecene	Column	LOD (mg/kg)	r^2	LOD (mg/kg)	r^2
DON	HP-5	0.007	0.9993	0.004	0.9964
	HP-1701	0.016	0.9982	0.007	0.9977
NIV	HP-5	0.017	0.9975	0.020	0.9962
	HP-1701	0.028	0.9900	0.039	0.9902
3-ADON	HP-5	0.010	0.9995	0.010	0.9990
	HP-1701	0.016	0.9974	0.012	0.9991
15-ADON	HP-5	0.011	0.9991	0.012	0.9987
	HP-1701	0.018	0.9963	0.014	0.9974

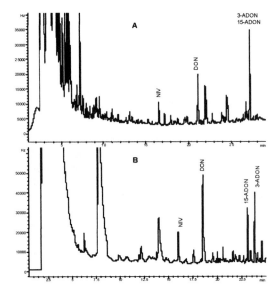

Fig. 1. Chromatograms of a blank wheat sample spiked with (A) 1.0 mg/kg of each DON, NIV, 3-ADON and 15-ADON, derivatized with PFPA, and separated by HP-5 column, and (B) 2.5 mg/kg of each of the same trichothecenes, derivatized with PFPA and separated by HP-1701 column. Clean-up: alumina-charcoal-C18 silica cartridge.

Differentiation between 3- and 15-ADON cannot be done with HP-5 column, because both compounds have the same retention time. To solve this problem, analysis was achieved with HP-1701 column, which provided different retention times for these compounds (Fig. 1). All the analyzed trichothecenes were separated with HP-1701 column, but the LODs were higher than those obtained with HP-5 column (Table 2).

3.2 OTA analysis

The chromatograms obtained by the method using lead hydroxyacetate showed more peaks than those obtained by IAC clean-up (Fig. 2). To avoid co-elution of OTA with beer components and to optimize peak performance, several acetonitrile-water mixtures were tested as mobile phases.

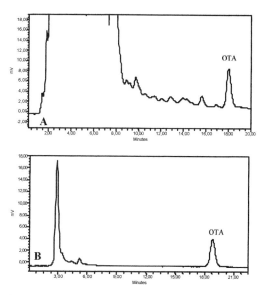

Fig.2. Chromatograms of a beer sample spiked with OTA (A) treated with lead hydroxyacetate and extracted with chloroform, and (B) treated with PEG 8000/NaHCO₃ and cleaned up with immunoaffinity column. Spiking level: 0.5 ng/ml. Conditions: excitation wavelength (330 nm); emission wavelength (460nm); mobile phase: acetonitrile-water (60:40, v/v) acidified at pH 3.0 with phosphoric acid as acid modifier at 1.4 ml/ min flow-rate.

The results observed with different mixtures of acetonitrile-water are shown in Table 3. Acetic acid used as a modifier proved worse than phosphoric acid with regard to OTA peak shape. The best results were obtained using acetonitrile-water 40:60 (v/v) acidified at pH 3.0 with a flow-rate of 1.4 ml/min in isocratic conditions.

Table 3. Assayed mobile phases for OTA separation by LC. Phosphoric acid was generally used as acid modifier.

Mixture	pH	Flow (ml/min)	Retention time (min)	OTA peak shape	Interference from matrix peaks
Acetonitrile:water 45:55 v/v	4.50	1.0	10.4	Acceptable	Yes
Acetonitrile:water 45:55 v/v	4.00ᵃ	1.0	6.2	Acceptable	Yes
Acetonitrile:water 45:55 v/v	3.55	1.2	16.1	Acceptable	Yes
Acetonitrile:water 40:60 v/v	3.55ᵃ	1.2	12.4	Acceptable	Yes
Acetonitrile:water 40:60 v/v	3.00	1.2	21.4	Acceptable	No
Acetonitrile:water 40:60 v/v	3.00	1.4	18.0	Acceptable	No
Acetonitrile:water 35:65 v/v	3.55	1.2	Aprox. 30	Broad with queue	-

ᵃ Acetic acid was used as acid modifier.

Addition of lead hydroxyacetate as clarification/precipitation agent helps to clean beer samples from bulk components, which leads to flat chromatographic baselines and lack of interfering neighbour peaks under appropriate LC conditions.

A calibration line was calculated using standard solutions. The OTA range was 0.05-10 ng of injected toxin (equivalent to 0.0025-0.5 ng/ml beer), and r^2 was 0.9998. The limit of detection of the method was 0.005 ng/ml beer (based on a 3:1 signal/noise ratio).

Table 4. Recovery data of the LC method using lead hydroxyacetate for analysis of OTA in beer (n=3).

OTA spiking level (ng/ml beer)	Mean recovery (%)	s	RSD (%)
0.50	99.8	o4.0	4.0
0.10	96.1	o6.0	6.2
0.05	94.7	o5.4	5.7
0.01	91.4	o6.1	6.7

The results of recovery experiments of the proposed analytical procedure are shown in Table 4. The recovery range was 91.4% - 99.8 % (average 95.5%) for spiking levels ranging 0.01 – 0.5 ng of OTA per ml of sample, respectively. On the basis of these data the method can be considered very good to be applied to OTA determination in beer. The alternative method using IAC clean-up provided worse recoveries (61.0% and 63.7% at 0.1 and 0.5 ng/ml spiking levels, respectively) according to a previous study [25]. Repeatability was also better working with the new procedure.

4   Conclusions

MycoSep 225 column and cartridges containing alumina-charcoal-C_{18} silica or alumina-charcoal-silica offered the best results for analysis of type B trichothecenes in wheat with similar reproducibility. Clean-up was performed in a simple step and was more effective than other assayed procedures. Although MycoSep 225 provided the best recoveries for DON, it is more expensive than 'made-in-laboratory' cartridges and provided lower recoveries for NIV.

HP-5 column shows low bleed and provided high sensitivity for the trichothecene derivatives. However, the use of HP-1701 column is needed to confirm the possible detection of 3- and 15-ADON. PFPA is preferred to HFBA as a derivatization reagent because of better stability against moisture, lower cost and similar sensitivity of trichothecene derivatives, except perhaps for DON.

A new method has been proposed for determination of ochratoxin A in beer. It uses reversed-phase LC separation and fluorescence detection. The main particularity is that many bulk components are precipitated by lead hydroxyacetate. The selected mobile phase is acetonitrile -water 40:60 (v/v, pH 3.0) using phosphoric acid as modifier.

The method is cheaper than others because SPE cartridges or expensive IAC columns are not needed. However, the performance is good. The method is sensitive, quite accurate as high recovery rates were obtained at low spiking levels, and repeatable.

Acknowledgements

The authors wish to thank financial support from the Spanish Government "Ministerio de Educación y Ciencia" (Project AGL-2001-2974-C05-01 and AGL 2004-07549-C05-02/ALI) and from the Valencian Government "Conselleria d'Empresa, Universitat i Ciència" (Project GV04B-111).

References

[1]   J. Olson, T. Börjesson, T. Lundstedt, and J. Schnürer, Int. J. Food Microbiol. **72**, 203 (2002).

[2]   D.B. Scott, Mycopathol. Mycol. Appl. **25**, 213 (1965).

[3]   P. Krogh, (ed.), Mycotoxins in food (Academic Press, London, 1987), p. 97.

[4]   J.C. Frisvad, and R.A. Samson, Stud. Mycol. **45**, 201 (2000).

[5]   M.M. Mossora, S. Adams, J.A.G. Roach, and M.W. Trucksess, J. AOAC Int. **79**, 1116 (1996).

[6]   D.R. Lauren, and R. Greenhalgh, J. AOAC Int., **70**, 479 (1987).

[7]   S.M. Croteau, D.B. Prelusky, and H.L. Trenholm, J. Agric. Food Chem. **42**, 928 (1994).

[8]   C. Hsueh, Y. Liu, and M.S. Freund, Anal. Chem. **71**, 4075 (1999).

[9]   L.K. Prom, R.D. Horsley, B.J. Steffenson, and P.B. Schwarz, J. Am. Soc. Chem. **57**, 60 (1999).

[10]   W. Langseth, A. Bernhoft, T. Rundberget, B. Kosiak, and M. Gareis, Mycopathologia **144**, 103 (1999).

[11]   J.J. Mateo, A. Llorens, R. Mateo, and M. Jiménez, J. Chromatogr. A. **918**, 99 (2001).

[12]   R.M. Black, R. J. Clarke, and R.W. Read, J. Chromatogr. **388**, 365 (1987).

[13]   R.C. Sinha, M.E. Savard, and R. Lau, J. Agric. Food Chem. **43**, 1740 (1995).

[14]   A.W. Schaafsma, R.W. Nicol, M.E. Savard, R.C. Sinha, L.M. Reid, and G. Rottinghaus, Mycopathologia **142**, 107 (1998).

[15]   P.A. Burdaspal, and T.M. Legarda, Alimentaria **98**, 115 (1998).

[16]   A. Visconti, M. Pascale, and G. Centonze, J. AOAC Int. **84**, 1818 (2001).

[17]   K. Jorgensen, Food Addit. Contam. **15**, 550 (1998).

[18]   G.S. Shephard, A. Fabiani, S. Stockenström, N. Mshicileli, and V. Sewram, J. Agric. Food Chem. **51**, 1102 (2003).

[19]   Food and Agriculture Organization of the United Nations: Worldwide regulations for mycotoxins 1995. A compendium Vol. 64 (FAO Food and Nutrition Paper, Rome, 1996), p. 6.

[20]   E.K.Tangny, S. Ponchaut, M. Madoux, R. Rozenberg, and Y. Larondelle, Food Addit. Contam. **19**, 1169 (2002).

[21]   I. Barna-Vetró, L. Solti, J. Teren, A. Gÿongÿosi, E. Szabo, and A. Wölfling, J. Agric. Food Chem. **44**, 4071 (1996).

[22]   K.F. Nielsen, and U. Thrane, J. Chromatogr. A. **929**, 75 (2001).

[23]   T. Tanaka, A. Yoneda, S. Inoue, Y. Sugiura, and Y. Ueno, J. Chromatogr. A. **882**, 23 (2000).

[24]   G.M. Wood, S. Patel, A.C. Entwisle, and A. Boenke, Food Addit. Contam. **13**, 519 (1996).

[25]   J.M Sáez, A. Medina, J.V. Gimeno-Adelantado, R. Mateo, and M. Jiménez, J. Chromatogr. A. **1029**, 125(2004).

DVC-FISH procedure to enumerate specific viable cells of *Lactobacillus delbrueckii* subsp. *bulgaricus* DN-100182

J. García[1], Y. Moreno[1], M. Carmen Collado[1], J. María Cobo[2] and M. Hernández[1]

[1]Departamento de Biotecnología. Universidad Politécnica de Valencia. Valencia, Spain
[2]RED INDE, Investigación Nutricional Danone España, Barcelona, Spain.

We have developed a DVC-FISH procedure to discriminate rapid and easily viable and non viable cells of *Lactobacillus delbrueckii subsp. bulgaricus* strain DN-100182. Samples containing cells were incubated in MRS broth with antibiotic and then sample was hybridizated with a specific rRNA oligonucleotide probe. Of the four antibiotics tested (novobiocin, nalidixic acid, pipemidic acid and ciprofloxacin), nalidixic acid was the most effective to DVC method and 7 hours of incubation was optimal. The number of viable cells was obtained by enumeration of specific hybridizated cells which were elongated at least 2 times their original size. Results show that this DVC-FISH procedure with the use of novobiocin is a rapid and culture-independent useful method to detect specifically viable *Lactobacillus delbrueckii subsp. bulgaricus* in different samples.

Keywords: FISH, DVC, *Lactobacillus delbrueckii subsp. bulgaricus,* probe, antibiotic
Abbreviation key: FISH = fluorescent *in situ* hybridization
DVC = direct viable count

1   Introduction

Probiotics have been defined as living organisms that, when included as a part of the diet, confer some favourable effect on the host [1]. Fermented milks have the property of containing these live bacteria belonging mainly to *Lactobacillus* and *Bifidobacterium genus*. It has been suggested that the exercise of the lactic acid bacteria (LAB) action as probiotics should arrive alive and in sufficient number to the intestine (approximately 10^6-10^7 microorganisms per ml), in order to appreciate their effects and to reach adhere, implant or multiply in the intestinal tract [2].

Plate counting is the classical method to determine the LAB viability. However this method has a number of disadvantages such as the time consuming and the underestimation of the number of viable bacteria due to the irregular distribution of the microorganisms in the sample and the lost of culturability under stress conditions.

There is a clear need for cultivation-independent methods to quantify the microbial communities on fermented foods [3]. Ribosomal RNA probe hybridization (FISH) has become widely adopted for detection and enumeration of specific bacterial groups in mixed populations [4]. This approach has been successfully used to quantify *Lactococcus and Leuconostoc in fermented foods* [3]; *Lactococcus, Enterococcus and Streptococcus in milk and Bifidobacterium* in fecal samples [5]. Although the rRNA content varies among cell types depending on growth rate and physiological state [6] it is not enough to determine the viability of a cell. Direct viable count method (DVC) discriminates viable and nonviable cells by direct microscopy [7]. Viable bacteria are able to elongate but their division are inhibited in presence of nutrients and a gyrase inhibitor. The combination of direct viable count (DVC) procedure which increases intracellular rRNA levels, with FISH performed on rRNA-targeted sequences could prove useful in detecting and identifying viable cells in mixed microbial communities [8].

The aim of this study was to develop a DVC-FISH procedure to enumerate specific viable LAB for a future application on fermented milks and in vivo assays following yogurt ingestions. Several antibiotics, different concentrations of each antibiotic and incubation times were tested to establish the optimal conditions to elongate the cells. A 23S rRNA probe was designed to hybridize with *Lactobacillus delbruecki* in FISH analysis and the specificity of this probe was tested. This improved procedure will be an effective and rapid method to determine viability of LAB in survival studies.

2   Materials and methods

2.1 Culture conditions

A *Lactobacillus delbrueckii* subsp. *bulgaricus* strain (DN-100182, Danone Vitapole) isolated from yogurt was used. Strain was grown in MRS broth (MERCK) under anaerobiosis conditions at 37°C. Following 24 hours of incubation, the culture was tested for DVC incubation.

2.2 Probe

FISH analysis was performed with a 23S rRNA oligonucleotide probe specific to the DN-100182 strain (LDE23-GCGTGTTCCRTCCTTAAGC-) designed by us. The specificity of LDE23 probe for *Lactobacillus delbrueckii* subsp. *bulgaricus* was confirmed by the gapped Probe Match at RDP II (Michigan State University) and BLAST search. LDE23 probe specificity was evaluated by whole-cell hybridization with different species of *Lactobacillus* and non-*Lactobacillus* species such as *Streptococcus thermophilus* (Table 1). Probe was synthesized and labelled by Tib Molbiol (Berlin, Germany) with CY3.

Table 1   Lactic acid bacteria strains used to test the probe specificity.

Species or subspecies	Strain	Hybridization with LDE23
Lactobacillus delbrueckii subsp. *bulgaricus*	DN-100182	+
Lactobacillus delbrueckii subsp. *delbrueckii*	CECT 286	-
Lactobacillus paracasei	CECT 4022	-
Lactobacillus brevis	CECT 4121	-
Lactobacillus acidophilus	CECT 903	-
Lactobacillus casei	CECT 475	-
Lactobacillus rhamnosus	CECT 278	-
Lactobacillus salivarius	CECT 4063	-
Streptococcus thermophilus	CECT 986	-

2.3 Hybridization

Samples were hybridized with 20% formamide at 46°C for 2 hours as previously described [9].
Hybridized samples were examined with an Olympus microscope BX50 equipped with a 100W mercury high-pressure bulb and set filters U-MWB, U-MWIB and U-MWIG. Colour

micrographs were taken with an Olympus DP 10 digital camera (Olympus Optical Co., Hamburg, Germany).

2.4 DVC-assay

The inhibitors of DNA gyrase nalidix acid, novobiocin, ciprofloxacin and pipemidic acid (Sigma Chemical Co.) were tested for the DVC assay. Different incubation times (0, 7, 24 hours) and antibiotic concentrations (1, 10, 100 and 1000 μg/ml) were assessed to obtain the optimal conditions. The stock antibiotics solutions were appropriately diluted in distilled water and added to MRS broth flasks.

A control sample was prepared with a broth containing no antibiotic. Samples were incubated in flasks under anaerobic conditions (anaerogen, Oxoid) at 37°C. An aliquot of broth was sampled at the moment of the addition of the *Lactobacillus delbrueckii* subsp. *bulgaricus* culture into the flask and after 7 and 24 hours of incubation and then processed for FISH analysis [4]. Briefly, cells were washed with phosphate-buffered saline and then fixed with 4% paraformaldehyde at 4°C.

2.5 Scanning electron microscopy (SEM)

For SEM, *Lactobacillus delbrueckii* subsp. *bulgaricus* cells were prepared by fixing in 0.1 M sodium phosphate buffer (PBS) (pH 7.2), containing 2.5% glutaraldehyde (Sigma Chemical Co.) and 2% osmium tetroxide (Sigma) at 4°C for 8 hours. The fixed cells were washed in 0.1 M PBS. This suspension was carefully dispensed onto the surface of 25mm 0.2 μm Poretics polycarbonate membranes (Sigma) and immersed in a liquid nitrogen bath. Membranes were coated with gold and examined by using a JEOL JSM-5410 (JEOL, Ltd., Tokyo, Japan) scanning electron microscopy operating at 20 kV.

2.6 Culture counts

Viability of bacteria was also tested by culture. 100 μl of different dilutions from each MRS flask sample were spread on MRS agar plates and incubated under anaerobic conditions at 37°C for 48 hours. Counts were made in duplicate.

3 Results and discussion

The LDE23 probe resulted to be specific to *Lactobacillus delbrueckii* subsp. *bulgaricus* when applied to FISH *in vitro* assays and directly to yogurt samples.

For the DVC procedure, bacteria must be incubated in a broth media with an optimal concentration of antibiotics which inhibit cell replication but allows other synthetic pathways to continue [10]. Therefore, in these conditions viable bacteria continue to metabolize nutrients and become elongated but not replicated.

Effectiveness and optimum concentrations were tested for each antibiotic: ciprofloxacin, nalidix acid, novobiocin and pipemidic acid (Table 2).

Table 2 Length of cells (μm) at different incubation times with several antibiotics.

Antibiotic	Conc. μg/ml	Length t 0		Length t 7h		Length t 24h	
		min	max	min	max	min	max
Nalidix acid	1	2	9	2	9	2	9
	10	2	9	5	32	5	20
	100	2	9	5	35	5	18
	1000	2	9	2	9	2	9
Novobiocin	1	2	9	2	10	2	12
	10	2	9	5	56	5	59
	100	2	9	5	50	4	52
	1000	2	9	2	10	2	9
Ciprofloxacin	1	2	9	3	9	3	9
	10	2	9	2	9	2	9
	100	2	9	3	29	3	15
	1000	2	9	2	9	2	9
Pipemidic acid	1	2	9	2	9	2	9
	10	2	9	2	9	2	9
	100	2	9	2	9	2	9
	1000	2	9	2	9	2	9

When pipemidic acid was used for the DVC technique, no increase in cell length was observed for any of the concentrations tested. Therefore, this procedure was ruled out.

Although bacteria which had been incubated with novobiocin (at concentrations ranging from 10 to 100 ≥g/ml) showed the largest increase in length (changing the maximum value from 9≥m to 59≥m; Fig. 1 and Fig. 2), a decrease in the total number of microorganisms was found after being incubated for 7 hours or even longer, both for FISH and for plate counts. Therefore as the use of this antibiotic leads to a bacteriocide effect, it was also discarded as the optimum one.

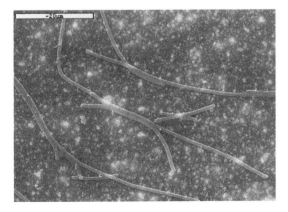

Fig. 1 Scanning electron microscopy after DVC with novobiocin (100 ≥g/ml).

776

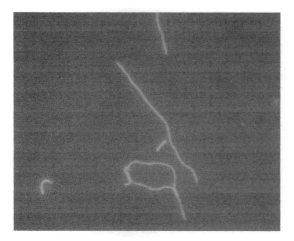

Fig. 2 Cell size differences between live and dead cells after DVC-FISH with novobiocin (10 ≥g/ml).

A similar bacteriocide effect was observed for ciprofloxacin, although less significant lengths were reached, being also rejected.

Since nalidixic acid also produced an outstanding increase in cell length (reaching the cells sizes up to 35≥m for concentrations between 10 and 100 ≥g/m; Fig 3 and Fig 4) and no reduction in the number of bacteria after 24 hours of incubation was produced, it could be concluded that this antibiotic is the most suitable when applying DVC technique for *Lactobacillus delbrue*ckii subsp. *bulgaricus.*

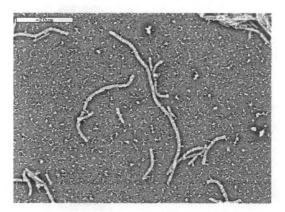

Fig. 3 Scanning electron microscopy after DVC with nalidixic acid (10 ≥g/ml).

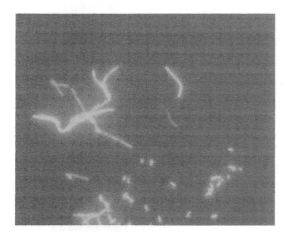

Fig. 4 Cell size differences between live and dead cells after DVC-FISH with nalidixic acid (10 ≥g/ml).

When effectiveness of the DVC-FISH technique was tested in yogurt using nalidixic acid as antibiotic, the elongation of *Lactobacillus delbrue*ckii subsp. *bulgaricus* obtained was similar to the elongation observed in the *in vitro* assay. Using plate counts and FISH (cells with a length at least twice the initial one) the total number of live cells was in the same log order in both cases. After an incubation period of 7 hours, nalidixic acid finished its action, allowing making a difference between live and dead cells (Fig 5).

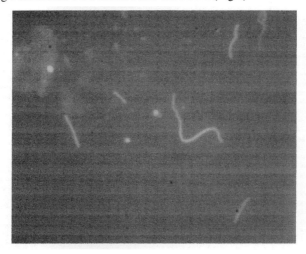

Fig. 5 Effect of the DVC-FISH technique tested in yogurt using nalidixic acid as antibiotic.

The fluorescent intensity signal of the hybridized cells following DVC treatment was stronger than before the treatment. This effect could be due to an increase of intracellular rRNA levels which has been verified by others authors [11].

778

DVC-FISH incubation with nalidixic acid resulted a useful tool to determine specifically the presence of viable Lactobacillus for *in vitro* and *in vivo* assays.

References

[1]   R. Fuller, Probiotics in human medicine. Gut **32**, 439-442 (1991).
[2]   Y. Bouhnik, Survie et effets chez l'homme des bactéries ingérées dans les laits fermentés. Lait **73**, 241-247 (1993).
[3]   F. Ampe , N. Omar, C. Moizan, C. Wacher and J. P, Guyot. Polyphasic Study of the Spatial Distribution of Microorganisms in Mexican Pozol, a Fermented Maize Dough, Demonstrates the Need for Cultivation-Independent Methods To Investigate Traditional Fermentations. Applied and Environmental Microbiology. Vol. **65**, 5464-5473 (1999).
[4]   R. Amann, W. Ludwig, and K.H. Schleifer. Phylogenetic identification and in situ detection of individual microbial cells without cultivation. Microbiol. Rev. **59**, 143-169 (1995).
[5]   A.C. Ouwehand, T. Kurvinen and P. Rissanen. Use of a probiotic Bifidobacterium in a dry food matrix, an in vivo study. Int. J. Food Microbiol. **95**,103-106 (2004).
[6]   A. Moter and U. B. Göbel. Fluorescence in situ hybridization (FISH) for direct visualization of microorganisms. J. Microbiol. Methods **41**, 85-112 (2000).
[7]   K. Kogure, U. Simidu and N. Taga. A tentative direct microscopic method for counting living marine bacteria. Can. J. Microbiol. **25**, 415-420 (1979).
[8]   S. Kalmbach, W. Manz, and U. Szewzyk. Isolation of new bacterial species from drinking water biofilms and proof of their in situ dominance with highly specific 16S rRNA probes. Appl. Environ. Microbiol. **63**, 4164-4170 (1997).
[9]   Y. Moreno, M.A. Ferrús, J.L. Alonso, A. Jiménez and J. Hernández. Use of fluorescent in situ hybridization to evidence the presence of *Helicobacter pylori* in water. Wat. Res. **37**, 2251-2256 (2003).
[10]   C. Buchrieser and C.W. Kaspar. An improved direct viable count for enumeration of bacteria in milk. Int. J. Food. Microbiol. **20**, 227-236 (1993).
[11]   J. Baudart, J. Coallier, P. Laurent and M. Prévost. Rapid and Sensitive Enumeration of Viable Diluted Cells of Members of the Family *Enterobacteriaceae* in Freshwater and Drinking Water. Applied and Environmental Microbiology, **68**, 5057-5063 (2002).

Fluorescence imaging and new fluorescence techniques applied to cellular pathopharmacology

Dalgis Mesa[1]*, Elli Kohen**, Joseph G. Hirschberg**, Roger Leblanc *

*Nanomethods Laboratory, Chemistry Annex, Department of Chemistry
University of Miami 1301 Memorial Drive Coral Gables, Florida 33146-0431 USA. **Department of Physics, University of Miami James L. Knight Physics Building
1320 Campo Sano Drive Coral Gables, FL 33146 USA.

[1]Corresponding author: e-mail: dalgis450@bellsouth.net , Phone: (786) 487-6066, (305) 260 0892

Fluorescence imaging is carried out on a variety of Cancer cell in culture (colon, breast, skin, neuroblastoma, and ovary). Vital Fluorescence Probes of mitochondria, endoplasmic reticulum, Golgi, lysosomes, nucleoli nucleolar channels are added, with single spectral band excitation.This is a first step towards multiwavelength fluorescence excitation by Fourier Interferometry. It is now planned to initiate work with a new family of fluorophores, i.e. Quantum Dots (QDs) which may supercede other fluorescence techniques. QDs are bright fluorophores with high quantum yields, narrow fluorescence emission bands, and high resistance to photobleaching; they can provide excitation of several different emission colors using a single excitation wavelength.

1 Introduction

The spectral fluorescence imaging methods applied to a variety of cancer cells or otherwise genetically deficient cells are starting to reveal the related organelle alterations and interactions [1,2]. Early studies indicate that in terms of cellular physiology the information obtained from intracellular metabolic probes would be significantly more valuable than the information derived form exogenous vital fluorescent probes [1,2] requires operating at the limits of detection sensitivity. The most informative natural probes are NAD(P)H and Flavins, which connect to practically all biochemical pathways of energy metabolism whether in the mitochondria, the cytoplasm, or the nucleus. With the CCD camera, in specific, ICCD (intensified CCD) interesting results have been obtained. The Lakowicz group has enabled to gently enhance the fluorescence intensity available for studies in bio-medicine by using radiative decay engineering in the presence of metallic surfaces in the medium containing the radiative fluorophores [3,4]. Molecular Probes (Eugene, Oregon, USA) has introduced Prolong Gold antifade reagent for protection from photobleaching. The results indicate good intracellular penetration of this fluorescence enhancer. Therefore, it is expected that the sensitivity barrier to detail studies of metabolic control and regulation or the regulation connected to cellular Pathopharmacology will be overcome by the above approach. It is now planned to initiate work with a new family of fluorophores, i.e. Quantum Dots (QDs) i.e bright fluorophores with high quantum yields, narrow fluorescence emission bands, and high resistance to photobleaching; they can provide excitation of several different emission colors using a single excitation wavelength [3]. QDs are suitable for detecting multiple ligands in a single experiment. These fluorophores can be used at all levels of structural resolution from micro to nano-levels; toxicity can be minimized by protective coating. In addition treatment with the single excitation wavelength fluorophore DiOC$_6$ (3) a Carbocyanine dye, has resulted in multi-organelle

780

localization (Golgi, lysosomes, nucleoli). These methods are in principle applicable to cytodiagnostics, cytoprognostic, and drug trials.

2 Materials and methods

Fluorescence imaging using a CCD camera equipped with an INTEGRATER (for operation in live and frame-accumulate) modes has been extended to a broad variety of extrinsic probes [2]:

Table 2.1

Cell Organelle	Probes
Mitochondria	DASPMI, rhodamine 123, tetramethylrhodamine methylesther (TMRM), TMRI, and mitochondria green.
Lysosomes	atebrine (quinacrine) and fluoro-bora or flurorogenic ptoboes of lysosomal hydrolases, i.e. nonylmethulumbelliferylglucose (UG9)
Golgi apparatus	NBD ceramide
Endoplasmic Reticulum	$DiOC_6$ (3)
Nuclear channels	$DiOC_6$ (3)
Cytoplasmic membrane	merocyanine

Table 2.2 Biological material.

Cells used	Expected results
Yeast	Membrane studies, mitochondrial organization, photo-irradiation, accumulation of Schiff-bases.
Chlamydomonas	Regulation of hydrogen production.
Malignant Cells colon, neuroblastoma, osteosarcoma, and skin cancer	Multi-organelle complex formation, patho-pharmacological evaluations, cyto diagnostics, sito prognostics, drug trials (preferably with NADPH and Flavin reductions)

In addition to probes in **Table 2.1** studies were initiated with the fluorescent nano-crystals called quantum dots [3]. One of the quantum dots used has a Cadmiun Selenide core and a Zinc Sulfide shell. The probe which is water soluble stabilized with Mercaptoacetic acid. The core particle size is around 2.5 nm. Initial trials were on yeast cells (zaccharomyces cereviseie). Within seconds the probe accumulated on the membrane of yeast cells and in peripheral granules with a ring like organization. At this stage, it is reasonable to conclude that we have no clear evidence of intracellular probe penetration except for accumulation on the cell membrane. The ultimate goal however, which is targeted should be probe penetration and attachment to cytoplasmic and eventually nuclear structures.

3   Results and discussion

Metabolic Studies

The metabolic studies depend on reliability of NAD(P)H imaging in response to reducing glycolitic substrates or Kreb Cycle substrates. There is the glycolitic response which has been described over a century ago by Lois Pasteur over a century ago. There are however, very few actual images of the mitochondrial NAD(P)H response to glucose challenge. Figure 1 (before and after glucose) illustrate such as response within mitochondrial clusters of an osteosarcoma cell in culture.

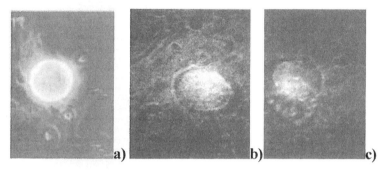

Fig. 1a. Phase image of an osteosarcoma cell in culture. The main characteristic is the gigantism of the nucleus.
Fig. 1b. Fluorescence Image of the same cell under UV excitation by 365 nm mercury line.The cell has been treated with 10micromolar rotenone which blocks the NADH-cytochrome B reductase.
Fig. 1c. Within ten seconds of glucose addition (five to ten milli-molar) glucose there is intense fluorescence of mitochondria-like corpuscles. This metabolic response is attributed to mitochondrial NAD(P)H.

Study on Yeast:

Figure 2 shows the fluorescence of a mitochondrial probe localized within yeast mitochondria organized at the periphery of the cell.

Fig. 2. Localization of DASPMI fluorochrome in yeast mitochondria. Alternatively diffuse fluorescence of Schiff bases may be obtained after 10-15 sec. continuous UV irradiation (not shown here). These shieff-bases are the product of lipids peroxidation induced by exposure to UV light.

Studies on Cancer Cells:

Figure 3 illustrates one of the major applications of cancer cell fluorescence imaging. These are further discussed in the conclusion explaining the diagnostics, prognostic and therapeutic potentials of the methods.

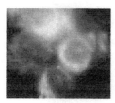

Fig. 3. Fluorescence image of cytoplasmic organelles in a cancer cell in culture (an example for undifferentiated colon, breast, and ovarian cancer).

Projected Instrumental Design:
Figure 4 describe the projected design aimed to enhance the potentials, sensitivity, and versatility of the methods.

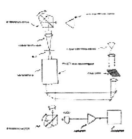

Fig. 4. Projected design for fluorescence excitation-emission spectral imaging using Fourier Interferometry. The excitation source can be simplified to a light emitting diode provided only the visible part of the light spectrum is used. An interferometer (Hirschberg's Pentaferometer) is placed on both the excitation and emission sides. A liquid crystal array (LCA) on the excitation side and microphone on the culture cuvette for combined photo- acoustic and fluorescence imaging [3,5].

Studies with Quantum Dots:
Figure 5 a and b show expected results using nano crystals of Cadmiun Selenide (QDs). Because of the very high quantum yields it has been possible to record bright images indicating nano-crystals accumulation on the membrane of yeast, using live recording with 32 milli-second frame scan (Fig. 6). To be really applicable intracellular penetration, organelle localization, or even intranuclear localization is searched for and should be realizable by attachment of peptides.

Fig. 5a. Bright dots within the nuclei of human gioblastoma cell culture are FISH (fluorescence in situ hybridization) signals from repetitive DNA sequences of chromosome seven.

Fig. 5b. Same as above in chromosomal images.Chromosomeimages showing localization of
FISH probe in glioma cell (From of Prof. P.E. MacKeever and el, Dept of Pathology Ann Arbor,
Michigan). It is expected that similar results will be obtained with Quantum Dots using a design like in Fig. 6.4 [3,6]

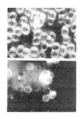

Fig. 6. Accumulation of Quantum dots at the surface of yeast membranes in the intact cell. At this time there is no evidence of intracellular penetration, but it is hoped that with addition of peptides to the quantum dots (nano-crystals of cadmium selenium) penetration ultimately will be achieved.

4 Conclusion

The overall long range goal, is to make this methods accessible to broad applications in cyto-diagnostics, cyto-prognostics, and drug trials. A two-pronged approach should yield promising results: on one side the endogenous metabolic probes and on the other side the highest quantum yield fluorescence probe, such as the quantum dots nano-crystals. The metabolic probes are applicable to what may be called systems cellular physiology, and the unifying concept can be developed for the regulation of metabolism from bacteria through hydrogen-producing algae, and mammalian cells.

5 References

[1]   Elli Kohen, et al (2002). Fluorescence probes in Oncology. Imperial College Press, London.
[2]   Elli Kohen, et al (2004). Atlas of Cell Organelles Fluorescence. CRC Press, Boca Raton.
[3]   Elli Kohen, et al. Advances in Phototoxicology and Implications in Cellular Pathopharmacology, Recent Res. Devel. Photochem. Photobio., 7(2004): **175-200**.
[4]   Lakowicz, J. R. (2001). Radiative Decay Engineering: Biophysical and biomedical Applications. Analytical Biochemistry 298, 1-24.
[5]   Hirschberg, J.G, and Kohen, E. (1999). Pentaferometer: a solid Sagnac interferometer, *Applied Optics*, 38:**136-138**.
[6]   McKeever PE (1996), *Cancer Gene. Cytogen.* 87: **41-47**.

Impact of *Bifidobacterium* animalis DN-173010 on Human Intestinal Microbiota by Fluorescent in situ Hybridization

Y. Moreno[1], M. C. Collado[1], J. García[1], J. M. Cobo[2], E. Hernández[1], M. Hernández[1]*

[1]Department of Biotechnology, Polytechnic University of Valencia, Camino de Vera 14, 46022 Valencia, Spain.
[2] Red INDE, Investigación Nutricional Danone España, Barcelona, Spain

The aim of this study was to obtain accurate data on the fate of ingested bifidobacteria in humans as a first step towards assessing the physiological importance of ingested bifidobacteria. We have developed a culture-independent protocol based on fluorescent *in situ* hybridization (FISH) for the identification and enumeration of bifidobacteria directly in fecal samples. An oligonucleotide probe specific to the 16S rRNA region of *B. animalis* DN-173010 was designed to identify this strain in fecal samples of healthy subjects throughout the course of a controlled feeding study by FISH technique. Fermented milk supplemented with *B. animalis* DN-173010 was administered to 12 healthy volunteers for four weeks. A significant increase in the number of *Bifidobacterium* cells from both the product and endogenous flora was observed in feces. Along the assay the number of bifidobacteria enumerated by culturing was 10 to 100-fold lower than by FISH technique. These results confirm the usefulness of 16S rRNA-based techniques for the direct and specific detection of bifidobacteria in fecal samples in few hours.

Keywords: *Bifidobacterium*, fermented milk, FISH, fecal samples
Abbreviation key: FISH = fluorescent in situ hybridization

1   Introduction

Some organisms used as probiotics are bacteria from the genus *Bifidobacterium,* which consists of gram-positive anaerobic microorganisms with a variety of rod morphologies that appear to be among the most prevalent microflora in the gastrointestinal tract of humans and animals [1–3]. The contribution of these bacteria to good health has been recognized for quite some time and has led to widespread use of bifidobacteria as probiotics for maintaining or improving human and animal health [4, 5]. It has been suggested that the action of probiotics should occur when they are alive and numerous enough in the intestine (approximately 10^6–10^7 microorganisms per ml) in order to appreciate their effects and to reach, adhere, implant or multiply in the intestinal tract [6]. However, Mechanisms responsible for the beneficial effects associated with bifidobacteria are not clearly understood [7]. Since there are inherent difficulties in obtaining definitive evidence for the proposed effects of ingesting exogenous bifidobacteria, there is a great deal of speculation about the possible prophylactic and therapeutic properties of foods containing bifidobacteria. One difficulty is the presence of endogenous bifidobacteria in the gastrointestinal tract and human feces, which complicates the task of differentiating these from ingested bifidobacteria and unequivocally demonstrating the survival of ingested bifidobacteria through the gastrointestinal tract.
Ribosomal rRNA probe hybridization without cultivation has been widely adopted for detection of specific bacterial groups in mixed populations [8] and to identify and enumerate different bacterial groups in fecal samples in few hours [9]. The aim of this study was to obtain accurate data on the fate of ingested bifidobacteria in humans as a first

step towards assessing the physiological importance of ingested bifidobacteria. Thus, an oligonucleotide probe specific to the 16S rRNA region of *B. animalis* DN-173010 was designed to identify product bifidobacteria by FISH technique directly in fecal samples of healthy subjects throughout the course of a controlled feeding study.

2   Materials and methods

2.1 Bacterial strains

Four reference strains, *Bifidobacterium infantis* ATCC15697, *Bifidobacterium longum* ATCC 15707, *Bifidobacterium bifidum* ATCC 2952, *Lactobacillus delbrueckii bulgaricus* ATCC 11842 and twenty strains of *Bifidobacterium animalis/lactis, Streptococcus thermophilus* and *Lactobacillus delbrueckii* isolated from different commercial dairy products were used to test the specificity of the probes. Bifidobacteria were cultured on TPY (Tryptone, Peptone and Yeast Extract; Scharlau-Chemie, Barcelona, Spain) agar plates and Lactobacilli were plated on MRS agar (Merck, Darmstad, Germany) and incubated under anaerobic conditions at 37°C for 48-72 hours.

2.2 Subjects and experimental schedule

The initial population consisted of 12 healthy volunteers (6 women and 6 men with ages ranging from 25 to 60 years-old) who did not include dairy products in their regular diet. The duration of the study was 12 weeks, which was divided into three periods: (1) 4 week with no ingestion of any fermented milk with bifidobacteria (control period); (2) 4 week with administration of fermented milk; (3) 4 week with no administration of bifidobacteria. During the bifidobacteria ingesting period, subjects consumed 250 ml of the assigned fermented milk daily. The number of viable bifidobacteria in 250 ml of fermented product was estimated by microscopy counts as 1.4×10^9 viable cells/ml. Throughout the entire experimental period the volunteers did not consume any liquid milk products or fermented dairy products prepared using lactic acid bacteria (LAB). The feces of an individual without product ingestion (sample L, negative control) and of another subject with continuous product ingestion for three months before starting the study (sample K, positive control) were monitored as controls.

2.3 Collection of fecal samples and plate counts

Fecal analysis was performed every week during the three periods. All samples were collected fresh in sterile plastic recipients, placed in sterile glass bottles, refrigerated, transported immediately to the laboratory, and processed without further delay. The fresh fecal samples were homogenized (1:10 wt/v) with PBS buffer (130 mM sodium chloride, 10 mM sodium phosphate, pH 7.2) .
Serial dilutions ranging from 10^{-2} to 10^{-8} from homogenized fecal samples were plated on the appropriate agar media. Bifidobacteria were enumerated on BFM Agar [10] plates and incubated anaerobically (Anaerogen , Oxoid) at 37°C for 72 h. Counts were made by triplicate, the values were averaged and bacterial concentrations were expressed as the $\log_{10}$ of the number of CFU/g wet weight of feces.

2.4 Oligonucleotide probes

A 16S rRNA sequence of *Bifidobacterium animalis* (BIFA1; 5′-CCAGCGTTCATCCTGAGCT-3′) retrieved from GeneBank/EMBL was selected as possible oligonucleotide probe to be used for specific hybridization of bacteria. The BLAST program (www.ncbi.nlm.nih.gov) was used to align and compare the sequence of the probe with other available sequences. The specificity of the probe was tested by whole cell in situ hybridization of several species belonging to genus *Bifidobacterium*, *Lactobacillus* and *Streptococcus* isolated from dairy products and from reference collections. The Bif668 genus specific probe was used to target all *Bifidobacterium* species [11]. The EUB 338 universal probe, complementary to a 16S rRNA region of the domain Bacteria [8], was used as a positive control. Probes were synthesized and labelled with fluorescein (FLUOS) and CY3 by Tib Molbiol (Berlin, Germany).

2.5 FISH conditions

One millilitre from an exponential culture of each bacteria used for specificity assay was centrifuged (1000 x g, at 4°C for 10 min), resuspended in PBS buffer, and fixed with three volumes of 4% paraformaldehyde at 4°C for 2 h. Subsequently, fixed samples were centrifuged, washed with PBS buffer and finally resuspended in 1:1 PBS/ethanol (vol/vol). For feces analysis, 1 g of each sample was homogenized in sterile PBS buffer. An aliquot of the mix was fixed as describe above and stored at –20°C. An aliquot of 5 μl fixed bacteria was placed on a gelatine-coated slide, air dried, dehydrated (50, 80, 100% ethanol) and hybridized as described by Moreno et al. [12]. To provide specific hybridization for the target organisms, a final concentration of formamide was established at 20 % in the hybridization buffer (0.9 M NaCl, 0.01% SDS, 20 mM Tris-HCl, pH 7.6). Slides were examined under epifluorescence Olympus BX50 microscope equipped with a set of U-MWIB and U-MWIG filters. Ten to twenty fields were counted for sample.

3   Results and discussion

3.1 Probes

Under stringent conditions, the BIFA1 probe was able to hybridize with the bifidobacteria isolated from the product (DN-173010), whereas none of the other bifidobacteria and LAB species tested showed fluorescent signal.

3.2 Fecal analysis

To enumerate all the different bifidobacteria contained in the feces of the 12 volunteers, we used the Bif662 genus-specific probe and the BIFA1 specific probe (Table 1). The remaining microbiota present in these samples could be monitored with the EUB338 probe. The total number of bifidobacteria before product ingestion ranged from 2.81×10^9 cells/ml to 3.6×10^7 cells/ml, depending on the individual (Table 1). According with the results of others authors [13, 14] the initial bifidobacteria counts in feces using plate counts were between one and two orders of magnitude less than those obtained by FISH (Table 1).

Table 1 Results obtained using cultivation and FISH techniques with Bif 662 (*Bifidobacterium* genus) and BIFA1 (*B. animalis* species) probes in control period (4 weeks without DN-173010 ingestion). Positive control ingested *B. animalis* during control period. [A] mean value (mean standard error).

		Control period			
Subjects		Log CFU/ml		Log cells/ml	
		Bifidobacteria (n=3)	Enterobacteria	Bifidobacteria Bif662 (n=20)	Bifidobacteria Bif a1 (n=20)
Individual A	Men	6.2 (0.08)[A]	6.2	9.03 (0.12)[A]	-
Individual B	Men	6.0 (0.02)	5.9	9.00 (0.99)	-
Individual C	Men	6.5 (0.01)	4.7	8.08 (1.87)	-
Individual D	Men	7.0 (0.01)	6.2	8.93 (0.42)	-
Individual E	Men	6.0 (0.02)	6.5	8.51 (0.28)	-
Individual F	Women	6.0 (0.02)	6.1	8.98 (0.65)	-
Individual G	Women	6.0 (0.01)	5.0	9.41 (1.07)	-
Individual H	Women	6.0 (0.02)	6.2	9.45 (1.38)	-
Individual I	Women	6.0 (0.03)	5.2	9.20 (0.94)	-
Individual J	Women	6.0 (0.03)	6.7	7.56 (0.39)	-
Control +	Men	7.7 (0.03)	7.3	8.53 (0.38)	8.68 (1.43)
Control -	Women	7.2 (0.08)	7.3	8.36 (1.87)	-

This shows that FISH is useful for detecting bacteria with or without culturable capacity, or damaged by an adverse circumstance such as transit through the gastric cavity. The fluctuation in the number of bifidobacteria due to product ingestion and the error due to the duplicate count of 10 fields from the same sample were determined.

Bifidobacteria counts using FISH technique during all the study are showed in Table 2. In general, the total bifidobacteria (endogenous and exogenous) population increased with product ingestion.

After two weeks of ingestion, the quantity of bifidobacteria from the product was ten times lower than the total bifidobacteria population (Table 2), except for the positive control, for which all bifidobacteria hybridized with both the Bif662 and the BIFA1 probes. Therefore, all bifidobacteria were procedent from the product. In these samples, we also observed a general increase in both exogenous and endogenous bifidobacteria.

After three weeks of ingestion, hybridization with BIFA1 showed that the number of exogenous bifidobacteria had increased ten times in all individuals. We observed competition between exogenous and endogenous bifidobacteria in the women's subgroup, with an inital increase in exogenous bifidobacteria leading to a temporarily slight reduction in endogenous bifidobacteria.

Table 2 Results obtained using in situ hybridization with Bif 662 (genus *Bifidobacterium*) and BIFA1 (species *B. animalis*) probes throughout the experiment. ^ mean value (mean standard error)

Subjects	1 week ingestion		2 weeks ingestion		3 weeks ingestion		4 weeks ingestion		2 weeks no ingestion		3 weeks no ingestion	
	Bif662	Bif a1	Bif662	Bif a1	Bif662	Bif a1	Bif662	Bif a1	Bif662	Bif a1	Bif662	Bif a1
A	9.44 (0.32)^	-	9.02 (1.00)^	8.66 (0.50)	9.55 (0.36)	8.85 (0.13)	10.24 (1.20)^	9.77 (0.46)	10.00 (0.93)	8.90 (0.27)	9.54 (0.96)^	8.06 (0.33)
B	9.57 (0.44)	-	9.59 (2.56)	8.28 (0.34)	11.00 (0.67)	9.91 (0.89)	10.50 (0.74)	9.62 (0.62)	10.48 (1.77)	8.69 (0.13)	-	-
C	9.39 (0.23)	-	9.23 (1.25)	8.67 (0.57)	10.61 (2.63)	10.14 (1.32)	10.77 (0.59)	9.49 (0.30)	10.42 (0.42)	8.86 (0.15)	9.04 (0.04)	6.83 (0.11)
D	9.18 (0.40)	-	9.63 (2.80)	8.74 (0.75)	10.86 (0.26)	10.60 (4.90)	10.32 (1.63)	9.56 (0.34)	9.90 (0.57)	8.56 (0.14)	8.80 (0.06)	7.46 (0.17)
E	9.57 (3.61)	-	10.16 (0.64)	8.64 (0.51)	11.09 (0.79)	9.81 (0.62)	11.51 (1.06)	9.68 (0.36)	10.87 (0.85)	8.31 (0.16)	10.89 (0.02)	8.23 (0.02)
F	10.20 (3.30)	-	10.67 (0.25)	9.48 (0.34)	10.34 (2.29)	10.05 (1.70)	10.65 (1.45)	9.97 (0.67)	10.33 (1.22)	8.61 (0.18)	10.60 (0.02)	8.19 (0.06)
G	10.25 (1.70)	-	10.82 (0.09)	9.41 (0.25)	10.78 (1.81)	9.79 (0.48)	10.87 (0.28)	9.36 (0.17)	11.10 (0.85)	8.61 (0.14)	10.99 (0.05)	8.61 (0.10)
H	10.54 (3.20)	-	10.78 (0.40)	9.52 (0.26)	10.54 (1.56)	9.79 (1.75)	10.90 (0.70)	9.84 (0.55)	10.14 (0.99)	9.19 (0.13)	11.08 (0.02)	8.89 (0.06)
I	-	-	9.78 (0.03)	8.52 (0.43)	9.09 (0.40)	8.79 (0.28)	9.89 (0.53)	9.27 (0.30)	10.01 (0.76)	8.49 (0.11)	10.09 (0.03)	8.31 (0.09)
J	9.21 (0.23)	-	8.48 (0.28)	7.95 (0.19)	9.87 (1.73)	9.74 (1.55)	8.44 (0.26)	8.19 (0.30)	-	-	-	-
+ Control	8.53 (0.94)	8.59 (0.11)	9.00 (1.64)	8.68 (0.83)	9.63 (0.86)	9.57 (0.70)	9.74 (0.49)	9.62 (0.39)	9.16 (0.47)	9.16 (0.47)	9.31 (0.08)	9.16 (0.07)
- Control	8.13 (0.33)	-	8.69 (1.17)	-	8.83 (1.56)	-	8.37 (1.70)	-	7.68 (0.49)	-	8.20 (0.50)	-

From the fourth week on, in the men's subgroup, the Bif662 probe indicated that there was an increase in the number of total bifidobacteria in 60% of the individuals, due to the exogenous bifidobacteria reached the maximun.

After two weeks without product ingestion, the probe counts revealed a progressive decrease in the number of exogenous bifidobacteria. In samples analyses after four weeks without product ingestion, neither exogenous nor endogenous bifidobacteria were detected by FISH or by plate counting in 10% of the samples from each subgroup (samples B and J), this maybe due to limitations of the methods (limit for FISH was approximately 10^5 bacteria/g) . These results are according with other authors who reported the presence of bifidobacteria in feces after oral administration and they were not able to detect bifidobacteria by FISH in some subjects [15]. Nevertheless, the endogenous bifidobacteria population appeared to remain stable throughout the study although in some cases the number of endogenous bifidobacteria at the end of the study was greater or smaller than at the beginning. This means that consumption of *B. animalis* DN-173010 contained in fermented milk product change the bifidobacteria composition in fecal samples.

Therefore, continued product ingestion is required to maintain the appropriate number of *B. animalis* DN-173010 to cause beneficial effects for the host, since they remain in the intestinal tract for approximately 3 weeks. This assay appears to show the beneficial effect of product bifidobacteria on endogenous microbiota recovery.

FISH technique applied to detect bifidobacteria in human fecal samples demonstrated that this group of bacteria could be enumerated at least as accurately as by the conventional cultivation technique which has been observed by other authors [11]. FISH technique has the potential to be used as a quick and sensitive tool for the detection of *Bifidobacterium* spp. in fecal samples.

References

[1]   B. Biavati, M. Vescovo, S. Torriani, V. Bottazzi, Bifidobacteria: history, ecology, physiology and applications, Ann. Microbiol. 50 (2000) 117-131.

[2]   V. Scardovi, in Bergey's manual of systematic bacteriology. P. H. A. Sneath, N. S. Mair, M. E. Sharpe, and J. G. Holt, (Eds.), Genus *Bifidobacterium*. Williams and Wilkins, Baltimore, 1986, Vol. 2. Pages 1418-1434.

[3]   G.L. Simon, and S.L. Gorbach, Intestinal flora in health and disease. Gastroenterology 86 (1984)174-193.

[4]   A.M.P. Gomes and F.X. Malcata, *Bifidobacterium* spp. and *Lactobacillus acidophilus*: biological, biochemical, technological and therapeutical properties relevant for use as probiotics. Trends Food Sci. Technol. 10 (1999)139-157.

[5]   C. Stanton, G. Gardiner, H. Meehan, K. Collins, G. Fitzgerald, P. B. Lynch and R. P. Ross. Market potential for probiotics. Am. J. Clin. Nutr. 73 (2001) 476-483.

[6]   J. A. Kurmann and J. L. Rasic, in Therapeutic properties of fermented milks. R. K. Robinson, (Ed.) The health potential of products containing bifidobacteria, Elsevier Applied Food Sciences, London, 1991, Pages 117-158.

[7]   M.E. Sanders, Summary of conclusions from a consensus panel of experts on health attributes of lactic cultures: significance to fluid milk products containing cultures. J. Dairy Sci. 76 (1993)1819-1828.

[8]   R.I. Amann, W. Ludwig and K.H. Schleifer, Phylogenetic identification and *in situ* detection of individual microbial cells without cultivation, Microbiol. Rev. 59 (1995)143-169.

[9]   P. Marteau, P. Pochart, J. Doré, C. Béra-Maillet, A. Bernalier, G. Corthier, Comparative study of bacterial groups within the human cecal and fecal microbiota. J. Appl. Environ Microbiol. 67 (2001) 4939-4942.

[10]   Y. Nebra, A. R. Blanch, A new selective medium for *Bifidobacterium* spp. Appl. Environ. Microbiol. 65 (1999) 5173-5176.

[11]   P. S. Langendijk, F. Schut, J. J. Gijsbert, G. C. Raangs, G. R. Kamphuis, M. H. F. Willeinson and G. W. Welling, Quantitative fluorescence *in situ* hybridization of *Bifidobacterium spp.* with genus-specific 16S rRNA-targeted probes and its application in fecal samples. Appl. Environ. Microbiol. 61 (1995) 3069-3075.

[12]   Y. Moreno, M.A. Ferrús, J.L. Alonso, A. Jiménez and J. Hernández. Use of fluorescent in situ hybridization to evidence the presence of *Helicobacter pylori* in water. Wat. Res. 37: 2251-2256. (2003).

[13]   A.H. Franks, H.J.M. Harmsen, G.C. Raangs, G.J. Jansen, F. Scut and G.W. Welling, Variations of bacterial populations in human feces measured by fluorescent in situ hybridization with group-specific 16S rRNA-targeted oligonucleotide probes. Appl. Environ. Microbiol. (1998) 64, 3336-3345.

[14]   H. J. M. Harmsen, G.R. Gibson, P. Elfferich, G.C. Raangs, A.C.M. Wildeboer-Veloo, A. Argaiz, M.B. Roberfroid and G.W. Welling, Comparison of viable cell counts and fluorescence in situ hybridization using specific rRNA-based probes for the quantification of human fecal bacteria. FEMS Microbiol. Lett. 183 (2000) 125-129.

[15]   A.C. Ouwehand, T. Kurvinen and P. Rissanen. Use of a probiotic Bifidobacterium in a dry food matrix, an in vivo study. Int. J. Food Microbiol. 95: 103-106. (2004).

Scanning electron microscope study of fish and rice flour coextrudates

Tumuluru Jaya Shankar[1] and Sukumar Bandyopadhyay[2*]

[1] Department of Process and Chemical Engineering, University College Cork, Ireland
[2] Agricultural and Food Engineering Department, Indian Institute of Technology, Kharagpur-721302, India

[*] Corresponding Author: e-mail: bandyo@agfe.iitkgp.ernet.in, Fax: +91-3222-82244; Tel: 91-3222-283102;

Microstructure of fish and rice flour coextrudates was observed using scanning electron microscope (SEM). Experiments were conducted based on rotatable experimental design and microstructure was studied of the products extruded at different combinations of extruder barrel temperature, extruder screw speed, and fish and moisture contents of the feed. SEM photographs revealed that a high extruder barrel temperature of 200° C effected the formation of porous and fibrous product due to starch gelatinization with network of holes whereas the product extruded at 100°C was more compact. A high barrel temperature of 200°C during extrusion cooking resulted in extensive protein denaturation and fiber formation and the fiber strands got cemented with starch material leading to cross-linking. Extruder screw speed of 110 rev/min helped in formation of smooth texture, whereas low extruder screw speed of 70 rev/min resulted in more fragmentations. High protein levels due to maximum fish content of 45 % favored greater fiber formation, better organization and orientation than that with low protein level. Feed moisture content of 60 and 20 % induced cracks in the texture, but cracks were more pronounced at lower moisture content indicating less cross linking and binding.

Keywords: Extrusion cooking; scanning electron microscope

1 Introduction

Fish mince or powder, blended and coextruded with rice flour, produces some promising snack food products with nutritional combinations [1, 2]. Extrusion cooking imparts characteristic texture to the product, which is a desirable functional property for use as cereal-based fish snacks and crackers. Texture of these products varies depending upon extrusion processing parameters. Extruder barrel temperature and screw speed coupled with moisture and protein level of the raw material are the four important determinants of texture. Food texture being a direct consequence of microstructure, an examination of product microstructure by Scanning electron microscope (SEM) has proven a satisfactory method for evaluation and prediction of textural properties. Several researchers studied SEM photographs of extruded starch based protein products [3–5]. However little information is available on microstructure of fish and rice flour coextrudates. The present work deals with SEM studies of extrudates prepared out of fish and rice flour blends to see the effect of process variables on texturization, cross-linking, cracks formations, fibrillation and orientation of starch and protein matrix.

2   Materials and methods

2.1 Extrusion

Tropical marine fish known as Bombay duck (Harpodon nehereus) and rice flour both made into powder, passing through 353 um mesh sieve, were used for preparing the extrudates. A five level central composite design was used to analyze the data. Table 1 gives the actual and coded levels and experiments were conducted based on rotatable experimental design [2]. The combinations of extruder barrel temperature, extruder screw speed, fish content and feed moisture content used for SEM study are indicated in Table 2.Calculated amount of fish and rice flour was mixed well, adjusted to a desired level of moisture content of the mixture by adding water, and was extruded. The extrudate strands were dried at 60–65°C for 2–2.5 h. The dry extrudate containing moisture content within 7--15 % (w.b) were kept in sealed polyethylene pouches and stored in airtight containers for the moisture content to equilibrate. The samples were periodically checked and found to vary between 7 and 9 % moisture content (w.b) and were used for further analysis [2].

Table 1 Actual and coded levels of experimental design

Independent variables		Coded levels				
		-2	-1	0	-1	2
Barrel temperature (°C)	X_1	100	12 5	150	175	200
Screw speed (rev/min)	X_2	70	80	90	100	110
Fish content (%, w/w)	X_3	5	15	25	35	45
Feed moisture content (%, v/w)	X_4	20	30	40	50	60

Table 2 Level combinations for SEM observations

Expt. No	Temperature, °C X_1	Screw speed, rev/min X_2	Fish content, % (w/w) X_3	Feed moisture content, % (v/w) X_4
Q	2	0	0	0
R	-2	0	0	0
S	0	2	0	0
T	0	-2	0	0
U	0	0	2	0
V	0	0	-2	0
W	0	0	0	2
X	0	0	0	-2

2.2 Scanning electron microscope (SEM)

Small quantity of dry extruded rods, dried under vacuum to moisture content 5-6 % (d.b) were cryo-fractured by liquid nitrogen immersion technique. This method consisted of dipping the extrudate rods in liquid nitrogen for 2-3 minutes. The frozen extruded rods were broken into 2-5 mm sizes, and cut into sections of sharp edges. The sections were collected and lyophilized to make them free of moisture. The samples were then coated with gold in vacuum using a sputter coater and examined at 20 kV with JEOL JSM-5800 scanning microscope (made in Japan). Each SEM photograph was observed at 1000 λ magnification.

3   Results and discussions

The SEM photographs were taken where one process variable was at its maximum and minimum values and the other three were at the center point of the rotatable experimental design. This helped to understand the effect of each individual process variable on microstructure within the range of maximum and minimum values.

3.1 Effect of temperature

Fig.1 (Q) shows the effect of extruder barrel temperature of 200°C. At 200°C maximum expansion ratio was achieved and the product became porous and fibrous with network of holes. Extrusion of previously moistened soy flour at temperature of 140–180°C resulted in network of holes after solidification at ambient temperatures [6]. Sudden drop of pressure and vaporization of superheated entrapped water results in porous and fibrous product. When the extrudate leaves the die, the starch embedded in the protein matrix quickly solidifies, retains partially open-celled structure and produces a low density product. The high temperature during extrusion cooking resulted in extensive protein denaturation and fiber formation and the fiber strands got cemented with starch material and leading to cross-linking. Fig.1 (R) shows the product extruded at 100°C where no holes were observed

Q)   R)

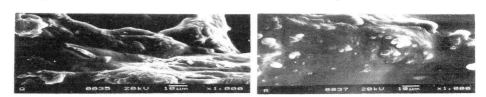

Fig. 1 SEM photographs (Q and R) of the extruded product at a barrel temperature of 200 and 100°C.

3.2 Effect of screw speed

Fig. 2 (S) at 110 rev/min there was quicker movement of the material inside the extruder giving less time for the material to gelatinize and expand, resulting in minimum holes and smooth structure. In Fig.2 (T) at 70 rev/min the material was in the barrel for longer time resulting in more fragmentations and small holes.

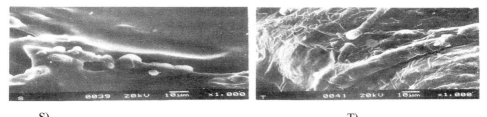

S)   T)

Fig. 2 SEM photographs (S and T) of the extruded product at a screw speed of 110 and 70 rev/min.

3.3 Effect of fish content

In case of Fig. 3 (U) a maximum fish content of 45 % resulted in fibrous compact matrix with minimum holes and reasonable expansion and greater homogeneity. Higher amount of protein present in extruder feed favored the formation of unbroken fibers better organization and orientation. Incase of Fig.3 (V) where the fish content was 5 % cracks were clearly visible in the photographs and structure was less compacted due to insufficient binding and cross linking at the time of extrusion.

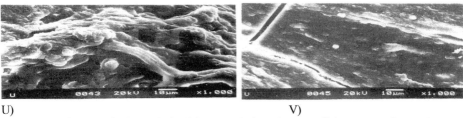

U)   V)

Fig. 3 SEM photographs (U and V) of the extruded product at a fish content of 45 and 5 %.

3.4 Effect of feed moisture content

Fig.4 (W and X) show the effect of feed moisture content at 60 and 20 % respectively. The holes were observed more with 60 % than that with 20 %. The increase in moisture content leads to formation of more holes. Low fat content and increased water content lowered the protein concentration, where the involvement of protein in the matrix formation got reduced. At feed moisture content of 60 and 20 % cracks were observed, but cracks were more pronounced at lower moisture content indicating less cross linking and binding. At higher moisture content the product appeared to be more multi layered and less compacted.

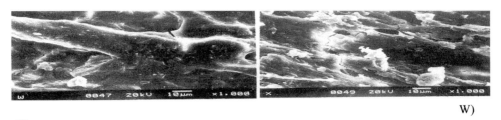

W)

X)

Fig. 4 SEM photographs (W and X) of the extruded product at a feed moisture content of 60 and 20 %.

4   Conclusions

High extruder barrel temperature of 200°C effected the formation of porous and fibrous product with network of holes leading to maximum expansion ratio. Extruder screw speed of 110 rev/min helped in formation of smooth texture, whereas low extruder screw speed of 70 rev/min resulted in fragmentations. High protein levels due to maximum fish content favored greater fiber formation, better organization and orientation than that with low protein levels. Feed moisture content of 60 and 20 % induced cracks in the texture, which were however more pronounced at 20 %.

References

[1]   S. K. Giri, and S. Bandyopadhyay, Effect of extrusion variables on extrudate characteristics of fish muscle- rice flour blend in a single screw extruder. J. Food Process. Pres. **24**, pp. 177-190 (2000).

[2]   T. J. Shankar and S. Bandyopadhyay, Optimization of extrusion process variables using a genetic algorithm, Trans IChemE, Part C, Food and Bioproducts process, **82(C2)**, pp. 143-150 (2004).

[3]   T. J. Maurice, and D. W. Stanely, Texture-structure relationships in texturized soy protein. iv. Influence of process variables on extrusion texturization. Can. Inst. Food Sci. Technol. J. **11**, pp. 1-6. (1972).

[4]   S. Bhattacharya, H. Das, and A. N. Bose, Effect of extrusion process variables on microstructure of blends of minced fish and wheat flour. J. Food Sci. Technol. **27(1)**, pp. 22-28 (1990)

[5]   P. L Harries, S. L Cuppett, and K. W. Lee, Scanning electron microscope study of maize gluten meal and soy coextrudates. Cereal Chem. **65 (3)**: pp. 228-232, (1988).

[6]   Harper, J.M, Extrusion texturization of foods. Food Technol. **40**, pp. 70-89, (1986).

Thin section and freeze-fracture electron microscopy of *Paracoccus denitrificans* bacterium

Ezzatollah Keyhani

Laboratory for Life Sciences, Saadat Abade, Sarve Sharghi 34, 19979 Tehran, IRAN
E-mail: keyhanius2002@yahoo.com; keyhanie@ibb.ut.ac.ir,
Phone: +98-21-207-4804, Fax: +98-21 640-4680

Paracoccus denitrificans thin sections showed its outer and inner (plasma) membranes and the periplasmic space between them filled with periplasm and the peptidoglycans layer. This space was usually homogenous but fibrillar material was occasionally seen. Moreover, plasma membrane evaginations towards the periplasmic space were seen as well as mesosomes and convoluted plasma membrane invaginations towards the periplasmic space. In the cytosol, a tubular fragment of roughly 24 nm was sometimes visible. Freeze-fracture of *Paracoccus denitrificans* showed the fractured plane through the outer and plasma membranes. The plasma membrane protoplasmic face showed numerous intramembrane particles which were random in some cases or aggregated in others. The plasma membrane exoplasmic face showed scarce intramembrane particles, randomly distributed. Both exoplasmic and protoplasmic surfaces of the outer membrane showed fewer intramembrane particles than the plasma membrane.

Keywords thin section, freeze-fracture, *Paracoccus denitrificans*

1   Introduction

Many bacteria living in soil, sewage or sludge possess a highly branched respiratory network. In *Paracoccus denitrificans*, three types of terminal oxidases have been described. The one that is most expressed at atmospheric oxygen concentration is the aa_3-type cytochrome c oxidase, which has a relatively low affinity for oxygen. Its expression level decreases with decreasing oxygen concentrations. The second type of oxidase is the ccb_3-type cytochrome c oxidase, which has a relatively high affinity for oxygen and which is increasingly synthesized at decreasing oxygen concentrations. The third type of oxidase is a bd-type quinol oxidase, which is the counterpart of the bo-type quinol oxidase in *Escherichia coli* and *Salmonella typhimurium* [1–5].

The morphology of bacteria has been the subject of many investigations [6–9]. However, even though *Paracoccus denitrificans* has been the object of thorough biochemical investigations, as for example on its respiratory chain, a comparative study between its fine structure as revealed by thin section and freeze-fracture has not been reported. This is the purpose of the present study.

2   Materials and methods

P. denitrificans (strain ATCC 17741) was cultured either in enriched or synthetic medium, harvested at the stationary phase and washed twice in phosphate buffer 0.1 M, pH 7.4 before processing for electron microscopy. Thin section and freeze-fracture electron microscopy were done according to [10–12].

2.1 Thin sections

A small pellet of bacteria was fixed with glutaraldehyde 2.5%, phosphate buffer 0.1 M, pH 7.4, for one hour, washed several times with phosphate buffer 0.1 M, pH 7.4, post–fixed with osmium tetroxide 1%, phosphate buffer 0.1 M, pH 7.4, dehydrated in ethanol and embedded in Epoxy resin [10].

2.2 Freeze-fracture

Paracoccus denitrificans cells were freeze-fractured after incubation in 30% glycerol for 5 to 30 minutes. A suspension of cells (approximately 3 μl) was transferred to a side-loading gold apposed specimen holder and rapidly quenched in liquid Freon 22 (chlorodifluoromethane) cooled to its freezing point by liquid nitrogen. Freeze-fracture was done by standard techniques according to reference [11, 12].

3   Results and discussion

Three structures separated the cytoplasmic region of *Paracoccus denitrificans* from the environment.

In thin cross section, the outermost structure was the outer membrane which was 3-nm thick. Anothersimilar membrane, also 3-nm thick, covering the cytoplasm, was the plasma membrane. Between these two membranes was a layer of material, referred to as the periplasmic region. The thickness of the periplasm was generally between 10 and 15 nm. Its aspect could be classified into two types, sometimes homogenous (Fig. 1A), and sometimes containing fibrillar material (Fig. 1B).

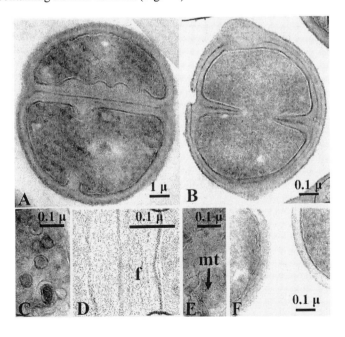

798

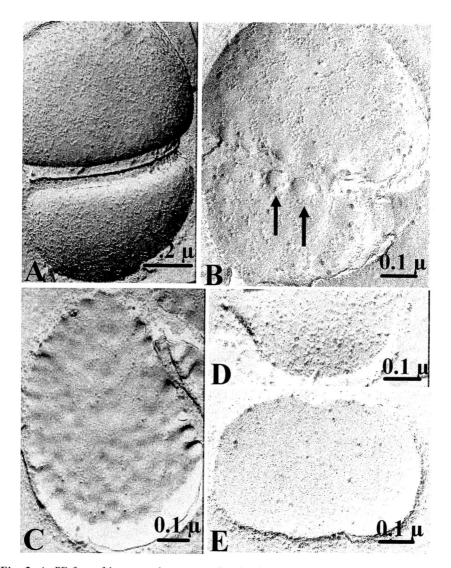

Fig. 2 A: PF face of inner membrane; note the abundant random distribution of intramembrane particles. B: PF face of inner membrane; arrows show surelevated blebs. C: EF face of inner membrane; note the depressions corresponding to surelevations in the PF face. D: EF face of outer membrane. E: PF face of outer membrane exhibiting rare and randomly distributed intramembrane particles.

The outer membrane was generally more visible and easy to delineate than the plasma membrane. The periplasmic region was, as indicated above, subjected to some structural variations. A detail of the fibrillar material sometimes seen is shown in Fig. 1D. On other occasions, one or two dense continuous dark lines (Fig. 1B) surrounding the bacterial cell were visible. This intermediate layer was not seen in many bacteria.

The plasma membrane occasionally appeared to be folded into the cytoplasm. These mesosomal structures were seen in any area of the plasma membrane. Another particularity of the plasma membrane in dividing cells was the formation of membrane evaginations towards the periplasmic space (Fig. 1F).

In the cytoplasmic region, three distinct structures were visible. The ribosomal granules were scattered throughout the cytoplasm. Another detectable structure was a fibrous material, generally centrally located in the cytoplasm, that corresponded to the chromosome of the bacterium. In addition, various spherical vesicles, of 51 nm diameter on the average, surrounded by a unit membrane, were also visible in the cytoplasm (Fig. 1C). A tubular structure of 24-nm thickness (Fig. 1E), presumably the remnant of flagella, was also seen in the cytoplasm.

Freeze-fracture of *Paracoccus denitrificans* (Fig. 2) showed the fractured plane through the inner and outer plasma membranes. The plasma membrane protoplasmic face showed numerous intramembrane particles either randomly dispersed (Fig. 2A) or aggregated (Fig. 2B). Spherical surrelevations of 50 nm were also visible at this surface (Fig. 2B, arrows). The exoplasmic surface of inner membrane showed numerous depressions corresponding to surrelevations of the protoplasmic face and scarce intramembrane particles distributed throughout the surface of the membrane. The exoplasmic surface of the outer membrane showed numerous intramembrane particles (Fig. 2D), while the protoplasmic surface of the outer membrane showed scarce intramembrane particles (Fig. 2E).

Our observations showed that two types of *Paracoccus denitrificans* were identified, both by thin section and freeze-fracture electron microscopy, with respect to the fine structure of their inner plasma membrane. One type was characterized by a smooth surface and the other one exhibited surrelevations at the surface of the membrane. This may correspond to two metabolic states of the bacterium grown to stationary phase in batch culture. However, biochemical characterization of the respiratory chain of any sample from the cultures, and the effect of various inhibitors upon it, were typical of *Paracoccus denitrificans*.

References

[1]   M. F. Otten, J. van der Oost, W. N. M. Reijnders, H. V. Westerhoff, B. Ludwig, and R. J. M. Van Spanning, J. Bacteriol., **183**, 7017 (2001).
[2]   M. F. Otten, D. M. Stork, W. N. M. Reijnders, H. V. Westerhoff, and R. J. M. Van Spanning, Eur. J. Biochem., **268**, 2486 (2001).
[3]   E. Keyhani, and D. Miani-Tehrani, Biochim. Biophys. Acta, **1506**, 1 (2001).
[4]   S. Jünemann, Biochim. Biophys. Acta, **1321**, 107 (1997).
[5]   M. R. Pudek, and P. D. Bragg, Arch. Biochem. Biophys., **174**, 546 (1976).
[6]   G. W. Claus, and L. E. Roth, J. Cell Biol., **20**, 217 (1964).
[7]   Z. Li, A. J. Clarke, and T. J. Beveridge, J. Bacteriol., **180**, 5478 (1998).
[8]   N. Nanninga, Microbiol. Molecul. Biol. Rev., **62**, 110 (1998).
[9]   T. J. Beveridge, J. Bacteriol., **181**, 4725 (1999).
[10] E. Keyhani, J. Microsc. (Paris), **9**, 63 (1970)
[11] E. Keyhani, J. Cell Sci., **46**, 289 (1980)
[12] E. Keyhani, in: Proceedings of the 11th EUREM Conference, Dublin, Ireland, 1998, pp. 568–569.

Subject Index *

---

*   The page numbers refer to the first page of the respecting article